Mathematics for Physics
and Physicists

Mathematics for Physics and Physicists

Walter APPEL

Translated by Emmanuel Kowalski

Princeton University Press
Princeton and Oxford

Published by Princeton University Press, 41 William Street,
Princeton, New Jersey 08540
In the United Kingdom: Princeton University Press, 3 Market Place,
Woodstock, Oxfordshire OX20 1SY

Originally published in French under the title *Mathématiques pour la physique... et les physiciens!*, copyright © 2001 by H&K Editions, Paris.

Translated from the French by Emmanuel Kowalski

Ouvrage publié avec le concours du Ministère français chargé de la culture – Centre national du livre.

[This book was published with the support of the French Ministry of Culture – Centre national du livre.]

Library of Congress Control Number: 2006935871

ISBN-13: 978-0-691-13102-3
ISBN-10: 0-691-13102-3

British Library Cataloging-in-Publication Data is available

Printed on acid-free paper.

pup.princeton.edu

Printed in the United States of America

10 9 8 7 6 5 4 3 2 1

Formatted with LaTeX 2_ε under a Linux environment.

Contents

Appendices

Tables

To Kirone MALLICK, Angel ALASTUEY and Michael KIESSLING,
for their friendship and our discussions – scientific or otherwise.

Thanks

Many are those who I wish to thank for this book: Jean-François Colombeau who was in charge of the course of Mathematics for the "Magistère des sciences de la matière" of the ENS Lyon before me; Michel Peyrard who then asked me to replace him; Véronique Terras and Jean Farago who were in charge of the exercise sessions during the three years when I taught this course.

Many friends and colleagues were kind enough to read and re-read one or more chapters: Julien Barré, Maxime Clusel, Thierry Dauxois, Kirone Mallick, Julien Michel, Antoine Naert, Catherine Pépin, Magali Ribaut, Erwan Saint-Loubert Bié; and especially Paul Pichaureau, TeX-guru, who gave many (very) sharp and (always) pertinent comments concerning typography and presentation as well as contents. I also wish to thank the Éditions H&K and in particular Sébastien Desreux, for his hard work and stimulating demands; Bérangère Condomines for her attentive reading.

I am also indebted to Jean-François Quint for many long and fascinating mathematical discussions.

This edition owes a lot to Emmanuel Kowalski, friend of many years, Professor at the University of Bordeaux, who led me to refine certain statements, correct others, and more generally helped to clarify some delicate points.

For sundry diverse reasons, no less important, I want to thank Craig Thompson, author of *Good-bye Chunky Rice* [89]and *Blankets* [90], who created the illustrations for Chapter 1; Jean-Paul Marchal, master typographer and maker of popular images in Épinal; Angel Alastuey who taught me so much in physics and mathematics; Frédéric, Laëtitia, Samuel, Koupaïa and Alanis Vivien for their support and friendship; Claude Garcia, for his *joie de vivre*, his *fideos*, his fine bottles, his knowledge of topology, his advice on Life, the Universe... and the Rest! without forgetting Anita, Alice and Hugo for their affection.

I must not forget the many readers who, since the first French edition, have communicated their remarks and corrections; in particular Jean-Julien Fleck, Marc Rezzouk, Françoise Cornu, Céline Chevalier and professors Jean Cousteix, Andreas de Vries, and François Thirioux; without forgetting all the students of the "Magistère des sciences de la matière" of ENS Lyon, promotions 1993 to 1999, who followed my classes and exercise sessions and who contributed, through their remarks and questions, to the elaboration and maturation of this book.

Et, bien sûr, les plus importants, Anne-Julia, Solveig et Anton – avec tout mon amour.

Although the text has been very consciously written, read, and proofread, errors, omissions or imprecisions may still remain. The author welcomes any remark, criticism, or correction that a reader may wish to communicate, care of the publisher, for instance, by means of the email address *Errare.humanum.est@h-k.fr*.

A list of *errata* will be kept updated on the web site

http://press.princeton.edu

A book's apology

WHY SHOULD A PHYSICIST STUDY MATHEMATICS? *There is a fairly fashionable current of thought that holds that the use of advanced mathematics is of little real use in physics, and goes sometimes as far as to say that knowing mathematics is already by itself harmful to true physics. However, I am and remain convinced that mathematics is still a precious source of insight, not only for students of physics, but also for researchers.*

Many only see mathematics as a tool – and of course, it is in part a tool, but they should be reminded that, as Galileo said, the book of Nature is written in the mathematician's language.[1] *Since Galileo and Newton, the greatest physicists give examples that knowing mathematics provides the means to understand precise physical notions, to use them more easily, to establish them on a sure foundation, and even more importantly, to discover new ones.*[2] *In addition to ensuring a certain rigor in reasoning which is indispensable in any scientific study, mathematics belongs to the natural lexicon of physicists. Even if the rules of proportion and the fundamentals of calculus are sufficient for most purposes, it is clear that a richer vocabulary can lead to much deeper insights. Imagine if Shakespeare or Joyce had only had a few hundred words to chose from!*

Is is therefore discouraging to sometimes hear physicists dismiss certain theories because "this is only mathematics." In fact, the two disciplines are so closely related that the most prestigious mathematical award, the Fields medal, was given to the physicist Edward WITTEN *in 1990, rewarding him for the remarkable mathematical discoveries that his ideas led to.*

How should you read this book? Or rather how should you not read *some parts?*

---

[1] *"Philosophy is written is this vast book which is forever open before our eyes – I mean, the Universe – but it can not be read until you have learned the tongue and understood the character in which it is written. It is written in the language of mathematics, and its characters are triangles, circles, and other geometric pictures, without which it is not humanly possible to understand a single word [...]."*

[2] I will only mention NEWTON (gravitation, differential and integral calculus), GAUSS (optics, magnetism, *all* the mathematics of his time, and quite a bit that was only understood much later), HAMILTON (mechanics, differential equations, algebra), HEAVISIDE (symbolic calculus, signal theory), GIBBS (thermodynamics, vector analysis) and of course EINSTEIN. One could write a much longer list. If Richard FEYNMAN presents a very "physical" description of his art in his excellent physics course [35], appealing to remarkably little formalism, it is nevertheless the fact that he was the master of the elaborate mathematics involved, as his research works show.

Since the reader may want to learn first certain specific topics among those present, here is a short description of the contents of this text:

- *The first chapter serves as a reminder of some simple facts concerning the fundamental notion of* convergence. *There shouldn't be much in the way of new mathematics there for anyone who has followed a "rigorous calculus" course, but there are many counter-examples to be aware of. Most importantly, a long section describes the traps and difficulties inherent in the process of exchanging two limits* in the setting of physical models. *It is not always obvious where, in physical reasoning, one has to exchange two* mathematical *limits, and many apparent paradoxes follow from this fact.*

- *The real beginning concerns the theory of integration, which is briefly presented in the form of the Lebesgue integral based on measure theory (Chapter 2). For many readers, this may be omitted in a first reading. Chapter 3 discusses the basic results and techniques of integral calculus.*

- *Chapters 4 to 6 present the theory of functions of a complex variable, with a number of applications:*
  - *the "residue method," which is an amazing tool for integral calculus;*
  - *some physical notions, such as* causality, *are very closely related to analyticity of functions on the complex plane (Section 13.4);*
  - *harmonic functions (such that $\triangle f = 0$) in two dimensions are linked to the real parts of holomorphic (analytic) functions (Chapter 5);*
  - *conformal maps (those which preserve angles) can be used to simplify boundary conditions in problems of hydrodynamics or electromagnetism (Chapter 6);*

- *Chapters 7 and 8 concern the theory of distributions ("generalized functions") and their applications to physics. These form a relatively self-contained subset of the book.*

- *Chapters 9 to 12 deal with Hilbert spaces, Fourier series, Fourier and Laplace transforms, which have too many physical applications to attempt a list. Chapter 13 develops some of those applications, and this chapter also requires complex analysis.*

- *Chapter 14 is a short (probably too short) introduction to the Dirac notation used in quantum mechanics:* kets $|\psi\rangle$ *and* bras $\langle\psi|$. *The notions of generalized eigenbasis and self-adjoint operators on Hilbert space are also discussed.*

- *Several precise physical problems are considered and solved in Chapter 15 by the method of Green functions. This method is usually omitted from textbooks on electromagnetism (where a solution is taken straight out of a magician's hat) or of field theory (where it is assumed that the method is known). I hope to fill a gap for students by presenting the necessary (and fairly simple) computations from beginning to end, using physicist's notation.*

- *Chapters 16 and 17 about tensor calculus and differential forms are also somewhat independent from the rest of the text. Those two chapters are only brief introductions to their respective topics.*
- *Chapter 18 has the modest goal of relating some notions of topology and group theory to the idea of* spin *in quantum mechanics.*
- *Probability theory, discussed in Chapters 19 to 21, is almost entirely absent from the standard physics curriculum, although the basic vocabulary and results of probability seem necessary to any physicist interested in theory (stochastic equations, Brownian motion, quantum mechanics and statistical mechanics all require probability theory) or experiments (Gaussian white noise, measurement errors, standard deviation of a data set...)*
- *Finally, a few appendices contain further reminders of elementary mathematical notions and the proofs of some interesting results, the length of which made their insertion in the main text problematic.*

Many physical applications, using mathematical tools with the usual notation of physics, are included in the text. They can be found by looking in the index at the items under "Physical applications."

Remarks on the biographical summaries

The short biographies of mathematicians which are interspersed in the text are taken from many sources:

- Bertrand HAUCHECORNE and Daniel SURATTEAU, *Des mathématiciens de A à Z*, Ellipses, 1996 (French).
- The web site of Saint-Andrews University (Scotland)

  www-history.mcs.st-andrews.ac.uk/history/Mathematicians

  and the web site of the University of Colorado at Boulder

  www.colorado.edu/education/DMP

  from which certain pictures are also taken.
- *The literary structure of scientific argument*, edited by Peter DEAR, University of Pennsylvania Press, 1991.
- Simon GINDIKIN, *Histoires de mathématiciens et de physiciens*, Cassini, 2000.
- *Encyclopædia Universalis*, Paris, 1990.

Translator's foreword

I am a mathematician and have now forgotten most of the little physics I learned in school (although I've probably picked up a little bit again by translating this book). I would like to mention here two more reasons to learn mathematics, and why this type of book is therefore very important.

First, physicists benefit from knowing mathematics (in addition to the reasons Walter mentioned) because, provided they immerse themselves in mathematics sufficiently to become fluent in its language, they will gain access to new *intuitions*. Intuitions are very different from any set of techniques, or tools, or methods, but they are just as indispensable for a researcher, and they are the hardest to come by.[3] A mathematician's intuitions are very different from those of a physicist, and to have both available is an enormous advantage.

The second argument is different, and may be subjective: physics is *hard*, much harder in some sense than mathematics. A very simple and fundamental physical problem may be all but impossible to solve because of the complexity (apparent or real) of Nature. But mathematicians *know* that a simple, well-formulated, natural mathematical problem (in some sense that is impossible to quantify!) *has a "simple" solution*. This solution may require inventing entirely new concepts, and may have to wait for a few hundred years before the idea comes to a brilliant mind, but it *is* there. What this means is that if you manage to put the physical problem in a very natural mathematical form, the guiding principles of mathematics may lead you to the solution. Dirac was certainly a physicist with a strong sense of such possibilities; this led him to discover antiparticles, for instance.

[3] Often, nothing will let you understand the intuition behind some important idea except, essentially, rediscovering by yourself the most crucial part of it.

Notation

Symbol	Meaning	Page
$\setminus$	without (example: $A \setminus B$)	
$\sqsupset$	"has for Laplace transform"	333
$\langle \cdot, \cdot \rangle$	duality bracket: $\langle a, x \rangle = a(x)$	
$(\cdot \vert \cdot)$	scalar product	251
$\vert\psi\rangle$, $\langle\psi\vert$	ket, bra	379,380
$\lfloor \cdot \rfloor$	integral part	
Ш	Dirac comb	186
$\mathbf{1}$	constant function $x \mapsto 1$	
$*$	convolution product	211
$\otimes$	tensor product	210, 439
$\wedge$	exterior product	464
∂	boundary (example: $\partial\Omega$, boundary of Ω)	
$\triangle$	laplacian	
$\square$	d'Alembertian	414
$\sim$	equivalence of sequences or functions	580
$\simeq$	isomorphic to (example: $E \simeq F$)	
$\approx$	approximate equality (in physics)	
$\stackrel{\text{def}}{=}$	equality defining the left-hand side	
$[a, b]$	closed interval : $\{x \in \mathbb{R}\ ;\ a \leqslant x \leqslant b\}$	
$]a, b[$	open interval : $\{x \in \mathbb{R}\ ;\ a < x < b\}$	
$[\![1, n]\!]$	$\{1, \ldots, n\}$	

Latin letter	Meaning	Page
$\mathscr{B}$	Borel σ-algebra	58
$\mathcal{B}(a\,;r)$, $\overline{\mathcal{B}}(a\,;r)$	open (closed) ball centered at a with radius r	
$\mathcal{B}il(E\times F,G)$	space of bilinear maps from $E\times F$ to G	
$\complement_\Omega A$	complement of A in Ω	512
$\overline{\mathbb{C}}$	$\mathbb{C}\cup\{\infty\}$	146
$\mathscr{C}(I,\mathbb{R})$	vector space of continuous real-valued functions on I	
$\mathscr{C}(a\,;r)$	circle centered at a with radius r	
$\mathrm{C}^{\underline{\mathrm{nt}}}$	any constant	
$\mathscr{D}$	vector space of test functions	182
$\mathscr{D}'$	vector space of distributions	183
$\mathscr{D}'_+$, $\mathscr{D}'_-$	vector space of distributions with bounded support on the left (right)	236
$\mathrm{diag}(\lambda_1,\ldots,\lambda_n)$	square matrix with coefficients $m_{ij}=\lambda_i\,\delta_{ij}$	
e	Neper's constant (e: electric charge)	
$(\mathbf{e}_\mu)_\mu$	basis of a vector space	433
$\mathsf{E}(X)$	expectation of the r.v. X	527
$\mathscr{F}[f]$, $\widetilde{f}$	Fourier transform of f	278
$\widehat{f}$	Laplace transform of f	333
$\check{f}$	transpose of f; $\check{f}(x)=f(-x)$	189
$f^{(n)}$	n-th derivative of f	
$\mathrm{GL}_n(\mathbb{K})$	group of invertible matrices of order n over $\mathbb{K}$	
$\mathscr{GL}(E)$	group of automorphisms of the vector space E	
H	Heaviside distribution or function	193
$\mathbb{K}$, $\mathbb{K}'$	$\mathbb{R}$ or $\mathbb{C}$	
ℓ^2	space of sequences $(u_n)_{n\in\mathbb{N}}$ such that $\sum\lvert u_n\rvert^2<+\infty$	261
L^1	space of integrable functions	280
L^2	space of square-integrable functions	262
$\mathscr{L}(E,F)$	space of linear maps from E to F	
$\mathscr{L}$	Lebesgue σ-algebra	60
$\mathfrak{M}_n(\mathbb{K})$	algebra of square matrices of order n over K	
$\mathfrak{P}(A)$	set of subsets of A	57
$\mathsf{P}(A)$	probability of the event A	514
$\overline{\mathbb{R}}$	$\mathbb{R}\cup\{-\infty,+\infty\}$	
$\mathrm{Re}(z),\mathrm{Im}(z)$	real or imaginary part of z	
$\mathscr{S}$	space of Schwartz functions	289
$\mathscr{S}'$	space of tempered distributions	300
$\mathfrak{S}$, $\mathfrak{S}_n$	group of permutations (of n elements)	26
${}^t\square$	transpose of a matrix : $({}^tA)_{ij}=A_{ji}$	
$\mathbf{u}$, $\mathbf{v}$, $\mathbf{w}$, $\mathbf{x}$	vectors in E	434
pv	principal value	188
X,Y,Z	random variables	521

Greek letter		
$(\boldsymbol{\alpha}^\mu)_\mu$	dual basis of the basis $(\mathbf{e}_\mu)_\mu$	436
Γ	Euler function	154
δ, δ'	Dirac distribution, its derivative	185, 196
δ^μ_ν, δ_{ij}	Kronecker symbol	445
$\varepsilon(\sigma)$	signature of the permutation σ	465
Λ^{*k}	space of exterior k-forms	464
Π	rectangle function	213, 279
$\sigma(X)$	standard deviation of the r.v. X	528
χ_A	characteristic function of A	63
$\boldsymbol{\omega}, \boldsymbol{\sigma}$	differential forms	436

Abbreviations		
(n).v.s.	(normed) vector space	
r.v.	random variable	521
i.r.v.	independent random variables	537

Chapter 1

Reminders: convergence of sequences and series

This first chapter, which is quite elementary, is essentially a survey of the notion of convergence of sequences and series. Readers who are very confortable with this concept may start reading the next chapter.

However, although the mathematical objects we discuss are well known in principle, they have some unexpected properties. We will see in particular that the order of summation may be crucial to the evaluation of the series, so that changing the order of summation may well change its sum.

We start this chapter by discussing two physical problems in which a limit process is *hidden*. Each leads to an apparent paradox, which can only be resolved when the underlying limit is explicitly brought to light.

1.1 The problem of limits in physics

1.1.a Two paradoxes involving kinetic energy

First paradox

Consider a truck with mass m driving at constant speed $v = 60$ mph on a perfectly straight stretch of highway (think of Montana). We assume that, friction being negligible, *the truck uses no gas to remain at constant speed.*

On the other hand, to accelerate, it must use (or guzzle) ℓ gallons of gas in order to increase its kinetic energy by an amount of 1 joule. This assumption, although it is imperfect, is physically acceptable because each gallon of gas yields the same amount of energy.

So, when the driver decides to increase its speed to reach $v' = 80$ mph, the quantity of gas required to do so is equal to the difference of kinetic energy, namely, it is

$$\ell(E_c' - E_c) = \frac{1}{2}\ell m(v'^2 - v^2) = \frac{1}{2}\ell m(6\,400 - 3\,600) = 1\,400 \times \ell m.$$

With $\ell \cdot m = \frac{1}{10\,000}\,\mathrm{J} \cdot \mathrm{mile}^{-2} \cdot \mathrm{h}^2$, say, this amounts to 0.14 gallon. Jolly good.

Now, let us watch the same scene of the truck accelerating, as observed by a highway patrolman, initially driving as fast as the truck $w = v = 60$ mph, but with a motorcycle which is unable to go faster.

The patrolman, having his college physics classes at his fingertips, argues as follows: "in my own galilean reference frame, the relative speed of the truck was previously $v^* = 0$ and is now $v^{*\prime} = 20$ mph. To do this, the amount of gas it has guzzled is equal to the difference in kinetic energies:

$$\ell(E_c^{*\prime} - E_c^*) = \frac{1}{2}\ell m\Big((v^{*\prime})^2 - (v^*)^2\Big) = \frac{1}{2}\ell m(400 - 0) = 200 \times \ell m,$$

or around 0.02 gallons."

There is here a clear problem, and one of the two observers must be wrong. Indeed, the galilean relativity principle states that all galilean reference frames are equivalent, and computing kinetic energy in the patrolman's reference frame is perfectly legitimate.

How is this paradox resolved?

We will come to the solution, but first here is another problem. The reader, before going on to read the solutions, is earnestly invited to think and try to solve the problem by herself.

Second paradox

Consider a highly elastic rubber ball in free fall as we first see it. At some point, it hits the ground, and we assume that *this is an elastic shock*.

Most high-school level books will describe the following argument: "assume that, at the instant $t = 0$ when the ball hits the ground, the speed of the ball is $v_1 = -10\ \mathrm{m{\cdot}s^{-1}}$. Since the shock is elastic, there is conservation of total energy before and after. Hence the speed of the ball after the rebound is $v_2 = -v_1$, or simply $+10\ \mathrm{m{\cdot}s^{-1}}$ going up."

This looks convincing enough. But it is not so impressive if seen from the point of view of an observer who is also moving down at constant speed $v_{\text{obs}} = v_1 = -10\ \mathrm{m{\cdot}s^{-1}}$. For this observer, the speed of the ball before the shock is $v_1^* = v_1 - v_{\text{obs}} = 0\ \mathrm{m{\cdot}s^{-1}}$, so it has zero kinetic energy. However, after rebounding, the speed of the ball is $v_2^* = v_2 - v_{\text{obs}} = 20\ \mathrm{m{\cdot}s^{-1}}$, and therefore it has nonzero kinetic energy! With the analogue of the reasoning above, one should still have found $v_2^* = v_1^* = 0$ (should the ball go through the ground?)

So there is something fishy in this argument also. It is important to remember that *the fact that the right answer is found in the first case does not imply that the argument that leads to the answer is itself correct.*

Readers who have solved the first paradox will find no difficulty in this second one.

Paradoxes resolved

Kinetic energy is of course not the same in every reference frame. But this is not so much the kinetic energy we are interested in; rather, we want the difference before and after the event described.

Let's go back to elementary mechanics. What happens, in two distinct reference frames, to a system of N solid bodies with initial speed $\boldsymbol{v}_i$ ($i = 1, \ldots, N$) and final speed $\boldsymbol{v}'_i$ after some shock?

In the first reference frame, the difference of kinetic energy is given by

$$\Delta E_c = \sum_{i=1}^{N} m_i(\boldsymbol{v}'^2_i - \boldsymbol{v}^2_i).$$

In a second reference frame, with relative speed $\boldsymbol{w}$ with respect to the first, the difference is equal to

$$\begin{aligned}\Delta E_c^* &= \sum_{i=1}^{N} m_i\Big((\boldsymbol{v}'_i - \boldsymbol{w})^2 - (\boldsymbol{v}_i - \boldsymbol{w})^2\Big)\\ &= \sum_{i=1}^{N} m_i(\boldsymbol{v}'^2_i - \boldsymbol{v}_i{}^2) - 2\boldsymbol{w}\cdot\left(\sum_{i=1}^{N} m_i(\boldsymbol{v}'_i - \boldsymbol{v}_i)\right) = \Delta E_c - 2\boldsymbol{w}\cdot\Delta\boldsymbol{P},\end{aligned}$$

(we use $*$ as exponents for any physical quantity expressed in the new reference frame), so that $\Delta E_c^* = \Delta E_c$ as long as *the total momentum is preserved* during the shock, in other words if $\Delta\boldsymbol{P} = \boldsymbol{0}$.

In the case of the truck and the patrolman, we did not really take the momentum into account. In fact, the truck can accelerate *because it "pushes back" the whole earth behind it!*

So, let us take up the computation with a terrestrial mass M, which is large but not infinite. We will take the limit $[M \to \infty]$ at the very end of the computation, and more precisely, we will let $[M/m \to \infty]$.

At the beginning of the "experiment," in the terrestrial reference frame, the speed of the truck is $\boldsymbol{v}$. At the end of the experiment, the speed is $\boldsymbol{v}'$. Earth, on the other hand, has original speed $\boldsymbol{V} = \boldsymbol{0}$, and if one remains in the same galilean reference frame, final speed $\boldsymbol{V}' = \frac{m}{M}(\boldsymbol{v} - \boldsymbol{v}')$ (because of conservation of total momentum).[1] The kinetic energy of the system at the beginning is then $\frac{1}{2}mv^2$ and at the end it is $\frac{1}{2}mv'^2 + \frac{1}{2}MV'^2$. So, the difference is given by

$$\Delta E_c = \frac{1}{2}m(v'^2 - v^2) + \frac{1}{2}\frac{m^2}{M}(v - v')^2 = \frac{1}{2}m(v'^2 - v^2)\left[1 + O\left(\frac{m}{M}\right)\right].$$

This is the amount of gas involved! So we see that, up to negligible terms, the first argument gives the right answer, namely, 0.14 gallons.

We now come back to the patrolman's frame, moving with speed w with respect to the terrestrial frame. The initial speed of the truck is $v^* = v - w$, and the final speed is $v'^* = v' - w$. The Earth has initial speed $V^* = -w$ and final speed $V'^* = -w + \frac{m}{M}'(v - v')$. The difference is now:

$$\begin{aligned}
\Delta E_c^* &= \frac{1}{2}m\left(v'^{*2} - v^{*2}\right) + \frac{1}{2}M\left(V'^{*2} - V^{*2}\right) \\
&= \frac{1}{2}m(v' - w)^2 - \frac{1}{2}m(v - w)^2 + \frac{1}{2}M\left[\frac{m}{M}(v - v') - w\right]^2 - \frac{1}{2}Mw^2 \\
&= \frac{1}{2}mv'^2 - \frac{1}{2}mv^2 + m(v - v')\cdot w + \frac{1}{2}\frac{m^2}{M}(v - v')^2 - m(v - v')\cdot w, \\
\Delta E_c^* &= \Delta E_c.
\end{aligned}$$

Hence the difference of kinetic energy is preserved, as we expected. So even in this other reference frame, a correct computation shows that the quantity of gas involved is the same as before.

The patrolman's mistake was to forget the *positive* term $-m(v - v')\cdot w$, corresponding to the difference of kinetic energy of the Earth in its galilean frame. This term does not tend to 0 as $[M/m \to \infty]$!

From the point of view of the patrolman's frame, 0.02 gallons are needed to accelerate the truck, and the remaining 0.12 gallons are needed to accelerate the Earth!

We can summarize this as a table, where T is the truck and E is the Earth.

	Initial speed	Final speed	E_c init.	E_c final	ΔE_c
T	v	v'	$\frac{1}{2}mv^2$	$\frac{1}{2}mv'^2$	$\frac{m}{2}(v'^2 - v^2)$
E	0	$\frac{m}{M}(v - v')$	0	$\frac{1}{2}\frac{m^2}{M}(v - v')^2$	$+\frac{m^2}{2M}(v - v')^2$
T*	$v - w$	$v' - w$	$\frac{1}{2}m(v - w)^2$	$\frac{1}{2}m(v' - w)^2$	$\frac{m}{2}(v'^2 - v^2)$
E*	$-w$	$\frac{m}{M}(v - v') - w$	$\frac{1}{2}Mw^2$	$\frac{M}{2}\left(\frac{m}{M}(v - v') - w\right)^2$	$+\frac{m^2}{2M}(v - v')^2$

[1] Note that the terrestrial reference frame is then not galilean, since the Earth started "moving" under the truck's impulsion.

(Illustration © Craig THOMPSON 2001)

Fig. 1.1 – Romeo, Juliet, and the boat on a lake.

The second paradox is resolved in the same manner: the Earth's rebound energy must be taken into account after the shock with the ball.

The interested reader will find another paradox, relating to optics, in Exercise 1.3 on page 43.

1.1.b Romeo, Juliet, and viscous fluids

Here is an example in mechanics where a function $f(x)$ is defined on $[0, +\infty[$, but $\lim_{x\to 0^+} f(x) \neq f(0)$.

Let us think of a summer afternoon, which Romeo and Juliet have dedicated to a pleasant boat outing on a beautiful lake. They are sitting on each side of their small boat, immobile over the waters. Since the atmosphere is conducive to charming murmurs, Romeo decides to go sit by Juliet.

Denote by M the mass of the boat and Juliet together, m that of Romeo, and L the length of the walk from one side of the boat to the other (see Figures 1.1 and 1.2).

Two cases may be considered: one where the friction of the boat on the lake is negligible (a perfect fluid), and one where it is given by the formula $f = -\eta v$, where f is the force exerted by the lake on the boat, v the speed of the boat on the water, and η a viscosity coefficient. We consider the problem only on the horizontal axis, so it is *one-dimensional.*

We want to compute how far the boat moves

1. in the case $\eta = 0$;

2. in the case $\eta \neq 0$.

Let ℓ be this distance.

The first case is very easy. Since no force is exerted in the horizontal plane on the system "boat + Romeo + Juliet," the center of gravity of this system

(Illustration © Craig THOMPSON 2001)

Fig. 1.2 – Romeo moved closer.

does not move during the experiment. Since Romeo travels the distance L relative to the boat, it is easy to deduce that the boat must cover, in the opposite direction, the distance

$$\ell = \frac{m}{m+M} L\,.$$

In the second case, let $x(t)$ denote the positive of the boat and $y(t)$ that of Romeo, relative to the Earth, not to the boat. The equation of movement for the center of gravity of the system is

$$M\ddot{x} + m\ddot{y} = -\eta\dot{x}.$$

We now integrate on both sides between $t = 0$ (before Romeo starts moving) and $t = +\infty$. Because of the friction, we know that as $[t \to +\infty]$, the speed of the boat goes to 0 (hence also the speed of Romeo, since he will have been long immobile with respect to the boat). Hence we have

$$(M\dot{x} + m\dot{y})\Big|_0^{+\infty} = 0 = -\eta\big(x(+\infty) - x(0)\big)$$

or $\eta\ell = 0$. Since $\eta \neq 0$, we have $\ell = 0$, whichever way Romeo moved to the other side. In partcular, if we take the limit when $\eta \to 0$, hoping to obtain the nonviscous case, we have:

$$\lim_{\substack{\eta\to 0\\ \eta>0}} \ell(\eta) = 0 \qquad \text{hence} \qquad \lim_{\substack{\eta\to 0\\ \eta>0}} \ell(\eta) \neq \ell(0).$$

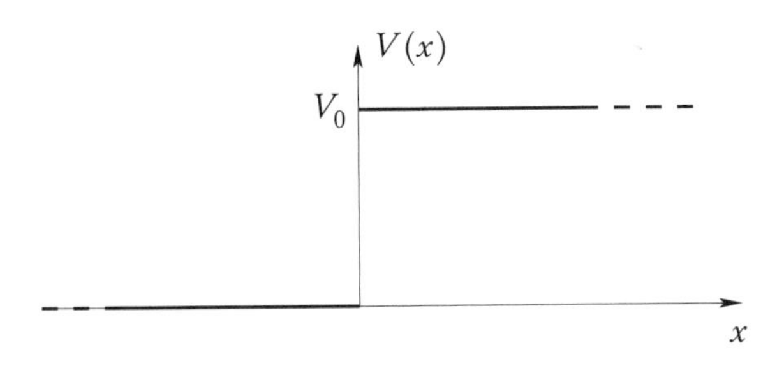

Fig. 1.3 – Potential wall $V(x) = V_0 H(x)$.

Conclusion: *The limit of a viscous fluid to a perfect fluid is singular.* It is not possible to formally take the limit when the viscosity tends to zero to obtain the situation for a perfect fluid. In particular, it is easier to model flows of nonviscous perfect fluids by "real" fluids which have large viscosity, because of turbulence phenomena which are more likely to intervene in fluids with small viscosity. The interested reader can look, for instance, at the book by Guyon, Hulin, and Petit [44].

Remark 1.1 The exact form $f = -\eta v$ of the friction term is crucial in this argument. If the force involves additional (nonlinear) terms, the result is completely different. Hence, if you try to perform this experiment in practice, it will probably not be conclusive, and the boat is not likely to come back to the same exact spot at the end.

1.1.c Potential wall in quantum mechanics

In this next physical example, there will again be a situation where we have a limit $\lim_{x\to 0} f(x) \neq f(0)$; however, the singularity arises here in fact because of a *second* variable, and the true problem is that we have a double limit which *does not commute*: $\lim_{x\to 0}\lim_{y\to 0} f(x,y) \neq \lim_{y\to 0}\lim_{x\to 0} f(x,y)$.

The problem considered is that of a quantum particle arriving at a potential wall. We look at a one-dimensional setting, with a potential of the type $V(x) = V_0 H(x)$, where H is the Heaviside function, that is, $H(x) = 0$ if $x < 0$ and $H(x) = 1$ if $x > 0$. The graph of this potential is represented in Figure 1.3.

A particle arrives from $x = -\infty$ in an energy state $E > V_0$; part of it is transmitted beyond the potential wall, and part of it is reflected back. We are interested in the reflection coefficient of this wave.

The incoming wave may be expressed, for negative values of x, as the sum of a progressive wave moving in the direction of increasing x and a reflected wave. For positive values of x, we have a transmitted wave in the direction of increasing x, but no component in the other direction. According to the Schrödinger equation, the wave function can therefore be written in the form

$$\varphi(x,t) = \psi(x) f(t),$$

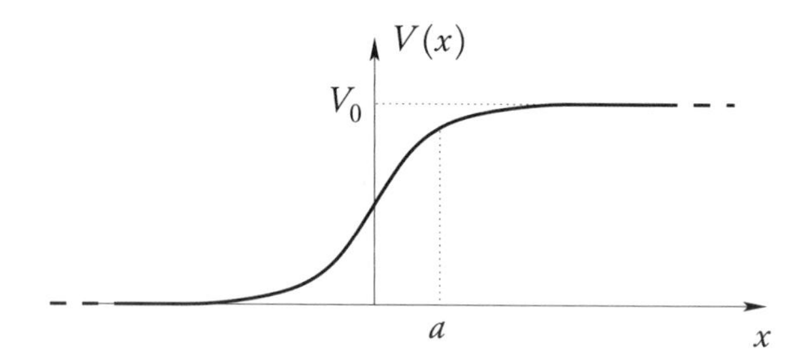

Fig. 1.4 – "Smoothed" potential $V(x) = V_0/(1 + \mathrm{e}^{-x/a})$, with $a > 0$.

where

$$\psi(x) = \begin{cases} \mathrm{e}^{\mathrm{i}kx} + B\,\mathrm{e}^{-\mathrm{i}kx} & \text{if } x < 0, \text{ with } k \stackrel{\text{def}}{=} \frac{\sqrt{2mE}}{\hbar}, \\ A\,\mathrm{e}^{\mathrm{i}k'x} & \text{if } x > 0, \text{ with } k' \stackrel{\text{def}}{=} \frac{\sqrt{2m(E-V_0)}}{\hbar}. \end{cases}$$

The function $f(t)$ is only a time-related phase factor and plays no role in what follows. The reflection coefficient of the wave is given by the ratio of the currents associated to ψ and is given by $R = 1 - \frac{k'}{k}|A|^2$ (see [20, 58]). There remains to find the value of A. To find it, it suffices to write the equation expressing the continuity of ψ and ψ' at $x = 0$, Since $\psi(0^+) = \psi(0^-)$, we have $1 + B = A$. And since $\psi'(0^+) = \psi'(0^-)$, we have $k(1 - B) = k'A$, and we deduce that $A = 2k/(k + k')$. The reflection coefficient is therefore equal to

$$R = 1 - \frac{k'}{k}|A|^2 = \left(\frac{k - k'}{k + k'}\right) = \left(\frac{\sqrt{E} - \sqrt{E - V_0}}{\sqrt{E} + \sqrt{E - V_0}}\right)^2. \tag{I.1}$$

Here comes the surprise: this expression (I.1) is independant of $\hbar$. In particular, the limit as $[\hbar \to 0]$ (which defines the "classical limit") yields a nonzero reflection coefficient, although we know that in classical mechanics a particle with energy E does not reflect against a potential wall with value $V_0 < E$![2] So, displaying explicitly the dependency of R on $\hbar$, we have:

$$\lim_{\substack{\hbar \to 0 \\ \hbar \neq 0}} R(\hbar) \neq 0 = R(0).$$

In fact, we have gone a bit too fast. We take into account the physical aspects of this story: the "classical limit" is certainly not the same as brutally writing "$\hbar \to 0$." Since Planck's constant is, as the name indicates, just a constant, this makes no sense. To take the limit $\hbar \to 0$ means that one arranges for the quantum dimensions of the system to be much smaller than all other dimensions. Here the quantum dimension is determined by the de

[2] The particle goes through the obstacle with probability 1.

Broglie wavelength of the particle, that is, $\lambda = \hbar/p$. What are the other lengths in this problem? Well, there are none! At least, the way we phrased it: because in fact, expressing the potential by means of the Heaviside function is rather *cavalier*. In reality, the potential must be continuous. We can replace it by an infinitely differentiable potential such as $V(x) = V_0/(1+\mathrm{e}^{-x/a})$, which increases, roughly speaking, on an interval of size $a > 0$ (see Figure 1.4). In the limit where $a \to 0$, the discontinuous Heaviside potential reappears.

Computing the reflection coefficient with this potential is done similarly, but of course the computations are more involved. We refer the reader to [58, chapter 25]. At the end of the day, the reflection coefficient is found to depend not only on $\hbar$, but also on a, and is given by

$$R(\hbar, a) = \left(\frac{\sinh a\pi(k-k')}{\sinh a\pi(k+k')} \right)^2.$$

($\hbar$ appears in the definition of $k = \sqrt{2mE}/\hbar$ and $k' = \sqrt{2m(E-V_0)}/\hbar$.) We then see clearly that *for fixed nonzero* a, the de Broglie wavelength of the particle may become infinitely small compared to a, and this defines the correct classical limit. Mathematically, we have

$$\forall a \neq 0 \qquad \lim_{\substack{\hbar \to 0 \\ \hbar \neq 0}} R(\hbar, a) = 0 \qquad \text{classical limit}$$

On the other hand, if we keep $\hbar$ fixed and let a to to 0, we are converging to the Heaviside potential and we find that

$$\forall \hbar \neq 0 \qquad \lim_{\substack{a \to 0 \\ a \neq 0}} R(\hbar, a) = \left(\frac{k-k'}{k+k'} \right)^2 = R(\hbar, 0).$$

So the two limits $[\hbar \to 0]$ and $[a \to 0]$ cannot be taken in an arbitrary order:

$$\lim_{\substack{\hbar \to 0 \\ \hbar \neq 0}} \lim_{\substack{a \to 0 \\ a \neq 0}} R(\hbar, a) = \left(\frac{k-k'}{k+k'} \right)^2 \qquad \text{but} \qquad \lim_{\substack{a \to 0 \\ a \neq 0}} \lim_{\substack{\hbar \to 0 \\ \hbar \neq 0}} R(\hbar, a) = 0.$$

To speak of $R(0,0)$ has *a priori* no physical sense.

1.1.d Semi-infinite filter behaving as waveguide

We consider the cicuit AB,

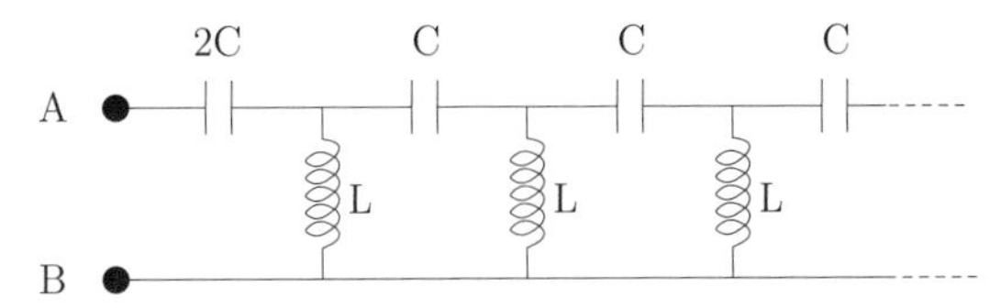

made up of a cascading sequence of "T" cells $(2C, L, 2C)$,

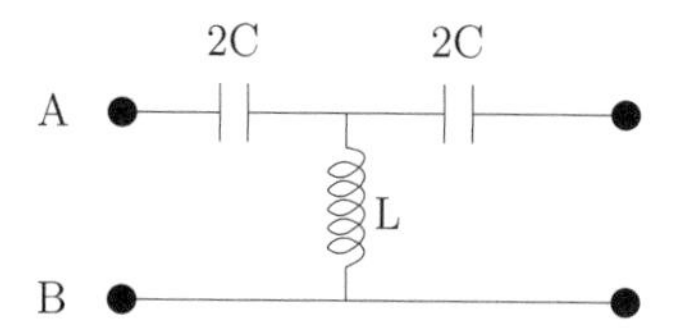

(the capacitors of two successive cells in series are equivalent with one capacitor with capacitance C). We want to know the total impedance of this circuit.

First, consider instead a circuit made with a finite sequence of n elementary cells, and let Z_n denote its impedance. Kirchhoff's laws imply the following recurrence relation between the values of Z_n:

$$Z_{n+1} = \frac{1}{2\mathrm{i}C\omega} + \frac{\mathrm{i}L\omega\left(\dfrac{1}{2\mathrm{i}C\omega} + Z_n\right)}{\mathrm{i}L\omega + \dfrac{1}{2\mathrm{i}C\omega} + Z_n}, \tag{1.2}$$

where ω is the angular frequency. In particular, note that if Z_n is purely imaginary, then so is Z_{n+1}. Since Z_1 is purely imaginary, it follows that

$$Z_n \in \mathrm{i}\mathbb{R} \qquad \text{for all } n \in \mathbb{N}.$$

We don't know if the sequence $(Z_n)_{n\in\mathbb{N}}$ converges. But one thing is certain: if this sequence $(Z_n)_{n\in\mathbb{N}}$ converges to some limit, this must be purely imaginary (the only possible real limit is zero).

Now, we compute the impedance of the infinite circuit, noting that this circuit AB is *strictly* equivalent to the following:

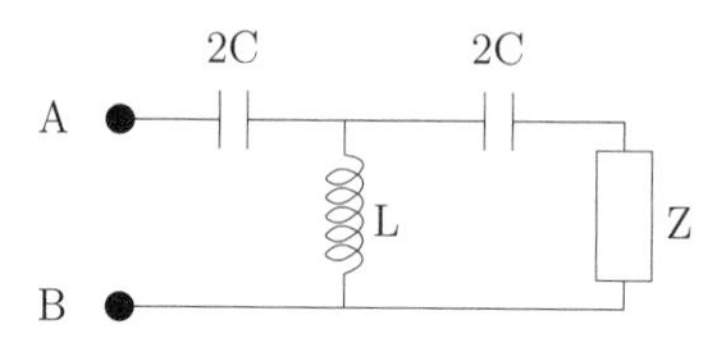

Hence we obtain an equation involving Z:

$$Z = \frac{1}{2\mathrm{i}C\omega} + \frac{\mathrm{i}L\omega\left(\dfrac{1}{2\mathrm{i}C\omega} + Z\right)}{\mathrm{i}L\omega + \dfrac{1}{2\mathrm{i}C\omega} + Z}. \tag{1.3}$$

Some computations yield a second-degree equation, with solutions given by

$$Z^2 = \frac{1}{C} \cdot \left(L - \frac{1}{4C\omega^2}\right).$$

We must therefore distinguish two cases:

• **If** $\omega < \omega_c = \dfrac{1}{2\sqrt{LC}}$, we have $Z^2 < 0$ and hence Z is purely imaginary of the form

$$Z = \pm i\sqrt{\frac{1}{4C^2\omega^2} - \frac{L}{C}}.$$

Remark 1.2 Mathematically, there is nothing more that can be said, and in particular there remains an uncertainty concerning the "sign" of Z.

However, this can also be determined by a physical argument: let ω tend to 0 (continuous regime). Then we have

$$Z(\omega) \underset{\omega\to 0^+}{\sim} \pm\frac{i}{2C\omega},$$

the modulus of which tends to infinity. This was to be expected: the equivalent circuit is open, and the first capacitor "cuts" the circuit. Physically, it is then natural to expect that, the first coil acting as a plain wire, the first capacitor will be dominant. Then

$$Z(\omega) \underset{\omega\to 0^+}{\sim} -\frac{i}{2C\omega}$$

(corresponding to the behavior of a single capacitor).

Thus the physically acceptable solution of the equation (1.3) is

$$Z = -i\sqrt{\frac{1}{4C^2\omega^2} - \frac{L}{C}}.$$

• **If** $\omega > \omega_c = \dfrac{1}{2\sqrt{LC}}$, then $Z^2 > 0$ and Z is therefore *real*:

$$Z = \pm\sqrt{\frac{L}{C} - \frac{1}{4C^2\omega^2}}.$$

Remark 1.3 Here also the sign of Z can be determined by physical arguments. The real part of an impedance (the "resistive part") is always *non-negative* in the case of a passive component, since it accounts for the dissipation of energy by the Joule effect. Only active components (such as operational amplifiers) can have negative resistance. Thus, the physically acceptable solution of equation (1.3) is

$$Z = +\sqrt{\frac{L}{C} - \frac{1}{4C^2\omega^2}}.$$

In this last case, there seems to be a paradox since Z cannot be the limit as $n \to +\infty$ of $(Z_n)_{n\in\mathbb{N}}$. Let's look at this more closely.

From the mathematical point of view, Equation (1.3) expresses nothing but the fact that Z is a "fixed point" for the induction relation (1.2). In other words, this is the equation we would have obtained from (1.2), by continuity, *if* we had known that the sequence $(Z_n)_{n\in\mathbb{N}}$ converges to a limit Z. However, there is no reason for the sequence $(Z_n)_{n\in\mathbb{N}}$ to converge.

Remark 1.4 *From the physical point of view*, the behavior of this infinite chain is rather surprising. How does resistive behavior arise from purely inductive or capacitative components? Where does energy dissipate? And where does it go?

In fact, there is no dissipation of energy in the sense of the Joule effect, but energy does disappear from the point of view of an operator "holding points A and B." More precisely, one can show that there is a flow of energy propagating from cell to cell. So at the beginning of the circuit, it looks like there is an "energy well" with fixed power consumption. Still, no energy disappears: an infinite chain can consume energy without accumulating it anywhere.[3] In the regime considered, this infinite chain corresponds to a waveguide.

We conclude this first section with a list of other physical situations where the problem of noncommuting limits arises:

- taking the "classical" (nonquantum) limit, as we have seen, is by no means a trivial matter; in addition, it may be in conflict with a "non-relativistic" limit (see, e.g., [6]), or with a "low temperature" limit;
- in plasma thermodynamics, the limit of infinite volume ($V \to \infty$) and the nonrelativistic limit ($c \to \infty$) are incompatible with the thermodynamic limit, since a characteristic time of return to equilibrium is $V^{1/3}/c$;
- in the classical theory of the electron, it is often reproached that such a classical electron, with zero naked mass, rotates too fast at the level of the equator (200 times the speed of light) for its magnetic moment and renormalized mass to conform to experimental data. A more careful calculation by Lorentz himself[4] gave about 10 times the speed of light at the equateur. But in fact, the limit $[m \to 0^+]$ requires care, and if done correctly, it *imposes* a limit $[v/c \to 1^-]$ to maintain a constant renormalized mass [7];
- another interesting example is an "infinite universe" limit and a "diluted universe" limit [56].

1.2 Sequences

1.2.a Sequences in a normed vector space

We consider in this section a normed vector space $(E, \|\cdot\|)$ and sequences of elements of E.[5]

---

[3] This is the principle of Hilbert's infinite hotel.

[4] Pointed out by Sin-Itiro Tomonaga [91].

[5] We recall the basic definitions concerning normed vector spaces in Appendix A.

DEFINITION 1.5 (Convergence of a sequence) Let $(E, \|\cdot\|)$ be a normed vector space and $(u_n)_{n\in\mathbb{N}}$ a sequence of elements of E, and let $\ell \in E$. The sequence $(u_n)_{n\in\mathbb{N}}$ **converges to ℓ** if, for any $\varepsilon > 0$, there exists an index starting from which u_n is at most at distance ε from ℓ:

$$\forall \varepsilon > 0 \quad \exists N \in \mathbb{N} \quad \forall n \in \mathbb{N} \qquad n \geqslant N \Longrightarrow \|u_n - \ell\| < \varepsilon.$$

Then ℓ is called the **limit of the sequence $(u_n)_{n\in\mathbb{N}}$**, and this is denoted

$$\ell = \lim_{n\to\infty} u_n \qquad \text{or} \qquad u_n \xrightarrow[n\to\infty]{} \ell.$$

DEFINITION 1.6 A sequence $(u_n)_{n\in\mathbb{N}}$ of real numbers **converges to $+\infty$** (resp. **to $-\infty$**) if, for any $M \in \mathbb{R}$, there exists an index N, starting from which all elements of the sequence are larger than M:

$$\forall M \in \mathbb{R} \quad \exists N \in \mathbb{N} \quad \forall n \in \mathbb{N} \qquad n \geqslant N \Longrightarrow u_n > M \qquad (\text{resp. } u_n < M).$$

In the case of a complex-valued sequence, a type of convergence to infinity, in modulus, still exists:

DEFINITION 1.7 A sequence $(z_n)_{n\in\mathbb{N}}$ of complex numbers **converges to infinity** if, for any $M \in \mathbb{R}$, there exists an index N, starting from which all elements of the sequence have modulus larger than M:

$$\forall M \in \mathbb{R} \quad \exists N \in \mathbb{N} \quad \forall n \in \mathbb{N} \qquad n \geqslant N \Longrightarrow |z_n| > M.$$

Remark 1.8 There is only "one direction to infinity" in $\mathbb{C}$. We will see a geometric interpretation of this fact in Section 5.4 on page 146.

Remark 1.9 The strict inequalities $\|u_n - \ell\| < \varepsilon$ (or $|u_n| > M$) in the definitions above (which, in more abstract language, amount to an emphasis on *open* subsets) may be replaced by $\|u_n - \ell\| \leqslant \varepsilon$, which are sometimes easier to handle. Because $\varepsilon > 0$ is arbitrary, this gives an equivalent definition of convergence.

1.2.b Cauchy sequences

It is often important to show that a sequence converges, without explicitly knowing the limit. Since the definition of convergence depends on the limit ℓ, it is not conveninent for this purpose.[6] In this case, the most common tool is the *Cauchy criterion*, which depends on the convergence "of elements of the sequence with respect to each other":

---

[6] As an exemple: how should one prove that the sequence $(u_n)_{n\in\mathbb{N}}$, with

$$u_n = \sum_{p=1}^{n} \frac{(-1)^{p+1}}{p^4},$$

converges? Probably not by guessing that the limit is $7\pi^4/720$.

DEFINITION 1.10 (Cauchy criterion) A sequence $(u_n)_{n\in\mathbb{N}}$ in a normed vector space **is a Cauchy sequence**, or **satisfies the Cauchy criterion**, if

$$\forall \varepsilon > 0 \quad \exists N \in \mathbb{N} \quad \forall p, q \in \mathbb{N} \qquad q > p \geqslant N \Longrightarrow \left\| u_p - u_q \right\| < \varepsilon$$

or, equivalently, if

$$\forall \varepsilon > 0 \quad \exists N \in \mathbb{N} \quad \forall p, k \in \mathbb{N} \qquad p \geqslant N \Longrightarrow \left\| u_{p+k} - u_p \right\| < \varepsilon.$$

A common technique used to prove that a sequence $(u_n)_{n\in\mathbb{N}}$ is a Cauchy sequence is therefore to find a sequence $(\alpha_p)_{p\in\mathbb{N}}$ of real numbers such that

$$\lim_{p\to\infty} \alpha_p = 0 \qquad \text{and} \qquad \forall p, k \in \mathbb{N} \quad \left\| u_{p+k} - u_p \right\| \leqslant \alpha_p.$$

PROPOSITION 1.11 *Any convergent sequence is a Cauchy sequence.*

This is a trivial consequence of the definitions. But we are of course interested in the converse. Starting from the Caucy criterion, we want to be able to conclude that a sequence converges – without, in particular, requiring the limit to be known beforehand. However, that is not always possible: there exist normed vector spaces E and Cauchy sequences in E which do not converge.

Example 1.12 Consider the set of rational numbers $\mathbb{Q}$. With the absolute value, it is a normed $\mathbb{Q}$-vector space. Consider then the sequence

$$u_0 = 3 \quad u_1 = 3.1 \quad u_2 = 3.14 \quad u_3 = 3.141 \quad u_4 = 3.1415 \quad u_5 = 3.14159\cdots$$

(you can guess the rest[7]...). This is a sequence of rationals, which is a Cauchy sequence (the distance between u_p and u_{p+k} is at most 10^{-p}). However, it does not converge in $\mathbb{Q}$, since its limit (in $\mathbb{R}$!) is π, which is a notoriously irrational number.

The space $\mathbb{Q}$ is not "nice" in the sense that it leaves a lot of room for Cauchy sequences to exist without converging *in* $\mathbb{Q}$. The mathematical terminology is that $\mathbb{Q}$ *is not complete.*

DEFINITION 1.13 (Complete vector space) A normed vector space $(E, \|\cdot\|)$ is **complete** if all Cauchy sequences in E are convergent.

THEOREM 1.14 *The spaces $\mathbb{R}$ and $\mathbb{C}$, and more generally all finite-dimensional real or complete normed vector spaces, are complete.*

PROOF

First case: It is first very simple to show that a Cauchy sequence $(u_n)_{n\in\mathbb{N}}$ of real numbers is bounded. Hence, according to the Bolzano-Weierstrass theorem (Theorem A.41, page 581), it has a convergent subsequence. But any Cauchy sequence which has a convergent subsequence is itself convergent (its limit being that of the subsequence), see Exercise 1.6 on page 43. Hence any Cauchy sequence in $\mathbb{R}$ is convergent.

Second case: Considering $\mathbb{C}$ as a normed real vector space of dimension 2, we can suppose that the base field is $\mathbb{R}$.

---

[7] This is simply $u_n = 10^{-n} \cdot \lfloor 10^n \pi \rfloor$ where $\lfloor \cdot \rfloor$ is the integral part function.

Consider a basis $\mathcal{B} = (b_1, \dots, b_d)$ of the vector space E. Then we deduce that E is complete from the case of the real numbers and the following two facts: (1) a sequence $(\boldsymbol{x}_n)_{n\in\mathbb{N}}$ of vectors, with coordinates $(x_n^1, \dots, x_n^d)$ in $\mathcal{B}$, converges in E if and only if each coordinate sequence $(x^k{}_n)_{n\in\mathbb{N}}$; and (2), if a sequence is a Cauchy sequence, then each coordinate is a Cauchy sequence.

Both facts can be checked immediately when the norm of E is defined by $\|\boldsymbol{x}\| = \max|x^k|$, and other norms reduce to this case since all norms are equivalent on E.

Example 1.15 The space L^2 of square integrable functions (in the sense of Lebesgue), with the norm $\|f\|_{L^2}^2 \stackrel{\text{def}}{=} \int_{\mathbb{R}} |f|^2$, is complete (see Chapter 9). This infinite-dimensional space is used very frequently in quantum mechanics.

Counterexample 1.16 Let $E = \mathbb{K}[X]$ be the space of polynomials with coefficients in K (and arbitrary degree). Let $P \in E$ be a polynomial, written as $P = \sum \alpha_n X^n$, and define its norm by $\|P\| \stackrel{\text{def}}{=} \max_{i\in\mathbb{N}} |\alpha_i|$. Then the normed vector space $(E, \|\cdot\|)$ is not complete (see Exercise 1.7 on page 43).

Here is an important example of the use of the Cauchy criterion: the fixed point theorem.

1.2.c The fixed point theorem

We are looking for solutions to an equation of the type

$$f(x) = x,$$

where $f : E \to E$ is an application defined on a normed vector space E, with values in E. Any element of E that satisfies this equation is called a **fixed point** of f.

DEFINITION 1.17 (Contraction) Let E be a normed vector space, U a subset of E. A map $f : U \to E$ is a **contraction** if there exists a real number $\rho \in [0, 1[$ such that $\|f(y) - f(x)\| \leqslant \rho \|y - x\|$ for all $x, y \in U$. In particular, f is continuous on U.

THEOREM 1.18 (Banach fixed point theorem) *Let E be a complete normed vector space, U a non-empty closed subset of E, and $f : U \to U$ a contraction. Then f has a unique fixed point.*

PROOF. Chose an arbitrary $u_0 \in U$, and define a sequence $(u_n)_{n\in\mathbb{N}}$ for $n \geqslant 0$ by induction by $u_{n+1} = f(u_n)$. Using the definition of contraction, an easy induction shows that we have

$$\|u_{p+1} - u_p\| \leqslant \rho^p \|u_1 - u_0\|$$

for any $p \geqslant 0$. Then a second induction on $k \geqslant 0$ shows that for all $p, k \in \mathbb{N}$, we have

$$\|u_{p+k} - u_p\| \leqslant (\rho^p + \cdots + \rho^{p+k-1}) \cdot \|u_1 - u_0\| \leqslant \frac{\rho^p}{1-\rho} \cdot \|u_1 - u_0\|,$$

and this proves that the sequence $(u_n)_{n\in\mathbb{N}}$ is a Cauchy sequence. Since the space E is complete, this sequences has a limit $a \in E$. Since U is closed, we have $a \in U$.

Now from the continuity of f and the relation $u_{n+1} = f(u_n)$, we deduce that $a = f(a)$. So this a is a fixed point of f. If b is an arbitrary fixed point, the inequality $\|a-b\| = \|f(a) - f(b)\| \leqslant \rho \|a-b\|$ proves that $\|a-b\| = 0$ and thus $a = b$, showing that the fixed point is unique.

Remark 1.19 Here is one reason why Banach's theorem is very important. Suppose we have a normed vector space E and a map $g : E \to E$, and we would like to solve an equation $g(x) = b$. This amounts to finding the fixed points of $f(x) = g(x) + x - b$, and we can hope that f may be a contraction, at least locally. This happens, for instance, in the case of the Newton method, if the function used is nice enough, and if a suitable (rough) approximation of a zero is known.

This is an extremely fruitful idea: one can prove this way the Cauchy-Lipschitz theorem concerning existence and unicity of solutions to a large class of differential equations; one can also study the existence of certain fractal sets (the von Koch snowflake, for instance), certain stability problems in dynamical systems, etc.

Not only does it follow from Banach's theorem that certain equations have solutions (and even better, unique solutions!), but the proof provides an *effective* way to find this solution by a successive approximations: it suffices to fix u_0 arbitrarily, and to define the sequence $(u_n)_{n\in\mathbb{N}}$ by means of the recurrence formula $u_{n+1} = f(u_n)$; then we know that this sequence converges to the fixed point a of f. Moreover, the convergence of the sequence of approximations is exponentially fast: the distance from the approximate solution u_n to the (unknown) solution a decays as fast as ρ^n. An example (the *Picard iteration*) is given in detail in Problem 1 on page 46.

1.2.d Double sequences

Let $(x_{n,k})_{(n,k)\in\mathbb{N}^2}$ be a double-indexed sequence of elements in a normed vector space E. We assume that the sequences made up of each row and each column converge, with limits as follows:

$$\begin{array}{cccccc} x_{11} & x_{12} & x_{13} & \cdots & \longrightarrow & A_1 \\ x_{21} & x_{22} & x_{23} & \cdots & \longrightarrow & A_2 \\ x_{31} & x_{32} & x_{33} & \cdots & \longrightarrow & A_3 \\ \vdots & \vdots & \vdots & & & \\ \downarrow & \downarrow & \downarrow & & & \\ B_1 & B_2 & B_3 & & & \end{array}$$

The question is now whether the sequences $(A_n)_{n\in\mathbb{N}}$ and $(B_k)_{k\in\mathbb{N}}$ themselves converge, and if that is the case, whether their limits are equal. In general, it turns out that the answer is "No." However, under certain conditions, if one sequence (say (A_n)) converges, then so does the other, and the limits are the same.

DEFINITION 1.20 A double sequence $(x_{n,k})_{n,k}$ **converges uniformly with respect to k to a sequence $(B_k)_{k\in\mathbb{N}}$ as $n \to \infty$** if

$$\forall \varepsilon > 0 \quad \exists N \in \mathbb{N} \quad \forall n \in \mathbb{N} \quad \forall k \in \mathbb{N} \qquad n \geqslant N \Longrightarrow \left|x_{n,k} - B_k\right| < \varepsilon.$$

In other words, there is convergence with respect to n for fixed k, but in such a way that the speed of convergence is *independent* of k; or one might say that "all values of k are similarly behaved."

Uniform convergence with respect to n toward the sequence $(A_n)_{n\in\mathbb{N}}$ is similarly defined.

THEOREM 1.21 (Double limit) *With notation as above, if the following three conditions hold:*

① *each row converges, and* $A_n = \lim_{k\to\infty} x_{n,k}$ *for all* $n \in \mathbb{N}$,

② *each column converges, and* $B_k = \lim_{n\to\infty} x_{n,k}$ *for all* $k \in \mathbb{N}$,

③ *the convergence is* ***uniform*** *either with respect to n or with respect to k;*

then the sequences $(A_n)_{n\in\mathbb{N}}$ *and* $(B_k)_{k\in\mathbb{N}}$ *converge,* $\lim_{n\to\infty} A_n = \lim_{k\to\infty} B_k = \ell$. *One says that* ***the double sequence*** $(x_{n,k})_{n,k}$ ***converges to the limit*** ℓ.

Be aware that the uniform convergence condition ③ is very important. The following examples gives an illustration: here both limits exist, but they are different.

$$\begin{array}{ccccccc} 1 & 0 & 0 & 0 & \cdots & \to & 0 \\ 1 & 1 & 0 & 0 & \cdots & \to & 0 \\ 1 & 1 & 1 & 0 & \cdots & \to & 0 \\ 1 & 1 & 1 & 1 & \cdots & \to & 0 \\ \vdots & \vdots & \vdots & \vdots & & \vdots & \\ \downarrow & \downarrow & \downarrow & \downarrow & & & \\ 1 & 1 & 1 & 1 & \cdots & & \end{array}$$

1.2.e Sequential definition of the limit of a function

DEFINITION 1.22 Let $f : \mathbb{K} \to \mathbb{K}'$ (where $\mathbb{K}, \mathbb{K}' = \mathbb{R}$ or $\mathbb{C}$ or any normed vector space), let $a \in \mathbb{K}$, and let $\ell \in \mathbb{K}'$. Then f **has the limit** ℓ**, or tends to** ℓ**, at the point** a if we have

$$\forall \varepsilon > 0 \quad \exists \eta > 0 \quad \forall z \in \mathbb{K} \qquad |z - a| < \eta \Longrightarrow \left|f(z) - \ell\right| < \varepsilon.$$

There are also limits at infinity and infinite limits, defined similarly:

DEFINITION 1.23 Let $f : \mathbb{K} \to \mathbb{K}'$ (where $\mathbb{K}, \mathbb{K}' = \mathbb{R}$ or $\mathbb{C}$). Let $\ell \in \mathbb{K}'$. Then f **tends to** ℓ **at** $+\infty$, resp. **at** $-\infty$ (in the case $\mathbb{K} = \mathbb{R}$), resp. **at infinity** (in the case $\mathbb{K} = \mathbb{C}$), if we have

$$\begin{aligned} &\forall \varepsilon > 0 \quad \exists A \in \mathbb{R} \quad \forall x \in \mathbb{R} \qquad x > A \Longrightarrow \left|f(x) - \ell\right| < \varepsilon, \\ \big[\text{resp.}\ &\forall \varepsilon > 0 \quad \exists A' \in \mathbb{R} \quad \forall x \in \mathbb{R} \qquad x < A' \Longrightarrow \left|f(x) - \ell\right| < \varepsilon \\ \text{resp.}\ &\forall \varepsilon > 0 \quad \exists A \in \mathbb{R}^+ \quad \forall z \in \mathbb{C} \qquad |z| > A \Longrightarrow \left|f(z) - \ell\right| < \varepsilon\big]. \end{aligned}$$

Similarly, a function $f : \mathbb{R} \to \mathbb{R}$ **tends to $+\infty$ at $+\infty$** if

$$\forall M > 0 \quad \exists A \in \mathbb{R} \quad \forall x \in \mathbb{R} \qquad x > A \Longrightarrow f(x) > M,$$

and finally a function $f : \mathbb{C} \to \mathbb{C}$ **tends to infinity at infinity** if

$$\forall M > 0 \quad \exists A \in \mathbb{R} \quad \forall z \in \mathbb{C} \qquad |z| > A \Longrightarrow \left|f(z)\right| > M.$$

In some cases, the definition of limit is refined by introducing a *punctured neighborhood*, i.e., looking at the values at points other than the point where the limit is considered:

DEFINITION 1.24 Let $f : \mathbb{K} - \{a\} \to \mathbb{K}'$ (with $\mathbb{K}, \mathbb{K}' = \mathbb{R}$ or $\mathbb{C}$ and $a \in \mathbb{K}$) and let $\ell \in \mathbb{K}'$. Then **$f(x)$ converges to ℓ in punctured neighborhoods of a** if

$$\forall \varepsilon > 0 \quad \exists \eta \in \mathbb{R} \quad \forall z \in \mathbb{K}$$
$$(z \neq a \text{ and } |z - a| < \eta) \Longrightarrow \left(\left|f(z) - \ell\right| < \varepsilon\right).$$

This is denoted

$$\ell = \lim_{\substack{z \to a \\ z \neq a}} f(z).$$

This definition has the advantage of being practically identical to the definition of convergence at infinity. It is often better adapted to the physical description of a problem, as seen in Examples 1.1.b and 1.1.c on page 5 and the following pages. A complication is that it reduces the applicability of the theorem of composition of limits.

THEOREM 1.25 (Sequential characterization of limits) *Let $f : \mathbb{K} \to \mathbb{K}'$ be a function, and let $a \in \overline{\mathbb{K}}$ and $\ell \in \overline{\mathbb{K}'}$. Then $f(x)$ converges to ℓ as x tends to a if and only if, for any convergent sequence $(x_n)_{n \in \mathbb{N}}$ with limit equal to a, the sequence $\big(f(x_n)\big)_{n \in \mathbb{N}}$ converges to ℓ.*

1.2.f Sequences of functions

Consider now the case of a sequence of functions $(f_n)_{n \in \mathbb{N}}$ each defined on a same subset X of $\mathbb{R}$ or $\mathbb{C}$ and taking values in $\mathbb{R}$ or $\mathbb{C}$. Denote by $\|\cdot\|_\infty$ the "supremum norm"[8]:

$$\|f\|_\infty = \sup_{x \in X} \left|f(x)\right|.$$

[8] In fact, it is not a norm on the space of all functions, but only on the subspace of bounded functions. We disregard this subtlety and consider here that $\|\cdot\|$ takes values in $\overline{\mathbb{R}^+} = \mathbb{R} \cup \{+\infty\}$.

The definition of convergence of real or complex sequences may be extended to functions in two ways: one is a local notion, called *simple*, or *pointwise*, *convergence*, and the other, more global, is *uniform convergence*.[9]

DEFINITION 1.26 (Simple convergence) Let $(f_n)_{n\in\mathbb{N}}$ be a sequence of functions, all defined on the same set X, which may be arbitrary. Then **the sequence $(f_n)_{n\in\mathbb{N}}$ converges simply** (or: **pointwise on X**) **to a function f** defined on X if, for any x in X, the sequence $\big(f_n(x)\big)_{n\in\mathbb{N}}$ converges to $f(x)$. This is denoted

$$f_n \xrightarrow{\text{cv.s.}} f.$$

DEFINITION 1.27 (Uniform convergence) Let $(f_n)_{n\in\mathbb{N}}$ be a sequence of functions, all defined on the same set X, which may be arbitrary. Then **the sequence $(f_n)_{n\in\mathbb{N}}$ converges uniformly to the function f** if

$$\forall \varepsilon > 0 \quad \exists N \in \mathbb{N} \quad \forall n \in \mathbb{N} \qquad n \geqslant N \Longrightarrow \|f_n - f\|_\infty < \varepsilon.$$

This is denoted $$f_n \xrightarrow{\text{cv.u.}} f.$$

In other words, in the case where X is a subset of $\mathbb{R}$, the graph of the function f is located inside a smaller and smaller band of constant width in which all the graphs of f_n must also be contained if n is large enough:

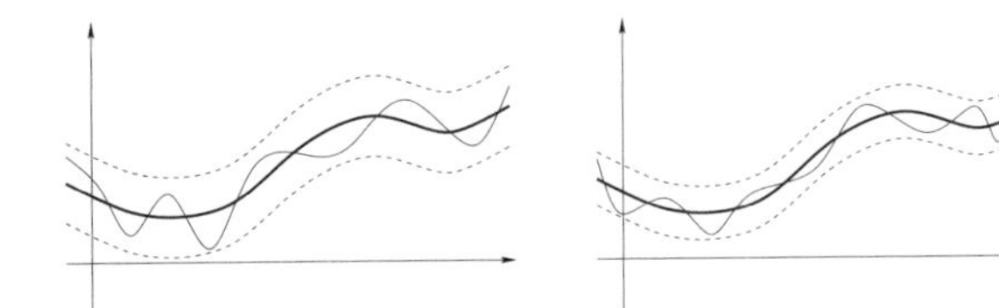
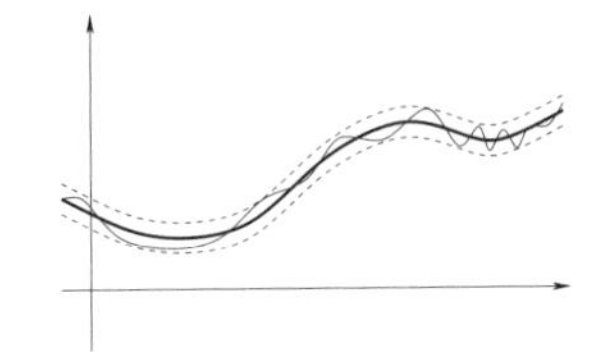

If we have functions $f_n : X \to E$, where $(E, \|\cdot\|)$ is a normed vector space, we similarly define pointwise and uniform convergence using convergence in E; for instance, $(f_n)_{n\in\mathbb{N}}$ converges uniformly to f if

$$\forall \varepsilon > 0 \quad \exists N \in \mathbb{N} \quad \forall n \in \mathbb{N} \qquad n \geqslant N \Longrightarrow \sup_{x\in X} \big\|f_n(x) - f(x)\big\| < \varepsilon.$$

Remark 1.28 Uniform convergence is an important theoretical mathematical notion. If one wishes to compute numerically a function f using successive approximations f_n (for instance, partial sums of a series expansion), then to get an error of size at most ε for the value $f(x_1)$ of f at some given point x_1, it suffices to find N_1 such that $\big|f_n(x_1) - f(x_1)\big| \leqslant \varepsilon$ for any $n \geqslant N_1$. Similarly, if the value of f at other points $x_2, x_3, \dots, x_p$ is needed, it will be enough to find corresponding integers $N_2, N_3, \dots, N_p$. However, if it is not known beforehand at which points the function will evaluated, it will be necessary to know an integer N such that $\big|f_n(x) - f(x)\big| \leqslant \varepsilon$ for all $n \geqslant N$ *and* for all $x \in \mathbb{R}$. Uniform convergence is then desirable.

[9] The concept of uniform convergence is due to George Stokes (see page 472) and Philipp Seidel (1821–1896), independently.

It is clear that uniform convergence implies pointwise convergence, but the converse is not true.

Example 1.29 Define a sequence $(f_n)_{n\geqslant 1}$ a functions on $\mathbb{R}$ by

$$f_n(x) = \begin{cases} nx & \text{if } x \in [0, 1/n], \\ 2 - nx & \text{if } x \in [1/n, 2/n], \\ 0 & \text{if } x \in [2/n, 1]. \end{cases}$$

The reader will have no trouble proving that $(f_n)_{n\geqslant 1}$ converges pointwise to the zero function. However, the convergence is not uniform, since we have $\|f_n - f\|_\infty = 1$ for all $n \geqslant 1$.

Example 1.30 The sequence of functions $f_n : \mathbb{R} \to \mathbb{R}$ defined for $n \geqslant 1$ by

$$f_n : x \longmapsto \sin\left(x + \frac{x}{n}\right)$$

converges uniformly to $f : x \mapsto \sin x$ on the interval $[0, 2\pi]$, and in particular it converges pointwise on this interval. However, although the sequence converges pointwise to the sine function on $\mathbb{R}$, *the convergence is not uniform* on all of $\mathbb{R}$. Indeed, for $n \geqslant 1$, we have

$$f_n\left(\frac{n\pi}{2}\right) = \sin\left(\frac{n\pi}{2} + \frac{\pi}{2}\right) \qquad \text{and} \qquad f\left(\frac{n\pi}{2}\right) = \sin\left(\frac{n\pi}{2}\right),$$

and those two values differ by 1 in absolute value. However, one can check that the convergence is uniform on any bounded segment in $\mathbb{R}$.

♦ **Exercise 1.1** Let $g(x) = \mathrm{e}^{-x^2}$ and $f_n(x) = g(x - n)$. Does the sequence $(f_n)_{n\in\mathbb{N}}$ converge pointwise on $\mathbb{R}$? Does it converge uniformly?

Remark 1.31 In the case where the functions f_n are defined on a subset of $\mathbb{R}$ with finite measure (for instance, a finite segment), a theorem of Egorov shows that pointwise convergence implies uniform convergence except on a set of arbitrarily small measure (for the definitions, see Chapter 2).

Remark 1.32 There are other ways of defining the convergence of a sequence of functions. In particular, when some norm is defined on a function space containing the functions f_n, it is possible do discuss convergence in the sense of this norm. Uniform convergence corresponds to the case of the $\|\cdot\|_\infty$ norm. In Chapter 9, we will also discuss the notion of *convergence in quadratic mean*, or convergence in L^2 norm, and *convergence in mean* or convergence in L^1 norm. In pre-Hilbert spaces, there also exists a *weak convergence*, or convergence in the sense of scalar product (which is not defined by a norm if the space is infinite-dimensional).

A major weakness of pointwise convergence is that it does not preserve continuity (see Exercise 1.10 on page 44), or limits in general (Exercise 1.12). Uniform convergence, on the other hand, does preserve those notions.

THEOREM 1.33 (Continuity of a limit) *Let $(f_n)_{n\in\mathbb{N}}$ be a sequence of functions defined on a subset D in $\mathbb{K}$ (or in a normed vector space), with values in an arbitrary normed vector space. Assume that the sequence $(f_n)_{n\in\mathbb{N}}$ converges uniformly to a function f.*

i) Let $a \in D$ be such that all functions f_n are continuous at the point a. Then f is also continuous at the point a.

ii) In particular, if each f_n is continuous on D, then the limit function f is also continuous on D.

This property extends to the case where a is not in D, but is a limit point of D. However, it is then necessary to reinforce the hypothesis to assume that the functions f_n have values in a complete normed vector space.

THEOREM 1.34 (Double limit) *Let D be a subset of $\mathbb{R}$ (or of a normed vector space) and let $x_0 \in \overline{D}$ be a limit point[10] of D. Let $(f_n)_{n\in\mathbb{N}}$ be a sequence of functions defined on D with values in a* complete *normed vector space E. Assume that, for all n, the function f_n has a limit as x tends to x_0. Denote $\ell_n = \lim_{x\to x_0} f_n(x)$.*

If $(f_n)_{n\in\mathbb{N}}$ converges uniformly to a function f, then

i) $f(x)$ has a limit as $x \to x_0$;

ii) $(\ell_n)_{n\in\mathbb{N}}$ has a limit as $n \to \infty$;

iii) the two limits are equal: $\lim_{x\to x_0} f(x) = \lim_{n\to\infty} \ell_n$.

In other words:
$$\lim_{x\to x_0} \lim_{n\to\infty} f_n(x) = \lim_{n\to\infty} \lim_{x\to x_0} f_n(x).$$

If we want a limit of differentiable functions to remain differentiable, stronger assumptions are needed:

THEOREM 1.35 (Differentiation of a sequence of functions) *Let I be an interval of $\mathbb{R}$ with non-empty interior, and let $(f_n)_{n\in\mathbb{N}}$ be a sequence of functions defined on I with values in $\mathbb{R}$, $\mathbb{C}$, or a normed vector space. Assume that the functions f_n are differentiable on I, and moreover that:*

i) the sequence $(f_n)_{n\in\mathbb{N}}$ converges pointwise to a function f;

ii) the sequence $(f'_n)_{n\in\mathbb{N}}$ converges uniformly to a function g.

Then f is differentiable on I and $f' = g$.

Remark 1.36 If the functions take values in $\mathbb{R}$ or $\mathbb{C}$ or more generally any *complete* normed vector space, it is possible to weaken the assumptions by asking, instead of (i), that the sequence $(f_n(x_0))$ converges at a single point $x_0 \in I$. Assumption (ii) remains identical, and the conclusion is the same: $(f_n)_{n\in\mathbb{N}}$ converges uniformly to a differentiable function with derivative equal to g.

Counterexample 1.37 The sequence of functions given by
$$f_n : x \longmapsto \frac{8}{\pi} \sum_{k=1}^{n} \frac{\sin^2(nx)}{4n^2 - 1}$$

---

[10] See Definition 4.52 on page 106; the simplest example is $D =]a, b]$ and $x_0 = a$.

converges uniformly to the function $f : x \mapsto |\sin x|$ (see Exercise 9.3 on page 270), but the sequence of derivatives does not converge uniformly. The previous theorem does not apply, and indeed, although each f_n is differentiable at 0, the limit f is not.

Remark 1.38 It happens naturally in some physical situations that a limit of a sequence of functions is not differentiable. In particular, in statistical thermodynamics, the state functions of a finite system are smooth. However, as the number of particles grows to infinity, discontinuities in the state functions or their derivatives may appear, leading to phase transitions.

Uniform convergence is also useful in another situation: when trying to exchange a limit (or a sum) and an integration process. However, in that situation, pointwise convergence is often sufficient, using the powerful tools of Lebesgue integration (see Chapter 2).

THEOREM 1.39 (Integration on a finite interval) *Let $(f_n)_{n\in\mathbb{N}}$ be a sequence of integrable functions (for instance, continuous functions), which converges uniformly to a function f on a finite closed interval $[a,b] \subset \mathbb{R}$. Then f is integrable on $[a,b]$ and we have*

$$\lim_{n\to\infty} \int_a^b f_n(x)\,\mathrm{d}x = \int_a^b f(x)\,\mathrm{d}x.$$

Example 1.40 This theorem is very useful, for instance, when dealing with a power series expansion which is known to converge uniformly on the open disc of convergence (see Theorem 1.66 on page 34). So, if we have $f(x) = \sum_{n=0}^{\infty} a_n x^n$ for $|x| < R$, then for any x such that $|x| < R$, we deduce that

$$\int_0^x f(s)\,\mathrm{d}s = \sum_{n=0}^{\infty} \frac{a_n}{n+1}\, x^{n+1}.$$

To establish that a sequence converges uniformly, in practice, it is necessary to compute $\|f_n - f\|_\infty$, or rather to bound this expression by a quantity which itself converges to 0. This is sometimes quite tricky, and it is therefore useful to know the following two results of Dini:[11]

THEOREM 1.41 (Dini) *Let K be a compact subset of $\mathbb{R}^k$, for instance, a closed ball. Let $(f_n)_{n\in\mathbb{N}}$ be an increasing sequence of continuous functions converging pointwise on K to a continuous function f. Then the sequence $(f_n)_{n\in\mathbb{N}}$ converges uniformly to f on K.*

THEOREM 1.42 (Dini) *Let $I = [a,b]$ be a compact interval in $\mathbb{R}$, and let $(f_n)_{n\in\mathbb{N}}$ be a sequence of increasing functions from I to $\mathbb{R}$ that converges pointwise on I to a continous function f. Then $(f_n)_{n\in\mathbb{N}}$ converges uniformly on I.*

---

[11] Ulisse DINI (1845–1918) studied in Pisa and Paris before taking a position in Pisa. His work concerned the theory of functions of a real variable, and he contributed to the early development of functional analysis.

Remark 1.43 Be careful to distinguish between an *increasing sequence* of functions and a sequence of *increasing functions*. The former is a sequence $(f_n)_{n\in\mathbb{N}}$ of real-valued functions such that $f_{n+1}(x) \geqslant f_n(x)$ for any $x \in \mathbb{R}$ and $n \in \mathbb{N}$. The latter is a sequence of real-valued functions defined on $K \subset \mathbb{R}$ such that for any $x, y \in K$: $x \leqslant y \Longrightarrow f_n(x) \leqslant f_n(y)$.

As mentioned briefly already, it is possible with Lebesgue's dominated convergence theorem to avoid requiring uniform convergence to exchange an integral and a limit, as in Theorem 1.39. See Chapters 2 and 3 for details on this theory.

1.3 Series

1.3.a Series in a normed vector space

We first recall the definition of convergence and absolute convergence of a series in a normed vector space.

DEFINITION 1.44 (Convergence of a series) Let $(a_n)_n$ be a sequence with values in a normed vector space. Let $(S_n)_{n\in\mathbb{N}}$ denote the **sequence of partial sums**

$$S_n \stackrel{\text{def}}{=} \sum_{k=0}^{n} a_k.$$

- The **series $\sum a_n$ converges, and its sum is equal to A** if the sequence $(S_n)_{n\in\mathbb{N}}$ converges to A. This is denoted

$$\sum_{n=0}^{\infty} a_n = A.$$

- The **series $\sum a_n$ converges absolutely** if the series $\sum \|a_n\|$ converges in $\mathbb{R}$.
- In particular, a series $\sum a_n$ of real or complex numbers converges absolutely if the series $\sum |a_n|$ converges in $\mathbb{R}$.

As in the case of sequences, there exists a Cauchy criterion for convergence of series[12]:

[12] This criterion was stated by Bernhard Bolzano (see page 581) in 1817. But Bolzano was isolated in Prague and little read. Cauchy presented this criterion, without proof and as an obvious fact, in his analysis course in 1821.

THEOREM 1.45 (Cauchy criterion) *If a series $\sum u_n$ with values in a normed vector space E converges, then it satisfies the Cauchy criterion:*

$$\forall \varepsilon > 0 \quad \exists N \in \mathbb{N} \quad \forall p, q \in \mathbb{N} \qquad (q > p \geqslant N) \Longrightarrow \left\| \sum_{n=p}^{q} u_n \right\| < \varepsilon,$$

or in other words:

$$\lim_{p,q \to \infty} \sum_{p \leqslant n \leqslant q} u_n = 0.$$

Conversely, any series which satisfies the Cauchy criterion and takes values in $\mathbb{R}$, $\mathbb{C}$, any finite-dimensional normed vector space, or more generally, any complete normed vector space, converges.

From this the following fundamental theorem is easily deduced:

THEOREM 1.46 *Any absolute convergent series $\sum a_n$ with values in a* complete *normed vector space is convergent.*

In particular, any absolutely convergent series of real or complex numbers is convergent.

PROOF. Let $\sum u_n$ be an absolutely convergent series. Although we can write

$$\left\| \sum_{n=0}^{k} u_n \right\| \leqslant \sum_{n=0}^{k} \|u_n\|,$$

nothing can be deduced from this, because the right-hand side does not tend to zero. But we can use the Cauchy critetion: for all p, $q \in \mathbb{N}$, we have of course

$$\left\| \sum_{n=p}^{q} u_n \right\| \leqslant \sum_{n=p}^{q} \|u_n\|,$$

and since $\sum \|u_n\|$ satisfies the Cauchy criterion, so does $\sum u_n$. Since u_n lies in a complete space by assumption, this means that the series $\sum u_n$ is indeed convergent.

1.3.b Doubly infinite series

In the theory of Fourier series, we will have to deal with formulas of the type

$$\int_0^1 |f(t)|^2 \, \mathrm{d}t = \sum_{n=-\infty}^{+\infty} |c_n|^2.$$

To give a precise meaning to the right-hand side, we must clarify the meaning of the convergence of a series indexed by integers in $\mathbb{Z}$ instead of $\mathbb{N}$.

DEFINITION 1.47 A **doubly infinite series** $\sum_{n \in \mathbb{Z}} a_n$, with a_n is a normed vector space, **converges** if $\sum a_n$ and $\sum a_{-n}$ are both convergent, the index ranging over $\mathbb{N}$ in each case. Then we denote

$$\sum_{n=-\infty}^{+\infty} a_n \stackrel{\text{def}}{=} \sum_{n=0}^{\infty} a_n + \sum_{n=1}^{\infty} a_{-n}$$

and say that this is the sum of $\sum_{n\in\mathbb{Z}} a_n$.

In other words, a series of complex numbers $\sum_{n\in\mathbb{Z}} a_n$ converges to ℓ if and only if, for any $\varepsilon > 0$, there exists $N > 0$ such that

$$\text{for any } i \geqslant N \text{ and } j \geqslant N, \qquad \left| \left(\sum_{n=-i}^{j} a_n \right) - \ell \right| < \varepsilon.$$

Remark 1.48 It is crucial to allow the upper and lower bounds i and j to be independent. In particular, if the limit of $\sum_{n=-k}^{k} a_n$ exists, it does not follow that the doubly infinite series $\sum_{n\in\mathbb{Z}} a_n$ converges.

For instance, take $a_n = 1/n$ for $n \neq 0$ and $a_0 = 0$. Then we have $\sum_{n=-k}^{k} a_n = 0$ for all k (and so this sequence does converge as k tends to infinity), but the series $\sum_{n\in\mathbb{Z}} a_n$ *diverges* according to the definition, because each of the series $\sum a_n$ and $\sum a_{-n}$ (over $n \geqslant 0$) is divergent.

1.3.c Convergence of a double series

As in Section 1.2.d, let $(a_{ij})_{i,j\in\mathbb{N}}$ be a family of real numbers indexed by two integers. For any $p, q \in \mathbb{N}$, we have

$$\sum_{i=1}^{p} \sum_{j=1}^{q} a_{ij} = \sum_{j=1}^{q} \sum_{i=1}^{p} a_{ij},$$

since each sum is finite. On the other hand, even if all series involved are convergent, it is not always the case that

$$\sum_{i=1}^{\infty} \sum_{j=1}^{\infty} a_{ij} \qquad \text{and} \qquad \sum_{j=1}^{\infty} \sum_{i=1}^{\infty} a_{ij},$$

as the following example shows:

$$(a_{ij}) = \begin{pmatrix} 1 & -1 & 0 & 0 & 0 & 0 & \dots \\ 0 & 1 & -1 & 0 & 0 & 0 & \dots \\ 0 & 0 & 1 & -1 & 0 & 0 & \dots \\ \vdots & \vdots & \ddots & \ddots & \ddots & \ddots & \end{pmatrix}$$

where we have (note that i is the row index and j is the column index):

$$\sum_{i=1}^{\infty} \sum_{j=1}^{\infty} a_{ij} = 0 \qquad \text{but} \qquad \sum_{j=1}^{\infty} \sum_{i=1}^{\infty} a_{ij} = 1.$$

We can find an even more striking example by putting

$$(a_{ij}) = \begin{pmatrix} 1 & -1 & 0 & 0 & 0 & 0 & \dots \\ 0 & 2 & -2 & 0 & 0 & 0 & \dots \\ 0 & 0 & 3 & -3 & 0 & 0 & \dots \\ \vdots & \vdots & \ddots & \ddots & \ddots & \ddots & \end{pmatrix}$$

in which case

$$\sum_{i=1}^{\infty}\sum_{j=1}^{\infty} a_{ij} = \sum_{i=1}^{\infty} 0 = 0 \qquad \text{but} \qquad \sum_{j=1}^{\infty}\sum_{i=1}^{\infty} a_{ij} = \sum_{j=1}^{\infty} 1 = +\infty.$$

1.3.d Conditionally convergent series, absolutely convergent series

DEFINITION 1.49 A series $\sum a_n$ with a_n in a normed vector space is **conditionnally convergent** if it is convergent but not absolutely convergent.

DEFINITION 1.50 We denote by $\mathfrak{S}$ the group of **permutations**, that is, the group of bijections from $\mathbb{N}$ to $\mathbb{N}$, and we denote by $\mathfrak{S}_n$ the finite group of permutations of the set $\{1, \ldots, n\}$.

DEFINITION 1.51 A **series $\sum x_n$ is commutatively convergent** if it is convergent with sum X, and for any permutation $\varphi \in \mathfrak{S}$, the rearranged series $\sum x_{\varphi(n)}$ converges to X.

Is a convergent series necessarily *commutatively* convergent? In other words, *is it legitimate to change arbitrarily the order of the terms of a convergent series?*

At first sight, it is very tempting to say "Yes," almost without thinking, since permuting terms in a finite sum has no effect on the result. The problem is that we have here an *infinite sum*, not a finite sum in the algebraic sense. So there is a limiting process involved, and we will see that this brings a very different picture: only *absolutely convergent* series will be *commutatively convergent*. So, if a series is conditionally convergent (convergent but not absolutely so), changing the order of the terms may alter the value of the sum – or even turn it into a divergent series.

THEOREM 1.52 *Let $\sum a_n$ be a conditionnally convergent series, with $a_n \in \mathbb{R}$. Then for any $\ell \in \overline{\mathbb{R}}$, there exists a permutation $\varphi \in \mathfrak{S}$ such that the rearranged series $\sum a_{\varphi(n)}$ converges to ℓ.*

In fact, for any $a, b \in \overline{\mathbb{R}}$, with $a \leqslant b$, there exists a permutation $\psi \in \mathfrak{S}$ such that the set of limit points of the sequence of rearranged partial sums $\left(\sum_{k=0}^{n} a_{\psi(k)}\right)_{n\in\mathbb{N}}$ is the interval $[a, b]$.

PROOF. We assume that $\sum a_n$ is conditionnally convergent.

► First remark: *there are infinitely many positive values* and *infinitely many negative values* of the terms a_n of the series. Let α_n denote the sequence of non-negative terms, in the order they occur, and let β_n denote the sequence of negative terms.

Here is an illustration:

u_n :	1	3	2	−4	−1	2	−1	0	2	...
	α_1	α_2	α_3	β_1	β_2	α_4	β_3	α_5	α_6	...

► Second remark: both series $\sum \alpha_n$ and $\sum \beta_n$ are divergent, their partial sums converging, respectively, to $+\infty$ and $-\infty$. Indeed, if both series were to converge, the

series $\sum a_n$ would be absolutely convergent, and if only one were to converge, then $\sum a_n$ would be divergent (as follows from considering the sequence of partial sums).
► Third remark: both sequences $(\alpha_n)_{n\in\mathbb{N}}$ and $(\beta_n)_{n\in\mathbb{N}}$ tend to 0 (since $(a_n)_{n\in\mathbb{N}}$ tends to 0, as a consequence of the convergence of $\sum a_n$).

Now consider $\ell \in \mathbb{R}$. Let S_n denote the sequence of sums of values of α and β which is constructed as follows. First, sum all consecutive values of α_n until their sum is larger than ℓ; call this sum S_1. Now add to S_1 all consecutive values of β_n until the resulting sum $S_1 + \beta_1 + \cdots$ is *smaller* than ℓ; call this sum S_2. Then start again adding from the remaining values of α_n until getting a value larger than ℓ, called S_3, and continue in this manner until the end of time.

Now notice that:

- Since at each step we add at least one value of α or one of β, it is clear that *all* values of α will be used sooner or later, as well as all values of β, that is, when all is said and done, all values of a_n will have been involved in one of the sums S_n.
- Since, at each step, the distance $|\ell - S_n|$ is at most equal to the absolute value of the last value of α or β considered, the distance from S_n to ℓ tends to 0 as n tends to infinity.

From this we deduce that the sequence (S_n) is a sequence of partial sums of a rearrangement of the series $\sum a_n$, and that it converges to ℓ. Hence this proves that *by simply changing the order of the terms, one may cause the series to converge to an arbitrary sum.*

Let now $a, b \in \overline{\mathbb{R}}$ with $a < b$ (the case $a = b$ being the one already considered).
• **If a and b are both finite**, we can play the same game of summation as before, but this time, at each step, we either sum values of α_n until we reach a value larger than b, or we sum values of β_n until the value is less than a.
• **If $b = +\infty$ and a is finite**, we sum from a to above $a+1$, then come back to below a, then sum until we are above $a+2$, come back below a, etc. Similarly if $a = -\infty$ and b is finite.
• **If $a = -\infty$ and $b = +\infty$**, start from 0 to go above 1, then go down until reaching below -2, then go back up until reaching above 3, etc.

Example 1.53 Consider the sequence $(a_n)_{n\in\mathbb{N}^*}$ with general term $a_n = (-1)^{n+1}/n$. It follows from the theory of power series (Taylor expansion of $\log(1+x)$) that the series $\sum a_n$ converges and has sum equal to $\log 2$. If we sum the same values a_n by taking one positive term followed by two negative terms, then the resulting series converges to $\frac{1}{2}\log 2$. Indeed, if $(S_n)_{n\in\mathbb{N}}$ and $(S'_n)_{n\in\mathbb{N}}$ denote the sequence of partial sums of the original and modified series, respectively, then for $n \in \mathbb{N}$ we have

$$S_{2n} = 1 - \tfrac{1}{2} + \tfrac{1}{3} - \tfrac{1}{4} + \cdots + \tfrac{1}{2n-1} - \tfrac{1}{2n}$$

and

$$S'_{3n} = \underbrace{1 - \tfrac{1}{2}}\, - \tfrac{1}{4} + \underbrace{\tfrac{1}{3} - \tfrac{1}{6}}\, - \tfrac{1}{8} + \cdots + \underbrace{\tfrac{1}{2n-1} - \tfrac{1}{4n-2}}\, - \tfrac{1}{4n}$$

$$= \tfrac{1}{2} - \tfrac{1}{4} + \tfrac{1}{6} - \tfrac{1}{8} + \cdots + \tfrac{1}{4n-2} - \tfrac{1}{4n} = \tfrac{1}{2} S_{2n}.$$

As an exercise, the reader can check that if one takes instead two positive terms followed by one negative terms, the resulting series converges with a value equal to $\frac{3}{2}\log 2$.

The following result shows that, on the other hand, one can rearrange at will the order of the terms of an absolutely convergent series.

THEOREM 1.54 *A series of complex numbers is commutatively convergent if and only if it is absolutely convergent.*

PROOF. Assume first that the terms of the series are real numbers. The theorem above shows that if $\sum a_n$ is commutatively convergent, it must be absolutely convergent. Conversely, assume the series is absolutely convergent and let ℓ denote its sum. Let ψ be any permutation. By convergence of the series, there exists $N \geqslant 1$ such that

$$\left|\sum_{k=1}^{n} a_k - \ell\right| < \varepsilon$$

for $n \geqslant N$. For each such $n \geqslant N$, there exists N' such that the set $\{\psi(1), \dots, \psi(N')\}$ contains $\{1, \dots, n\}$ (it suffices that N' be larger than the maximum of the images of 1, $\dots$, N by the inverse permutation ψ^{-1}). Then for any $m \geqslant N'$, we have

$$\left|\sum_{k=1}^{m} a_{\psi(k)} - \ell\right| \leqslant \left|\sum_{k=1}^{n} a_k - \ell\right| + \sum_{k>n} |a_k| \leqslant \varepsilon + \sum_{k>n} |a_k|,$$

since the set $\{\psi(1), \dots, \psi(m)\}$ contains $\{1, \dots, n\}$, and possibly additional values which are all larger than n. The absolute convergence makes its appearance now: the last sum on the right is the remainder for the convergent series $\sum |a_k|$, and for $n \geqslant N''$ it is therefore itself smaller than ε. Since, given ε, we can take $n = N''$ and find the value N' from it, such that

$$\left|\sum_{k=1}^{m} a_{\psi(k)} - \ell\right| \leqslant 2\varepsilon$$

for $m \geqslant N'$, and so we have proved that the rearranged series converges with sum equal to ℓ.

If the terms of the series are complex numbers, it suffices to apply the result for real series to the series of real and imaginary parts.

The possibility of rearranging at will the order of summation explains the importance of absolutely convergent series[13].

Remark 1.55 In statistical mechanics, there are so-called *diagrammatic* methods to compute the values of certains quantities, such as pressure or mean-energy, at equilibrium. Those methods are based on rearrangements of the terms of certain series, summing "by packets" in particular. Those methods are particularly useful when the original series is not absolutely convergent. This means that all results obtained in this manner must be treated carefully, if not suspiciously. They belong to the gray area of *exact* (at least, this is what everyone believes!) results, but which are not *rigorous*. (It is of course much more difficult to obtain results which can be judged with mathematical standards of rigor; the reader is invited to read the beautiful papers [63, 64] for convincing illustrations.)

[13] Peter Gustav LEJEUNE-DIRICHLET showed in 1837 that a convergent series with non-negative terms is commutatively convergent. In 1854, Bernhard RIEMANN wrote three papers in order to obtain a position at the university of Göttingen. In one of them, he describes commutatively convergent series in the general case. However, another paper was selected, concerning the foundations of geometry.

1.3.e Series of functions

We can define pointwise and uniform convergence of series of functions just as was done for sequences of functions.

DEFINITION 1.56 (Pointwise convergence) Let X be an arbitrary set, $(E, \|\cdot\|)$ a normed vector space. A series $\sum f_n$ of functions $f_n : X \to E$ **converges pointwise to a function $F : X \to E$** if, for any $x \in X$, the series $\sum f_n(x)$ converges to $F(x)$ in E, that is, if

$$\forall \varepsilon > 0 \quad \forall x \in X \quad \exists N \in \mathbb{N} \quad \forall n \in \mathbb{N} \qquad n \geqslant N \Longrightarrow \left\| \sum_{k=1}^{n} f_k(x) - F(x) \right\| < \varepsilon.$$

The function F is called the **pointwise**, or **simple, limit** of the series $\sum f_n$, and this is denoted $\sum f_n \xrightarrow{\text{cv.s.}} F$.

DEFINITION 1.57 (Uniform convergence) Let X be an arbitrary set, $(E, \|\cdot\|)$ a normed vector space. A series $\sum f_n$ of functions $f_n : X \to E$ **converges uniformly to a function $F : X \to E$** if the sequence of partial sums of the series converges uniformly to F, that is, if

$$\forall \varepsilon > 0 \quad \exists N \in \mathbb{N} \quad \forall x \in X \quad \forall n \in \mathbb{N} \qquad n \geqslant N \Longrightarrow \left\| \sum_{k=1}^{n} f_k(x) - F(x) \right\| < \varepsilon.$$

This is denoted $\sum f_n \xrightarrow{\text{cv.u.}} F$. This amounts to

$$\lim_{n\to\infty} \left\| \sum_{k=1}^{n} f_k - F \right\|_\infty = 0 \qquad \text{where} \qquad \|g\|_\infty = \sup_{x\in X} \|g(x)\|.$$

DEFINITION 1.58 (Absolute convergence) Let X be an arbitrary set, $(E, \|\cdot\|)$ a normed vector space. A series $\sum f_n$ of functions $f_n : X \to E$ **converges absolutely** if the series $\sum \|f_n\|_\infty$ converges, where

$$\|f_n\|_\infty = \sup_{x\in X} \|f_n(x)\|.$$

The following theorem is the most commonly used to prove uniform convergence of a series of functions:

THEOREM 1.59 *Any absolutely convergent series with values in a* complete *normed vector space is uniformly convergent, and hence pointwise convergent.*

Corresponding to the continuity and differentiability results for sequences of functions, we have:

THEOREM 1.60 (Continuity and differentiability of a series of functions) *Let D be a subset of $\mathbb{R}$ or of a normed vector space. Let $(f_n)_{n\in\mathbb{N}}$ be a sequence of functions $f_n : D \to E$, where $(E, \|\cdot\|)$ is some normed vector space, for instance, $\mathbb{R}$ or $\mathbb{C}$. Assume that the series $\sum f_n$ converges pointwise to a function F.*

i) If each f_n is continuous on D, and if the series $\sum f_n$ converges uniformly on D, then F is continuous on D.

ii) If D is an interval of $\mathbb{R}$, each f_n is differentiable on D, and the series $\sum f_n'$ converges uniformly, then F is differentiable and we have

$$F' = \sum_{n=0}^{\infty} f_n'.$$

1.4 Power series, analytic functions

Quite often, physicists encounter series expansions of some function. These expansions may have different origins:

- the superposition of many phenomena (as in the Fabry-Perot interferometer);
- perturbative expansions, when exact computations are too difficult to perform (e.g., hydrodynamics, semiclassical expansions, weakly relativistic expansions, series in astronomy, quantum electrodynamics, etc.);
- sometimes the exact evalution of a function which expresses some physical quantity is impossible; a *numerical* evaluation may then be performed using Taylor series expansions, Fourier series, infinite product expansions, or asymptotic expansions.

We first recall various forms of the Taylor formula. The general idea is that there is an approximate expression

$$f(x) \approx f(a) + (x-a)f'(a) + \frac{(x-a)^2}{2!}f''(a) + \cdots + \frac{(x-a)^k}{k!}f^{(k)}(a)$$

for a function f which is at least k times differentiable on an interval J, with values in some normed vector space $(E, \|\cdot\|)$, and for a given point $a \in J$, where x lies is some neighborhood of a.

The question is to make precise the meaning of the symbol "$\approx$"!

Define $R_k(x)$ to be the difference between $f(x)$ and the sum on the right-hand side of the above expression; in other words, we have

$$f(x) = \sum_{n=0}^{k} \frac{(x-a)^n}{n!} f^{(n)}(a) + R_k(x) = T_k(x) + R_k(x)$$

Brook TAYLOR (1685–1731), English mathematician, was a student at Cambridge, then member and secretary of the prestigious Royal Society, a venerable institution dedicated to the advancement of Science. He wrote the famous formula

$$f(a+\varepsilon)=f(a)+\varepsilon f'(a)+\tfrac{\varepsilon^2}{2}f''(a)+\cdots$$

without considering the issue of convergence. He was also interested in the physical and mathematical aspects of vibrating strings.

by definition. Of course, we hope that the **Taylor remainder** $R_k(x)$ is a "small quantity," so that we may approximate the value of $f(x)$ by the value of the **Taylor polynomial of order** k **at** x, that is, by $T_k(x)$. There are different ways in which this remainder may become small:

- one may let x tend to a (for a fixed value of k);
- or let k tend to infinity (for a fixed value of x).

The Taylor-Lagrange and Taylor-Young formulas are relevant for the first case, while the second belongs to the theory of power series.

1.4.a Taylor formulas

THEOREM 1.61 (Taylor formula with integral remainder) *Let J be an interval of $\mathbb{R}$, and $(E,\|\cdot\|)$ a normed vector space. Let $f : J \to E$ be a function of $\mathscr{C}^k$ class on J, which is piecewise of $\mathscr{C}^{k+1}$ class on J. For any a and $x \in J$, we have*

$$f(x)=\sum_{n=0}^{k}\frac{(x-a)^n}{n!}f^{(n)}(a)+\int_a^x\frac{(x-t)^k}{k!}f^{(k+1)}(t)\,\mathrm{d}t.$$

THEOREM 1.62 (Taylor-Lagrange formula) *Let $f : J \to \mathbb{R}$ be a* real-valued *function of $\mathscr{C}^k$ class on an interval J of $\mathbb{R}$, which is $k+1$ times differentiable in the interior of J. Let $a \in J$. Then, for any $x \in J$, there exists $\theta \in\,]0,1[$ such that*

$$f(x)=\sum_{n=0}^{k}\frac{(x-a)^n}{n!}f^{(n)}(a)+\frac{(x-a)^{k+1}}{(k+1)!}f^{(k+1)}\big(a+\theta(x-a)\big).$$

Remark 1.63 This formula is only valid for real-valued functions. However, the following corollary is also true for functions with complex values, or functions with values in a normed vector space.

COROLLARY 1.64 (Taylor-Lagrange inequality) *Let $f : J \to E$ be a function of $\mathscr{C}^k$ class on an interval J of $\mathbb{R}$, with values in a normed vector space E. Assume f is $k+1$ times differentiable in the interior of J. Let $a \in J$. Then for any $x \in J$ we have*

$$\left\| f(x) - \sum_{n=0}^{k} \frac{(x-a)^n}{n!} f^{(n)}(a) \right\|_E \leqslant \frac{|x-a|^{k+1}}{(k+1)!} \sup_{t \in J} \left\| f^{(k+1)}(t) \right\|_E.$$

THEOREM 1.65 (Taylor-Young formula) *Let f be a function which is k times differentiable on an interval J of $\mathbb{R}$, with values in a normed vector space E. Let $a \in J$. Then we have*

$$f(x) - \sum_{n=0}^{k} \frac{(x-a)^n}{n!} f^{(n)}(a) = \underset{x \to a}{\mathrm{o}} \left((x-a)^k \right).$$

1.4.b Some numerical illustrations

Suppose we want to compute numerically some values of the inverse tangent function arctan, which is of course infinitely differentiable on $\mathbb{R}$. It is easy to compute the values of the successive derivatives of this function at 0, and we can write down explicitly the Taylor polynomial at 0 of arbitrary order: this gives the expression

$$\arctan x = \sum_{n=0}^{k} \frac{(-1)^n}{2n+1} x^{2n+1} + R_k(x),$$

for the Taylor formula of order $2n+1$ (notice that only odd powers of x appear, because the inverse tangent function is odd).

If we represent graphically those polynomials with $k = 0, 1, 4, 36$ (i.e., of order 1, 5, 9, and 18, respectively), with the graph of the function itself for comparison, we obtain the following:

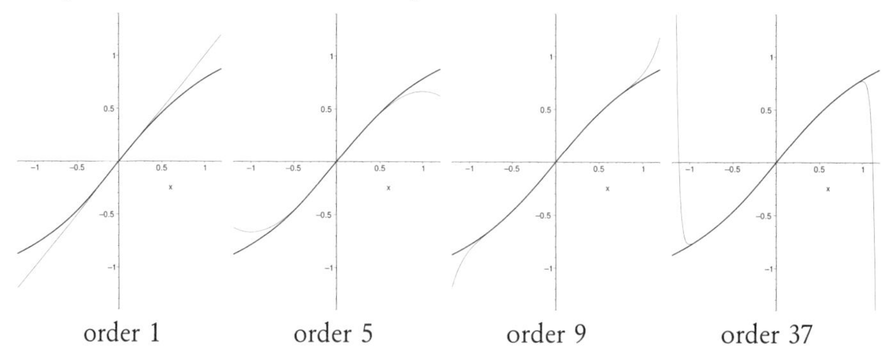

order 1 order 5 order 9 order 37

The following facts appear:

- on each graph (i.e., for fixed k), the Taylor polynomial and the inverse tangent functions get closer and closer together as x approaches 0;

- for a fixed real number $x \in [-1, 1]$ (for instance 0.8), the values at x of the Taylor polynomial of increasing degree get closer and closer to the value of the function as k increases;
- on the other hand, for a real number x such that $|x| > 1$, disaster strikes: the larger k is, the further away to $\arctan x$ is the value of the Taylor polynomial!

The first observation is simply a consequence of the Taylor-Young formula. The other two deserve more attention. It seems that the sequence $(T_k)_{k\in\mathbb{N}}$ of the Taylor polynomials converges on $[-1, 1]$ and diverges outside.[14] However, *the function* arctan *is perfectly well-defined, and very regular, at the point* $x = 1$; it does not seem that anything special should happen there. In fact, it is possible to write down the Taylor expansion centered at $a = 1$ instead of $a = 0$ (this is a somewhat tedious computation[15]), and (using approximations of the same order as before), we obtain the following graphs:

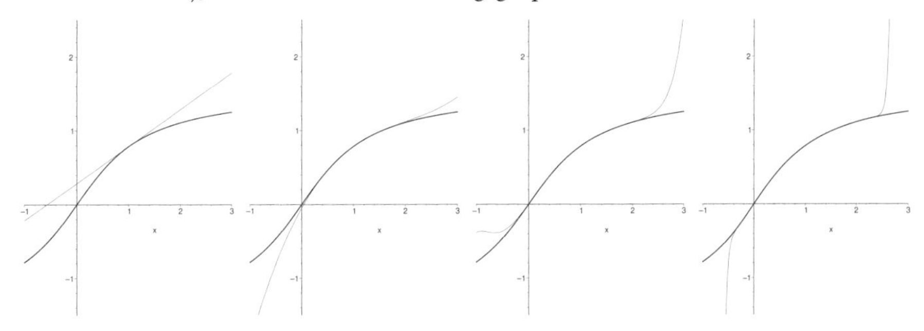

We can see the same three basic facts, except that convergence seems to be restricted now to the interval $\left[1 - \sqrt{2}, 1 + \sqrt{2}\right]$.

In order to understand why such intervals occur, it is necessary to dwell further on the theory of power series (see below) and especially on holomorphic functions of a complex variable (in particular, Theorem 4.40 on page 101). We will only state here that the function arctan can be continued naturally to a function on the complex plane ($\arctan z_0$ is defined as the value of the integral of the function $1/(1+z^2)$ on a certain path[16] joining the origin to z_0. The function thus obtained is well-defined, independently of the chosen path, up to an integral multiple of π[17] and is a well-defined function on $\mathbb{C}$ minus the two single points where $1 + z^2$ vanishes, namely i and $-$i. Then, one shows that for such a function, the sequence of Taylor polynomials centered

---

[14] To be honest, it is difficult to ascertain from the graphs above if the interval to consider is $[-1, 1]$ or $]-1, 1[$, for instance. The general theory of series shows that $(T_n(x))_{n\in\mathbb{N}}$ converges quickly if $|x| < 1$ and very slowly if $|x| = 1$.

[15] The n-th coefficient of the polynomial is $(-1)^{n+1}\sin(n\pi/4)2^{-n/2}/n$ and the constant term is $\pi/4$.

[16] This notion of integral on a path is defined by the formula (4.2) page 94.

[17] This is a consequence of the residue theorem 4.81 on page 115.

at a converges on the open disc centered at a with radius equal to the distance from a to the closest singulirity (hence the radius is $|1-0|=1$ in the first case of Taylor expansions at $a=0$, and is $|1-\mathrm{i}|=\sqrt{2}$ in the second case).

1.4.c Radius of convergence of a power series

A **power series centered at z_0** is any series of functions of the type

$$z \longmapsto \sum_{n=0}^{\infty} a_n (z-z_0)^n,$$

where $(a_n)_{n\in\mathbb{N}}$ is a given sequence of real or complex numbers, which are sometimes called the *coefficients* of the power series.

THEOREM 1.66 (Radius of convergence) *Let $\sum a_n(z-z_0)^n$ be a power series centered at z_0. The **radius of convergence** is the element in $\overline{\mathbb{R}}^+$ defined by*

$$R \stackrel{\text{def}}{=} \sup\left\{t \in \mathbb{R}^+ \,;\, (a_n t^n)_{n\in\mathbb{N}} \text{ is bounded}\right\}.$$

The power series converges absolutely and uniformly on any compact subset in the disc $\mathcal{B}(z_0\,;R) \stackrel{\text{def}}{=} \left\{z \in \mathbb{C} \,;\, |z-z_0| < R\right\}$, in the complex plane $\mathbb{C}$, and it diverges for any $z \in \mathbb{C}$ such that $|z| > r$. For $|z| = r$, the series may be convergent, conditionally convergent, or divergent at z.

Note that "absolute convergence" here refers to absolute convergence as a series of functions, which is stronger than absolute convergence for every z involved: in other words, for any compact set $D \subset \mathcal{B}(z_0\,;R)$, we have

$$\sum_{n=0}^{\infty} \sup_{z\in D} |a_n z^n| < +\infty.$$

Example 1.67 The power series $-\log(1-z) = \sum_{n=1}^{\infty} z^n/n$ converges for any $z \in \mathbb{C}$ such that $|z| < 1$ and diverges if $|z| > 1$ (the radius of convergence is $R = 1$). Moreover, this series is divergent at $z = 1$, but conditionnally convergent at $z = -1$ (by the alternate series test), and more generally, it is conditionnally convergent at $z = \mathrm{e}^{\mathrm{i}\theta}$ for any $\theta \notin 2\pi\mathbb{Z}$ (this can be shown using the Abel transformation, also known as "summation by parts").

DEFINITION 1.68 (Power series expansion) Let Ω be an open subset in $\mathbb{R}$ or $\mathbb{C}$. A function $f : \Omega \to \mathbb{C}$ defined on Ω **has a power series expansion centered at some $z_0 \in \Omega$** if there exist an open subset $V \subset \Omega$ containing z_0 and a sequence $(a_n)_{n\in\mathbb{N}}$ of complex numbers such that

$$\forall z \in V \qquad f(z) = \sum_{n=0}^{\infty} a_n (z-z_0)^n.$$

The radius of convergence of a power series depends only weakly on the precise values of the coefficients, so, for instance, if $F = P/Q$ is a rational function with no pole in $\mathbb{N}$, the power series $\sum F(n)\, a_n\, z^n$ and $\sum a_n\, z^n$ have the same radius of convergence. From this and Theorem 1.60, it follows in particular that a power series can be differentiated term by term inside the disc of convergence:

THEOREM 1.69 (Derivative of a power series) *Let J be an open subset of $\mathbb{R}$, $x_0 \in J$, and $f : J \to \mathbb{C}$ a function which has a power series expansion centered at x_0:*

$$f(x) = \sum_{n=0}^{\infty} a_n (x - x_0)^n.$$

Let $R > 0$ be the radius of convergence of this power series. Then f is infinitely differentiable on the open interval $]x_0 - R, x_0 + R[$, and each derivative has a power series expansion on this interval, which is obtained by repeated term by term differentiation, that is, we have

$$f'(x) = \sum_{n=1}^{\infty} n a_n\, (x - x_0)^{n-1} \qquad \textit{and} \qquad f^{(k)}(x) = \sum_{n=k}^{\infty} \frac{n!}{(n-k)!}\, a_n (x - x_0)^{n-k}$$

for any $k \in \mathbb{N}$. Hence the n-th coefficient of the power series $f(x)$ can be expressed as

$$a_n = \frac{f^{(n)}(x_0)}{n!}.$$

Remark 1.70 The power series $\sum \big(f^{(n)}(x_0)/n!\big) \cdot (x - x_0)^n$ is the **Taylor series of f at x_0**. On any compact subset inside the open interval of convergence, it is the uniform limit of the sequence of Taylor polynomials.

Remark 1.71 In Chapter 4, this result will be extended to power series of one complex variable (Theorem 4.40 on page 101).

1.4.d Analytic functions

Consider a function that may be expended into a power series in a neighborhood V of a point z_0, so that for $z \in V$, we have

$$f(z) = \sum_{n=0}^{\infty} a_n (z - z_0)^n.$$

Given such a $z \in V$, a natural question is the following: *may f also be expended into a power series centered at z?*

Indeed, it might seem possible *a priori* that f can be expanded in power series only around z_0, and around no other point. However, this is not the case:

DEFINITION 1.72 (Analytic function) A function $f : \Omega \to \mathbb{C}$ defined on an open subset Ω of $\mathbb{C}$ or $\mathbb{R}$ is **analytic** on Ω if, for any $z_0 \in \Omega$, f has a power series expansion centered at z_0.

Note that the radius of convergence of the power series may (and often does!) vary with the point z_0.

THEOREM 1.73 *Let $\sum a_n z^n$ be a power series with positive radius of convergence $R > 0$, and let f denote the sum of this power series on $\mathcal{B}(0\,;R)$. Then the function f is analytic on $\mathcal{B}(0\,;R)$.*

Example 1.74 The function $f : x \mapsto 1/(1-x)$ has the power series expansion

$$f(x) = \sum_{n=0}^{\infty} x^n$$

around 0, with radius of convergence equal to 1. Hence, for any $x_0 \in \left]-1,1\right[$, there exists a power series expansion centered at x_0 (obviously with different coefficients). This can be made explicit: let $h \in \mathcal{B}(0\,;|1-x_0|)$, then with $x = x_0 + h$, we have

$$f(x) = f(x_0+h) = \frac{1}{1-(x_0+h)} = \frac{1}{1-x_0} \cdot \frac{1}{1-h/(1-x_0)} = \sum_{n=0}^{\infty} \frac{(x-x_0)^n}{(1-x_0)^{n+1}}.$$

Remark 1.75 (Convergence of Taylor expansions) Let $f : U \to \mathbb{C}$ be a function defined on an open subset U of $\mathbb{R}$. Under what conditions is f analytic?[18] There are two obvious necessary conditions:

- f is infinitely differentiable on U;
- for any $x_0 \in U$, there exists an open disc $\mathcal{B}(x_0, r)$ such that the series $\sum \frac{1}{n!} f^{(n)}(x_0)(x-x_0)^n$ converges for any $x \in \mathcal{B}(x_0, r)$.

However, those two conditions are *not* sufficient. The following classical counter-example shows this: let

$$f(x) \stackrel{\text{def}}{=} \exp\left(-1/x^2\right) \quad \text{if } x \neq 0, \qquad f(0) = 0.$$

It may be shown[19] that f is indeed of $\mathscr{C}^\infty$ and that each derivative of f at 0 vanishes, which ensures (!) the convergence of the Taylor series everywhere. But since the function vanishes only at $x = 0$, it is clear that the Taylor series does not converge to f on any open subset, hence f is not analytic.

It is therefore important not to use the terminology "analytic" where "infinitely differentiable" is intended. This is a confusion that it still quite frequent in scientific literature.

The Taylor formulas may be used to prove that a function is analytic. If the sequence $(R_n)_{n\in\mathbb{N}}$ of remainders for a function f converges uniformly to 0 on a neighborhood of $a \in \mathbb{R}$, then the function f is analytic on this neighborhood. To show this, one may use the integral expression of the remainder terms in the Taylor formula. A slightly different but useful approach is to prove that both the function under consideration and its Taylor series (which must be shown to have positive radius of convergence) satisfy the same differential equation, with the corresponding initial conditions; then f is analytic because of the unicity of solutions to a Cauchy problem.

Also, it is useful to remember that if f and g have power series expansions centered at z_0, then so do $f+g$ and fg. And if $f(z_0) \neq 0$, the function $1/f$ also has a power series expansion centered at z_0.

---

[18] The same question, for a function of a complex variable, turns out to have a completely different, and much simpler, answer: if f is differentiable – in the complex sense – on the open set of definition, then it is always analytic. See Chapter 4.

[19] By induction, proving that $f^{(n)}(x)$ is for $x \neq 0$ of the form $x \mapsto Q_n(x) f(x)$, for some rational function Q_n.

1.5

A quick look at asymptotic and divergent series

1.5.a Asymptotic series

DEFINITION 1.76 (Asymptotic expansion) Let F be a function of a real or complex variable z, defined for all z with $|z|$ large enough. The function F **has an asymptotic expansion** if there exists a sequence $(a_n)_{n\in\mathbb{N}}$ of complex numbers such that

$$\lim_{z\to\infty} z^N \left\{ F(z) - \sum_{n=0}^{N} \frac{a_n}{z^n} \right\} = 0$$

for any positive integer N. This is denoted

$$F(z) \underset{z\to\infty}{\sim} \sum_{n=0}^{\infty} \frac{a_n}{z^n}. \tag{1.4}$$

The definition means that the expansion (1.4) is a good approximation for large values of z. Indeed, if we only consider the first twenty terms of the series, for instance, we see that the sum of those approximates $f(z)$ "to order $1/z^{20}$ at least" when $[z \to \infty]$.

However, it frequently turns out that *for fixed* z, the behavior of the series in (1.4) is quite bad as $N \to \infty$. In particular, the series may be divergent. This phenomenon was pointed out and studied in detail by Henri Poincaré in the case of asymptotic series used in astronomy, at the beginning of the twentieth century [70].

How can a divergent asymptotic series still be used? Since $\sum a_n/z^n$ is asymptotic to F, if there is some R such that F is continuous for $|z| \geqslant R$, then we see that there exist constants C_1, C_2,... such that

$$\left| F(z) - \sum_{n=0}^{N} \frac{a_n}{z^n} \right| \leqslant \frac{C_N}{|z|^{N+1}} \qquad \text{for } N \in \mathbb{N} \text{ and } |z| \geqslant R.$$

For fixed z, we can look for the value of N such that the right-hand side of this inequality is minimal, and truncate the asymptotic series at this point. Of course, *we do not obtain* $F(z)$ *with infinite precision*. But in many cases the actual precision increases with $|z|$, as described in the next section, and may be pretty good.

It is also possible to speak of **asymptotic expansion** as $z \to 0$, which corresponds to the existence of a sequence $(a_n)_{n\in\mathbb{N}}$ such that

$$\lim_{z\to 0} \frac{1}{z^N} \left\{ f(z) - \sum_{n=0}^{N} a_n z^n \right\} = 0,$$

which is denoted

$$f(z) \underset{z\to 0}{\sim} \sum_{n=0}^{\infty} a_n z^n. \tag{1.5}$$

Remark 1.77 If it exists, an asymptotic expansion of a function is unique, but there may be two different functions with the same asymptotic expansion! For instance, e^{-x} and e^{-x^2} both have asymptotic expansions with $a_n = 0$ for all n as $x \to +\infty$.

A physical example is given by quantum electrodynamics. This quantum theory of electromagnetic interactions gives physical results in the form of series in powers of the coupling constant $\alpha = e^2/\hbar c \approx 1/137$ (the *Sommerfeld fine structure constant*), which means that a *perturbative expansion in* α is performed:

As shown by Dyson [32], when studying a physical quantity F we can expect to find a perturbative series of the following type (with the normalization $\hbar = c = 1$):

$$F(e^2) = F(\alpha) = \sum_{n=0}^{\infty} f_n\, \alpha^n.$$

Since the value of α is fixed by Nature, with a value given by experiments, only the truncated series can give a physical result if the series is divergent. The truncation must be performed around the 137-th term, which means that we still expect a very precise result – certainly more precise, by far, than anything the most precise experiment will ever give! However, if $F(e^2)$ is not analytic at $e = 0$, the question is raised whether the asymptotic expansion considered gives access to F uniquely or not.

Studying asymptotic series is in itself a difficult task. Their implications in physics (notably field theory) are at the heart of current research [61].

1.5.b Divergent series and asymptotic expansions

Since EULER, CAUCHY (page 88), and especially POINCARÉ (page 475), it has been realized that divergent series may be very useful in physics. As seen in the previous section, they appear naturally in computations of asymptotic series.

As a general rule, convergent series are used to prove numerical or functional identities (between power series, Fourier series, etc). Thus the series may be used instead of the value of their sum at any time in a computation. Another remark is that, from the *computational* viewpoint, some series are more interesting than others, because they converge faster. For example, we have the following two identities for $\log 2$:

$$\log 2 = 1 - \frac{1}{2} + \frac{1}{3} - \frac{1}{4} + \cdots + \frac{(-1)^{n+1}}{n} + \cdots$$

$$-\log\frac{1}{2} = \log 2 = \frac{1}{2} + \frac{1}{2^2\cdot 2} + \frac{1}{2^3\cdot 3} + \cdots + \frac{1}{2^n\cdot n} + \cdots$$

The second of those (which comes from expanding $x \mapsto \log(1-x)$ in power series at $x = 1/2$) converges much faster than the first (which results from a

Leonhard EULER (1707–1783), a Swiss mathematician, an exceptional teacher, obtained a position at the Academy of Sciences of Saint Petersburg thanks to Nicolas and Daniel BERNOULLI when he was only twenty. He also spent some years in Berlin, but came back to Russia toward the end of his life, and died there at seventy-six (while drinking tea). His works are uncountable! We owe him the notations e and i and he imposed the use of π that was introduced by Jones in 1706. Other notations due to Euler are sin, cos, tang, cot, sec, and cosec. He also introduced the use of complex exponents, showed that $e^{ix} = \cos x + i \sin x$, and was particularly fond of the formula $e^{i\pi} + 1 = 0$. He defined the function Γ, which extends the factorial function from integers to $\mathbb{C} \setminus (-\mathbb{N})$, and used the Riemann zeta function for real values of the variable. No stone of the mathematical garden of his time was left unturned by Euler; let us only add the Euler angles in mechanics and the Euler equation in fluid mechanics.

similar expansion at $x = -1$, hence on the *boundary* of the disc of convergence[20]).

While studying problems of celestial mechanics, Poincaré realized that the meaning of "convergent series" was not the same for mathematicians, with rigor in mind, or astronomers, interested in efficiency:

> Geometers, preoccupied with rigorousness and often indifferent to the length of the inextricable computations that they conceive, with no idea of implementing them in practice, say that a series is *convergent* when the sum of its terms tends to some well-defined limit, however slowly the first terms might diminish. Astronomers, on the contrary, are used to saying that a series converges when the twenty first terms, for instance, diminish very quickly, even though the next terms may well increase indefinitely. Thus, to take a simple example, consider the two series with general terms $\dfrac{1000^n}{1 \cdot 2 \cdot 3 \cdots n}$ and $\dfrac{1 \cdot 2 \cdot 3 \cdots n}{1000^n}$.
>
> Geometers will say that the first series converges, and even that it converges

[20] The number of terms necessary to approximate $\log 2$ within 10^{-6}, for instance, can be estimated quite precisely for both series. Using the Leibniz test for alternating sums, the remainder of the first series is seen to satisfy

$$|R_n| \leqslant |u_{n+1}| = \frac{1}{n+1},$$

and this is the right order of magnitude (a pretty good estimate is in fact $R_n \approx 1/2n$). If we want $|R_n|$ to be less than 10^{-6}, it suffices to take $n = 10^6$ terms. This is a very slow convergence. The remainder of the second series, on the other hand, can be estimated by the remainder of a geometric series:

$$R'_n = \sum_{k=n+1}^{\infty} \frac{1}{n \cdot 2^n} \leqslant \sum_{k=n+1}^{\infty} \frac{1}{2^k} = \frac{1}{2^n}.$$

Hence twenty terms or so are enough to approximate $\log 2$ within 10^{-6} using this expansion (since $2^{20} \approx 10^6$).

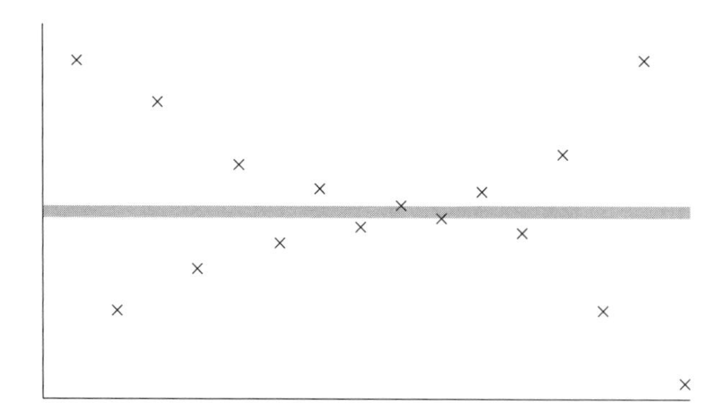

Fig. 1.5 – The precise value of $f(x)$ is always found between two successive values of the partial sums of the serie $\sum f_k(x)$. Hence, it is inside the gray strip.

> rapidly, [...] but they will see the second as divergent.
>
> Astronomers, on the contrary, will see the first as divergent [...] and the second as convergent. [70]

How is it possible to speak of convergence of a series which is really divergent? We look at this question using a famous example, the **Euler series.** Let

$$\forall x > 0 \qquad f(x) = \int_0^{+\infty} \frac{\mathrm{e}^{-t/x}}{1+t}\,\mathrm{d}t,$$

and say we wish to study the behavior of f for small values of x. A first idea is to expand $1/(1+t)$ as $\sum (-1)^k\, t^k$ and exchange the sum and the integral if permitted. Substitute $y = t/x$ and then integrate by parts; a simple induction then shows that

$$\int_0^{+\infty} t^k\, \mathrm{e}^{-t/x}\,\mathrm{d}t = k!\, x^{k+1},$$

and since the $\sum (-1)^k\, k!\, x^{k+1}$ is obviously divergent for any nonzero value of x, this first idea is a lamentable failure.

To avoid this problem, it is possible to truncate the power-series expansion of the denominator, and write

$$\frac{1}{1+t} = \sum_{k=0}^{n-1} (-1)^k\, t^k + \frac{(-1)^n\, t^n}{1+t},$$

from which we derive an expression for f of the type $f = f_n + R_n$, where

$$f_n(x) = x - x^2 + 2!\, x^3 - 3!\, x^4 + \cdots + (-1)^{n-1}(n-1)!\, x^n \tag{1.6}$$

and

$$R_n(x) = (-1)^n \int_0^{+\infty} \frac{t^n\, \mathrm{e}^{-t/x}}{1+t}\,\mathrm{d}t.$$

Since $(1+t)^{-1} \leqslant 1$, the remainder satisfies $\left|R_n(x)\right| \leqslant n!\, x^{n+1}$, which means that $R_n(x)$ is of absolute value smaller than the first omitted term; moreover,

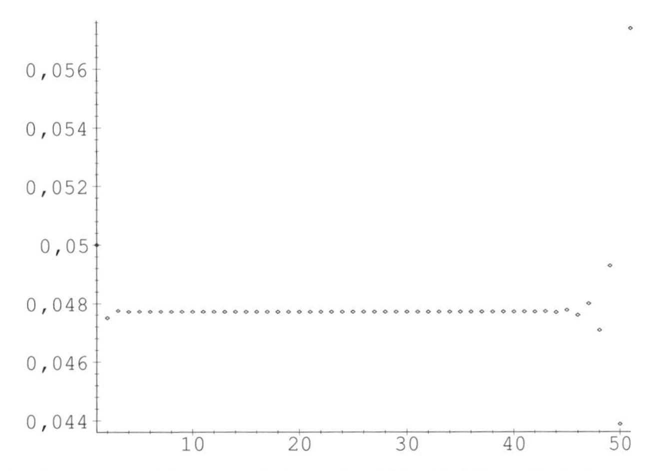

Fig. 1.6 – The first 50 partial sums of the series $\sum(-1)^{k-1}(k-1)!\,x^k$ for $x = 1/20$. Notice that, starting from $k = 44$, the series diverges rapidly. The best precision is obtained for $n = 20$, and gives $f(x)$ with an error roughly of size $2 \cdot 10^{-8}$.

they are of the same sign. It follows (see the proof of the alternating series test) that

$$f_{2n}(x) < f(x) < f_{2n+1}(x), \tag{1.7}$$

although, in contrast with the case of alternating series with terms converging to 0, the general term here $(-1)^n n!\,x^{n+1}$ diverges. Hence it is not possible to deduce from (1.7) an *arbitrarily precise* approximation of $f(x)$. However, *if x is small, we can still get a very good approximation*, as we now explain.

Fix a positive value of x. There exists an index N_0 such that the distance $|f_{2n+1}(x) - f_{2n}(x)|$ is smallest (the ratio between consecutive terms is equal to nx, so this value of n is in fact $N_0 = \lfloor 1/x \rfloor$). This means that, if we look at the first N_0 values, the series "seems to converge," before it starts blowing up. It is interesting to remark that the "convergence" of the first N_0 terms is *exponentially fast*, since the minimal distance $|f_{N+1}(x) - f_N(x)|$ is roughly given by

$$N!\,x^N \approx N!\,N^{-N} \sim \sqrt{2\pi/x}\,\mathrm{e}^{-1/x}$$

(using the Stirling formula, see Exercise 5.4 on page 154.) Thus, if we wish to know the value of $f(x)$ for a "small" value of x, and if a precision of the order of $\sqrt{2\pi/x}\,\mathrm{e}^{-1/x}$ suffices, it is possible to use the divergent asymptotic series (1.6), by computing and summing the terms *up to the smallest term* (see Figure 1.5). For instance, we obtain for $x = 1/50$ a precision roughly equal to $6 \cdot 10^{-20}$, which is perfectly sufficient for most physical applications! (see Figure 1.6.)

For a given value of x, on the other hand, the asymptotic series does not allow any improvement on the precision.[21] But the convergence is so fast that

[21] For instance, in quantum field theory, the asymptotic series in terms of α has a limited

Sir George Biddel AIRY (1801–1892), English astronomer, is known in particular for discovering the theory of diffraction rings. He determined approximately the solar apex, the direction toward which the sun and the solar system seem to be directed, in the Hercules region. He was also interested in geology. He was director of the Royal Observatory and took part in the controversy concerning priority for the discovery of Neptune (the French pushing the claim of Le Verrier while the English defended Adams).
The picture here represents a fake stamp, painted directly on the envelope, representing a contemporary caricature of Sir Airy; the post office was bluffed and stamped and delivered the letter.

it makes it possible to do some computations which are out of reach of a standard method! And what Poincaré remarked is, in fact, a fairly general rule: *divergent series converge, in general, much more rapidly than convergent series.*

In 1857, George Stokes was studying the Airy integral [3]

$$\mathrm{Ai}(z) \stackrel{\text{def}}{=} \frac{1}{\pi}\int_0^{+\infty} \cos\left(\frac{t^3}{3} + zt\right) \mathrm{d}t,$$

which appears in the computations of caustics. The goal was to find zeros of this function, and compare the with "experimental" zeros (corresponding to dark bands in any optics figure, which had been measured with great precision, at least as far as the first twenty-five). Airy himself, using a convergent series expansion of Ai at 0, managed fairly easily to compute the position of the first band, and with considerable difficulty, found the second one. In fact, his mathematically convergent expansion was "divergent" in the sense of astronomers (all the more so as one gets farther away from the origin). Stokes used instead the "devilish" method of divergent series[22] and, after bypassing some nontrivial difficulties (linked, in particular, to complex integration), obtained *all* the hands[23] with a precision of 10^{-4}!

Remark 1.78 There are other well-known techniques to give a sense to the sum of (some) divergent series. The interested reader may read the classic book of Émile Borel [13], the first part of which at least is very readable. Concerning asymptotic expansions, see [72].

precision since α is fixed (equal to $1/137$ approximately) and cannot be made to tend to zero. This suggests that quantum field theory, in its current perturbative form, will one day be replaced by another theory. Of course, as long as a precision to 10^{-100} is "enough"...

[22] Niels ABEL wrote in 1826 that divergent series are "the Devil's invention, and it is shameful to base any proof of any kind on them. By using them, one can get from them whatever result is sought: they have done much evil and caused many paradoxes" (letter to his professor Holmboë).

[23] Only the first is less precise, because it is too small and Stokes used an asymptotic expansion at $+\infty$.

EXERCISES

Physical "paradoxes"

♦ **Exercise 1.2 (Electrical energy)** Consider an electric circuit consisting of two identical capacitors in series, with capacitance C and resistance R. Suppose that for $t \leqslant 0$, the circuit is open, one of the capacitors carries the charge Q, and the other has no charge. At $t = 0$, the circuit is closed, and is left to evolve freely. What is the state of equilibrium for this circuit? What is the energy of the system at $t = 0$? What is the energy as $t \to +\infty$? Show that the missing energy depends only on R. What happened to this energy?

Now assume that $R = 0$. What is the energy of the system at any arbitrary t? What is the limit of this energy as $t \to +\infty$? Do you have any comments?

♦ **Exercise 1.3 (A paradox in optics)** We know that two distinct sources of monochromatic light do not create a clear interference picture in an experiment with Young slits. As the distance between the sources increases, we first see a contrast decrease in the interference picture. This is called a **defect of spatial coherence**.

Hence, a famous experiment gives a measurement of the angular distance between two components of a double star by the observation of the first disappearance of the interference fringes when slowly moving two Young slits apart.

This experiment works very well with monochromatic light. However, if we define two monochromatic sources S_1 and S_2 mathematically, each emits a signal proportional to $e^{2i\pi\nu t}$, and there should be no problem of spatial coherence.

Perform the computation properly. A computation in optics *always* starts with amplitudes (possibly, one may show that the crossed terms cancel out in average, and do the computations with intensity only). Here, the cross terms are fine, and never disappear. In other words, this shows that two different monochromatic light sources are always perfectly coherent.

But experiment shows the opposite: a defect of spatial coherence. How can this be explained?

♦ **Exercise 1.4** In the rubber ball paradox of page 2, give an interpretation of the variation of kinetic energy of the ball, in the moving reference frame, in terms of the work of the force during the rebound. The shock may be modeled by a very large force lasting a very short amount of time, or one can use the formalism of distributions (see Chapter 7).

Sequences and series

♦ **Exercise 1.5** It is known that $\mathbb{Q}$, and hence also $\mathbb{Q} \cap [0,1]$, is countable. Let $(x_n)_{n\in\mathbb{N}}$ be a sequence of rational numbers such that $\mathbb{Q} \cap [0,1] = \{x_n \ ; \ n \in \mathbb{N}\}$. Show that the sequence $(x_n)_{n\in\mathbb{N}}$ diverges.

♦ **Exercise 1.6** In an arbitrary normed vector space, show that a Cauchy sequence which has a convergent subsequence is convergent.

♦ **Exercise 1.7** Show that the space $\mathbb{K}[X]$ of polynomials with coefficients in $\mathbb{K}$ is not complete with the norm given by

$$P = \sum_{i=1}^{n} \alpha_i X^i \longmapsto \|P\| \stackrel{\text{def}}{=} \max_{1\leqslant i\leqslant n} |\alpha_i| .$$

♦ **Exercise 1.8 (Fixed point)** Let $a, b \in \mathbb{R}$ be real numbers with $a < b$, and let $f : [a,b] \to [a,b]$ be a continuous function with a fixed point ℓ. Assume that there exists a real number

λ and an interval $V = [\ell - \varepsilon, \ell + \varepsilon]$ around ℓ, contained in $[a, b]$ and stable under f (i.e., $f(x) \in V$ if $x \in V$), such that

$$\forall x \in V \qquad \left|f(x) - f(\ell)\right| \leqslant \lambda \left|x - \ell\right|^2.$$

i) Let $\alpha \in V$ be such that $\lambda(\alpha - \ell) < 1$. Let u be the sequence defined by induction by

$$u_0 = \alpha, \qquad u_{n+1} = f(u_n) \quad \text{for all } n \in \mathbb{N}.$$

Show that $(u_n)_{n\in\mathbb{N}}$ converges to ℓ.

ii) Show in addition that $|u_n - \ell| \leqslant \lambda^{-1} \cdot \big(\lambda(\alpha - \ell)\big)^{2^n}$.

♦ **Exercise 1.9** Let

$$f_n(x) = \begin{cases} 2n^3 x & \text{if } 0 \leqslant x \leqslant 1/2n, \\ n^2 - 2n^3\,(x - 1/2n) & \text{if } 1/2n \leqslant x \leqslant 1/n, \\ 0 & \text{if } 1/n \leqslant x \leqslant 1, \end{cases}$$

for $n \in \mathbb{N}$ and $x \in [0, 1]$. Plot a graph of f, and compute

$$\lim_{n\to\infty} \int_0^1 f_n(x)\,\mathrm{d}x \qquad \text{and} \qquad \int_0^1 \lim_{n\to\infty} f_n(x)\,\mathrm{d}x.$$

♦ **Exercise 1.10** Let $(f_n)_{n\in\mathbb{N}}$ be a sequence of functions converging simply to a function f. If each f_n is increasing, show that f is also increasing. Show that the same stability holds for the properties "f_n is convex" and "f_n is k-Lipschitz." Show that, on the other hand, it is possible that each f_n is continuous, but f is not (take $f_n(x) = \sin^{2n} x$).

♦ **Exercise 1.11** Let φ_n be the function defined on $[-1, 1]$ by

$$\varphi_n(x) = \int_0^x \left(1 - \mathrm{e}^{-1/nt^2}\right) \mathrm{d}t$$

for $n \in \mathbb{N}^*$.

Show that φ_n is infinitely differentiable, and that the sequence $(\varphi_n)_{n\in\mathbb{N}^*}$ converges uniformly on $[-1, 1]$. What is its limit?

Let $\varepsilon > 0$ be given. Show that for any $p \in \mathbb{N}$, there exists a map Ψ_p from $[-1, 1]$ into $\mathbb{R}$, infinitely differentiable, such that

i) $\Psi_p^{(k)}(0) = 0$ for $k \neq p$, and $\Psi_p^{(p)}(0) = 1$.

ii) for $k \leqslant p - 1$ and $x \in [-1, 1]$, $\quad \left|\Psi_p^{(k)}(x)\right| \leqslant \varepsilon$.

Now let $(a_n)_{n\in\mathbb{N}}$ be an arbitrary sequence of real numbers. Construct an infinitely differentiable map f from $[-1, 1]$ to $\mathbb{R}$ such that $f^{(n)}(0) = a_n$ for all $n \in \mathbb{N}$.

♦ **Exercise 1.12 (Slightly surprising exercise)** Construct a series of functions $\sum f_n$ defined on $\mathbb{R}$, which converges pointwise to a sum $F(x)$, the convergence being uniform on any finite interval of $\mathbb{R}$, and which, moreover, satisfies:

$$\forall n \in \mathbb{N} \qquad \lim_{x\to+\infty} f_n(x) = +\infty$$

but

$$\lim_{x\to+\infty} F(x) = -\infty.$$

♦ **Exercise 1.13** Consider a power series centered at the point $a \in \mathbb{C}$, given by

$$f(z) \stackrel{\text{def}}{=} \sum_{n=0}^{\infty} c_n (z-a)^n.$$

Let R denote its radius of convergence, and assume $R > 0$.

i) Prove the **Cauchy formula**: for any $n \in \mathbb{N}$ and any $r \in]0, R[$, we have

$$c_n = \frac{1}{2\pi\, r^n} \int_0^{2\pi} f(a + r\,\mathrm{e}^{\mathrm{i}\theta})\,\mathrm{e}^{-\mathrm{i}n\theta}\,\mathrm{d}\theta.$$

ii) Prove the **Gutzmer formula**: for $r \in]0, R[$, we have

$$\sum_{n=0}^{\infty} |c_n|^2\, r^{2n} = \frac{1}{2\pi} \int_0^{2\pi} \left| f(a + r\,\mathrm{e}^{\mathrm{i}\theta}) \right|^2 \mathrm{d}\theta.$$

iii) Prove that if $R = +\infty$, in which case the sum $f(z)$ of the power series is said to be an *entire function*, and if moreover f is bounded on $\mathbb{C}$, then f is constant (this is Liouville's theorem, which is due to Cauchy).

iv) Is the sine function a counter-example to the previous result?

♦ **Exercise 1.14** Let f be a function of $\mathscr{C}^\infty$ class defined on an open set $\Omega \subset \mathbb{R}$. Show that f is analytic if and only if, for any $x_0 \in \Omega$, there are a neighborhood $\mathscr{V}$ of x_0 and positive real numbers M and t such that

$$\forall x \in \mathscr{V} \quad \forall p \in \mathbb{N} \qquad \left| \frac{f^{(p)}(x)}{p!} \right| \leqslant M\, t^p.$$

Function of two variables

♦ **Exercise 1.15** Let $f : \mathbb{R}^2 \to \mathbb{R}$ be a function of two real variables. This exercise gives examples showing that the limits

$$\lim_{x\to 0} \lim_{y\to 0} f(x,y) \qquad \lim_{y\to 0} \lim_{x\to 0} f(x,y) \qquad \text{and} \qquad \lim_{(x,y)\to(0,0)} f(x,y)$$

are "independent": each may exist without the other two existing, and they may exist without being equal.

i) Let $f(x,y) = \begin{cases} \dfrac{xy}{x^2+y^2} & \text{if } x^2+y^2 \neq 0, \\ 0 & \text{if } x = y = 0. \end{cases}$

Show that the limits $\lim_{x\to 0} \lim_{y\to 0} f(x,y)$ and $\lim_{y\to 0} \lim_{x\to 0} f(x,y)$ both exist, but that the limit[24] $\lim_{(x,y)\to(0,0)} f(x,y)$ is not defined.

---

[24] The limit of f as the *pair* (x,y) tends to a value $(a,b) \in \mathbb{R}^2$ is defined using any of the natural norms on $\mathbb{R}^2$, for instance the norm $\|(x,y)\|_\infty = \max\big(|x|, |y|\big)$ or the euclidean norm $\|(x,y)\|_2 \stackrel{\text{def}}{=} \sqrt{x^2+y^2}$, which are equivalent. Thus, we have

$$\lim_{\substack{(x,y)\to(a,b) \\ (x,y)\neq(a,b)}} f(x,y) = \ell$$

if and only if

$$\forall \varepsilon > 0 \quad \exists \eta > 0 \qquad \big((x,y) \neq (a,b) \text{ and } \left\|(x-a, y-b)\right\|_\infty < \eta\big) \Longrightarrow \big(|f(x,y) - \ell| \leqslant \varepsilon\big).$$

ii) Let $f(x,y) = \begin{cases} y + x\sin(1/y) & \text{if } y \neq 0, \\ 0 & \text{if } y = 0. \end{cases}$

Show that both limits $\lim\limits_{(x,y)\to(0,0)} f(x,y)$ and $\lim\limits_{y\to 0}\lim\limits_{x\to 0} f(x,y)$ exist, but on the other hand $\lim\limits_{x\to 0}\lim\limits_{y\to 0} f(x,y)$ does not exist.

iii) Let $f(x,y) = \begin{cases} \dfrac{xy}{x^2+y^2} + y\sin\left(\dfrac{1}{x}\right) & \text{if } x \neq 0, \\ 0 & \text{if } x = 0. \end{cases}$

Show that $\lim\limits_{\substack{x\to 0 \\ x\neq 0}}\lim\limits_{\substack{y\to 0 \\ y\neq 0}} f(x,y)$ exists. Show that neither $\lim\limits_{(x,y)\to(0,0)} f(x,y)$, nor $\lim\limits_{\substack{y\to 0 \\ y\neq 0}}\lim\limits_{x\to 0} f(x,y)$ exist.

iv) Let $f(x,y) = \begin{cases} \dfrac{x^2-y^2}{x^2+y^2} & \text{if } x^2+y^2 \neq 0, \\ 0 & \text{if } x = y = 0. \end{cases}$

Show that the limits $\lim\limits_{x\to 0}\lim\limits_{y\to 0} f(x,y)$ and $\lim\limits_{y\to 0}\lim\limits_{x\to 0} f(x,y)$ both exist, but are different.

PROBLEM

♦ **Problem 1 (Solving differential equations)** The goal of this problem is to illustrate, in a special case, the Cauchy-Lipschitz theorem that ensures the existence and unicity of the solution to a differential equation with a given initial condition.

In this problem, I is an interval $[0,a]$ with $a > 0$, and we are interested in the nonlinear differential equation

$$y' = \frac{ty}{1+y^2} \tag{E}$$

with the initial condition

$$y(0) = 1. \tag{CI}$$

The system of two equations (E) + (CI) is called the **Cauchy problem**. In what follows, E denotes the space $\mathscr{C}(I,\mathbb{R})$ of real-valued continuous functions defined on I, with the norm $\|f\|_\infty = \sup\limits_{t\in I} |f(t)|$.

i) Let $(f_n)_{n\in\mathbb{N}}$ be a Cauchy sequence in $(E, \|\cdot\|_\infty)$.

(a) Show that for any $x \in I$ the sequence $(f_n(x))_{n\in\mathbb{N}}$ converges in $\mathbb{R}$. For $x \in I$, we let $f(x) \stackrel{\text{def}}{=} \lim\limits_{n\to\infty} f_n(x)$.

(b) Show that $(f_n)_{n\in\mathbb{N}}$ converges uniformly f on I.

(c) Show that the function $f : I \to \mathbb{R}$ is continuous.

(d) Deduce from this that $(E, \|\cdot\|_\infty)$ is a complete normed vector space.

ii) For any $f \in E$, define a function $\Phi(f)$ by the formula

$$\Phi(f) : I \longrightarrow \mathbb{R},$$
$$t \longmapsto \Phi(f)(t) = 1 + \int_0^t \frac{u\,f(u)}{1+\big(f(u)\big)^2}\,du.$$

Show that the functions $f \in E$ which are solutions of the Cauchy problem (E) + (CI) are exactly the fixed points of Φ.

iii) Show that the function $x \longmapsto \dfrac{x}{1+x^2}$ is 1-Lipschitz on $\mathbb{R}$, i.e.,

$$\left| \frac{y}{1+y^2} - \frac{x}{1+x^2} \right| \leqslant |y-x| . \tag{1.8}$$

iv) Show that Φ is a contracting map if a is sufficiently small.

v) Show that there exists a unique solution to the Cauchy problem. Give an explicit iterative method to solve the system numerically (**Picard iterations**).

Remark 1.79 In general, all this detailed work need not be done: the Cauchy-Lipschitz theorem states that for any continous function $\psi(x, y)$ which is locally Lipschitz with respect to the second variable, the Cauchy problem

$$y' = \psi(t, y),$$

has a unique maximal solution (i.e., a solution defined on a maximal interval).

SOLUTIONS

♦ **Solution of exercise 1.2.** The energy of the circuit at the beginning of the experiment is the energy contained in the charged capacitor, namely $E = Q^2/2C$. At equilibrium, when $[t \to \infty]$, no current flows, and the charge of each capacitor is $Q/2$ (it is possible to write down the necessary differential equations and solve them to check this). Thus the final energy is $E' = 2(Q/2)^2/C = E/2$. The energy which is dissipated by the Joule effect (computed by the integral $\int_0^{+\infty} Ri^2(t)\,dt$, where $t \mapsto i(t)$ is the current flowing through the circuit at time t) is of course equal to $E - E'$, and does not depend on R.

However, if $R = 0$, one observes oscillations of charge in each capacitor. The total energy of the system is conserved (it is not possible to compute it from relations in a quasi-stationary regime; one must take magnetic fields into account!). In particular, as $[t \to +\infty]$, the initial energy is recovered. The explanation for this apparent contradiction is similar to what happened for Romeo and Juliet: the time to reach equilibrium is of order $2/RC$ and tends to infinity as $[R \to 0]$. This is a typical situation where the limits $[R \to 0]$ and $[t \to +\infty]$ do not commute.

Finally, if we carry the computations even farther, it is possible to take into account the electromagnetic radiation due to the variations of the electric and magnetic fields. There is again some loss of energy, and for $[t \to +\infty]$, the final energy $E - E' = E/2$ is recovered.

♦ **Solution of exercise 1.3.** Light sources are never purely monochromatic; otherwise there would indeed be no spatial coherence problem. What happens is that light is emitted in wave packets, and the spectrum of the source necessarily has a certain width $\Delta\lambda > 0$ (in a typical example, this is order of magnitude $\Delta\nu = 10^{14}\ \text{s}^{-1}$, corresponding to a coherence length of a few microns for a standard light-bulb; the coherence length of a small He-Ne laser is around thirty centimeters, and that of a monomode laser can be several miles). All computations must be done first with $\Delta\lambda \neq 0$ before taking a limit $\Delta\lambda \to 0$. Thus, surprisingly, spatial coherence is *also* a matter of temporal coherence. This is often hidden, with the motto being "since the sources are not coherent, I must work by summing intensities instead of amplitudes."

In fact, when considering an interference figure, one must always sum amplitudes, and then (this may be a memory from your optics course, or an occasion to read Born and Wolf [14]) perform a *time average* over a period Δt, which may be very small, but not too much (depending on the receptor; the eyes are pretty bad in this respect, an electronic receptor is better, but none can have $\Delta t = 0$).

The delicate issue is to be careful of a product $\Delta t \cdot \Delta\lambda$. If you come to believe (wrongly!) that two purely monochromatic sources interfere without any spatial coherence defect, this means that you have assumed $\Delta t \cdot \Delta\lambda = 0$. To see the spatial coherence issue arise, one must keep $\Delta\lambda$ large enough so that $\Delta t \cdot \Delta\lambda$ cannot be neglected in the computation.

♦ **Solution of exercise 1.7.** Let $P_n = \sum_{k=1}^{n} X^k/k$. It is easy to see that $(P_n)_{n\in\mathbb{N}}$ is a Cauchy sequence: for any integers k and p, we have $\|P_{p+k} - P_p\| = \frac{1}{p+1}$.

However, this sequence does not converge in $\mathbb{K}[X]$; indeed, if it were to converge to a limit L, we would have $L \in \mathbb{K}[X]$, and all coefficients of L of degree large enough ($\geqslant N$, say) would be zero, which implies that $\|P_k - L\| \geqslant 1/(\deg L + 1)$ if $k \geqslant N$, contradicting that $(P_n)_{n\in\mathbb{N}}$ converges to L.

This example shows that $\mathbb{K}[X]$ is not complete.

♦ **Solution of exercise 1.9.** For any $n \in \mathbb{N}$, we have $\int f_n = n/2$, and for any $x \in [0,1]$, the sequence $\big(f_n(x)\big)_n$ tends to 0, showing that

$$\lim_{n\to\infty} \int f_n(x)\,\mathrm{d}x = +\infty \qquad \text{whereas} \qquad \int \Big(\lim_{n\to\infty} f_n(x)\Big)\,\mathrm{d}x = 0.$$

♦ **Solution of exercise 1.11.** The sequence $(\varphi_n)_{n\in\mathbb{N}}$ converges uniformly to 0.

Notice also the property

$$\varphi_n(0) = 0, \qquad \varphi_n'(0) = 1, \qquad \varphi_n^{(k)}(0) = 0 \qquad \forall k \geqslant 2.$$

For given $\varepsilon > 0$, it suffices to define Ψ_p as the $(p-1)$-st primitive of φ_N, where N is sufficiently large so that $\sup_{x\in[-1,1]} \left|\psi_p^{(p-1)}(x)\right| \leqslant \varepsilon$. Here, each primitive is selected to be the one vanishing at 0 (i.e., the integral from 0 to x of the previous one). It is easy to see that the successive derivatives of this function satisfy the require condition, and the last property follows from the construction.

Now let $(a_n)_{n\in\mathbb{N}}$ be an arbitrary sequence of real numbers. For all $n \in \mathbb{N}$, one can apply the previous construction to find Ψ_n such that

$$\sup_{x\in[-1,1]} \left|\psi_n^{(n-1)}(x)\right| \leqslant \frac{1}{2^n \cdot \max(1, |a_n|)}.$$

It is then immediate that the series $\sum a_n \Psi_n$ converges uniformly to a function f having all desired properties.

Of course, the function f thus constructed is by no means unique: one may add a term $\alpha(\varphi_n' - 1)$, where $\alpha \in \mathbb{R}$, without changing the values of the derivatives at 0.

♦ **Solution of exercise 1.12.** Let

$$f_n(x) = -\frac{x^{4n-1}}{(4n-1)!} + \frac{x^{4n+1}}{(4n+1)!}$$

for $n \geqslant 1$; the series $\sum f_n$ converges to $F(x) = \sin x - x$.

♦ **Solution of exercise 1.13**

i) The power series for $f(a + r\,\mathrm{e}^{\mathrm{i}\theta})$ may be integrated term by term because of its absolute convergence in the disc centered at a of radius $r < R$. Since we have

$$\int_0^{2\pi} \mathrm{e}^{\mathrm{i}(k-n)\theta}\,\mathrm{d}\theta = \delta_{kn} = \begin{cases} 1 & \text{if } k = n, \\ 0 & \text{otherwise,} \end{cases}$$

the stated formula follows.

ii) Similarly, expand $\left|f(a + r\,e^{i\theta})\right|^2$ as a product of two series and integrate term by term. Most contributions cancel out using the formula above, and only the terms $|c_n|^2\, r^{2n}$ remain.

iii) If f is bounded on $\mathbb{C}$, we have $|c_n\, r^n| \leqslant \|f\|_\infty$. Letting $r \to +\infty$, it follows that $c_n = 0$ for $n \geqslant 1$, which means that f is constant.

iv) The function sin is *not* bounded on $\mathbb{C}$! Indeed, we have for instance $\lim_{x\to+\infty} \left|\sin(ix)\right| = +\infty$. So there is no trouble.

♦ **Solution of problem 1**

i) (a) Let $x \in I$. For any $p, q \in \mathbb{N}$, we have

$$\left|f_p(x) - f_q(x)\right| \leqslant \sup_{y\in I} \left|f_p(y) - f_q(y)\right| = \left\|f_p - f_q\right\|_\infty,$$

and this proves that the sequence $\big(f_n(x)\big)_{n\in\mathbb{N}}$ is a Cauchy sequence in $\mathbb{R}$, so it converges.

(b) Let $\varepsilon > 0$ be fixed. There exists N such that $\left\|f_p - f_q\right\|_\infty \leqslant \varepsilon$ for all $p > q > N$. Let $x \in I$. We then have

$$\left|f_p(x) - f_q(x)\right| \leqslant \varepsilon \qquad \text{for any } p > q > N,$$

and since this holds for all p, we may fix q and let $p \to \infty$. We obtain

$$\left|f(x) - f_q(x)\right| \leqslant \varepsilon \qquad \text{for all } q > N.$$

This bound holds independently of $x \in I$. Thus we have shown that

$$\left\|f - f_q\right\|_\infty \leqslant \varepsilon \qquad \text{for any } q \geqslant N.$$

Finally, this being true for any $\varepsilon > 0$, it follows that the sequence $(f_n)_{n\in\mathbb{N}}$ converges uniformly to f.

Remark: *At this point, we haven't proved that there is convergence in the normed vector space* $\big(E, \|\cdot\|_\infty\big)$. *It remains to show that the limit f is in E, that is, that f is continuous. This follows from Theorem 1.33, but we recall the proof.*

(c) Let $x \in I$; we now show that f is continuous at x.

Let $\varepsilon > 0$ be fixed. From the preceding question, there exists an integer N such that $\|f_n - f\|_\infty \leqslant \varepsilon$ for all $n \geqslant N$, and in particular $\|f_N - f\|_\infty \leqslant \varepsilon$. Since f_N is an element of E, it is continuous. So there exists $\eta > 0$ such that

$$\forall y \in I \qquad |y - x| \leqslant \eta \Longrightarrow \left|f_N(y) - f_N(x)\right| \leqslant \varepsilon.$$

Using the triangle inequality, we deduce from this that for all $y \in I$ such that $|x - y| \leqslant \eta$, we have

$$\left|f(y) - f(x)\right| \leqslant \left|f(y) - f_N(y)\right| + \left|f_N(y) - f_N(x)\right| + \left|f_N(x) - f(x)\right| \leqslant 3\varepsilon.$$

This proves the continuity of f at x, and since x is arbitrary, this proves that f is continuous on I, and hence is an element of E.

(d) For any Cauchy sequence $(f_n)_{n\in\mathbb{N}}$ in E, the previous questions show that $(f_n)_{n\in\mathbb{N}}$ converges in E. Hence the space $\big(E, \|\cdot\|_\infty\big)$ is complete.

ii) Let f be a fixed point of Φ. Then we have

$$\forall t \in I \qquad f'(t) = \big(\Phi(f)\big)'(t) = \frac{t\,f(t)}{1 + \big(f(t)\big)^2}.$$

Moreover, it is easy to see that $f(0) = \Phi(f)(0) = 1$.

Conversely, let f be a solution of the Cauchy problem $(E)+(CI)$. Then $\Phi(f)$ is differentiable and we have

$$\forall t \in I \qquad \Phi(f)'(t) = \frac{t\,f(t)}{1+\big(f(t)\big)^2} = f'(t).$$

The functions f and $\Phi(f)$ have the same derivative on I, and moreover satisfy

$$\Phi(f)(0) = 1 \qquad \text{and} \qquad f(0) = 1.$$

It follows that $\Phi(f) = f$.

iii) Looking at the derivative $g' : x \mapsto \dfrac{1-x^2}{(1+x^2)^2}$ we see that $|g'(x)| \leqslant 1$ for all $x \in \mathbb{R}$. The mean-value theorem then proves that g is 1-Lipschitz, as stated.

iv) Let $f, g \in E$. Then we have

$$\begin{aligned}
\big\|\Phi(f) - \Phi(g)\big\|_\infty &= \sup_{t\in I}\left|\int_0^t \left(\frac{u\,f(u)}{1+\big(f(u)\big)^2} - \frac{u\,g(u)}{1+\big(g(u)\big)^2}\right) \mathrm{d}u\right| \\
&\leqslant \sup_{t\in I}\int_0^t u\left|\frac{f(u)}{1+\big(f(u)\big)^2} - \frac{g(u)}{1+\big(g(u)\big)^2}\right| \mathrm{d}u \\
&\leqslant \int_0^a u\left|\frac{f(u)}{1+\big(f(u)\big)^2} - \frac{g(u)}{1+\big(g(u)\big)^2}\right| \mathrm{d}u
\end{aligned}$$

by positivity. Using the inequality of the previous question, we get

$$\big\|\Phi(f) - \Phi(g)\big\|_\infty \leqslant \int_0^a u\big|f(u) - g(u)\big|\,\mathrm{d}u \leqslant \|f-g\|_\infty \int_0^a u\,\mathrm{d}u = \frac{a^2}{2}\,\|f-g\|_\infty\,.$$

This is true for any $f, g \in E$, and hence Φ is $(a^2/2)$-Lipschitz; if $0 \leqslant a < \sqrt{2}$, this map Φ is a contraction.

v) According to the fixed-point theorem, the previous results show that Φ has a unique fixed point in E.

According to Question ii), this means that there exists a unique solution of the Cauchy Problem $(E)+(CI)$ on an interval $[0,a]$ for $a < \sqrt{2}$.

To approximate the solution numerically, it is possible to select an arbitrary function f_0 (for instance, simply $f_0 = 0$), and construct the sequence $(f_n)_{n\in\mathbb{N}}$ defined by $f_{n+1} = \Phi(f_n)$ for $n \geqslant 0$. This requires computing (numerically) some integrals, which is a fairly straightforward matter (numerical integration is usually numerically stable: errors do not accumulate in general[25]). The speed of convergence of the sequence $(f_n)_{n\in\mathbb{N}}$ to the solution f of the Cauchy problem is exponential: with $I = [0,1]$, the distance (from the norm on E) between f_n and f is divided by 2 (at least) after each iterative step. It is therefore possible to expect a good numerical approximation after few iterations (the precision after ten steps is of the order of $\|f_0 - f\|_\infty/1000$ since $2^{10} = 1024$).

[25] On the other hand, numerical differentiation tends to be much more delicate.

Chapter 2

Measure theory and the Lebesgue integral

This chapter is devoted to the theory of integration. After a survey of the Riemann integral and a discussion of its limitations, we present the Lebesgue integral. This requires a brief discussion of basic measure theory, which is important for probability theory as well.

Note that the main theorems and techniques of integral calculus (change of variable, justification of exchanges of limits and integrals) are discussed in the next chapter.

2.1 The integral according to Mr. Riemann

2.1.a Riemann sums

One possible way of defining the integral of a function f on a finite interval $[a, b]$ is the following: start by subdividing the interval in n parts, which are more or less of the same size $\approx (b - a)/n$, by choosing real numbers

$$a = x_0 < x_1 < x_2 < \cdots < x_n = b ;$$

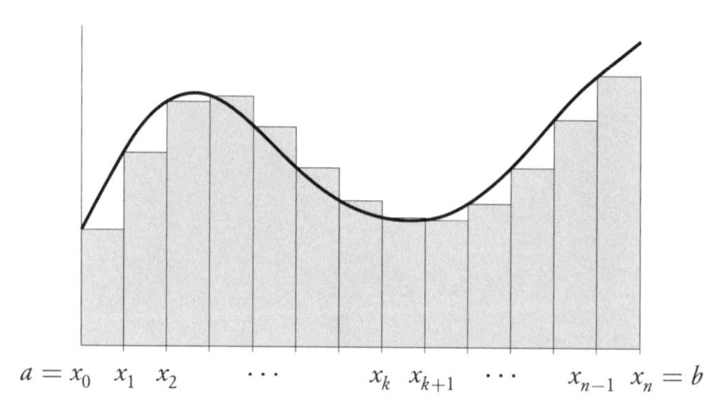

Fig. 2.1 – The interval $[a, b]$ is here partitioned uniformly, with n subintervals of constant length $(b - a)/n$.

then "approximate" the function f by a "step" function which is constant on each interval of the subdivision, and takes there the same value as f at the beginning of the interval:

The sum of the areas of the small rectangles is then equal to

$$S_n = \sum_{k=1}^{n} (x_k - x_{k-1}) \cdot f(x_{k-1}),$$

and it may be expected (or hoped) that as n goes to infinity, this value will converge to some limit. Riemann's result is that this happens if the approximations used for f improve steadily as n increases. This means that the values taken by f on an interval $[x, x + \varepsilon]$ must be very close to $f(x)$ when ε is small. This is a continuity assumption for f. Here is the precise result.

DEFINITION 2.1 (Subdivision) A **subdivision** of the interval $[a, b]$ is any tuple of real numbers of the type

$$\sigma_n = (x_i)_{i=0,\dots,n},$$

where
$$a = x_0 < x_1 < x_2 < \cdots < x_n = b.$$

The **step** of the subdivision σ_n is the length of the largest subinterval:

$$\pi(\sigma) = \max_{i \in [\![0, n-1]\!]} \left| x_{i+1} - x_i \right|.$$

A **marked subdivision** of $[a, b]$ is a pair

$$\sigma'_n = \left((x_i)_{i=0,\dots,n}, (\xi_k)_{k=1,\dots,n} \right),$$

where $(x_i)_{i=0,\dots,n}$ is a subdivision of $[a, b]$ and $\xi_k \in \left[x_{k-1}, x_k \right]$ for all k.

DEFINITION 2.2 (Riemann sums) Let $\sigma'_n = \left((x_i)_i, (\xi_k)_k \right)$ be a marked subdivision of $[a, b]$ and let $f : [a, b] \to \mathbb{K}$ be a function with values in $\mathbb{K} = \mathbb{R}$

or $\mathbb{C}$. The **Riemann sum** of f associated to σ'_n is

$$S(f,\sigma'_n) = \sum_{k=1}^{n}(x_k - x_{k-1}) \cdot f(\xi_k).$$

The following result provides a rigorous definition of the integral of a continuous function on a finite interval:

THEOREM 2.3 (and definition of the integral) *Let $f : [a,b] \to \mathbb{K}$ be a continuous function. For any sequence $(\sigma'_n)_{n\in\mathbb{N}^*}$ of marked subdivisions of $[a,b]$, such that the step of the subdivisions tends to 0 as n tends to infinity, the sequence of associated Riemann sums converges. Moreover, the limit of this sequence is independent of the chosen sequence of marked subdivisions. The* ***(Riemann) integral***[1] ***of the function*** f *is defined to be this common limit:*

$$\int_a^b f(x)\,\mathrm{d}x = \int_a^b f = \lim_{n\to\infty} S(f,\sigma'_n)$$

where $(\sigma'_n)_n$ satisfies $\lim_{n\to\infty} \pi(\sigma'_n) = 0$.

This result and definition may easily be extended to piecewise continuous functions,[2] and to even larger classes of functions, by means of a small change in the construction used. Namely, define a **step function on $[a,b]$** to be a function φ which is constant on the subintervals of a subdivision

$$\sigma = (a_0, \ldots, a_n)$$

of $[a,b]$. The integral of a step function φ associated to σ and taking the value y_i on $]a_{i-1}, a_i[$ is obviously equal to

$$\int_a^b \varphi = \sum_{i=1}^{n}(a_i - a_{i-1})\,y_i.$$

For any function $f : [a,b] \to \mathbb{R}$, define

$$I_{-}(f) = \sup\left\{\int \varphi\ ;\ \varphi \text{ step function with } \varphi \leqslant f\right\}$$

and

[1] The notation "d" for differentiation, "$\mathrm{d}y/\mathrm{d}x$", "$\int$" (which originally represented an "s," the initial of *summa*), are due to Gottfried Leibniz (1646–1716), who also invented the name "function" and popularized the use of the equal sign "=" and of a dot to represent multiplication.

[2] Recall that $f : [a,b] \to \mathbb{K}$ is **piecewise continuous** if there exist an integer n and real numbers

$$a = a_0 < a_1 < \cdots < a_n = b$$

such that, for all $i \in [\![0, n-1]\!]$, the function f restricted to $]a_i, a_{i+1}[$ is continuous and has finite limits as x tends to a_i from above, and to a_{i+1} from below.

$$I_+(f) = \inf\{\textstyle\int \varphi \,;\ \varphi \text{ step function with } \varphi \geqslant f\}.$$

Then f is **integrable in the sense of Riemann, or Riemann-integrable**, if we have $I_-(f) = I_+(f)$.

A more delicate issue is to extend the definition to functions which are defined on an arbitrary interval (not necessarily closed or bounded), such as $]\alpha, \beta]$ or $[\alpha, +\infty[$.

2.1.b Limitations of Riemann's definition

- Some easily defined functions cannot be approximated by step functions, and the Riemann sums for those functions do not converge to a common value. This is the case, for instance, for the function $d : [0,1] \to \mathbb{R}$, which is defined by $d(x) = 0$ if x is irrational, and $d(x) = 1$ if x is rational. This function is therefore not Riemann-integrable.

- The definition turns out to be the very devil to deal with as soon as questions of convergence of sequences of integrals appear. In particular, to justify that the integral of the sum of a series is the sum of the series of integrals, one needs very stringent conditions.

- It also turns out that if we want to deal with a *complete* space of functions with the norm $\int_a^b |f(t)| dt$, not even the most liberal definition of Riemann integral is sufficient. The smallest complete space containing $\mathscr{C}([a,b])$ is much larger than the space of Riemann-integrable functions; it is the space L^1 defined in the next section.

2.2 The integral according to Mr. Lebesgue

To solve the problems related to Riemann's definition of integrable function, the idea is to provide an alternate definition for which a much larger class of functions will be integrable.[3] In particular, this space of integrable functions, denoted L^1, will turn out to be complete.

---

[3] From the point of view of trying to obtain a complete space with the norm

$$\|f\|_1 = \int_a^b |f(x)| \,\mathrm{d}x,$$

it is always possible to "complete" a metric space, in particular, a normed vector space; more precisely, given a normed vector space $(E, \|\cdot\|)$, there exists a complete normed vector space $(F, \|\cdot\|')$ such that E is isometric to a dense subspace E' of F. Using this isometry, E may be seen as a dense subspace of a complete space. The general construction of F is somewhat abstract (F is the set of Cauchy sequences in E, modulo the equivalence relation $u \sim v$, if and

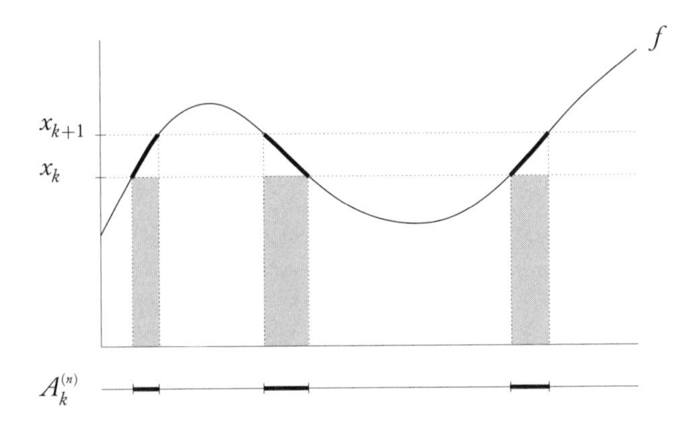

Fig. 2.2 – Construction of the sets $A_k^{(n)} = f^{-1}\big(\,[x_k, x_{k+1}]\,\big)$, where $x_k = k/2^n$, with the value of $\mu\big(A_k^{(n)}\big) \cdot k/2^n$.

2.2.a Principle of the method

Lebesgue's idea to define the integral of a real-valued function is the following: for various possible values (denoted α) of the function, we consider the "size" of the set of points where f takes the value α (or is suitably close to α). Multiply the size of this set by α, and repeat with other values, tallying the total sum thus obtained. Taking a limit in a suitable manner, this sum converges to the integral of the function.[4]

In other words, instead of a "vertical" dissection as in Riemann's method, one has to perform an "horizontal" dissection as follows. Assume first that f takes non-negative values. Subdivide the set $\mathbb{R}^+$ of values in infinitely many intervals, say of size $1/2^n$. For each subinterval of the type $\left[\frac{k}{2^n}, \frac{k+1}{2^n}\right[$, consider the set of real numbers x such that $f(x)$ belongs to this interval (see Figure 2.2):

$$A_k^{(n)} = f^{-1}\left(\left[\frac{k}{2^n}, \frac{k+1}{2^n}\right[\right).$$

Denote by $\mu\big(A_k^{(n)}\big)$ the "size" of this set (temporarily; this crucial notion of "size" will be made precise later), and then define

$$S_n(f) = \sum_{k=0}^{\infty} \mu\big(A_k^{(n)}\big) \cdot \frac{k}{2^n}.$$

only if $u - v$ is a sequence converging to zero), but it is very powerful. In particular, it provides one construction of the set $\mathbb{R}$ of real numbers, starting with Cauchy sequences in $\mathbb{Q}$.

[4] Henri Lebesgue used the following analogy: to sum up a certain amount of money, we can – instead of adding up the values of all coins as they come up one by one – start by putting together all dimes, all nickels, all \$1 bills, and then count the number of each stack, multiply it by the common value of each coin or bill in the stack, and finally add up everything.

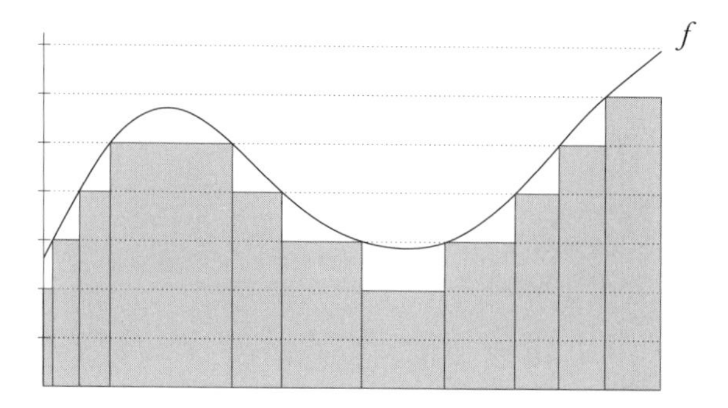

Fig. 2.3 – Construction of the integral of f.

It is quite easy to check that the sequence $\left(S_n(f)\right)_n$ is increasing (see Figure 2.4).

Hence it converges in $\overline{\mathbb{R}^+} = \mathbb{R}^+ \cup \{+\infty\}$, and the limit is denoted $\int f$. The function f is said to be integrable (in the sense of Lebesgue) if the limit is finite.[5]

This is the principle. As can be seen, the issue of evaluating the "size" of the sets $A_k^{(n)}$ has been left open. This very important point is the subject of the next sections.

Remark 2.4 Before going farther, we briefly explain what is gained from this new method:

- the set of integrable functions is vast, beyond human imagination;[6]
- the set of Lebesgue-integrable functions (not yet precisely defined), denoted L^1 or, to be precise, $L^1(\mathbb{R})$ or $L^1[a,b]$, is *complete* for the norm $\|f\| = \int |f|$;
- the theorems of the Lebesgue integral calculus are infinitely simpler and more powerful (in particular, concerning the inversion of limits and integrals, or differentiation under the integral sign), even when applied to very regular functions.

2.2.b Borel subsets

We now take up the question of measuring the "size" of a subset A of $\mathbb{R}$. As we will see, it turns out to be impossible to assign a measure to *all* subsets of $\mathbb{R}$. It is necessary to restrict which subsets are considered; they are called "measurable sets." Since it is natural to expect that the union $A \cup B$ of two

[5] The exact definition below will be slightly different, but equivalent; it will not involve a particular choice of intervals (bounded by numbers of the form $k/2^n$) to subdivide $\mathbb{R}^+$.

[6] Much larger than the set of Riemann-integrable functions, in particular, because even very irregular functions become integrable. It is true that some Riemann-integrable functions are not Lebesgue integrable, and that this causes trouble, but they are typically functions which create all the difficulties of the Riemann theory.

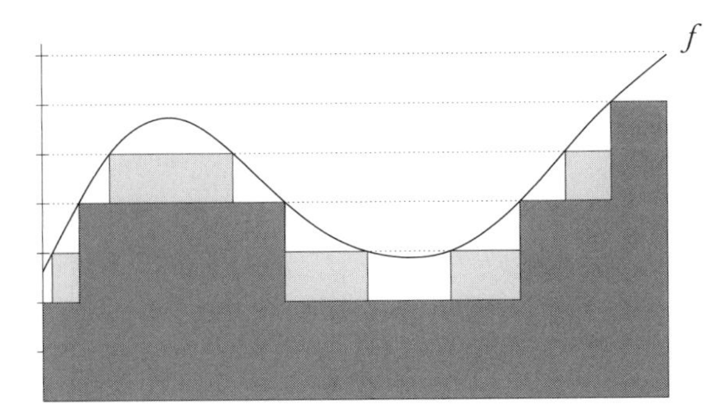

Fig. 2.4 – $S_{n-1}(f)$ (dark gray) is less than $S_n(f)$ (light gray).

subsets which can be measured should also be measurable, and since analysis quickly requires some infinite unions, the following definition is reasonable:

DEFINITION 2.5 Let X be an arbitrary set. The **set of subsets of X** is denoted $\mathfrak{P}(X)$. A **σ-algebra**[7] on X is any subset $\mathscr{T} \subset \mathfrak{P}(X)$ of the set of subsets of X such that the following conditions hold:

(T1) $X \in \mathscr{T}$ and $\varnothing \in \mathscr{T}$;

(T2) for all $A \in \mathscr{T}$, the complement $X \setminus A$ is also in $\mathscr{T}$;

(T3) if $(A_n)_{n \in \mathbb{N}}$ is a sequence of elements of $\mathscr{T}$, then $\bigcup_{n=0}^{\infty} A_n \in \mathscr{T}$.

Hence a σ-algebra $\mathscr{T}$ is *stable* by complement and countable union. Moreover, since the complement of the union of a family of sets is the intersection of their complements, $\mathscr{T}$ is also stable by countable intersections.

Example 2.6 $\mathfrak{P}(X)$ is a σ-algebra (the largest σ-algebra on X); $\{\varnothing, X\}$ is also a σ-algebra (the smallest). For $a \in X$, $\{\varnothing, X, \{a\}, X \setminus \{a\}\}$ is a σ-algebra.

Consider a subset $\mathscr{X}$ of $\mathfrak{P}(X)$ (hence a set, the elements of which are themselves subsets of X). The intersection of all σ-algebras containing $\mathscr{X}$ is still a σ-algebra, and of course it is contained in any σ-algebra containing $\mathscr{X}$. This is therefore the smallest σ-algebra containing $\mathscr{X}$. This leads to the following definition:

DEFINITION 2.7 Let $\mathscr{X} \subset \mathfrak{P}(X)$. The **$\sigma$-algebra generated** by $\mathscr{X}$ is the smallest σ-algebra containing $\mathscr{X}$. It is denoted $\mathscr{T}(\mathscr{X})$. For $A \subset X$, the **σ-algebra generated by A** is the σ-algebra $\mathscr{T}(A)$ generated by $X = \{A\}$.

[7] An **algebra** is a set of subsets stable under complement and *finite* union. The "σ" prefix indicates the countable operations which are permitted.

Émile BOREL (1871–1956), was ranked first at the entrance examination of both the École Polytechnique and the École Normale Supérieure. He chose the latter school and devoted himself to mathematics; to Marguerite, daughter of Paul Appell (also a famous writer under the pseudonym Camille MARBO); and to politics (in the footsteps of PAINLEVÉ). In 1941, he was jailed in Fresnes because of his opposition to the regime of Vichy, and became a member of the resistance as soon as he was freed. In 1956, he received the first Gold Medal from the CNRS. He founded the Institut Henri-Poincaré. Among his most important works, we can mention that he studied the measurability of sets, and in particular defined the sets of reals of measure zero and the Borel sets, on which Lebesgue measure is well defined. He also worked in probability theory and mathematical physics (the kinetic theory of gas). His student Henri LEBESGUE (see page 61) used his results to develop his theory of integration.

Example 2.8 For $a \in X$, the *σ-algebra generated by* $\{a\}$ is $\{\varnothing, X, \{a\}, X \setminus \{a\}\}$.

DEFINITION 2.9 (Borel sets) The **Borel σ-algebra** is the σ-algebra on $\mathbb{R}$ generated by the open subsets of $\mathbb{R}$. It is denoted $\mathscr{B}(\mathbb{R})$. Its elements are called **Borel sets.**

It is easy to check that the Borel σ-algebra is also the σ-algebra generated by intervals of $\mathbb{R}$ of the type $]-\infty, a]$, where a ranges over all real numbers.[8] In particular, the following sets are Borel subsets of $\mathbb{R}$:

1. all open sets;
2. all closed sets (since their complement is open);
3. all countable unions and intersections of open and closed sets.

However, there are many others! The Borel σ-algebra is *much more* complex and resists any simple-minded exhaustive description.

2.2.c Lebesgue measure

It is unfortunately beyond the scope of this book to describe the construction of the Lebesgue measure in all details. Interested readers are invited to look, for instance, in Rudin's book [76] or in Burk's very clear course [16].

DEFINITION 2.10 A **measurable space** is a pair $(X, \mathscr{T})$, where X is an arbitrary set and $\mathscr{T}$ is a σ-algebra on X. Elements of $\mathscr{T}$ are called **measurable sets.**

DEFINITION 2.11 Let $(X, \mathscr{T})$ be a measurable space. A **measure on $(X, \mathscr{T})$**, or **on X**, by abuse of notation, is a map μ defined on the σ-algebra $\mathscr{T}$, with values in $\overline{\mathbb{R}}^+ \stackrel{\text{def}}{=} \mathbb{R}^+ \cup \{+\infty\}$, such that

[8] This is of great importance in probability theory, where it is used to reduce the study of a random variable to the study of its distribution function.

(MES1) $\mu(\varnothing) = 0$;

(MES2) for any *countable* family $(A_n)_{n\in\mathbb{N}}$ of *pairwise disjoint* measurable sets in $\mathscr{T}$, we have

$$\mu\Bigl(\bigcup_{n=0}^{\infty} A_i\Bigr) = \sum_{i=0}^{\infty} \mu(A_i).$$

The second property is called the σ-additivity of the measure.

A **measure space**[9] is a triple $(X, \mathscr{T}, \mu)$, where $(X, \mathscr{T})$ is a measurable space and μ is a measure on $(X, \mathscr{T})$.

Remark 2.12 Since a measure takes values in $\overline{\mathbb{R}}^+ \stackrel{\text{def}}{=} \mathbb{R}^+ \cup \{+\infty\}$, it is necessary to specify the addition rule involving $+\infty$, namely,

$$x + \infty = +\infty \qquad \text{for all } x \in \overline{\mathbb{R}}^+.$$

Note that the usual property "$(x+z = y+z) \Longrightarrow (x = y)$" is no longer valid in full generality (e.g., $2 + \infty = 3 + \infty$); however, it holds if $z \neq +\infty$.

The following theorem, due to Lebesgue, shows that there exists a measure on the Borel σ-algebra which extends the length of an interval.

THEOREM 2.13 (The Lebesgue measure on Borel sets) *The pair* $(\mathbb{R}, \mathscr{B})$ *is a measurable set and there exists on* $(\mathbb{R}, \mathscr{B})$ *a unique measure* μ *such that*

$$\mu\bigl([a, b]\bigr) = |b - a| \qquad \textit{for all } a, b \in \mathbb{R}.$$

The idea for the construction of the Lebesgue measure is described in Sidebar 1 on the following page.

The Lebesgue measure μ is translation-invariant (i.e., translating a measurable set does not change its measure), and is **diffuse** (that is, for any finite or countable set Δ, we have $\mu(\Delta) = 0$). In particular, $\mu(\mathbb{N}) = \mu(\mathbb{Z}) = \mu(\mathbb{Q}) = 0$. Note that the converse is false: for instance, the Cantor triadic set is not countable, yet it is of measure zero (see Exercise 2.5 on page 68).

2.2.d The Lebesgue σ-algebra

The Lebesgue measure on Borel sets is sufficient for many applications (in particular, it is commonly used in probability theory).

Nevertheless, it is interesting to know that the construction may be refined. Consider a Borel set A with measure zero. There may exist subsets $B \subset A$ which are *not* of measure zero, simply because B is not a Borel set, and hence is not measurable. But it seems intuitively obvious that a subset of a set of measure zero should also be considered negligible and be of measure zero (in

---

[9] Note that very similar definitions will appear in Chapter 19, concerning *probability spaces*.

Sidebar 1 (Lebesgue measure for Borel sets)

Consider the "length" map ℓ *which associates its length* $b-a$ *to any finite interval* $[a,b]$. *We try to* extend *this definition to an arbitrary subset of* $\mathbb{R}$.

Let $A \in \mathfrak{P}(\mathbb{R})$ *be an* arbitrary *subset of* $\mathbb{R}$.

Let $\mathscr{E}(A)$ *be the set of all sequences* $\mathscr{A} = (A_n)_{n\in\mathbb{N}}$ *of intervals such that* A *is contained in* $\bigcup_{i=0}^{\infty} A_i$. *For any* $\mathscr{A} = (A_n)_{n\in\mathbb{N}} \in \mathscr{E}(A)$, *let*

$$\lambda(\mathscr{A}) \stackrel{\text{def}}{=} \sum_{i=0}^{\infty} \ell(A_i),$$

and define the ***Lebesgue exterior measure*** *of* A *as follows:*

$$\lambda^*(A) \stackrel{\text{def}}{=} \inf_{\mathscr{A}\in\mathscr{E}(A)} \lambda(\mathscr{A}).$$

Unfortunately, this "exterior measure" is not *a measure as defined previously: the property of* σ*-additivity (*MES2*) is not satisfied (this would contradict the result of Sidebar 2).*

On the other hand, one can show that the restriction of λ^* *to the Borel* σ*-algebra*

$$\begin{aligned} \mu : \mathscr{B} &\longrightarrow \mathbb{R} \\ A &\longmapsto \mu(A) = \lambda^*(A) \end{aligned}$$

is a measure (this is by no means obvious or easy to prove). It is the ***Lebesgue measure for Borel sets.***

particular, measurable). In fact, it is possible to define a σ-algebra containing all Borel sets, and all subsets of Borel sets with Lebesgue measure zero, and to extend the Lebesgue measure to this σ-algebra. This σ-algebra is called the **Lebesgue σ-algebra** and is denoted $\mathscr{L}$. We have therefore $\mathscr{B} \subset \mathscr{L}$. (The construction of $\mathscr{L}$ and of the required extension of the Lebesgue measure are again beyond the scope of this book.) We thus have:

PROPOSITION 2.14 *Let* A *be an element of the Lebesgue* σ*-algebra such that* $\mu(A) = 0$. *Then any subset* $B \subset A$ *is in the Lebesgue* σ*-algebra and satisfies* $\mu(B) = 0$.

This property is called *completeness* of the measure space $(\mathbb{R}, \mathscr{L}, \mu)$ or of the Lebesgue σ-algebra.[10] The Borel σ-algebra is not complete, which justifies the introduction of the larger Lebesgue σ-algebra.

Remark 2.15 *In the remainder of this and the next chapter,* we will consider the Lebesgue measure on the Lebesgue σ-algebra, and a measurable set will be an element of $\mathscr{L}$. Note that in practice (for a physicist) there is really no difference between the Lebesgue and Borel σ-algebras.

[10] This is not the same notion as completeness of a topological space.

Henri LEBESGUE (1875–1941), son of a typographical worker, entered the École Normale Supérieure, where he followed the courses of Émile BOREL. In 1901, he published in his thesis the theory of the Lebesgue measure, generalizing previous work of Borel (and also of Camille JORDAN and Giuseppe PEANO). The following year, he developed his theory of integration, and used it to study series of functions and Fourier series. He proved that a bounded function is integrable in the sense of Riemann if and only if it is continuous except on a negligible set. In his later years, Lebesgue turned against analysis and would only discuss elementary geometry.

2.2.e Negligible sets

DEFINITION 2.16 A subset $N \subset X$ of a measure space $(X, \mathscr{T}, \mu)$ is **negligible** if it is contained in a measurable set $N' \in \mathscr{T}$ such that $\mu(N') = 0$.

In particular, a subset N of $\mathbb{R}$ is **negligible** if and only if it is in the Lebesgue σ-algebra and is of Lebesgue measure zero.

Remark 2.17 In terms of the simpler Borel σ-algebra, a subset N of $\mathbb{R}$ is negligible if and only if there exists a Borel set N' containing N which is of Lebesgue measure zero. But N itself is not necessarily a Borel set.

Example 2.18 The set $\mathbb{Q}$ of rational numbers is negligible since, using σ-additivity, we have

$$\mu(\mathbb{Q}) = \sum_{r \in \mathbb{Q}} \mu(\{r\}) = \sum_{r \in \mathbb{Q}} 0 = 0.$$

DEFINITION 2.19 A property P defined for elements of a measure space $(X, \mathscr{T}, \mu)$ is **true for almost all $x \in X$** or **holds almost everywhere**, abbreviated **holds a.e.**, if it is true for all $x \in X$ except those in a negligible set.

Example 2.20 The Heaviside function

$$H : x \longmapsto \begin{cases} 0 & \text{if } x < 0, \\ 1 & \text{if } x \geqslant 0, \end{cases}$$

is differentiable almost everywhere.

The Dirichlet function

$$D : x \longmapsto \begin{cases} 1 & \text{if } x \in \mathbb{Q}, \\ 0 & \text{if } x \in \mathbb{R} \setminus \mathbb{Q}, \end{cases}$$

is zero almost everywhere.

DEFINITION 2.21 Let $(X, \mathscr{T})$ and $(Y, \mathscr{T}')$ be measurable spaces. A function $f : X \to Y$ is **measurable** if, for any measurable subset $B \subset Y$, the set

$$f^{-1}(B) \stackrel{\text{def}}{=} \{x \in X \; ; \; f(x) \in B\}$$

of X is measurable, that is, if $f^{-1}(B) \in \mathscr{T}$ for all $B \in \mathscr{T}'$.

This definition should be compared with the definition of continuity of a function on a topological space (see Definition A.3 on page 574).

Remark 2.22 *In practice*, all functions defined on $\mathbb{R}$ or $\mathbb{R}^d$ that a physicist (or almost any mathematician) has to deal with are measurable, and it is usually pointless to bother checking this. Why is that the case? The reason is that it is *difficult* to exhibit a nonmeasurable function (or even a single nonmeasurable set); it is *necessary* to invoke at some point the Axiom of Choice (see Sidebar 2), which is not physically likely to happen. The famous Banach-Tarski paradox shows that nonmeasurable sets may have very weird properties.[11]

2.2.f Lebesgue measure on $\mathbb{R}^n$

Part of the power of measure theory as defined by Lebesgue is that it extends fairly easily to product spaces, and in particular to $\mathbb{R}^n$. First, one can consider on $\mathbb{R} \times \mathbb{R}$ the σ-algebra generated by sets of the type $A \times B$, where $(A, B) \in \mathscr{B}(\mathbb{R})^2$. This σ-algebra is denoted $\mathscr{B}(\mathbb{R}) \otimes \mathscr{B}(\mathbb{R})$; in fact, it can be shown to coincide with the Borel σ-algebra[12] of $\mathbb{R}^2$ with the usual topology. It is then possible to prove that there exists a unique measure denoted $\pi = \mu \otimes \mu$ on $\mathscr{B}(\mathbb{R}) \otimes \mathscr{B}(\mathbb{R})$ which is given on rectangles by the natural definition

$$\pi(A \times B) = \mu(A)\, \mu(B)$$

for A, $B \in \mathscr{B}(\mathbb{R})$. This measure is called the **product measure**, or the Lebesgue measure on $\mathbb{R}^2$. The construction may be repeated, yielding a Borel σ-algebra and a Lebesgue measure on $\mathbb{R}^n$ for all $n \geqslant 1$.

2.2.g Definition of the Lebesgue integral

Here we give a general definition of the Lebesgue integral on an arbitrary measure space $(X, \mathscr{T}, \mu)$, which – in a first reading – the reader may assume is $\mathbb{R}$ with the Lebesgue measure. It yields the same results as the construction sketched in Section 2.2.a, but the added generality is convenient to prove the important results of the Lebesgue integral calculus.

The first step is to define the integral for a non-negative measurable simple function (i.e., a function taking only finitely many values). This definition

[11] Consider a solid ball $\mathcal{B}$ of radius 1 in $\mathbb{R}^3$. There exists a way of partitioning the ball in five disjoint pieces $\mathcal{B} = \mathcal{B}_1 \cup \cdots \cup \mathcal{B}_5$ such that, after translating and rotating the pieces $\mathcal{B}_i$ suitably – without dilation of any kind – the first two and the last three make up *two* distinct balls *of the same radius*, not missing a single point:

It seems that this result flies in the face of the principle of "conservation of volume" during a translation or a rotation. But that is the point: the pieces $\mathcal{B}_i$ (or at least some of them) are *not* measurable, and so the notion of "volume" of $\mathcal{B}_i$ makes no sense; see [74].

[12] In the sense of the σ-algebra generated by open sets.

is then extended to non-negative functions, and finally to (some) real- or complex-valued measurable functions.

DEFINITION 2.23 (Simple function) Let X be an arbitrary set, $A \subset X$ a subset of X. The **characteristic function** of A is the function $\chi_A : X \to \{0,1\}$ such that $\chi_A(x) = 1$ if $x \in A$ and $\chi_A(x) = 0$ otherwise.

Let $(X, \mathscr{T})$ be a measurable space. A function $f : X \to \overline{\mathbb{R}}$ is a **simple function** if there exist finitely many disjoint measurable sets $A_1, \ldots, A_n$ in $\mathscr{T}$ and elements $\alpha_1, \ldots, \alpha_n$ in $\overline{\mathbb{R}}$ such that

$$f = \sum_{i=1}^{n} \alpha_i \, \chi_{A_i}.$$

Now if $(X, \mathscr{T}, \mu)$ is a measure space and f is a *non-negative* simple function on X, the **integral of f with respect to the measure μ** is defined by

$$\int f \, d\mu \stackrel{\text{def}}{=} \sum_{i=1}^{n} \alpha_i \, \mu(A_i),$$

which is either a non-negative real number or $+\infty$. In this definition, the convention $\infty \times 0 = 0$ is used if either $\alpha_i = \infty$ or $\mu(A_i) = \infty$. If the integral of f is finite, then f is **integrable**.

Remark 2.24 In the case where $X = \mathbb{R}$ and μ is the Lebesgue measure, the following notation is also used for the integral of f with respect to μ:

$$\int f \, d\mu = \int_{\mathbb{R}} f \, d\mu = \int f(x) \, dx = \int_{-\infty}^{+\infty} f(x) \, dx.$$

Example 2.25 Let D be the Dirichlet function, which is 1 for rational numbers and 0 for irrational numbers. Then D is a simple function (since $\mathbb{Q}$ is measurable), and

$$\int_{[0,1]} (1 - D) \, d\mu = 1 \quad \text{whereas} \quad \int_{\mathbb{R}} (1 - D) \, d\mu = +\infty.$$

The definition is extended now to any measurable non-negative function.

DEFINITION 2.26 (Integral of a non-negative function) Let $(X, \mathscr{T}, \mu)$ be a measure space and let $f : X \to \overline{\mathbb{R}}^+$ be a measurable function with non-negative values. The **integral of f** is defined by

$$\int f \, d\mu \stackrel{\text{def}}{=} \sup \left\{ \int g \, d\mu \; ; \; g \text{ simple such that } 0 \leqslant g \leqslant f \right\}.$$

It is either a non-negative real number or $+\infty$. If it is finite, then f is **integrable**.

Finally, to deal with real-valued measurable functions f, the decomposition of f into its non-negative and non-positive parts is used: The **non-negative part** of a function f on X is the function $f^+ \stackrel{\text{def}}{=} \max(f, 0)$, and the **non-positive part** of f is the function $f^- \stackrel{\text{def}}{=} \max(-f, 0)$. We have $f = f^+ - f^-$ and the functions f^+ and f^- are non-negative measurable functions.

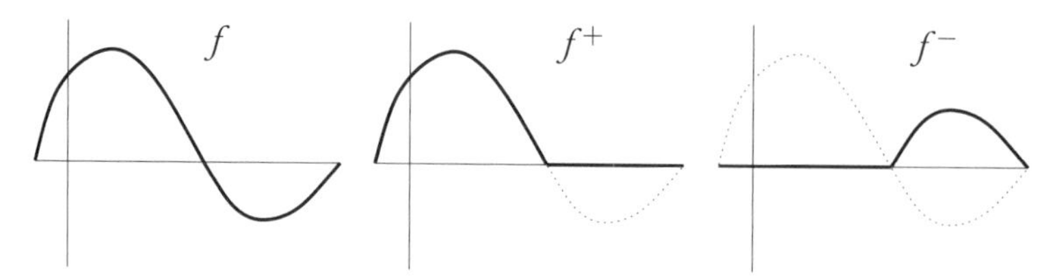

DEFINITION 2.27 (Integrability and integral of a function) Let $(X, \mathcal{T}, \mu)$ be a measure space and let $f : X \mapsto \overline{\mathbb{R}}$ be a measurable function, $E \subset X$ a measurable set. Then **f is integrable on E with respect to μ** or **f is μ-integrable**, if $\int_E f^+ \, d\mu$ and $\int_E f^- \, d\mu$ are both finite, or equivalently if the non-negative function $|f|$ is integrable on E. The **integral of f on E with respect to μ** is then

$$\int_E f \, d\mu \stackrel{\text{def}}{=} \int_E f^+ \, d\mu - \int_E f^- \, d\mu.$$

If f is a measurable complex-valued function, one looks separately at the real and imaginary parts $\text{Re}(f)$ and $\text{Im}(f)$. If both are integrable on X with respect to μ, then so is f and we have

$$\int f \, d\mu = \int \text{Re}(f) \, d\mu + i \int \text{Im}(f) \, d\mu.$$

In practice, the definitions above are justified by the fact that any measurable function f is the pointwise limit of a sequence $(f_n)_{n \in \mathbb{N}}$ of simple functions. If, moreover, f is bounded, it is a *uniform* limit of simple functions.[13]

As it should, the Lebesgue integral enjoys the same formal properties as the Riemann integral. In particular, the following properties hold:

PROPOSITION 2.28 *Let $(X, \mathcal{T}, \mu)$ be a measure space, and let f, g be measurable functions on X, with values in $\mathbb{R}$ or $\mathbb{C}$. Then we have:*

i) if f and g are integrable with respect to μ, then for any complex numbers α, $\beta \in \mathbb{C}$, the function $\alpha f + \beta g$ is integrable and

[13] This should be compared with the Riemann definition of integral, where step functions are used. It turns out that there are relatively few functions which are uniform limits of step functions (because step functions, constant on intervals, are much more "rigid" than simple functions can be, defined as they are using arbitrary measurable sets).

$$\int (\alpha f + \beta g)\,\mathrm{d}\mu = \alpha \int f\,\mathrm{d}\mu + \beta \int g\,\mathrm{d}\mu;$$

ii) *if f and g are integrable and real-valued, and if $f \leqslant g$, then $\int f\,\mathrm{d}\mu \leqslant \int g\,\mathrm{d}\mu$;*

iii) *if f is integrable, then $\left|\int f\,\mathrm{d}\mu\right| \leqslant \int |f|\,\mathrm{d}\mu$;*

iv) *if g is non-negative and integrable and if $|f| \leqslant g$, then f is integrable and $\int |f|\,\mathrm{d}\mu \leqslant \int g\,\mathrm{d}\mu$.*

Finally, under some suitable assumptions,[14] for any two measure spaces $(X, \mathscr{T}, \mu)$ and $(Y, \mathscr{T}', \nu)$, the product measure space $(X \times Y, \mathscr{T} \otimes \mathscr{T}', \mu \otimes \nu)$ is defined in such a way that Fubini's formula holds: for any integrable function f on $X \times Y$, we have

$$\begin{aligned}\iint_{X\times Y} f\,\mathrm{d}(\mu\otimes\nu) &= \int_X \left(\int_Y f(x,y)\,\mathrm{d}\nu(y)\right)\mathrm{d}\mu(x)\\ &= \int_Y \left(\int_X f(x,y)\,\mathrm{d}\mu(x)\right)\mathrm{d}\nu(y)\end{aligned}$$

(an implicit fact here is that the integral with respect to one variable is a measurable function of the other variable, so that all integrals which appear make sense). In other words, the integral may be computed in any order (first over X then Y and conversely). Moreover, a measurable function f defined on $X \times Y$ is integrable if either

$$\int_X \left|\int_Y f(x,y)\,\mathrm{d}\nu(y)\right| \mathrm{d}\mu(x) \quad \text{or} \quad \int_Y \left|\int_X f(x,y)\,\mathrm{d}\mu(x)\right| \mathrm{d}\nu(y)$$

is finite.

All this applies in particular to integrals on $\mathbb{R}^2$ with respect to the Lebesgue measure $\mu \otimes \mu$, and by induction on $\mathbb{R}^d$ for $d \geqslant 1$. The Fubini formula becomes

$$\begin{aligned}\iint_{\mathbb{R}^2} f(x,y)\,\mathrm{d}x\,\mathrm{d}y &= \int_{\mathbb{R}} \left(\int_{\mathbb{R}} f(x,y)\,\mathrm{d}y\right)\mathrm{d}x\\ &= \int_{\mathbb{R}} \left(\int_{\mathbb{R}} f(x,y)\,\mathrm{d}x\right)\mathrm{d}y.\end{aligned}$$

The following notation is also commonly used for integrals on $\mathbb{R}^2$:

$$\begin{aligned}\iint_{\mathbb{R}^2} f\,\mathrm{d}^2\mu = \iint_{\mathbb{R}^2} f\,\mathrm{d}(\mu\otimes\mu) &= \iint_{\mathbb{R}^2} f(x,y)\,\mathrm{d}x\,\mathrm{d}y\\ &= \iint_{\mathbb{R}^2} f(x,y)\,\mathrm{d}\mu(x)\,\mathrm{d}\mu(y).\end{aligned}$$

[14] Always satisfied in applications.

2.2.h Functions zero almost everywhere, space L^1 of integrable functions

PROPOSITION 2.29 *Let $(X, \mathscr{T}, \mu)$ be a measure space, for instance, $X = \mathbb{R}$ with the Lebesgue σ-algebra and measure. Let $f : X \to \mathbb{R}^+ \cup \{+\infty\}$ be a non-negative measurable function. Then we have*

$$\int f \, \mathrm{d}\mu = 0 \iff f \text{ is zero almost everywhere.}$$

PROOF. See Exercise 2.7 on page 69.

In particular, if the values of an integrable function f are changed on a set of measure zero, the integral of f does not change. We have

$$f = g \quad \text{a.e.} \quad \Longrightarrow \quad \int f \, \mathrm{d}\mu = \int g \, \mathrm{d}\mu.$$

In the chapter concerning distributions, we will see that two functions which differ only on a set of measure zero should not be considered to be really different. Hence it is common to say that such functions are equal, instead of merely[15] "equal almost everywhere."

DEFINITION 2.30 (L^1 spaces) Let $(X, \mathscr{T}, \mu)$ be a measure space. The space $L^1(X, \mu)$ or simply $L^1(X)$ if no confusion concerning μ can arise, is the vector space of μ-integrable complex-valued functions on X, up to equality almost everywhere.

In particular, $\boldsymbol{L^1(\mathbb{R})}$ or simply $\boldsymbol{L^1}$ denotes the vector space of functions integrable on $\mathbb{R}$ with respect to Lebesgue measure, up to equality almost everywhere.

For any interval $[a, b]$ of $\mathbb{R}$, $\boldsymbol{L^1[a, b]}$ is the space of functions on $[a, b]$ integrable with respect to the restriction of Lebesgue measure to $[a, b]$. This integrability condition can also be expressed as saying that the function on $\mathbb{R}$ obtained by extending f by 0 outside $[a, b]$ is in $L^1(\mathbb{R})$.

The following important theorem justifies all the previous theory:

THEOREM 2.31 (Riesz-Fischer) *Let $(X, \mathscr{T}, \mu)$ be a measure space. Then the space $L^1(X, \mu)$ of complex-valued integrable functions, with the norm $\|f\|_1 = \int |f| \, d\mu$, is a complete normed vector space. In particular, the spaces $L^1(\mathbb{R})$ and $L^1[a, b]$ are complete normed vector spaces.*

---

[15] This is not simply an abuse of notation. Mathematically, what is done is take the set of integrable functions, and take its *quotient* modulo the *equivalence relation* "being equal almost everywhere." This means that an element in $L^1(X)$ is in fact an *equivalence class* of functions. This notion will come back during the course.

Moreover, the space $\mathscr{C}([a,b])$ of continuous functions on $[a,b]$ is dense in the space $L^1[a,b]$, which is therefore the completion of $\mathscr{C}([a,b])$, and similarly for the space $\mathscr{C}(\mathbb{R})$ of continuous functions on $\mathbb{R}$ and $L^1(\mathbb{R})$.

2.2.i And today?

Lebesgue's theory of integration became very quickly the standard approach at the beginning of the twentieth century. However, some other constructions remain useful, such as that of Stieltjes[16] (see Sidebar 8). Generalizations of the integral, such as the Itō stochastic integral,[17] or recently the work of Cécile DeWitt-Morette and Pierre Cartier on the Feynman path integrals [53], have also appeared. There is also a "pedagogical"[18] definition of an integral which combines advantages of the Lebesgue and Riemann integrals, due to Kurzweil and Henstock (see, e.g., [93]).

[16] The so-called Riemann-Stieltjes integral defines integration on $\mathbb{R}$ with respect to a "Borel measure"; they can be used to avoid appealing to distributions in some situations, and are still of great use in probabiliy theory.

[17] This concerns Brownian motion in particular.

[18] This is not the only reason it is interesting.

EXERCISES

♦ **Exercise 2.1** Let $d \in \mathbb{N}^*$. Prove or disprove (with a counterexample) the following statements:

i) An open set in $\mathbb{R}^d$ is bounded if and only if its Lebesgue measure is finite.

ii) A Borel set in $\mathbb{R}^d$ has positive Lebesgue measure if and only if it contains a non-empty open set.

iii) For any integer p and $0 < p < d$, the set $\mathbb{R}^p \times \{0\}^{d-p}$ has Lebesgue measure zero in $\mathbb{R}^d$.

♦ **Exercise 2.2** Find the σ-algebras on $\mathbb{R}$ generated by a single element.

♦ **Exercise 2.3** Show that each of the following subsets generate the Borel σ-algebra $\mathscr{B}(\mathbb{R})$:

1. closed sets in $\mathbb{R}$;
2. intervals of the type $]a, b]$;
3. intervals of the type $]-\infty, b]$.

Show that $\mathscr{B}(\mathbb{R}^d)$ is generated by

1. closed sets in $\mathbb{R}^d$;
2. "bricks," that is, sets of the type $\{x \in \mathbb{R}^d \mid a_i < x_i \leqslant b_i,\ i = 1, \ldots, d\}$;
3. closed half-spaces $\{x \in \mathbb{R}^d \mid x_i \leqslant b\}$, where $1 \leqslant i \leqslant d$ and $b \in \mathbb{R}$.

♦ **Exercise 2.4** Let $(X, \mathcal{T}, \mu)$ be a measure space. Prove the following properties:

i) $\mu(\varnothing) = 0$;

ii) μ is increasing: if $A, B \in \mathcal{T}$ and $A \subset B$, then we have $\mu(A) \leqslant \mu(B)$;

iii) let $(A_n)_{n\in\mathbb{N}}$ be an increasing sequence of measurable sets in $\mathcal{T}$, and let $A = \bigcup_n A_n$. Then we have $\mu(A_n) \xrightarrow[n\to\infty]{} \mu(A)$;

iv) let $(A_n)_{n\in\mathbb{N}}$ be a decreasing sequence of measurable sets in $\mathcal{T}$ such that $\mu(A_1) < +\infty$, and let $A = \bigcap_n A_n$. Then we have $\mu(A_n) \xrightarrow[n\to\infty]{} \mu(A)$. Show also that this is not necessarily true if $\mu(A_1) = +\infty$;

v) let $(A_n)_{n\in\mathbb{N}}$ be any sequence of measurable sets in $\mathcal{T}$. Then we have

$$\mu\Big(\bigcup_n A_n\Big) \leqslant \sum_n \mu(A_n) \qquad (\sigma\text{-subadditivity of the measure}).$$

♦ **Exercise 2.5** In this exercise, we denote by $\mathscr{O}$ the set of open subsets in $\mathbb{R}$ for the usual topology, and (as usual) by $\mathscr{B}$ and $\mathscr{L}$ the Borel σ-algebra and the Lebesgue σ-algebra. We have $\mathscr{O} \subset \mathscr{B} \subset \mathscr{L} \subset \mathfrak{P}(\mathbb{R})$, and we will show that all inclusions are strict.

i) Show that $\mathscr{O} \subsetneq \mathscr{B}$.

ii) Let C be the Cantor triadic set, defined by

$$C = \bigcap_{p\geqslant 1} F_p, \qquad \text{where} \qquad F_p \stackrel{\text{def}}{=} \Big\{x \in \mathbb{R}\,;\ \exists (k_n)_{n\in\mathbb{N}} \in E_p,\ x = \sum_{n\geqslant 1} k_n/3^n\Big\},$$

$$E_p \stackrel{\text{def}}{=} \Big\{(k_n)_{n\in\mathbb{N}} \in \{0,1,2\}^{\mathbb{N}}\,;\ \forall n \leqslant p,\ k_n \neq 1\Big\}.$$

a) Show that C is a compact set with $\mu(C) = 0$.

b) Show that C is in one-to-one correspondance with $\{0,1\}^{\mathbb{N}}$, and in one-to-one correspondence with $\mathbb{R}$.

c) Show that $\mathscr{L}$ is in one-to-one correspondance with $\mathfrak{P}(\mathbb{R})$.

d) One can show that there exists an injective map from $\mathscr{B}$ to $\mathbb{R}$. Deduce that the second inclusion is strict.

iii) Use Sidebar 2 on the following page to conclude that the last inclusion is also strict.

♦ **Exercise 2.6** Let f be a map from $\mathbb{R}$ to $\mathbb{R}$. Is it true that the two statements below are equivalent?

i) f is almost everywhere continuous;

ii) there exists a continuous function g such that $f = g$ almost everywhere.

♦ **Exercise 2.7** Let f be a measurable function on a measure space $(X, \mathscr{T}, \mu)$, such that $f(x) > 0$ for $x \in A$ with $\mu(A) > 0$. Show that $\int f \, d\mu > 0$.

Deduce from this a proof of Theorem 2.29 on page 66.

♦ **Exercise 2.8 (Egorov's theorem)** Let $(X, \mathscr{T}, \mu)$ be a finite measure space, that is, a measure space such that $\mu(X) < +\infty$ (for instance, $X = [a, b]$ with the Lebesgue measure). Let $(f_n)_{n\in\mathbb{N}}$ be a sequence of complex-valued measurable functions on X *converging pointwise* to a function f.

For $n \in \mathbb{N}$ and $k \in \mathbb{N}^*$, let

$$E_n^{(k)} \stackrel{\text{def}}{=} \bigcap_{p \geqslant n} \left\{ x \in X \; ; \; \left| f_p(x) - f(x) \right| \leqslant \frac{1}{k} \right\}.$$

For fixed k, show that the sequence $\left(E_n^{(k)}\right)_{n\geqslant 1}$ is *increasing* (for inclusion) and that the measure of$E_n^{(k)}$tends to $\mu(X)$ as n tends to infinity. Deduce the following theorem:

THEOREM 2.32 (Egorov) *Let $(X, \mathscr{T}, \mu)$ be a finite measure space. Let $(f_n)_{n\in\mathbb{N}}$ be a sequence of complex-valued measurable functions on X converging pointwise to a function f. Then, for any $\varepsilon > 0$, there exists a measurable set A_ε with measure $\mu(A_\varepsilon) < \varepsilon$ such that $(f_n)_{n\in\mathbb{N}}$ converges uniformly to f on the complement $X \setminus A_\varepsilon$.*

This theorem shows that, on a finite measure space, pointwise convergence is not so different from uniform convergence: one can find a set of measure arbitrarily close to the measure of X such that the convergence is uniform on this set.

Sidebar 2 (A nonmeasurable set) *We describe here the construction of a subset of $\mathbb{R}$ which is not measurable (for the Lebesgue σ-algebra).*

Consider the following relation on $\mathbb{R}$: for $x, y \in \mathbb{R}$, we say that $x \sim y$ if and only if $(x-y) \in \mathbb{Q}$.

This relation "$\sim$" is of course an equivalence relation.[a]

We now appeal to the Axiom of Choice of set theory[b] *to define a subset $E \subset [0,1]$ containing* exactly one *element in each equivalence class of $\sim$*[c]*. Note that the use of the Axiom of Choice means that this set E is not defined constructively: we know it exists, but it is impossible to give an explicit definition of it.*

We now claim that E is not in the Lebesgue σ-algebra.

To prove this, for any $r \in \mathbb{Q}$, let $E_r = \{x + r \,;\, x \in E\}$.

If $r, s \in \mathbb{Q}$ and $r \neq s$, the sets E_r and E_s are disjoint. Indeed, if there exists $x \in E_r \cap E_s$, then by definition there exist $y, z \in E$ such that $x = y + r = z + s$, which implies that $z - y = r - s$ is a rational number, and so $z \sim y$. But E contains a single *element in each equivalence class, so this means that $y = z$ and hence $r = s$.*

Assuming now that E is measurable, let $\alpha = \mu(E)$. Then for each $r \in \mathbb{Q}$, E_r is measurable with measure $\mu(E_r) = \alpha$ (because it is a translate of E).

Let
$$S \stackrel{\text{def}}{=} \bigcup_{r \in \mathbb{Q} \cap [-1,1]} E_r.$$

As a countable union of measurable sets, this set is itself measurable. If $\alpha \neq 0$, then $\mu(S) = +\infty$ since S is a disjoint union, so we can use σ-additivity to compute its measure. But clearly $S \subset [-1,2]$ so $\mu(S) \leqslant 3$. The assumption $\alpha > 0$ is therefore untenable, and we must have $\alpha = 0$. Then $\mu(S) = 0$. But we can also easily check that

$$[0,1] \subset S \subset [-1,2].$$

This implies that $\mu(S) \geqslant 1$, contradicting this last assumption.

It follows that E is not measurable.

---

[a] This means that, for all x, y, z :

- $x \sim x$;
- $x \sim y$ if and only if $y \sim x$;
- if $x \sim y$ and $y \sim z$, then $x \sim z$.

[b] The Axiom of Choice states for any non-empty set A, there exists a map $f : \mathfrak{P}(A) \to A$, called a **choice function**, such that $f(X) \in X$ for any non-empty subset $X \subset A$. In other words, the choice function *selects*, in an arbitrary way, an element in any non-empty subset of X. It is interesting to know that a more restricted form of the Axiom of Choice (the Axiom of Countable Choice) is compatible with the assertion that any subset of $\mathbb{R}$ is Lebesgue-measurable. (But is not sufficient to prove this statement). The interested reader may look in books such as [51] for a very intuitive approach, or [45] which is very clear.

[c] This means that for any $x \in \mathbb{R}$, there is a *unique* $y \in E$ with $x \sim y$.

SOLUTIONS

♦ **Solution of exercise 2.1**

i) Wrong. Take, for instance, $d = 1$ and the unbounded open set

$$U = \bigcup_{k \geqslant 1} \left] k, k + \frac{1}{2^k} \right[$$

which has Lebesgue measure $\mu(U) = \sum_{k \geqslant 1} 2^{-k} = 1$.

ii) Wrong. For $d = 1$, consider for example the Borel set $(\mathbb{R} \setminus \mathbb{Q}) \cap [0, 1]$, which has measure 1 but does not contain any non-empty open set.

iii) True. It is easy to describe for any $\varepsilon > 0$ a subset of $\mathbb{R}^d$ containing $\mathbb{R}^p \times \{0\}^{d-p}$ with measure less than ε.

♦ **Solution of exercise 2.6.** Wrong. Let $f = D$ be the Dirichlet function, equal to 1 on $\mathbb{Q}$ and zero on $\mathbb{R} \setminus \mathbb{Q}$. Then f is nowhere continuous, but $f = 0$ almost everywhere.

♦ **Solution of exercise 2.7.** For any $n \geqslant 1$, let

$$E_n \stackrel{\text{def}}{=} \{x \in X \; ; \; f(x) > 1/n\}.$$

The sequence $(E_n)_{n \in \mathbb{N}^*}$ is an increasing sequence of measurable sets, and its union is $\{x \in X \; ; \; f(x) > 0\}$, which has positive measure since it contains A. Using the σ-additivity, there exists n_0 such that E_{n_0} has positive measure. Then the integral of f is at least equal to $\mu(E_{n_0})/n_0 > 0$.

♦ **Solution of exercise 2.8.** For any $k \in \mathbb{N}$, we have $\mu\big(E_n^{(k)}\big) \xrightarrow[n \to \infty]{} \mu(X)$. It is then possible to construct a strictly increasing sequence $(n_k)_{k \in \mathbb{N}}$ such that $\mu\big(X - E_{n_k}^{(k)}\big) < \varepsilon/2^k$ for all $k \in \mathbb{N}$, and then conclude.

Chapter 3

Integral calculus

3.1

Integrability in practice

A physicist is much more likely to encounter "concrete" functions than abstract functions for which integrability might depend on subtle theoretical arguments. Hence it is important to remember the usual tricks that can be used to prove that a function is integrable.

The standard method involves two steps:

- first, find some general criteria proving that some "standard" functions are integrable;
- second, prove and use comparison theorems, which show how to reduce the integrability of a "complicated" function to the integrability of one of the standard ones.

3.1.a Standard functions

The situation we consider is that of functions f defined on a half-open interval $[a,b[$, where b may be either a real number or $+\infty$, and where the functions involved are assumed to be integrable on any interval $[a,c]$ where $a<c$ is a real number (for instance, f may be a continuous function on $[a,b[$). This is a fairly general situation, and integrability on $\mathbb{R}$ may be reduced to integrability on $[0,+\infty[$ and $]-\infty,0]$.

PROPOSITION 3.1 (Integrability of a non-negative function) *Let a be a real and let $b \in \mathbb{R} \cup \{\pm\infty\}$. Let $f : [a,b[\rightarrow \mathbb{R}^+$ be a non-negative measurable function which has a primitive F. Then f is integrable on $[a,b[$ if and only if F has a finite limit in b.*

Example 3.2 The function log is integrable on $]0,1]$. The functions $x \mapsto e^{-ax}$ are integrable on $[0,+\infty[$ if and only if $a > 0$.

PROPOSITION 3.3 (Riemann functions t^α) *The function $f_\alpha : t \mapsto t^\alpha$ with $\alpha \in \mathbb{R}$ is integrable on $[1,+\infty[$ if and only if $\alpha < -1$. It is integrable on $]0,1]$ if and only if $\alpha > -1$.*

PROPOSITION 3.4 (Bertrand functions $t^\alpha \log^\beta t$) *The function $t \mapsto t^\alpha(\log t)^\beta$ with α and β real numbers is integrable on $[1,+\infty[$ if and only if*

$$\alpha < -1 \qquad \text{or} \qquad (\alpha = 1 \quad \text{and} \quad \beta < -1).$$

3.1.b Comparison theorems

Once we know some integrable functions, the following simple result gives information concerning the integrability of many more functions.

PROPOSITION 3.5 (Comparison theorem) *Let $g : [a,b[\rightarrow \mathbb{C}$ be an integrable function, and let $f : [a,b[\rightarrow \mathbb{C}$ be a measurable function. Then*

i) if $|f| \leqslant |g|$, then f is integrable on $[a,b[$;

ii) if $f = \underset{b}{O}(g)$ or if $f = \underset{b}{o}(g)$ and if f is integrable on any interval $[a,c]$ with $c < b$, then f is integrable on $[a,b[$;

iii) if g is non-negative, $f \underset{b}{\sim} g$, and f is integrable on any interval $[a,c]$ with $c < b$, then f is integrable on $[a,b[$.

♦ **Exercise 3.1** Study the integrability of the following functions (in increasing order of difficulty):

i) $t \mapsto \cos(t)\log(\tan t)$ on $]0,\pi/2[$;

ii) $t \mapsto (\sin t)/t$ on $]0,+\infty[$;

iii) $t \mapsto 1/(1+t^\alpha \sin^2 t)$ on $[0,+\infty[$, where $\alpha > 0$.

PROPOSITION 3.6 (Asymptotic comparison) *Let g be a non-negative measurable function on $[a,b[$, and let f be a measurable function on $[a,b[$. Assume that g is not integrable and that f and g are both integrable on any interval $[a,c]$ with $c < b$. Then asymptotic comparison relations between f and g extend to their integrals as follows:*

$$f \underset{b}{\sim} g \implies \int_a^x f \underset{x\to b}{\sim} \int_a^x g \qquad \text{and} \qquad f = \underset{b}{o}(g) \implies \int_a^x f = \underset{x\to b}{o}\left(\int_a^x g\right).$$

If g and h are integrable on $[a,b[$, then asymptotic comparison relations between f and h extend to the "remainders" as follows:

$$f \underset{b}{\sim} h \implies \int_x^b f \underset{x\to b}{\sim} \int_x^b h \qquad \text{and} \qquad f = \underset{b}{\mathrm{o}}(h) \implies \int_x^b f = \underset{x\to b}{\mathrm{o}}\left(\int_x^b h\right).$$

♦ **Exercise 3.2** Show that

$$\int_1^{+\infty} \frac{\mathrm{e}^{-xt}}{\sqrt{1+t^2}}\,\mathrm{d}t \underset{x\to 0^+}{\sim} -\log x.$$

3.2 Exchanging integrals and limits or series

Even when dealing with "concrete" functions, Lebesgue's theory is extremely useful because of the power of its general statements about exchanging limits and integrals. The main theorem is Lebesgue's dominated convergence theorem.

THEOREM 3.7 (Lebesgue's dominated convergence theorem) *Let $(X, \mathscr{T}, \mu)$ be a measure space, for instance, $\mathbb{R}$ or $\mathbb{R}^d$ with the Borel σ-algebra and the Lebesgue measure. Let $(f_n)_{n\in\mathbb{N}}$ be a sequence of complex-valued measurable functions on X. Assume that $(f_n)_{n\in\mathbb{N}}$ converges pointwise almost everywhere to a function $f : X \to \mathbb{R}$, and moreover assume that there exists a μ-integrable function $g : X \to \mathbb{R}^+$ such that $|f_n| \leqslant g$ almost everywhere on X, for all $n \in \mathbb{N}$. Then*

i) f is μ-integrable;

ii) for any measurable subset $A \subset X$, we have

$$\lim_{n\to\infty} \int_A f_n \,\mathrm{d}\mu = \int_A f\,\mathrm{d}\mu.$$

This result is commonly used both to prove the integrability of a function obtained by a limit process, and as a way to compute the integral of such functions.

Example 3.8 Suppose we want to compute $\displaystyle\lim_{n\to\infty} \frac{1}{\pi}\int_{-\infty}^{+\infty} \frac{n\,\mathrm{e}^{-x^2}\cos x}{1+n^2x^2}\,\mathrm{d}x$. By the change of variable $y = nx$, we are led to

$$I_n = \frac{1}{\pi}\int_{-\infty}^{+\infty} \frac{\mathrm{e}^{-y^2/n^2}\cos(y/n)}{1+y^2}\,\mathrm{d}y.$$

The sequence of functions $f_n : y \mapsto \mathrm{e}^{-y^2/n^2}\cos(y/n)/(1+y^2)$ converges pointwise on $\mathbb{R}$ to the function $f : x \mapsto 1/(1+x^2)$. Moreover, $|f_n| \leqslant f$, which is integrable on $\mathbb{R}$ (by comparison with $t \mapsto 1/t^2$ at $\pm\infty$).

Therefore, by Lebesgue's theorem, the limit exists and is equal to

$$\frac{1}{\pi}\int_{-\infty}^{+\infty} \frac{1}{1+x^2}\,\mathrm{d}x = \frac{1}{\pi}\big[\arctan x\big]_{-\infty}^{+\infty} = 1.$$

Another useful result does not require the "domination" assumption, but instead is restricted to increasing sequences. This condition is more restrictive, but often very easy to check, in particular for series of non-negative terms.

THEOREM 3.9 (Beppo Levi's monotone convergence theorem) *Let $(X, \mathscr{T}, \mu)$ be a measure space, for instance, $\mathbb{R}$ or $\mathbb{R}^d$ with the Borel σ-algebra and the Lebesgue measure. Let $(f_n)_n$ be a sequence of non-negative measurable functions on X. Assume that the sequences $\big(f_n(x)\big)_n$ are increasing for all $x \in X$. Then the function $f : X \to \mathbb{R} \cup \{+\infty\}$, defined by*

$$f(x) = \lim_{n\to\infty} f_n(x) \qquad \text{for all } x \in X,$$

is measurable and we have

$$\lim_{n\to\infty} \int f_n \,\mathrm{d}\mu = \int f \,\mathrm{d}\mu$$

(this quantity may be either finite or infinite).

This result is often used as a first step to prove that a function is integrable.

Here finally is the version of the dominated convergence theorem adapted to a series of functions:

THEOREM 3.10 (Term by term integration of a series) *Let $(X, \mathscr{T}, \mu)$ be a measure space, for instance, $\mathbb{R}$ or $\mathbb{R}^d$ with the Borel σ-algebra and the Lebesgue measure. Let $\sum f_n$ be a series of complex-valued measurable functions on X such that*

$$\sum_{n=0}^{\infty} \int |f_n| \,\mathrm{d}\mu < +\infty.$$

Then the series of functions $\sum f_n$ converges absolutely almost everywhere on X and its sum is μ-integrable. Moreover, its integral may be computed by term by term integration: we have

$$\int \sum_{n=0}^{\infty} f_n \,\mathrm{d}\mu = \sum_{n=0}^{\infty} \int f_n \,\mathrm{d}\mu.$$

3.3

Integrals with parameters

3.3.a Continuity of functions defined by integrals

Let $(X, \mathscr{T}; \mu)$ be a measure space. Assume we have a function of two variables $f : [a, b] \times X \to \mathbb{R}$ or $\mathbb{C}$. Assuming the integrals make sense, one can define a function by integrating over X for fixed values of the parameter $x \in [a, b]$: let

$$I(x) \stackrel{\text{def}}{=} \int_{\mathbb{R}} f(x,t)\,\mathrm{d}t \qquad \text{for } x \in [a,b].$$

A natural question is to determine what properties of *regularity* the integral $I(x)$ will inherit from properties of f.

The two main results concern the continuity or derivability of f. Here again, the Lebesgue theory gives very convenient answers:

THEOREM 3.11 (Continuity of functions defined by integrals) *Let $x_0 \in [a, b]$. Assume that for almost all $t \in X$, the function $x \mapsto f(x,t)$ is continuous at x_0, and that there exists a μ-integrable function $g : X \to \mathbb{R}$ such that*

$$\left|f(x,t)\right| \leqslant g(t) \quad \textit{for almost all } t \in X,$$

for all x in a neighborhood of x_0. Then $x \mapsto \int_{\mathbb{R}} f(x,t)\,\mathrm{d}t$ is continuous at x_0.

PROOF. This is a simple consequence of the dominated convergence theorem. Indeed, let $(x_n)_{n\in\mathbb{N}}$ be any sequence of real numbers converging to x_0. For n large enough, the values of x_n lie in the neighborhood of x_0 given by the statement of the theorem. Then for almost all $t \in X$, we have $\lim_{n\to\infty} f(x_n, t) = f(x_0, t)$ by continuity, and, moreover, $\left|f(x_n,t)\right| \leqslant g(t)$ for all t. From the dominated convergence theorem, we deduce

$$\lim_{n\to\infty} I(x_n) = \lim_{n\to\infty} \int_{\mathbb{R}} f(x_n,t)\,\mathrm{d}t = \int_{\mathbb{R}} f(x_0,t)\,\mathrm{d}t = \int_{\mathbb{R}} \lim_{n\to\infty} f(x_n,t)\,\mathrm{d}t, = I(x_0),$$

which, being valid for an arbitrary sequence $(x_n)_{n\in\mathbb{N}}$, is the statement we wanted to prove.

Counterexample 3.12 It is of course natural to expect that the conclusion of Theorem 3.11 should be true in most cases. Here is an example where, in the absence of the "domination" property, the continuity of a function defined by an integral with parameters does not hold.

Let f be the function on $\mathbb{R} \times \mathbb{R}$ defined by

$$f(x,t) = \begin{cases} \exp\left[-\dfrac{\pi}{2}\left(t - \dfrac{1}{x}\right)^2\right] & \text{if } x \neq 0, \\ 0 & \text{if } x = 0. \end{cases}$$

For all $t \in \mathbb{R}$, the function $x \mapsto f(x,t)$ is continuous on $\mathbb{R}$, and in particular at 0.

Let $I(x) = \int f(x,t)dt$.

For any fixed nonzero x, the function $t \longmapsto f(x,t)$ is a gaussian centered at $1/x$, so its integral is easy to compute; it is in fact equal to 1. But for $x = 0$, the function $t \longmapsto f(0,t)$ is identically zero, so its integral is also zero. Hence we have

$$I(0) = 0 \qquad \text{but} \qquad \forall x \in \mathbb{R}^*, \quad I(x) = 1.$$

This shows that the function I is not continuous at zero.

3.3.b Differentiating under the integral sign

THEOREM 3.13 (Differentiability of functions defined by integrals) *Let $x_0 \in [a,b]$. Assume that there exists a neighborhood $\mathscr{V}$ of x_0 such that for $x \in \mathscr{V}$, the function $x \longmapsto f(x,t)$ is differentiable at x_0 for almost all $t \in X$, and $t \longmapsto f(x,t)$ is integrable. Assume, moreover, that the derivatives are dominated by an integrable function g, that is, there exists an integrable function $g : X \to \mathbb{R}$ such that*

$$\left| \frac{\partial f}{\partial x}(x,t) \right| \leqslant g(t) \qquad \textit{for all } x \in \mathscr{V} \textit{ and almost all } t \in X.$$

Then the function $x \longmapsto I(x) = \int f(x,t)\,\mathrm{d}t$ is differentiable at x_0, and moreover we have

$$\frac{\mathrm{d}}{\mathrm{d}x}\left(\int_{\mathbb{R}} f(x,t)\,\mathrm{d}t \right)\Bigg|_{x=x_0} = \int_{\mathbb{R}} \frac{\partial f}{\partial x}(x_0,t)\,\mathrm{d}t.$$

Those theorems are very useful in particular when dealing with the theory of the Fourier transform.

Counterexample 3.14 Here again, the assumption of domination cannot be dispensed with. For instance, let

$$F(x) = \int_0^{+\infty} \frac{\cos(xt)}{1+t^2}\,\mathrm{d}t;$$

one can then show (see Exercise 3.10 on page 84) that

$$F(x) = \frac{\pi}{2}\,\mathrm{e}^{-|x|}$$

for all $x \in \mathbb{R}$. This function is continuous on $\mathbb{R}$, but is not differentiable at 0.

3.3.c Case of parameters appearing in the integration range

THEOREM 3.15 *With the same assumption as in Theorem 3.13, let moreover $v : \mathscr{V} \to \mathbb{R}$ be a function differentiable at x_0. Let*

$$I(x) = \int_0^{v(x)} f(x,t)\,\mathrm{d}t.$$

Then I is differentiable at x_0 and we have

$$\frac{\mathrm{d}}{\mathrm{d}x}\left(\int_0^{v(x)} f(x,t)\,\mathrm{d}t \right)\Bigg|_{x=x_0} = \int_0^{v(x)} \frac{\partial f}{\partial x}(x_0,t)\,\mathrm{d}t + v'(x)\cdot f\big(x_0, v(x_0)\big).$$

The derivative of this function defined by an integral involves two terms: one obtained by keeping the limits of integration constant, and differentiating under the integral sign; and the second obtained by differentiating the integral with varying boundary.

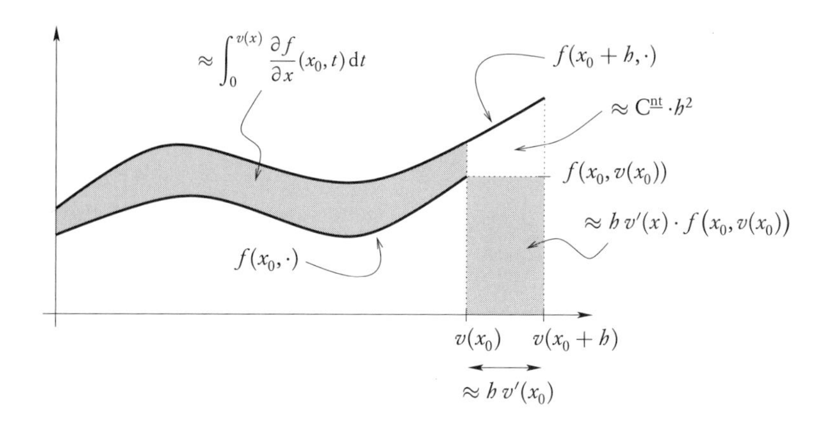

3.4

Double and multiple integrals

For completeness and ease of reference, we recall the theorem of Fubini mentioned at the end of the previous chapter.

THEOREM 3.16 (Fubini-Lebesgue) *Let $(X, \mathscr{T}, \mu)$ and $(X', \mathscr{T}', \nu)$ be measure spaces, each of which is the union of a sequence of sets with finite measure, for instance, $X = Y = \mathbb{R}$ with the Lebesgue measure. Let $f : X \times Y \to \mathbb{R}$ be a mesurable function on $X \times Y$ for the product σ-algebra. Then:*

i) f is integrable on $X \times Y$ with respect to the product measure $\mu \otimes \nu$ if and only if one of the following integrals is finite:

$$\int \left(\int \left| f(x, y) \right| \mathrm{d}y \right) \mathrm{d}x \qquad \text{or} \qquad \int \left(\int \left| f(x, y) \right| \mathrm{d}x \right) \mathrm{d}y.$$

ii) If f is integrable on $X \times Y$, then for almost all $x \in X$, the function $y \mapsto f(x, y)$ is ν-integrable on Y, and similarly when exchanging the role of x and y.

iii) If f is integrable on $X \times Y$, the function defined for almost all $x \in X$ by

$$x \longmapsto I(x) \stackrel{\text{def}}{=} \int_Y f(x, y)\,\mathrm{d}\nu(y)$$

A child prodigy, the Venetian Guido Fubini (1879–1943) found, in nineth grade, a formula for π as a series converging faster than anything known at the time. At the Scuola normale superiore di Pisa, he was student of Dini (see page 22). He then held positions at Pisa and Torino. Mussolini's fascist regime forbade him to teach, and he went into exile to Paris and then to Princeton University in the United States, where he finished his career. In addition to his work on the Lebesgue integral, he proved important results in functional analysis, differential geometry, and projective geometry.

is μ-integrable on X and we have

$$\int_X I(x)\,\mathrm{d}\mu(x) = \iint_{X\times Y} f\,\mathrm{d}(\mu\otimes\nu)(x,y).$$

Remark 3.17 Since X and Y play symmetric roles, we also have

$$\iint_{X\times Y} f = \int_Y \left(\int_X f(x,y)\,\mathrm{d}\mu(x)\right)\mathrm{d}\nu(y) = \int_X \left(\int_Y f(x,y)\,\mathrm{d}\nu(y)\right)\mathrm{d}\mu(x).$$

For non-negative functions, we have a converse:

THEOREM 3.18 (Fubini-Tonelli) *Let $(X,\mathscr{T},\mu)$ and $(X',\mathscr{T}',\nu)$ be measure spaces, each of which is the union of a sequence of sets with finite measure, for instance, $X = Y = \mathbb{R}$ with the Lebesgue measure. Let $f : X\times Y \to \mathbb{R}$ be a non-negative measurable function. Then the integral of f on $X\times Y$ with respect to the product measure $\mu\otimes\nu$, which is either a non-negative real number or $+\infty$, is given by the Fubini formula*

$$\begin{aligned}\iint_{X\times Y} f(x,y)\,\mathrm{d}(\mu\otimes\nu)(x,y) &= \int_Y \left(\int_X f(x,y)\,\mathrm{d}\mu(x)\right)\mathrm{d}\nu(y)\\ &= \int_X \left(\int_Y f(x,y)\,\mathrm{d}\nu(y)\right)\mathrm{d}\mu(x).\end{aligned}$$

In particular, if the following two conditions hold:

i) for almost all $x\in X$, the function $y\mapsto f(x,y)$ is integrable on Y;

ii) the function defined for almost all $x\in X$ by $x\mapsto I(x) \stackrel{\text{def}}{=} \int_{\mathbb{R}} f(x,y)\,\mathrm{d}y$ is μ-integrable,

then f is integrable on $X\times Y$ with respect to the product measure $\mu\otimes\nu$.

3.5 Change of variables

The last important point of integral calculus is the change of variable formula in multiple integrals. Suppose we wish to perform a change of variable in 3 dimensions

$$(x,y,z) \longmapsto (u,v,w) = \varphi(x,y,z),$$

(and analogously in n dimensions); then the *jacobian* of this change of variable is defined as follows:

DEFINITION 3.19 Let Ω and Ω' be two open sets in $\mathbb{R}^3$. Let φ be a bijection from Ω to Ω', which is continuous and has continuous differential, as well as its inverse. Denote $(u,v,w) = \varphi(x,y,z)$. The **jacobian matrix** of φ is the matrix of partial derivatives

$$\operatorname{Jac}_\varphi(x,y,z) \stackrel{\text{def}}{=} \frac{\mathrm{D}(u,v,w)}{\mathrm{D}(x,y,z)} \stackrel{\text{def}}{=} \begin{pmatrix} \dfrac{\partial u}{\partial x} & \dfrac{\partial u}{\partial y} & \dfrac{\partial u}{\partial z} \\ \dfrac{\partial v}{\partial x} & \dfrac{\partial v}{\partial y} & \dfrac{\partial v}{\partial z} \\ \dfrac{\partial w}{\partial x} & \dfrac{\partial w}{\partial y} & \dfrac{\partial w}{\partial z} \end{pmatrix}.$$

This is a continuous function on Ω. The **jacobian** of the change of variable is the determinant of the jacobian matrix:

$$J_\varphi(x,y,z) = \det \operatorname{Jac}_\varphi(x,y,z).$$

It is also a continuous function on Ω.

The generalization of this definition to change of variables in $\mathbb{R}^n$ is immediate.

THEOREM 3.20 (Change of variables) *Let U and V be open sets in $\mathbb{R}^n$. Let $\varphi : U \to V$ be a bijection which is continuous, has continuous differential, and is such that the jacobian of φ does not vanish on U. Then for any function $f : V \to \mathbb{R}^p$, we have*

$$f \text{ is integrable on } V \iff (f \circ \varphi)\, J_\varphi \text{ is integrable on } U$$

and moreover

$$\int_V f(u_1,\ldots,u_n)\,\mathrm{d}u_1\ldots\mathrm{d}u_n = \int_U f\circ\varphi(x_1,\ldots,x_n)\,\big|\,J_\varphi(x_1,\ldots,x_n)\big|\,\mathrm{d}x_1\ldots\mathrm{d}x_n.$$

Here integration and integrability are with respect to the Lebesgue measure on $\mathbb{R}^n$, and mean that each component of f is integrable.

As an example, take the change of variable in polar coordinates in the plane $\mathbb{R}^2$, denoted $\varphi : (r,\theta) \mapsto (x,y)$, which we may assume to be defined on the open set $]0,+\infty[\, \times\,]-\pi,\pi[$. We can write

$$\begin{cases} x = r\cos\theta, \\ y = r\sin\theta, \end{cases} \qquad \text{or} \qquad \begin{cases} r = \sqrt{x^2+y^2}, \\ \theta = 2\arctan\left(\dfrac{y}{x+\sqrt{x^2+y^2}}\right). \end{cases}$$

(Tricky question: why shouldn't we write $\theta = \arctan(y/x)$ as is often seen? Why do most people write it anyway?)

The jacobian of this change of variable is

$$\left|\frac{\mathrm{D}(x,y)}{\mathrm{D}(r,\theta)}\right| = \begin{vmatrix} \dfrac{\partial x}{\partial r} & \dfrac{\partial y}{\partial r} \\ \dfrac{\partial x}{\partial \theta} & \dfrac{\partial y}{\partial \theta} \end{vmatrix} = \begin{vmatrix} \cos\theta & \sin\theta \\ -r\sin\theta & r\cos\theta \end{vmatrix} = r,$$

and we recover the well-known formula:

$$\iint f(x,y)\,\mathrm{d}x\,\mathrm{d}y = \int_{-\pi}^{\pi}\int_{-\infty}^{+\infty} f(r,\theta)\,r\,\mathrm{d}r\,\mathrm{d}\theta.$$

Remark 3.21 We might be interested in the inverse transformation. The computation of $\frac{\mathrm{D}(r,\theta)}{\mathrm{D}(x,y)}$ directly from the formula is rather involved, but a classical result of differential calculus avoids this computation: we have

$$\frac{\mathrm{D}(r,\theta)}{\mathrm{D}(x,y)} \cdot \frac{\mathrm{D}(x,y)}{\mathrm{D}(r,\theta)} = 1,$$

and this result may be generalized to arbitrary locally bijective differentiable maps in any dimension.

Remark 3.22 In one dimension, the change of variable formula is

$$\int_{\mathbb{R}} f(u)\,\mathrm{d}u = \int_{\mathbb{R}} f\big(\varphi(x)\big)\,|\varphi'(x)|\,\mathrm{d}x.$$

It is important not to forget the absolute value in the change of variable, for instance,

$$\int_{\mathbb{R}} f(x)\,\mathrm{d}x = \int_{\mathbb{R}} f(-y) \times |-1|\,\mathrm{d}y = \int_{\mathbb{R}} f(-y)\,\mathrm{d}y.$$

Sometimes absolute values are omitted, and one writes, for instance, $\mathrm{d}y = -\mathrm{d}x$. In this case, the range of integration is considered to be *oriented*, and the orientation is reversed when the change of variable is performed if it corresponds to a decreasing function $\varphi : x \mapsto y = \varphi(x)$. Thus we have

$$\int_{-\infty}^{+\infty} f(x)\,\mathrm{d}x = \int_{+\infty}^{-\infty} f(-y)(-\mathrm{d}y) = \int_{-\infty}^{+\infty} f(-y)\,\mathrm{d}y,$$

with the minus sign cancelling the effect of interchanging the bounds of integration.

EXERCISES

♦ **Exercise 3.3** Show that

$$\int_1^x \frac{|\sin t|}{t}\,\mathrm{d}t \underset{x\to+\infty}{\sim} \frac{2}{\pi}\log x.$$

♦ **Exercise 3.4** Check that the jacobian for the change of variables from cartesian coordinates to spherical coordinates $(x, y, z) \mapsto (r, \theta, \varphi)$ in $\mathbb{R}^3$ is $J = r^2 \sin\theta$.

♦ **Exercise 3.5** Find (and justify!) the following limits:

$$\lim_{n\to\infty} \int_0^1 \frac{n^{3/2}x}{1+n^2x^2}\,\mathrm{d}x \qquad \text{and} \qquad \lim_{n\to\infty} \int_0^\infty \frac{|\sin x|^{1/n}}{1+x^2}\,\mathrm{d}x.$$

♦ **Exercise 3.6** Let $f(\nu)$ be the distribution function of energy emitted by a black body, as a function of the frequency. Recall that, by definition, the total amount of energy with frequency between ν_0 and ν_1 is given by the integral

$$\int_{\nu_0}^{\nu_1} f(\nu)\,\mathrm{d}\nu.$$

Max Planck showed at the end of the nineteenth century that the distribution function $f(\nu)$ is given by the formula (which now bears his name)

$$f(\nu) = \frac{A\nu^3}{\mathrm{e}^{2\pi\beta\hbar\nu} - 1},$$

where A is a constant, $\beta = 1/k_{\mathrm{B}}T$ is the inverse temperature, and $\hbar$ is the Planck constant.

What is the distribution function $g(\lambda)$ of the energy as function of the *wavelength* $\lambda \stackrel{\text{def}}{=} c/\nu$? For which frequency $\nu_{\max}$ is $f(\nu)$ maximal? What is the wavelength for which $g(\lambda)$ is maximal? Why is it that $\lambda_{\max} \neq c/\nu_{\max}$?

♦ **Exercise 3.7** Show that the function $x \mapsto \arctan(x) - \arctan(x-1)$ is integrable on $\mathbb{R}$ (with respect to Lebesgue measure, of course) and show that

$$\int_{-\infty}^{+\infty} \big[\arctan(x) - \arctan(x-1)\big]\,\mathrm{d}x = \pi.$$

Generalize this formula.

♦ **Exercise 3.8 (Counterexample to Fubini's theorem)** For $(x, y) \in [0, 1]^2$, define

$$f(x, y) = \frac{\operatorname{sgn}(y - x)}{\max(x, y)^2}.$$

i) Compute $\int_0^1 f(x, y)\,\mathrm{d}x$ for all $y \in [0, 1]$.
   Deduce from this the value of $\int_0^1 \left(\int_0^1 f(x, y)\,\mathrm{d}x\right)\mathrm{d}y$.

ii) Similarly, compute $\int_0^1 \left(\int_0^1 f(x, y)\,\mathrm{d}y\right)\mathrm{d}x$.

iii) Any comments?

♦ **Exercise 3.9 (Integrability of a radial function)** Let f be a (measurable) function on $\mathbb{R}^3$, such that we have

$$f(x,y,z) \underset{r\to 0}{\sim} r^\alpha$$

for some real number α, where $r \stackrel{\text{def}}{=} \sqrt{x^2+y^2+z^2}$. Assuming that f is integrable on any bounded region in $\mathbb{R}^3 - \{0\}$, discuss the integrability of f "at the origin," depending on the value of α. Generalize this to functions on $\mathbb{R}^d$.

♦ **Exercise 3.10 (Fourier transform of a Lorentzian function)** Let

$$F(x) = \int_{-\infty}^{+\infty} \frac{\cos ux}{1+u^2}\,\mathrm{d}u \qquad \text{for all } x \in \mathbb{R}.$$

i) Where is f defined? Is it continuous on its domain of definition?

ii) Let $g(x) = \displaystyle\int_{-\infty}^{+\infty} \frac{x\cos t}{x^2+t^2}\,\mathrm{d}t$. What is the relation between F and g?

iii) Deduce from the previous question a differential equation satisfied by g. (Start by comparing
$$\frac{\partial^2}{\partial x^2}\left(\frac{x}{x^2+t^2}\right) \qquad \text{and} \qquad \frac{\partial^2}{\partial t^2}\left(\frac{x}{x^2+t^2}\right).$$
Solve the differential equation, and deduce the value of F.)

SOLUTIONS

♦ **Solution of exercise 3.4.** Let (r, θ, φ) denote spherical coordinates (where θ is latitude and φ longitude). We have

$$x = r\cos\theta\cos\varphi, \qquad y = r\cos\theta\sin\varphi, \qquad z = r\sin\theta,$$

and therefore the jacobian is

$$J = \begin{vmatrix} \cos\theta\cos\varphi & \cos\theta\sin\varphi & \sin\theta \\ r\sin\theta\sin\varphi & r\sin\theta\sin\varphi & -r\cos\theta \\ -r\cos\theta\sin\varphi & r\cos\theta\cos\varphi & 0 \end{vmatrix} = r^2\sin\theta.$$

♦ **Solution of exercise 3.6.** The function $g(\lambda)$ has the property that the amount of energy radiated by frequencies in the interval $[\nu_0, \nu_1]$, corresponding to wavelengths λ_0 and λ_1, is equal to $\displaystyle\int_{\lambda_1}^{\lambda_0} g(\lambda)\,\mathrm{d}\lambda$. The change of variable $\nu = c/\lambda$ leads to

$$\int_{\lambda_1}^{\lambda_0} g(\lambda)\,\mathrm{d}\lambda = \int_{\nu_0}^{\nu_1} g(c/\nu)\left|\frac{\mathrm{d}\lambda}{\mathrm{d}\nu}\right|\mathrm{d}\nu = \int_{\nu_0}^{\nu_1} g(c/\nu)\frac{c}{\nu^2}\,\mathrm{d}\nu,$$

and this shows that

$$f(\nu) = c\,\frac{g(c/\nu)}{\nu^2}.$$

Note that there is no reason for the functions $\nu \mapsto f(\nu)$ and $\nu \mapsto \nu^2 f(\nu)$ to reach a maximum at the same point. In fact, rather than the functions f and g themselves, it is the differential forms (or measures) $f(\nu)\,\mathrm{d}\nu$ and $g(\lambda)\,\mathrm{d}\lambda$ which have physical meaning.

♦ **Solution of exercise 3.7.** Integrability comes from the inequalities

$$\frac{1}{1+(x-1)^2} \leqslant \arctan(x) - \arctan(x-1) \leqslant \frac{1}{1+x^2},$$

which are consequences of the mean value theorem. To compute the value of the integral, let F denote a primitive of arctan. Then we have

$$\begin{aligned}\int_{-M}^{M} \big[\arctan(x) - \arctan(x-1)\big]\,\mathrm{d}x &= F(M) - F(M-1) - F(-M) + F(-M-1) \\ &= \arctan(\xi_M) - \arctan(\xi'_M),\end{aligned}$$

where $\xi_M \in [M-1, M]$ and $\xi'_M \in [-M-1, -M]$ are given by the mean value theorem again. The result stated follows immediately.

♦ **Solution of exercise 3.10.**

i) The uniform upper bound $u \mapsto \frac{1}{1+u^2}$ and the continuity theorem show that F is defined and continuous on $\mathbb{R}$. It is obvious that $F(x) \leqslant \pi$, with equality for $x = 0$.

ii) By the change of variable $t = xy$, we have for $x > 0$

$$g(x) = \int_{-\infty}^{+\infty} \frac{\cos(xy)}{1+y^2}\,\mathrm{d}y = F(x).$$

It follows that $g(x)$ has a limit as x tends to zero on the right, namely,

$$g(0^+) = F(0) = \pi.$$

Similarly, for $x < 0$ we have $g(x) = -F(x)$ and hence $g(0^-) = -\pi$. Since the definition gives $g(0) = 0$, we see that the function g is not continuous at 0.

iii) A simple computation shows that the two derivatives mentioned in the statement of the exercise are equal. From this, we deduce that

$$\frac{\partial^2}{\partial x^2} g(x) = \int_{-\infty}^{+\infty} \cos t \frac{\partial^2}{\partial x^2} \left(\frac{x}{x^2 + t^2} \right) \mathrm{d}t = -\int_{-\infty}^{+\infty} \cos t \frac{\partial^2}{\partial t^2} \left(\frac{x}{x^2 + t^2} \right) \mathrm{d}t$$

$$= \int_{-\infty}^{+\infty} \cos t \left(\frac{x}{x^2 + t^2} \right) \mathrm{d}t,$$

using two integrations by parts. It follows that g is a solution of the differential equation

$$g''(x) - g(x) = 0.$$

The general solution of this equation is of the form

$$g(x) = \alpha_{\pm} \mathrm{e}^{x} + \beta_{\pm} \mathrm{e}^{-x},$$

where (α_+, β_+) are real constants (which give the value of g on $\mathbb{R}^{*+}$), and (α_-, β_-) similarly on $\mathbb{R}^{*-}$.

Since g is bounded, the coefficient α_+ of e^t on $\mathbb{R}^+$ is zero, and the coefficient β_- of e^{-t} on $\mathbb{R}^-$ also. Since g is odd and $g(0^+) = \pi$, we finally find that

$$g(x) = \pi \operatorname{sgn}(x) \mathrm{e}^{-|x|}.$$

Since $F(x) = \operatorname{sgn}(x) \cdot g(x)$ and $F(0) = \pi$, we finally find that

$$\forall x \in \mathbb{R} \qquad F(x) = \pi \, \mathrm{e}^{-|x|}.$$

Chapter 4

Complex Analysis I

This chapter deals with the theory of complex-valued functions of one complex variable. It introduces the notion of a *holomorphic function*, which is a function $f : \Omega \to \mathbb{C}$ defined and *differentiable* on an open subset of $\mathbb{C}$. We will see that the (weak) assumption of differentiability implies, in strong contrast with the real case, the (much stronger) consequence that f is infinitely differentiable. We will then study functions with *singularities* at isolated points and which are holomorphic except at these points; we will see that their study has important applications to the computation of many integrals and sums, notably for the computation of the Fourier transforms that occur in physics. In Chapter 5, we will see how techniques of *conformal analysis* provide elegant solutions for certain problems of physics in two dimensions, in particular, problems of electrostatics and of incompressible fluid mechanics, but also in the theory of diffusion and in particle physics (see also Chapter 15).

4.1 Holomorphic functions

Whereas differentiability in $\mathbb{R}$ is a relatively weak constraint,[1] we will see during the course of this chapter that differentiability in terms of a complex variable implies by contrast many properties and "rigidifies" the situation, in a sense that will soon be made precise.

[1] A function f, defined on an open interval I of $\mathbb{R}$ and with values in $\mathbb{R}$ or $\mathbb{C}$ may be differentiable at all points of I without, for instance, the derivative being continuous. For instance, if we set $f(0) = 0$ and $f(x) = x^2 \sin(1/x)$ for all nonzero x, then f is differentiable at all points of $\mathbb{R}$, $f'(0) = 0$, but f' is not continuous at 0 (there is no limit of f' at 0).

The life of Baron Augustin-Louis CAUCHY (1789–1857) is quite eventful since Cauchy, for political reasons (as a supporter of the French monarchy), went into exile in Italy and then Prague. He was a long-time teacher at the École Polytechnique. His work is extremely wide-ranging, but it is to analysis that Cauchy brought a fundamental contribution: a new emphasis on rigorous arguments. He was one of the first to worry about *proving* the convergence of various processes (in integral calculus and in the theory of differential equations); he worked on determinants and introduced the notation of matrices with double indices. In complex analysis, his contributions are fundamental, as shown by the many theorems bearing his name.

4.1.a Definitions

Let us start by defining differentiability on the complex plane. Let Ω be an open subset of the complex[2] plane $\mathbb{C}$, and let $f : \Omega \to \mathbb{C}$ be a complex-valued function of a complex variable. Let $z_0 \in \Omega$ be given.

DEFINITION 4.1 The function $f : \Omega \to \mathbb{C}$ is said to be **differentiable at** $z_0 \in \mathbb{C}$ if

$$f'(z_0) \stackrel{\text{def}}{=} \lim_{z \to z_0} \frac{f(z) - f(z_0)}{z - z_0}$$

exists, i.e., if there exists a complex number, denoted $f'(z_0)$ such that

$$\left| f(z) - f(z_0) - f'(z_0) \times (z - z_0) \right| = \mathrm{o}(z - z_0) \quad [z \to z_0].$$

One may ask whether this definition is not simply equivalent to differentiability in $\mathbb{R}^2$, after identifying the set of complex numbers and the real plane (see sidebar 3 on page 134 for a reminder on basic facts about differentiable functions on $\mathbb{R}^2$). The following theorem, due to Cauchy, shows that this is not the case and that $\mathbb{C}$-differentiability is stronger.

THEOREM 4.2 (Cauchy-Riemann equations) *Let $f : \Omega \to \mathbb{C}$ be a complex-valued function of one complex variable; denote by $\widetilde{f}$ the complex-valued function on $\mathbb{R}^2$ naturally associated to f, that is, the function*

$$\begin{aligned} \widetilde{f} : \ \mathbb{R}^2 &\longrightarrow \mathbb{C}, \\ (x, y) &\longmapsto \widetilde{f}(x, y) \stackrel{\text{def}}{=} f(x + \mathrm{i}y). \end{aligned}$$

[2] While we are speaking of this, we can recall that the study of the complex plane owes much to the works of Rafaele BOMBELLI (1526–1573) (yes, so early!), who used i to solve algebraic equations (he called it "di meno," that is, "*[root]* of minus *[one]*"), and of Jean-Robert ARGAND (1768–1822) who gave the interpretation of $\mathbb{C}$ as a geometric plane. The complex plane is sometimes called "the Argand plane." It is also ARGAND who introduced the *modulus* for a complex number. The notation "i," replacing the older notation $\sqrt{-1}$, is due to Euler (see page 39).

Then the function f is $\mathbb{C}$-differentiable at the point $z_0 = x_0 + \mathrm{i}\, y_0 \in \Omega$ if and only if

$$\left.\begin{array}{l} \tilde{f} \text{ is } \mathbb{R}^2\text{-differentiable} \\ \dfrac{\partial \tilde{f}}{\partial x}(x_0, y_0) = -\mathrm{i}\dfrac{\partial \tilde{f}}{\partial y}(x_0, y_0). \end{array}\right\} \qquad \textit{Cauchy-Riemann equation.}$$

Proof

• Assume that f is differentiable in the complex sense at a point z_0. We can define a linear form $\mathrm{d}\tilde{f}(z_0) : \mathbb{R}^2 \to \mathbb{C}$ in the following manner:

$$\mathrm{d}\tilde{f}(z_0)\big((k,l)\big) \stackrel{\text{def}}{=} f'(z_0) \times \big(k + \mathrm{i}\, l\big).$$

We have then, according to the definition of differentiability of f at z_0,

$$\left|\tilde{f}(x,y) - \tilde{f}(x_0,y_0) - \mathrm{d}\tilde{f}(x_0,y_0)\big((x-x_0),(y-y_0)\big)\right| = o(z - z_0),$$

which shows that $\tilde{f}$ is $\mathbb{R}^2$-differentiable at the point (x_0, y_0). Moreover, since

$$\mathrm{d}\tilde{f}(x_0,y_0)(k,l) = f'(z_0)k + \mathrm{i} f'(z_0) l,$$

it follows that

$$\mathrm{d}f(z_0) = f'(z_0)\,\mathrm{d}x + \mathrm{i} f'(z_0)\,\mathrm{d}y.$$

Comparing with the formula

$$\mathrm{d}\tilde{f} = \frac{\partial \tilde{f}}{\partial x}\,\mathrm{d}x + \frac{\partial \tilde{f}}{\partial y}\,\mathrm{d}y$$

(see page 134), we get

$$\frac{\partial f}{\partial x} = f'(z_0) \qquad \text{and} \qquad \frac{\partial f}{\partial y} = \mathrm{i} f'(z_0),$$

and therefore the required equality.

• Conversely: if $\tilde{f}$ is $\mathbb{R}^2$-differentiable and satisfies $\partial\tilde{f}/\partial x = -\mathrm{i}\partial\tilde{f}/\partial y$, then

$$\begin{aligned} \mathrm{d}\tilde{f}(x_0,y_0)\big((k,l)\big) &= \left[\frac{\partial \tilde{f}}{\partial x}\,\mathrm{d}x + \frac{\partial \tilde{f}}{\partial y}\,\mathrm{d}y\right]\big((k,l)\big) = \frac{\partial \tilde{f}}{\partial x}\left[\mathrm{d}x + \mathrm{i}\,\mathrm{d}y\right]\big((k,l)\big) \\ &= \frac{\partial \tilde{f}}{\partial x} \times (k + \mathrm{i} l), \end{aligned}$$

which shows, by the definition of differentiability of $\tilde{f}$, that

$$\left| f(z) - f(z_0) - \frac{\partial f}{\partial x} \times (z - z_0) \right| = o(z - z_0),$$

and thus that f is differentiable and that its derivative is equal to $f'(z_0) = \frac{\partial \tilde{f}}{\partial x}(z_0)$.

We henceforth write f instead of $\tilde{f}$, except in special circumstances.

In the following, we will very often write a complex-valued function in the form

$$f = P + \mathrm{i}Q \qquad \text{with} \quad P = \operatorname{Re}(f), \quad Q = \operatorname{Im}(f).$$

The relation obtained in the preceding theorem allows us to write

THEOREM 4.3 (Cauchy-Riemann equations) *Let $f : \Omega \to \mathbb{C}$ be a function differentiable at $z_0 \in \Omega$. Then we have the following equations:*

$$\left.\begin{aligned} \left.\frac{\partial P}{\partial x}\right|_{(x_0,y_0)} &= \left.\frac{\partial Q}{\partial y}\right|_{(x_0,y_0)} \\ \left.\frac{\partial Q}{\partial x}\right|_{(x_0,y_0)} &= -\left.\frac{\partial P}{\partial y}\right|_{(x_0,y_0)} \end{aligned}\right\} \quad \textit{Cauchy-Riemann equations.}$$

DEFINITION 4.4 Let Ω be an open subset of $\mathbb{C}$. A function $f : \Omega \to \mathbb{C}$ is **holomorphic**[3] **at $z_0 \in \mathbb{C}$** if it is differentiable at z_0, and it is holomorphic on Ω if it is holomorphic at every point of Ω. We will denote by $\mathscr{H}(\Omega)$ the set of all functions holomorphic on Ω.

If A is a non-empty subset of $\mathbb{C}$, a function f defined on A is **holomorphic on A** if there exist an open neighborhood Ω' of A and a holomorphic function g on Ω' such that $g(z) = f(z)$ for $z \in A$.

The following exercise (which is elementary) is an example of the "rigidity" of holomorphic functions that we mentioned before.

DEFINITION 4.5 A **domain** of $\mathbb{C}$ is a *connected*[4] open subset of $\mathbb{C}$.

♦ **Exercise 4.1** Show that if $f : \Omega \to \mathbb{C}$ is a function holomorphic on a domain of $\mathbb{C}$, the following properties are equivalent:

(i) f is constant on Ω;

(ii) $P = \text{Re}(f)$ is constant on Ω;

(iii) $Q = \text{Im}(f)$ is constant on Ω;

(iv) $|f|$ is constant on Ω;

(v) $\bar{f}$ is holomorphic on Ω.

(Solution on page 129)

4.1.b Examples

Example 4.6 The functions $z \mapsto z^2$, $z \mapsto e^z$, the polynomials in z, are holomorphic on $\mathbb{C}$.

Example 4.7 The function $z \mapsto 1/z$ is holomorphic on $\mathbb{C}^*$.

On the other hand, the function $f : z \mapsto \bar{z}$ is not differentiable at any point. Indeed, let us show, for instance, that f is not differentiable at 0. For

[3] The name *holomorphic* was introduced by Jean-Claude BOUQUET (1819–1885) and his colleague Charles BRIOT (1817–1882). Some texts use equivalently the words "analytic" or "regular."

[4] A topological space is said to be **connected** if it is not the union of two disjoint open sets; in other words, it is a set which is "in one piece"; see Appendix A.

this, we compute

$$\frac{f(z) - f(0)}{z - 0} = \frac{\bar{z}}{z}.$$

If now we let z follow a path included in $\mathbb{R}$ approaching 0, this quotient is equal to 1 thoughout the path, whereas if we let z follow a path included in the set of purely imaginary numbers, we find a constant value equal to -1; this shows that *the quantity of interest has no limit as* $z \to 0$, and thus that f is not differentiable at 0. If we add a constant, we see that it is not differentiable at z_0 for any z_0.

Example 4.8 The function $z \mapsto |z|$ is nowhere holomorphic, and similarly for the functions $z \mapsto \text{Re}(z)$ and $z \mapsto \text{Im}(z)$.

As a general rule, holomorphic functions will be those which depend "on z only" (and of course sufficiently smooth), whereas those functions "that depend not only on z but also on $\bar{z}$" will not be holomorphic, as for instance $|z| = \sqrt{z \cdot \bar{z}}$ or $\text{Re}(z) = \frac{z+\bar{z}}{2}$. One may say that the variables z and $\bar{z}$ are treated as if they were *independent* variables.[5]

4.1.c The operators $\partial/\partial z$ and $\partial/\partial \bar{z}$

We consider the plane $\mathbb{R}^2$, which we will freely identify with $\mathbb{C}$.

DEFINITION 4.9 We will denote by $\mathrm{d}x$ the linear form which associates its first coordinate to any vector (k, l) in $\mathbb{R}^2$. Similarly, we denote by $\mathrm{d}y$ the linear form which associates its second coordinate to any vector (k, l) in $\mathbb{R}^2$:

$$\mathrm{d}x(k, l) = k \qquad \mathrm{d}y(k, l) = l.$$

We will write, moreover, $\mathrm{d}z \stackrel{\text{def}}{=} \mathrm{d}x + \mathrm{i}\,\mathrm{d}y$ and $\mathrm{d}\bar{z} \stackrel{\text{def}}{=} \mathrm{d}x - \mathrm{i}\,\mathrm{d}y$.

♦ **Exercise 4.2** Show that the differential form $\mathrm{d}z = \mathrm{d}x + \mathrm{i}\mathrm{d}y$ is the identity ($\mathrm{d}z = \mathrm{Id}_{\mathbb{C}}$) and that the differential form $\mathrm{d}\bar{z}$ is the complex conjugation.

For any function $f : \Omega \to \mathbb{C}$, $\mathbb{R}^2$-differentiable (and therefore not necessarily differentiable in the complex sense), we have (check sidebar 3 on page 134):

$$\mathrm{d}f = \frac{\partial f}{\partial x}\mathrm{d}x + \frac{\partial f}{\partial y}\mathrm{d}y.$$

The functions $z = x + \mathrm{i}y$ and $\bar{z} = x - \mathrm{i}y$ are $\mathbb{R}^2$-differentiable on $\mathbb{R}^2 \simeq \mathbb{C}$; moreover, $\mathrm{d}z = \mathrm{d}x + \mathrm{i}\,\mathrm{d}y$ and $\mathrm{d}\bar{z} = \mathrm{d}x - \mathrm{i}\,\mathrm{d}y$. We deduce that $\mathrm{d}x = \frac{1}{2}(\mathrm{d}z + \mathrm{d}\bar{z})$

---

[5] For those readers interested in theoretical physics, this treatment "as independent variables" can be found also in field theory, for instance in the case of Grassmann variables.

and $\mathrm{d}y = \frac{1}{2\mathrm{i}}(\mathrm{d}z - \mathrm{d}\bar{z})$, which allows us to write:

$$\begin{aligned}\mathrm{d}f &= \frac{1}{2}\frac{\partial f}{\partial x}(\mathrm{d}z + \mathrm{d}\bar{z}) + \frac{\mathrm{i}}{2}\frac{\partial f}{\partial y}(\mathrm{d}\bar{z} - \mathrm{d}z) \\ &= \frac{1}{2}\left(\frac{\partial f}{\partial x} - \mathrm{i}\frac{\partial f}{\partial y}\right)\mathrm{d}z + \frac{1}{2}\left(\frac{\partial f}{\partial x} + \mathrm{i}\frac{\partial f}{\partial y}\right)\mathrm{d}\bar{z}.\end{aligned}$$

This leads us to the following definition:

DEFINITION 4.10 We introduce the differential operators

$$\frac{\partial}{\partial z} \stackrel{\text{def}}{=} \frac{1}{2}\left(\frac{\partial}{\partial x} - \mathrm{i}\frac{\partial}{\partial y}\right) \quad \text{and} \quad \frac{\partial}{\partial \bar{z}} \stackrel{\text{def}}{=} \frac{1}{2}\left(\frac{\partial}{\partial x} + \mathrm{i}\frac{\partial}{\partial y}\right).$$

We will sometimes denote them ∂ and $\overline{\partial}$, when there is no ambiguity.

With these operators, the differential of a function $\mathbb{R}^2$-differentiable can therefore be written in the form

$$\mathrm{d}f = \frac{\partial f}{\partial z}\,\mathrm{d}z + \frac{\partial f}{\partial \bar{z}}\,\mathrm{d}\bar{z}.$$

We can then write another form of Theorem 4.2:

THEOREM 4.11 (Cauchy-Riemann equations) *If f is $\mathbb{C}$-differentiable, that is, holomorphic, at z_0, we have*

$$\frac{\partial f}{\partial z}(z_0) = f'(z_0) \qquad \textit{and} \qquad \frac{\partial f}{\partial \bar{z}}(z_0) = 0.$$

This last relation is none other than the Cauchy-Riemann condition. *One may write also the two preceding equations as*

$$\frac{\partial f}{\partial z}(z_0) = \frac{\partial f}{\partial x}(z_0) = -\mathrm{i}\,\frac{\partial f}{\partial y}(z_0).$$

Example 4.12 The function $f : z \mapsto \mathrm{Re}(z) = \frac{z+\bar{z}}{2}$ is not holomorphic because $\frac{\partial f}{\partial \bar{z}} = \frac{1}{2}$ at every point.

Remark 4.13 We could also change our point of view and look at functions that can be expressed in terms of $\bar{z}$ only:

DEFINITION 4.14 A function $f : \Omega \to \mathbb{C}$ is **antiholomorphic** if it satisfies

$$\frac{\partial f}{\partial z} = 0$$

for all point in Ω.

Remark 4.15 Be careful not to argue that "$\partial f/\partial z = 0$ at all points, so the function is constant," which is absurd. The function $z \mapsto \bar{z}$ does satisfy the condition, and therefore is antiholomorphic, but it is certainly not constant! The quantity $\partial f/\partial z$ is equal to $f'(z)$ **only** for a holomorphic function. For an arbitrary function, it does not have such a clear meaning.

4.2 Cauchy's theorem

In this section, we will describe various versions of a fundamental result of complex analysis, which says that, under very general conditions, *the integral of a holomorphic function around a closed path is equal to zero.* We will start by making precise what a path integral is, and then present a simple version of Cauchy's theorem (with proof), followed by a more general version (proof omitted).

4.2.a Path integration

We now need a few definitions: *argument*, *path*, and *winding number of a path around a point.*

DEFINITION 4.16 (Argument) Let Ω be an open subset of $\mathbb{C}^* = \mathbb{C} \setminus \{0\}$ and let $\theta : \Omega \to \mathbb{R}$ be a continuous function such that for $z \in \Omega$ we have

$$e^{i\theta(z)} = \frac{z}{|z|}.$$

Such a function θ is called a **continuous determination of the argument**. If Ω is connected, then two continuous determinations of the argument differ by a constant integral multiple of 2π. There exists a unique continuous determination of the argument

$$\theta : \mathbb{C} \setminus \mathbb{R}^- \longrightarrow]-\pi, \pi[$$

which satisfies $\theta(x) = 0$ for any real number $x > 0$. It is called the **principal determination of the argument** and is denoted "Arg."

Remark 4.17 One should be aware that a continuous determination of the argument does not necessarily exist for a given open subset. In particular, one cannot find one for the open subset $\mathbb{C}^*$ itself, since starting from argument 0 on the positive real axis and making a complete circle around the origin, one comes back to this half-axis with argument equal to 2π, so that continuity of the argument is not possible.

DEFINITION 4.18 (Curve, path) A **curve** in an open subset Ω in the complex plane is a continuous map from $[0, 1]$ into Ω (one may also allow other sets of definitions and take an arbitrary real segment $[a,b]$ with $a < b$). If $\gamma : [0,1] \to \Omega$ is a curve, we will denote by $\widetilde{\gamma}$ its image in the plane; for simplicity, we will also often write simply γ without risk of confusion. A curve γ is **closed** if $\gamma(0) = \gamma(1)$. It is **simple** if the image doesn't "intersect itself," which means that $\gamma(s) = \gamma(t)$ only if $s = t$ or, in the case of a closed curve, if $s = 0$, $t = 1$ (or the opposite). A **path** is a curve which is piecewise continuously differentiable.

DEFINITION 4.19 (Equivalent paths) Two paths

$$\gamma : [a,b] \to \mathbb{C} \qquad \text{and} \qquad \gamma^* : [a^*,b^*] \to \mathbb{C}$$

are called **equivalent** if their images coincide and the corresponding curves have the same orientation: $\gamma([a,b]) = \gamma^*([a^*,b^*])$, with $\gamma(a) = \gamma^*(a^*)$, $\gamma(b) = \gamma^*(b)$. In other words, there exists a **re-parameterization** of the path γ^*, preserving the orientation, which transforms it into γ, which means a bijection $u : [a,b] \to [a^*,b^*]$ which is continuous, piecewise differentiable and strictly increasing, and such that $\gamma(x) = \gamma^*\big(u(x)\big)$ for all $x \in [a,b]$.

Example 4.20 Let $\gamma : \left| \begin{array}{l} [0,2\pi] \longrightarrow \mathbb{C} \\ t \longmapsto e^{it} \end{array} \right.$ and $\gamma^* : \left| \begin{array}{l} [0,\pi/2] \longrightarrow \mathbb{C} \\ t \longmapsto e^{-2\pi i \cos t}. \end{array} \right.$

These define equivalent paths, with image equal to the unit circle.

If $\gamma : [a,b] \to \Omega$ is a path, we have of course the formula

$$\gamma(b) - \gamma(a) = \int_a^b \gamma'(t)\,dt.$$

On the other hand, if $f : \Omega \to \mathbb{R}$ is a *real-valued* function, twice continuously differentiable, then denoting by γ_1 and γ_2 the real and imaginary parts of γ, we have

$$f\big(\gamma(b)\big) - f\big(\gamma(a)\big) = \int_a^b \left(\frac{\partial f}{\partial x}\big(\gamma(t)\big)\frac{d\gamma_1}{dt} + \frac{\partial f}{\partial y}\big(\gamma(t)\big)\frac{d\gamma_2}{dt} \right) dt, \qquad (4.1)$$

the right-hand side of which is nothing else, in two dimensions, than

$$\int_\gamma \nabla f \cdot d\boldsymbol{\ell}.$$

One must be aware that this formula holds only for a *real-valued* function.

DEFINITION 4.21 (Integral on a path) Let $f : \Omega \to \mathbb{C}$ be an arbitrary (measurable) function (not necessarily holomorphic), and γ a path with image contained in Ω. The **integral of f on the path γ** is given, if the right-hand side exists, by

$$\int_\gamma f(z)\,dz \stackrel{\text{def}}{=} \int_a^b f\big(\gamma(t)\big)\gamma'(t)\,dt. \qquad (4.2)$$

The path γ is called **the integration contour**.

One checks (using the change of variable formula) that this definition does not depend on the parameterization of the path γ, that is, if γ^* is a path equivalent to γ, the right-hand integral is simultaneously defined and gives the same result with γ or γ^*. We then have the following theorem:

THEOREM 4.22 *Let f be a holomorphic function on Ω and $\gamma : [a,b] \to \mathbb{C}$ a path with image contained in Ω. Then*

$$f(\gamma(b)) - f(\gamma(a)) = \int_\gamma \frac{\partial f}{\partial z}(z)\,\mathrm{d}z. \tag{4.3}$$

PROOF. We write $f = f_1 + \mathrm{i} f_2$ and then for each of the *real-valued* functions f_1 and f_2 we appeal to formula (4.1). Formula (4.2), applied to $\partial f/\partial z$, and the Cauchy conditions then imply the result.

Remark 4.23 This formula is only valid if f is a *holomorphic* function. If we take the example of the function $f : z \mapsto \bar{z}$, which is not holomlorphic, we have $\partial f/\partial z = 0$ at all points of $\mathbb{C}$. Formula (4.3) does not apply, since otherwise it would follow that f is identically zero! An extension of the formula, valid for any complex function (holomorphic or not) is given in Theorem 4.51 on page 106 (the Green formula).

DEFINITION 4.24 The **length** of a path γ is the quantity

$$\mathscr{L}_\gamma \stackrel{\text{def}}{=} \int_a^b |\gamma'(t)|\,\mathrm{d}t.$$

We then have the following intuitive property: the integral of a function along a path is at most the length of the path multiplied by the supremum of the modulus of the function.

PROPOSITION 4.25 *If f is holomorphic on Ω and if γ is a path contained in Ω, then*

$$\left| \int_\gamma f(z)\,\mathrm{d}z \right| \leqslant \left(\sup_{z\in\gamma} |f(z)| \right) \times \mathscr{L}_\gamma.$$

4.2.b Integrals along a circle

Some integrals can be computed easily using a suitable parameterization of the path; this is the case for instance for integrals along circles.

Consider the circle $\mathscr{C}(a\,;R)$ centered at the point a with radius R, with positive (counterclockwise) orientation; one can use the parameterization given by

$$\gamma : [0, 2\pi] \longrightarrow \mathbb{C}, \qquad \theta \longmapsto a + R\,\mathrm{e}^{\mathrm{i}\theta},$$

and, therefore, we have $z = a + R\,\mathrm{e}^{\mathrm{i}\theta}$ and $\mathrm{d}z = \mathrm{i}R\,\mathrm{e}^{\mathrm{i}\theta}$.

PROPOSITION 4.26 (Integral along a circle) *Let $a \in \mathbb{C}$ and $R > 0$. Let f be a function defined on an open subset Ω in $\mathbb{C}$ containing the circle $\gamma = \mathscr{C}(a\,;R)$. Then*

$$\int_\gamma f(z)\,\mathrm{d}z = \int_0^{2\pi} f(a + R\mathrm{e}^{\mathrm{i}\theta})\,\mathrm{i}R\,\mathrm{e}^{\mathrm{i}\theta}\,\mathrm{d}\theta.$$

As an example, let us compute the integral of the function $z \mapsto 1/z$ on the unit circle. The path can be represented by $\gamma : [0, 2\pi] \to \mathbb{C}$, $\theta \mapsto z = \mathrm{e}^{\mathrm{i}\theta}$. We find

$$\int_\gamma \frac{\mathrm{d}z}{z} = \int_0^{2\pi} \frac{\mathrm{i}\mathrm{e}^{\mathrm{i}\theta}}{\mathrm{e}^{\mathrm{i}\theta}}\,\mathrm{d}\theta = 2\mathrm{i}\pi. \tag{4.4}$$

4.2.c Winding number

THEOREM 4.27 (and definition) *Let γ be a closed path in $\mathbb{C}$, U the complement of the image of γ in the complex plane. The* ***winding number of γ around the point*** z *is given by*

$$\operatorname{Ind}_\gamma(z) \stackrel{\text{def}}{=} \frac{1}{2\pi\mathrm{i}} \int \frac{\mathrm{d}\zeta}{\zeta - z} \qquad \textit{for all } z \in U. \tag{4.5}$$

Then $\operatorname{Ind}_\gamma(z)$ is an integer, which is constant as a function of z on each connected component of U and is equal to zero on the unbounded connected component of U.

PROOF. See Appendix D.

Example 4.28 A small picture is more eloquent than a long speech.

The picture on the right shows the various values of the function $\operatorname{Ind}_\gamma$ for a curve in the complex plane. One sees that the meaning of the value of the function $\operatorname{Ind}_\gamma$ is very simple: for any point not on γ, it gives the number of times the curve *turns around* z, this number being counted algebraically (taking orientation into account). Being in the unbounded connected component of U means simply being entirely outside the loop.

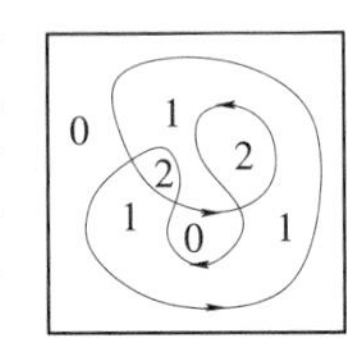

♦ **Exercise 4.3** Show by a direct computation that if γ is a circle with center a and radius r, positively oriented, then

$$\operatorname{Ind}_\gamma(z) = \begin{cases} 1 & \text{if } |z-a| < r, \\ 0 & \text{if } |z-a| > r. \end{cases}$$

4.2.d Various forms of Cauchy's theorem

We are now going to show that if $f \in \mathscr{H}(\Omega)$, where Ω is open and "has no holes," and if γ is a path in Ω, then the integral of f along γ is zero.

We start with the following simple result:

PROPOSITION 4.29 *Let Ω be an open set and γ a* closed *path in Ω. If $F \in \mathscr{H}(\Omega)$ and F' is continuous, then $\int_\gamma F'(z)\,\mathrm{d}z = 0$.*

PROOF. Indeed, $\int_\gamma F'(z)\,\mathrm{d}z = F\big(\gamma(b)\big) - F\big(\gamma(a)\big) = 0$ since γ is a closed path so $\gamma(b) = \gamma(a)$.

In particular we deduce the following result.

THEOREM 4.30
For $\left\{ \begin{array}{l} \textit{any closed path } \gamma \textit{ and any } n \in \mathbb{N} \\ \textit{or} \\ \textit{any closed path } \gamma,\ 0 \notin \gamma \textit{ and } n = -2,\ -3, \ldots \end{array} \right\}$, *we have* $\displaystyle\int_\gamma z^n\,\mathrm{d}z = 0.$

PROOF. For any $n \in \mathbb{Z}$, $n \neq -1$, the function $z \mapsto z^n$ is the derivative of

$$z \longmapsto \frac{z^{n+1}}{(n+1)},$$

which is holomorphic on $\mathbb{C}$ if $n \geqslant 0$ or on $\mathbb{C} \setminus \{0\}$ if $n \leqslant -2$. The preceding proposition gives the conclusion.

This property is very important, and will be used to prove a number of results later on. An important point: the (excluded) case $n = -1$ is very special. We have shown (see formula (4.4)) that the integral of $1/z$ along at least one closed path is not zero. This comes from the fact that the function $z \mapsto 1/z$ is *the only one* among the functions $z \mapsto 1/z^n$ which is not the derivative of a function holomorphic on $\mathbb{C}^*$. It is true that we will define a "logarithm function" in the complex plane, with derivative equal to $z \mapsto 1/z$, but we will see (Section 5.1.a on page 135) that this "complex logarithm" can *never* be defined on the whole of $\mathbb{C}^*$, but only on smaller domains where it is not possible for a closed path to turn completely around 0.

Here is now a first version, somewhat restricted, of Cauchy's theorem.

PROPOSITION 4.31 (Cauchy's theorem for triangles) *Let Δ be a closed triangle in the plane (by this is meant a "filled" triangle, with boundary $\partial\Delta$ consisting of three line segments), entirely contained in an open set Ω. Let $p \in \Omega$ and let $f \in \mathscr{H}\big(\Omega \setminus \{p\}\big)$ be a function continuous on Ω and holomorphic on Ω except possibly at the point p. Then*

$$\int_{\partial\Delta} f(z)\,\mathrm{d}z = 0.$$

PROOF. The proof, elementary but clever, is well worthy of attention. Because of its length, we have put it in Appendix D.

Remark 4.32 We have added some flexibility to the statement by allowing f not to be holomorphic at one point (at most) in the triangle. In fact, this extra generality is illusory. Indeed, we will see further on that if a function is holomorphic in an open subset minus a single point, but is continuous on the whole open set, then in fact it is holomorphic everywhere on the open set.

The following version is much more useful.

THEOREM 4.33 (Cauchy's theorem for convex sets) *Let Ω be an open convex subset of the plane and let $f \in \mathscr{H}(\Omega)$ be a function holomorphic on Ω. Then, for any closed path γ contained in Ω,*

$$\int_\gamma f(z)\,\mathrm{d}z = 0.$$

PROOF. Fix a point $a \in \Omega$ and, for all $z \in \Omega$, define, using the convexity property,

$$F(z) \stackrel{\text{def}}{=} \int_{[a,z]} f(\zeta)\,\mathrm{d}\zeta,$$

where we have denoted

$$[a,z] \stackrel{\text{def}}{=} \big\{(1-\theta)a + \theta z\ ;\ \theta \in [0,1]\big\}.$$

We now show that F is a primitive of f.

□ Let z_0 be a point in Ω. For any $z \in \Omega$, the triangle $\Delta \stackrel{\text{def}}{=} \langle a, z, z_0\rangle$ is contained in Ω. Applying Cauchy's Theorem for a triangle, we get that $\int_{[z_0,z]} f(z)\,\mathrm{d}z = F(z) - F(z_0)$, hence

$$\frac{F(z)-F(z_0)}{z-z_0} - f(z_0) = \frac{1}{z-z_0}\int_{[z_0,z]} \big[f(\zeta) - f(z_0)\big]\,\mathrm{d}z \xrightarrow[z\to z_0]{} 0$$

using the continuity of f. The function F is therefore differentiable at z_0 and its derivative is given by $F'(z_0) = f(z_0)$. □.

We have proved that $f = F'$. Proposition 4.29 then ensures that the integral of f on any closed path is zero.

Finally, there exists an even more general version of the theorem, which is optimal in some sense. It requires the notion of a *simply connected set* (see Definition A.10 on page 575 in Appendix A). The proof of this is unfortunately more delicate. However, the curious reader will be able to find it in any good book devoted to complex analysis [43, 59].

THEOREM 4.34 (Cauchy's theorem for simply connected sets) *Let Ω be a simply connected open subset of the plane and let $f \in \mathscr{H}(\Omega)$. Then for any closed γ inside Ω,*

$$\int_\gamma f(z)\,\mathrm{d}z = 0.$$

Remark 4.35 In fact, one can show that if $f \in \mathscr{H}(\Omega)$, where Ω is an open simply connected set (as for instance a convex set), then f admits a primitive on Ω. The vanishing of the integral then follows as before.

In the case of an open set with holes, one can use a generalization of Cauchy's theorem. It is easy to see that such an open set can be dissected in a manner similar to what is indicated in Figure 4.1. One takes the convention that the boundary of the holes is given the negative (clockwise) orientation, and the exterior boundary of the open set is given the positive orientation. The integral of a holomorphic function along the boundary of the open set, with this orientation, is zero.

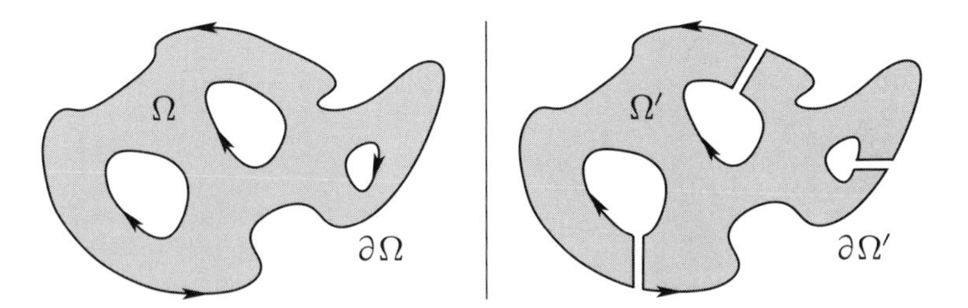

Fig. 4.1 – One can dissect an open set having one or more "holes" to make it simply connected. The integral on the boundary $\partial\Omega'$ of the new open set thus defined is zero since Ω' is simply connected, which shows that the integral on $\partial\Omega$ (with the orientation chosen as described on the left) is also zero.

4.2.e Application

One consequence of Cauchy's theorem is the following: one can move the integration contour of an holomorphic function continuously without changing the value of the integral, under the condition at least that the function be holomorphic at all the points which are swept by the path during its deformation.

THEOREM 4.36 *Let Ω be an open subset of $\mathbb{C}$, $f \in \mathscr{H}(\Omega)$, and γ a path in Ω. If γ can be deformed continuously to a path γ' in Ω, then* $\int_\gamma f(z)\,\mathrm{d}z = \int_{\gamma'} f(z)\,\mathrm{d}z$.

Proof. The proof is somewhat delicate, but one can give a very intuitive graphical description of the idea. Indeed, one moves from the contour γ in Figure 4.2 (a) to the contour γ' of Figure 4.2 (b) by adding the contour of Figure 4.2 (c). But the latter does not contain any "hole" (since γ is deformed *continuously* to γ'); therefore the integral of f on the last contour is zero.

4.3 Properties of holomorphic functions

4.3.a The Cauchy formula and applications

THEOREM 4.37 (Cauchy's formula in a convex set) *Let γ be a closed path in a convex open set Ω and let $f \in \mathscr{H}(\Omega)$. If $z \in \Omega$ and $z \notin \gamma$, we have*

$$f(z) \times \operatorname{Ind}_\gamma(z) = \frac{1}{2\pi\mathrm{i}} \int_\gamma \frac{f(\xi)}{\xi - z}\,\mathrm{d}\xi.$$

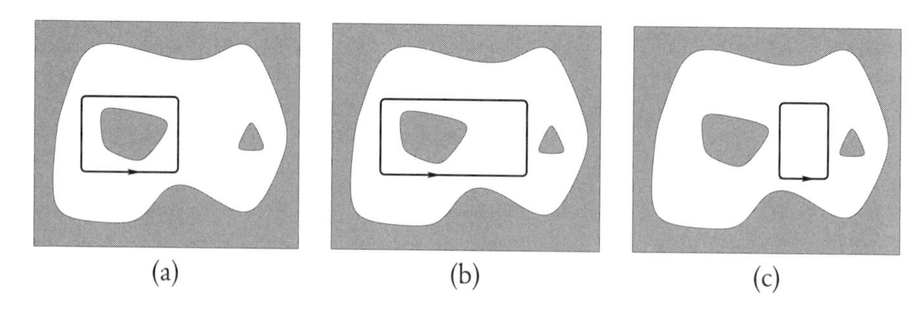

Fig. 4.2 – The three contours γ, γ' et $\gamma' - \gamma$.

In particular, if γ turns once around z counterclockwise, then

$$f(z) = \frac{1}{2\pi i} \int_\gamma \frac{f(\xi)}{\xi - z} \, d\xi.$$

PROOF. Let $z \in \Omega$, $z \notin \gamma$ and define a function g by

$$g(\xi) \stackrel{\text{def}}{=} \begin{cases} \frac{f(\xi)-f(z)}{\xi - z} & \text{if } \xi \in \Omega,\ \xi \neq z, \\ f'(z) & \text{if } \xi = z. \end{cases}$$

Then $g \in \mathscr{H}(\Omega \setminus \{z\})$ and is continuous at z; it is then possible to show (see Theorem 4.56) that g is holomorphic on Ω. So, by Cauchy's theorem for a convex set, $\int_\gamma g(\xi) \, d\xi = 0$, hence

$$\begin{aligned} 0 &= -\int_\gamma \frac{f(\xi)}{\xi - z} \, d\xi + \int_\gamma \frac{d\xi}{\xi - z} \times f(z) \\ &= -\int_\gamma \frac{f(\xi)}{\xi - z} \, d\xi + 2\pi i \operatorname{Ind}_\gamma(z) \, f(z) \end{aligned}$$

by definition of the winding number of a path around a point.

Remark 4.38 This result remains valid for a simply connected set.

Let us apply this theorem to the case, already mentioned, where the path of integration is a circle centered at z_0 with radius r: $\gamma = \mathscr{C}(0; r)$. One can then parameterize γ by $z = z_0 + re^{i\theta}$ and we have

$$\begin{aligned} f(z_0) = \frac{1}{2\pi i} \int_\gamma \frac{f(z)}{z - z_0} \, dz &= \frac{1}{2\pi i} \int_{-\pi}^{\pi} \frac{f(z_0 + re^{i\theta})}{r\, e^{i\theta}} ri e^{i\theta} \, d\theta \\ &= \frac{1}{2\pi} \int_{-\pi}^{\pi} f(z_0 + re^{i\theta}) \, d\theta, \end{aligned}$$

which allows us to write:

THEOREM 4.39 (Mean value property) *If Ω is an open subset of $\mathbb{C}$ and if f is holomorphic on Ω, then for any $z_0 \in \Omega$ and any $r \in \mathbb{R}$ such that $\overline{\mathcal{B}}(z_0\,;\,r) \subset \Omega$, we* *have*

$$f(z_0) = \frac{1}{2\pi} \int_{-\pi}^{\pi} f(z_0 + r e^{i\theta}) \, d\theta.$$

This theorem links the value at a point of a holomorphic function with the average of its values on an arbitrary circle centered at this point, which explains the name.

Here now is one the most fundamental results from the theoretical viewpoint; it also has practical implications, as we will see in Chapter 15 about Green functions.

THEOREM 4.40 (Cauchy formulas) *Let Ω be an open subset of $\mathbb{C}$ and let $f : \Omega \to \mathbb{C}$ be an arbitratry function. We have*

$$f \in \mathscr{H}(\Omega) \iff f \text{ is analytic on } \Omega.$$

Moreover, for any $n \in \mathbb{N}$ and for any $a \in \Omega$,

$$f^{(n)}(a) = \frac{n!}{2\pi i} \int_\gamma \frac{f(z)}{(z-a)^{n+1}} \, dz,$$

where γ is any path inside Ω such that $\text{Ind}_\gamma(a) = 1$.

Recall that f is analytic on an open subset Ω if and only if, at any point of Ω, f admits locally a power series expansion converging in an open disc with non-zero radius. The proof will use Cauchy's Formula:

PROOF. ▶ **Let us prove the implication $\Leftarrow$.** If the power series $f(z) = \sum_{n=0}^{\infty} c_n (z-a)^n$ converges on an open ball with radius $r > 0$ around the point a, then according to Theorem 1.69, the series $\sum n c_n (z-a)^{n-1}$ converges on the same open ball. Without loss of generality, we can work with $a = 0$. Denote

$$g(z) \stackrel{\text{def}}{=} \sum_{n=0}^{\infty} n c_n z^{n-1}, \qquad z \in \mathcal{B}(0\,;r).$$

We now show that we have indeed $f' = g$ on $\mathcal{B}(0\,;r)$.

□ Let $w \in \mathcal{B}(0\,;r)$. Choose a real numer ρ such that $|w| < \rho < r$. Let $z \neq w$ be a complex number. We have

$$\frac{f(z) - f(w)}{z - w} - g(z) = \sum_{n=1}^{\infty} c_n \left[\frac{z^n - w^n}{z - w} - n w^{n-1} \right].$$

The term between brackets is zero for $n = 1$. For $n \geqslant 2$, it is equal to

$$\left[\frac{z^n - w^n}{z - w} - n w^{n-1} \right] = (z - w) \sum_{k=1}^{n-1} k w^{k-1} z^{n-k-1}.$$

(To prove this formula, it suffices to express the right-hand side using the well-known relation $1 - x^n = (1-x)\sum_{k=0}^{n-1} x^k$, putting $x = z/w$ or $x = w/z$.) Hence, if $|z| < \rho$, the modulus of the term between brackets is at most equal to $n(n-1)\rho^{n-2}/2$, and

$$\left| \frac{f(z) - f(w)}{z - w} - g(z) \right| \leqslant |z - w| \sum_{n=1}^{\infty} n^2 \left| c_n \right| \rho^{n-2}.$$

But this last series is convergent (Theorem 1.69). Letting z tend to w, we find that f is differentiable at w and that $g(w) = f'(w)$. □

So it follows that f is holomorphic on $\mathcal{B}(a\,;r)$, and its derivative admits a power series expansion of the type $f'(z) = \sum_{n\geqslant 1} nc_n(z-a)^{n-1}$. By induction one can even show that the successive derivatives of f are given by the formula in Theorem 1.69 on page 35, which is therefore valid for complex variables as well as for real variables.

► **Let us prove the implication ⇒.** We assume then that $f \in \mathscr{H}(\Omega)$.

□ Let $a \in \Omega$, and let us show that f admits a power series expansion around a. Since Ω is open, one can find a real number $R > 0$ such that $\mathcal{B}(a\,;R) \subset \Omega$. Let γ be a circle centered at a and with radius $r < R$. We have (Theorem 4.37)

$$f(z) = \frac{1}{2\pi i}\int_\gamma \frac{f(\xi)}{\xi - z}\,d\xi \qquad z \in \mathcal{B}(a\,;r).$$

But, for any $\xi \in \gamma$, we have

$$\left|\frac{z-a}{\xi - a}\right| \leqslant \frac{|z-a|}{r} < 1;$$

hence the geometric series

$$\sum_{n=0}^{\infty} \frac{(z-a)^n}{(\xi-a)^{n+1}} = \frac{1}{\xi - z}$$

converges *uniformly with respect to* ξ on γ, and this for any $z \in \mathcal{B}(a\,;r)$, which allows us to exchange the series and the integral and to write

$$f(z) = \frac{1}{2\pi i}\int_\gamma \sum_{n=0}^{\infty} \frac{(z-a)^n}{(\xi-a)^{n+1}} f(\xi)\,d\xi = \frac{1}{2\pi i}\sum_{n=0}^{\infty}(z-a)^n \int \frac{f(\xi)}{(\xi-a)^{n+1}}\,d\xi$$
$$= \sum_{n=0}^{\infty} c_n\,(z-a)^n,$$

with

$$c_n \stackrel{\text{def}}{=} \frac{1}{2\pi i}\int \frac{f(\xi)}{(\xi-a)^{n+1}}\,d\xi.$$

So we have shown that f admits a power series expansion around a. Moreover we now (see the general theory of power series) that in fact we have $c_n = f^{(n)}(a)/n!$, and this gives the formula of the theorem. □

In conclusion, f is indeed analytic on Ω and the formula is proved.

An immediate consequence of this theorem is:

THEOREM 4.41 *Any function holomorphic on an open set Ω is infinitely differentiable on Ω.*

We can now pause to review our trail so far. We have first defined a holomorphic function on Ω as a function differentiable at any point of Ω. Having a single derivative at any point turns out to be sufficient to prove that, in fact, f has derivatives of all order, which is already a very strong result. Even stronger is that f is *analytic* on Ω, that is, that it can be expanded in power series around every point. And the proof shows something even beyond this: is we take a point $z \in \Omega$, and any open ball centered at z entirely

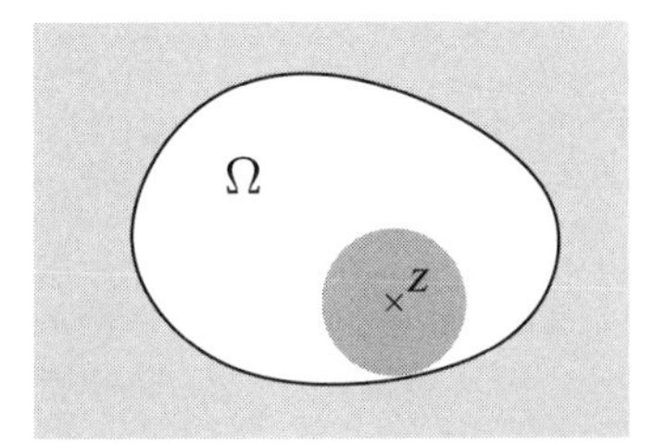

Fig. 4.3 – If f is holomorphic on Ω, then f has a power series expansion converging on the gray open ball (the open ball with largest radius contained in Ω).

contained in Ω, then f can be expanded in a power series that converges on this ball[6] (see Figure 4.3). In particular, we have the following theorem.

THEOREM 4.42 *Let f be a function holomorphic on the whole of $\mathbb{C}$. Then f admits around any point a power series expansion that converges on the whole complex plane.*

Example 4.43 The function $z \mapsto \exp(z)$ has, around 0, the power series expansion given by

$$e^z = \sum_{n=0}^{\infty} \frac{z^n}{n!}$$

with infinite radius of convergence. Around a point $z_0 \in \mathbb{C}$, the function has (another) expansion

$$e^z = \sum_{n=0}^{\infty} \frac{e^{z_0}(z - z_0)^n}{n!}$$

which converges also on the whole of $\mathbb{C}$.

DEFINITION 4.44 An **entire function** is any function which is holomorphic on the whole complex plane.

The following theorem is due to Cauchy, but bears the name of Liouville[7]:

THEOREM 4.45 (Liouville) *Any bounded entire function is constant.*

This theorem tells us that any entire function which is not constant must necessarily "tend to infinity" somewhere. This "somewhere" is of course "at infinity" in the complex plane, but one must not think that it means that "$f(z)$ tends to infinity when $|z|$ tends to infinity." For instance, if we consider the function $z \mapsto \exp(z)$, we see that if we let z tend to infinity on the left

[6] There can of course be convergence even outside this ball; this brings the possibility of analytic continuation (see Section 5.3 on page 144).

[7] Not because it was thought that Cauchy already had *far too many* theorems to his name but because, Liouville having quoted it during a lecture, a German mathematician attributed the result to him by mistake; as sometimes happens, the mistake took hold.

Joseph LIOUVILLE (1809–1882), son of a soldier and student at the École Polytechnique, became professor there at the age of 29, a member of the Académie des Sciences at 30 and a professor at the Collège de France at 42. A powerful and eclectic mind, Liouville studied with equal ease geometry (surfaces with constant curvature), number theory (constructing the first transcendental number, which bears his name), mechanics (three-body problem, celestial dynamics,...) analysis,... and he became actively involved in the diffusion of mathematics, creating an important journal, the *Journal de mathématiques pures et appliquées*, in 1836.

side of the plane (that is, by letting the real part of z go to $-\infty$), then $\exp(z)$ tends to 0. It is only if the real part of z tends to $+\infty$ that $|e^z|$ tends to infinity.

PROOF OF LIOUVILLE'S THEOREM. We start by recalling the following result:

LEMMA 4.46 *If f is an entire function such that $f'(z) \equiv 0$ on $\mathbb{C}$, then f is constant.*

According to the Cauchy formulas (Theorem 4.40), if γ is a contour turning once around z in the positive direction, we have

$$\frac{1}{2\pi i}\int_\gamma \frac{f(\zeta)}{(\zeta - z)^2}\,d\zeta = f'(z).$$

If we assume that f is bounded, there exists $M \in \mathbb{R}$ such that $|f(z)| < M$ for any $z \in \mathbb{C}$.

Let then $z \in \mathbb{C}$, and let us show that $f'(z) = 0$.

□ Let $\varepsilon > 0$. Consider the circle γ centered at z and with arbitrary radius $R > 0$. We have

$$|f'(z)| \leqslant \frac{1}{2\pi i}\left|\int_\gamma \frac{f(\zeta)}{(\zeta - z)^2}\,d\zeta\right| \leqslant \frac{1}{2\pi}\int_\gamma \frac{|f(\zeta)|}{|\zeta - z|^2}\,d\zeta \leqslant \frac{M}{R}.$$

If we select the radius R so that $R \geqslant M/\varepsilon$ (which is always possible since the function f is defined on $\mathbb{C}$), we obtain therefore $|f'(z)| \leqslant \varepsilon$. □

As a consequence, we have found that $f'(z) = 0$. This being true for any $z \in \mathbb{C}$, the function f is constant.

4.3.b Maximum modulus principle

THEOREM 4.47 (Maximum modulus principle) *Let Ω be a domain in $\mathbb{C}$, f a function holomorphic on Ω, and $a \in \Omega$. Then, either f is constant, or f has no local maximum at the point a, which means that any neighborhood of a contains points b such that $|f(a)| < |f(b)|$.*

What does this result mean? At any point in the open set of definition, the modulus of the holomorphic function cannot have a local maximum. It can have a minimum: it suffices to consider $z \mapsto z$ which a local (in fact, global) minimum at $z = 0$. But we can never find a maximum.

COROLLARY 4.48 *Let $f \in \mathscr{H}(\Omega)$ and let V be a bounded open subset of Ω, such that the closure $\overline{V}$ of V is contained in Ω. Then the supremum of f on V is achieved on the boundary of V, which means that there exists $w \in \partial V$ such that*

$$\left|f(w)\right| = \sup_{z \in V} \left|f(z)\right|.$$

We will see later that this theorem also applies to real harmonic functions (such as the electrostatic potential in a vacuum), which will have important physical consequences (see Chapter 5; it is in fact there that we will give, page 141, a graphical interpretation of the maximum modulus principle).

4.3.c Other theorems

We write down for information a theorem which is the converse of Cauchy's theorem for triangles:

THEOREM 4.49 (Morera) *Let f be a complex-valued function, continuous on an open subset Ω of $\mathbb{C}$, such that*

$$\int_{\partial \Delta} f(z)\,\mathrm{d}z = 0$$

for any closed triangle $\Delta \subset \Omega$. Then $f \in \mathscr{H}(\Omega)$.

This theorem has mostly historical importance, and some interest for pure mathematicians, but is of little use in physics. It is only stated for the reader's general knowledge.

On the other hand, here is a theorem which is much more useful, due to Green[8]:

THEOREM 4.50 (Green-Riemann) *Let P and Q be functions from $\mathbb{R}^2$ to $\mathbb{R}$ or $\mathbb{C}$, $\mathbb{R}^2$-differentiable, and let K be a sufficiently regular compact subset.[9] Then we have*

$$\int_{\partial K} (P\,\mathrm{d}x + Q\,\mathrm{d}y) = \iint_K \left(\frac{\partial Q}{\partial x} - \frac{\partial P}{\partial y} \right) \mathrm{d}x\,\mathrm{d}y.$$

[8] George Green (1793–1841), English mathematician, was born and died in Nottingham, where he wasn't sheriff, but rather miller (the mill is still functional) and self-taught. He entered Cambridge at 30, where he got his doctorate four years later. He studied potential theory, proved the Green-Riemann formula, and developed an English school of mathematical physics, followed by Thompson (discoverer of the electron), Stokes (see page 472), Rayleigh, and Maxwell.

[9] The boundary of K, denoted ∂K, must be of class $\mathscr{C}^1$ and, for any point $z \in \partial K$, there must exist a neighborhood $\mathscr{V}_z$ of z homeomorphic to the unit open ball $\mathcal{B}(0;1)$, such that $\mathscr{V}_z \cap K$ corresponds by the homeomorphism with the upper half-plane. This avoids a number of pathologies.

This result can be stated in the language of functions on $\mathbb{C}$ in the following manner.

THEOREM 4.51 (Green) *Let $F(z,\bar{z})$ be a function of the variables z and $\bar{z}$, admitting continuous partial derivatives with respect to each of them, and let K be a sufficiently regular compact subset of $\mathbb{C}$. Then we have*

$$\int_{\partial K} F(z,\bar{z})\,\mathrm{d}z = 2\mathrm{i} \iint_K \frac{\partial F}{\partial \bar{z}}\,\mathrm{d}A,$$

where $\mathrm{d}A$ denotes the area integration element in the complex plane. One can also write F as a function of x and y, which gives us (with a slight abuse of notation)

$$\int_{\partial K} F(x,y)\,\mathrm{d}z = 2\mathrm{i} \iint_K \frac{\partial F}{\partial \bar{z}}\,\mathrm{d}x\,\mathrm{d}y.$$

PROOF. It is enough to write $F = P + \mathrm{i}Q$ and to apply the formula from Theorem 4.50 to the real and to the imaginary parts of $\int_{\partial K} F\,\mathrm{d}z$.

The Green-Riemann formula should be compared with the Stokes formula[10] for differential forms, given page 473.

4.3.d Classification of zero sets of holomorphic functions

The rigidity of a holomorphic function is such that, if one knows the values of f in a very small subset of the open set of definition Ω, then these data are theoretically sufficient to know f on the whole of Ω (see Exercise 4.1 on page 90). What amount of information actually suffices to determine f uniquely? The answer is given by the following theorem: a sequence of points having one accumulation point in Ω is enough.

First we recall the definition of an accumulation point.

DEFINITION 4.52 Let Z be a subset of $\mathbb{C}$. A point $c \in Z$ is an **accumulation point** if there exists a sequence $(z_n)_{n\in\mathbb{N}}$ of elements of $Z \setminus \{c\}$ such that $z_n \to c$.

(It is important to remark that the definition imposes that $z_n \neq c$ for all $n \in \mathbb{N}$; without this condition, any point would be an accumulation point, a constant sequence equal to c converging always to c.)

THEOREM 4.53 (Classification of zero sets) *Let Ω be a domain of $\mathbb{C}$ (recall that this means a* connected *open set) and let $f \in \mathscr{H}(\Omega)$. Define the zero set of f by*

$$Z(f) \stackrel{\text{def}}{=} \{a \in \Omega\,;\ f(a) = 0\}.$$

[10] An approach to complex analysis based on differential forms is presented by Dolbeault in [29]. It is a more abstract viewpoint at first sight, but it is very rich.

If $Z(f)$ has at least one accumulation point in Ω, then $Z(f) = \Omega$, that is, the function is identically zero on Ω.

If $Z(f)$ has no accumulation point, then for any $a \in Z(f)$, there exist a unique positive integer $m(a)$ and a function $g \in \mathscr{H}(\Omega)$ which does not vanish at a, such that one can write f in the factorized form

$$f(z) = g(z) \times (z - a)^{m(a)} \qquad \text{for all } z \in \Omega.$$

Moreover, $Z(f)$ is at most countable.

DEFINITION 4.54 The integer $m(a)$ defined in the preceding theorem is called the **order of the zero** of f at the point a.

Interpretation

Suppose we have two holomorphic functions f_1 and f_2, both defined on the same domain Ω, and that we know that $f_1(z) = f_2(z)$ for any z belonging to a certain set Z. Then if Z has at least one accumulation point in Ω, the functions f_1 and f_2 are necessarily equal on the whole set Ω.

Thus, if we have a function holomorphic on $\mathbb{C}$ and if we know its values for instance on $\mathbb{R}^+$, then it is – theoretically – entirely determined. In particular, if f is zero on $\mathbb{R}^+$, then it is zero everywhere.

Note that Ω is not simply open, but it must also be *connected*. Indeed, if Ω were the union of two disjoint open subsets, the rigidity of the function would not be able to "bridge the gap" between the two open subsets, and the behavior of f on one of the two open subsets would be entirely uncorrelated with its behavior on the other.

Counterexample 4.55 Assume we know a holomorphic function f at the points $z_n \stackrel{\text{def}}{=} 1/n$, for all $n \in \mathbb{N}^*$, and that in fact $f(z_n) = n$. In this case, f is not uniquely determined. In fact, both functions $f_1(z) = 1/z$ and $f_2(z) = \exp(2\pi i/z)/z$ satisfy $f_1(z_n) = f_2(z_n) = f(z)$ for all $n \in \mathbb{N}$. What gives? Here, because $f(z) \to \infty$ when z tends to 0, f is not holomorphic at 0; the accumulation point, which is precisely 0, does not belong to Ω and the theorem cannot be applied to conclude.

COROLLARY 4.55.1 *Let Ω be a connected open subset and assume that the open ball $\mathcal{B} = \mathcal{B}(z_0\,;\,r)$ with $r > 0$ is contained in Ω. If $f \in \mathscr{H}(\Omega)$ and if f is zero on $\mathcal{B}$, then $f \equiv 0$ on Ω.*

COROLLARY 4.55.2 *Let $\Omega \subset \mathbb{C}$ be a connected open subset, and let f and g be holomorphic on Ω and such that $fg \equiv 0$ on Ω. Then either $f \equiv 0$ on Ω or $g \equiv 0$ on Ω.*

PROOF. Let us suppose that $f \not\equiv 0$; let then z_0 be a point of Ω such that $f(z_0) \neq 0$. Since f is continuous, there exists a positive real number $r > 0$ such that $f(z) \neq 0$ for all points in $\mathcal{B} \stackrel{\text{def}}{=} \mathcal{B}(z_0\,;\,r)$. Hence $g \equiv 0$ on $\mathcal{B}$ and Corollary 4.55.1 allows us to conclude that $g \equiv 0$ on Ω.

COROLLARY 4.55.3 *Let f and g be entire functions. If $f(x) = g(x)$ for all $x \in \mathbb{R}$, then $f \equiv g$.*

These theorems also show that functional relations which hold on $\mathbb{R}$ remain valid on $\mathbb{C}$. For instance, we have

$$\forall x \in \mathbb{R} \qquad \cos^2 x + \sin^2 x = 1.$$

The function $z \mapsto \cos^2 z + \sin^2 z - 1$ being holomorpic (it is defined using the squares of power series with infinite radius of convergence, and is therefore holomorphic on $\mathbb{C}$) and being identically zero on $\mathbb{R}$, it must be zero on $\mathbb{C}$; we therefore have

$$\forall z \in \mathbb{C} \qquad \cos^2 z + \sin^2 z = 1.$$

4.4 Singularities of a function

Certain functions are only holomorphic on an open subset minus one or a few points, for instance the function $z \mapsto 1/z$, which belongs to the set $\mathscr{H}(\mathbb{C} \setminus \{0\})$. It is in general those functions which are useful in physics. The points at which f is not holomorphic are called *singularities* and have a physical significance in many problems.[11]

4.4.a Classification of singularities

There are three kinds of singularities: *artificial* (or removable) singularities, *poles*, and *essential singularities*. The first case is not very interesting. As the name suggests, it is not a "true" singularity. It corresponds to the case of a bounded f. We have the following theorem due to Riemann (see page 138).

THEOREM 4.56 *Let a be a point in $\mathbb{C}$. Let f be a function defined on an open neighborhood Ω of a, except at a, such that f is holomorphic and bounded on $\Omega \setminus \{a\}$. Then $\lim_{z\to a} f(z)$ exists and the function f can be continued to a function f defined on Ω by putting*

$$f(a) \stackrel{\text{def}}{=} \lim_{\zeta \to a} f(\zeta).$$

The function f so defined is holomorphic *on the whole of Ω.*

---

[11] Thus, in linear response theory, the response function will be, depending on conventions, analytic (say) in the complex upper half-plane, but will have poles in the lower half-plane. Those poles correspond to the energies of the various modes. In particle physics, they will be characteristic of the *mass* of an excitation (particle) and its *half-life*.

DEFINITION 4.57 If a is a singularity of f and f is bounded in a neighborhood of a, then a is called a **removable singularity**, or an **artificial singularity**.

Consequently, if f has an artificial singularity, it must have been badly written to begin with. Consider, for instance, $z \mapsto (4z^2 - 1)/(2z + 1)$, which has an artificial singularity at -1; we should instead have written the function as $z \mapsto 2z - 1$, which would then have been defined on the whole of $\mathbb{C}$. All artificial singularities are not so "visible," however, as shown by the next example.

Example 4.58 The most important example of an artificial singularity is the following. Let f be a function holomorphic on a domain $\Omega \subset \mathbb{C}$. Let $z_0 \in \Omega$ and define

$$g(z) \stackrel{\text{def}}{=} \frac{f(z) - f(z_0)}{z - z_0} \qquad \forall z \in \Omega \setminus \{z_0\}.$$

Then g is holomorphic on $\Omega \setminus \{z_0\}$ and has an artificial singularity at z_0. Indeed, if we extend g by continuity by putting

$$g(z) \stackrel{\text{def}}{=} \begin{cases} \dfrac{f(z) - f(z_0)}{z - z_0} & \text{if } z \neq z_0, \\ f'(z_0) & \text{if } z = z_0, \end{cases}$$

then we obtain a function which is holomorphic on Ω.

Other singularities are classified in two categories.

THEOREM 4.59 (Casorati-Weierstrass) *Let $a \in \Omega$ be an isolated singularity of a function $f \in \mathscr{H}(\Omega \setminus \{a\})$. If a is not a removable singularity, then only two possibilities can occur:*

a) $\lim_{z \to a} |f(z)|$ *exists and is equal to* $+\infty$ *;*

b) for any $r > 0$ such that $\mathcal{B}(a\,; r) \subset \Omega$, the image of $\mathcal{B}(a\,; r) \setminus \{a\}$ by f is dense in $\mathbb{C}$.

Proof. Assume there exists $r > 0$ such that the image of $\mathcal{B}(a\,; r)$ is not dense in $\mathbb{C}$. Then, for some $\lambda \in \mathbb{C}$ and some $\varepsilon > 0$, we have

$$|f(z) - \lambda| > \varepsilon \qquad \text{for all } z \in \big(\mathcal{B}(a\,; r) \setminus \{a\}\big).$$

Consider now the function g, defined on $\mathcal{B}(a\,; r) \setminus \{a\}$ by $g(z) \stackrel{\text{def}}{=} 1/\big(f(z) - \lambda\big)$. Then g is holomorphic on $\mathcal{B}(a\,; r) \setminus \{a\}$ and, moreover, $|g(z)| < 1/\varepsilon$ for all $z \in \mathcal{B}(a\,; r) \setminus \{a\}$. According to the removable singularity theorem, we can continue the function g to a function $\hat{g} : \mathcal{B}(a\,; r) \to \mathbb{C}$. This new function $\hat{g}$ is then nonzero everywhere, except possibly at a. For all $z \neq a$, we have

$$f(z) = \lambda + \frac{1}{\hat{g}(z)}. \tag{$*$}$$

If $\hat{g}(a)$ were not 0, the right-hand side of $(*)$ would be holomorphic on $\mathcal{B}(a\,; r)$ and f would have an artificial singularity at a, which we assumed was not the case. Hence $\hat{g}(a) = 0$ and $\lim_{z \to a} |f(z)| = +\infty$.

DEFINITION 4.60 (Poles) If $f \in \mathscr{H}\big(\Omega \setminus \{a\}\big)$ and if

$$\lim_{z \to a} \big|f(z)\big| = +\infty,$$

then f admits a **pole** at a, and a is called a **pole** of f.

DEFINITION 4.61 (Essential singularity) If $f \in \mathscr{H}\big(\Omega \setminus \{a\}\big)$ and if $\big|f(z)\big|$ has no limit as z tends to a, then f is said to have an **essential singularity** at a.

Poles and essential singularities are of a very different nature. Whereas functions behave quite "reasonably" in the neighborhood of a pole, they are very wild in the neighborhood of an essential singularity.

Example 4.62 Consider the function $f : z \mapsto \exp(1/z)$. Then, in any neighborhood of 0, the function f takes *any* value in $\mathbb{C}$ except 0 and oscillates so fast, in fact, that it reaches any of these complex values infinitely often. This can be seen as follows. Let λ be an arbitrary nonzero complex number. There exists a complex number w such that $e^{w} = \lambda$. If we now put $z_n \stackrel{\text{def}}{=} 1/(w + 2\pi i n)$, then $z_n \xrightarrow[n\to\infty]{} 0$ and $f(z_n) = \exp(1/z_n) = \lambda$ for all $n \in \mathbb{N}$.

4.4.b Meromorphic functions

DEFINITION 4.63 A subset S of $\mathbb{C}$ is **locally finite** if, for any closed disk $\mathscr{D}$, the subset $\mathscr{D} \cap S$ is finite. If Ω is an open set in $\mathbb{C}$ and if $S \subset \Omega$, S is **locally finite in Ω** if and only if for any closed disque $\mathscr{D}$ contained in Ω, the set $\mathscr{D} \cap S$ is finite.

In practice, to say that S is locally finite in $\mathbb{C}$ boils down to saying that either S is finite, or S is countable and can be written $S = \{z_n \,;\, n \in \mathbb{N}\}$ with $\lim\limits_{n\to\infty} |z_n| = +\infty$.

Example 4.64 Let f be a function holomorphic on an open set Ω and not identically zero. Then the set of zeros of f is locally finite in Ω.

Example 4.65 The set $\{1/n \,;\, n \in \mathbb{N}^*\}$ is locally finite in the open ball $\mathcal{B}(1\,;\,1)$ (which does not contain 0), but not in $\mathbb{C}$.

DEFINITION 4.66 Let Ω be an open subset of $\mathbb{C}$ and $S \subset \Omega$ a subset of Ω. A function $f : \Omega \setminus S \to \mathbb{C}$ is a **meromorphic function** on Ω if and only if

i) S is a closed subset of Ω and is locally finite in Ω;

ii) f is holomorphic on $\Omega \setminus S$ (which is then necessarily closed);

iii) for any $z \in S$, f has a pole at z.

Example 4.67 The function $z \mapsto 1/z$ is meromorphic on $\mathbb{C}$. The set of its singularities is the singleton $\{0\}$.

PROPOSITION 4.68 *If Ω is an open subset of $\mathbb{C}$ and if $f : \Omega \to \mathbb{C}$ is holomorphic and not identically zero, then the function*

$$F(z) \stackrel{\text{def}}{=} \frac{1}{f(z)} \qquad \text{for all } z \in U \setminus Z(f)$$

is a meromorphic function.

PROOF. The function F is holomorphic on $\Omega \setminus Z(f)$ and, since $Z(f)$ is locally finite, according to Theorem 4.53 on page 106, it suffices to check that for any point $a \in Z(f)$, we have $\lim_{z \to a} |F(z)| = +\infty$, which is immediate by continuity.

Example 4.69 If P and Q are coprime polynomials with complex coefficients (i.e., if they have no common zero in $\mathbb{C}$), with $Q \neq 0$, the rational function P/Q is meromorphic on $\mathbb{C}$. The set of its poles is the set of zeros of Q, which is locally finite (in fact finite) because Q is holomorphic.

Remark 4.70 Two other kinds of singularities are sometimes defined:

Branch points appear when a function cannot be defined on a simply connected open set, as, for instance, the complex logarithm or other *multivalued* functions (see Section 5.1.a on page 135).

Singularities at infinity are defined by making the substitution $w = 1/z$. One obtains a new function $F : w \mapsto F(w) = f(1/z)$; then f has a singularity at infinity if F has a singularity at 0, those singularities being by definition of the same kind (see page 146).

4.5 Laurent series

4.5.a Introduction and definition

We have just seen that a function holomorphic on $\Omega \setminus z_0$ could have singularities of two different kinds at z_0: an essential singularity or a pole. We can shed some light on this distinction in the following manner. If f can be written in the form of the following series (which is not a power series, since coefficients with negative indices occur):

$$f(z) = \sum_{n \in \mathbb{Z}} a_n (z - z_0)^n,$$

then

- f has a pole at z_0 if there are only *finitely many* nonzero coefficients with negative indices;
- f has an essential singularity at z_0 if there are infinitely many.

The next theorem stipulates that f usually admits such an expansion.

THEOREM 4.71 (Laurent series) *Let f be a function holomorphic on the annulus $\rho_1 < |z - z_0| < \rho_2$, where $0 \leqslant \rho_1 < \rho_2 \leqslant +\infty$. Then f admits a unique expansion, in this annulus, of the type*

$$f(z) = \sum_{n=-\infty}^{+\infty} a_n (z - z_0)^n.$$

Moreover, the coefficients a_n are given by

$$a_n = \frac{1}{2i\pi} \int_{\mathscr{C}} \frac{f(\zeta)}{(\zeta - z_0)^{n+1}} \, d\zeta,$$

where $\mathscr{C}$ is any contour inside the annulus turning once around z_0 in the positive direction (i.e., $\operatorname{Ind}_{\mathscr{C}}(z_0) = 1$).

Remark 4.72 The annulus considered may also be "infinite," as, for instance, $\mathbb{C}^* = \mathbb{C} \setminus \{z_0\}$.

This motivates the following definition, due to Laurent[12]:

DEFINITION 4.73 The preceding expansion is called the **Laurent series expansion of f**. If there are only finitely many nonzero coefficients with negative index in this expansion, then the absolute value of the smallest index with nonzero coefficient is called the **order of the pole** at z_0. A pole of order 1 is also called a **simple pole**.

PROOF OF THEOREM 4.71. Without loss of generality, we may assume that z_0 in the proof. We cut the annulus in the way shown in the figure below. The interior circle, with radius ρ_1, is taken with negative orientation, and the exterior circle with positive orientation. The two parallel segments on the figure are very close, and we will take the limit where they become identical at the end of the proof. Denote by γ the contour thus defined, and by γ_1 and γ_2 the two circles with *positive* orientation. Denote moreover by Ω the interior of the contour γ, that is, the open annulus minus a small strip. This open set is not convex, but it is still *simply connected*. One can therefore apply Cauchy's formula on Ω (or, more precisely, on a simply connected neighborhood of Ω containing γ).

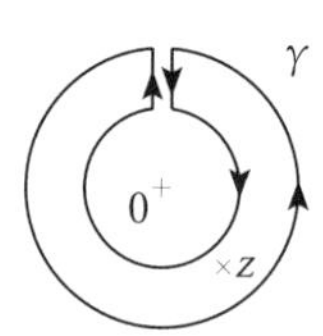

We have then, for any point z in the annulus, by Cauchy's formula applied to f and γ:

$$f(z) = \frac{1}{2\pi i} \int_{\gamma} \frac{f(\zeta)}{\zeta - z} \, d\zeta = \frac{1}{2\pi i} \int_{\gamma_1} \frac{f(\zeta)}{\zeta - z} \, d\zeta - \frac{1}{2\pi i} \int_{\gamma_2} \frac{f(\zeta)}{\zeta - z} \, d\zeta$$

[12] Pierre LAURENT, (1813–1853), a former student at the École Polytechnique, was a hydraulic engineer and mathematician.

in the limit where both segments coincide. It is enough then to write that, for any $\zeta \in \gamma_1$, we have $|\zeta - z| < 1$ and hence

$$\frac{1}{\zeta - z} = \frac{1}{\zeta} \times \frac{1}{1 - z/\zeta} = \sum_{n=0}^{\infty} \frac{z^n}{\zeta^{n+1}},$$

whereas for $\zeta \in \gamma_2$, we have $|z/\zeta| < 1$ and

$$\frac{1}{\zeta - z} = \frac{1}{z} \times \frac{1}{1 - \zeta/z} = -\sum_{n=-\infty}^{-1} \frac{z^n}{\zeta^{n+1}}.$$

This proves the first formula stated. Moreover, the value of a_n is given by the integral on γ_1 or γ_2 depending on whether n is positive or negative. The integral on any intermediate curve gives – of course – the same result.

Remark 4.74 We know that a power series has a *radius of convergence*, that is, if $\sum a_n z^n$ converges for a certain $z_0 \in \mathbb{C}$, then it converges for any $z \in \mathbb{C}$ with $|z| < |z_0|$. We have, for Laurent series, a similar result: if $\sum_{n=-\infty}^{+\infty} a_n z^n$ converges for z_1 and z_2 with $|z_1| < |z_2|$, then it converges for any $z \in \mathbb{C}$ such that $|z_1| < |z| < |z_2|$. Similarly, for a given Laurent series, there exist two real numbers ρ_1 and ρ_2 (with ρ_2 possibly $+\infty$) such that the Laurent series converges uniformly and absolutely in the interior of the open annulus $\rho_1 < |z| < \rho_2$ and diverges outside. This result is elementary and does not require the theory of holomorphic functions.

Remark 4.75 The proof of the theorem also shows that f can be written as $f = f_1 + f_2$, where f_1 is holomorphic in the disc $|z| < \rho_2$ and f_2 is holomorphic in the domain $|z| > \rho_1$. If one asks, in addition, that $f_1(z) \xrightarrow[z\to 0]{} 0$, then this decomposition is unique.

4.5.b Examples of Laurent series

In the case where a function f has only one singularity, for instance,

$$f(z) \stackrel{\text{def}}{=} \frac{1}{z-1},$$

the Laurent series may be defined on the (infinite) "annulus"

$$\mathcal{D}_1 \stackrel{\text{def}}{=} \{z \in \mathbb{C}\,;\ |z| > 1\}.$$

Indeed, one can write

$$f(z) = \frac{1}{z-1} = \frac{1}{z} \frac{1}{1 - \frac{1}{z}} = \sum_{n=1}^{\infty} \frac{1}{z^n} \tag{4.6}$$

for $|z| > 1$. Note that this series has infinitely many negative indices, but *beware*, this does not mean that there is an essential singularity! Indeed, this infinite series comes from having performed the expansion in an annulus around 0, while the pole is at 1. A Laurent series expansion around $z = 1$ gives of course

$$f(z) = \frac{1}{z-1}$$

with no other term. Remark that this Laurent series expansion (4.6) is valid only in the "annulus" $|z| > 1$, outside of which the series diverges rather trivially.

This same function f can be expanded on the annulus $0 < |z| < 1$ by

$$f(z) = -\sum_{n=0}^{\infty} z^n.$$

When a function has more than one singularity, the annulus must remain between the poles; for example, the function $g(z) = 1/(z-1)(z+2)$ admits a Laurent series expansion on the annulus $1 < |z| < 2$. Note that it also admits a (different) Laurent series expansion on the annulus $2 < |z|$, and a third one on $0 < |z| < 1$ (this last expansion without any nonzero coefficient with negative index, since the function g is very cosily holomorphic at 0).

4.5.c The Residue theorem

DEFINITION 4.76 An open subset $\Omega \subset \mathbb{C}$ is **holomorphically simply connected** if, for any holomorphic function $f : \Omega \to \mathbb{C}$, there exists a holomorphic function $F : \Omega \to \mathbb{C}$ such that $F' = f$.

We will admit the following results:

LEMMA 4.77 *An open subset $\Omega \subset \mathbb{C}$ is holomorphically simply connected if and only if, for any holomorphic function $f : \Omega \to \mathbb{C}$ and for any closed path γ inside Ω, we have*

$$\int_\gamma f(z)\,\mathrm{d}z = 0.$$

LEMMA 4.78 *An open subset $\Omega \subset \mathbb{C}$ is holomorphically simply connected if and only if it is simply connected.*

Consequently, we will drop the appellation "holomorphically simply connected" from now on. The notion of simple connectedness acquires a richer meaning because of the preceding lemma however.

DEFINITION 4.79 (Residue) Let f be a function that can be expanded in Laurent series in an annulus $\rho_1 < |z-a| < \rho_2$:

$$f(z) = \sum_{n\in\mathbb{Z}} a_n (z-a)^n \qquad \text{for } \rho_1 < |z-a| < \rho_2.$$

The coefficient a_{-1} (corresponding to the term with $1/(z-a)$) is called the **residue** of f at a and is denoted

$$a_{-1} = \operatorname{Res}(f; a).$$

In particular, if $f \in \mathscr{H}(\Omega \setminus \{a\})$ and f has a pole at a, we can write

$$f(z) = g(z) + \sum_{n=1}^{N} \frac{a_{-n}}{(z-a)^n},$$

with g holomorphic on Ω, and a_{-1} is the residue of f at a.

We will see in the next section how to compute the residue of a function at a given point.

THEOREM 4.80 *Let f be a function holomorphic on the annulus $\rho_1 < |z-a| < \rho_2$. Let γ be a closed path contained inside this annulus. Then we have*

$$\frac{1}{2\pi i}\int_\gamma f(z)\,dz = \operatorname{Ind}_\gamma(a) \times \operatorname{Res}(f; a). \tag{4.7}$$

PROOF. We can indeed write (taking $a = 0$)

$$f(z) = \frac{a_{-1}}{z} + g(z), \qquad \text{where} \quad g(z) = \sum_{n\neq -1} a_n z^n.$$

Since the series $\sum_{n\neq -1} a_n z^n$ converges, as well as the series $\sum_{n\neq -1} a_n z^{n+1}/(n+1)$, it follows that the function g has a holomorphic primitive. Hence $\int_\gamma g(z)\,dz = 0$ and we get

$$\int_\gamma f(z)\,dz = a_{-1}\int_\gamma \frac{dz}{z} = 2\pi i\, a_{-1} \operatorname{Ind}_\gamma(z),$$

which is the required formula.

Another formulation of this theorem, which will be of more use in practice, is as follows:

THEOREM 4.81 (Residue theorem) *Let $\Omega \subset \mathbb{C}$ be a simply connected open set and let $P_1, \dots, P_n$ be distinct points in Ω. Let $f \in \mathscr{H}(\Omega \setminus \{P_1, \dots, P_n\})$ and let γ be a closed path contained in Ω which does not contain any of the points $P_1, \dots, P_n$. Then we have*

$$\int_\gamma f(z)\,dz = 2\pi i \sum_{i=1}^{n} \operatorname{Res}(f; P_i) \operatorname{Ind}_\gamma(P_i).$$

Why is the Residue theorem useful?

We can make clear immediately that its interest, for a physicist, is mostly computational. Then it is, in many cases, irreplaceable. It relates a path integral in the complex plane with a sum which depends only on the behavior of the function near *finitely many* points. It is therefore possible to compute the integral around a closed path (see Section 4.6.d) or, through a small trick, an integral over $\mathbb{R}$ (see Sections 4.6.b and 4.6.c). Conversely, the sum of certain series can be identified with an infinite sum of residues, which is linked to an integral over a closed path; by deforming this closed path, one may then relate this integral to a *finite* sum of residues (see Section 4.6.e).

But before going into this, we must see how to compute in practice the residue of a function at a given point.

4.5.d Practical computations of residues

In this short section, we will see how to compute in practice the residue of a function. In almost all cases we will have to consider a function $f \in \mathscr{H}(\Omega \setminus \{a\})$, where Ω is a neighborhood of a point $a \in \mathbb{C}$. In the case where f admits a simple pole at a, namely, if one can express this function in the form $f(z) = b_{-1}/(z-a) + g(z)$, où $g \in \mathscr{H}(\Omega)$, then we have

$$b_{-1} = \text{Res}\,(f;a) = \lim_{z\to a}\,(z-a)\,f(z). \tag{4.8}$$

How can one know if f has a simple pole at a and not a higher order pole? If one has no special intuition in the matter, one can try to use the preceding formula: if it works, this means the pole was simple; if it explodes, the pole was of higher order![13]

Consider, for example, the case where f is a function that has a pole of order 2 at a. Then we have

$$f(z) = \frac{b_{-2}}{(z-a)^2} + \frac{b_{-1}}{(z-a)} + g(z) \qquad \text{with } g \in \mathscr{H}(\Omega). \tag{4.9}$$

Formula (4.8) then gives indeed an infinite value in this situation. How, then, can one obtain b_{-1} without being annoyed by the "more divergent" part? If we multiply equation (4.9) by $(z-a)^2$, we get

$$(z-a)^2 f(z) = b_{-2} + b_{-1}(z-a) + (z-a)^2 g(z).$$

Differentiate with respect to z:

$$\frac{\mathrm{d}}{\mathrm{d}z}\big((z-a)^2 f(x)\big) = b_{-1} + 2(z-a)\,g(z) + (z-a)^2\,g'(z).$$

By taking the limit $[z \to a]$, we obtain exactly b_{-1} and this without explosion. As a general rule, for a pole of order k, one can use the following formula, the proof of which is immediate:

THEOREM 4.82 *Let $f \in \mathscr{H}(\Omega \setminus \{a\})$ be such that f has a pole of order k at a. Then the residue of f at a is given by*

$$\text{Res}\,(f;a) = \frac{1}{(k-1)!}\lim_{z\to a}\frac{\mathrm{d}^{k-1}}{\mathrm{d}z^{k-1}}\Big\{(z-a)^k f(z)\Big\}. \tag{4.10}$$

Example 4.83 Let a be a nonzero complex number. Consider the function $f(z) = 1/(z^2+a^2)$, which is meromorphic on the complex plane and has two poles at the points $+\mathrm{i}a$ and $-\mathrm{i}a$. These poles are simple and we have

$$\begin{aligned}\text{Res}\,(f;\mathrm{i}a) = \lim_{z\to \mathrm{i}a}(z-\mathrm{i}a)\,f(z) &= \lim_{z\to \mathrm{i}a}\frac{(z-\mathrm{i}a)}{(z-\mathrm{i}a)(z+\mathrm{i}a)}\\ &= \lim_{z\to \mathrm{i}a}\frac{1}{(z+\mathrm{i}a)} = \frac{1}{2\mathrm{i}a}.\end{aligned}$$

[13] I agree that this is a dangerous method, but as long as it is restricted to mathematics and is not applied to chemistry, things will be fine.

4.6

Applications to the computation of horrifying integrals or ghastly sums

The goal of this section is to show how the knowledge of Cauchy's theorems and residues can help computing some integrals which are otherwise quite difficult to evaluate explicitly. In certain cases one can also evaluate the sum of a *series* using similar techniques.

The idea, when trying to compute an integral on $\mathbb{R}$, is to start with only a piece of the integral, on an interval $[-R, R]$, then close this path in the complex plane, most often by a semicircle, checking that the added piece does not contribute to the integral (at least in a suitable limit), and compute this new path integral by the Residue theorem. Before presenting examples, we must establish conditions under which the "extra" piece of the integral will indeed be negligible.

4.6.a Jordan's lemmas

Jordan's lemmas are little theorems which are constantly useful when trying to compute integrals using the method of residues.

The first of Jordan's lemmas is useful for the computation of arbitrary integrals.

THEOREM 4.84 (First Jordan lemma) *Let $f : \mathbb{C} \to \mathbb{C}$ be a continuous function in the sector*

$$\mathscr{S} \stackrel{\text{def}}{=} \left\{ r\,\mathrm{e}^{\mathrm{i}\theta};\ r > 0 \text{ and } 0 \leqslant \theta_1 \leqslant \theta \leqslant \theta_2 \leqslant \pi \right\},$$

such that $\lim\limits_{|z|\to 0} z\,f(z) = 0$ *(resp.* $\lim\limits_{|z|\to\infty} z\,f(z) = 0$*).*

Denote $\gamma(r) = \left\{ r\,\mathrm{e}^{\mathrm{i}\theta}; \theta_1 \leqslant \theta \leqslant \theta_2 \right\}$. *Then we have*

$$\lim_{r\to 0^+} \int_{\gamma(r)} f(z)\,\mathrm{d}z = 0 \qquad \left(\textit{resp.} \lim_{r\to +\infty} \int_{\gamma(r)} f(z)\,\mathrm{d}z = 0\right).$$

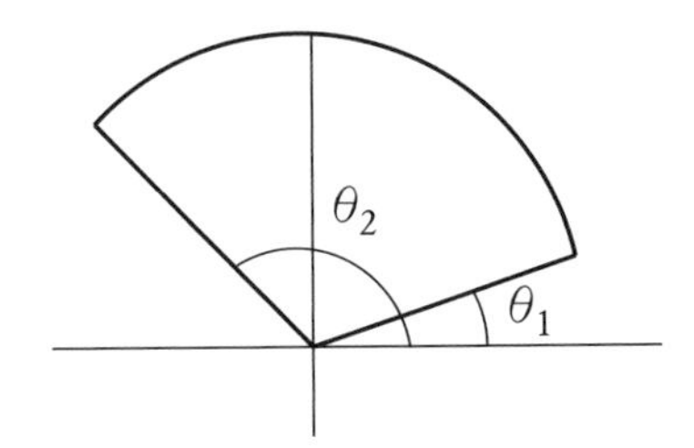

The second lemma is more specifically useful to compute Fourier transforms.

THEOREM 4.85 (Second Jordan lemma) *Let $f : \mathbb{C} \to \mathbb{C}$ be a continuous function on the sector*

$$\mathscr{S} = \left\{ r\,\mathrm{e}^{\mathrm{i}\theta};\ r \geqslant 0 \text{ and } 0 \leqslant \theta_1 \leqslant \theta \leqslant \theta_2 \leqslant \pi \right\},$$

such that $\lim\limits_{\substack{z\to\infty \\ z\in\mathscr{S}}} f(z) = 0$. *Then*

$$\lim_{r\to+\infty} \int_{\gamma(r)} f(z)\,\mathrm{e}^{\mathrm{i}z}\,\mathrm{d}z = 0.$$

PROOF. Consider the case (the most difficult) where the sector is $0 \leqslant \theta \leqslant \pi$. Remarking that

$$\left| \mathrm{e}^{\mathrm{i}r\mathrm{e}^{\mathrm{i}\theta}} \right| = \mathrm{e}^{-r\sin\theta},$$

we can bound the integral under consideration by

$$\begin{aligned}
\left| \int_{\gamma(r)} f(z)\,\mathrm{e}^{\mathrm{i}z}\,\mathrm{d}z \right| &\leqslant \int_0^{\pi} \left| f(r\,\mathrm{e}^{\mathrm{i}\theta}) \right| r\,\mathrm{e}^{-r\sin\theta}\,\mathrm{d}\theta \\
&\leqslant \int_0^{\pi/2} \left(\left| f(r\,\mathrm{e}^{\mathrm{i}\theta}) \right| + \left| f(r\,\mathrm{e}^{\mathrm{i}(\pi-\theta)}) \right| \right) r\,\mathrm{e}^{-r\sin\theta}\,\mathrm{d}\theta.
\end{aligned}$$

Let $\varepsilon > 0$. Since f is a function that tends to 0 at infinity, there exists $R > 0$ such that, for any $r \geqslant R$ and for any $\theta \in [0, \pi]$, we have $\left| f(r\,\mathrm{e}^{\mathrm{i}\theta}) \right| \leqslant \varepsilon$. In particular, for any $t \geqslant R$, we have

$$\left| \int_{\gamma(r)} f(z)\,\mathrm{e}^{\mathrm{i}z}\,\mathrm{d}z \right| \leqslant 2\varepsilon \int_0^{\pi/2} r\,\mathrm{e}^{-r\sin\theta}\,\mathrm{d}\theta.$$

But for any $\theta \in \left[0, \frac{\pi}{2}\right]$, we have $0 \leqslant \frac{2\theta}{\pi} \leqslant \sin\theta$. We deduce that

$$\int_0^{\pi/2} r\,\mathrm{e}^{-r\sin\theta} \leqslant \int_0^{\pi/2} r\,\mathrm{e}^{-2r\theta/\pi}\,\mathrm{d}\theta = \frac{\pi}{2}(1 - \mathrm{e}^{-r}) \leqslant \frac{\pi}{2},$$

which shows that for any $r \geqslant R$, we have

$$\left| \int_{\gamma(r)} f(z)\,\mathrm{e}^{\mathrm{i}z}\,\mathrm{d}z \right| \leqslant \pi\varepsilon.$$

4.6.b Integrals on $\mathbb{R}$ of a rational function

We consider the case where F is a rational function (i.e., a function of the form $F(x) = P(x)/Q(x)$, where P and Q are two polynomials, which we assume have no common zeros), and we wish to evaluate $\int_{\mathbb{R}} F(x)\,\mathrm{d}x$. We will assume here that the function F has no real pole, i.e., the polynomial Q does not vanish on $\mathbb{R}$ (the situation where Q vanishes will be treated in Chapter 8, during the study of Cauchy's principal values, page 223).

The nephew of the symbolist painter Puvis de Chavanne, Camille JORDAN (1838–1922), born in Lyons, was first in the entrance exam for the École Polytechnique at 17, then entered the École des Mines and became engineer in charge of Parisian quarries. He taught at the École Polytechnique, then entered Collège de France and the Académie des sciences. He studied linear algebra, group theory (he introduced the terminology *abelian groups* and studied cristallography; among his students were Felix Klein and Sophus Lie), geometry, and Fourier analysis, and made some of the first steps in measure theory. He introduced some fundamental topological concepts (homotopy). His pedagogical influence was enormous.

We will also assume that the integral in question is indeed convergent, which is the case if the degrees of P and Q satisfy $\deg Q \geqslant \deg P + 2$.

We start by extending the domain of the real variable to the complex plane, then truncate the integral on $\mathbb{R}$ to an integral on $[-R, R]$, with $R > 0$, and close the path of integration by means of a semicircle $\mathscr{C}_R$ of radius R as in Figure 4.4.

Denote by $\gamma(R) = [-R, R] \cup \mathscr{C}_R$ the contour thus defined. According to the assumptions, we can use the first of Jordan's lemmas 4.84, with the consequence that $\int_{\mathscr{C}_R} F(z)\,\mathrm{d}z \xrightarrow[R\to\infty]{} 0$ and hence

$$\lim_{R\to+\infty} \int_{\gamma(R)} F(z)\,\mathrm{d}z = \int_{-\infty}^{+\infty} F(x)\,\mathrm{d}x.$$

Moreover, when R goes to infinity, the contour $\gamma(R)$ ends up containing all the poles of F in the upper half-plane, i.e., for R large enough, we have

$$\int_{\gamma(R)} F(z)\,\mathrm{d}z = 2\pi\mathrm{i} \sum_{\substack{a \text{ pole in} \\ \text{upper half-plane}}} \operatorname{Res}(F\,;\,a),$$

which gives, in the limit $R \to +\infty$, the desired integral

$$\boxed{\int_{-\infty}^{+\infty} F(x)\,\mathrm{d}x = 2\pi\mathrm{i} \sum_{\substack{a \text{ pole in} \\ \text{upper half-plane}}} \operatorname{Res}(F\,;\,a).} \tag{4.11}$$

Remark 4.86 We might, of course, have used a different contour in the argument to close the segment $[-R, R]$ in the lower half-plane. Then we would have had to compute the residues in the *lower* half-plane. But then one would have had to be careful that with the orientation chosen the winding number of the contour around each pole would be equal to -1 and not -1, since the contour would have been oriented negatively.

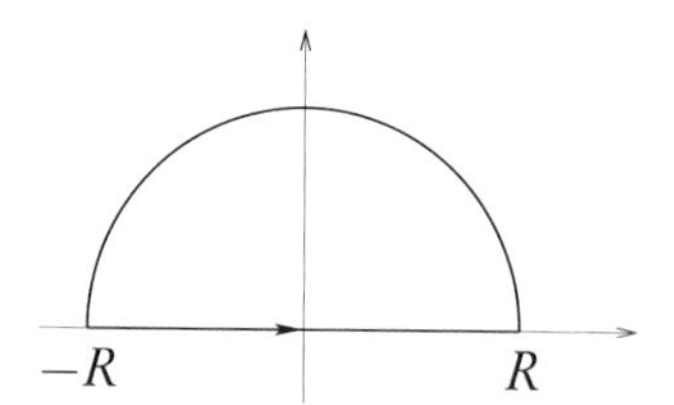

Fig. 4.4 – Integration path for a rational function.

4.6.c Fourier integrals

We are looking here at evaluating integrals of the type

$$\mathscr{I} = \int_{-\infty}^{+\infty} f(x)\,\mathrm{e}^{\mathrm{i}kx}\,\mathrm{d}x \qquad \text{with } k \in \mathbb{R},$$

which occur very often in practice, since they correspond to the value of the Fourier transform of f evaluated at k. One can use the same type of argument and contour as the one used in Section 4.6.b, using this time the second of Jordan's lemmas, but one must *be careful of the sign of k!* Indeed, the integral on the upper semicircle

$$\mathscr{C}^{+} \stackrel{\text{def}}{=} \left\{R\,\mathrm{e}^{\mathrm{i}\theta}\ ;\ \theta \in [0,\pi]\right\} = \tag{4.12}$$

−R R

of $f(z)\,\mathrm{e}^{\mathrm{i}kz}$ only tends to 0 when r tends to infinity when the real part of $\mathrm{i}kz$ is *negative*; if $z \in \mathscr{C}^{+}$, we must then have k *positive*. If k is negative, we must rather perform the integration along the lower semicircle

$$\mathscr{C}^{-} \stackrel{\text{def}}{=} \left\{R\,\mathrm{e}^{-\mathrm{i}\theta}\ ;\ \theta \in [0,\pi]\right\} = \tag{4.13}$$

−R R

Example 4.87 We try to compute

$$\mathscr{I} \stackrel{\text{def}}{=} \int_{-\infty}^{\infty} \frac{\cos(kx)}{x^2+a^2}\,\mathrm{d}x,$$

where a is a strictly positive real number. We start by writing the cosine in exponential form. One way to do it is to use $\cos(kx) = (\mathrm{e}^{\mathrm{i}kx} + \mathrm{e}^{-\mathrm{i}kx})/2$, but here it is simpler to notice that

$$\mathscr{I} = \operatorname{Re}\left(\int_{-\infty}^{\infty} \frac{\mathrm{e}^{\mathrm{i}kx}}{x^2+a^2}\,\mathrm{d}x\right).$$

For this last integral, we must now distinguish the two cases $k > 0$ and $k < 0$.

① When $k > 0$, we take the contour (4.12) and we obtain

$$\mathscr{I} = \lim_{R\to\infty} \int_{\gamma^+(R)} \frac{\mathrm{e}^{\mathrm{i}kz}}{z^2 + a^2}\,\mathrm{d}z = 2\pi\mathrm{i}\,\mathrm{Res}\left(\frac{\mathrm{e}^{\mathrm{i}kz}}{z^2 + a^2}\,;\,\mathrm{i}a\right),$$

since the only poles of the function being integrated are $\mathrm{i}a$ and $-\mathrm{i}a$. These poles are simple (because $1/(z^2 + a^2) = 1/(z + \mathrm{i}a)(z - \mathrm{i}a)$), and we compute easily

$$\mathrm{Res}\left(\frac{\mathrm{e}^{\mathrm{i}kz}}{z^2 + a^2}\,;\,\mathrm{i}a\right) = \lim_{z\to\mathrm{i}a} (z - \mathrm{i}a)\frac{\mathrm{e}^{\mathrm{i}kz}}{z^2 + a^2} = \lim_{z\to\mathrm{i}a} \frac{\mathrm{e}^{\mathrm{i}kz}}{z - \mathrm{i}a} = \frac{\mathrm{e}^{-ka}}{2\mathrm{i}a},$$

which gives

$$\mathscr{I} = \frac{\pi}{a}\,\mathrm{e}^{-ka}.$$

② When $k < 0$, we must take the contour (4.13) and we get

$$\mathscr{I} = \lim_{R\to\infty} \int_{\gamma^-(R)} \frac{\mathrm{e}^{\mathrm{i}kz}}{z^2 + a^2}\,\mathrm{d}z = -2\pi\mathrm{i}\,\mathrm{Res}\left(\frac{\mathrm{e}^{\mathrm{i}kz}}{z^2 + a^2}\,;\,-\mathrm{i}a\right).$$

We then evaluate the residue at $-\mathrm{i}a$, to obtain

$$\mathscr{I} = \frac{\pi}{a}\,\mathrm{e}^{ka}.$$

To conclude, we have, independently of the sign of k (and even if $k = 0$)

$$\int_{-\infty}^{\infty} \frac{\cos kx}{x^2 + a^2}\,\mathrm{d}x = \frac{\pi}{a}\,\mathrm{e}^{-|k|a}.$$

4.6.d Integral on the unit circle of a rational function

We are trying here to compute an integral of the type

$$\mathscr{K} = \int_0^{2\pi} R(\sin t, \cos t)\,\mathrm{d}t,$$

where $R(x, y)$ is a rational function which has no pole on the unit circle ($x^2 + y^2 = 1$). Putting, naturally, $z = \mathrm{e}^{\mathrm{i}t}$, we obtain the relations

$$\mathrm{d}z = \mathrm{i}\mathrm{e}^{\mathrm{i}t}\,\mathrm{d}t, \quad \sin t = \frac{\mathrm{e}^{\mathrm{i}t} - \mathrm{e}^{-\mathrm{i}t}}{2\mathrm{i}} = \frac{1}{2\mathrm{i}}\left(z - \frac{1}{z}\right), \quad \cos t = \frac{1}{2}\left(z + \frac{1}{z}\right).$$

The integral becomes

$$\mathscr{K} = \int_{\mathscr{C}} R\left(\frac{z - 1/z}{2\mathrm{i}}, \frac{z + 1/z}{2}\right)\frac{\mathrm{d}z}{\mathrm{i}z},$$

where $\mathscr{C}$ is the unit circle with the positive orientation. An immediate application of the Residue theorem shows that

$$\boxed{\mathscr{K} = 2\pi\mathrm{i} \sum_{a \text{ pole} \in \mathcal{B}(0\,;\,1)} \mathrm{Res}\left[\frac{1}{\mathrm{i}z}R\left(\frac{1}{2\mathrm{i}}\left(z - \frac{1}{z}\right), \frac{1}{2}\left(z + \frac{1}{z}\right)\right); a\right]} \tag{4.14}$$

Example 4.88 We want to compute the following integral:

$$I = \int_0^{2\pi} \frac{\mathrm{d}t}{a + \sin t}, \qquad a \in \mathbb{R},\ a > 1.$$

Put $R(x, y) = 1/(a + x)$. Then we have

$$I = 2\pi \mathrm{i} \sum_{a \text{ pôle} \in \mathcal{B}(0;\,1)} \operatorname{Res}\left(\frac{1}{\mathrm{i}z} \frac{1}{a + \frac{1}{2\mathrm{i}}\left(z - \frac{1}{z}\right)}\ ;\ a\right).$$

Define now $f(z) \stackrel{\text{def}}{=} 2/(z^2 + 2\mathrm{i}az - 1)$, and it only remains to find the poles of f. These are the solutions of the equation $z^2 + 2\mathrm{i}az - 1 = 0$, and so they are the points $\zeta = -\mathrm{i}a + \mathrm{i}\sqrt{a^2 - 1}$ and $\zeta' = -\mathrm{i}a - \mathrm{i}\sqrt{a^2 - 1}$. They are located as shown in the figure below

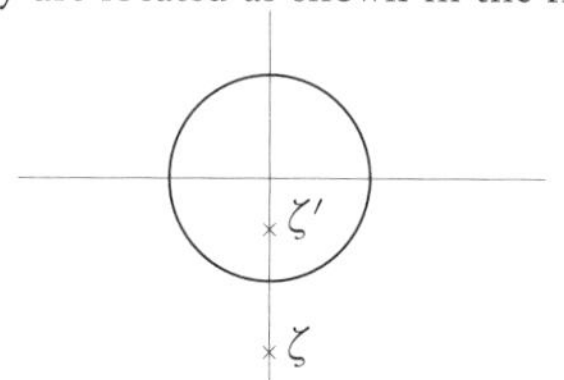

Only the pole ζ contributes to the integral, it is a simple pole with residue $-\mathrm{i}/\sqrt{a^2 - 1}$ and the integral is therefore equal to $I = 2\pi/\sqrt{a^2 - 1}$.

4.6.e Computation of infinite sums

Consider a series of the form

$$S = \sum_{n \in \mathbb{Z}} f(n),$$

where f is a given function, decaying sufficiently rapidly for the sum to exist, and with no pole on the real axis. One can check easily that

$$f(n) = \operatorname{Res}\big(\pi f(z) \operatorname{cotan} \pi z\ ;\ z = n\big),$$

and so

$$S = \sum_{n \in \mathbb{Z}} f(n) = \sum_{n \in \mathbb{Z}} \operatorname{Res}\big(\pi f(z) \operatorname{cotan}(\pi z)\ ;\ z = n\big).$$

The partial sum from $-N$ to N is therefore given by the integral of f along the following contour[14] (where each circle is taken with positive orientation):

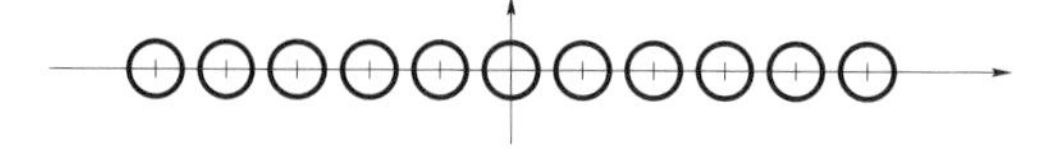

which we can deform (adding contributions which cancel each other in the integral) into

[14] Which is not a contour properly speaking since it consists of multiple pieces, but this is without importance, the integral on this "contour" being defined as the sum of the integrals on each circle.

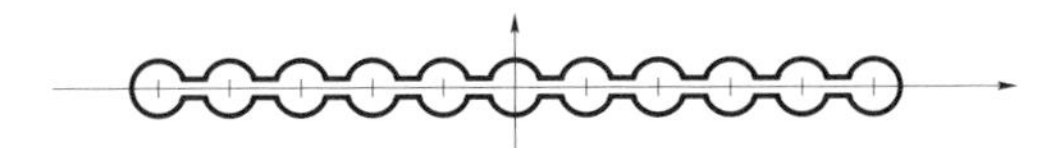

and by continuity into

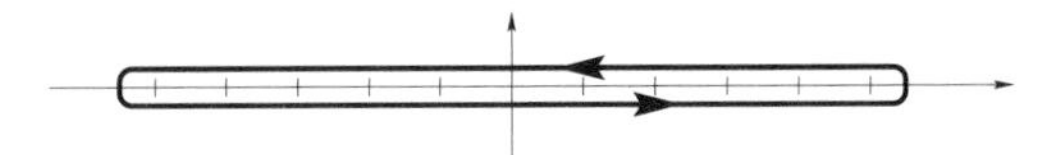

In the limit where N tends to infinity, and if the function f decays to 0, the contour becomes simply

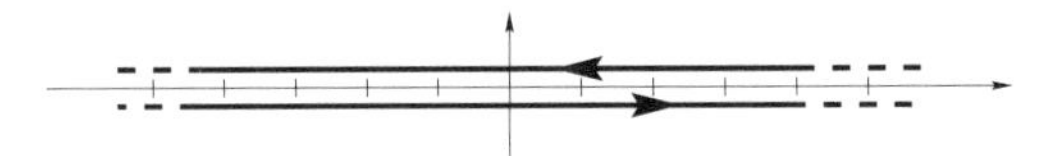

Finally, this last integral can be computed by the method of residues, by using Jordan's Lemmas and closing each of the lines by semicircles. There only remains to evaluate the residues of the function f outside the real axis.

Example 4.89 Suppose we want to evaluate the sum

$$S = \sum_{n \in \mathbb{Z}} \frac{1}{n^2 + a^2},$$

where a is a nonzero real number. We will then define

$$f(z) \stackrel{\text{def}}{=} \frac{1}{z^2 + a^2} \pi \operatorname{cotan}(\pi z),$$

and S is the sum of the residues of this function at all the integers if $a \neq 0$. We will therefore assume $a > 0$. The reasoning explained above shows that this sum is equal to the opposite of the sum of the residues of f at nonreal poles. These are located at $z = \pm \mathrm{i}a$, and we have

$$\operatorname{Res}(f\,;\, z = \pm \mathrm{i}a) = \frac{1}{2\mathrm{i}a} \pi \operatorname{cotan}(\mathrm{i}\pi a) = -\frac{\pi}{2a} \coth(\pi a),$$

hence the result

$$S = \frac{\pi}{a} \coth(\pi a). \tag{4.15}$$

One can note that, if a tends to 0, we have $S \sim 1/a^2$, which is the expected behavior (the term $n = 0$ of the sum is $1/a^2$). In the same manner, we have

$$\sum_{n=1}^{\infty} \frac{1}{n^2 + a^2} = \frac{\pi}{2a} \coth(\pi a) - \frac{1}{2a^2}.$$

The method must obviously be tailored to each case, as the next example shows:

Example 4.90 We now try to evaluate the sum

$$T = \sum_{n=1}^{\infty} \frac{(-1)^n}{n^4}.$$

Instead of using the function cotan, it is interesting to use the function $z \mapsto 1/\sin \pi z$, which has poles at every integer and for which the residue at $n \in \mathbb{N}$ is precisely equal to $(-1)^n/\pi$.

Unfortunately, the function $z \longmapsto 1/z^4$ has a pole at 0. This pole must then be treated separately. We start by noticing that

$$T = \frac{T'}{2} \qquad \text{with} \quad T' = \sum_{\substack{n=-\infty \\ n\neq 0}}^{\infty} \frac{(-1)^n}{n^4},$$

and we deduce that the sum T' is given by $1/2\mathrm{i}$ times the integral on the following contour of the function $z \longmapsto 1/z^4 \sin(\pi z)$

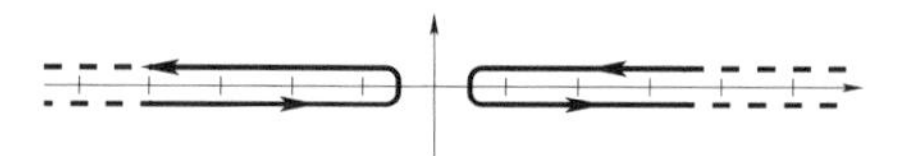

which we then close (appealing to Jordan's second lemma) into

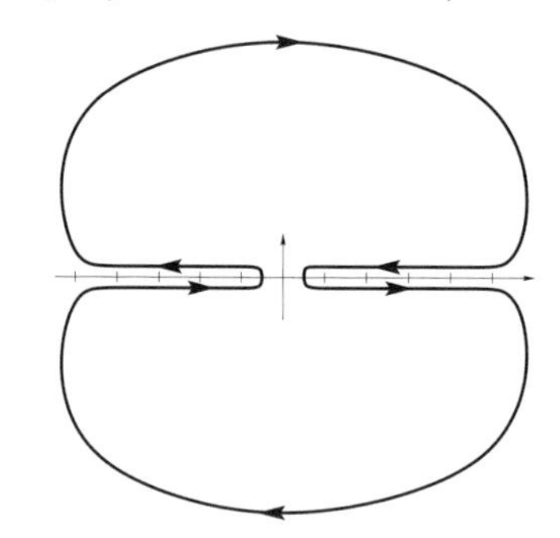

before deforming it by continuity into

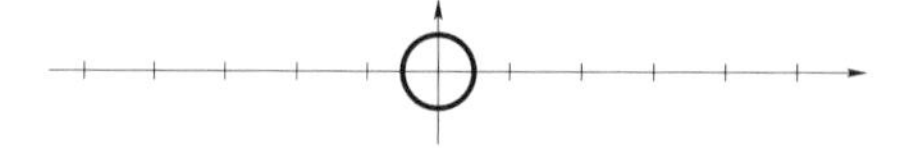

which gives

$$T' = -\operatorname{Res}\left(\frac{1}{z^4 \sin z}\ ;\ z = 0\right),$$

the minus sign coming from the fact that the last path is taken with negative orientation. There only remains to evaluate this residue, using, for instance, a Taylor expansion of $1/\sin \pi z$:

$$\begin{aligned}
\sin z &= z - \frac{z^2}{6} + \frac{z^5}{120} + \cdots, \\
\frac{1}{z^4 \sin \pi z} &= \frac{1}{z^5}\left(1 - \frac{\pi^2 z^2}{6} + \frac{\pi^4 z^4}{120} + \cdots\right)^{-1} \\
&= \frac{1}{z^5}\left(1 + \frac{\pi^2 z^2}{6} + \frac{14\pi^4 z^4}{720} + \cdots\right),
\end{aligned}$$

which shows that the residue of $1/z^4 \sin \pi z$ at 0 is $\frac{14\pi^4}{720}$ and that

$$\sum_{n=1}^{\infty} \frac{(-1)^n}{n^4} = \frac{-7\pi^4}{720}.$$

EXERCISES

♦ **Exercise 4.4** Show that if f is holomorphic on an open subset $\Omega \subset \mathbb{C}$, then

$$\triangle\left(|f|^2\right) = 4\left|\frac{\partial f}{\partial z}\right|^2.$$

♦ **Exercise 4.5** Let f and g be continuously differentiable functions on $\mathbb{R}^2$ identified with $\mathbb{C}$ (i.e., functions such that the partial derivatives $\partial f/\partial x$ and $\partial f/\partial y$ exist and are continuous). Show that the following chain rules hold:

$$\frac{\partial}{\partial z}(f \circ g) = \frac{\partial f}{\partial z}\frac{\partial g}{\partial z} + \frac{\partial f}{\partial \bar{z}}\frac{\partial \bar{g}}{\partial z},$$

and
$$\frac{\partial}{\partial \bar{z}}(f \circ g) = \frac{\partial f}{\partial z}\frac{\partial g}{\partial \bar{z}} + \frac{\partial f}{\partial \bar{z}}\frac{\partial \bar{g}}{\partial \bar{z}}.$$

♦ **Exercise 4.6** Let U be a connected open set and let $f : U \to \mathbb{C}$ be a function such that f^2 and f^3 are holomorphic. Show that f is holomorphic on U.

Path integration

♦ **Exercise 4.7** Consider, in $\mathbb{C}$, the segments $\mathcal{T}_1 = [-\mathrm{i}, 6\mathrm{i}]$, $\mathcal{T}_2 = [-\mathrm{i}, 2+6\mathrm{i}]$ and $\mathcal{T}_3 = [2+5\mathrm{i}, 6\mathrm{i}]$. Define now a function $f : \mathbb{C} \to \mathbb{C}$ by $f(x+\mathrm{i}y) = x^2 - 3\mathrm{i}y$. Compute explicitly $\int_{\mathcal{T}_1} f(z)\,\mathrm{d}z$ and $\int_{\mathcal{T}_2 \cup \mathcal{T}_3} f(z)\,\mathrm{d}z$. Conclude.

♦ **Exercise 4.8** Denoting by $\mathscr{T}$ the circular arc joining 3 to $\mathrm{i}\sqrt{3}$ with equation

$$(x-1)^2 + y^2 = 4,$$

compute $\int_{\mathscr{T}} \mathrm{d}z/z^2$.

♦ **Exercise 4.9** Recall that $\int_{-\infty}^{+\infty} \mathrm{e}^{-x^2}\,\mathrm{d}x = \sqrt{\pi}$. One then wishes to compute, for any nonzero $a \in \mathbb{R}$, the integral

$$\int_{-\infty}^{+\infty} \mathrm{e}^{-x^2} \cos(ax)\,\mathrm{d}x.$$

Let $f(z) = \mathrm{e}^{-z^2}$ for any $z \in \mathbb{C}$, and denote by $\mathcal{T}$ the rectangle

$$\mathcal{T} = [-R, R, R + \mathrm{i}a/2, -R + \mathrm{i}a/2].$$

Compute the integral $\int_{\mathcal{T}} f(z)\,\mathrm{d}z$ and conclude by letting R go to infinity.

Integrals using residues

♦ **Exercise 4.10** Compute the residue at 0 of $\sin(z)/z^2$.

♦ **Exercise 4.11** Prove the convergence of the following integrals and compute their values:

$$\int_{-\infty}^{+\infty} \frac{(1+x)\sin 2x}{x^2+2x+2}\,\mathrm{d}x \qquad \text{and} \qquad \int_{-\infty}^{+\infty} \frac{\mathrm{d}x}{(1+x^2)^n} \qquad \text{with } n \in \mathbb{N}^*$$

♦ **Exercise 4.12 (Fresnel integral)** Does the Fresnel integral

$$\mathscr{I} \stackrel{\text{def}}{=} \int_{-\infty}^{+\infty} \mathrm{e}^{\mathrm{i}x^2}\,\mathrm{d}x$$

exist in the sense of Lebesgue?

We nevertheless try to assign a meaning to this integral. Show that

$$\mathscr{I}(R,R') \stackrel{\text{def}}{=} \int_{-R}^{R'} \mathrm{e}^{\mathrm{i}x^2}\,\mathrm{d}x$$

exists and has a limit as R and R' go to $+\infty$, and compute this limit, using a suitable contour and remembering the following value of the gaussian integral:

$$\int_{-\infty}^{+\infty} \mathrm{e}^{-x^2}\,\mathrm{d}x = \sqrt{\pi}.$$

♦ **Exercise 4.13 (Fourier transform of the gaussian)** Let $\nu \in \mathbb{R}$. Compute

$$\int_{-\infty}^{+\infty} \mathrm{e}^{-\pi x^2}\mathrm{e}^{-2\mathrm{i}\pi\nu x}\,\mathrm{d}x,$$

which is the **Fourier transform** of the gaussian $x \mapsto \mathrm{e}^{-\pi x^2}$ at the point ν. One may use Exercise 4.9.

(Another method will be given in Exercise 10.1 on page 295, using the Fourier transform.)

♦ **Exercise 4.14** Let $\mathcal{A}$ be a *finite* subset of $\mathbb{C}$ containing no positive real number. Let $\mu \in \mathbb{C} \setminus \mathbb{Z}$. Let finally $f : \mathbb{C} \to \mathbb{C}$ be a function meromorphic on $\mathbb{C}$ and holomorphic on $\mathbb{C} \setminus \mathcal{A}$, such that

$$\lim_{\substack{r\to 0 \\ \text{or } r\to+\infty}} I_r = 0 \qquad \text{with} \qquad I_r \stackrel{\text{def}}{=} \int_0^{2\pi} \left|f(r\,\mathrm{e}^{\mathrm{i}\theta})\right| r^{-\operatorname{Re}\mu+1}\,\mathrm{d}\theta.$$

Show that we have then

$$\int_0^{+\infty} f(x)x^{-\mu}\,\mathrm{d}x = \frac{2\mathrm{i}\pi}{1-\mathrm{e}^{2\mathrm{i}\pi\mu}} \sum_{a\in\mathcal{A}} \operatorname{Res}\left(f(z)\mathrm{e}^{-\mu L(z)}\,;\,a\right),$$

where L denotes the complex logarithm defined on $\mathbb{C} \setminus \mathbb{R}^+$ by

$$L(z) \stackrel{\text{def}}{=} \log|z| + \mathrm{i}\theta(z) \qquad \text{with } \theta(z) \in\,]0,2\pi[\,.$$

♦ **Exercise 4.15** Let f be a function holomorphic in the open disc $\mathscr{D}(0;R)$ with $R > 1$. Denote by γ the unit circle: $\gamma \stackrel{\text{def}}{=} \{\mathrm{e}^{\mathrm{i}\theta}\,;\,\theta \in [0,2\pi[\}$.

Computing in two different ways the integrals

$$I = \int_\gamma \left(2 + (z + {}^1\!/_z)\right)\frac{f(z)}{z}\,\mathrm{d}z, \qquad J = \int_\gamma \left(2 - (z + {}^1\!/_z)\right)\frac{f(z)}{z}\,\mathrm{d}z,$$

show that

$$\begin{cases} \dfrac{2}{\pi}\displaystyle\int_0^{2\pi} f(\mathrm{e}^{\mathrm{i}\theta})\cos^2\left({}^\theta\!/_2\right)\,\mathrm{d}\theta = 2f(0) + f'(0), \\[2ex] \dfrac{2}{\pi}\displaystyle\int_0^{2\pi} f(\mathrm{e}^{\mathrm{i}\theta})\sin^2\left({}^\theta\!/_2\right)\,\mathrm{d}\theta = 2f(0) - f'(0). \end{cases}$$

♦ **Exercise 4.16** Compute the value of

$$\int_0^{+\infty} \frac{\mathrm{d}x}{x^6+1}.$$

On may use the classical contour or, more cleverly, a sector with angle $\pi/3$.

♦ **Exercise 4.17** Compute

$$\int_0^{\infty} \frac{\log x}{1+x^3}\,\mathrm{d}x.$$

♦ **Exercise 4.18** Compute

$$\int_0^{\infty} \frac{\cos\left(\frac{\pi x}{2}\right)}{x^2-1}\,\mathrm{d}x.$$

♦ **Exercise 4.19** Show that

$$\int_0^{+\infty} \sin x^2\,\mathrm{d}x = \int_0^{+\infty} \cos x^2\,\mathrm{d}x = \frac{1}{2}\sqrt{\frac{\pi}{2}}.$$

HINT: Use the following contour: from 0 to R on the real axis, then an eighth of a circle up to $R\,\mathrm{e}^{\mathrm{i}\pi/4}$, and then back to the starting point along the imaginary axis.

♦ **Exercise 4.20** Show that

$$\int_0^{+\infty} \frac{\cosh ax}{\cosh x}\,\mathrm{d}x = \frac{\pi}{2\cos\left(\frac{\pi}{2}a\right)}, \qquad \text{where } |a|<1.$$

HINT: Look for the poles of $F(z) = \dfrac{\mathrm{e}^{az}}{\cosh z}$, and consider the contour integral on the rectangle with vertices at $[-R, R, R+\mathrm{i}\pi, -R+\mathrm{i}\pi]$.

♦ **Exercise 4.21** Show that if R is a rational function without poles on the half-axis of positive real numbers $\mathbb{R}^+$, if $\alpha \in]0,1[$ and if $R(x)/x^\alpha$ is integrable on $\mathbb{R}^+$, then

$$\left(1-\mathrm{e}^{-2\mathrm{i}\pi\alpha}\right) \times \int_0^{+\infty} \frac{R(x)}{x^\alpha}\,\mathrm{d}x = 2\mathrm{i}\pi \sum_{\substack{\text{poles other}\\ \text{than } z=0}} \operatorname{Res}\left(\frac{R(z)}{z^\alpha}\right),$$

where the function $z \mapsto z^\alpha$ is defined by $\rho\,\mathrm{e}^{\mathrm{i}\theta} \mapsto \rho^\alpha\,\mathrm{e}^{\mathrm{i}\alpha\theta}$ for all $\theta \in [0, 2\pi[$.

HINT: Use the following contour and take the limits $[R \to +\infty]$ and $[\varepsilon \to 0]$.

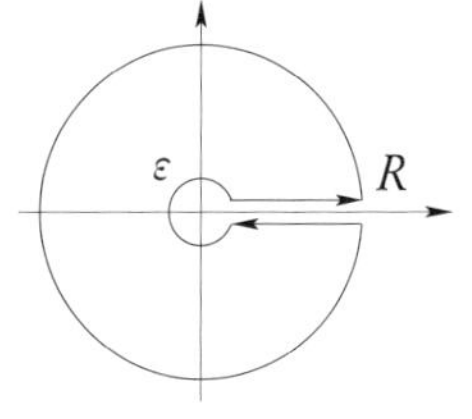

♦ **Exercise 4.22** Show that if $\alpha \in]0,1[$, then

$$\int_0^{+\infty} \frac{\mathrm{d}x}{x^\alpha(1+x)} = \frac{\pi}{\sin\pi\alpha}.$$

♦ **Exercise 4.23** Let a be a positive real number. Compute then

$$\lim_{\varepsilon \to 0^+} \int_{-\infty}^{+\infty} \frac{\mathrm{d}x}{(x^2 - a^2 - \mathrm{i}\varepsilon)}.$$

HINT: Take the following contour and the limit $[R \to +\infty]$.

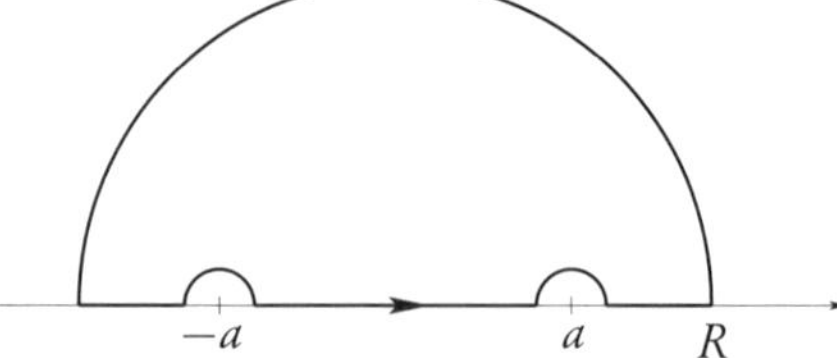

Sums using residues

♦ **Exercise 4.24** Compute the following two series:

$$S = \sum_{n \in \mathbb{N}^*} \frac{1}{n^4 + a^4} \qquad \text{and} \qquad T = \sum_{n \in \mathbb{N}^*} \frac{n^2}{n^4 + a^4} \qquad \text{with } a \in \mathbb{R},\ a \geqslant 0.$$

♦ **Exercise 4.25 (Ramanujan's formula)** Using the function $z \mapsto \operatorname{cotan} \pi z$, deduce, by means of a clever contour, that

$$\frac{\coth \pi}{1^7} + \frac{\coth 2\pi}{2^7} + \cdots + \frac{\coth n\pi}{n^7} + \cdots = \frac{19\pi^7}{56\,700}.$$

Note: This beautiful formula is due to the great indian mathematician Ramanujan.[15] But Ramanujan did not use the calculus of residues.

PROBLEM

♦ **Problem 2 Application to finite temperature field theory** — In quantum statistical mechanics, some quantities at equilibrium (such as the density of a gas, its pressure, its average energy) are computed using sums over discrete frequencies (called **Matsubara frequencies**). We consider here a simple case where these sums are easily computable using the method of residues.

For instance, if one considers a gas of free (i.e., noninteracting) electrons (and positrons[16]), denote by μ the chemical potential of the electrons; the positrons have chemical potential equal to $-\mu$. The temperature being equal to T, put $\beta = 1/k_{\mathrm{B}} T$, where k_{B} is the Boltzmann constant.

The free charge density, in momentum space, is given [55, 60] by

$$N(p) \stackrel{\text{def}}{=} \sum_{l} \frac{4(\mathrm{i}\beta\omega_l + \beta\mu)}{(\mathrm{i}\beta\omega_l + \beta\mu)^2 - \beta^2 E_p^2},$$

[15] Srinivasa RAMANUJAN (1887–1920) was probably the most romantic character in the mathematical pantheon. An Indian mathematician of genius, but lacking in formal instruction, he was discovered by G. H. HARDY, to whom he had mailed page after page covered with formulas each more incredible than the last. The reader may read *Ramanujan, an Indian Mathematician* by Hardy, in the book [46]. One will also find there the famous anecdote of the number "1729."

[16] In relativistic quantum mechanics, one cannot have a description of matter which is purely electronic, for instance; positrons necessarily occur, as was discovered by Dirac [49].

with $$\omega_l = \frac{(2l+1)\pi}{\beta} \qquad \text{and} \qquad E_p \stackrel{\text{def}}{=} (m^2c^4 + p^2c^2)^{1/2}.$$

1. Put this sum in the form of a sum of residues of a well-chosen function $f(z)$. Make a drawing of the complex plane with the corresponding poles.
2. Does the chosen function admit any other poles?
3. By considering the decay of the function $f(z)$ at infinity, show that the sum of the residues vanishes.
4. Show then that we have

$$\begin{aligned} N(\boldsymbol{p}) &= -\tanh\left(\frac{\beta E_p - \beta\mu}{2}\right) + \tanh\left(\frac{\beta E_p + \beta\mu}{2}\right) \\ &= 2N_F^{\text{el}}(\boldsymbol{p}) - 2N_F^{\text{pos}}(\boldsymbol{p}), \end{aligned}$$

where we use the electronic and positronic Fermi distributions

$$N_F^{\text{el}}(\boldsymbol{p}) = \frac{1}{\mathrm{e}^{\beta(E_p-\mu)}+1} \qquad \text{and} \qquad N_F^{\text{pos}}(\boldsymbol{p}) = \frac{1}{\mathrm{e}^{\beta(E_p+\mu)}+1}.$$

Interpret these two terms.

SOLUTIONS

♦ **Solution of exercise 4.1 on page 90**. Put $f = P + \mathrm{i}Q$.

We start by remarking that property (ii) implies all the others. Let us show the equivalence of (ii) and (iii). Suppose, for example, that $P = \mathrm{C}^{\text{nt}}$. Then, at any point,

$$\frac{\partial P}{\partial x} = \frac{\partial P}{\partial y} = 0,$$

which, by the Cauchy equations, implies that

$$\frac{\partial Q}{\partial y} = \frac{\partial Q}{\partial x} = 0$$

and hence that $Q = \mathrm{C}^{\text{nt}}$. One deduces that (ii) and (iii) are equivalent. Moreover, (ii) and (iii), together, imply property (i). Thus we have shown the equivalence of (i), (ii), and (iii).

Let us show that (iv) implies (iii) – this will prove also that (iv) implies (i) and (ii). Assume then property (iv). We can also assume that f is not the zero function (which is an obvious case). Then f has no zero at all, and so the sum $P^2 + Q^2 = \mathrm{C}^{\text{nt}}$ is never zero. We have

$$0 = \frac{\partial\,|f|^2}{\partial x} = 2P\frac{\partial P}{\partial x} + 2Q\frac{\partial Q}{\partial x} \stackrel{\text{Cauchy}}{=} 2P\frac{\partial Q}{\partial y} + 2Q\frac{\partial Q}{\partial x}$$

and

$$0 = \frac{\partial\,|f|^2}{\partial y} = 2P\frac{\partial P}{\partial y} + 2Q\frac{\partial Q}{\partial y} = -2P\frac{\partial Q}{\partial x} + 2Q\frac{\partial Q}{\partial y}.$$

The two preceeding equations can therefore be written in the form

$$\begin{pmatrix} Q & P \\ -P & Q \end{pmatrix} \begin{pmatrix} \partial Q/\partial x \\ \partial Q/\partial y \end{pmatrix} = 0.$$

But the determinant of this matrix is $P^2 + Q^2$ which is everywhere non-zero. So we deduce that, at any point, we have $\frac{\partial Q}{\partial x} = \frac{\partial Q}{\partial y} = 0$ and, consequently, that $Q = \mathrm{C}^{\text{nt}}$.

There only remains to show that (v) implies (ii). Assume that both f and $\overline{f}$ are holomorphic; then the Cauchy equations allow us to write

$$(f \in \mathscr{H}) \Longrightarrow \frac{\partial P}{\partial x} = \frac{\partial Q}{\partial y} \qquad \text{and} \qquad (\overline{f} \in \mathscr{H}) \Longrightarrow \frac{\partial P}{\partial x} = \frac{\partial(-Q)}{\partial y},$$

and hence $\partial P/\partial x = 0$; similarly $\partial P/\partial y = 0$, so that $P = \text{C}^{\underline{\text{nt}}}$. Which, taking into account the remarks above, completes the proof.

♦ **Solution of exercise 4.4.** One checks first that the laplacian can be written $\triangle = 4\partial\overline{\partial}$. Then write $\triangle |f|^2 = 4\partial\overline{\partial}(f\overline{f}) = 4\partial\left[(\overline{\partial} f)\overline{f} + f(\overline{\partial}\,\overline{f})\right]$. Since f is holomorphic, we have $\overline{\partial} f = 0$ hence $\triangle |f|^2 = 4\partial\left[f\overline{\partial}\,\overline{f}\right] = 4\partial f \cdot \overline{\partial}\,\overline{f}$ since $\overline{\partial}\,\overline{f}$ is an *antiholomorphic* function; thus $\partial\overline{\partial}\,\overline{f} = 0$. To conclude, one finds

$$\triangle |f|^2 = 4\partial f \cdot \overline{\partial f} = 4\,|\partial f|^2 .$$

♦ **Solution of exercise 4.6.** Disregarding the trivial case where $f = 0$, write then $f = f^3/f^2$ at any point where f does not vanish. But since f^2 is holomorphic, its zeros are isolated, and therefore those of f also, since they are the same. Denote by Z the set of these zeros. Then f is holomorphic on $\Omega \setminus Z$ and bounded in the neighborhood of each point in Z, which are therefore artificial singularities. Hence, if we extend by continuity the function $g = f^3/f^2$ (by putting $g(z) = 0$ for any $x \in Z$), the resulting function g is holomorphic by Theorem 4.56 on page 108, and moreover $g = f$ by construction.

♦ **Solution of exercise 4.8.** The function $z \mapsto (1/z^2)$ being holomorphic on the open set $\Omega = \{z \in \mathbb{C}\ ;\ \text{Im}(z) > \text{Re}(z)\}$ which contains the contour $\mathscr{T}$, it admits a primitive on Ω, which is evidently $z \mapsto -1/z$. Then

$$\int_{\mathscr{T}} \frac{1}{z^2}\,\mathrm{d}z = \left[-\frac{1}{z}\right]_3^{\mathrm{i}\sqrt{3}} = \frac{1}{3} + \frac{\mathrm{i}}{\sqrt{3}}.$$

♦ **Solution of exercise 4.9.** Show that the integrals on the vertical segments tend to 0 when $[R \to \infty]$. Show moreover that

$$\int_{-R}^{R} \mathrm{e}^{-(x+\mathrm{i}a/2)^2}\,\mathrm{d}x = \mathrm{e}^{a^2}\int_{-R}^{R} \mathrm{e}^{-x^2}\cos ax\,\mathrm{d}x$$

by parity. The integral considered is equal to $\sqrt{\pi}\mathrm{e}^{-a^2}$.

♦ **Solution of exercise 4.10.** The Laurent series expansion of the sine function coincides with its Taylor series expansion and is given by

$$\sin z = z - \frac{z^3}{3!} + \frac{z^5}{5!} - \cdots; \qquad \text{hence} \qquad \frac{\sin z}{z^2} = \frac{1}{z} - \frac{z}{3!} + \frac{z^3}{5!} - \cdots,$$

and the residue is the coefficient of $1/z$, namely 1.

♦ **Solution of exercise 4.11.** For the first integral, put

$$f(z) = \frac{(1+z)\,\mathrm{e}^{2\mathrm{i}z}}{z^2 + 2z + 2}$$

and integrate by closing the contour from above (with justification!). The only residue inside the contour is the one in $z = 1 + \mathrm{i}$. The desired integral is the imaginary part of $\int_\gamma f(z)\,\mathrm{d}z = \mathrm{i}\pi\mathrm{e}^{-2-2\mathrm{i}}$, which is $(\pi\cos 2)/\mathrm{e}^2$.

For the second integral, close the contour from above, using Jordan's second lemma, with

$$f(z) = \frac{1}{(1+z^2)^n} = \frac{1}{(z-\mathrm{i})^2\,(z+\mathrm{i})^n}.$$

Then the desired integral is equal to $2i\pi \operatorname{Res}(f\,;\,i)$ (or, if the contour is closed from below, to $-2i\pi \operatorname{Res}(f\,;\,-i)$). But, according to formula (4.10), page 116, since i is a pole of order n,

$$\operatorname{Res}(f\,;\,i) = \frac{1}{(n-1)!}\frac{d^{n-1}}{dz^{n-1}}\frac{1}{(z+i)^n}\bigg|_{z=i}.$$

An easy induction shows that

$$\frac{d^{n-1}}{dz^{n-1}}\frac{1}{(z+i)^n} = \frac{(-n)\cdot(-n-1)\cdots(-2n+2)}{(z+i)^{2n-1}} = \frac{(-1)^{n-1}}{(z+i)^{2n-1}}\frac{(2n-2)!}{(n-1)!},$$

which leads, after short algebraic manipulations, to

$$\int_{-\infty}^{+\infty}\frac{dz}{(1+z^2)^n} = \frac{\pi}{2^{2n-2}}\frac{(2n-2)!}{\big[(n-1)!\big]^2}.$$

◆ **Solution of exercise 4.12.** The Fresnel integral is not defined in the Lebesgue sense since the modulus of e^{ix^2} is constant and nonzero, and therefore not integrable. However, the integral is defined as an "improper integral," which means that the integral $\mathscr{I}(R,R')$ admits a limit as R and R' (separately) tend to infinity. Consider now the contour

$$\mathscr{C}(R,R') \stackrel{\text{def}}{=}$$

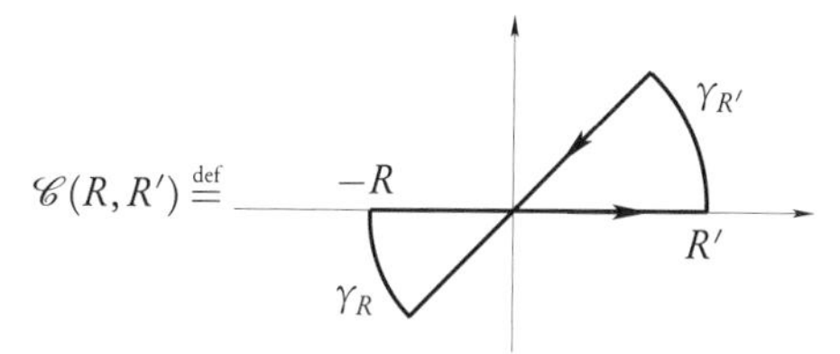

The integral on the upper eighth arc $\gamma_{R'}$ is, by a change of variable, equal to

$$\int_{\gamma_{R'}} e^{iz^2}\,dz = \int_{C'} e^{i\zeta}\frac{d\zeta}{\sqrt{\zeta}},$$

where C' is the upper *quarter*-circle with radius R'^2, and this last integral does tend to 0 by Jordan's second lemma. The second eighth of the circle is treated in the same manner. Since the integral on $\mathscr{C}(R,R')$ vanishes (the function is holomorphic), the Fresnel integral is equal to an integral on the line segment with angle $\pi/4$:

$$\lim_{R,R'\to+\infty}\left(\int_{-R}^{R'} e^{ix^2}\,dx - \int_{-R}^{R'} e^{i(e^{i\pi/4}x)^2}e^{i\pi/4}\,dx\right) = 0,$$

hence
$$\int_{-\infty}^{+\infty} e^{ix^2}\,dx = e^{i\pi/4}\int_{-\infty}^{+\infty} e^{-x^2}\,dx = \frac{1+i}{\sqrt{2}}\sqrt{\pi}.$$

◆ **Solution of exercise 4.13.** Denote by $F(\nu)$ the integral under consideration. It can be put in the form

$$F(\nu) = e^{-\pi\nu^2}\int e^{-\pi(x+i\nu)^2}\,dx.$$

Denote by γ_ν the line with imaginary part ν, hence parallel to the real axis. Then, using Cauchy's theorem for the function $z \mapsto e^{-\pi z^2}$, which is holomorphic on $\mathbb{C}$, show as in exercise 4.9, that

$$\int_{\gamma_\nu} e^{-\pi z^2}\,dz = \int_{-\infty}^{+\infty} e^{-\pi x^2}\,dx = 1,$$

which gives $F(\nu) = e^{-\pi\nu^2}$.

♦ **Solution of exercise 4.16.** Consider the contour $\gamma(R)$ made from the line segments $[0, R]$, $\left[0, R\,\mathrm{e}^{\mathrm{i}\pi/3}\right]$ and the part of the circle $\mathscr{C}(0; R)$ with argument varying from 0 and $\pi/3$. Let g be the function $g : z \mapsto z^6 + 1$. Then the function $f = 1/g$ has only one pole inside this contour, at the point $\zeta = \exp(\mathrm{i}\pi/6)$. Check then (using the first Jordan lemma and a few algberaic computations) that

$$\int_{\gamma(R)} f(z)\,\mathrm{d}z \xrightarrow[R\to+\infty]{} \left(1 - \mathrm{e}^{\mathrm{i}\pi/3}\right) \int_0^{+\infty} \frac{\mathrm{d}x}{x^6 + 1}.$$

Using the residue theorem, it follows therefore that

$$\int_0^{+\infty} \frac{\mathrm{d}x}{x^6 + 1} = \frac{2\mathrm{i}\pi \operatorname{Res}(f\,;\,\zeta)}{1 - \mathrm{e}^{\mathrm{i}\pi/3}}.$$

To compute the preceeding residue, one may notice that

$$\begin{aligned} \operatorname{Res}(f\,;\,\zeta) = \lim_{z\to\zeta} \frac{z - \zeta}{g(z)} &= \lim_{z\to\zeta} \frac{z - \zeta}{g(z) - g(\zeta)} \qquad \text{because } g(\zeta) = \zeta^6 - 1 = 0 \\ &= \frac{1}{g'(\zeta)} = \frac{\mathrm{e}^{-5\pi/6}}{6}. \end{aligned}$$

Thus we get

$$\int_0^{+\infty} \frac{\mathrm{d}x}{x^6 + 1} = \frac{\mathrm{i}\pi}{3} \frac{\mathrm{e}^{-5\pi/6}}{1 - \mathrm{e}^{\mathrm{i}\pi/3}} = \frac{\pi}{3}.$$

♦ **Solution of exercise 4.17.** Define

$$f(z) = \frac{\log z}{1 + z^3},$$

where the logarithm function is defined on $\mathbb{C} \setminus \mathbb{R}^-$, for instance. Integrate on the contour going from ε to R (for $\varepsilon > 0$), followed by an arc with angle $2\pi/3$, then by a line segment toward the origin, but finished by another arc of the circle with radius ε. Show that the integrals on the circles tend to 0 when R tends to infinity and ε to 0. The only residue inside the contour is from the pole at $z = \mathrm{e}^{\mathrm{i}\pi/3}$ and is equal to $-\frac{\pi}{9}\mathrm{e}^{-\mathrm{i}\pi/6}$. Compute the integral on the second side in terms of the desired integral and in terms of

$$\int_0^{+\infty} \frac{1}{1 + x^3}\,\mathrm{d}x = \frac{2\pi}{3\sqrt{3}}$$

(which can be computed using the partial fraction expansion of $1/(1 + x^3)$, for instance).

One finds then

$$\int_0^{+\infty} \frac{\log x}{1 + x^3}\,\mathrm{d}x = -\frac{2}{27}\pi^2.$$

♦ **Solution of exercise 4.18.** Show that the function is integrable using a Taylor expansion at 1 and -1. Integrate on a contour which consists of the line segment from $-R$ to R, avoiding the poles at 1 and -1 by half-circles above the real axis with small radius ε, and close the contour by a half-circle of radius R. Show that the integral on the large circle of

$$f(z) = \frac{\mathrm{e}^{\mathrm{i}\pi z/2}}{z^2 - 1}$$

tends to 0 and compute the integrals on the small circles (which do not vanish; their sum is equal to π in the limit where $[\varepsilon \to 0]$).

Hence one finds that $\displaystyle\int_{-\infty}^{+\infty} f(x)\,\mathrm{d}x = \pi$ and the desired integral has value $-\pi/2$.

♦ **Solution of exercise 4.24.** Notice that

$$T + \mathrm{i}a^2 S = \sum_{n=1}^{\infty} \frac{1}{(n^2 - \mathrm{i}a^2)}.$$

Use then the result given by equation (4.15), page 123, and take the real and imaginary parts, respectively.

♦ **Solution of exercise 4.25.** Let

$$f(z) = \pi \frac{\operatorname{cotan}(\pi z)\coth(\pi z)}{z^7}.$$

This function has poles at n and $\mathrm{i}n$, for any $n \in \mathbb{Z}$. The residue at $n \neq 0$ is simply $\coth(\pi n)/n^7$, and the residue at $\mathrm{i}n \neq 0$ is $\operatorname{cotan}(\mathrm{i}\pi n)/(\mathrm{i}n)^7 = \coth(\pi n)/n^7$, which gives us that the integral of f on the square with center at 0 and edges parallel to the axis at distance $R = (n + \frac{1}{2})\pi$ (which does not pass through any pole) is

$$I_n = 2\mathrm{i}\pi \sum_{\substack{k=-n \\ k\neq 0}}^{k=n} \frac{2\coth \pi k}{k^7} + 2\mathrm{i}\pi \operatorname{Res}(f\,;\,0).$$

Since (by an adaptation to this situation of Jordan's second lemma) I_n tends to 0 when n tends to infinity, we deduce that

$$S = \sum_{k=1}^{n} \frac{\coth \pi k}{k^7} = -\frac{1}{4}\operatorname{Res}(f\,;\,0).$$

It suffices now to expand $\pi \coth \pi z \operatorname{cotan} \pi z$ to order 6 around 0 (by hand for the most courageous, with a computer for the others):

$$\pi \coth \pi z \operatorname{cotan} \pi z = \frac{1}{\pi z^2} - \frac{7\pi^3}{45} z^2 - \frac{19\pi^7}{14\,175} z^6 + O(z^{10})$$

which indicates that the residue of f at 0 is $-19\pi^7/14\,175$, and the formula stated follows.

Sidebar 3 (Differentiability of a function on $\mathbb{R}^2$) *Consider*

$$f: \quad \mathbb{R}^2 \longrightarrow \mathbb{R}^2, \qquad (x,y) \longmapsto f(x,y),$$

a function of two real variables with values in $\mathbb{R}^2$. Denote its coordinates by $f = (f_1, f_2)$; thus the functions f_1 and f_2 take values in $\mathbb{R}$.

*The function f is **differentiable** at a point $(x_0, y_0) \in \mathbb{R}^2$ if there exists a linear map $\Theta \in \mathscr{L}(\mathbb{R}^2)$ which satisfies*

$$f(x_0 + h, y_0 + k) = f(x_0, y_0) + \Theta(h,k) + o(h,k) \qquad [(h,k) \to (0,0)].$$

*Such a linear map Θ is called the **differential of f at the point (x_0, y_0)**, and is denoted $\mathrm{d}f_{(x_0,y_0)}$. It can be represented as a 2×2 matrix:*

$$\operatorname{mat}\left(\mathrm{d}f_{(x_0,y_0)}\right) = \begin{pmatrix} \dfrac{\partial f_x}{\partial x} & \dfrac{\partial f_x}{\partial y} \\ \dfrac{\partial f_y}{\partial x} & \dfrac{\partial f_y}{\partial y} \end{pmatrix},$$

which means that

$$\begin{aligned} \begin{pmatrix} f_x(x_0 + h, y_0 + k) \\ f_y(x_0 + h, y_0 + k) \end{pmatrix} &= \begin{pmatrix} f_x(x_0, y_0) \\ f_y(x_0, y_0) \end{pmatrix} + \begin{pmatrix} \dfrac{\partial f_x}{\partial x} & \dfrac{\partial f_x}{\partial y} \\ \dfrac{\partial f_y}{\partial x} & \dfrac{\partial f_y}{\partial y} \end{pmatrix} \cdot \begin{pmatrix} h \\ k \end{pmatrix} + o(h,k) \\ &= \begin{pmatrix} f_x(x_0, y_0) \\ f_y(x_0, y_0) \end{pmatrix} + \begin{pmatrix} \dfrac{\partial f_x}{\partial x} \cdot h + \dfrac{\partial f_x}{\partial y} \cdot k \\ \dfrac{\partial f_y}{\partial x} \cdot h + \dfrac{\partial f_y}{\partial y} \cdot k \end{pmatrix} + o(h,k). \end{aligned}$$

*The map which associates $\mathrm{d}f_{(x_0,y_0)}$ to (x_0, y_0) is called the **differential** of f and is denoted $\mathrm{d}f$.*

It is customary to write $\mathrm{d}x : (h,k) \mapsto h$ and $\mathrm{d}y : (h,k) \mapsto k$, and consequently the differential $\mathrm{d}f$ can also be written, in the complex notation $f = f_x + \mathrm{i} f_y$, as

$$\mathrm{d}f = \frac{\partial f}{\partial x}\,\mathrm{d}x + \frac{\partial f}{\partial y}\,\mathrm{d}y. \qquad (*)$$

(Notice that we began with a vector function (f_x, f_y) and ended with a scalar function $f_x + i f_y$; therefore the matrix of $\mathrm{d}f$ is a simple line 1×2 and no more a 2×2 square matrix. It is now the matrix of a linear form.)

If f is differentiable, then its partial derivatives are well-defined. The converse is not true, but still, it the partial derivatives of f are defined and continuous on an open set $\Omega \subset \mathbb{R}^2$, then f is indeed differentiable, and its differential $\mathrm{d}f$ satisfies the equation $(*)$.

Chapter 5

Complex Analysis II

5.1

Complex logarithm; multivalued functions

5.1.a The complex logarithms

We seek to extend the real logarithm function $x \longmapsto \log x$, defined on $\mathbb{R}^{+*}$, to the complex plane – or at least to a part, as large as possible, of the complex plane.

The most natural idea is to come back to one of the possible definitions of the real logarithm: it is the primitive of $x \longmapsto 1/x$ which vanishes at $x = 1$. Let us therefore try to "primitivize" the function $z \longmapsto 1/z$ and so put, for any $z \in \mathbb{C}^*$,

$$L(z) \stackrel{\text{def}}{=} \int_{\gamma(z)} \frac{1}{\zeta}\,\mathrm{d}\zeta, \tag{5.1}$$

where $\gamma(z)$ is a path in the complex plane joining 1 to z, not going through 0 (so that the integral exists). Does equation (5.1) really define a function? The answer is "yes" if any choice of the path γ leads to the same value of $L(z)$, and "no" otherwise. However, two paths γ_1 and γ_2, joining 1 to z, but going on "opposite sides" of the point 0, give different results (see Figure 5.1); if we call γ the path consisting of γ_1 followed by $-\gamma_2$, the integral of $1/\zeta$ around γ is equal to $2\pi\mathrm{i}$ times the residue at 0 (which is 1), times the number of turns of γ around 0; so, in the end, $\int_\gamma (1/\xi)\,\mathrm{d}\xi = 2\pi\mathrm{i}$.

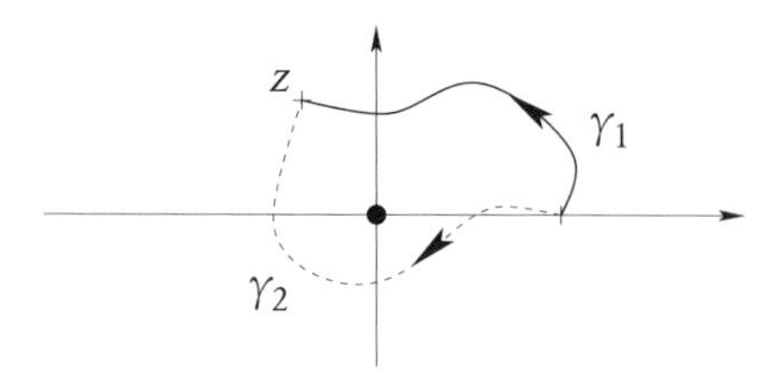

Fig. 5.1 – It is not possible to integrate the function $z \mapsto 1/z$ on the complex plane, since the result depends on the chosen path. Thus, the integrals on the solid contour γ_1 and the dashed contour γ_2 differ by $2\pi i$.

Consequently, it is not possible to define the complex logarithm on the punctured plane $\mathbb{C}^* = \mathbb{C} \setminus \{0\}$.

Still, we may restrict our ambitions a bit and only consider the plane from which a half-line has been removed, namely $\mathbb{C} \setminus \mathbb{R}^-$.

PROPOSITION 5.1 *The complex plane minus the half-line of negative real numbers* $\mathbb{C} \setminus \mathbb{R}^-$ *is simply connected.*

A consequence of this theorem is that, on this simply connected set, all closed paths are homotopic[1] to a point, and the relation (5.1) *does* define a function correctly.

DEFINITION 5.2 The function $L(z)$ defined by

$$L(z) \stackrel{\text{def}}{=} \int_{\gamma(z)} \frac{1}{\zeta} \, d\zeta, \tag{5.2}$$

where $\gamma(z)$ is a path inside $\mathbb{C} \setminus \mathbb{R}^-$ joining 1 to z, is called the **principal complex logarithm.**

Notice that one could just as well remove *another* half-line from the complex plane, for instance, the half-line of positive real numbers, or one of the half-lines of the purely imaginary axis. Even more, it would be possible to take an arbitrary simple curve $\mathscr{C}$ joining 0 to infinity; then, on the complement $\mathbb{C} \setminus \mathscr{C}$, which is simply connected, one can define a new logarithm function.

DEFINITION 5.3 A **cut** is any simple curve joining 0 to infinity in the complex plane. To each cut $\mathscr{C}$ is associated a **logarithm** function $\ell_{\mathscr{C}}$ on $\mathbb{C} \setminus \mathscr{C}$. The point 0 is called a **branch point.**

The various logarithm functions are of course related to each other by the following result:

---

[1] I.e., a closed path can be contracted continuously to a point, see page 575.

THEOREM 5.4 *Let ℓ and ℓ' be two complex logarithms, defined on $\mathbb{C} \setminus \mathscr{C}$ and $\mathbb{C} \setminus \mathscr{C}'$, respectively. Then, for any $z \in \mathbb{C} \setminus \{\mathscr{C} \cup \mathscr{C}'\}$, the difference $\ell(z) - \ell'(z)$ is an integral multiple of $2\pi\mathrm{i}$.*

This difference is constant on each connected component of $\mathbb{C} \setminus \{\mathscr{C} \cup \mathscr{C}'\}$.

In addition, the principal logarithm satisfies the following property:

PROPOSITION 5.5 *If $z \in \mathbb{C}$ is written $z = \rho\,\mathrm{e}^{\mathrm{i}\theta}$, with $\theta \in \left]-\pi, \pi\right[$ and $\rho \in \mathbb{R}^{+*}$, then $L(z) = \log \rho + \mathrm{i}\theta$.*

Example 5.6 Let $\ell(z)$ be the complex logarithm defined by the cut $\mathbb{R}^+$. The reader will easily check that, if $z = \rho\,\mathrm{e}^{\mathrm{i}\theta}$ with $\theta \in]0, 2\pi[$ this time and $\rho \in \mathbb{R}^{+*}$, then $\ell(z) = \log \rho + \mathrm{i}\theta$.

5.1.b The square root function

The logarithm is not the only function that poses problems when one tries to extend it from the (half) real line to the complex plane. The "square root"[2] function is another.

Indeed, if $z = \rho\,\mathrm{e}^{\mathrm{i}\theta}$, one may be tempted to define the square root of z by $\sqrt{z} \stackrel{\text{def}}{=} \sqrt{\rho}\,\mathrm{e}^{\mathrm{i}\theta/2}$. But a problem arises: the argument θ is only defined up to 2π, and so $\theta/2$ is only defined up to a multiple of π; thus, there is ambiguity on the sign of $\sqrt{z}$.

It is therefore necessary to specify which determination of the argument will be used to define the square root function. To this end, we introduce again a *cut* and choose a continuous determination of the argument (which always exists on a simply connected open set).

In the following, each time the complex square root function is mentioned, it will be necessary to indicate, first which is the cut, and second, which is the determination of the argument considered.

Example 5.7 If the principal determination of the argument $\theta = \operatorname{Arg} z \in \left]-\pi, \pi\right[$ is chosen, one can defined $z \longmapsto \sqrt{z} \stackrel{\text{def}}{=} \sqrt{\rho}\,\mathrm{e}^{\mathrm{i}\theta/2}$ and the restriction of "this" square root function to the real axis is the usual real square root.

By chosing $\mathrm{i}\mathbb{R}^-$ as the cut, one can define a square root function on the whole real axis.

5.1.c Multivalued functions, Riemann surfaces

There is a trick to avoid having to introduce a cut for functions like the logarithm or the square root. Physicists call these in general *multivalued functions*, and mathematicians speak of *functions defined on a Riemann surface*.

We start with this second point of view.

The main idea is to replace the punctured complex plane (i.e., without 0), on which the definition of a holomorphic logarithm or square root was

[2] The notation "$\sqrt{\ }$" to designate the square root (this symbol is a deformation of the letter "r") was introduced by the German mathematician Christoff Rudolff (1500–1545).

Georg Friedrich Bernhard RIEMANN (1826–1866), German mathematician of genius, brought extraordinary results in almost all fields: differential geometry (riemannian manifolds, riemannian geometry), complex analysis (Riemann surfaces, Riemann's mapping theorem, the Riemann zeta function in relation to number theory, the nontrivial zeros of which are famously conjectured to lie on the line of real part $\frac{1}{2}$), integration (Riemann integral), Fourier analysis (Riemann-Lebesgue theorem), differential calculus (Green-Riemann theorem), series,... He also contributed to numerous physical problems (heat and light propagation, magnetism, hydrodynamics, acoustics, etc.).

unsuccessful, by a larger domain which is simply connected.

Denote by $\mathscr{H}$ the helix-shaped surface which is constructed as follows. Take an infinite collection of planes, stacked vertically. Each of these planes is then cut, for instance along the positive real axis, and the upper part of the cut of one plane is then *glued* to the the lower part of the cut in the plane immediately above:

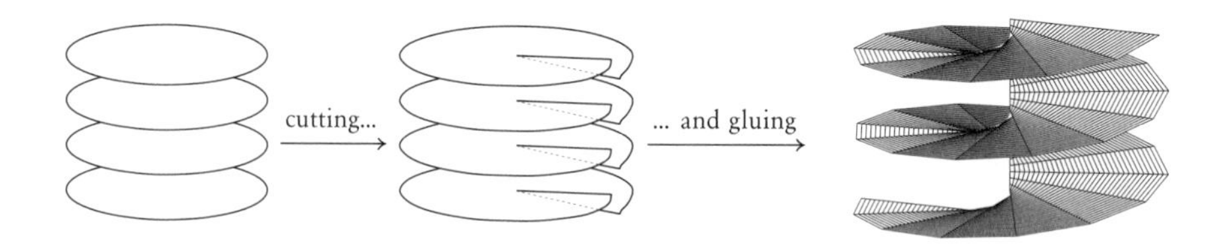

(The third figure is what is called a **helicoid** in geometry. Here, only the topology of the surface concerns us. The helicoid can be found in spiral staircases. Close to the axis of the staircase, the steps are very steep; this is why masons use an axis with nonzero radius.)

This surface is called the **Riemann surface of the logarithm function.** Each of the points on this surface can be mapped to a point on the complex plane, simply by projecting vertically. Two points of the helicoid precisely vertical to each other are therefore often *identified to the same complex number.* They are distinguished only by being on a different "floor." Note that walking on the helicoid and turning once around the axis (i.e., the origin), means that the floor changes, as in a parking garage.

PROPOSITION 5.8 *$\mathscr{H}$ is simply connected.*

It it then possible to define a logarithm function on $\mathscr{H}$ by integrating the function $\zeta \mapsto 1/\zeta$ along a path on $\mathscr{H}$ joining 1 to z ($1/\zeta$ denoting the inverse of the complex number identified with a point of the Riemann surface). It is easy to see that this path integral makes sense and, because of this proposition, it does not depend on the choice of the path from 1 to z. This gives a well-defined logarithm function $L : \mathscr{H} \rightarrow \mathbb{C}$ such that $\exp(L(z)) = z$ for all points in $\mathscr{H}$.

The point of view of a physicist is somewhat different. Instead of considering $\mathscr{H}$, she is happy with $\mathbb{C} \setminus \{0\}$, but says that the logarithm function is **multivalued**, that is, for a given $z \in \mathbb{C} \setminus \{0\}$, the function $L(z)$ takes not a single value, but rather a whole sequence of values; the one that should be chosen depends on the path taken to reach the point z. The function L is then holomorphic on any simply connected subset of $\mathbb{C} \setminus \{0\}$ (and the classical local results, such as the Taylor formulas, are applicable). It is also said that the logarithm function has many **branches** (corresponding to a choice of the floor of $\mathscr{H}$).

This viewpoint can seem quite shocking to the adepts of mathematical rigor, but it has its advantages and is particularly intuitive.

Next, let us construct the Riemann surface associated to the square-root function. It is clear that by extending the square-root function by continuity on $\mathbb{C} \setminus \{0\}$, the sign of the function changes after one turn around the origin, but comes back to its initial value after turning *twice*. Hence, we take two copies of the complex plane, and cut them before gluing again in the following manner:

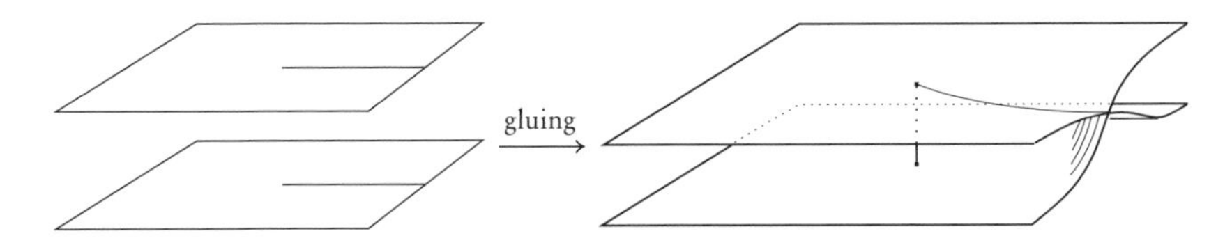

When the square-root function has a given value on one of the sheets, it has the opposite value on the equivalent point of the other sheet. However, it should be noticed that this surface is not simply connected.

5.2

Harmonic functions

5.2.a Definitions

DEFINITION 5.9 (Harmonic function) A function $u : \mathbb{R}^2 \rightarrow \mathbb{R}$, defined on an open set Ω of the plane, is **harmonic** on Ω if its **laplacian** is identically zero, that is, if it satisfies

$$\triangle u = \frac{\partial^2 u}{\partial x^2} + \frac{\partial^2 u}{\partial y^2} = 0 \qquad \text{at any point of } \Omega.$$

We will see, in this section, that a number of theorems proved for holomorphic functions remain true for harmonic functions, notably the mean value theorem, and the maximum principle. This will have important consequences in physics, where there is an abundance of harmonic functions.

First, what is the link between holomorphic functions and harmonic functions?

If $f \in \mathscr{H}(\Omega)$ is a holomorphic function, it can be expressed in the form $f = u + \mathrm{i}v$, where u and v are real-valued functions; it is then easy to show that u and v are harmonic. Indeed, f being $\mathscr{C}^\infty$, the functions u and v are also $\mathscr{C}^\infty$, so

$$\triangle u = \frac{\partial^2 u}{\partial x^2} + \frac{\partial^2 u}{\partial y^2} = \frac{\partial}{\partial x}\frac{\partial u}{\partial x} + \frac{\partial}{\partial y}\frac{\partial u}{\partial y} \overset{\text{Cauchy}}{=} \frac{\partial}{\partial x}\frac{\partial v}{\partial y} - \frac{\partial}{\partial y}\frac{\partial v}{\partial x} \overset{\text{Schwarz}}{=} 0$$

from the Cauchy-Riemann equations on the one hand, and Schwarz's theorem concerning mixed derivatives of functions of class $\mathscr{C}^2$ on the other hand. The argument for v is similar. To summarize:

THEOREM 5.10 *The real and the imaginary part of a holomorphic function are harmonic.*

Now, we introduce the notion of harmonicity for a complex-valued function. If f is $\mathscr{C}^2$ we have

$$\triangle f = \frac{\partial^2 f}{\partial x^2} + \frac{\partial^2 f}{\partial y^2} = \left(\frac{\partial}{\partial x} - \mathrm{i}\frac{\partial}{\partial y}\right)\left(\frac{\partial}{\partial x} + \mathrm{i}\frac{\partial}{\partial y}\right) f = 4\,\frac{\partial}{\partial z}\frac{\partial}{\partial \bar{z}} f = 4\,\partial\bar{\partial} f.$$

DEFINITION 5.11 A complex-valued function f defined on an open subset Ω of the plane is **harmonic** if it satisfies

$$\triangle f = 0, \qquad \text{i.e.,} \qquad \partial\bar{\partial} f = 0.$$

5.2.b Properties

PROPOSITION 5.12 *A complex-valued function f is harmonic if and only if its real and imaginary parts are harmonic.*

PROOF. If u and v are harmonic, then $u + \mathrm{i}v$ is certainly also harmonic since the operators ∂ and $\bar{\partial}$ are $\mathbb{C}$-linear.

Conversely, if f is harmonic, it suffices to take the real part of the relation $\triangle f = 0$ to obtain $\triangle u = 0$, and similarly for the imaginary part.

Hence, in particular, any holomorphic function is harmonic. What about the converse? It cannot hold in general because there need be no relation whatsoever between the real and imaginary parts of a complex-valued harmonic function, whereas there is a very subtle one for holomorphic functions (as seen from the Cauchy-Riemann equations). However, there is *some* converse for the real (or imaginary) part only:

THEOREM 5.13 *Any real-valued harmonic funcion is* locally *the real part (resp. the imaginary part) of a holomorphic function.*

What is meant by "locally" here? Using only the methods of differential equations, one can show that:

PROPOSITION 5.14 *Let $\mathcal{B}$ be an open ball and $f : \mathcal{B} \to \mathbb{R}$ be a harmonic function. There exists a harmonic function $v : \mathcal{B} \to \mathbb{R}$ such that $F = u + \mathrm{i}v$ is holomorphic on $\mathcal{B}$.*

The astute reader will undoubtedly have already guessed what prevents the extension of this theorem to any open subset Ω: if Ω has "holes," the function v may change value when following its extensions around the hole. The same phenomenon will be seen in the section about analytic continuation, and it has already been visible in Cauchy's theorem. On the other hand, if u is harmonic on a *simply connected* open subset, then u is indeed the real part of a function holomorphic on the whole of Ω.

COROLLARY 5.15 *If $u : \Omega \to \mathbb{C}$ is harmonic, then it is of class $\mathscr{C}^\infty$.*

PROOF. It is enough to show that for any $z \in \Omega$, there is some open ball $\mathcal{B}$ containing z such that u restricted to $\mathcal{B}$ is $\mathscr{C}^\infty$. But on $\mathcal{B}$ there is a holomorphic function f such that $u = \mathrm{Re}(f)$ on $\mathcal{B}$; in particular, f is $\mathscr{C}^\infty$ on $\mathcal{B}$, hence so is its real part *in the sense of $\mathbb{R}^2$-differentiability.*

The following theorems are also deduced from Proposition 5.14.

THEOREM 5.16 (Mean value theorem) *Let Ω be an open subset of $\mathbb{R}^2 \simeq \mathbb{C}$ and let u be harmonic on Ω. For all $z_0 \in \Omega$ and for all $r \in \mathbb{R}$ such that $\overline{\mathcal{B}}(z_0\,;\,r) \subset \Omega$, we have*

$$u(z_0) = \frac{1}{2\pi} \int_{-\pi}^{\pi} u(z_0 + r\,\mathrm{e}^{\mathrm{i}\theta})\,\mathrm{d}\theta,$$

or, in real notation (on $\mathbb{R}^2$),

$$u(x_0, y_0) = \frac{1}{2\pi} \int_{-\pi}^{\pi} u(x_0 + r\cos\theta, y_0 + r\sin\theta)\,\mathrm{d}\theta.$$

THEOREM 5.17 (Maximum principle) *Let Ω be a domain in $\mathbb{C}$, $u : \Omega \to \mathbb{R}$ a harmonic function and $a \in \Omega$. Then either u is constant on Ω or a is not a local maximum of u, that is, any neighborhood of a contains a point b such that $u(b) < u(a)$.*

Similarly, if u is not constant, it does not have a local minimum either.

For harmonic functions, it is possible to explain the maximum principle graphically. Indeed, if z is a local extremum (minimum or maximum), and is not constant, the function u must have the following aspect:

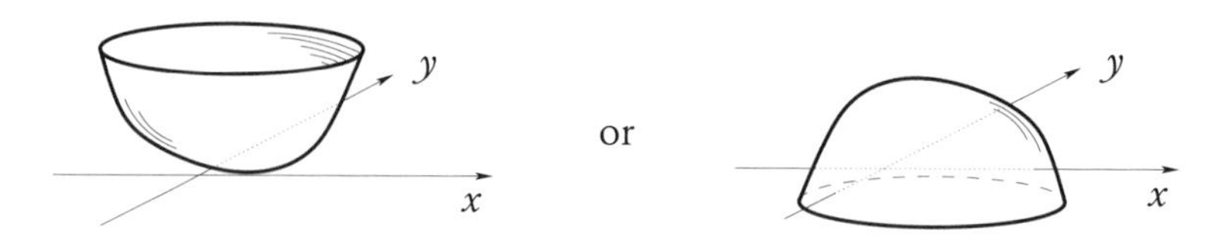

But, in the first case, it is transparent that $\partial^2 u/\partial x^2$ must be positive, *as well as* $\partial^2 u/\partial y^2$, and so the laplacian of u cannot be zero. In the second case, those same partial derivatives are negative, and the laplacian cannot be zero either.

A harmonic function must have one of the two partial derivatives negative and the other positive, which gives the graph the shape of a saddle or a mountain pass or a potato chip[3]:

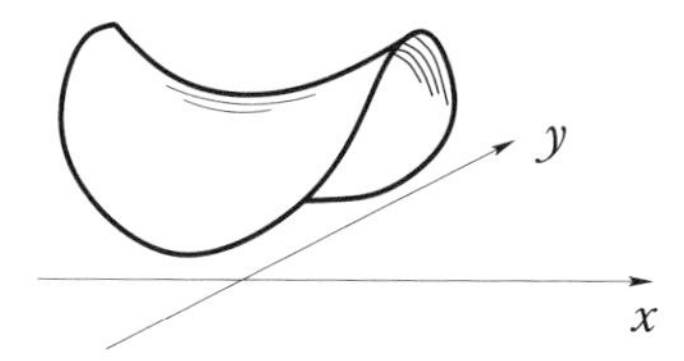

Such a surface obviously has no local extremum, as any bicyclist will tell you!

5.2.c A trick to find f knowing u

Consider a function $u : \Omega \to \mathbb{R}$ which is harmonic on a simply connected open set Ω. To find a function $v : \Omega \to \mathbb{R}$ which is also harmonic and such that $u + iv$ is holomorphic, we have to solve the system

$$\begin{cases} \dfrac{\partial u}{\partial x} = \dfrac{\partial v}{\partial y} \\ \dfrac{\partial u}{\partial y} = -\dfrac{\partial v}{\partial x}. \end{cases} \qquad \text{for any } z \in \Omega.$$

There is a classical method: integrating the second equation, we write

$$v(x,0) = -\int_0^x \frac{\partial u}{\partial y}(x',0)\,dx',$$

then

$$v(x,y) = v(x,0) + \int_0^y \frac{\partial u}{\partial x}(x,y')\,dy'$$

[3] A question for physicists: why is it that a potato chip has a hyperbolic geometry? Another question: why is it also true of the leaves of certain plants, and more spectacularly, for many varieties of iris flowers?

(integrating the first equation).

In fact, a "trick" exists to find easily the holomorphic function f with real part equal to a given harmonic function u, avoiding the approach with differential equations. Therefore let u be a harmonic function on a simply connected open set $\Omega \subset \mathbb{C}$. We are looking for a holomorphic function $f : \Omega \to \mathbb{C}$ such that $u = \text{Re}(f)$. Hence in particular

$$u(x, y) = \frac{f(z) + \overline{f(z)}}{2} \qquad \text{with } z = x + \mathrm{i}\, y$$

namely,

$$u(x, y) = \frac{f(x + \mathrm{i}\, y) + \overline{f(x + \mathrm{i}\, y)}}{2} \tag{$*$}$$

We are considering the function $u(x, y)$ as a function of two real variables. Let us extend it to a function of two complex variables $\widetilde{u}(z, z')$. For instance, if $u(x, y) = 3x + 5xy$, we define $\widetilde{u}(z, z') = 3z + 5zz'$. Now put

$$g(z) \stackrel{\text{def}}{=} 2\widetilde{u}\left(\frac{z}{2}, \frac{z}{2\mathrm{i}}\right). \tag{$**$}$$

Then, using formulas ($*$) and ($**$), we get

$$\begin{aligned} g(z) = 2\widetilde{u}\left(\frac{z}{2}, \frac{z}{2\mathrm{i}}\right) &= \frac{2}{2}\left\{f\left(\frac{z}{2} + \mathrm{i}\frac{z}{2\mathrm{i}}\right) + \overline{f}\left(\frac{z}{2} - \mathrm{i}\frac{z}{2\mathrm{i}}\right)\right\} \\ &= f(z) + \overline{f}(0), \end{aligned} \tag{5.3}$$

so that $f(z)$ is, up to a purely imaginary constant, $g(z) - g(0)/2$, and hence

$$f(z) = g(z) - u(0, 0) + \mathrm{i}\, \mathrm{C}^{\underline{\mathrm{nt}}}.$$

To establish formula (5.3), we used the fact that the function $z \mapsto \overline{f(z)}$ can also be expressed as a certain function $\widetilde{f}$ of the variable $\bar{z}$, and more precisely as $\overline{f}(\bar{z})$. For instance, if $f(z) = 3z + \mathrm{i}z^2$, then $\overline{f(z)} = 3\bar{z} - \mathrm{i}\bar{z}^2 = \overline{f}(\bar{z})$.

Now, isn't all this a little bit fishy? We have considered in the course of this reasoning, that the functions (u, f,...) behaved as "operators" performing certain operations on their variables (raise it to a power, multiply by a constant, take the cosine...), and that they could just as well perform them on a complex variable if they could do so on a real one, or on $\bar{z}$ if they knew how to operate on z. This is not at all the usual way to see a function! However, the expansions of f and u in power series give a rigorous justification of all those manipulations. They also give the limit of the method: it is necessary that the functions have power series expansions valid on the whole set Ω considered. Thus, this technique works if Ω is an open ball (or a connected and simply connected open subset, by analytic continuation), but not on a domain which is not simply connected, an annulus, for example.[4]

[4] I wish to thank Michael Kiessling for teaching me this little-known trick.

Example 5.18 Let $u(x, y) = 3x^2 - 3y^2 + 5xy + 2$, which satisfies $\triangle u = 0$. Then we have $\widetilde{u}(z, \bar{z}) = 3z^2 - 3\bar{z}^2 + 5z\bar{z} + 2$ and $2\widetilde{u}(z/2, z/2\mathrm{i}) = 3z^2 + 5z^2/2\mathrm{i} + 4 = f(z) + \overline{f}(0)$, which gives in particular that $2\mathrm{Re}\, f(0) = 4$, hence $f(0) = 2 + \mathrm{i} \times \mathrm{C}^{\mathrm{nt}}$. Hence we have $f(z) = 3z^2 + 5z^2/2\mathrm{i} + 2$ (up to a purely imaginary constant) and one checks that the real part of f is indeed equal to u.

Example 5.19 Consider the function $u(x, y) = \mathrm{e}^{-x} \sin y$. It is easy to see that this function is harmonic. Then we have $g(z) = 2\mathrm{e}^{-z/2} \sin(z/2\mathrm{i}) = -2\mathrm{i}\mathrm{e}^{-z/2} \sinh(z/2)$. Since $g(0) = 0$, we have $f(z) = g(z) + \mathrm{i}\, \mathrm{C}^{\mathrm{nt}}$, hence

$$\mathrm{e}^{-x} \sin y = \mathrm{Re}\left[2\mathrm{i}\mathrm{e}^{-z/2} \sinh\left(\frac{z}{2}\right)\right].$$

5.3 Analytic continuation

Assume a function f is known on an open subset Ω by a power series expansion that converges on Ω. We would like to know it is possible to *extend* the definition of f to a larger set.

Take a convenient example: we define the function f on the open ball $\Omega = \mathcal{B}(0; 1)$ by

$$f(x) = \sum_{n=0}^{\infty} z^n. \tag{5.4}$$

We know that this power series is indeed convergent on Ω, but divergent for $|z| > 1$. However, if we start from the point $z_0 = \frac{1}{2} + \frac{2\mathrm{i}}{3}$, for instance, we can find a new power series which converges to f on a certain neighborhood Ω' of z_0:

$$f(z) = \sum_{n=0}^{\infty} a_n (z - z_0)^n \qquad \text{on a neighborhood of } z_0. \tag{5.5}$$

But – surprise! – this neighborhood may well extend beyond Ω, as is shown in Figure 5.2 on the facing page.

We can then choose a path γ and try to find, for any point z of γ, a disk of convergence of a power series centered at z coinciding with f on Ω. This is what is called an **analytic continuation along a path** (see Figure 5.3).

This means that it has been possible to *continue* the function f to certain points of the complex plane where the original series (5.4) was not convergent. On the intersection $\Omega' \cap \Omega$, the two series (5.4) and (5.5) coincide, because of the "rigidity" of holomorphic functions.

Is this rigidity sufficient to ensure that this analytic continuation is unique?

Well, yes and no.

If we try to continue a function analytically by going around a singularity, the continuation thus obtained may depend on which side of the singularity

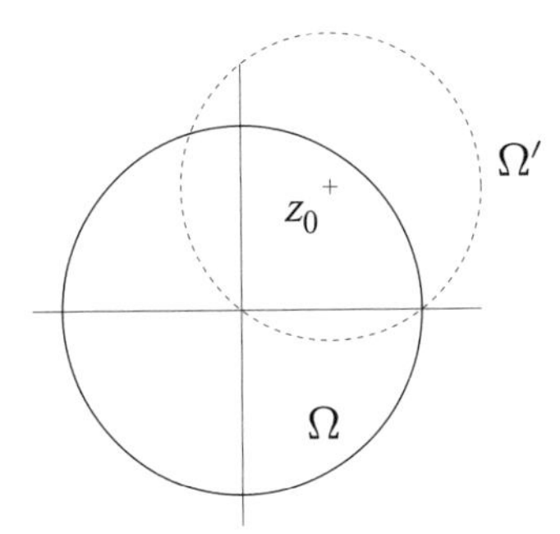

Fig. 5.2 – Analytic continuation of f on an open set Ω' extending beyond Ω.

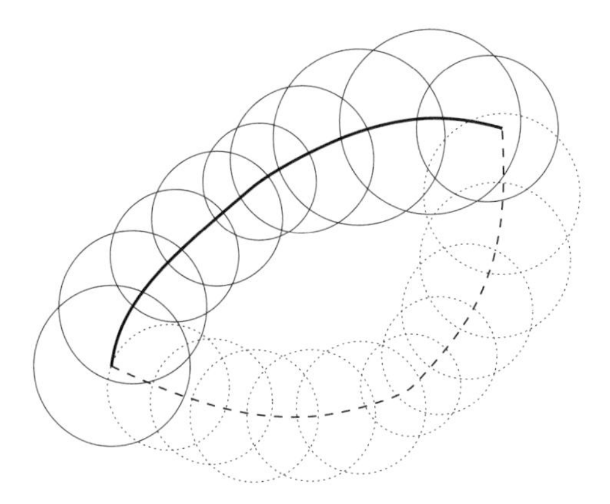

Fig. 5.3 – Continuation along a path γ with successive disks, and second continuation along a path γ'.

the chosen path takes. For instance, in the case of the function f defined by $f(z) = \sum_{n=1}^{\infty}(-1)^{n+1} z^n/n$, which is equal to $z \mapsto \log(1+z)$ on $\Omega = \mathcal{B}(0\,;\,1)$, there exists a singularity at the point $z = -1$. Two continuations, defined by paths running above or under the point $z = 1$, will differ by $2\pi i$.

On the other hand, the following uniqueness result holds:

THEOREM 5.20 *Let f be a function holomorphic on a domain Ω_0. Let $\Omega_1, \dots, \Omega_n$ be domains in $\mathbb{C}$ such that $\bigcup_{i=1}^{n} \Omega_i$ is connected and* simply connected *and contains Ω_0. If there exists an analytic continuation of f to $\bigcup_{i=1}^{n} \Omega_i$, then this analytic continuation is unique.*

5.4

Singularities at infinity

It is possible to compactify the complex plane by adding a single point at infinity.

DEFINITION 5.21 (Riemann sphere) We denote by $\overline{\mathbb{C}} \stackrel{\text{def}}{=} \mathbb{C} \cup \{\infty\}$ the set consisting of the complex plane together with an additional point "infinity."

The point at infinity corresponds to the limit "$|z| \to +\infty$" in $\mathbb{C}$. Since we add a single point, independent of the argument of z, this means that all the "directions" to infinity are identified. Then we are led to represent this extended complex plane in the form of a sphere,[5] called the **Riemann sphere**. This can also be obtained, for example, by stereographic projection (see Figure 5.4 on the next page).

DEFINITION 5.22 (Residue at infinity) Let f be a function meromorphic on the complex plane. Define a function F by $F(z) = f(1/z)$. The function f is said to have a **pole at infinity** if F has a pole at 0, and f is said to have an **essential singularity at infinity** if F has an essential singularity at 0.

The **residue at infinity** of f, denoted $\operatorname{Res}(f;\infty)$, is the quantity

$$\operatorname{Res}(f\,;\,\infty) = \operatorname{Res}\left(-\frac{1}{z^2}F(z)\,;\,0\right).$$

[5] One can understand why "a plane with all directions to infinity identified is a sphere" by taking a round tablecloth and grasping in one hand the whole circumference (the points of which are "infinitely far" from the center for a physicist). One obtains a pouch, which is akin to a sphere.

The open sets in $\overline{\mathbb{C}}$ are the usual open sets in $\mathbb{C}$ and the complements of compact subsets in $\mathbb{C}$. The topology obtained is the usual topology for the sphere, in which the point ∞ is like any other point.

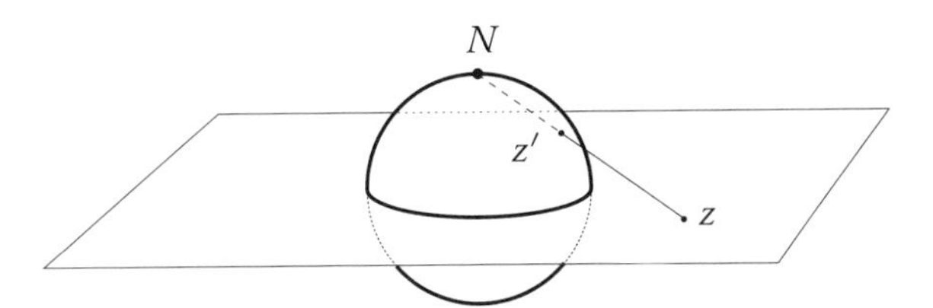

Fig. 5.4 – The Riemann sphere, obtained by stereographic projection. If z is a point in the complex plane $\mathbb{C}$, the line (Nz) meets the sphere centered at 0 with radius 1 in a single point z', the image of z on the Riemann sphere. The equator of the sphere is the image of the circle centered at the origin with radius 1. The point N corresponds to the point at infinity (and to "going to infinity" in all directions).

The entire functions which have a pole at infinity are (only) the polynomials.

Remark 5.23 The definition of the residue at infinity is justified by the fact that it is not really the function f which matters for the residue but rather the **differential form** $f(z)\,\mathrm{d}z$ (which is the natural object occuring in the theory of path integration; see Chapter 17). By putting $w = 1/z$, it follows that

$$f(z)\,\mathrm{d}z = -\frac{1}{w^2} f\left(\frac{1}{w}\right)\mathrm{d}w = -\frac{1}{w^2} F(w)\,\mathrm{d}w.$$

The residue theorem can then be extended to the Riemann sphere:

THEOREM 5.24 (Generalized residue theorem) *Let Ω be an open subset of $\overline{\mathbb{C}}$ (the topology of which is that of a sphere in space by stereographic projection) and K a closed subset with oriented boundary $\gamma = \partial\Omega$ which is made of one or more closed paths in $\mathbb{C}$ (hence* not *passing through the point at infinity). Let f be a function holomorphic on Ω, except possibly at finitely many points. Denote by z_k the singularities of f inside K; then*

$$\int_\gamma f(z)\,\mathrm{d}z = 2\pi\mathrm{i}\sum_k \operatorname{Res}(f\,;\,z_k),$$

where the sum ranges over the singularities, including possibly the point at infinity.

Proof. It suffices to treat the case where the point at infinity is in the interior of K. Assume that all the finite singularities (in $\mathbb{C}$) are contained in the ball $\mathcal{B}(0;R)$. The set $\{|z| > R\}$ is then a neighborhood of ∞, which we may assume is contained in K (increasing R if need be). We then define $K' = K \setminus \{z \in \mathbb{C}\,;\,|z| > R\}$, which has a boundary given by

$$\partial K' = \gamma \cup \mathscr{C}(0;R)$$

(with positive orientation). Applying the residue theorem to K', we find

$$\int_\gamma f(z)\,\mathrm{d}z + \int_{|z|=R} f(z)\,\mathrm{d}z = 2\pi\mathrm{i}\sum_{\substack{\text{finite}\\ \text{singularities}}} \operatorname{Res}(f\,;\,z_k).$$

Thus we obtain

$$\int_{|z|=R} f(z)\,\mathrm{d}z = \int_{|w|=1/R} \frac{1}{w^2}\, f\Big(\frac{1}{w}\Big)\,\mathrm{d}w = -2\pi\mathrm{i}\,\mathrm{Res}\,(f\,;\,\infty).$$

(Note that the change of variable $w = 1/z$ transforms a positively oriented contour to a negatively oriented one.)

The stated formula follows.

Example 5.25 The function $f : z \longmapsto 1/z$ has a pole at infinity, with residue equal to -1. If we let $\gamma = \mathscr{C}(5\,;1)$, a contour which does not contain any singularity in $\mathbb{C}$, we have

$$\int_\gamma f(z)\,\mathrm{d}z = 0.$$

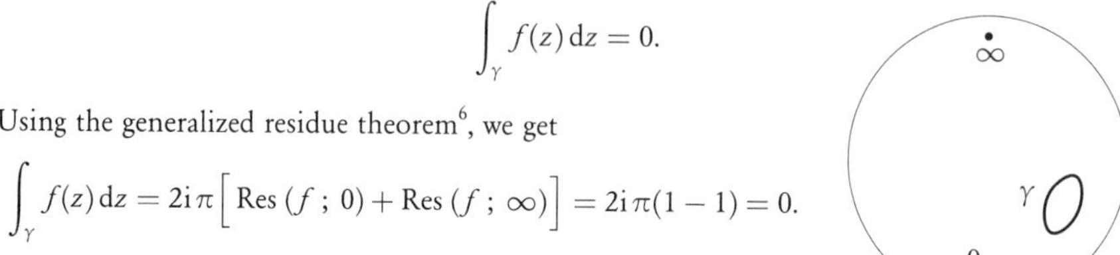

Using the generalized residue theorem[6], we get

$$\int_\gamma f(z)\,\mathrm{d}z = 2\mathrm{i}\pi\Big[\mathrm{Res}\,(f\,;\,0) + \mathrm{Res}\,(f\,;\,\infty)\Big] = 2\mathrm{i}\pi(1-1) = 0.$$

5.5 The saddle point method

The saddle point method is a computational "recipe" developed by Debye[7] to compute integrals of the type

$$I(\alpha) \stackrel{\text{def}}{=} \int_\gamma \mathrm{e}^{-\alpha f(z)}\,\mathrm{d}z.$$

Here, z is a real or complex variable.[8] The goal is to find an approximation of $I(\alpha)$ in the limit when α goes to $+\infty$, assuming that the function f admits a local minimum.

---

[6] With respect to the closed set K which is the "outside" of γ on the picture. In fact, the notions of the parts "inside" and "outside" a curve do not make sense on the Riemann sphere. Which side is the "outside" of the equator on Earth? The northern hemisphere or the southern hemisphere?

[7] Petrus Josephus Wilhelmus Debye (1884–1966), a Dutch physicist, born in Maastricht, studied the specific heat of solids (Debye model), ionic solutions (Debye potential, see problem 5 p. 325, Debye-Hückel method), among other things. He received the Nobel prize in chemistry in 1936.

[8] There exist generalizations of this method for functions of vector variables, even for fields, which are used in quantum field theory.

5.5.a The general saddle point method

Consider, more generally, an integral of the type

$$I_\alpha \stackrel{\text{def}}{=} \int_\gamma e^{-f_\alpha(z)}\,dz, \tag{5.6}$$

where α is a *complex* parameter and γ is a path in $\mathbb{C}$. We assume that f_α is a function analytic on an open subset containing γ, and we want to evaluate this integral in the limit where $|\alpha| \to \infty$.

The first idea is to evaluate the integral in the parts of the domain of integration where the *real part* of f_α is minimal, since it is in those regions that the modulus of $\exp^{-f_\alpha}$ is largest and so they should bring the main contribution to the integral. Unfortunately, in those regions, the *imaginary part* of f may very well vary and, therefore, e^{-f_α} will oscillate, possibly very rapidly. These oscillations may well destroy a large part of the integral by compensations, and the suggested method will not be justified.

The idea of Debye is then to *deform* the contour γ into an invented path γ_α satisfying the following properties:

i) along γ_α, $\operatorname{Im}\big[f_\alpha(z)\big]$ is constant;

ii) there exists a point z_α on γ_α such that

$$\left.\frac{df_\alpha(z)}{dz}\right|_{z_\alpha} = 0;$$

iii) the real part of f_α has a local minimum at $z = z_\alpha$;

iv) moreover, this minimum becomes "steeper and steeper";

v) finally, the initial path γ can be deformed continuously into γ_α (i.e., during the deformation, it must not pass through any of the singularities of f_α).

DEFINITION 5.26 A point of $\mathbb{C}$ satisfying Conditions ii) and iii) above is called a **saddle point** for f_α.

The condition $\operatorname{Im}\big[f_\alpha(z)\big] = C^{\underline{\text{nt}}}$ gives, in general, the equation for the curve γ_α. Under this condition, the phase of the exponential is constant, and the computation reduces to a real integral.

Consider now the behavior of the function f_α around the saddle point, if it exists. Suppose therefore that

$$\left.\frac{df_\alpha}{dz}\right|_{z=z_\alpha} = 0 \qquad \text{and} \qquad \left.\frac{d^2 f_\alpha}{dz^2}\right|_{z=z_\alpha} \neq 0.$$

Then we can write

$$f_\alpha(z) = f_\alpha(z_\alpha) + \frac{1}{2}(z - z_\alpha)^2 \left.\frac{\mathrm{d}^2 f_\alpha}{\mathrm{d}z^2}\right|_{z=z_\alpha} + \mathrm{o}(z - z_\alpha)^2. \tag{5.7}$$

Denoting

$$\left.\frac{\mathrm{d}^2 f_\alpha}{\mathrm{d}z^2}\right|_{z=z_\alpha} = \rho\,\mathrm{e}^{\mathrm{i}\theta} \qquad \text{and} \qquad z - z_\alpha = r\,\mathrm{e}^{\mathrm{i}\varphi},$$

we have, as a first approximation,

$$\mathrm{Re}\{f_\alpha(z) - f_\alpha(z_\alpha)\} = r^2 \rho \cos(2\varphi + \theta) + \mathrm{o}(r^2),$$
$$\mathrm{Im}\{f_\alpha(z) - f_\alpha(z_\alpha)\} = r^2 \rho \sin(2\varphi + \theta) + \mathrm{o}(r^2).$$

Hence we see that, at the saddle point z_α, the angle φ between the path γ_α and the real axis should be such that $\sin(2\varphi + \theta) = 0$, which leaves two possible directions, orthogonal to each other (solid lines in Figure 5.5). The two directions characterized by $\cos(2\varphi + \theta) = 0$ are those where the real part of $\{f_\alpha(z) - f_\alpha(z_\alpha)\}$ is constant. Between those lines, the real part is either positive or negative, depending on the sign of $\cos(2\varphi + \theta)$, which shows that we are indeed dealing with a *saddle point*, or a "mountain pass" (see Figure 5.5).

The two orthogonal directions where the imaginary part of $f_\alpha(z) - f_\alpha(z_\alpha)$ vanishes (to second order) are precisely the directions where the real part of $f_\alpha(z) - f_\alpha(z_\alpha)$ varies most quickly, either increasing or decreasing.

Consequently, we must choose, at the saddle point, the direction characterized by $\sin(2\varphi + \theta) = 0$ and such that, along this direction, $\mathrm{Re}(f_\alpha)$ has a local minimum at z_α, and not a local maximum.

We have seen that, because of the properties of holomorphic functions, it is usually possible to find a path along which the oscillations of e^{-f_α} have disappeared and where, moreover, f_α will have a local minimum as pronounced as possible. This justifies the other common name of the method: the **method of steepest descent**.

There only remains to evaluate the integral along the path γ_α. We introduce a new variable t parameterizing the tangent line to γ_α at z_α, so that we can write

$$f_\alpha(z) = f_\alpha(z_\alpha) + \frac{t^2}{2} + \mathrm{o}(z - z_\alpha)^2,$$

and, from (5.7), we have

$$t = (z - z_\alpha)\sqrt{\frac{\mathrm{d}^2 f_\alpha}{\mathrm{d}z^2}}$$

(the sign of the square root being chosen in such a way that its real part is positive). We now rewrite formula (5.6) in terms of this parameter t, which gives, noticing that $\mathrm{d}z = \big(\mathrm{d}z(t)/\mathrm{d}t\big)\,\mathrm{d}t$,

$$I_\alpha \approx \mathrm{e}^{-f_\alpha(z_\alpha)} \int_{-\infty}^{+\infty} \mathrm{e}^{-t^2} \frac{\mathrm{d}z(t)}{\mathrm{d}t}\,\mathrm{d}t.$$

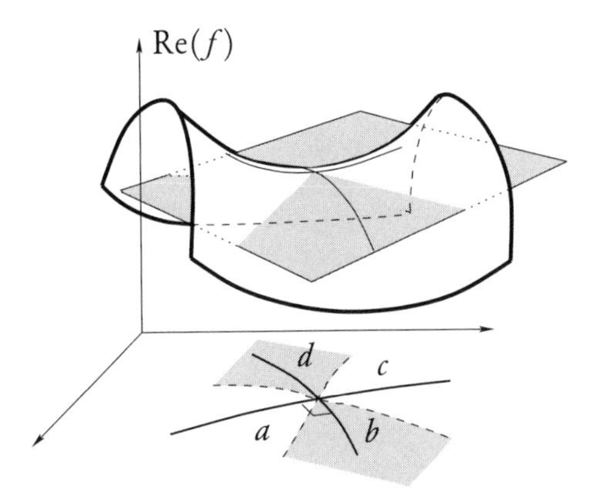

Fig. 5.5 – The two solid curves are perpendicular and locally determined by the condition $\sin(2\varphi + \theta) = 0$. These are not only the curves such that $\mathrm{Im}(f_\alpha)$ is constant, but also those where $\mathrm{Re}(f_\alpha)$ varies most rapidly. The two dotted lines separate the regions a, b, c and d. In regions a and c, $\mathrm{Re} f_\alpha(z) > \mathrm{Re} f_\alpha(z_\alpha)$, whereas in regions b and d, $\mathrm{Re} f_\alpha(z) < \mathrm{Re} f_\alpha(z_\alpha)$. The chosen path therefore goes through regions a and c.

Since the function $\mathrm{d}f_\alpha/\mathrm{d}z$ varies slowly in general, compared to the gaussian term, we can consider that $\mathrm{d}z/\mathrm{d}t = (\mathrm{d}^2 f_\alpha/\mathrm{d}z^2)^{-1/2}$ is constant, and evaluate the remaining gaussian integral, which gives

$$\boxed{I_\alpha \approx \mathrm{e}^{-f_\alpha(z_\alpha)} \left\{ \frac{2\pi}{\mathrm{d}^2 f_\alpha/\mathrm{d}z^2\big|_{z=z_\alpha}} \right\}^{1/2}.} \tag{5.8}$$

This formula can be rigorously justified, case by case, by computing the non-gaussian contributions to the integral and comparing them with the main term we just computed. The computation must be done by expanding f_α to higher order. One is then led to compute gaussian moments (which is easy) to derive an asymptotic expansion of I_α to higher order. A condition to ensure that the higher orders do not interfere is that the ratio between

$$\left\{ \frac{\mathrm{d}^p f_\alpha}{\mathrm{d}x^p}\bigg|_{x=x_\alpha} \right\}^{1/p} \qquad \text{and} \qquad \left\{ \frac{\mathrm{d}^2 f_\alpha}{\mathrm{d}x^p}\bigg|_{x=x_\alpha} \right\}^{1/2}$$

tends to 0 when α tends to infinity. This condition is necessary, but not sufficient.

5.5.b The real saddle point method

There exists a particularly simple case of the problem when f is a *real-valued* function. Then it is not possible to work in the complex plane, but a gaussian integral computation leads to a similar result.

Consider an integral of the shape

$$I_\alpha = \int_{\mathbb{R}} \mathrm{e}^{-f_\alpha(x)}\,\mathrm{d}x,$$

where f_α is a function at least twice continuously differentiable, depending on a parameter α, and with the following property:

$$\begin{cases} \text{for every } \alpha,\ f_\alpha \text{ has a unique } \textit{minimum}^{9} \text{ at a point } x_\alpha, \text{ denoted } m_\alpha = f_\alpha(x_\alpha); \text{ moreover, this minimum is more and} \\ \text{more "narrow"; i.e., if we denote } M_\alpha \stackrel{\text{def}}{=} \mathrm{d}^2 f_\alpha/\mathrm{d}x^2\big|_{x=x_\alpha}, \text{ we} \\ \text{have } M_\alpha > 0 \text{ and } M_\alpha \longrightarrow +\infty \text{ as } \alpha \to \infty. \end{cases} \tag{5.9}$$

The dominating contribution to the integral is evidently given by the region of x close to the local minimum. So the idea is to expand the function f_α around this minimum. If it suffices to expand to order 2 (the first derivative is zero because the point is a local minimum), then the computation reduces to a gaussian integral. This approximation should be all the more justified as we take the limit $[\alpha \to \infty]$.

We thus write

$$f_\alpha(x) = m_\alpha + \frac{1}{2}M_\alpha(x - x_\alpha)^2 + \mathrm{o}(x - x_\alpha)^2,$$

and as a first approximation estimate the integral by

$$I_\alpha \approx \mathrm{e}^{-m_\alpha} \int \mathrm{e}^{-\frac{1}{2}M_\alpha(x-x_\alpha)^2}\,\mathrm{d}x.$$

We recognize here a gaussian integral, so the result is

$$\boxed{I_\alpha \approx \mathrm{e}^{-m_\alpha}\sqrt{\frac{2\pi}{M_\alpha}} = \mathrm{e}^{-m_\alpha}\sqrt{\frac{2\pi}{\mathrm{d}^2 f_\alpha/\mathrm{d}x^2\big|_{x=x_\alpha}}}.}$$

The first factor is simply the value of the exponential function at its maximum, and the second factor expresses the "characteristic width" of this maximum, proportional to $1/\sqrt{M_\alpha}$. This result should be compared to the result of (5.8).

---

[9] If f_α has several local minima, each one contributes to the integral.

Remark 5.27 Sometimes, an integral of the same type as before arises, but with an imaginary exponential, that is, of the type

$$I_\alpha = \int_{\mathbb{R}} e^{i f_\alpha(x)}\,dx,$$

where f_α is a *real-valued* function.

In this case, if $f_\alpha(x_0)$ varies fast enough as α tends to infinity for a given x_0, we expect that the neighborhood of this point will only have a small contribution to the integral, because of destructive interferences. This phenomenon can be understood "by hand" by the following rough reasoning: if we expand around x_0

$$f_\alpha(x) = f_\alpha(x_0) + (x - x_0)\left.\frac{df_\alpha}{dx}\right|_{x=x_0} + o(x - x_0),$$

and if $df_\alpha/dx \neq 0$, we have, as a first approximation, an integral of the type

$$M \int e^{ik(x-x_0)}\,dx, \qquad \text{with} \quad k \stackrel{\text{def}}{=} \left.\frac{df_\alpha}{dx}\right|_{x=x_0},$$

which is zero if $k \neq 0$.

We must therefore look for extremums of f_α to find important contributions, which are called for obvious reasons **stationary phase points**; assuming there is a unique such point x_α, a computation very similar to the previous one for the real saddle point method gives the result

$$I_\alpha \approx e^{i f_\alpha(x_\alpha)} \left[\frac{2\pi i}{d^2 f_\alpha/dx^2|_{x=x_\alpha}}\right]^{1/2}.$$

This case is also called the **stationary phase method**.

Remark 5.28 The saddle point method is not just a trick for physicists. It is very often used in pure mathematics, analytic number theory in particular; for instance, it occurs in the proof of the irrationality of one of the numbers $\zeta(5), \zeta(7), \ldots, \zeta(21)$ by Tanguy Rivoal [73]! (Recall that

$$\zeta(s) = \sum_{n \geqslant 1} \frac{1}{n^s}$$

for $\mathrm{Re}(s) > 1$).

EXERCISES

♦ **Exercise 5.1** Define

$$u(x, y) = e^x\,(x \cos y + \mu y \sin y).$$

Find μ so that u is the *real* part of a holomorphic function f (identifying the plane $\mathbb{R}^2$ and the complex plane). What is the function f?

♦ **Exercise 5.2** Similarly, find all the holomorphic functions with real part equal to

$$P(x, y) = \frac{x(1+x) + y^2}{(1+x)^2 + y^2}.$$

♦ **Exercise 5.3** Show that, in two dimensions, a cavity in a conductor, devoid of charge, has potential identically zero.

♦ **Exercise 5.4 (Stirling formula)** Using the Euler Γ function defined on the complex half-plane $\text{Re}(z) > 0$ by the integral representation

$$\Gamma(z) \stackrel{\text{def}}{=} \int_0^{+\infty} e^{-x} x^{z-1} \, dx,$$

and using the fact that $\Gamma(n+1) = n!$ for all $n \in \mathbb{N}$, show, by the complex saddle point method, that

$$\Gamma(n+1) = n! \sim n^n e^{-n} \sqrt{2\pi n}.$$

This formula, established by Abraham DE MOIVRE, is called the **Stirling formula**.

SOLUTIONS

♦ **Solution of exercise 5.3.** Denote by Ω the domain defined by the cavity; Ω is simply connected. The electrosatic potential satisfies $\varphi(z) = 0$ at any point $z \in \partial\Omega$. This potential is also a harmonic function, hence achieves its minimum *and* its maximum on the boundary $\partial\Omega$. Consequently, we have $\varphi(z) = 0$ for any $z \in \Omega$.

♦ **Solution of exercise 5.4.** The first thing to do is to find a suitable contour. To start with, write

$$\Gamma(\alpha+1) = \int_0^{+\infty} e^{-x-\alpha \log x} \, dx,$$

then look for the points where $df_\alpha/dz = 0$, with $f_\alpha(z) = z - \alpha \log z$. We find $z_\alpha = \alpha$, and $d^2 f_\alpha/dz^2|_{z_\alpha} = 1/\alpha$, which is a positive real number. The two lines for which, locally, the imaginary part of f_α is constant are the real axis and the line with equation $z = \alpha + iy$. It is the first which corresponds to a local minimum of f at $z_\alpha = \alpha$; therefore the real axis is the most suitable contour.

Hence we evaluate the integral according to the formula (5.8), which gives

$$\Gamma(\alpha+1) \approx e^{-\alpha-\alpha\log\alpha} \sqrt{2\pi\alpha} \qquad [\alpha \to \infty].$$

One can check explicitly that the non-gaussian contributions (those given by the terms of order larger than 2 in the expansion of f_α) are inverse powers of α. Hence

$$\Gamma(\alpha+1) = \alpha^\alpha e^{-\alpha} \sqrt{2\pi\alpha}\bigl(1 + O(1/\alpha)\bigr),$$

and in particular, for integral values of α,

$$\Gamma(n+1) = n! \sim n^n e^{-n} \sqrt{2\pi n}.$$

Notice that this approximation is very quickly quite precise; for $n = 5$, the exact value is $5! = 120$ and the Stirling formula gives approximately 118, which is an error of about 2%.

Even more impressive, the relative error from applying the Stirling formula for $n = 100$ is about 0,08%... (note that this error is nevertheless very large in absolute value: it is roughly of size 10^{155}, whereas 100! is about 10^{158}).

Remark 5.29 If one seeks the asymptotic behavior of $\Gamma(z)$ for large z not necessarily real, the computation is more or less the same, except that the path γ is not the real axis any more. However, the result of the computation is identical (the result is valid uniformly in any region where the argument $\theta \in]-\pi, \pi]$ of z satisfies $-\pi + \varepsilon < \theta < \pi - \varepsilon$ for some $\varepsilon > 0$).

Chapter 6

Conformal maps

6.1
Conformal maps

6.1.a Preliminaries

Consider a change of variable $(x, y) \mapsto (u, v) = \big(u(x, y), v(x, y)\big)$ in the plane $\mathbb{R}^2$, identified with $\mathbb{C}$. This change of variable really only deserves the name if f is locally bijective (i.e., one-to-one); this is the case if the jacobian of the map is nonzero (then so is the jacobian of the inverse map):

$$\left|\frac{D(u,v)}{D(x,y)}\right| = \begin{vmatrix} \dfrac{\partial u}{\partial x} & \dfrac{\partial u}{\partial y} \\ \dfrac{\partial v}{\partial x} & \dfrac{\partial v}{\partial y} \end{vmatrix} \neq 0 \qquad \text{and} \qquad \left|\frac{D(x,y)}{D(u,v)}\right| = \begin{vmatrix} \dfrac{\partial x}{\partial u} & \dfrac{\partial x}{\partial v} \\ \dfrac{\partial y}{\partial u} & \dfrac{\partial y}{\partial v} \end{vmatrix} \neq 0.$$

THEOREM 6.1 *In a complex change of variable*

$$z = x + \mathrm{i}y \longmapsto w = f(z) = u + \mathrm{i}v,$$

and if f is holomorphic, *then the jacobian of the map is equal to*

$$J_f(z) = \left|\frac{D(u,v)}{D(x,y)}\right| = \left|f'(z)\right|^2.$$

PROOF. Indeed, we have $f'(z) = \dfrac{\partial u}{\partial x} + i\dfrac{\partial v}{\partial x}$ and hence, by the Cauchy-Riemann relations,

$$|f'(z)|^2 = \left(\frac{\partial u}{\partial x}\right)^2 + \left(\frac{\partial v}{\partial x}\right)^2 = \frac{\partial u}{\partial x}\frac{\partial v}{\partial y} - \frac{\partial v}{\partial x}\frac{\partial u}{\partial y} = J_f(z).$$

DEFINITION 6.2 A **conformal map** or **conformal transformation** of an open subset $\Omega \subset \mathbb{R}^2$ into another open subset $\Omega' \subset \mathbb{R}^2$ is any map $f : \Omega \to \Omega'$, locally bijective, that preserves angles and orientation.

THEOREM 6.3 *Any conformal map is given by a holomorphic function f such that the derivative of f does not vanish.*

This justifies the next definition:

DEFINITION 6.4 A **conformal transformation** or **conformal map** of an open subset $\Omega \subset \mathbb{C}$ into another open subset $\Omega' \subset \mathbb{C}$ is any holomorphic function $f : \Omega \to \Omega'$ such that $f'(z) \neq 0$ for all $z \in \Omega$.

PROOF THAT THE DEFINITIONS ARE EQUIVALENT. We will denote in general $w = f(z)$. Consider, in the complex plane, two line segments γ_1 and γ_2 contained inside the set Ω where f is defined, and intersecting at a point z_0 in Ω. Denote by γ_1' and γ_2' their images by f.

We want to show that if the angle between γ_1 and γ_2 is equal to θ, then the same holds for their images, which means that the angle between the tangent lines to γ_1' and γ_2' at $w_0 = f(z_0)$ is also equal to θ.

Consider a point $z \in \gamma_1$ close to z_0. Its image $w = f(z)$ satisfies

$$\lim_{z\to z_0} \frac{w - w_0}{z - z_0} = f'(z_0),$$

and hence
$$\lim_{z\to z_0} \operatorname{Arg}(w - w_0) - \operatorname{Arg}(z - z_0) = \operatorname{Arg} f'(z_0),$$

which shows that the angle between the curve γ_1' and the real axis is equal to the angle between the original segment γ_1 and the real axis, plus the angle $\alpha = \operatorname{Arg} f'(z_0)$ (which is well defined because $f'(z) \neq 0$).

Similarly, the angle between the image curve γ_2' and the real axis is equal to that between the segment γ_2 and the real axis, plus the same α.

Therefore, the angle between the two image curves is the same as that between the two line segments, namely, θ.

Another way to see this is as follows: the tangent vectors of the curves are transformed according to the rule $\boldsymbol{v}' = \mathrm{d}f_{z_0}\,\boldsymbol{v}$. But the differential of f (when f is seen as a map from $\mathbb{R}^2$ to $\mathbb{R}^2$) is of the form

$$\mathrm{d}f_{z_0} = \begin{pmatrix} \dfrac{\partial P}{\partial x} & \dfrac{\partial P}{\partial y} \\[2ex] \dfrac{\partial Q}{\partial x} & \dfrac{\partial Q}{\partial y} \end{pmatrix} = |f'(z_0)| \begin{pmatrix} \cos\alpha & -\sin\alpha \\ \sin\alpha & \cos\alpha \end{pmatrix}, \tag{6.1}$$

where α is the argument of $f'(z_0)$. This is the matrix of a rotation composed with a homothety, that is, a similitude.

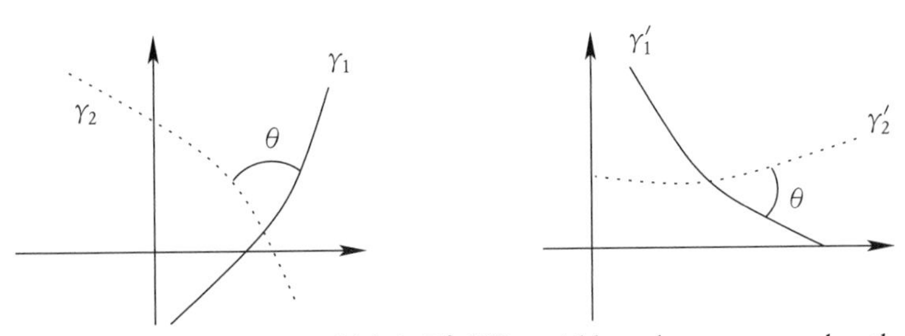

Conversely, if f is a map which is $\mathbb{R}^2$-differentiable and preserves angles, then at any point $\mathrm{d}f$ is an endomorphism of $\mathbb{R}^2$ which preserves angles. Since f also preserves orientation, its determinant is positive, so $\mathrm{d}f$ is a similitude, and its matrix is exactly as in equation (6.1). The Cauchy-Riemann equations are immediate consequences.

Remark 6.5 An *antiholomorphic* map also preserves angles, but it reverses the orientation.

6.1.b The Riemann mapping theorem

DEFINITION 6.6 Two open subsets U and V of $\mathbb{C}$ are **homeomorphic** if there exists a bijection $f : U \to V$ such that f and its inverse f^{-1} are both continuous. Such a map is called a **homeomorphism** of U into V.

Remark 6.7 In fact, in this case the continuity of f^{-1} is an unnecessary condition, because it turns out to be implied by the continuity of f, the fact that f is bijective, and that U and V are open. However, this is quite a delicate result.

Example 6.8 The complex plane $\mathbb{C}$ is homeomorphic to the (interior of the) unit disc, through the homeomorphism

$$f : z \longmapsto \frac{z}{1+|z|}.$$

See also Exercise 6.2 on page 174.

DEFINITION 6.9 Two open sets U and V are **conformally equivalent** if there exists an injective conformal application $f : U \to \mathbb{C}$ such that $V = f(U)$.

THEOREM 6.10 (Riemann mapping theorem) *Let U be any open subset of $\mathbb{C}$ such that $U \neq \mathbb{C}$. If U is homeomorphic to the unit disc $D = \mathcal{B}(0; 1)$, then U is conformally equivalent to D.*

In general, two open sets which are homeomorphic are *not* conformally equivalent. For instance, two annuli $r_1 < |z| < r_2$ and $r_1' < |z| < r_2'$ are conformally equivalent if and only if $r_2'/r_1' = r_2/r_1$.

So the Riemann mapping theorem is a very strong result, since *any connected and simply connected open set*, except $\mathbb{C}$, can be transformed, via a conformal map, into the unit disc.

To illustrate, consider in Figure 6.1 on the next page some open subset U of the complex plane. Riemann's theorem guarantees the existence of a conformal map $w = f(z)$ such that $\mathcal{B}(0; 1) = f(U)$. The boundary ∂U of U is mapped to the unit disc $\mathscr{C}(0;1)$. (This seems natural, but to make

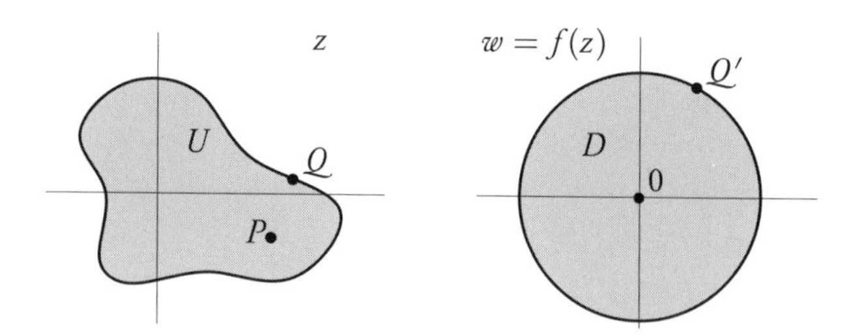

Fig. 6.1 – The open subset U being homeomorphic to the unit disc, it is conformally equivalent to it. There exists therefore a conformal map $w = f(z)$ such that the image of U by f is D. One can ask that $f(P) = 0$ and $f(Q) = Q'$.

sense of it we need to know in general that f extends to the boundary of U. That this is the case is shown by Carathéodory and Osgood, provided this boundary is a simple closed continuous curve; in the explicit examples below, this extension will be obvious.) Moreover, it is possible to impose additional conditions on f, for instance, by asking that a particular point $P \in U$ be mapped to the center of the unit disc: $f(P) = 0$, and by asking also that a point $Q \in \partial U$ be mapped to a given point on the unit circle.

It is important to note that, while Riemann's theorem tells us that the conformal representation of U exists theoretically, it does not tell *how* to find it. Physicists like Joukovski[1] found clever conformal representations of various domains in the plane; these are useful to compute easily (in terms of computer time) the effects of a fluid flowing around those domains.

6.1.c Examples of conformal maps

The following examples of conformal maps are given without proof. The reader is invited to check their properties by herself.

Representation of a half-plane into a disc

Let z_0 be any point in the open upper half-plane. Let

$$w = f(z) = \frac{z - z_0}{z - \bar{z}_0}.$$

The upper half-plane is then transformed bijectively into the interior $\mathcal{B}(0\,;1)$ of the unit circle. The point z_0 is mapped to 0. The real points x tending to either $+\infty$ or $-\infty$ are sent to the same limit 1. As z runs over the real

[1] Nicolaï JOUKOVSKI (1847–1921), a Russian physicist, studied hydrodynamics and aerodynamics. He discovered a conformal map $z \longmapsto \frac{1}{2}(z + a^2/z)$, which transforms the circles going through a and $-a$ into a profile similar to an airplane wing.

axis in increasing order, w goes around the unit circle in the counterclockwise direction, starting and ending at the point $w = 1$.

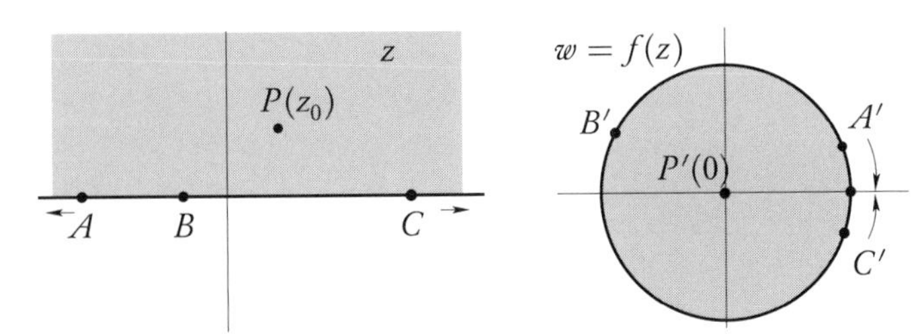

Infinite strip of width a

Using the complex exponential function, we can transform an infinite strip of width a into the upper half-plane, by putting

$$w = f(z) = e^{\pi z/a}.$$

The points on the real axis and the line $\mathrm{Im}(z) = a$ in the limit $\mathrm{Re}(z) \to -\infty$ both converge to 0, while in the limit $\mathrm{Re}(z) \to +\infty$, the real points converge to $+\infty$ and those with height a converge to $-\infty$ (on the real axis).

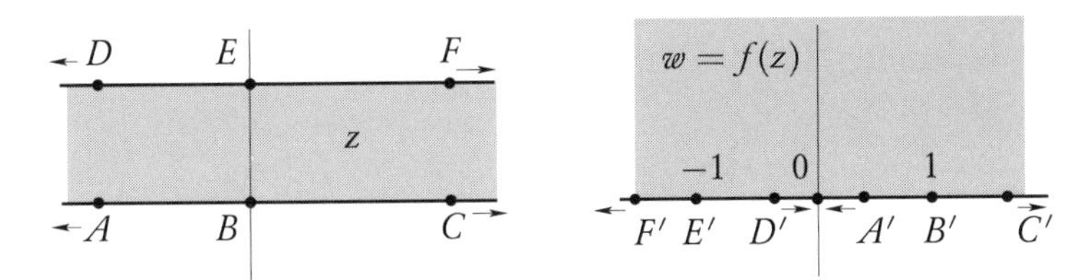

Half-plane outside a half-circle

The upper half-plane minus a half-circle is mapped to the upper half-plane by putting

$$w = f(z) = \frac{a}{2}\left(z + \frac{1}{z}\right),$$

where a is an arbitrary real number. This transformation is not conformal at the points -1 and 1 (the derivative vanishes), which explains that the right angles between the half-circle and the real axis are transformed into "flat" angles (equal to π).

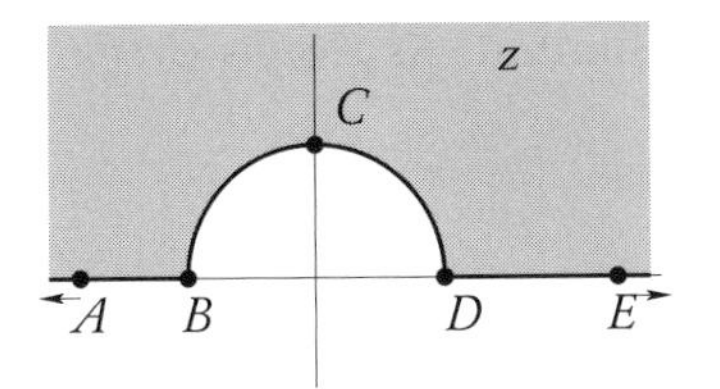

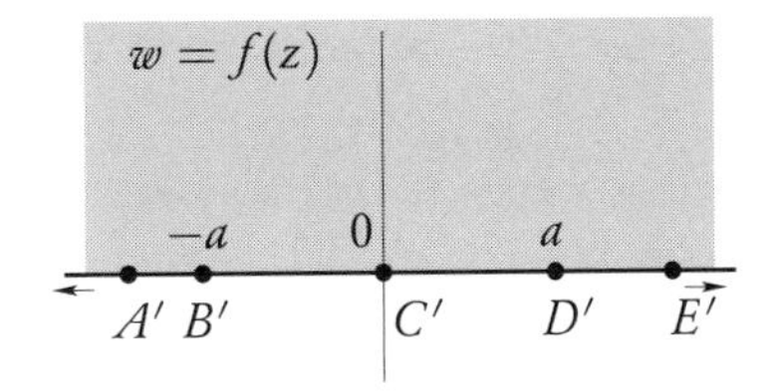

Half-circle

The unit upper half-circle is mapped into the upper half-plane by putting

$$w = f(z) = \left(\frac{1+z}{1-z}\right)^2.$$

The point -1 (C in the figure) is mapped to 0 while the point 1 (A) is mapped to infinity on the real axis (either $+\infty$ or $-\infty$, depending on whether one approaches A on the real axis or on the boundary of the half-circle).

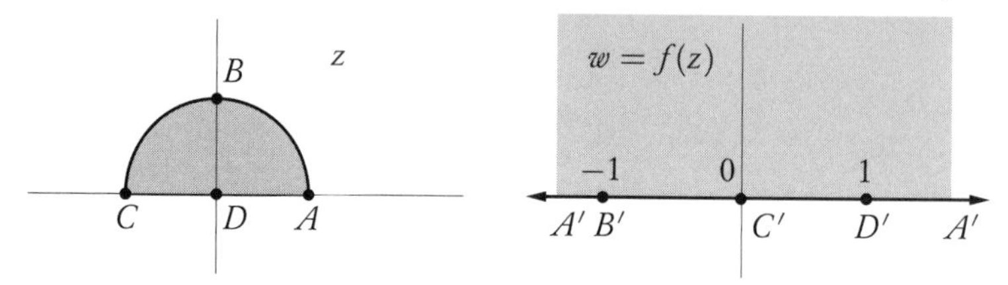

Some conformal mappings into the unit disc

The exterior of the unit disc is mapped conformally to the interior by simply putting $w = 1/z$.

Also the exterior of an ellipse with major axis equal to $2\cosh(\alpha)$ and minor axis equal to $2\sinh(\alpha)$ is mapped to the interior of the unit disc by putting

$$w = f(z) = \frac{1}{2}\left(z\,e^{-\alpha} + \frac{1}{z}\,e^{\alpha}\right).$$

Other examples

The angular sector $\{r\,e^{i\theta}\,;\ 0 < r < 1,\ 0 < \theta < \frac{\pi}{2}\}$ (a quarter-disc) is mapped into the upper half-disc by putting $w = z^2$.

Similarly, the angular sector of angle π/k (with $k \in \mathbb{R}$, $k \geqslant \frac{1}{2}$) is mapped into the upper half-plane by putting $w = z^k$.

The upper half-disc in mapped to a semi-infinite strip (infinite to the left) by putting $w = \log z$ with the logarithm defined by a cut in the lower half-plane.

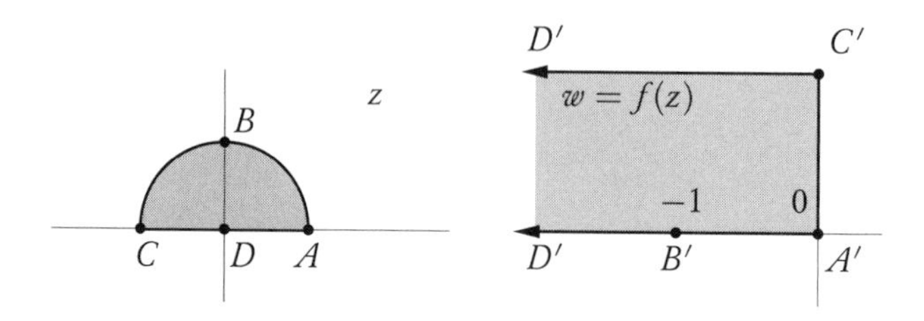

The same transformation maps the upper half-plane into an infinite strip.

The German mathematician Hermann SCHWARZ (1843–1921) succeeded his teacher Weierstrass at the University of Berlin in 1892. He worked on conformal mappings, potential theory (proving the existence of a solution to the Dirichlet problem in 1870), and partial differential equations.

6.1.d The Schwarz-Christoffel transformation

We are now looking for a conformal transformation that will map a domain bounded by a polygonal line into a simpler domain, for instance, the upper half-plane (or the unit disc, but we already know how to go from the unit disc to the upper half-plane). For this, we use the following result:

THEOREM 6.11 (Schwarz-Christoffel transformation) *Let $n \in \mathbb{N}$ be an integer, $x_1, \dots, x_n$ real numbers, and $\alpha_1, \dots, \alpha_n$ real numbers between 0 and 2π. Let finally $a \in \mathbb{C}$ be an arbitrary complex number. The map given by*

$$\frac{\mathrm{d}w}{\mathrm{d}z} = a(z - x_1)^{\alpha_1/\pi - 1}(z - x_2)^{\alpha_2/\pi - 1} \cdots (z - x_n)^{\alpha_n/\pi - 1} \tag{6.2}$$

transforms the real axis into a polygonal curve with successive angles $(\alpha_1, \dots, \alpha_n)$ and transforms the upper half-plane into the domain bounded by this curve (see Figure 6.2 on the following page).

Formula (6.2) provides only a local expression of the map. By integrating along a complex path, we obtain the equivalent formula

$$w = f(z) = a \int_{\gamma(z)} \prod_{i=1}^{n} (z' - x_i)^{\frac{\alpha_i}{\pi} - 1}\, \mathrm{d}z' + w_0,$$

where $w_0 \in \mathbb{C}$ is an arbitrary constant and $\gamma(z)$ is a path coming from infinity, contained in the open upper half-plane and ending at z. Since the upper half-plane is simply connected and the only poles of the function to integrate are on the real axis, the result does not depend on the actual path we choose.

Remark 6.12

① Changing w_0 amounts simply to translating the image of the map.

② Changing the value of the constant a gives the possibility to make rotations (by changing the argument of a) or homotheties (by changing its module).

③ By imposing the condition

$$\sum_i \left(\frac{\alpha_i}{\pi} - 1\right) = -2,$$

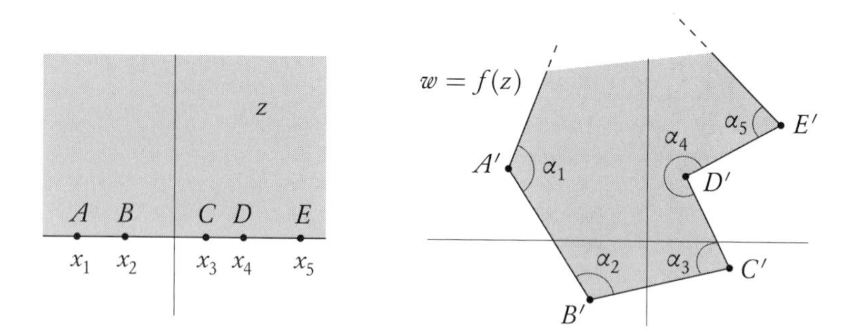

Fig. 6.2 – The Schwarz-Christoffel transform maps the open upper half-plane into the domain Ω bounded by the polygonal curve with successive angles $\alpha_1, \ldots, \alpha_n$.

the polygonal line is closed and we obtain a closed polygon. An infinite polygon (when the condition does not hold) is a limiting case of closed polygons.

④ If the problem to solve is to find a conformal mapping transforming the upper half-plane into a polygonal domain (either closed or open), it is possible to choose arbitrarily three of the real numbers $(x_1, \ldots, x_n)$. The others are then imposed by the various quantities in the picture.

⑤ It is often convenient to move one of the x_i (for instance, x_1 or x_n) to infinity, which simplifies the computations.

Example 6.13 We describe how to use the Schwarz-Christoffel transform to map conformally an infinite half-strip into the upper half-plane.

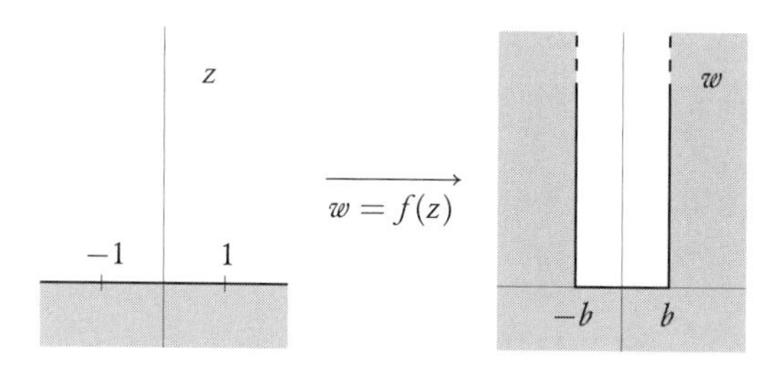

The two angles in the polygonal line are equal to $\frac{\pi}{2}$ and we fix the choice of $x_1 = -1$ and $x_2 = 1$. So we need to solve

$$\frac{\mathrm{d}w}{\mathrm{d}z} = a(z+1)^{-\frac{1}{2}}(z-1)^{-\frac{1}{2}} = \frac{a}{\sqrt{z^2-1}} = \frac{\alpha}{\sqrt{1-z^2}},$$

where a is a constant to be determined and $\alpha = -\mathrm{i}a$. We therefore put

$$w = f(z) = \alpha \arcsin(z) + B,$$

and to have $f(1) = b$ and $f(-1) = -b$, we must take $B = 0$ and $\alpha = 2b/\pi$. Hence the solution is

$$w = \frac{2b}{\pi} \arcsin(z), \qquad \text{i.e.,} \qquad z = \sin\left(\frac{\pi w}{2b}\right).$$

Elwin CHRISTOFFEL (1829–1900), born in Montjoie Aachen (present-day Monshau) in Germany, from a family of clothes merchants, studied in Berlin from 1850, where he had in particular Dirichlet as a teacher (see page 171). He defended his doctoral on the flow of electricity in homogeneous bodies, then taught there before ending his career in the new University of Strasbourg. Shy, irritable, asocial, Christoffel was nevertheless a brilliant mathematician who studied Riemannian geometry, introducing the *Christoffel symbols*, as well as conformal mappings.

6.2
Applications to potential theory

(This section may be omitted for a first reading; it is preferable to study it with some knowledge of Green functions, which we will present later.)

Conformal transformations are useful to solve a number of problems of the type "Poisson equation with boundary condition," for instance, the problem of the diffusion of heat in a solid, in a stationary regime, with the temperature fixed on the boundary of the solid; or the problem of the electrostatic field created by a charge distribution in the presence of conductors. The main idea is to transform the "boundary" of the problem, which may be complicated, into a simpler boundary, by a conformal map. In particular, the Dirichlet problem for a half-plane or for a circle (see page 170) can be solved, and this makes it possible to solve it, at least in principle, for more complicated boundaries.

THEOREM 6.14 *Let $\Omega \subset \mathbb{R}^2$ and $\Omega' \subset \mathbb{R}^2$ be two domains of $\mathbb{R}^2 \simeq \mathbb{C}$ and f a* bijective *conformal map from Ω into Ω'. Let $\varphi : \Omega \to \mathbb{R}$ be a real-valued function, with laplacian denoted by $\triangle_{(x,y)}\varphi = \frac{\partial^2\varphi}{\partial x^2} + \frac{\partial^2\varphi}{\partial y^2}$. Moreover, denote $(u,v) \stackrel{\text{def}}{=} f(x,y)$ and consider the image of by the change of variable f:*

$$\Phi(u,v) \stackrel{\text{def}}{=} \varphi\Big(f^{-1}(u,v)\Big)$$

(in other words, $\Phi = \varphi \circ f^{-1}$). Then, if we denote $\triangle_{(u,v)}\Phi \stackrel{\text{def}}{=} \frac{\partial^2\Phi}{\partial u^2} + \frac{\partial^2\Phi}{\partial v^2}$ the laplacian of Φ in the new coordinates, we have

$$\triangle_{(x,y)}\varphi = \left|f'(z)\right|^2 \triangle_{(u,v)}\,\Phi.$$

PROOF. This is an elementary but tiresome computation. It suffices to use the chain rule formulas

$$\frac{\partial}{\partial u} = \frac{\partial x}{\partial u}\frac{\partial}{\partial x} + \frac{\partial y}{\partial u}\frac{\partial}{\partial y} \qquad \text{and} \qquad \frac{\partial}{\partial v} = \frac{\partial x}{\partial v}\frac{\partial}{\partial x} + \frac{\partial y}{\partial v}\frac{\partial}{\partial y}$$

for partial derivatives, together with the Cauchy-Riemann relations.

COROLLARY 6.15 *The property that a function is harmonic is invariant by conformal transformation.*

Hence, suppose we have to solve a problem like the Dirichlet problem, namely, we must find a harmonic function φ on a domain domain $\Omega \in \mathbb{R}^2$ such that $\varphi(z)_{|\partial\Omega} = f(z)$, where $f : \partial\Omega \to \mathbb{C}$ is a given function. If we know how to transform Ω by conformal mapping into a half-circle or a half-plane, we can solve the associated Dirichlet problem and then perform the inverse change of variable, which will preserve the harmonicity of the solution.

However, in general the goal is to solve not a Laplace equation $\triangle\varphi = 0$, but rather a Poisson equation

$$\triangle\varphi(x, y) = \delta(x - x_0, y - y_0).$$

It is therefore necessary to first determine how this equation is transformed by a conformal mapping. In a dilation (change of scale) $x \longmapsto ax$ in the argument of a Dirac distribution (see Chapter 7), a factor $1/|a|^d$ appears, where d is the dimension. Here, we deal with two dimensions and, locally, a conformal transformation is simply a similitude, with homothety factor given by $|f'(z)|$. We have therefore the following result (see Theorem 7.45 on page 199 for a generalization when f is not bijective):

THEOREM 6.16 *Let $\Omega \subset \mathbb{R}^2$ and $\Omega' \subset \mathbb{R}^2$ be two domains of $\mathbb{R}^2 \simeq \mathbb{C}$ and let f be a bijective conformal mapping of Ω into Ω'. Then for $(x_0, y_0) \in \Omega$, we have*

$$\delta\big(f(x, y) - f(x_0, y_0)\big) = \frac{1}{|f'(x_0, y_0)|^2}\delta(x - x_0, y - y_0).$$

As a consequence, we have the following result:

THEOREM 6.17 *Let $\Omega \subset \mathbb{R}^2$ and $\Omega' \subset \mathbb{R}^2$ be two domains of $\mathbb{R}^2 \simeq \mathbb{C}$ and let f be a bijective conformal mapping of Ω into Ω'. Let $\varphi : \Omega \to \mathbb{R}$ be a real-valued function and let Φ be the image of φ by the change of variable f. Then, if φ satisfies the Poisson equation*

$$\triangle_{(x,y)}\varphi = \delta(x - x_0, y - y_0),$$

its image by the conformal map f also satisfies the Poisson equation in the new coordinates:

$$\triangle_{(u,v)}\Phi = \delta\big(f(x, y) - f(x_0, y_0)\big).$$

6.2.a Application to electrostatics

We will give an example of application of those theorems to solve the Dirichlet problem in the situation corresponding to electrostatics. We consider, in a two-dimensional space, a conductor which leaves a "hole" shaped like an infinite half-strip, denoted Ω (see Figure 6.3).

Now, we would like to compute the potential created by an arbitrary distribution of charge $\rho(x,y)$. For this, because of the linearity of the electrostatic Poisson equation ($\triangle\varphi = -\rho/\varepsilon_0$), we will see (Chapter 7) that it suffices to find the potential $G(x,y\,;x_0,y_0)$ created at the point (x,y) by a *Dirac distribution* at (x_0,y_0), which satisfies

$$(C)\quad \begin{cases} \triangle G(x,y\,;x_0,y_0) = \delta(x-x_0,y-y_0) \quad \text{where} \quad \triangle = \dfrac{\partial^2}{\partial x^2} + \dfrac{\partial^2}{\partial y^2}, \\ G(x,y\,;x_0,y_0) = 0 \quad \text{for any } (x,y)\in\partial\Omega \text{ and } (x_0,y_0)\in\Omega. \end{cases}$$

Once this function is found, the solution to the problem is given by[2]

$$\varphi(x,y) = -\frac{1}{\varepsilon_0}\iint_\Omega \rho(x',y')\,G(x,y\,;x',y')\,\mathrm{d}x'\,\mathrm{d}y'.$$

One should note that the function $G(x,y\,;x',y')$ *is not invariant under translations* and therefore it does not depend only on $x-x'$ and $y-y'$, because of the presence of the conductor. This function is called the **Green function** for the problem.

To find this function $G(x,y\,;x',y')$, we start by introducing complex variables $z = x+\mathrm{i}y$ and $w = u+\mathrm{i}v$; then we perform the conformal transformation $w = \sin(\pi z/2b)$ which, as we have seen, transforms Ω into the upper half-plane (see Figure 6.3). Then we look for the Green function of the new problem, $\mathscr{G}(w\,;w')$, which satisfies, according to Theorem 6.17,

$$(C')\quad \begin{cases} \triangle_{(u,v)}\mathscr{G}(u,v\,;u',v') = \delta(u-u',v-v'), \\ \qquad\mathscr{G}(u,v\,;u',v') = 0 \qquad \forall(u,v)\in f(\partial\Omega), \quad \forall(u',v')\in f(\Omega), \end{cases}$$

where the image of the boundary $\partial\Omega$ is the real axis $f(\partial\Omega) = \mathbb{R}$. In other words, we are looking for the electrostatic response of a simpler system (a half-plane) in the presence of an electric charge.

To solve this last problem, one can use the method of virtual images (which is justified, for instance, by means of the Fourier or Laplace transform): the potential created by a charge at w' in the neighborhood of a conductor in the lower half-plane is the sum of the "free" potentials[3] created, first, by the particle at w' and, second, by another particle, with the opposite charge, placed symmetrically of the first with respect to the real axis (namely, at $\overline{w}'$).

---

[2] The physicist's "proof" being that the quantity $-\varepsilon_0^{-1}\rho(x',y')\,G(x,y\,;x',y')\,\mathrm{d}x'\,\mathrm{d}y'$ is the potential created at (x,y) by the elementary charge $\rho(x',y')\,\mathrm{d}x'\,\mathrm{d}y'$.

[3] I.e., of the Green functions tending to 0 at infinity.

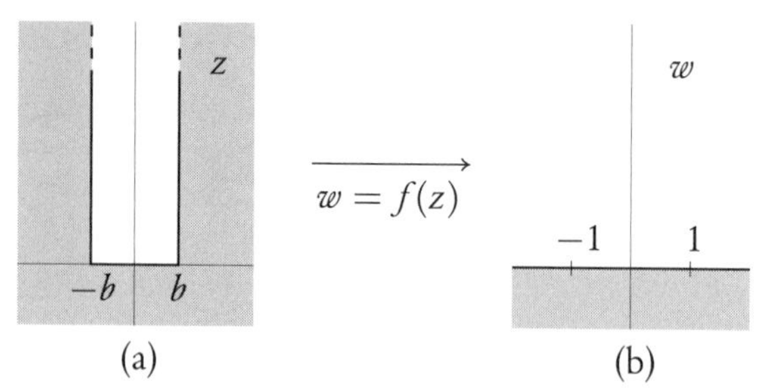

(a) (b)

Fig. 6.3 – (a) The domain Ω left free by the conductor is the set of points $z = x + \mathrm{i}y$ with $y > 0$ and $-b < x < b$. The Dirichlet conditions are that the potential is constant, for instance, zero, on $\partial\Omega$. (b) The image $f(\Omega)$ of the domain Ω is the upper half-plane.

However, we will see in the chapter concerning distributions that the Coulomb potential[4] in two dimensions is simply

$$\Phi(w) = -\frac{q}{2\pi\varepsilon_0} \log |w - w'| .$$

$\mathscr{G}$ is thus equal to

$$\mathscr{G}(w\,;w') = \frac{q}{2\pi} \log \left| \frac{w - w'}{w - \overline{w}'} \right| .$$

The Green function of the initial problem is obtained by putting, in the last formula, $w = \sin(\pi z/2b)$, which gives

$$G(z\,;z_0) = \frac{q}{2\pi} \log \left| \frac{\sin \dfrac{\pi z}{2b} - \sin \dfrac{\pi z_0}{2b}}{\sin \dfrac{\pi z}{2b} - \sin \dfrac{\pi \overline{z_0}}{2b}} \right| ,$$

or, by expanding in real form $\sin(a + \mathrm{i}b) = \sin a \cosh b + \mathrm{i} \cos a \sinh b$,

$$G(x, y\,; x_0, y_0) = \frac{q}{4\pi}$$
$$\times \log \left| \frac{\left(\sin \frac{\pi x}{2b} \cosh \frac{\pi y}{2b} - \sin \frac{\pi x_0}{2b} \cosh \frac{\pi y_0}{2b}\right)^2 + \left(\cos \frac{\pi x}{2b} \sinh \frac{\pi y}{2b} - \cos \frac{\pi x_0}{2b} \sinh \frac{\pi y_0}{2b}\right)^2}{\left(\sin \frac{\pi x}{2b} \cosh \frac{\pi y}{2b} - \sin \frac{\pi x_0}{2b} \cosh \frac{\pi y_0}{2b}\right)^2 + \left(\cos \frac{\pi x}{2b} \sinh \frac{\pi y}{2b} + \cos \frac{\pi x_0}{2b} \sinh \frac{\pi y_0}{2b}\right)^2} \right| .$$

Which, as the reader will notice, is not properly speaking something very intuitive. However, it is remarkable how this computation has been done so *easily* and without difficulty. This is the strength of conformal mappings.

[4] The **Coulomb potential** in any dimension is a potential satisfying the Laplace equation $\triangle\varphi = -\delta/\varepsilon_0$; Theorem 7.54 on page 208 shows that the three-dimensional Coulomb potential is given by $\boldsymbol{r} \mapsto 1/4\pi\varepsilon_0 r$, whereas in two dimensions Theorem 7.55 on page 208 shows that it is $\boldsymbol{r} \mapsto -(\log r)/2\pi\varepsilon_0$.

6.2.b Application to hydrodynamics

Consider the bidimensional flow of a perfect fluid (i.e., an *incompressible*, *non-viscous*, and *irrotational* fluid). Then (see, for instance, the excellent book by Guyon et al. [44]) the fluid velocity is obtained locally, at any point, from a potential φ, in the sense that the x and y coordinates of the velocity are given by

$$v_x = \frac{\partial \varphi}{\partial x} \qquad \text{and} \qquad v_y = \frac{\partial \varphi}{\partial y}$$

This potential is *harmonic* since $\operatorname{div} \boldsymbol{v} = 0$ is equivalent to $\triangle \varphi = 0$. Solving the flow problem amounts to finding φ with the right boundary conditions (normal velocity equal to zero on all impenetrable surfaces). This is called the **Neumann problem.**[5]

Remark 6.18 The velocity is obtained from a *global* potential if the domain under consideration $\Omega \subset \mathbb{C}$ is *simply connected.* Otherwise, it may not be possible to find a global potential; it is then necessary to "cut" Ω into simply connected pieces, and try, after solving the problem on each component, to glue the solutions back together. In what follows, we assume that Ω is indeed simply connected.

DEFINITION 6.19 The **complex velocity of the fluid** is the complex quantity given by

$$\mathscr{V} \stackrel{\text{def}}{=} v_x + \mathrm{i} v_y.$$

Since φ is harmonic, we know (Theorem 5.13) that there exists a harmonic function $\psi : \Omega \to \mathbb{R}$ which is conjugate to φ, that is, such that $\Phi \stackrel{\text{def}}{=} \varphi + \mathrm{i}\psi$ is holomorphic on Ω. Then the complex derivative of Φ is the conjugate of the complex velocity:

$$\Phi'(z) = \frac{\partial \Phi}{\partial z} = \frac{\partial \Phi}{\partial x} = \frac{\partial \varphi}{\partial x} + \mathrm{i}\frac{\partial \psi}{\partial x} \stackrel{\text{Cauchy-Riemann}}{=} \frac{\partial \varphi}{\partial x} - \mathrm{i}\frac{\partial \varphi}{\partial y} = v_x - \mathrm{i} v_y = \overline{\mathscr{V}(z)}.$$

The complex velocity is therefore an *antiholomorphic* function.

DEFINITION 6.20 The holomorphic function Φ is called the **complex potential.** A *stagnation point* for the flow Φ is any point $z_0 \in \Omega$ where the complex velocity vanishes: $\Phi'(z_0) = 0$.

Consider, for instance, the flow of a perfect fluid in a domain $\mathscr{D}$ in $\mathbb{R}^2$ (identified with the complex plane $\mathbb{C}$), delimited by the ground and by a wall

---

[5] Carl Gottfried Neumann (1832–1925), German mathematician and physicist, was the son of the physicist Franz Ernst Neumann, who introduced the notion of potential in electromagnetism.

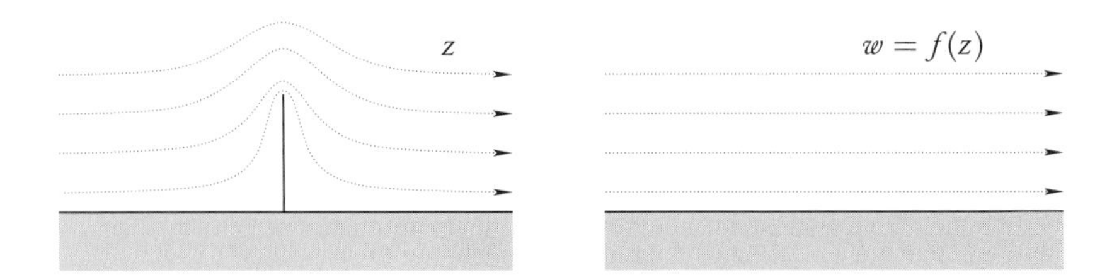

Fig. 6.4 – On the left: the wind over a wall. The white region is $\mathscr{D}$. On the right: the wind over the ground, without a wall. The white region is $\mathscr{D}'$.

of height a (see Figure 6.4, left). We will, by a conformal transformation, map $\mathscr{D}$ into the upper half-plane

$$\mathscr{D}' \stackrel{\text{def}}{=} \{z \in \mathbb{C} \, ; \, \operatorname{Im} z > 0\}$$

(see Figure 6.4, right). For this, it is enough to put $f(z) = \sqrt{z^2 + a^2} = w$.

Indeed, if we decompose the transformation f in three steps

$$f = (z \mapsto \sqrt{z}) \circ (z \mapsto z + a^2) \circ (z \mapsto z^2),$$

where the function $z \mapsto \sqrt{z}$ is taken with a cut along the positive real axis, then the domain $\mathscr{D}$ is transformed successively as follows:

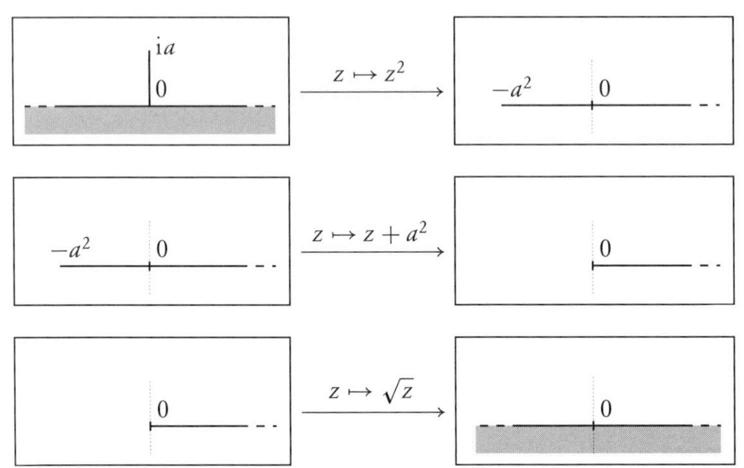

We can now deduce from this the flow in $\mathscr{D}$ with the boundary conditions[6]:

$$\lim_{\operatorname{Re}(z) \to \pm\infty} \boldsymbol{v}(z) = (v_\infty, 0).$$

[6] In the general case, one can proceed as follows. Search for the flow in the form

$$\Omega(z) = v_\infty z + g(z),$$

where g is a holomorphic function which describes a perturbation from the homogeneous flow $\Omega_0(z) = v_\infty z$, such that $\lim_{|z| \to \infty} g'(z) = 0$ so that the limit conditions are respected.

After the conformal mapping $w = f(z)$, the flow still obeys the same boundary conditions, since $f'(z) \sim 1$ as $|z|$ tends to infinity in the upper half-plane. The desired flow in the upper half-plane *without a wall* is easy to find: it a a uniform flow (the wind over the plain...), which can be described by the complex potential

$$\Theta(w) = v_\infty \times w.$$

Coming back to the variable w:

$$\Phi(z) = \Theta(w) = \Theta\left(\sqrt{z^2 + a^2}\right) = v_\infty \times \sqrt{z^2 + a^2}.$$

Two special points appear: the velocity of the flow is infinite on the top of the wall, at $z = \mathrm{i}a$ (this is why the wind "whistles" when going over a pointed obstacle: locally supersonic velocities are reached); it is zero "at the foot of the wall" (around $z = 0$). These two points are of course the two points where the transformation is no longer conformal[7]; it is therefore possible that singularities appear in this manner, which did not appear in the flow on the upper half-plane.

6.2.c Potential theory, lightning rods, and percolation

From the previous example we can understand the principle of the lightning rod. If we denote by $\varphi(\boldsymbol{r})$ the electrostatic potential, then close to a pointed object (like the wall above, or an acute angle in the definition of the domain), we can expect that φ' will diverge. Identifying $\mathbb{C}$ and $\mathbb{R}^2$, it is the *gradient* of φ that will diverge, namely, the electric field. But we know that in the air an electric field of intensity larger than some critical value, the "breakdown threshold," will cause a spontaneous electric discharge.[8] Saint Elmo's fire, well known to mariners, has the same origin.

But this is not yet the most remarkable thing. Consider a problem of random walk of a particle in an open domain $\Omega \subset \mathbb{C}$, assuming that, when the particle meets the *boundary* of Ω, it will stop. (It is possible to make numerical simulations of this problem.) What will happen after many particles have performed this process? One may think that the particles will accumulate on the boundary $\partial\Omega$, which will therefore grow, more or less regularly as time goes. In fact, this is not what happens. If the boundary $\partial\Omega$ has "cusps," they will attract more particles and, hence, will grow faster than the rest of the boundary. If, at a given time and due to the effect of chance, a "cusp" is created (by accumulation of some particles) on a "smooth" part of $\partial\Omega$, then this cusp will, in turn, attract more particles. In a word: the fluctuations of the boundary become unstable.

[7] Indeed, at the foot of the wall, we have $f'(z) = 0$, whereas around the top, $f'(z)$ diverges.

[8] Of course, the rough explanation given only concerns lightning rods in a two-dimensional world. In three dimensions, another argument must be found, because the one given uses specific properties of $\mathbb{C}$ and not of $\mathbb{R}^2$.

How to explain this phenomenon? If we denote by $p(z,t)$ the probability density for the particle to be at position z at time t, one can show that this probability satisfies an equation of *diffusion*, which is the manifestation on a macroscopic scale of the extremely erratic behavior of microscopic particles (the so-called **Brownian** motion). The diffusion equation takes the form

$$\left(\frac{\partial}{\partial t} - \varkappa \triangle\right) p(z,t) = 0$$

at any point of Ω. We will therefore have an equation of the same type if we perform a conformal transformation (independent of time) on the domain Ω. Once more, the cusps will attract more particles and we will observe a growth phenomenon.[9] (Here it is the flow of particles that diverges.)

6.3

Dirichlet problem and Poisson kernel

The Dirichlet problem is the following: the values of a harmonic function on the boundary $\partial\Omega$ of a domain Ω in the plane are assumed to be known, and we want to find the values of the function at any point in Ω.

For instance, one may be interested in the case of a metallic disc,[10] which is a heat conductor, in a stationary state, for which we impose a temperature T_0 on one half-circle, and a temperature T_1 on the other half-circle. The question is now to determine the temperature at any point inside the disc.

More precisely, if Ω is a domain (hence, a connected open set) of the complex plane $\mathbb{C}$, if $\partial\Omega$ denotes its boundary and if $u_0 : \partial\Omega \to \mathbb{C}$ is a *continuous* function, we are looking for a $u : \overline{\Omega} \to \mathbb{C}$ of $\mathscr{C}^2$ class, such that

$$\begin{cases} u_{|\partial\Omega} = u_0, \\ \triangle u = 0 \quad \text{in } \Omega. \end{cases}$$

A first important case is when the domain Ω is an open ball. Indeed, according to Riemann's mapping theorem, one can in principle reduce to this case whenever Ω is homeomorphic to an open ball (and not the whole of $\mathbb{C}$). In this situation, we have a first useful theorem:

THEOREM 6.21 *Let u be a function harmonic on a neighborhood of $\overline{\mathcal{B}}(0\,;\,1)$. Then, for any point $a \in \mathcal{B}(0;\,1)$, we have*

$$u(a) = \frac{1}{2\pi}\int_0^{2\pi} u(\mathrm{e}^{\mathrm{i}\psi})\,\frac{1-|a|^2}{\left|a-\mathrm{e}^{\mathrm{i}\psi}\right|^2}\,\mathrm{d}\psi.$$

[9] In particular, fractal growth may be observed by a mechanism similar to this one.

[10] Or, which amounts to the same thing, a metallic cylinder with infinite length, on which we assume translation invariance.

Peter Gustav LEJEUNE-DIRICHLET (1805–1859), German mathematician, was professor successively in Breslau, Berlin, and Göttingen, where he succeeded his teacher Gauss (see page 557). He was also the brother-in-law of Jacobi and a friend of Fourier. He studied trigonometric series and gave a condition for the convergence of a Fourier series at a given point (see Theorem 9.46 on page 268). His other works concern the theory of functions in analysis and number theory.

To prove this theorem, we start by stating a technical lemma which tells us, in substance, "anything that can be proved for the central point of the open disc $\mathcal{B}(0\,;1)$ can also be proved for an arbitrary point of the disc."

LEMMA 6.22 *Let $a \in \mathcal{B}(0\,;1)$. Denote $\varphi_a(z) \stackrel{\text{def}}{=} \frac{z-a}{1-\bar{a}z}$. Then $\varphi_a \in \mathscr{H}(\mathcal{B}')$, where $\mathcal{B}'$ is some neighborhood of $\mathcal{B}(0\,;1)$ (for instance $\mathcal{B}' = \mathcal{B}(0\,;|a|^{-1})$) and satisfies*

i) φ_a is bijective on $\mathcal{B}'$;

ii) $\varphi_a|_{\mathcal{B}(0;1)}$ is a bijection from $\mathcal{B}(0\,;1)$ into itself;

iii) $\varphi_a^{-1} = \varphi_{-a}$;

iv) $\varphi_a(a) = 0$.

In other words, φ_a transforms the disc into itself, but moves the point a to the center in a holomorphic manner.

PROOF OF THEOREM 6.21. We use the mean value property for harmonic functions:

$$u(a) = u \circ \varphi_{-a}(0) = \frac{1}{2\pi}\int_0^{2\pi} u\big(\varphi_{-a}(e^{i\theta})\big)\,d\theta = \frac{1}{2i\pi}\int_{\partial\mathcal{B}} \frac{u\big(\varphi_{-a}(\zeta)\big)}{\zeta}\,d\zeta.$$

With the change of variable $\xi = \varphi_{-a}(\zeta)$, we obtain $\zeta = \varphi_a(\xi)$,

$$\frac{d\zeta}{d\xi} = \varphi_a'(\xi) = \frac{1-|a|^2}{(1-\bar{a}\xi)^2},$$

and

$$u(a) = \frac{1}{2i\pi}\int_{\partial\mathcal{B}} \frac{u(\xi)}{\varphi_a(\xi)}\,\varphi_a'(\xi)\,d\xi = \frac{1}{2i\pi}\int_0^{2\pi} u(e^{i\psi})\,\frac{1-|a|^2}{|a-e^{i\psi}|^2}\,d\psi.$$

In Theorem 6.21, put $a = r\,e^{i\theta}$; there follows the equivalent formula:

$$u\big(r\,e^{i\theta}\big) = \int_0^{2\pi} u(e^{i\psi})\,P(r,\theta-\psi)\,d\psi,$$

where P is the Poisson kernel:

Siméon Denis POISSON (1781–1840) lost his father at fifteen, and was sent by his mother to study medicine in Paris to earn his living. Seeing patients die under his care, he decided (accidentally, having first chosen the wrong classroom!) to dedicate himself to mathematics. He was received first at the École Polytechnique at eighteen. Then honors came until the end of his life (he was even made a baron by Louis XVIII). In mathematics, his interests ranged over almost all of analysis: integration, Fourier series, probability, and, especially, mathematical physics, celestial mechanics and electrostatics in particular. A member of the Royal Council of Public Education, he tried to develop the teaching of mathematics.

DEFINITION 6.23 The **Poisson kernel** is the function $P : [0,1[\times\mathbb{R} \to \mathbb{R}$ defined by

$$P(r,\theta) = \frac{1}{2\pi}\,\frac{1-r^2}{1+r^2-2r\cos\theta} = \frac{1}{2\pi}\,\frac{1-r^2}{\left|r-\mathrm{e}^{\mathrm{i}\theta}\right|^2}.$$

Remark 6.24 In the case of holomorphic functions, the Cauchy formula not only reproduces holomorphic functions, but also *creates* them: if f is a given function on the circle $\mathscr{C} = \partial\mathcal{B}(0;1)$, assumed to be merely continuous, then

$$F(z) \stackrel{\text{def}}{=} \frac{1}{2\mathrm{i}\pi}\int_{\mathscr{C}} \frac{f(\zeta)}{\zeta - z}\,\mathrm{d}\zeta$$

is a holomorphic function on $\mathcal{B}(0;1)$.

However, there isn't a very strong link between f and F, contrary to what one might expect. For instance, for $z_0 \in \mathscr{C}$, nothing forces that $f(z_0) = \lim_{z\to z_0} F(z)$. One may, for instance, take $f(z) = \bar{z}$ (which, by the way, coincides on the unit circle with the function $g(z) = 1/z$). The preceding formula leads to $F(z) = 0$ for any $z \in \mathcal{B}(0;1)$, as the reader can amuse herself to prove.

On the other hand, for harmonic functions, there is a formula to create a harmonic function from boundary values, with a continuity property, using the Poisson kernel. This is the object of the next theorem.

THEOREM 6.25 (Dirichlet problem for a disc) *Let* $f : \partial\mathcal{B}(0;1) \to \mathbb{R}$ *be a* continuous *function defined on the unit circle. Put*

$$u(z) = \begin{cases} \dfrac{1}{2\pi}\displaystyle\int_0^{2\pi} f(\mathrm{e}^{\mathrm{i}\psi})\,\frac{1-|z|^2}{\left|z-\mathrm{e}^{\mathrm{i}\psi}\right|^2}\,\mathrm{d}\psi & \text{if } z \in \mathcal{B}(0;1), \\ f(z) & \text{if } z \in \partial\mathcal{B}(0;1). \end{cases}$$

Then u is continuous on $\overline{\mathcal{B}}(0;1)$ and harmonic on $\mathcal{B}(0;1)$.

If f is has discontinuities, then u is harmonic on $\mathcal{B}(0;1)$ and continuous in each point of the unit circle in which f happens to be continous.

PROOF. The proof of this theorem is omitted. Still, we remark that it involves a sequence of Dirac functions given by the Poisson kernel for $[r \to 1]$. After reading

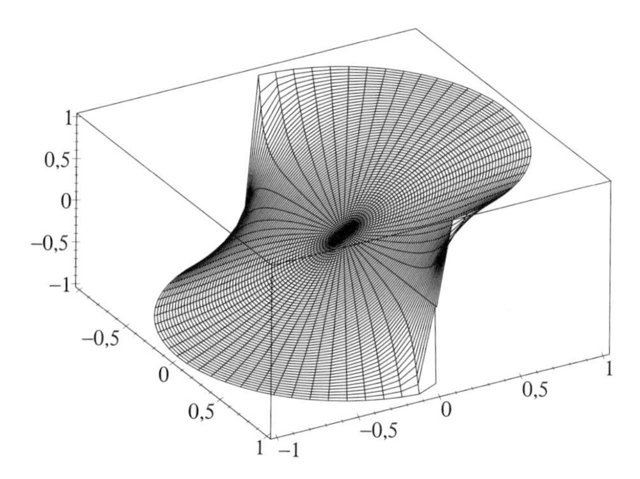

Fig. 6.5 – The temperature is equal to -1 on the lower half-circle, and to $+1$ on the upper half-circle; it is a harmonic function on the disc. Here it is represented on the vertical axis, and is deduced from the preceding theorem.

chapter 8 on distributions, and in particular section 8.2.b on page 231, the reader is invited to try to formalize the proof of this theorem.

The Dirichlet problem has many applications in physics. Indeed, there are many equations of "harmonic" type: electromagnetism, hydrodynamics, heat theory, and so on. The example already mentioned – find the temperature at any point in a disc where the temperature is imposed to be T_0 on the upper half-circle and T_1 on the lower half-circle – is illustrated in Figure 6.5. After integration, one finds (in polar coordinates)

$$T(r,\theta) = \frac{2}{\pi} \arctan\left(\frac{2r\sin\theta}{1-r^2}\right).$$

In Exercises 6.7 and 6.8, two other examples of solutions of the Dirichlet problem are given, for a half-plane and for a strip, respectively.

EXERCISES

♦ **Exercise 6.1** Let $U : \mathbb{R}^2 \to \mathbb{R}^2$ be a twice differentiable function. Show that if $f : \mathbb{C} \to \mathbb{C}$ is holomorphic, then denoting $f : (x + \mathrm{i}y) \mapsto (X + \mathrm{i}Y)$, we have

$$\triangle_{(x,y)} U(x, y) = \left|f'(x + \mathrm{i}y)\right|^2 \triangle_{(X,Y)} \widetilde{U}(X, Y),$$

with $\widetilde{U}(X, Y) \stackrel{\text{def}}{=} U \circ f^{-1}(X, Y)$.

♦ **Exercise 6.2** Show that any two annuli are homeomorphic (one can use, for instance, a purely radial function).

Complex flows

♦ **Exercise 6.3** Determine the flow of the wind in a two-dimensional domain delimited by two half-lines meeting at an angle α (for instance, one of the half-lines represents the horizontal ground, and the other a mountainside with uniform slope). Use the following boundary conditions: the wind must have velocity with modulus v and flow parallel to the boundaries, infinitely far away from the intersection of the lines.

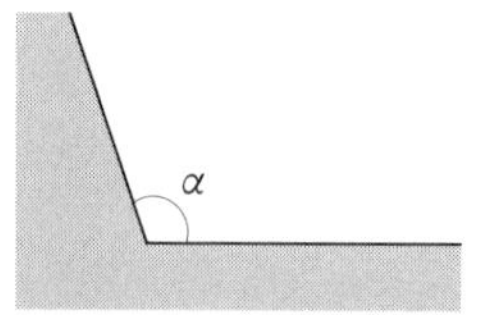

Assume that the flow is irrotational, incompressible, and stationary. In particular, what is the velocity of the wind at the point where the slope changes? Consider the various possibilities depending on the value of $\alpha \in [0, \pi]$.

♦ **Exercise 6.4** Let k be a real number. Interpret the complex flows given by

$$\Omega_1(z) = k \log(z - a) \qquad \text{and} \qquad \Omega_2(z) = \mathrm{i}k \log(z - a).$$

♦ **Exercise 6.5** We wish to solve the problem of the flow of a Newtonian fluid in three dimensions around a cylinder with infinite length and circular section. If only the solutions which are *invariant* by translation along the axis are sought, the problem is reduced to the study of the flow of a fluid in two dimensions around a disc.

We assume that the velocity of the fluid, far from the obstacle, is uniform. We look for the solution as a complex potential Φ written in the form

$$\Phi = V_0 z + G(z),$$

where the term $V_0 z$ is the solution unperturbed by the disc and where $G(z)$ is a perturbation, such that

$$G'(z) \xrightarrow[|z| \to +\infty]{} 0.$$

Check that the solution obtained is indeed compatible with the boundary conditions (the velocity of the fluid is tangent to the disc), and find the stagnation points of the fluid.

Dirichlet problem

♦ **Exercise 6.6** We wish to know the temperature of a sheet in the shape of an infinite half-plane, represented in the complex plane by the closed region

$$\mathscr{D} = \{z \in \mathbb{C}\,;\ \operatorname{Im}(z) \geqslant 0\}.$$

We assume that the temperature on the boundary of the sheet is given by

$$T(x) = \begin{cases} T_0 & \text{if } x < -1, \\ T_1 & \text{if } |x| < 1, \\ T_2 & \text{if } x > 1. \end{cases}$$

If $z \in \mathscr{D}$, we denote by θ_1 and θ_2 the angles defined in the picture below:

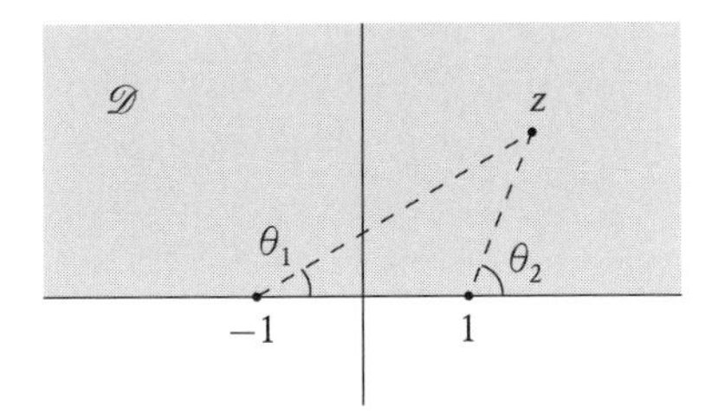

Show that any function $z \mapsto \alpha\theta_1 + \beta\theta_2 + \gamma$ (with $\alpha, \beta, \gamma \in \mathbb{R}$) is harmonic. Deduce from this the solution to the Dirichlet problem.

♦ **Exercise 6.7 Dirichlet problem for a half-plane:** consider the domain

$$\mathscr{D} = \{z \in \mathbb{C}\,;\ \operatorname{Im} z > 0\}.$$

We want to find u of $\mathscr{C}^2$ class defined on $\mathscr{D}$ with values in $\mathbb{R}$, admitting a continuous extension to $\overline{\mathscr{D}}$ and such that

$$\begin{cases} \triangle u = 0 & \text{in } \mathscr{D}, \\ u_{|\partial\mathscr{D}} = u_0 & \text{given.} \end{cases}$$

Find, using the Cauchy formula, an integral formula for $u(z)$ when $\operatorname{Im} z > 0$.

HINT: Show, before concluding, that if $z = x + \mathrm{i}y$ with $y > 0$ and if f is holomorphic on $\mathscr{D}$, then

$$f(z) = \frac{1}{\pi}\int_{-\infty}^{+\infty} \frac{y\, f(\eta)}{(\eta - x)^2 + y^2}\, \mathrm{d}\eta.$$

♦ **Exercise 6.8 Dirichlet problem for a strip:** we are seeking the solution of the Dirichlet problem for a strip

$$\mathscr{B} = \{z \in \mathbb{C}\,;\ 0 < \operatorname{Im} z < 1\}.$$

In other words, we are looking for a function u of $\mathscr{C}^2$ class defined on $\mathscr{B}$ with values in $\mathbb{R}$, admitting a continuous extension to $\overline{\mathscr{B}}$, and such that

$$\begin{cases} \triangle u = 0 & \text{in } \mathscr{B}, \\ u(x) = u_0(x) & \forall x \in \mathbb{R}, \\ u(x + i) = u_1(x) & \forall x \in \mathbb{R}. \end{cases}$$

Show that if $u \stackrel{\text{def}}{=} \operatorname{Re} f$ (with f holomorphic) is a solution of the problem, then we have

$$f(z) = \frac{\mathrm{i}}{2}\int_{-\infty}^{+\infty} u_0(t)\coth\left(\frac{\pi(t-z)}{2}\right)\mathrm{d}t + \frac{\mathrm{i}}{2}\int_{-\infty}^{+\infty} u_1(t)\tanh\left(\frac{\pi(t-z)}{2}\right)\mathrm{d}t.$$

SOLUTIONS

♦ **Solution of exercise 6.2.** Writing as usual $z = re^{i\theta}$ and denoting by $r_1 < r < r_2$ the first annulus and by $r'_1 < r < r'_2$ the second one, consider the map

$$\begin{cases} r \longmapsto \left(\dfrac{r'_2 - r'_1}{r_2 - r_1}\right)(r - r_1) + r'_1, \\ \theta \longmapsto \theta. \end{cases}$$

This application is indeed an homeomorphism from the first annulus onto the second (the transformation of the radius is affine). One can check that this is not a holomorphic map, except if the two annuli are homothetic to each other, in which case the formula given is the corresponding homothety.

♦ **Solution of exercise 6.3.** We can transform the domain $\mathscr{D}$ into the upper half-plane by means of the following conformal map:

$$z \longmapsto w = z^{\pi/\alpha}.$$

In the domain $\mathscr{D}' = \{z \in \mathbb{C}\,;\ \operatorname{Im}(z) > 0\}$, the flow of the wind is given by a complex potential $\Theta(w) = vw$; one checks indeed that the velocity field that derives from this is uniform and parallel to the ground (since $v \in \mathbb{R}$). The function Θ being holomorphic, its real part $\operatorname{Re}\Theta$, which is the real velocity potential, is harmonic: $\triangle(\operatorname{Re}\Theta) = 0$. Coming back to the initial domain, we put

$$\Phi(z) = \Theta\big(w(z)\big) = \Theta\big(z^{\pi/\alpha}\big) = v \times z^{\pi/\alpha},$$

which therefore satisfies $\triangle(\operatorname{Re}\Phi) = 0$ (because of Corollary 6.15) and has the required boundary conditions. The velocity of he wind is given by the gradient of $\operatorname{Re}\Phi$.

At $z = 0$, the mapping $z \mapsto w$ is not conformal. If $\alpha < \pi$, the derivative

$$\frac{\mathrm{d}w}{\mathrm{d}z} = \frac{\pi}{\alpha}\, z^{\pi/\alpha - 1}$$

tends to 0 (and so does the velocity of the wind), which shows that at the bottom of a V-shaped valley one is protected from the transversal wind (but not from the longitudinal wind, as people from Lyons know, who are victim of the North-South wind, whereas the East-West wind is almost absent). On the other hand, if $\alpha > \pi$, this derivative tends to infinity, which shows that on the crest of a mountain, the transverse wind is important.

♦ **Solution of exercise 6.4.** Ω_1 corresponds to the flow around a source or a well (depending on the sign of k), and Ω_2 corresponds to a whirlpool, as one can see by picturing the velocity field.

♦ **Solution of exercise 6.5.** Rotating the entire system if necessary, we may assume that the velocity at infinity is parallel to the real axis. Therefore we take $V_0 \in \mathbb{R}$ and look for a flow symmetric with respect to the real axis.

We know a conformal transformation that maps the half-plane minus a half-disc into the upper half-plane; it is given by $z = z + a^2/z$, where a is the radius of the disc that is removed. The circle $\mathscr{C}(0\,;a)$ is mapped to the segment $[-2a, 2a]$.

There only remains to find a free flow in the new domain: it is given trivially by

$$\Theta(w) = V_0 w.$$

The solution of the problem is therefore

$$\Phi(z) = V_0\left(z + \frac{a^2}{z}\right).$$

Notice that it is of the stated perturbed form, with $G(z) = V_0 a^2/z$.

One can show that, if $|z| = a$, then $\overline{\Phi'(z)}$ is tangent to the circle (i.e., $\Phi'(z)$ is orthogonal to the circle).

The stagnation points of the fluid are given by $\Phi'(z) = 0$, which gives $z = \pm a$.

♦ **Solution of exercise 6.6.** Notice that the function which is proposed is the imaginary part of

$$z \longmapsto \alpha\, L(z+1) + \beta\, L(z-1) + \gamma,$$

where L is the holomorphic logarithm defined on $\mathbb{C}$ with a cut along the lower half-plane. We look for the constants α, β, and γ by noticing that on the boundary of the sheet, $T = T_0$ for $\theta_1 = \theta_2 = \pi$, $T = T_1$ for $\theta_1 = 0$ and $\theta_2 = \pi$, and $T = T_2$ for $\theta_1 = \theta_2 = 0$. We obtain then

$$\begin{aligned} T &= \frac{T_0 - T_1}{\pi}\theta_1 + \frac{T_1 - T_2}{\pi}\theta_2 + T_2 \\ &= \frac{T_0 - T_1}{\pi} \arctan\left(\frac{y}{x+1}\right) + \frac{T_1 - T_2}{\pi} \arctan\left(\frac{y}{x-1}\right) + T_2. \end{aligned}$$

Chapter 7

Distributions I

7.1 Physical approach

7.1.a The problem of distribution of charge

We know that a point-like particle with electric charge q, placed at a point $\boldsymbol{r}_0$ in the usual space $\mathbb{R}^3$, produces, at any point $\boldsymbol{r}$ in $\mathbb{R}^3 \setminus \{\boldsymbol{r}_0\}$, an electrostatic field

$$\boldsymbol{E}(\boldsymbol{r}) = \frac{q}{4\pi\varepsilon_0} \frac{\boldsymbol{r} - \boldsymbol{r}_0}{\|\boldsymbol{r} - \boldsymbol{r}_0\|^3}. \tag{$*$}$$

If more charges are present, the linearity of the Maxwell equations ensures that the fields created by all the charges is the sum of those created by each charge. However, when working in *macroscopic* scale, it is sometimes better to describe the distribution of charges in *continuous* form; it is then modeled by a function $\rho : \mathbb{R}^3 \to \mathbb{R}$ which associates, to each point, the density of electric charge at this point. The interpretation of the function ρ is the following: if $\boldsymbol{r}_0 \in \mathbb{R}^3$ and if $\mathrm{d}^3\boldsymbol{r} = \mathrm{d}x\,\mathrm{d}y\,\mathrm{d}z$ is an elementary volume around $\boldsymbol{r}_0$, then

$$\rho(\boldsymbol{r}_0)\,\mathrm{d}x\,\mathrm{d}y\,\mathrm{d}z$$

represents the total charge contained in the elementary volume $\mathrm{d}x\,\mathrm{d}y\,\mathrm{d}z$.

The electric field at a point $\boldsymbol{r}$ is then (assuming the integral exists):

$$\boldsymbol{E}(\boldsymbol{r}) = \frac{1}{4\pi\varepsilon_0} \iiint_{\mathbb{R}^3} \rho(\boldsymbol{r}') \frac{\boldsymbol{r} - \boldsymbol{r}'}{\|\boldsymbol{r} - \boldsymbol{r}'\|^3} \, \mathrm{d}^3\boldsymbol{r}', \qquad (**)$$

and the total charge contained in a volume $\mathscr{V} \subset \mathbb{R}^3$ is

$$Q = \iiint_{\mathscr{V}} \rho(\boldsymbol{r}) \, \mathrm{d}^3\boldsymbol{r}.$$

Can we reconcile the two expressions $(*)$ and $(**)$ for the point-like charge and the density of charge respectively? In other words, can we write both equations in a uniform way?

To express the point-like charge with the "continuous" viewpoint, we would need a function $\delta_{\boldsymbol{r}_0}$ describing a point-like unit charge located at $\boldsymbol{r}_0$. This function would thus be zero on $\mathbb{R}^3 \setminus \{\boldsymbol{r}_0\}$; moreover, when integrated over a volume $\mathscr{V}$, we would obtain

$$\iiint_{\mathscr{V}} \delta_{\boldsymbol{r}_0}(\boldsymbol{r}) \, \mathrm{d}^3\boldsymbol{r} = \begin{cases} 1 & \text{if } \boldsymbol{r}_0 \in \mathscr{V}, \\ 0 & \text{if } \boldsymbol{r}_0 \notin \mathscr{V}. \end{cases}$$

Generalizing, for any continuous function $f : \mathbb{R}^3 \to \mathbb{R}$, we would need to have

$$\iiint_{\mathscr{V}} f(\boldsymbol{r}) \, \delta_{\boldsymbol{r}_0}(\boldsymbol{r}) \, \mathrm{d}^3\boldsymbol{r} = f(\boldsymbol{r}_0). \qquad (7.1)$$

But, according to the theory of integration, since $\delta_{\boldsymbol{r}_0}$ is zero almost everywhere, the integral (7.1) is necessarily zero also. *A "Dirac function" cannot exist.*

One may, following Dirac, use a sequence of functions with constant integral equal to 1, positive, and that "concentrate" around 0, as for instance:

$$\delta_n(x) = \begin{cases} n & \text{if } |x| \leqslant 1/2n, \\ 0 & \text{if } |x| > 1/2n, \end{cases}$$

or its equivalent in three dimensions

$$D_n(\boldsymbol{r}) = \begin{cases} 3n^3/4\pi & \text{if } \|\boldsymbol{r}\| \leqslant 1/2n, \\ 0 & \text{if } \|\boldsymbol{r}\| > 1/2n, \end{cases}$$

Every time the temptation arises to use the "δ-function," the sequence of functions $(\delta_n)_{n\in\mathbb{N}}$ (resp. $(D_n)_{n\in\mathbb{N}}$) can be used instead, and *at the very end of the computation* one must take the limit $[n \to \infty]$. Thus, $\delta_{\boldsymbol{r}_0}$ is replaced in (7.1) by $D_n(\boldsymbol{r} - \boldsymbol{r}_0)$, and the formula is written

$$\iiint_{\mathscr{V}} f(\boldsymbol{r}) \, \delta_{\boldsymbol{r}_0}(\boldsymbol{r}) \, \mathrm{d}^3\boldsymbol{r} = \lim_{n\to\infty} \iiint_{\mathscr{V}} f(\boldsymbol{r}) \, D_n(\boldsymbol{r} - \boldsymbol{r}_0) \, \mathrm{d}^3\boldsymbol{r} = f(\boldsymbol{r}_0).$$

The reader can check that this formula is valid for any function f continuous at $\boldsymbol{r}_0$ (see for instance a fairly complete discussion of this point of view in the book by Cohen-Tannoudji et al. [20, appendix II]). However, this procedure is fairly heavy. This is what motivated Laurent SCHWARTZ (and, independently, on the other side of the Iron Curtain, Israël GELFAND [38]) to create the **theory of distributions** (resp. of **generalized functions**), which gives a rigorous meaning to the "δ-function" and justifies the preceding technique (see Theorem 8.18 on page 232).

In the same manner we will want to describe not only the distribution of point-like charges, but also a charge supported on a surface or a curve, so that, if we denote by $\rho(\boldsymbol{r})$ the function giving the distribution of charges, we can compute the total charge contained within the volume $\mathscr{V}$ with the formula

$$\begin{aligned} Q(\mathscr{V}) &= \iiint_{\mathscr{V}} \rho(\boldsymbol{r})\,\mathrm{d}^3\boldsymbol{r} \\ &= \sum_{\substack{\text{charges } i \text{ in} \\ \text{the volume } \mathscr{V}}} q_i + \int_{\mathscr{L}\cap\mathscr{V}} \mu\,\mathrm{d}\ell + \iint_{\mathscr{S}\cap\mathscr{V}} \sigma\,\mathrm{d}^2 s + \iiint_{\mathscr{V}} \rho_{\text{vol.}}(\boldsymbol{r})\,\mathrm{d}^3\boldsymbol{r}, \end{aligned}$$

where $\rho_{\text{vol.}}$ is the volumic density of charge, σ the surface distribution on the surface $\mathscr{S}$, μ the linear distribution on the curve $\mathscr{L}$, and q_i the point-like charges.

7.1.b The problem of momentum and forces during an elastic shock

In this second example, we will look at the problem of an elastic shock between two objects.

Consider a game of squash. Assuming that at a given time the ball reaches a wall (orthogonally to the surface, to simplify) with velocity v_0, and rebounds. The ball gets "squashed" a little bit, so that the contact lasts for a period of time Δt which is nonzero; then it rebounds with velocity $-v_0$. Hence the graph of the speed of the ball as a function of time has the following shape:

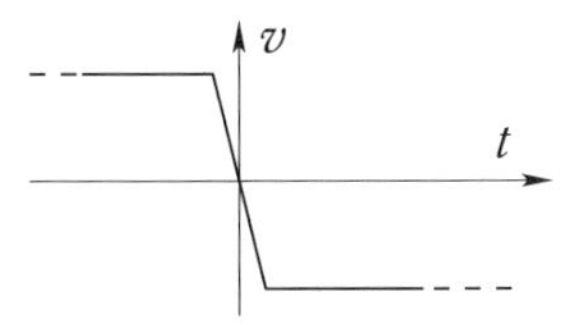

Newton's law stipulates that, during the movement, the force f exerted on the ball is such that $f = m\dot{v}$; it is therefore proportional to the derivative of the function graphed above. Now if we wish to model in a simple way a "hard" collision (where the "squashing" of the ball is not taken into account, for instance, as in a game of "pétanque"), the graph of the speed becomes

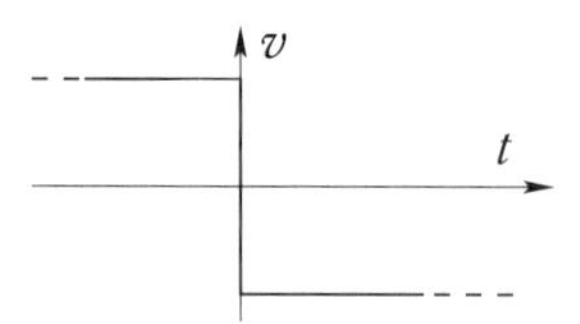

and the force exerted should still be proportional to the derivative of this function; in a word, f should be zero for any $t \neq 0$ and satisfy

$$\frac{1}{m}\int_{-\infty}^{+\infty} f(t)\,\mathrm{d}t = v(+\infty) - v(-\infty) = -2v_0.$$

Once again, neither the integral nor the previous derivative can be treated in the usual sense of functions.

7.2

Definitions and examples of distributions

We will now present a mathematical tool "δ" which, applied to a continuous function f, gives its value at 0, a relation which will be written

$$\delta(f) = f(0) \qquad \text{or} \qquad \langle \delta, f \rangle = f(0).$$

The notation $\langle \delta, f \rangle$ is simply a convenient equivalent to $\delta(f)$ so this means that δ is a "machine that acts on functions," associating them with a number. Now the rule above happens to be *linear* as a function of f; so it is natural to apply it to elements of a certain vector space of functions (that may still be subject to some choice). This object δ is thus what is called **a complex linear form** or **linear functional** on a space of functions.[1] So we have

$$\begin{aligned} \delta : \text{space of functions} &\longrightarrow \mathbb{C}, \\ \text{function} &\longmapsto \text{number}. \end{aligned}$$

There remains to make precise on which space of functions the functional δ (and other useful functionals) will be defined.

DEFINITION 7.1 (Test function) Let $n \geqslant 1$. The **test space**, denoted $\mathscr{D}(\mathbb{R}^n)$, is the vector space of functions φ from $\mathbb{R}^n$ into $\mathbb{C}$, which are of class $\mathscr{C}^\infty$ and have bounded support (i.e., they vanish outside some ball; in the case $n = 1$, they vanish outside a bounded interval).

A **test function** is any function $\varphi \in \mathscr{D}(\mathbb{R}^n)$.

[1] Bourbaki presents the operation of integration as a linear form on a space of functions. According to L. Schwartz, this abstract approach seems to have played a role in the genesis of the theory of distributions.

When the value of n is clear or irrelevant in the context, we will simply denote $\mathscr{D}$ instead of $\mathscr{D}(\mathbb{R}^n)$.

Example 7.2 There are an abundance of test functions, but it is not so easy to find one. (In particular because they are not analytic on $\mathbb{R}^n$ – can you see why?)

As an exercise, the reader will show that for $a, b \in \mathbb{R}$ with $a < b$, the function

$$\varphi(x) = \begin{cases} \exp\left(\dfrac{1}{(x-a)(x-b)}\right) & \text{if } x \in \left]a, b\right[, \\ 0 & \text{if } x \in \left]-\infty, a\right] \cup \left[b, +\infty\right[\end{cases}$$

is a test function on $\mathbb{R}$.

DEFINITION 7.3 A **distribution** on $\mathbb{R}^n$ is any continuous linear functional defined on $\mathscr{D}(\mathbb{R}^n)$. The distributions form a vector space called the **topological dual** of $\mathscr{D}(\mathbb{R}^n)$, also called the **space of distributions** and denoted $\mathscr{D}'(\mathbb{R}^n)$ or $\mathscr{D}'$.

For a distribution $T \in \mathscr{D}'$ and a test function $\varphi \in \mathscr{D}$, the value of T at φ will usually be denoted not $T(\varphi)$ but instead $\langle T, \varphi \rangle$. Thus, for any $T \in \mathscr{D}'$ and any $\varphi \in \mathscr{D}$, $\langle T, \varphi \rangle$ is a complex number.

Remark 7.4 This definition, simple in appearance, deserves some comments.

Why so many constraints in the definition of the functions of $\mathscr{D}$? The reason is simple: the topological dual of $\mathscr{D}$ gets "bigger," or "richer," if $\mathscr{D}$ is "small." The restriction of the space test $\mathscr{D}$ to a very restricted subset of functions produces a space of distribution which is very large.[2] (This is therefore the *opposite phenomenon* from what might have been expected from the experience in finite dimensions.[3])

The $\mathscr{C}^\infty$ regularity is necessary but the condition of bounded support can be relaxed, and there is a space of functions larger than $\mathscr{D}$ which is still small enough that most interesting distributions belong to its dual. It is the *Schwartz space* $\mathscr{S}$, which will be defined on page 289.

A distribution T is, by definition, a *continuous* linear functional on $\mathscr{D}$; this means that for any sequence $(\varphi_n)_{n\in\mathbb{N}}$ of test functions that converges to a test function $\varphi \in \mathscr{D}$, the sequence of complex numbers $(\langle T, \varphi_n \rangle)_{n\in\mathbb{N}}$ must converge to the value $\langle T, \varphi \rangle$. To make this precise we must first specify precisely what is meant by "a sequence of test functions $(\varphi_n)_{n\in\mathbb{N}}$ converging to φ."

DEFINITION 7.5 (Convergence in $\mathscr{D}$) A sequence of test functions $(\varphi_n)_{n\in\mathbb{N}}$ in $\mathscr{D}$ **converges in $\mathscr{D}$** to a function $\varphi \in \mathscr{D}$ if

- the supports of the functions φ_n are contained in a fixed bounded subset, independant of n;

---

[2] For those still unconvinced: we may define a linear functional on $\mathscr{D}$, which to any test function φ associates $\int \varphi(t)\, \mathrm{d}t$. However, if we had taken as $\mathscr{D}$ a space which is *too large*, for instance, the space L^1_{loc} of *locally* integrable functions (see Definition 7.8 on the next page), then this functional would not be well-defined since the integral $\int f(t)\, \mathrm{d}t$ may have been divergent, for instance, when $f(t) \equiv 1$. The functional $\varphi \mapsto \int \varphi(t)\, \mathrm{d}t$ is thus well-defined on the vector space $\mathscr{D}$ of test functions, but not on the space of locally integrable functions. Hence it belongs to $\mathscr{D}'$ but not to $(L^1_{\text{loc}})'$.

[3] Recall that in finite dimensions the dual space is of the same dimension as the starting space. Thus, the bigger the vector space, the bigger its dual.

- all the partial derivatives of all order of the φ_n converge uniformly to the corresponding partial derivative of φ: for $n = 1$, for instance, this means

$$\varphi_n^{(p)} \xrightarrow{\text{cv.u.}} \varphi^{(p)} \qquad \text{for any } p \in \mathbb{N}.$$

Now the continuity of a linear functional T defined on $\mathscr{D}$ can be defined in the following natural manner:

DEFINITION 7.6 (Continuity of a linear functional) A linear functional $T : \mathscr{D} \to \mathbb{C}$ is *continuous* if and only if:

$$\begin{cases} \text{for any sequence } (\varphi_n)_{n\in\mathbb{N}} \text{ of test functions} \\ \text{in } \mathscr{D}, \\ \text{if } (\varphi_n)_{n\in\mathbb{N}} \text{ converges in } \mathscr{D} \text{ to } \varphi, \\ \text{the sequence } \langle T, \varphi_n \rangle \text{ converges in } \mathbb{C} \text{ to} \\ \langle T, \varphi \rangle. \end{cases}$$

Example 7.7 We can now show that the functional δ defined by $\delta : \varphi \mapsto \varphi(0)$ is continuous.

Let $(\varphi_n)_{n\in\mathbb{N}}$ be a sequence of test functions, converging to φ in $\mathscr{D}$. Hence $\varphi_n \xrightarrow{\text{cv.u.}} \varphi$, which implies the simple convergence of $(\varphi_n)_n$ to φ; in particular, $\varphi_n(0)$ tends to $\varphi(0)$ in $\mathbb{C}$. This shows that $\langle \delta, \varphi_n \rangle$ tends to $\langle \delta, \varphi \rangle$ in $\mathbb{C}$, which establishes the continuity of δ.

7.2.a Regular distributions

Distributions can be seen as a generalization of the notion of functions or, more precisely, of *locally integrable functions.*

DEFINITION 7.8 A measurable function $f : \mathbb{R}^n \to \mathbb{C}$ is **locally integrable** if, for any compact $K \subset \mathbb{R}^n$, the function $f \cdot \chi_K$ is integrable on K. The space of locally integrable functions on $\mathbb{R}^n$ is denoted $L^1_{\text{loc}}(\mathbb{R}^n)$.

Functions which are locally integrable but not integrable are those which have integrability problems at infinity.

Example 7.9 The function $x \mapsto 1/\sqrt{|x|}$ is locally integrable (but not integrable on $\mathbb{R}$, for it decreases too slowly). More generally, any function in $L^1(\mathbb{R}^n)$, $L^2(\mathbb{R}^n)$, or $L^p(\mathbb{R}^n)$ with $p \geqslant 1$, is locally integrable.

Any locally integrable function defines then a distribution by means of the following theorem:

THEOREM 7.10 (Regular distributions) *For any locally integrable function f, there is an associated distribution, also denoted f, defined by*

$$\forall \varphi \in \mathscr{D} \qquad \langle f, \varphi \rangle \stackrel{\text{def}}{=} \int f(x)\, \varphi(x)\, \mathrm{d}x.$$

Such a distribution is called the **regular distribution** *associated to the locally integrable function f.*

PROOF. It suffices to show that the map $\varphi \mapsto \int f(x)\,\varphi(x)\,\mathrm{d}x$ is indeed linear (which is obvious) and continuous.

Let $(\varphi_n)_{n\in\mathbb{N}}$ be a sequence of test functions converging to $\varphi \in \mathscr{D}$. There exists a bounded closed ball B such that B contains the supports of *all* the φ_n. Since the function f is locally integrable, we can define $M = \int_B |f|$ and this M is a finite real number.

We then have $\left|\int f(\varphi_n - \varphi)\right| \leqslant M \int |\varphi_n - \varphi| \leqslant M\,V\,\|\varphi_n - \varphi\|_\infty$, where V is the finite volume of the ball B. But $\|\varphi_n - \varphi\|_\infty$ tends to 0 by the definition 7.5 and, therefore, $\langle f, \varphi_n - \varphi\rangle$ tends to 0. By linearity, $\langle f, \varphi_n\rangle$ tends to $\langle f, \varphi\rangle$.

Remark 7.11 If the functions f and g are equal almost everywhere, the distributions f and g will be equal, that is, $\langle f, \varphi\rangle = \langle g, \varphi\rangle$ for any $\varphi \in \mathscr{D}$. The converse is also true: if two regular distributions are equal as functionals, then the corresponding functions are equal almost everywhere. The reader is invited to construct a proof of this fact using the following property: any integrable function with compact support can be approximated in the mean by functions in $\mathscr{D}$. So the notion of distribution is in fact an extension of the notion of *classes of locally integrable functions equal almost everywhere.*

Example 7.12 Denote by $\mathbf{1}$ the constant function $\mathbf{1} : x \mapsto 1$ (defined on $\mathbb{R}$). The regular distribution associated to $\mathbf{1}$ is thus the map

$$\mathbf{1} : \mathscr{D} \longrightarrow \mathbb{C},$$

$$\varphi \longmapsto \langle \mathbf{1}, \varphi\rangle = \int_{-\infty}^{+\infty} \varphi(x)\,\mathrm{d}x.$$

Regular distributions will have considerable importance throughout this chapter, since they are those which "look the most like" classical functions. They will be used more than once to discover a new definition when time will come to generalize to distributions certain concepts associated to functions (such as the derivative, the Fourier transform, etc.).

7.2.b Singular distributions

We now define the Dirac distribution:

DEFINITION 7.13 The **Dirac distribution** is the distribution which, to any function φ in $\mathscr{D}(\mathbb{R})$, associates its value at 0:

$$\langle \delta, \varphi\rangle \stackrel{\text{def}}{=} \varphi(0) \qquad \forall \varphi \in \mathscr{D}.$$

For $a \in \mathbb{R}$, we define similarly the **Dirac distribution centered at a**, denoted δ_a, by its action on any test function:

$$\langle \delta_a, \varphi\rangle \stackrel{\text{def}}{=} \varphi(a) \qquad \forall \varphi \in \mathscr{D}.$$

Frequently (this will be justified page 189), δ_a will instead be denoted $\delta(x-a)$.

The preceding definition is easily generalized to the case of many dimensions:

The English physicist and mathematician Paul Adrien Maurice DIRAC (1902–1984) was one of the great founders of quantum mechanics. One of his major achievements was to give, for the first time, a correct description of a quantum relativistic electron *(Dirac equation)*, interpreting the solutions with negative energy that were embarrassing physicists as *antiparticles*. He also showed how the mathematical theory of groups could bring precious information concerning particle physics. According to Dirac, the mathematical beauty of a physical theory should be of paramount importance.

DEFINITION 7.14 (Dirac Distribution on $\mathbb{R}^n$) Let $\boldsymbol{r}_0 \in \mathbb{R}^n$ be a point in $\mathbb{R}^n$. The **Dirac distribution at $\boldsymbol{r}_0$** is the distribution defined by its action on any test function in $\mathscr{D}(\mathbb{R}^n)$:

$$\left\langle \delta_{\boldsymbol{r}_0}, \varphi \right\rangle \stackrel{\text{def}}{=} \varphi(\boldsymbol{r}_0) \qquad \forall \varphi \in \mathscr{D}.$$

As before, it will often be denoted $\delta(\boldsymbol{r} - \boldsymbol{r}_0)$.

Remark 7.15 We have thus introduced a notation that may at first lead to some confusion: if T is a distribution, $T(\boldsymbol{r})$ means not the value of T at the point $\boldsymbol{r}$, which would *make no sense*,[4] but the distribution T itself; this serves as reminder that the variable of the function on which T acts is denoted $\boldsymbol{r}$.

DEFINITION 7.16 The **Dirac comb** is the distribution **Ш** (pronounced "sha"[5]) defined by its action on any test functions $\varphi \in \mathscr{D}(\mathbb{R})$:

$$\langle Ш, \varphi \rangle \stackrel{\text{def}}{=} \sum_{n=-\infty}^{+\infty} \varphi(n),$$

that is,

$$Ш = \sum_{n=-\infty}^{+\infty} \delta_n \quad \text{or} \quad Ш(x) \stackrel{\text{def}}{=} \sum_{n=-\infty}^{+\infty} \delta(x-n).$$

Notice that the distribution Ш is well defined, because the sum $\sum_n \varphi(n)$ is in fact *finite* (φ being, by definition, with bounded support).

The continuity is left as an (easy) exercise for the reader.

The δ distribution is represented graphically by a vertical arrow, as here:

---

[4] This bears repeating again at the risk of being boring: a distribution is *not* a function, but a linear form on a space of functions.

[5] This is a Cyrillic alphabet letter, corresponding to the sound "sh."

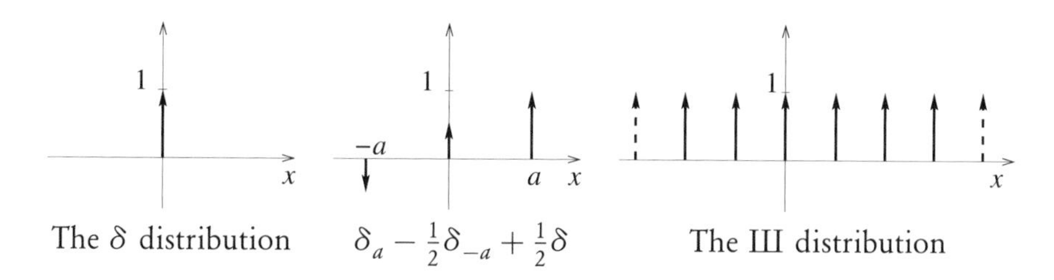

The δ distribution $\qquad \delta_a - \frac{1}{2}\delta_{-a} + \frac{1}{2}\delta \qquad$ The Ш distribution

7.2.c Support of a distribution

Two distributions S and T are **equal** if $\langle S, \varphi \rangle = \langle T, \varphi \rangle$ for any $\varphi \in \mathscr{D}$. They are **equal on an open subset Ω in $\mathbb{R}^n$** if $\langle S, \varphi \rangle = \langle T, \varphi \rangle$ for any $\varphi \in \mathscr{D}$ such that the support of φ is contained in Ω (i.e., for any $\varphi \in \mathscr{D}$ which is zero outside Ω).

Example 7.17 Let H be the Heaviside function. The regular distributions $\mathbf{1}$ and H are equal on $]0, +\infty[$. Indeed, for any function $\varphi \in \mathscr{D}$ which vanishes on $\mathbb{R}^-$, we have

$$\langle \mathbf{1}, \varphi \rangle = \int_{-\infty}^{+\infty} 1 \cdot \varphi(x)\,\mathrm{d}x = \int_{0}^{+\infty} \varphi(x)\,\mathrm{d}x = \langle H, \varphi \rangle .$$

Similarly, Ш and δ are equal on $]-\frac{1}{2}, \frac{1}{2}[$.

Consider now all the open subsets Ω on which a given distribution T is zero (i.e., is equal to the zero distribution on Ω). The union of these open sets is itself an open set. One can then show (but this is not obvious and we will omit it here, see [80] for instance) that this is the largest open set on which T is zero. Its complement, which is therefore closed, is called the **support of the distribution T**, and is denoted $\operatorname{supp}(T)$.

Example 7.18 The support of the Dirac distribution centered at 0 is $\operatorname{supp}(\delta) = \{0\}$, and more generally, for $a \in \mathbb{R}$, we have $\operatorname{supp}(\delta_a) = \{a\}$.

The support of the Dirac comb is $\operatorname{supp}(Ш) = \mathbb{Z}$.

7.2.d Other examples

PROPOSITION 7.19 *Let $\varphi \in \mathscr{D}(\mathbb{R})$ be a test function. The quantity*

$$\int_{-\infty}^{-\eta} \frac{\varphi(x)}{x}\,\mathrm{d}x + \int_{+\eta}^{+\infty} \frac{\varphi(x)}{x}\,\mathrm{d}x$$

has a finite limit as $[\eta \to 0^+]$. Moreover, the map

$$\varphi \longmapsto \lim_{\eta \to 0^+} \int_{|x|>\eta} \frac{\varphi(x)}{x}\,\mathrm{d}x \tag{7.2}$$

is linear and continuous on $\mathscr{D}(\mathbb{R})$.

PROOF. Because of its length, the proof of this theorem is given in Appendix D, page 600.

DEFINITION 7.20 The **Cauchy principal value of $1/x$** is the distribution on $\mathbb{R}$ defined by its action on any φ given by

$$\left\langle \operatorname{pv}\frac{1}{x}, \varphi \right\rangle = \operatorname{pv}\int \frac{\varphi(x)}{x}\,\mathrm{d}x \stackrel{\text{def}}{=} \lim_{\eta\to 0^+} \int_{|x|>\eta} \frac{\varphi(x)}{x}\,\mathrm{d}x.$$

It should be emphasized that this integral is defined *symmetrically* around the origin. It is this symmetry which provides the existence of the limit defining the distribution. The distribution $\operatorname{pv} 1/x$ can also be seen as the derivative of the regular distribution associated to the locally integrable function $x \mapsto \log|x|$ (see Exercise 8.4 on page 241).

Other properties and examples of physical applications of the Cauchy principal value will be given in the next chapter. In addition, a generalization of the method leading to the definition of a distribution from the function $x \mapsto 1/x$ (which is not locally integrable) is given in Exercises 8.5 and 8.6 on page 241.

7.3 Elementary properties. Operations

7.3.a Operations on distributions

We wish to define certain operations on distributions, such as translations, scaling, derivation, and so on. For this purpose, the method will be consistently the same:

> *Consider how those operations are defined for a locally integrable function; write them, in the language of distributions, for the associated regular distribution; then generalize this to an arbitrary distribution.*

For instance, for any locally integrable function f on $\mathbb{R}^n$ and any $a \in \mathbb{R}^n$, we define the **translate** f_a of the function f by

$$\forall x \in \mathbb{R} \qquad f_a(x) \stackrel{\text{def}}{=} f(x-a).$$

In other words, the graph (in $\mathbb{R}^{n+1}$) of the function f has been translated by the vector a:

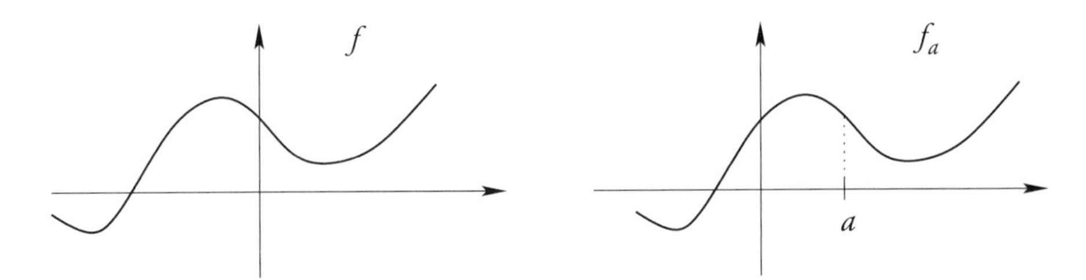

The regular distribution associated to this function, which is also denoted f_a, therefore satisfies

$$\langle f_a, \varphi \rangle = \int f(x-a)\,\varphi(x)\,\mathrm{d}x = \int f(y)\,\varphi(y+a)\,\mathrm{d}y = \langle f, \varphi_{-a} \rangle .$$

This can be generalized to an arbitrary distribution T in the following manner:

DEFINITION 7.21 The **translate** of the distribution T by a, denoted T_a or $T(x-a)$, is defined by

$$\langle T_a, \varphi \rangle \stackrel{\text{def}}{=} \langle T, \varphi_{-a} \rangle \qquad \text{for any } \varphi \in \mathscr{D}(\mathbb{R}^n),$$

which is also written

$$\big\langle T(x-a), \varphi(x) \big\rangle \stackrel{\text{def}}{=} \big\langle T(x), \varphi(x+a) \big\rangle .$$

The reader can easily check that the number $\langle T, \varphi_{-a} \rangle$ exists and that the map $\varphi \longmapsto \langle T, \varphi_{-a} \rangle$ is indeed linear and continuous.

Remark 7.22 As in Remark 7.15, the notation $T(x-a)$ is simply synonymous with T_a; similarly, with an abuse of language, $\varphi(x-a)$ sometimes designates (for physicists more than for mathematicians) the function $t \mapsto \varphi(t-a)$ and not the value of φ at $x-a$.

Example 7.23 Let $a \in \mathbb{R}$. The translate by a of the Cauchy principal value is

$$\operatorname{pv} \frac{1}{x-a} \; : \; \varphi \longmapsto \lim_{\eta \to 0^+} \left[\int_{-\infty}^{a-\eta} \frac{\varphi(x)}{x-a}\,\mathrm{d}x + \int_{a+\eta}^{+\infty} \frac{\varphi(x)}{x-a}\,\mathrm{d}x \right].$$

By analogous reasoning, we define the "transpose" of a distribution. For a function $f : \mathbb{R}^n \to \mathbb{C}$, we denote by $\check{f}$ the **transpose** of f, which is the function $x \mapsto \check{f}(x) \stackrel{\text{def}}{=} f(-x)$. If f is locally integrable, then so is $\check{f}$; hence both f and $\check{f}$ define regular distributions. Moreover, for any $\varphi \in \mathscr{D}(\mathbb{R}^n)$, we have

$$\langle \check{f}, \varphi \rangle = \int \check{f}(x)\,\varphi(x)\,\mathrm{d}x = \int f(-x)\,\varphi(x)\,\mathrm{d}x \stackrel{x=-t}{=} \int f(t)\,\varphi(-t)\,\mathrm{d}t = \langle f, \check{\varphi} \rangle .$$

This justifies the following definition:

DEFINITION 7.24 (Transpose) The **transpose** of a distribution T on $\mathbb{R}^n$, denoted $\check{T}$ or $T(-x)$, is defined by

$$\langle \check{T}, \varphi \rangle \stackrel{\text{def}}{=} \langle T, \check{\varphi} \rangle ,$$

which can also be written:

$$\big\langle T(-x), \varphi(x) \big\rangle \stackrel{\text{def}}{=} \big\langle T(x), \varphi(-x) \big\rangle$$

for any $\varphi \in \mathscr{D}(\mathbb{R}^n)$.

Similarly, for any $a \in \mathbb{R}^*$, a change of scale defines the **dilation** by a of a function f by $x \mapsto f(ax)$. The regular distribution associated to the dilation of a function satisfies

$$\big\langle f(ax), \varphi(x) \big\rangle = \int f(ax)\, \varphi(x)\, \mathrm{d}x = \int f(x)\, \varphi\Big(\frac{x}{a}\Big)\, \frac{\mathrm{d}x}{|a|}$$
$$= \frac{1}{|a|} \Big\langle f(x), \varphi\Big(\frac{x}{a}\Big) \Big\rangle .$$

This leads to the following generalization:

DEFINITION 7.25 (Dilation) Let $a \in \mathbb{R}^*$. The **dilation** of a distribution T by the factor a, denoted $T(ax)$, is defined by

$$\big\langle T(ax), \varphi(x) \big\rangle \stackrel{\text{def}}{=} \frac{1}{|a|} \Big\langle T(x), \varphi\Big(\frac{x}{a}\Big) \Big\rangle \qquad \text{for any } \varphi \in \mathscr{D}(\mathbb{R}).$$

Dilations of a distribution on $\mathbb{R}^n$ are defined in exactly the same manner. However, the jacobian associated to the change of variable $y = ax$ is $|a|^n$ instead of $|a|$, which gives the dilation relations

$$\boxed{\big\langle T(a\boldsymbol{x}), \varphi(\boldsymbol{x}) \big\rangle = \frac{1}{|a|^n} \Big\langle T(\boldsymbol{x}), \varphi\Big(\frac{\boldsymbol{x}}{a}\Big) \Big\rangle} \qquad \text{on } \mathbb{R}^n.$$

There is, unfortunately, no general definition of the product of two distributions.[6] Already, if f and g are two locally integrable functions (hence defining regular distributions), their product is not necessarily locally integrable.[7] However, if ψ is a function of class $\mathscr{C}^\infty$, then for any $\varphi \in \mathscr{D}$, the product $\psi\varphi$ is a test function (still being $\mathscr{C}^\infty$ and with bounded support); if f is locally integrable, then ψf is also locally integrable. From the point of view of distributions, we can write

$$\langle \psi f, \varphi \rangle = \int \big(\psi(x) f(x)\big)\, \varphi(x)\, \mathrm{d}x = \int f(x)\, \big(\psi(x)\, \varphi(x)\big)\, \mathrm{d}x = \langle f, \psi\varphi \rangle .$$

---

[6] In particular, it is easy to notice that taking the square of the Dirac distribution raises arduous questions. However, it turns out to be possible to define a product in a *larger* space than the space of distributions, the space of generalized distributions of Colombeau [22, 23].

[7] Simply take $f(x) = g(x) = 1/\sqrt{x}$.

This leads us to state the following definition:

DEFINITION 7.26 (Product of a distribution by a $\mathscr{C}^\infty$ function) Let $T \in \mathscr{D}'$ be a distribution and let ψ be a function of class $\mathscr{C}^\infty$. The **product** $\psi T \in \mathscr{D}'$ is defined by

$$\langle \psi T, \varphi \rangle \stackrel{\text{def}}{=} \langle T, \psi\varphi \rangle \qquad \text{for any } \varphi \in \mathscr{D}.$$

By applying this definition to the Dirac distribution, the following fundamental result follows:

THEOREM 7.27 *Let ψ be a function of $\mathscr{C}^\infty$ class. Then we have*

$$\boxed{\psi(x)\,\delta(x) = \psi(0)\,\delta(x)} \qquad \text{and in particular} \qquad \boxed{x\,\delta(x) = 0.}$$

Similarly, we have the relation

$$\psi(x)\,\text{Ш}(x) = \sum_{n\in\mathbb{Z}} \psi(n)\,\delta(x-n).$$

PROOF. If ψ is of $\mathscr{C}^\infty$ class, then by definition

$$\langle \psi\delta, \varphi \rangle = \langle \delta, \psi\varphi \rangle = \psi(0)\,\varphi(0) = \psi(0)\,\langle \delta, \varphi \rangle = \big\langle \psi(0)\,\delta, \varphi \big\rangle.$$

Since any test function has bounded support, the last formula follows by simple linearity.

It is important to notice that, for instance, the product $H\delta$ makes no sense here!

THEOREM 7.28 *The equation $x\,T(x) = 0$, with unknown a distribution T, admits the multiples of the Dirac distribution as solutions, and they are the only solutions:*

$$x\,T(x) = 0 \iff \big(T = \alpha\delta, \ \text{with } \alpha \in \mathbb{C}\big).$$

Remark 7.29 If ψ is merely continuous, the product ψT cannot be defined for an arbitrary distribution T, but the product $\psi\delta$ can in fact still be defined. Similarly, if ψ is of $\mathscr{C}^1$ class (in a neighborhood of 0), the product $\psi\delta'$ can be defined (see the next section).

7.3.b Derivative of a distribution

The next goal is to define derivation in the space of distributions. As before, consider a locally integrable function $f \in L^1_{\text{loc}}$, and assume moreover that it is differentiable and its derivative is also locally integrable. Then the regular distribution associated to f' is given by

$$\langle f', \varphi \rangle = \int f'(x)\,\varphi(x)\,\mathrm{d}x = -\int f(x)\,\varphi'(x)\,\mathrm{d}x = -\langle f, \varphi' \rangle.$$

The passage to the second integral is done by integrating by parts; the boundary terms vanish because φ has bounded support (this is one reason for the "very restrictive" choice of the space of test functions). This justifies the following generalization:

DEFINITION 7.30 The **derivative of a distribution** $T \in \mathscr{D}(\mathbb{R})'$ is defined by

$$\langle T', \varphi \rangle \stackrel{\text{def}}{=} - \langle T, \varphi' \rangle \qquad \text{for any } \varphi \in \mathscr{D}(\mathbb{R}). \tag{7.3}$$

Similarly, the higher order derivatives are defined by induction by

$$\left\langle T^{(m)}, \varphi \right\rangle \stackrel{\text{def}}{=} (-1)^m \left\langle T, \varphi^{(m)} \right\rangle \qquad \text{for any } \varphi \in \mathscr{D} \text{ and any } m \in \mathbb{N}.$$

Example 7.31 The function $x \mapsto \log|x|$ is locally integrable and therefore defines a regular distribution. However, its derivative (in the usual sense of functions) $x \mapsto 1/x$ is not locally integrable; therefore it does not define a (regular) distribution.

However, if the derivation is performed directly in the sense of distributions, $\big[\log|x|\big]'$ is a distribution (it is shown in Exercise 8.4 on page 241 that it is the Cauchy principal value $\text{pv}\,\frac{1}{x}$).

If working on $\mathbb{R}^n$, a similar computation justifies the definition of partial derivatives of a distribution $T \in \mathscr{D}'(\mathbb{R}^n)$ by

$$\left\langle \frac{\partial T}{\partial x_i}, \varphi \right\rangle \stackrel{\text{def}}{=} - \left\langle T, \frac{\partial \varphi}{\partial x_i} \right\rangle$$

for $\varphi \in \mathscr{D}(\mathbb{R}^n)$ and $1 \leqslant i \leqslant n$. Partial derivatives of higher order are obtained by successive applications of these rules: for instance,

$$\left\langle \frac{\partial^3 T}{\partial x_1^2 \partial x_2}, \varphi \right\rangle = (-1)^3 \left\langle T, \frac{\partial^3 \varphi}{\partial x_1^2 \partial x_2} \right\rangle.$$

In particular, on $\mathbb{R}^3$, we have

$$\left\langle \frac{\partial T}{\partial x}, \varphi \right\rangle = - \left\langle T, \frac{\partial \varphi}{\partial x} \right\rangle, \qquad \left\langle \frac{\partial T}{\partial y}, \varphi \right\rangle = - \left\langle T, \frac{\partial \varphi}{\partial y} \right\rangle,$$

$$\left\langle \frac{\partial T}{\partial z}, \varphi \right\rangle = - \left\langle T, \frac{\partial \varphi}{\partial z} \right\rangle.$$

Similarly, the laplacian being a differential operator of order 2, we define:

DEFINITION 7.32 The **laplacian** of a distribution on $\mathbb{R}^n$ is given by

$$\langle \triangle T, \varphi \rangle \stackrel{\text{def}}{=} \langle T, \triangle \varphi \rangle \qquad \text{for any } \varphi \in \mathscr{D}(\mathbb{R}^n).$$

THEOREM 7.33 (Differentiability of distributions) *Any distribution on $\mathbb{R}^n$ admits partial derivatives of arbitrary order. Any partial derivative of a distribution is a distribution.*

COROLLARY 7.33.1 *Any locally integrable function on $\mathbb{R}$ defining a distribution, it is infinitely differentiable in the sense of distributions:*

$$\forall f \in L^1_{\text{loc}}(\mathbb{R}) \quad \forall k \in \mathbb{N} \qquad f^{(k)} \in \mathscr{D}'(\mathbb{R}).$$

At the age of sixteen, Oliver HEAVISIDE (1850–1925), born in a very poor family, had to leave school. At eighteen, he was hired by Charles Wheatstone, the inventor of the telegraph. He then studied the differential equations which govern electric signals and in particular the "square root of the derivative" (he wrote $\sqrt{\mathrm{d}/\mathrm{d}t}\, H(t) = 1/\sqrt{\pi t}$), which gave apoplexies to the mathematicians of Cambridge.[9] One of his papers occasioned the severe disapproval of the Royal Society. He defended vector analysis (Maxwell was in favor of the use of quaternions instead of vectors), symbolic calculus, and the use of divergent series, and had faith in methods "which work" before being rigorously justified.

When Laurent Schwartz presented the theory of distributions, this result seemed particularly marvelous to the audience: all distributions are infinitely differentiable and their successive derivatives are themselves distributions.[8] Thus, if we consider a locally integrable function, however irregular, its associated distribution is still infinitely differentiable. This is one of the key advantages of the theory of distributions.

7.4 Dirac and its derivatives

7.4.a The Heaviside distribution

A particularly important example is the Heaviside distribution, which we will use constantly.

DEFINITION 7.34 The **Heaviside function** is the function $H : \mathbb{R} \to \mathbb{R}$ defined by

$$H(x) = \begin{cases} 0 & \text{if } x < 0, \\ 1/2 & \text{if } x = 0, \\ 1 & \text{if } x > 0. \end{cases}$$

[8] Mathematicians had been aware for less than a century of the fact that a continuous function could be very far from differentiable (since there exist, for instance, functions continuous on all of $\mathbb{R}$ which are nowhere differentiable, as shown by Peano, Weierstrass, and others).

[9] Enamored of rigor from a recent date only and, therefore, as inflexible as any new convert. The reader interested in the "battle" between Oliver Heaviside and the rigorists of Cambridge can read with profit the article [48].

DEFINITION 7.35 The **Heaviside distribution** is the regular distribution associated to the Heaviside function. It is thus defined by

$$\forall \varphi \in \mathscr{D}(\mathbb{R}) \qquad \langle H, \varphi \rangle \stackrel{\text{def}}{=} \int_0^{+\infty} \varphi(x)\,\mathrm{d}x.$$

The derivative, in the sense of distributions, of H is then by definition

$$\langle H', \varphi \rangle = -\langle H, \varphi' \rangle = -\int_0^{+\infty} \varphi'(x)\,\mathrm{d}x = -\left[\varphi\right]_0^{+\infty} = \varphi(0),$$

from which we deduce

THEOREM 7.36 *The derivative of the Heaviside distribution is the Dirac distribution*

$$\boxed{H' = \delta} \qquad \textit{in the sense of distributions.}$$

7.4.b Multidimensional Dirac distributions

In many cases, the physicist is working in $\mathbb{R}^3$ or $\mathbb{R}^4$. We will define here the notions of point-like, linear, and surface Dirac distributions, which occur frequently in electromagnetism, for instance.

Point-like Dirac distribution

Recall that in three dimensions the Dirac distribution is defined by

$$\langle \delta^{(3)}, \varphi \rangle \stackrel{\text{def}}{=} \varphi(\mathbf{0}).$$

The common notation $\delta^{(3)}(\boldsymbol{r}) = \delta(x)\,\delta(y)\,\delta(z)$ will be explained in Section 7.6.b. Most authors use δ instead of $\delta^{(3)}$ since, in general, there is no possible confusion. Thus a point-like charge in electrostatics is represented by a distribution $\rho(\boldsymbol{r}) = q\,\delta(\boldsymbol{r} - \boldsymbol{a})$. Note that, when it acts on a space variable, the three-dimensional Dirac distribution has the dimension of *the inverse of a volume*:

$$[\delta] = \frac{1}{[L]^3},$$

which implies that $\rho(\boldsymbol{r})$ has the dimension Q/L^3 of a volum density of charge.

Surface Dirac distribution

To describe a surface carrying a uniform charge, we use the surface Dirac distribution:

DEFINITION 7.37 Let $\mathscr{S}$ be a smooth surface in $\mathbb{R}^3$. The (normalized) **Dirac surface distribution** on $\mathscr{S}$ is given by its action on any test function $\varphi \in \mathscr{D}(\mathbb{R}^3)$:

$$\langle \delta_{\mathscr{S}}, \varphi \rangle \stackrel{\text{def}}{=} \iint_{\mathscr{S}} \varphi \, \mathrm{d}^2 \boldsymbol{s},$$

where $\mathrm{d}^2 \boldsymbol{s}$ is the integration element over the surface $\mathscr{S}$. In electromagnetism, a uniform surface density σ_0 on $\mathscr{S}$ is described by the distribution $\sigma_0 \delta_{\mathscr{S}}$. This can be extended to nonuniform surface distribution, since we know how to multiply a distribution by a function of $\mathscr{C}^\infty$ class. So a distribution of charge $\sigma(\boldsymbol{r})$ of $\mathscr{C}^\infty$ class on a surface $\mathscr{S}$ is represented by $\sigma(\boldsymbol{r})\,\delta_{\mathscr{S}}$.

Example 7.38 In $\mathbb{R}^3$, for $R > 0$, we denote by $\delta\big(\|\boldsymbol{x}\| - R\big)$ the Dirac surface distribution $\delta_{\mathscr{S}}$, where $\mathscr{S}$ is the sphere $\mathscr{S} = \big\{\boldsymbol{x} \in \mathbb{R}^3 \,;\, \|\boldsymbol{x}\| = R\big\}$.

It should be noted that when working with spacial variables in $\mathbb{R}^3$, the surface Dirac distribution has the dimension of *the inverse of length*:

$$[\delta_{\mathscr{S}}] = \frac{1}{[L]},$$

which can be recovered by writing (in a heuristic manner, indicated by large quotes)

$$\langle \delta_{\mathscr{S}}, \varphi \rangle = \text{“} \iiint \delta_{\mathscr{S}}(\boldsymbol{r})\, \varphi(\boldsymbol{r})\, \mathrm{d}^3 \boldsymbol{r} \text{”} = \iint_{\mathscr{S}} \varphi(\boldsymbol{r})\, \mathrm{d}^2 \boldsymbol{s}.$$

♦ **Exercise 7.1** Let $\mathscr{S}$ be a surface in $\mathbb{R}^3$ and let $a \in \mathbb{R}$. Denote by $\mathscr{S}'$ the surface obtained from $\mathscr{S}$ by homothety with coefficient $1/a$. Show that

$$\delta_{\mathscr{S}}(a\boldsymbol{r}) = \frac{1}{|a|} \delta_{\mathscr{S}'}(\boldsymbol{r}).$$

◊ **Solution**: Let $\varphi \in \mathscr{D}(\mathbb{R}^3)$. Then we have

$$\begin{aligned}\big\langle \delta_{\mathscr{S}}(a\boldsymbol{r}), \varphi \big\rangle &\stackrel{\text{def}}{=} \frac{1}{|a|^3} \Big\langle \delta_{\mathscr{S}}, \varphi\Big(\frac{\boldsymbol{r}}{a}\Big) \Big\rangle = \frac{1}{|a|^3} \iint_{\mathscr{S}} \varphi\Big(\frac{\boldsymbol{r}}{a}\Big)\, \mathrm{d}^2 \boldsymbol{r} \quad \text{(definition)} \\ &= \frac{1}{|a|^{3-2}} \iint_{\mathscr{S}'} \varphi(\boldsymbol{x})\, \mathrm{d}^2 \boldsymbol{x} = \frac{1}{|a|} \langle \delta_{\mathscr{S}'}, \varphi \rangle .\end{aligned}$$

Curvilinear Dirac distribution

By the same method, a density of charge on a curve is defined using the curvilinear Dirac distribution:

DEFINITION 7.39 Let $\mathscr{L}$ be a curve in $\mathbb{R}^3$. The unit **curvilinear Dirac distribution** on $\mathscr{L}$ is defined by its action on any test function $\varphi \in \mathscr{D}(\mathbb{R}^3)$ given by

$$\langle \delta_{\mathscr{L}}, \varphi \rangle \stackrel{\text{def}}{=} \int_{\mathscr{L}} \varphi \, \mathrm{d}\ell,$$

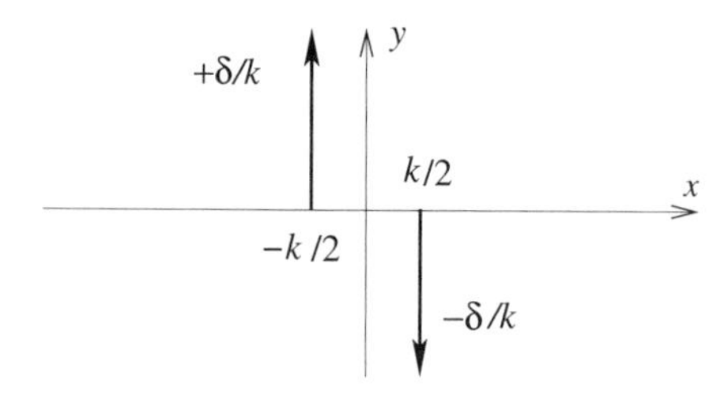

Fig. 7.1 – The distribution δ' can be seen as two Dirac peaks with opposite signs infinitely close to each other – that is, a dipole.

where $\mathrm{d}\ell$ is the integration element over $\mathscr{L}$. In electromagnetism, a density of charge $c(\boldsymbol{r})$ supported on the curve $\mathscr{L}$ is represented by the distribution $c(\boldsymbol{r})\,\delta_{\mathscr{L}}$.

When working with spacial variables in $\mathbb{R}^3$, the curvilinear Dirac distribution has the dimension of the *inverse of a surface*:

$$[\delta_{\mathscr{L}}] = \frac{1}{[L]^2}.$$

7.4.c The distribution δ'

As seen in the defining formula (7.3), the derivative δ' of the Dirac distribution is defined by

$$\langle \delta', \varphi \rangle \stackrel{\text{def}}{=} -\varphi'(0) \qquad \text{for any test function } \varphi \in \mathscr{D}.$$

Note that if φ is a test function, then

$$\begin{aligned}\langle \delta', \varphi \rangle = -\varphi'(0) &= -\lim_{k\to 0} \frac{\varphi\left(\frac{k}{2}\right) - \varphi\left(-\frac{k}{2}\right)}{k} \\ &= -\lim_{k\to 0} \left\langle \frac{1}{k}\left[\delta\left(x-\frac{k}{2}\right) - \delta\left(x+\frac{k}{2}\right)\right], \varphi \right\rangle,\end{aligned}$$

from which we conclude that (note the signs!)

$$\delta'(x) = \lim_{k\to 0} \frac{1}{k}\left[\delta\left(x+\frac{k}{2}\right) - \delta\left(x-\frac{k}{2}\right)\right].$$

This formula is interpreted as follows. The distribution represents a positive charge $1/k$ and a negative charge $-1/k$, situated at a distance equal to k, in the limit where $[k \to 0]$ (see Figure 7.1). Hence the distribution δ' represents a **dipole**, aligned on the horizontal axis $(\mathscr{O}x)$ with dipole moment -1, hence oriented toward *negative* values of x.

Remark 7.40 This expression requires a notion of convergence in the space $\mathscr{D}'$, which will be introduced in Section 8.2.a on page 230.

Can this be generalized to the three-dimensional case? In that situation, we cannot simply take the derivative, but we must indicate *the direction* in which to differentiate.

DEFINITION 7.41 In $\mathbb{R}^3$, the distributions δ'_x, δ'_y, and δ'_z, are defined by

$$\langle \delta'_x, \varphi \rangle \stackrel{\text{def}}{=} -\varphi'_x(\mathbf{0}) = -\frac{\partial \varphi}{\partial x}(\mathbf{0}), \qquad \langle \delta'_y, \varphi \rangle = -\varphi'_y(\mathbf{0}) = -\frac{\partial \varphi}{\partial y}(\mathbf{0}),$$

$$\langle \delta'_z, \varphi \rangle = -\varphi'_z(\mathbf{0}) = -\frac{\partial \varphi}{\partial z}(\mathbf{0}).$$

We denote by $\boldsymbol{\delta}'$ the vector $(\delta'_x, \delta'_y, \delta'_z)$.

Let's use these results to compute the electrostatic potential created by an elecrostatic dipole. The potential created at a point $\boldsymbol{s}$ by a regular distribution of charges ρ is

$$V(\boldsymbol{s}) = \frac{1}{4\pi\varepsilon_0} \iiint_{\mathbb{R}^3} \frac{\rho(\boldsymbol{r})}{\|\boldsymbol{r} - \boldsymbol{s}\|} \, \mathrm{d}^3\boldsymbol{r},$$

which can be written

$$V(\boldsymbol{s}) = \frac{1}{4\pi\varepsilon_0} \left\langle \rho(\boldsymbol{r}), \frac{1}{\|\boldsymbol{r} - \boldsymbol{s}\|} \right\rangle.$$

Now we extend this relation to other types of distributions, such as point-like, curvilinear, surface, or dipolar distributions.

PROPOSITION 7.42 *The potential created at a point $\boldsymbol{s}$ by the distribution of charges $\rho(\boldsymbol{r})$ is*

$$V(\boldsymbol{s}) = \frac{1}{4\pi\varepsilon_0} \left\langle \rho(\boldsymbol{r}), \frac{1}{\|\boldsymbol{r} - \boldsymbol{s}\|} \right\rangle. \tag{7.4}$$

Remark 7.43 The function $\boldsymbol{s} \mapsto 1/\|\boldsymbol{r} - \boldsymbol{s}\|$ is not of $\mathscr{C}^\infty$ class, nor with bounded support.

As far as the support is concerned, we can work around the problem by imposing a sufficiently fast decay at infinity of the distribution of charges. For the problem of the singularity of $1/\|\boldsymbol{r} - \boldsymbol{s}\|$ at $\boldsymbol{s} = \boldsymbol{r}$, we note that the function still remains integrable since the volume element is $r^2 \sin\theta \, \mathrm{d}r \, \mathrm{d}\theta \, \mathrm{d}\varphi$; this is enough for the expression (7.4) to make sense in the case of most distributions ρ of physical interest.

Consider now the dipolar distribution

$$T = -\boldsymbol{\delta}' \cdot \boldsymbol{P} = -P_x\delta'_x - P_y\delta'_y - P_z\delta'_z,$$

where we have denoted by $\boldsymbol{P}$ the dipole moment of the source. Then the potential is given by

$$\begin{aligned} V(\boldsymbol{s}) &= \left\langle T(\boldsymbol{r}), \frac{1}{\|\boldsymbol{r}-\boldsymbol{s}\|} \right\rangle \\ &= \left\langle -P_x\delta'_x - P_y\delta'_y - P_z\delta'_z, \frac{1}{\|\boldsymbol{r}-\boldsymbol{s}\|} \right\rangle \\ &= \left[P_x \frac{\partial}{\partial x}\frac{1}{\|\boldsymbol{r}-\boldsymbol{s}\|} + P_y \frac{\partial}{\partial y}\frac{1}{\|\boldsymbol{r}-\boldsymbol{s}\|} + P_z \frac{\partial}{\partial z}\frac{1}{\|\boldsymbol{r}-\boldsymbol{s}\|} \right]_{\boldsymbol{r}=\boldsymbol{0}}, \end{aligned}$$

namely,

$$V(\boldsymbol{s}) = P_x \frac{s_x}{\|\boldsymbol{s}\|^3} + P_y \frac{s_y}{\|\boldsymbol{s}\|^3} + P_z \frac{s_z}{\|\boldsymbol{s}\|^3} = \frac{\boldsymbol{P}\cdot\boldsymbol{s}}{\|\boldsymbol{s}\|^3},$$

which is consistent with the result expected [50, p. 138].

THEOREM 7.44 *In the usual space $\mathbb{R}^3$, an electrostatic dipole with dipole moment equal to $\boldsymbol{P}$ is described by the distribution $-\boldsymbol{\delta}' \cdot \boldsymbol{P}$.*

♦ **Exercise 7.2** Show that, in dimension one, dilating δ' by a factor $a \neq 0$ yields

$$\boxed{\delta'(ax) = \frac{1}{a\,|a|}\,\delta'(x)}$$

(Solution page 245)

7.4.d Composition of δ with a function

The goal of this section is to give a meaning to the distribution $\delta\big(f(x)\big)$, where f is a "sufficiently regular" function. This amounts to a change of variable in a distribution and is therefore a generalization of the notions of translation, dilation, and transposition.

Consider first the special case of a regular distribution and a change of variable $f : \mathbb{R} \to \mathbb{R}$ which is differentiable and bijective.

Let g be a locally integrable function. The function $g \circ f$ being still locally integrable, it defines a distribution and we have

$$\langle g \circ f, \varphi \rangle = \int g\big(f(x)\big)\,\varphi(x)\,\mathrm{d}x = \int g(y)\,\varphi\big(f^{-1}(y)\big)\,\frac{\mathrm{d}y}{\left|f'\big(f^{-1}(y)\big)\right|},$$

since, in the substitution $y = f(x)$, the absolute value of the jacobian is equal to

$$\left|\frac{\mathrm{d}y}{\mathrm{d}x}\right| = \left|f'(x)\right| = \left|f'\big(f^{-1}(y)\big)\right|.$$

By analogy, for a distribution T, we are led to define the distribution $T \circ f$ as follows:

$$\langle T \circ f, \varphi \rangle \stackrel{\text{def}}{=} \left\langle \frac{T}{\left|f' \circ f^{-1}\right|}, \varphi \circ f^{-1} \right\rangle.$$

Thus, the distribution $\delta\big(f(x)\big)$ satisfies

$$\langle \delta\big(f(x)\big), \varphi \rangle = \left\langle \frac{\delta}{\left|f' \circ f^{-1}\right|}, \varphi \circ f^{-1} \right\rangle = \frac{\varphi\big(f^{-1}(0)\big)}{\left|f'\big(f^{-1}(0)\big)\right|},$$

for any $\varphi \in \mathscr{D}$, which can also be written

$$\delta\big(f(x)\big) = \frac{1}{\left|f'(y_0)\right|}\delta(y - y_0),$$

where y_0 is the unique real number such that $f(y_0) = 0$. Note that it is necessary that $f'(y_0) \neq 0$: the function f must not vanish at the same time as its derivative.

We can then generalize to the case of a function f not necessarily bijective – but still differentiable – by writing (within quotes!)

$$\langle \delta \circ f, \varphi \rangle = \text{“}\int_{-\infty}^{+\infty} \delta\big(f(x)\big)\, \varphi(x)\, \mathrm{d}x\text{”} = \sum_{x \,|\, f(x)=0} \frac{1}{\left|f'(x)\right|}\, \varphi(x).$$

To summarize:

THEOREM 7.45 *Let f be a differentiable function which has only isolated zeros. Denote by $Z(f)$ the set of these zeros: $Z(f) = \{y \in \mathbb{R}\ ;\ f(y) = 0\}$. Assume that the derivative f' of f does not vanish at any of the points in $Z(f)$. Then we have*

$$\delta\big(f(x)\big) = \sum_{y \in Z(f)} \frac{1}{\left|f'(y)\right|}\delta(x - y).$$

A classical application of this result to special relativity is proposed in Exercise 8.9 on page 242.

7.4.e Charge and current densities

In this section, we study the application of the preceding change of variable formula, in a simple physical situation: the Lorentz transformation of a current four-vector.

In electromagnetism, the state of the system of charges is described by data consisting of

- the charge density $(\boldsymbol{x}, t) \mapsto \rho(\boldsymbol{x}, t)$;
- the current density $(\boldsymbol{x}, t) \mapsto \boldsymbol{j}(\boldsymbol{x}, t)$.

In the case of a *point-like* particle with charge q, if we denote by $t \mapsto \boldsymbol{R}(t)$ its position and by $t \mapsto \boldsymbol{V}(t)$ its velocity, with $\boldsymbol{V}(t) = \mathrm{d}\boldsymbol{R}(t)/\mathrm{d}t$, these densities are given by

$$\rho(\boldsymbol{x}, t) = q\, \delta^{(3)}\big(\boldsymbol{x} - \boldsymbol{R}(t)\big) \qquad \text{and} \qquad \boldsymbol{j}(\boldsymbol{x}, t) = q\boldsymbol{V}(t)\, \delta^{(3)}\big(\boldsymbol{x} - \boldsymbol{R}(t)\big).$$

Moreover, the densities of charge and current form, from the point of view of special relativity, a **four-vector**. In order to distinguish them, we will denote three-dimensional vectors in the form "$\boldsymbol{j}$" and four-vectors in the form "$\mathsf{j} = (\rho, \boldsymbol{j}/c)$".

During a change of galilean reference frame, characterized by a velocity $\boldsymbol{v}$ (or, in nondimensional form, $\boldsymbol{\beta} = \boldsymbol{v}/c$), these quantities are transformed according to the rule

$$(\rho', \boldsymbol{j}'/c) = \Lambda(\boldsymbol{\beta}) \cdot (\rho, \boldsymbol{j}/c), \quad \text{i.e.,} \quad \mathsf{j}' = \Lambda(\boldsymbol{\beta}) \cdot \mathsf{j} \quad \text{with} \quad \mathsf{j} \stackrel{\text{def}}{=} (\rho, \boldsymbol{j}/c), \tag{7.5}$$

where $\Lambda(\boldsymbol{\beta})$ is the matrix characterizing a Lorentz transformation. To simplify, we consider the case of a Lorentz transformation along the axis $(\mathcal{O}x)$, characterized by the velocity $\boldsymbol{\beta} = \beta \mathbf{e}_x$. Denote by $\mathsf{x} = (ct, \boldsymbol{x}) = (x^0, x^2, x^2, x^3)$ the coordinates in the original reference frame $\mathscr{R}$ and by $\mathsf{x}' = (ct', \boldsymbol{x}') = (x'^0, x'^1, x'^2, x'^3)$ the coordinates in the reference frame $\mathscr{R}'$ with velocity $\boldsymbol{\beta}$ relative to $\mathscr{R}$. The Lorentz transformation of the coordinates is then given by

$$\begin{pmatrix} ct' \\ x' \\ y' \\ z' \end{pmatrix} = \begin{pmatrix} \gamma & -\beta\gamma & & \\ -\beta\gamma & \gamma & & \\ & & 1 & \\ & & & 1 \end{pmatrix} \cdot \begin{pmatrix} ct \\ x \\ y \\ z \end{pmatrix}, \qquad \text{with } \gamma \stackrel{\text{def}}{=} \frac{1}{\sqrt{1-\beta^2}},$$

which we will write[10] as $\mathsf{x}' = \Lambda(\boldsymbol{\beta}) \cdot \mathsf{x}$. Expanding, we obtain

$$\begin{cases} t' = \gamma t - \gamma\beta x/c = \gamma(t - vx/c^2), \\ x' = -\gamma\beta ct + \gamma x = \gamma(x - vt), \\ y' = y, \\ z' = z, \end{cases} \quad \text{and} \quad \begin{cases} t = \gamma(t' + vx'/c^2), \\ x = \gamma(x' + vt'), \\ y = y', \\ z = z'. \end{cases}$$

Although it is almost always written in the form (7.5), the transformation law of the current four-vector is more precisely given by

$$\mathsf{j}(\mathsf{x}') = \Lambda(\boldsymbol{\beta}) \cdot \mathsf{j}\big(\Lambda(\boldsymbol{\beta})^{-1} \cdot \mathsf{x}'\big) = \Lambda(\boldsymbol{\beta}) \cdot \mathsf{j}(\mathsf{x}),$$

or, in the case under consideration

$$\rho'(\boldsymbol{x}', t) = \gamma\big(\rho(\boldsymbol{x}, t) - \tfrac{1}{c}\beta j_x(\boldsymbol{x}, t)\big), \qquad j'(\boldsymbol{x}', t) = \gamma\big(j_x(\boldsymbol{x}, t) - c\beta\rho(\boldsymbol{x}, t)\big). \tag{7.6}$$

The case of a particle at rest

Consider a simple case to begin with, where the particle is at rest in the laboratory reference frame $\mathscr{R}$.

[10] Of course, tensorial notation can also be used to write $x'^{\mu} = \Lambda^{\mu}{}_{\nu} x^{\nu}$.

Then $\mathsf{j}(\mathsf{x}) = q\big(\delta^{(3)}(\boldsymbol{x}), \mathbf{0}\big)$. Looking back at equation (7.6), and denoting by $\mathscr{L}^{-1}(\boldsymbol{x}') = \gamma(x' + vt')$ the spacial part of the inverse transformation $\Lambda(\boldsymbol{\beta})^{-1}$, we obtain

$$\big(\rho'(\boldsymbol{x}', t), \tfrac{1}{c}\, \boldsymbol{j}'(\boldsymbol{r}', t')\big) = q\Big(\gamma\, \delta^{(3)}\big(\mathscr{L}^{-1}(\boldsymbol{x})\big), -\boldsymbol{\beta}\gamma\, \delta^{(3)}\big(\mathscr{L}^{-1}(\boldsymbol{x}')\big)\Big).$$

Notice then that

$$\begin{aligned}\gamma\, \delta^{(3)}\big(\mathscr{L}^{-1}(\boldsymbol{x}')\big) &= \gamma\, \delta^{(3)}\big(\gamma(x' + vt'), y', z'\big) = \gamma\, \delta\big(\gamma(x' + vt')\big)\, \delta(y')\, \delta(z') \\ &= \delta(x' + vt')\, \delta(y')\, \delta(z') \qquad\qquad \text{(change of variable)} \\ &= \delta^{(3)}(\boldsymbol{x}' + \boldsymbol{v}t)\end{aligned}$$

which gives, as might be expected

$$\rho'(\boldsymbol{x}', t') = q\, \delta^{(3)}(\boldsymbol{x}' + \boldsymbol{v}t), \qquad \boldsymbol{j}'(\boldsymbol{x}', t) = -q\boldsymbol{v}\, \delta^{(3)}(\boldsymbol{x}' + \boldsymbol{v}t).$$

So we see that the Dirac distribution has some kind of invariance property with respect to Lorentz transformations: it does not acquire a "factor γ," contrary to what equation (7.6) could suggest. This is a happy fact, since the total charge in space is given by

$$Q' = \iiint_{\mathbb{R}^3} \rho'(\boldsymbol{x}', t)\, \mathrm{d}^3\boldsymbol{x}' = q,$$

and not γq!

This result generalizes of course to the case of a particle with an arbitrary motion compatible with special relativity.

7.5

Derivation of a discontinuous function

7.5.a Derivation of a function discontinuous at a point

We saw that any locally integrable function is differentiable in the sense of distributions. Hence, we can differentiate the Heaviside function and find that $H' = \delta$. If we had taken the derivative of H in the usual sense of functions, we would have obtained a function zero everywhere except at 0, where it is not defined. A primitive of this derivative would then have given the zero function, not H.

In other words, the following diagram cannot be "closed":

$$\begin{array}{ccc} \text{function } H & \xrightarrow{\text{derivation}} & 0 \\ \downarrow & & \not\downarrow \; \textit{no!} \\ \text{distribution } H & \xrightarrow{\text{derivation}} & \delta \end{array}$$

The same will happen for any function with an isolated discontinuity at a point: its derivative will exist in the sense of distributions, which differs from the usual derivative (if it exists) by a Dirac peak with height equal to the jump of the function.

The following notation will be used:

- f is the function being studied, or its associated distribution;
- $\{f'\}$ will be the *regular distribution* associated to the usual derivative of f in the sense of functions:

$$\langle \{f'\}, \varphi \rangle = \int_{-\infty}^{+\infty} f'(x)\, \varphi(x)\, \mathrm{d}x \; ;$$

  this is of course a distribution; the derivative of f in the sense of functions is a function which is not defined everywhere (in particular, it is not at the points of discontinuity of f), but the associated regular distribution is well-defined as long as f' is defined *almost everywhere* (and locally integrable);
- f' will be the derivative of the distribution f, defined as above by

$$\langle f', \varphi \rangle = -\langle f, \varphi' \rangle \; ;$$

  this is also a distribution.

Example 7.46 With this notation, the derivative in the sense of distributions of the Heaviside distribution H is $H' = \delta$, but the distribution associated to the derivative of H taken in the sense of functions is $\{H'\} = 0$.

Let f be a function which is piecewise of $\mathscr{C}^1$ class. Denote by a_i the points of discontinuity of f (we assume there are finitely many) and by $\sigma_i^{(0)}$ the jump of f at a_i: $\sigma_i^{(0)} = f(a_i^+) - f(a_i^-)$. One can then write f as the sum of a *continuous* function g and multiples of Heaviside functions (see Figure 7.2 on the next page):

$$f(x) = g(x) + \sum_i \sigma_i^{(0)}\, H(x - a_i).$$

Since the usual derivatives of the functions f and g are equal almost everywhere, we have $\{f'\} = \{g'\}$. However, remembering that $H' = \delta$, we obtain, in the sense of distributions

THEOREM 7.47 (Derivative of a discontinuous function) *Let f be a function which is piecewise $\mathscr{C}^1$. Then, with the notation above, we have*

$$f' = \{f'\} + \sum_i \sigma_i^{(0)}\, \delta(x - a_i). \tag{7.7}$$

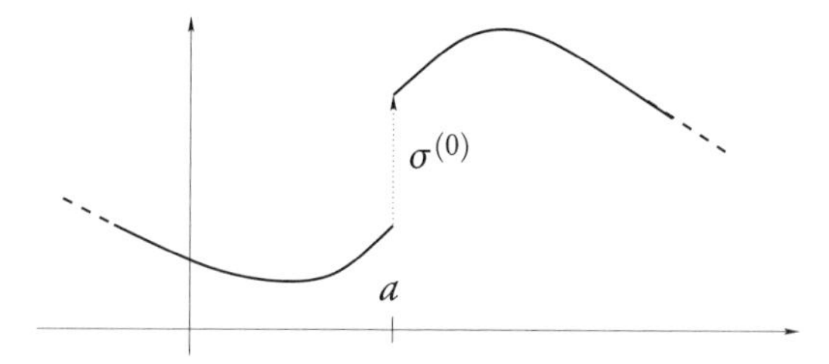

Fig. 7.2 – Example of a function with a discontinuity at a with jump equal to $\sigma^{(0)}$.

To lighten the notation, we will henceforth write equation (7.7) more compactly as follows:

$$f' = \{f'\} + \sigma^{(0)}\,\delta\ ;$$

all the discontinuities of f are *implicitly* taken into account. Similarly, if we denote by $\sigma^{(1)}$ any discontinuity of the derivative of the function f, $\sigma^{(2)}$ any discontinuity of its second derivative, and so on, then we will have the following result, with similar conventions:

THEOREM 7.48 (Successive derivatives of a piecewise $\mathscr{C}^\infty$ function) *Let f be a piecewise $\mathscr{C}^\infty$ function. Then we have*

$$f'' = \{f''\} + \sigma^{(1)}\,\delta + \sigma^{(0)}\,\delta',$$

and more generally

$$f^{(m)} = \left\{f^{(m)}\right\} + \sigma^{(m-1)}\,\delta + \cdots + \sigma^{(0)}\,\delta^{(m-1)} \qquad \forall m \in \mathbb{N}.$$

Example 7.49 Consider the function $E : x \longmapsto E(x) =$ "integral part of x." Then $\{E'\} = 0$, whereas, in the sense of distributions, we have $E' = Ш$.

♦ **Exercise 7.3** Compute the successive derivatives of the function $f : x \longmapsto |x|$.

◇ **Solution**: Since f is continuous, its derivative in the sense of distributions is the distribution associated to its usual derivative. But the derivative of f is the function $x \longmapsto \operatorname{sgn}(x)$ for any $x \neq 0$, which is defined everywhere except at $x = 0$. The associated regular distribution, denoted $\{f'\}$, is the distribution "sgn(x)" (meaning "sign of x") defined by

$$\langle\, \operatorname{sgn}(x), \varphi(x)\rangle = \int_0^{+\infty} \big(\varphi(x) - \varphi(-x)\big)\,\mathrm{d}x.$$

Therefore, the derivative of f in the sense of distributions is $f' = \{f'\} = \operatorname{sgn}$.

The function $x \longmapsto \operatorname{sgn}(x)$ has a discontinuity at 0 with jump equal to 2. Moreover, its usual derivative is zero almost everywhere (it is undefined at 0 and zero everywhere else), which shows that

$$\{\operatorname{sgn}'\} = 0 \qquad \text{and} \qquad f'' = \operatorname{sgn}' = 2\delta.$$

We deduce that the next derivatives of f are given by

$$f^{(k)} = 2\delta^{(k-2)} \qquad \text{for any } k \geqslant 2.$$

7.5.b Derivative of a function with discontinuity along a surface $\mathscr{S}$

Let $\mathscr{S}$ be a smooth surface in $\mathbb{R}^3$, and let $f : \mathbb{R}^3 \to \mathbb{C}$ be a function which is piecewise of $\mathscr{C}^1$ class, with a discontinuity along the surface $\mathscr{S}$.

We denote by $x_1 = x$, $x_2 = y$, and $x_3 = z$ the coordinates and assume the surface $\mathscr{S}$ is oriented. For $i = 1, 2, 3$, we denote by θ_i the angle between the exterior normal $\boldsymbol{n}$ to the surface $\mathscr{S}$ and the axis $(\mathscr{O}x_i)$. Arguing as in the previous section, the following theorem will follow:

THEOREM 7.50 (Derivatives of a function discontinuous along a surface) *Let f be a function with a discontinuity along the surface $\mathscr{S}$. Denoting by $\sigma^{(0)}$ the "jump" function $\mathscr{S}$, equal to the value of f just* outside *minus the value just* inside, *we have*

$$\frac{\partial f}{\partial x_i} = \left\{ \frac{\partial f}{\partial x_i} \right\} + \sigma^{(0)} \cos \theta_i \, \delta_{\mathscr{S}}. \tag{7.8}$$

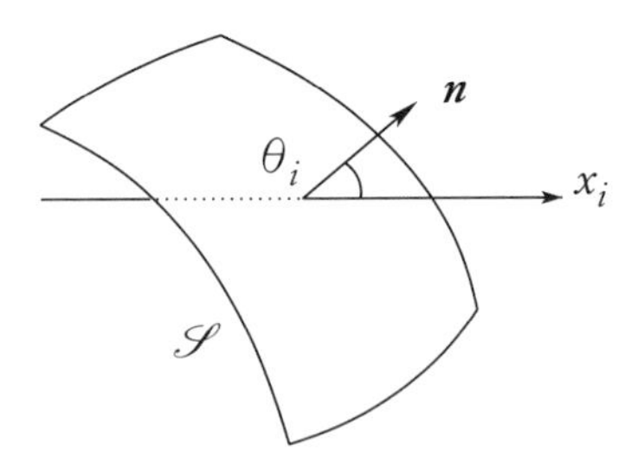

PROOF. Assume that f is of class $\mathscr{C}^1$ on $\mathbb{R}^3 \setminus \mathscr{S}$ and that its first derivatives admit a limit on each side of $\mathscr{S}$. We now evaluate the action of $\partial f / \partial x$ on an arbitrary test function $\varphi \in \mathscr{D}(\mathbb{R}^3)$:

$$\left\langle \frac{\partial f}{\partial x}, \varphi \right\rangle = -\left\langle f, \frac{\partial \varphi}{\partial x} \right\rangle = -\iiint f \frac{\partial \varphi}{\partial x} \, \mathrm{d}x \, \mathrm{d}y \, \mathrm{d}z$$
$$= -\iint \left(\int f(x, y, z) \frac{\partial \varphi}{\partial x} \, \mathrm{d}x \right) \mathrm{d}y \, \mathrm{d}z$$

for a smooth surface. Fix now y^* and z^* and denote by x^* the real number such that $(x^*, y^*, z^*) \in \mathscr{S}$ (generalizing to the case where more than one real number satisfies this property is straightforward). In addition, put $h(x) = f(x, y^*, z^*)$. Then we have

$$\int f(x, y^*, z^*) \frac{\partial \varphi}{\partial x} \, \mathrm{d}x = \int h(x) \frac{\partial \varphi}{\partial x} \, \mathrm{d}x = \left\langle h, \frac{\partial \varphi}{\partial x} \right\rangle = -\left\langle \frac{\partial h}{\partial x}, \varphi \right\rangle,$$

and we can now use Theorem 7.47, which gives

$$\left\langle \frac{\partial h}{\partial x}, \varphi \right\rangle = \left\langle \left\{ \frac{\partial h}{\partial x} \right\} + \sigma^{(0)}(x^*, y^*, z^*) \, \delta(x - x^*), \varphi \right\rangle$$
$$= \int \frac{\partial h}{\partial x} \varphi \, \mathrm{d}x + \sigma^{(0)}(x^*, y^*, z^*) \, \varphi(x^*, y^*, z^*),$$

where we have denoted by $\sigma^{(0)}(x^*, y^*, z^*)$ the jump of f across the surface $\mathscr{S}$ at the point (x^*, y^*, z^*).

But, integrating now with respect to y^* and z^*, we remark that

$$\iint_{\mathscr{S}} \sigma^{(0)}\varphi \, \mathrm{d}y \, \mathrm{d}z = \iint_{\mathscr{S}} \sigma^{(0)}\varphi \, \cos\theta_x \, \mathrm{d}^2 s,$$

since $\mathrm{d}^2 s \cos\theta_x = \mathrm{d}y \, \mathrm{d}y$, as can be seen in the figure.

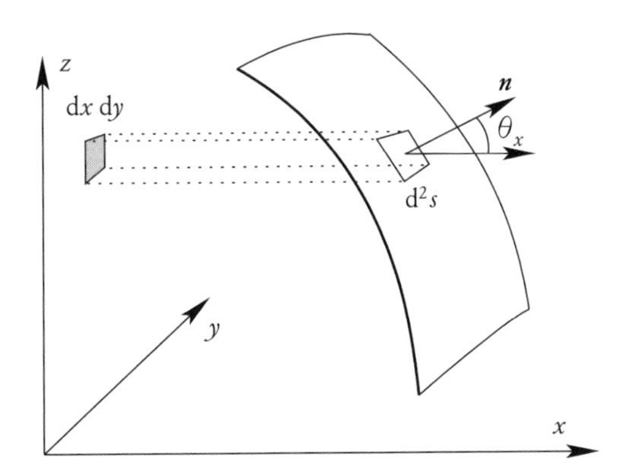

To conclude, we write

$$\iint_{\mathscr{S}} \sigma^{(0)}\varphi \, \cos\theta_x \, \mathrm{d}^2 s = \left\langle \sigma^{(0)} \cos\theta_x \, \delta_{\mathscr{S}}, \varphi \right\rangle.$$

Putting these results together, we obtain

$$\left\langle \frac{\partial f}{\partial x}, \varphi \right\rangle = \left\langle \sigma^{(0)} \cos\theta_x \, \delta_{\mathscr{S}}, \varphi \right\rangle + \iiint \frac{\partial f}{\partial x} \varphi \, \mathrm{d}x \, \mathrm{d}y \, \mathrm{d}z$$

for any $\varphi \in \mathscr{D}(\mathbb{R}^3)$. The same argument performed with the partial derivatives with respect to y or z leads to the stated formula (7.8).

Now we apply this result to a function f differentiable in a volume $\mathscr{V}$ delimitied by a *closed* surface $\mathscr{S} = \partial\mathscr{V}$, and equal to zero outside this volume. Taking the test function $\varphi(x) \equiv 1$, we get[11]

$$\iiint_{\mathscr{V}} \frac{\partial f}{\partial x_i} \, \mathrm{d}v - \iint_{\mathscr{S}} f \cos\theta_i \, \mathrm{d}s = 0 \qquad \text{for } i = 1, 2, 3,$$

or, combining the three formulas for $i = 1, 2, 3$ in vector form, and noticing that $\boldsymbol{n} = (\cos\theta_x, \cos\theta_y, \cos\theta_z)$

$$\boxed{\iiint_{\mathscr{V}} \mathbf{grad}\, f \, \mathrm{d}^3 v = \iint_{\mathscr{S}} f \, \boldsymbol{n} \, \mathrm{d}^2 s} \qquad \textit{Green-Ostrogradski formula.}$$

[11] More precisely, we should take a test function φ equal to 1 on $\mathscr{V}$, but with bounded support. If $\mathscr{V}$ is bounded, such a function always exists.

The formula (7.8) may also be applied to the components of a vector function $f = (f_1, f_2, f_3)$; then, after summing over indices, we get

$$\iiint_{\mathscr{V}} \sum_i \frac{\partial f_i}{\partial x_i} \, \mathrm{d}^3 v - \iint_{\mathscr{S}} \sum_i f_i \cos\theta_i \, \mathrm{d}^2 s = 0,$$

or equivalently

$$\boxed{\iiint_{\mathscr{V}} \operatorname{div} f \, \mathrm{d}^3 v = \iint_{\mathscr{S}} (f \cdot \boldsymbol{n}) \, \mathrm{d}^2 s} \qquad \textit{Green-Ostrogradski formula.}$$

7.5.c Laplacian of a function discontinuous along a surface $\mathscr{S}$

By applying the results of the previous section, it is possible to compute, in the sense of distributions, the laplacian of a function which is discontinuous along a smooth surface.

THEOREM 7.51 (Laplacian of a discontinuous function) *Let $\mathscr{S}$ be a smooth surface and f a function of $\mathscr{C}^\infty$ class except on the surface $\mathscr{S}$. Denote by $\sigma^{(0)}$ the "jump" of f across the surface $\mathscr{S}$. Denote moreover by $\frac{\partial f}{\partial \boldsymbol{n}}$ the derivative of f in the normal direction $\boldsymbol{n}$, defined by*

$$\frac{\partial f}{\partial \boldsymbol{n}} \stackrel{\text{def}}{=} \sum_i \frac{\partial f}{\partial x_i} \cos\theta_i = (\textbf{grad}\, f) \cdot \boldsymbol{n},$$

and let $\sigma_n^{(1)}$ be the jump of this function. Then we have

$$\triangle f = \sigma^{(0)} \, \delta'_n + \sigma_n^{(1)} \, \delta_{\mathscr{S}} + \{\triangle f\}, \tag{7.9}$$

where δ'_n is the normal derivative of δ: $\delta'_n = \sum_i \delta'_{x_i} \cos\theta_i = \boldsymbol{\delta}' \cdot \boldsymbol{n}$.

Remark 7.52 The action of the distribution δ'_n on any test function φ is therefore given by

$$\langle \delta'_n, \varphi \rangle = -\left\langle \delta_{\mathscr{S}}, \frac{\partial \varphi}{\partial \boldsymbol{n}} \right\rangle.$$

THEOREM 7.53 (Green formula) *Let f be a function twice differentiable in a volume $\mathscr{V}$ bounded by a surface $\mathscr{S}$, and let φ be a test function. Denoting by $\boldsymbol{n}$ the exterior normal to $\mathscr{S}$, we have*

$$\iiint_{\mathscr{V}} (f \triangle \varphi - \varphi \triangle f) \, \mathrm{d}^3 v = \iint_{\mathscr{S}} \left(f \frac{\partial \varphi}{\partial \boldsymbol{n}} - \varphi \frac{\partial f}{\partial \boldsymbol{n}} \right) \mathrm{d}^2 s. \tag{7.10}$$

PROOF. Extend f by putting $f(\boldsymbol{r}) = 0$ for any point outside $\mathscr{V}$.

Note that the jump of f accross the surface $\mathscr{S}$ at a point $\boldsymbol{r}$ is $-f(\boldsymbol{r})$ when crossing in the exterior direction, and similarly for the other jump functions involved.

Then apply directly the formula (7.9), noticing that

$$\langle \triangle f, \varphi \rangle = \iiint_{\mathscr{V}} f \triangle \varphi \, \mathrm{d}^3 v, \qquad \langle \sigma^{(0)} \delta'_{\boldsymbol{n}}, \varphi \rangle = \iint_{\mathscr{S}} f \frac{\partial \varphi}{\partial \boldsymbol{n}} \, \mathrm{d}^2 s,$$

$$\langle \sigma_{\boldsymbol{n}}^{(1)} \delta_{\mathscr{S}}, \varphi \rangle = \iint_{\mathscr{S}} \frac{-\partial f}{\partial \boldsymbol{n}} \, \mathrm{d}^2 s, \qquad \langle \{\triangle f\}, \varphi \rangle = \iiint_{\mathscr{V}} \varphi \triangle f \, \mathrm{d}^3 v.$$

7.5.d Application: laplacian of $1/r$ in 3-space

We continue working in $\mathbb{R}^3$. If f is a function on a subset of $\mathbb{R}^3$ which is twice differentiable and radial, its laplacian in the sense of functions can be computed by the formula

$$\triangle f = \frac{1}{r} \frac{\partial^2}{\partial r^2}(r f).$$

Hence, in the sense of functions, we have $\triangle \left(\frac{1}{r}\right) = 0$, except possibly at $r = 0$.

We now try to compute this laplacian in the sense of distributions.

For any $\varepsilon > 0$, define

$$f_\varepsilon(r) = \begin{cases} 1/r & \text{for } r > \varepsilon, \\ 0 & \text{for } r < \varepsilon. \end{cases}$$

Then apply the formula

$$\triangle f = \sigma^{(0)} \delta'_{\boldsymbol{n}} + \sigma_{\boldsymbol{n}}^{(1)} \delta_{\mathscr{S}} + \{\triangle f\}$$

to the *function* f_ε. For this, denote by $\mathscr{S}_\varepsilon$ the sphere $r = \varepsilon$ and by $\mathscr{V}_\varepsilon$ the volume interior to this surface. The jumps of f_ε and $\partial f_\varepsilon / \partial \boldsymbol{n}$ accross the surface $\mathscr{S}$ *oriented toward the exterior* are therefore equal to $1/r$ and $-1/r^2$, respectively. We thus find

$$\begin{aligned} \langle \triangle f_\varepsilon, \varphi \rangle &= \iiint f_\varepsilon \triangle \varphi \, \mathrm{d}^3 \boldsymbol{r} && \text{(by definition)} \\ &= \left\langle \sigma^{(0)} \delta'_{\boldsymbol{n}} + \sigma_{\boldsymbol{n}}^{(1)} \delta_{\mathscr{S}} + \{\triangle f_\varepsilon\}, \varphi \right\rangle && \text{but } \{\triangle f\} = 0 \\ &= \iint_{\mathscr{S}_\varepsilon} \frac{1}{\varepsilon} \left(-\frac{\partial \varphi}{\partial \boldsymbol{n}} \right) \mathrm{d}^2 s + \iiint_{\mathscr{V}_\varepsilon} \left(-\frac{1}{\varepsilon^2} \right) \varphi(\boldsymbol{r}) \, \mathrm{d}^3 \boldsymbol{r}. \end{aligned}$$

Each term in the last expression has a limit as $[\varepsilon \to 0]$, respectively, 0 (since the derivatives of φ are bounded and the surface of integration has area of order of magnitude ε^2) and

$$\lim_{\varepsilon \to 0^+} \iiint_{\mathscr{V}_\varepsilon} \left(-\frac{1}{\varepsilon^2} \right) \varphi(\boldsymbol{r}) \, \mathrm{d}^3 \boldsymbol{r} = -4\pi \, \varphi(\boldsymbol{0})$$

by continuity of φ (the factor 4π arises as the surface area of $\mathscr{S}_\varepsilon$ divided by ε^2).

Notice now that

$$\left\langle \triangle \frac{1}{r}, \varphi \right\rangle = \iiint_{\mathbb{R}^3} \frac{1}{r} \triangle \varphi(\boldsymbol{r}) \, \mathrm{d}^3\boldsymbol{r}$$

$$= \lim_{\varepsilon \to 0^+} \iiint_{\mathbb{R}^3} f_\varepsilon(\boldsymbol{r}) \triangle \varphi(\boldsymbol{r}) \, \mathrm{d}^3\boldsymbol{r} = \lim_{\varepsilon \to 0^+} \langle \triangle f_\varepsilon, \varphi \rangle ,$$

since the integrand on the left is an integrable function (because the volume element is given by $r^2 \sin\theta \, \mathrm{d}r \, \mathrm{d}\theta \, \mathrm{d}\varphi$). Putting everything together, we find that

$$\left\langle \triangle \frac{1}{r}, \varphi \right\rangle = \iiint_{\mathbb{R}^3} \frac{1}{r} \triangle \varphi(\boldsymbol{r}) \, \mathrm{d}^3\boldsymbol{r} = -4\pi \, \varphi(\boldsymbol{0}).$$

In other words, we have proved the following theorem:

THEOREM 7.54 *The laplacian of the radial function* $f : \boldsymbol{r} \mapsto \frac{1}{r} = \frac{1}{\|\boldsymbol{r}\|}$ *is given by*

$$\triangle \left(\frac{1}{r} \right) = -4\pi \, \delta.$$

It is this equation which, applied to electrostatics, will give us the Poisson law (see Section 7.6.g on page 216).

In the same manner, one can show that

THEOREM 7.55 *In* $\mathbb{R}^2$*, the laplacian of* $\boldsymbol{r} \mapsto \log \|\boldsymbol{r}\|$ *is*

$$\triangle \big(\log |r| \big) = 2\pi \, \delta. \tag{7.11}$$

Similarly, one gets (see [81]) :

PROPOSITION 7.56 *For* $n \geqslant 3$*, we have in* $\mathbb{R}^n$ *the formula*

$$\triangle \left(\frac{1}{r^{n-2}} \right) = -(n-2) S_n \, \delta,$$

where S_n *is the "area" of the unit sphere in* $\mathbb{R}^n$*, namely*

$$S_n = \frac{(2\pi)^{n/2}}{\Gamma(n/2)}.$$

7.6

Convolution

7.6.a The tensor product of two functions

DEFINITION 7.57 (Tensor product of functions) Let f and g be two functions defined on $\mathbb{R}$. The **direct product** of f and g (also called the **tensor product**) is the function $h : \mathbb{R}^2 \to \mathbb{R}$ defined by $h(x, y) = f(x)\,g(y)$ for any $x, y \in \mathbb{R}$. It will be denoted $h = f \otimes g$.

This definition is generalized in the obvious manner to the product of $f : \mathbb{R}^p \to \mathbb{C}$ by $g : \mathbb{R}^n \to \mathbb{C}$.

A somewhat grandiloquent word (tensor product!!) for a very simple thing, in the end. But despite the apparent triviality of this notion, the tensor product of functions will be very useful.

Example 7.58 Let $\Pi(x) = 1$ for $|x| < 1/2$ and $\Pi(x) = 0$ elsewhere. The tensor product of the functions H and Π is given by the following graphical representation, where the gray area corresponds to the points where $H \otimes \Pi$ takes the value 1, and the white area to those where it takes the value 0:

$H \otimes \Pi =$

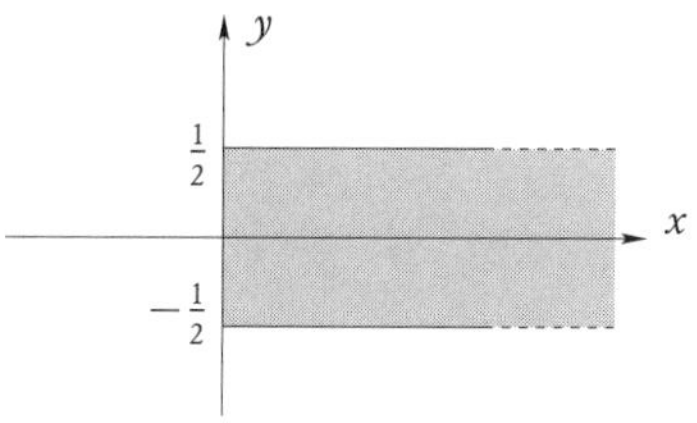

or $H \otimes \Pi(x, y) = H(x)\,\Pi(y)$.

7.6.b The tensor product of distributions

Let's now see how to generalize the tensor product to distributions. For this, as usual, we consider the distribution associated to the tensor product of two locally integrable functions. If $f : \mathbb{R}^p \to \mathbb{C}$ and $g : \mathbb{R}^n \to \mathbb{C}$ are locally integrable, and if we write $h : \mathbb{R}^{p+n} \to \mathbb{C}$ as their tensor product, then for any test function $\varphi \in \mathscr{D}(\mathbb{R}^{p+n})$ we have

$$\begin{aligned}\langle h, \varphi\rangle = \big\langle f(x)\,g(y), \varphi(x, y)\big\rangle &= \iint f(x)\,g(y)\,\varphi(x, y)\,\mathrm{d}x\,\mathrm{d}y \\ &= \int f(x)\left(\int g(y)\,\varphi(x, y)\,\mathrm{d}y\right)\mathrm{d}x = \Big\langle f(x), \big\langle g(y), \varphi(x, y)\big\rangle\Big\rangle.\end{aligned}$$

Take now a distribution $S(x)$ on $\mathbb{R}^p$ and a distribution $T(y)$ on $\mathbb{R}^n$. The function

$$x \longmapsto \theta(x) \stackrel{\text{def}}{=} \big\langle T(y), \varphi(x, y)\big\rangle$$

is a test function on $\mathbb{R}^n$; we can therefore let S act on it and, as for regular distributions, we define the distribution $S(x)\,T(y)$ by

$$\big\langle S(x)\,T(y), \varphi(x,y)\big\rangle \stackrel{\text{def}}{=} \Big\langle S(x), \big\langle T(y), \varphi(x,y)\big\rangle\Big\rangle.$$

DEFINITION 7.59 (Tensor product of distributions) Let S and T be two distributions. The **direct product**, or **tensor product**, of the distributions S and T is the distribution $S(x)\,T(y)$ defined on the space of test functions on $\mathbb{R}^p \times \mathbb{R}^n$ by

$$\big\langle S(x)\,T(y), \varphi(x,y)\big\rangle \stackrel{\text{def}}{=} \Big\langle S(x), \big\langle T(y), \varphi(x,y)\big\rangle\Big\rangle.$$

It is denoted $S \otimes T$, or $S(x)\,T(y)$.

The tensor product is of course not commutative: $S \otimes T \neq T \otimes S$.

Remark 7.60 Be carefull not to mistake $S(x)\,T(y)$ with a product of distributions, which is not well-defined in general (see note 6 on page 190).

Example 7.61 The Dirac distribution in two dimensions can be defined as the direct product of two Dirac distributions in one dimension:

$$\delta^{(2)}(\boldsymbol{r}) = \delta^{(2)}(x,y) = \delta(x)\,\delta(y).$$

This can be illustrated graphically as follows:

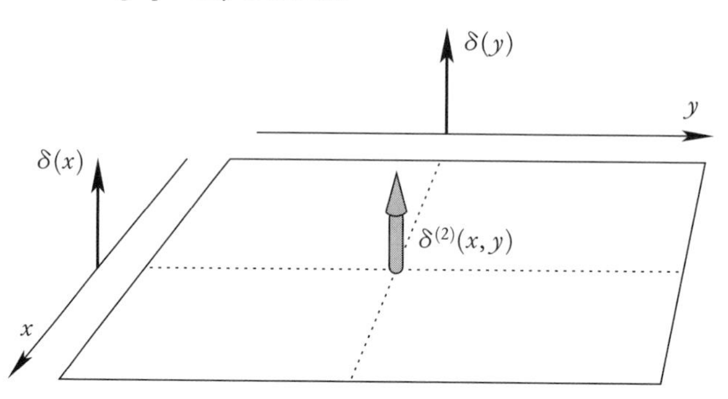

Similarly, in three dimensions, we have $\delta^{(3)}(\boldsymbol{x}) = [\delta \otimes \delta \otimes \delta](x,y,z) = \delta(x)\,\delta(y)\,\delta(z)$.

Example 7.62 In the same way, the tensor product $\delta \otimes H$ can be expressed by the graph

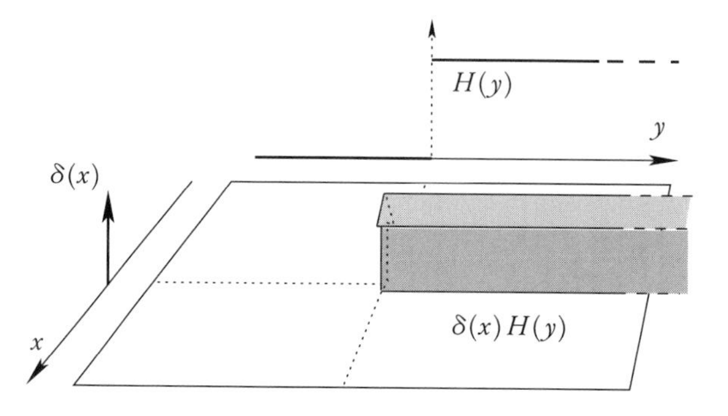

This distribution on $\mathbb{R}^2$ acts on a test function $\varphi \in \mathscr{D}(\mathbb{R}^2)$ by

$$\langle \delta(x)\,H(y), \varphi(x,y) \rangle = \int_0^{+\infty} \varphi(0,y)\,\mathrm{d}y.$$

Caution is that sometimes the constant function is omitted from the notation of a tensor product $f \otimes \mathbf{1}$. For instance, in two dimensions, the distribution denoted $\delta(x)$ is in reality equal to $\delta(x)\,\mathbf{1}(y)$ and acts therefore by the formula

$$\langle \delta(x), \varphi(x,y) \rangle = \langle \delta(x)\,\mathbf{1}(y), \varphi(x,y) \rangle = \int_{-\infty}^{+\infty} \varphi(0,y)\,\mathrm{d}y.$$

This should not be confused with $\delta(\boldsymbol{r}) = \delta(x,y) = \delta(x)\,\delta(y)$.

Example 7.63 (Special relativity) In the setting of Section 7.4.e, the density of charge $\rho(\mathbf{x}) = q\,\delta^{(3)}(\boldsymbol{x})$ can also be written

$$\rho = q\ \mathbf{1} \otimes \delta^{(3)},$$

since

$$\left(\mathbf{1} \otimes \delta^{(3)}\right)(ct, \boldsymbol{x}) = \mathbf{1}(ct)\,\delta^{(3)}(\boldsymbol{x}) = \delta^{(3)}(\boldsymbol{x}).$$

7.6.c Convolution of two functions

DEFINITION 7.64 (Convolution of functions) Let f and g be two locally integrable functions. Their convolution, or convolution product, is the function h defined by

$$h(x) \stackrel{\text{def}}{=} \int f(t)\,g(x-t)\,\mathrm{d}t,$$

where this is well-defined. It is denoted

$$h = f * g$$

or less rigorously $h(x) = f(x) * g(x)$. Note that the convolution of two functions does not always exist.

Example 7.65 The convolution of the Heaviside function with itself is given by

$$H * H(x) = \int_{-\infty}^{+\infty} H(t)\,H(x-t)\,\mathrm{d}t = x\,H(x).$$

Conversely, the convolution of H with $x \longmapsto 1/\sqrt{|x|}$ is not defined.

♦ **Exercise 7.4** Show that the convolution product $*$ is commutative, that is, $h = f*g = g*f$ whenever one of the two convolutions is defined.

♦ **Exercise 7.5** Let $a, b \in \mathbb{R}^+$ be real numbers such that $a \neq b$. Compute the convolution of $x \longmapsto \mathrm{e}^{-|ax|}$ with $x \longmapsto \mathrm{e}^{-|bx|}$.

(Solutions page 245)

It is possible to interpret graphically the convolution of two functions. Indeed, denoting by $h = f \otimes g$ their tensor product, we have

$$f * g(x) = \int f(t)\, g(x-t)\, \mathrm{d}t = \int h(t, x-t)\, \mathrm{d}t,$$

which means that one integrates the function $h : \mathbb{R}^2 \to \mathbb{C}$ on the path formed by the line $\mathscr{D}_x$ with slope -1 passing through the point $(0, x)$.

Example 7.66 Let $f = \Pi$ and $g(x) = \Pi(x - \frac{1}{2})$, where Π is the "rectangle" function (see page 213). Put $h = f \otimes g$ and $k = f * g$. Then the value of $k(x)$ at a point $x \in \mathbb{R}$ is given by the integral of h on the line $\mathscr{D}_x$ below. The gray area corresponds to those points where $h(x, y) = 1$, and the white area to those where $h(x, y) = 0$.

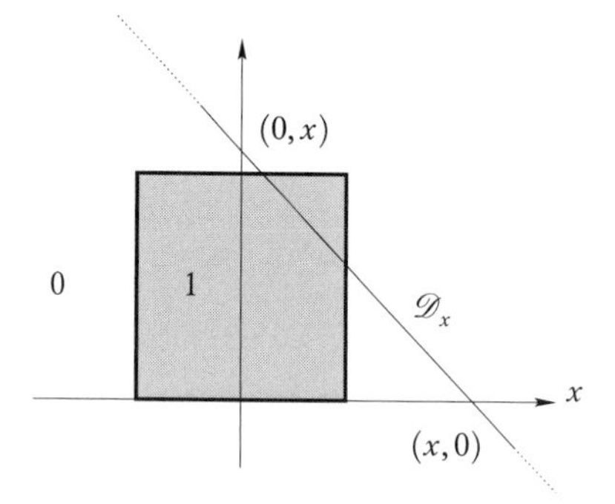

If the diagonal line $\mathscr{D}_x$ intersects the support of $f \otimes g$ in a finite segment, then the convolution of f by g is well defined at the point x.

Thanks to this interpretation, we have therefore proved the following result:

THEOREM 7.67 *The convolution ot two locally integrable functions f and g exists whenever one at least of the following conditions holds:*

i) the functions f and g both have bounded support;

ii) the functions f and g are both zero for sufficiently large x (they have bounded support "on the right");

iii) the functions f and g are both zero for sufficiently small (negative) x (they have bounded support "on the left").

Those conditions are of course sufficient, but not necessary.

♦ **Exercise 7.6** Show graphically that $H * H$ is well defined, but not $H * \check{H}$ (where, as defined previously, $\check{H}(x) = H(-x)$).

7.6.d "Fuzzy" measurement

The "rectangle" function Π is defined by

$$\Pi(x) = \begin{cases} 1 & \text{if } |x| < \frac{1}{2} \\ 0 & \text{if } |x| > \frac{1}{2}, \end{cases}$$

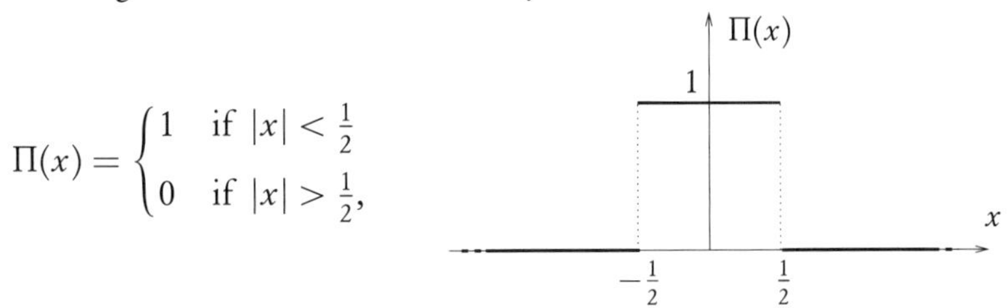

Let now $a > 0$ be fixed. Denoting by f_a the convolution of f by a similar rectangle with width a and height $1/a$, we have

$$f_a(x) = f(x) * \frac{1}{a}\Pi\left(\frac{x}{a}\right) = \frac{1}{a}\int_{x-a/2}^{x+a/2} f(t)\,\mathrm{d}t,$$

which is simply the average of the function f on an interval around x of length a. It is then possible to model simply an imperfect measurement of the function f, for which the uncertainty is of size a. Then details of size $\ell \ll a$ disappear, whereas those of size $\ell \gg a$ remain clearly visible.

This model can be refined by convolution of f by a gaussian or a lorentzian (see Example 279), for instance.

Example 7.68 A spectral ray has the shape of a lorentzian (a classical result of atomic physics), but what is observed experimentally is slightly different. Indeed, the atoms of the emitting gas are not at rest, but their velocities are distributed according to a certain distribution, given by a "maxwellian" function (this is the name given to the gaussian by physicists, because of the *Maxwell-Boltzmann* distribution). Due to the Doppler effect, the wavelengths emitted by the atoms are shifted by an amount proportional to the speed; therefore the spectral distribution observed is the convolution of the original shape (the lorentzian) by a gaussian, with variance proportional to the temperature.

It will be seen[12] that the sequence of functions $\big(n\Pi(nx)\big)_{n\in\mathbb{N}}$ converges (in a sense that will be made precise in Definition 8.12 on page 230) to the Dirac distribution δ. Thus, it can be expected that, in the limit $[n \to \infty]$, since the convolution does not change anything (the "precision" of measurement is infinite), we have $f * \delta = f$; thus we expect that δ will be a unit element for the convolution product.

To show this precisely, we first have to extend the convolution $*$ to the setting of distributions.

[12] See Exercise 8.12 on page 242.

7.6.e Convolution of distributions

By the same method as before, we wish to extend the convolution product to distributions, so that it coincides with the convolution of functions in the case of regular distributions. Now, for two locally integrable functions f and g (such that the convolution $f * g$ is defined), we have for any $\varphi \in \mathscr{D}(\mathbb{R})$:

$$\begin{aligned}\langle f * g, \varphi \rangle &= \int (f * g)(t)\, \varphi(t)\, \mathrm{d}t \\ &= \int \varphi(t) \left(\int f(s)\, g(t-s)\, \mathrm{d}s \right) \mathrm{d}t \\ &= \iint f(x)\, g(y)\, \varphi(x+y)\, \mathrm{d}x\, \mathrm{d}y \qquad (x = s, \quad y = t - s) \\ &= \big\langle f(x)\, g(y), \varphi(x+y) \big\rangle,\end{aligned}$$

since the jacobian in this change of variable is equal to $|\, J\,| = 1$.

DEFINITION 7.69 (Convolution of distributions) Let S and T be two distributions. The **convolution** of S and T is defined by its action on any test function $\varphi \in \mathscr{D}$ given by

$$\langle S * T, \varphi \rangle \stackrel{\text{def}}{=} \big\langle S(x)\, T(y), \varphi(x+y) \big\rangle.$$

The convolution of distributions does not always exist! The general conditions for its existence are difficult to write down. However, we may note that, as in the case of functions, the following result holds:

THEOREM 7.70 *The convolution of distributions with support bounded on the left (resp. bounded on the right) always exists.*

The convolution of an arbitrary distribution with a distribution with bounded support always exists.

Indeed, if $x \longmapsto \varphi(x)$ is a nonzero test function, the map $(x, y) \longmapsto \varphi(x+y)$ does not have bounded support, that is, it is not a test function on $\mathbb{R}^2$! If the support of $x \longmapsto \varphi(x)$ is a segment $[a, b]$, then the support of $(x, y) \longmapsto \varphi(x+y)$ is a strip as represented below:

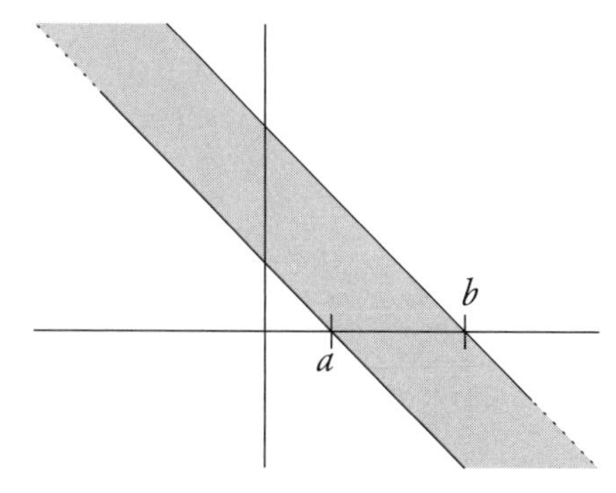

In particular, if we denote by Ω the support of S and by Ω' the support of T, then the convolution of S and T has a sense if the intersection of an arbitrary strip of this type with the cartesian product $\Omega \times \Omega'$ is always bounded. The reader can easily check that this condition holds in all the situations of the previous theorem.

PROPOSITION 7.71 *Let S and T be two distributions. Assume that the convolution $S * T$ exists. Then $T * S$ exists also and $S * T = T * S$ (*commutativity *of the convolution).*

Similarly, if S, T and U are three distributions, then

$$S * (T * U) = (S * T) * U = S * T * U$$

*if $S * T$, $T * U$, and $S * U$ make sense (*associativity *of the convolution product).*

Counterexample 7.72 Let $\mathbf{1}$ denote as usual the constant function $\mathbf{1} : x \longmapsto 1$. This function is locally integrable and therefore defines a regular distribution, Moreover, the reader can check that $(\mathbf{1} * \delta') = 0$, hence $(\mathbf{1} * \delta') * H = 0$. But, on the other hand, $(\delta' * H) = \delta$, and hence $\mathbf{1} * (\delta' * H) = \mathbf{1}$, which shows that

$$(\mathbf{1} * \delta') * H \neq \mathbf{1} * (\delta' * H).$$

This can be explained by the fact that $\mathbf{1} * H$ does not exist.

7.6.f Applications

Let f and g be two integrable functions. They define regular distributions, and, moreover (as easily checked), $f * \mathbf{1}$ and $g * \mathbf{1}$ exist and are in fact constant and equal to $\mathbf{1} * f = \int f(x)\,dx$, $\mathbf{1} * g = \int g(x)\,dx$, respectively. Assume moreover that $f * g$ exists. Then, writing $\mathbf{1} * [f * g] = [\mathbf{1} * f] * g$, we deduce the following result:

THEOREM 7.73 *Let f and g be two integrable functions such that the convolution of f and g exists. Then we have*

$$\int [f * g](x)\,dx = \left(\int f(x)\,dx\right) \cdot \left(\int g(x)\,dx\right).$$

Remark 7.74 The convolution product is the continuous equivalent of the Cauchy product of absolutely convergent series. Recall that the Cauchy product of the series $\sum a_n$ and $\sum b_n$ is the series $\sum w_n$ such that

$$w_n = \sum_{k=0}^{n} a_k\, b_{n-k}.$$

By analogy, one may write $w = a * b$. Then we have

$$\sum_{n=0}^{\infty} (a * b)_n = \left(\sum_{n=0}^{\infty} a_n\right) \cdot \left(\sum_{n=0}^{\infty} b_n\right).$$

Using the definition of convolution, it is easy to show the following relations:

THEOREM 7.75 *Let T be a distribution. Then we have*

$$\delta * T = T * \delta = T.$$

For any $a \in \mathbb{R}$, the translate of T by a can be expressed as the convolution

$$\delta(x-a) * T(x) = T(x) * \delta(x-a) = T(x-a),$$

and the derivatives of T can be expressed as

$$\delta' * T = T * \delta' = T' \quad \text{and} \quad \delta^{(m)} * T = T * \delta^{(m)} = T^{(m)} \quad \text{for any } m \in \mathbb{N}.$$

PROOF

- Let $T \in \mathscr{D}'$. Then, for any $\varphi \in \mathscr{D}$, we have

$$\begin{aligned}\langle T * \delta, \varphi \rangle &= \langle T(x)\,\delta(y), \varphi(x+y) \rangle \\ &= \Big\langle T(x), \langle \delta(y), \varphi(x+y) \rangle \Big\rangle = \langle T(x), \varphi(x) \rangle = \langle T, x \rangle,\end{aligned}$$

which shows that $T * \delta = T$. Hence we also have $\delta * T = T$.

- Let $a \in \mathbb{R}$. Then

$$\langle T * \delta_a, \varphi \rangle = \Big\langle T(x), \langle \delta(y-a), \varphi(x+y) \rangle \Big\rangle = \langle T(x), \varphi(x+a) \rangle = \langle T(x-a), \varphi(x) \rangle,$$

which shows that $T * \delta_a = T_a$.

- Finally, for any $\varphi \in \mathscr{D}$,

$$\langle T * \delta', \varphi \rangle = \Big\langle T(x), \langle \delta'(y), \varphi(x+y) \rangle \Big\rangle = \langle T(x), -\varphi'(x) \rangle = \langle T', \varphi \rangle.$$

It follows that $T * \delta' = \delta' * T = T'$.

In addition, if $T = R * S$, then $T' = \delta' * T = \delta' * R * S = R' * S$. Similarly, $T' = T * \delta' = R * S'$, which proves the following theorem:

THEOREM 7.76 *To compute the derivative of a convolution, it suffices to take the derivative one of the two factors and take the convolution with the other; in other words, for any $R, S \in \mathscr{D}'$, we have*

$$[R * S]' = R' * S = R * S'.$$

when $R * S$ exists. In the same manner, in three dimensions, for any $S, T \in \mathscr{D}'(\mathbb{R}^3)$ such that $S * T$ exists, we have

$$\triangle(S * T) = \triangle S * T = S * \triangle T.$$

7.6.g The Poisson equation

We have admitted that the electrostatic potential V created by a distribution of charge ρ is given by

$$V(\boldsymbol{r}) = \frac{1}{4\pi\varepsilon_0} \left\langle \rho(\boldsymbol{r}'), \frac{1}{\|\boldsymbol{r}' - \boldsymbol{r}\|} \right\rangle = \frac{1}{4\pi\varepsilon_0} \left[\rho * \frac{1}{\|\boldsymbol{r}\|} \right]$$

(if this expression makes sense), where $\|\boldsymbol{r}\|$ denotes, with a slight abuse of notation, the function $\boldsymbol{r} \longmapsto \|\boldsymbol{r}\|$. It is then easy to compute, in the sense of distributions, the laplacian of the potential. Indeed, it suffices to compute the laplacian of one of the two factors in this convolution. Since we know that $\triangle(1/r) = -4\pi\delta$, we obtain

$$\triangle V = \frac{1}{4\pi\varepsilon_0} \triangle \left[\rho * \frac{1}{\|\boldsymbol{r}\|}\right] = \frac{1}{4\pi\varepsilon_0} \left[\rho * \triangle \frac{1}{\|\boldsymbol{r}\|}\right] = \frac{1}{4\pi\varepsilon_0} \rho * (-4\pi\delta)$$
$$= -\frac{\rho}{\varepsilon_0}.$$

Thus we have recovered a classical result:

THEOREM 7.77 (Poisson equation) *The laplacian of the electrostatic potential created by a distribution of charge ρ is*

$$\triangle V = -\frac{\rho}{\varepsilon_0}.$$

7.7
Physical interpretation of convolution operators

We are going to see now that many physical systems, in particular measuring equipment can be represented by convolution operators. Suppose we have given a physical system which, when excited by an input signal E depending on time, produces an output signal, denoted $S = O(E)$. We make the assumptions that the operator O that transforms E into S is

- linear;
- continuous;
- invariant under translations, that is, the following diagram **commutes** (this means that composing the arrows from the upper left corner to the bottom right corner in the two possible ways has the same result):

$$\begin{array}{ccc} E(t) & \xrightarrow{O} & S(t) \\ {\scriptstyle\text{translation}}\big\downarrow & & \big\downarrow{\scriptstyle\text{translation}} \\ E(t-a) & \xrightarrow{O} & S(t-a) \end{array}$$

THEOREM 7.78 *If the three conditions stated above hold, then the operator O is a convolution operator, that is, there exists $R \in \mathscr{D}'$ such that*

$$S = O(E) = E * R = R * E$$

for any E.

The converse is of course true.

Remark 7.79 The case of an excitation depending on time is very simple. If, on the other hand, we consider an optical system where the source is an exterior object and where the output is measured on a photo-detector, the variable corresponding to the measurement (the coordinates, in centimeters, on the photographic plate) and the variable corresponding to the source (for instance, the angular coordinates of a celestial object) are diffent. One must then perform a change of variable to be able to write a convolution relation between the input and output signals.

Knowing the distribution R makes it possible to predict the response of the system to an arbitary excitation. The distribution R is called the **impulse response** because it corresponds to the output to an elementary δ input:

$$\delta \xrightarrow{O} O(\delta) = \delta * R = R.$$

Note that often, in physics, the operator linking $S(t)$ and $E(t)$ is a differential operator.[13] One can then write $D * S = E$ where D is a distribution which is a combination of derivatives of the unit δ.[14] The distribution R, being the output corresponding to a Dirac peak, therefore satisfies

$$R * D(t) = D * R(t) = \underbrace{\delta(t)}_{E(t)}.$$

Since δ is the neutral element for the convolution, the distribution R is thus the *inverse* of D for the convolution product. This point of view is developed in Section 8.4 on page 238. In general, this operator D is given by physical considerations; we will describe below the technique of Green functions (Chapter 15), where the purely technical difficulty is to compute, if it exists, the convolution inverse of D.

Remark 7.80 Experimentally, it is not always easy to send a Dirac function as input. It is doable in optics (use a star), but rather delicate in electricity. It is sometimes simpler to send a signal close to a Heaviside function. The response to such an excitation, which is called the **step response**, is then $S = H * R$. If we differentiate this, we obtain $S' = H' * R = \delta * R = R$, which shows the following result:

THEOREM 7.81 *The impulse response is the derivative of the step response.*

---

[13] Consider, for instance, a particle with mass m and position $x(t)$ on the real axis, subject to a force depending only of time $F(t)$. Take, for example, the excitation $E(t) = \frac{1}{m}F(t)$ and the response $S(t) = x(t)$. The evolution equation of the system is $S''(t) = E(t)$.

[14] In the previous example, one can take $D = \delta''$, since $\delta'' * S = S''$.

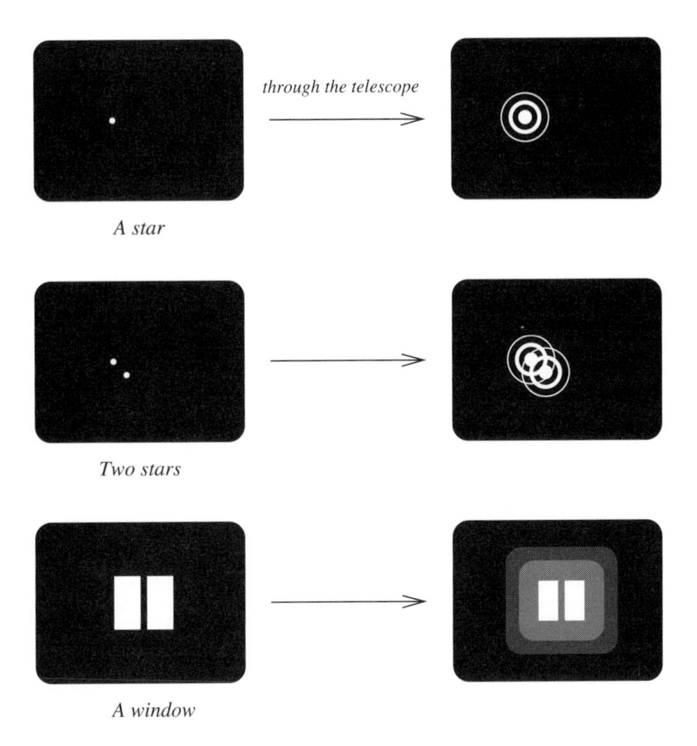

Fig. 7.3 – Some objects to look at with a telescope.

If T is sufficiently smooth, this differentiation can be performed numerically; however, differentiation is often hard to implement numerically. This is the price to pay.

DEFINITION 7.82 (Causal system) A system represented by a temporal convolution operator (i.e., by a time variable) is called a **causal system** if its impulse response $R(t)$ is zero on $\mathbb{R}^{*-}$ (which means that $\langle R, \varphi \rangle = 0$ for any test function φ which is zero for $t \geqslant 0$).

In a causal system, the effect of a signal cannot come before its cause (which is usually expected in physics).

Example 7.83 (in optics) Look at the sky with a telescope. Each star is a point-like light source, modeled by a Dirac δ. Represent the *light intensity* of a star centered in the telescope by a distribution $a\delta$; then the image of this star after passing through the telescope is given by a function of the intensity $a\psi(x, y)$, showing the diffraction spot due to the finite size of the mirror and to the Newton rings.[15] Any *incoherent* luminous object (which is indispensable to ensure linearity with respect to the light intensity functions), with intensity function in the coordinates (x, y) equal to $F(x, y)$, will give an image $F(x, y) * \psi(x, y)$ (see Figure 7.3).

[15] The function ψ is studied in Chapter 11, page 320.

7.8

Discrete convolution

It should be noted that it is quite possible to define a **discrete convolution** to model the "fuzziness" of a digitalized picture. Such a picture is given no longer by a continuous function, but by a sequence of values $n \longmapsto x_n$ with $n \in [\![1, N]\!]$ (in one dimension) or $(m, n) \longmapsto x_{mn}$ with $(m, n) \in [\![1, N]\!] \times [\![1, M]\!]$ (in two dimensions). Each of these discrete numerical values defines a *pixel*.

In the one-dimensional case, we are given values α_n for $n \in [\![1, N]\!]$ (and we put $\alpha_n = 0$ outside of this interval), and the convolution is defined by

$$y = \alpha * x \qquad \text{with } y_n = \sum_{k=-\infty}^{+\infty} \alpha_k \, x_{n-k},$$

the support of y being then equal to $[\![1 - M, N + M]\!]$. In two dimensions, the corresponding formula is

$$y = \alpha * x \qquad \text{with } y_{mn} = \sum_{k=-\infty}^{+\infty} \sum_{\ell=-\infty}^{+\infty} \alpha_{k\ell} \, x_{m-k,n-\ell}.$$

Photo-processing softwares (such as Photoshop, or the `GIMP`, which is Free – as in Free Speech – software) provide such convolutions. An example of the result is given in Figure 7.4 on the facing page.

The discrete convolution may be written as a matrix operation, with the advantages and inconveniences of such a representation. The inversion of this operation is done with the inverse matrix when it exists. We refer the reader desirous to learn more to a book about signal theory.

–

The exercises for this chapter are found at the end of Chapter 8, page 241.

Fig. 7.4 – A square in Prague, before and after discrete convolution.

Laurent Schwartz, born in Paris in 1915, son of the first Jewish surgeon in the Paris hospitals, grandson of a rabbi, was a student at the École Normale Supérieure. During World War II, he went to Toulouse and then to Clermont-Ferrand, before fleeing to Grenoble in 1943. In 1944, he barely escaped a German raid. After the war, he taught at the University of Nancy, where the main part of the Bourbaki group was located. He came back to Paris in 1953 and obtained a position at the École Polytechnique in 1959.

He worked in particular in functional analysis (under the influcence of Dieudonné) and, in 1944, created the theory of distributions. Because of these tremendously important works, he was awarded the Fields Medal in 1950. He showed how to use this theory in a physical context.

Finally, one cannot omit, side by side with the mathematician, the political militant, the pacifist, who raised his voice against the crimes of the French state during the Algerian war and fought for the independence of Algeria [82].

Laurent Schwartz passed away on July 4, 2002.

Chapter 8

Distributions II

In this chapter, we will first discuss in detail a particular distribution which is very useful in physics: the "Cauchy principal value" distribution. Notably, we will derive the famous formula

$$\frac{1}{x \pm \mathrm{i}\varepsilon} = \mathrm{pv}\,\frac{1}{x} \mp \mathrm{i}\pi\delta,$$

which appears in optics, statistical mechanics, and quantum mechanics, as well as in field theory. We will also treat the topology on the space of distributions and introduce the notion of convolution algebra, which will lead us to the notion of Green function. Finally, we will show how to solve in one stroke a differential equation with the consideration of initial conditions for the solution.

8.1 Cauchy principal value

8.1.a Definition

The function $x \mapsto \frac{1}{x}$ does not define a regular distribution, since it is not integrable in the neighborhood of $x = 0$. On the other hand, one can define, for $\varphi \in \mathscr{D}(\mathbb{R})$, the limit

$$\mathrm{pv}\int_{-\infty}^{+\infty} \frac{\varphi(x)}{x}\,\mathrm{d}x \stackrel{\text{def}}{=} \lim_{\varepsilon \to 0^+} \int_{|x| \geqslant \varepsilon} \frac{\varphi(x)}{x}\,\mathrm{d}x$$

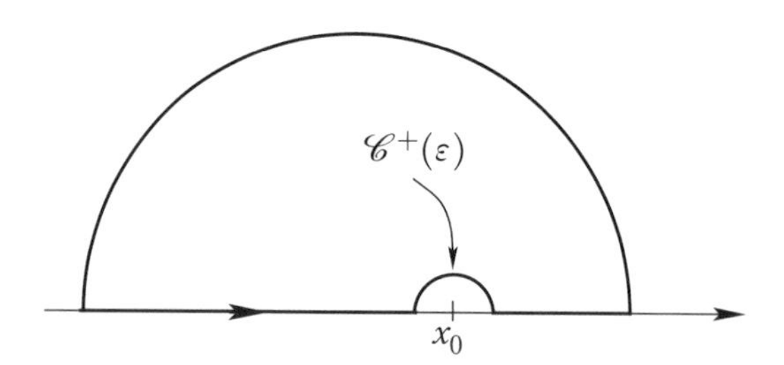

Fig. 8.1 – The contour γ^+.

which provides the definition of the distribution pv $\frac{1}{x}$, given by

$$\left\langle \text{pv}\,\frac{1}{x}, \varphi \right\rangle = \text{pv} \int_{-\infty}^{+\infty} \frac{\varphi(x)}{x}\,\mathrm{d}x \stackrel{\text{def}}{=} \lim_{\varepsilon \to 0^+} \left[\int_{-\infty}^{-\varepsilon} \frac{\varphi(x)}{x}\,\mathrm{d}x + \int_{\varepsilon}^{\infty} \frac{\varphi(x)}{x}\,\mathrm{d}x \right].$$

This is then generalized as follows:

DEFINITION 8.1 Let $x_0 \in \mathbb{R}$. The **Cauchy principal value** pv $\frac{1}{x-x_0}$ is defined as the distribution on $\mathbb{R}$ given by

$$\left\langle \text{pv}\,\frac{1}{x-x_0}, \varphi \right\rangle = \text{pv} \int \frac{\varphi(x)}{x-x_0}\,\mathrm{d}x \stackrel{\text{def}}{=} \lim_{\varepsilon \to 0^+} \int_{|x-x_0|>\varepsilon} \frac{\varphi(x)}{x-x_0}\,\mathrm{d}x.$$

8.1.b Application to the computation of certain integrals

The situation that we are considering is the computation of an integral of the type

$$\int_{-\infty}^{+\infty} \frac{f(x)}{x-x_0}\,\mathrm{d}x, \qquad \text{with} \quad f : \mathbb{R} \longrightarrow \mathbb{C}.$$

This integral is not correctly defined without additional precision; indeed, if $f(x_0) \neq 0$, the function $x \mapsto f(x)/(x-x_0)$ is not Lebesgue integrable. One possibility is to define the integral *in the sense of the principal value.*

We will assume that f is the restriction to the real axis of a function meromorphic on $\mathbb{C}$ and holomorphic at x_0. Moreover, we will assume that f decays sufficiently fast at infinity to permit the use of the method of residues. For instance, we will require that this decay holds at least in the upper half-plane.

Now define a contour γ^+ as in Figure 8.1: denote by $\mathscr{C}^+(\varepsilon)$ the small half-circle with radius ε centered at x_0 and above the real axis. Letting the radius of the large circle go to infinity (and assuming the corresponding integral goes to 0), we have then

$$\int_{\gamma^+} \frac{f(z)}{z-x_0}\,\mathrm{d}z = \int_{-\infty}^{-\varepsilon} \frac{f(x)}{x-x_0}\,\mathrm{d}x + \int_{\varepsilon}^{+\infty} \frac{f(x)}{x-x_0}\,\mathrm{d}x + \int_{\mathscr{C}^+} \frac{f(z)}{z-x_0}\,\mathrm{d}z \;;$$

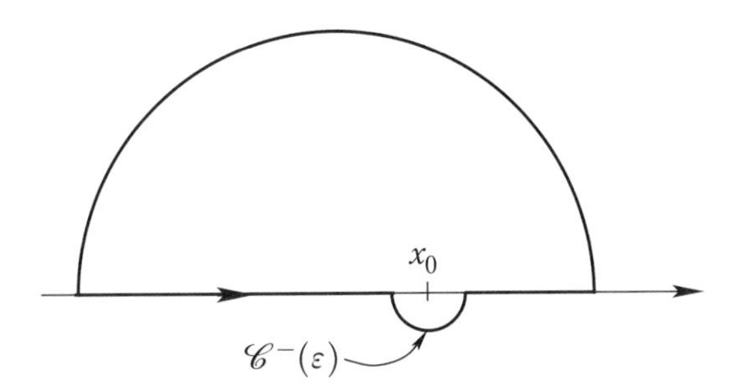

Fig. 8.2 – The contour γ^-.

but the last integral is equal to

$$\int_{\pi}^{0} \frac{f(\varepsilon\,\mathrm{e}^{\mathrm{i}\theta})}{\varepsilon\,\mathrm{e}^{\mathrm{i}\theta}}\,\mathrm{i}\varepsilon\,\mathrm{e}^{\mathrm{i}\theta}\,\mathrm{d}\theta = -\mathrm{i}\pi f(x_0) + \mathrm{O}(\varepsilon) \qquad [\varepsilon \to 0^+].$$

Thus, taking the limit $[\varepsilon \to 0^+]$, we obtain

$$2\mathrm{i}\pi \sum_{p\in\mathscr{H}} \mathrm{Res}\left(\frac{f(z)}{z-x_0}\,;\,p\right) = \mathrm{pv}\int_{-\infty}^{+\infty} \frac{f(x)}{x-x_0}\,\mathrm{d}x - \mathrm{i}\pi f(x_0), \tag{8.1}$$

where the sum $\sum_{p\in\mathscr{H}}$ is the sum over the poles of f located in the open upper half-plane $\mathscr{H} \stackrel{\text{def}}{=} \{z \in \mathbb{C}\,;\, \mathrm{Im}(z) > 0\}$.

So, to deal with the integral on a contour on which a pole is located, it was necessary to deform this contour slightly – in this case, by making a small detour around the pole to avoid it. One may wonder what would have happened, had another detour been chosen. So take the contour γ^- defined as in Figure 8.2. We then have

$$\begin{aligned}\int_{\gamma^-} \frac{f(z)}{z-x_0}\,\mathrm{d}z &= \mathrm{pv}\int_{-\infty}^{+\infty} \frac{f(x)}{x-x_0}\,\mathrm{d}x + \int_{\mathscr{C}^-} \frac{f(z)}{z-x_0}\,\mathrm{d}z \\ &= \mathrm{pv}\int_{-\infty}^{+\infty} \frac{f(x)}{x-x_0}\,\mathrm{d}x + \mathrm{i}\pi f(x_0).\end{aligned}$$

Since the integral on the left-hand side is equal to $2\pi\mathrm{i}$ times the sum of the residues located in the upper half-plane, plus the residue at x_0, which is equal to $2\pi\mathrm{i} f(x_0)$, the final formula is the same as (8.1).

8.1.c Feynman's notation

When evaluating an integral on a contour that passes through a pole, it must be specified on which side the pole the pole will be avoided (the result depends only on the side that is chosen, not on the precise shape of the detour).

Rather than going around the pole, Richard FEYNMAN proposed modifying the initial problem by moving the pole slightly off the real axis while keeping the original contour intact. Suppose the function considered has a pole on the real axis. Then one can write the symbolic equivalences

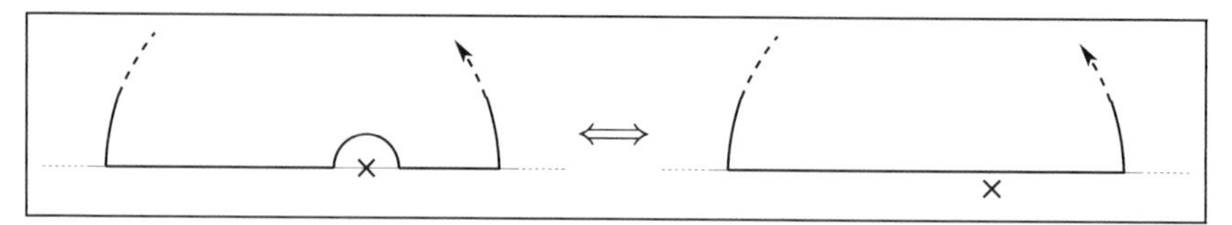

and

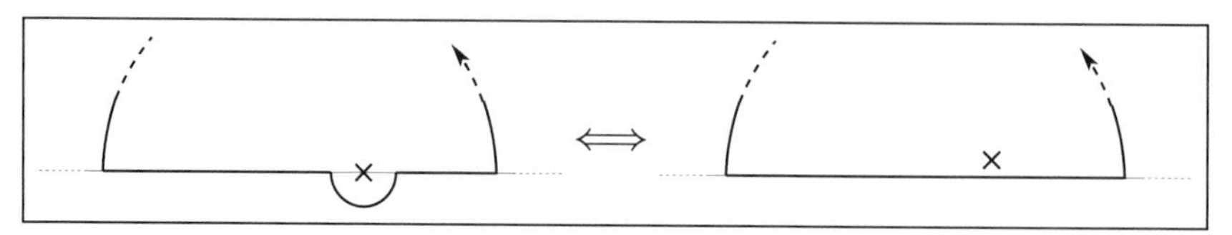

which means that the pole has been moved in the first case to $x_0 - \mathrm{i}\varepsilon$ and in the second case to $x_0 + \mathrm{i}\varepsilon$, and then the limit $[\varepsilon \to 0^+]$ is taken at the end of the computations.

To see clearly what this means, consider the equalities

$$\int_{\gamma^+} \frac{f(z)}{z - x_0}\,\mathrm{d}z = \lim_{\varepsilon \to 0^+} \int_{\gamma^+} \frac{f(z)}{z - x_0 + \mathrm{i}\varepsilon}\,\mathrm{d}z = \lim_{\varepsilon \to 0^+} \int_{\gamma} \frac{f(z)}{z - x_0 + \mathrm{i}\varepsilon}\,\mathrm{d}z,$$

where γ is the undeformed contour. Indeed, the first equality is a consequence of continuity under the integral sign, and the second equality comes from Cauchy's theorem, which, for any given $\varepsilon > 0$, justifies deforming γ^+ into γ, since no pole obstructs this operation. Thus we get, performing implicitly the operation " $\lim_{\varepsilon \to 0^+}$,"

$$\int_{\gamma^+} \frac{f(z)}{z - x_0}\,\mathrm{d}z = \int_{-\infty}^{+\infty} \frac{f(x)}{x - x_0 + \mathrm{i}\varepsilon}\,\mathrm{d}x = \mathrm{pv}\int_{-\infty}^{+\infty} \frac{f(x)}{x - x_0}\,\mathrm{d}x - \mathrm{i}\pi f(x_0),$$

or, in a rather more compact fashion

$$\left\langle \frac{1}{x - x_0 + \mathrm{i}\varepsilon}, f \right\rangle = \left\langle \mathrm{pv}\frac{1}{x - x_0}, f \right\rangle - \mathrm{i}\pi \langle \delta(x - x_0), f \rangle.$$

Symbolically, we write:

$$\boxed{\frac{1}{x - x_0 + \mathrm{i}\varepsilon} = \mathrm{pv}\left(\frac{1}{x - x_0}\right) - \mathrm{i}\pi\,\delta(x - x_0),}$$

and this formula must be applied to a meromorphic function f which is holomorphic on the real axis.

Similarly, we have the following result:

$$\boxed{\frac{1}{x - x_0 - \mathrm{i}\varepsilon} = \mathrm{pv}\left(\frac{1}{x - x_0}\right) + \mathrm{i}\pi\,\delta(x - x_0).}$$

Notice – and this will be convenient in certain computations – that

$$\text{pv}\,\frac{1}{x-x_0} = \text{Re}\left(\frac{1}{x-x_0 \pm \mathrm{i}\varepsilon}\right).$$

8.1.d Kramers-Kronig relations

Let F be a function from $\mathbb{C}$ into $\mathbb{C}$ which is meromorphic on $\mathbb{C}$ but holomorphic on the closed upper half-plane, which means that *all the poles of F have strictly negative imaginary part.* Consider now the contour $\mathscr{C}$ described below, with an arbitrarily large radius:

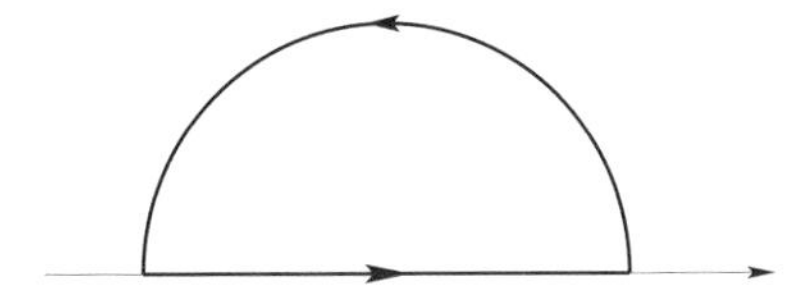

From Cauchy's formula, we have

$$\frac{1}{2\pi\mathrm{i}}\int_{\mathscr{C}} \frac{F(\zeta)}{\zeta - z}\,\mathrm{d}\zeta = \begin{cases} F(z) & \text{if } \mathrm{Im}(z) > 0, \\ 0 & \text{if } \mathrm{Im}(z) < 0. \end{cases}$$

In this formula, put $z = x + \mathrm{i}\varepsilon$ with $x \in \mathbb{R}$ and $\varepsilon > 0$:

$$F(x+\mathrm{i}\varepsilon) = \frac{1}{2\pi\mathrm{i}}\int_{-\infty}^{+\infty} \frac{F(x')}{x'-x-\mathrm{i}\varepsilon}\,\mathrm{d}x'.$$

Assume that the integral on the half-circle tends to 0 when the radius tends to infinity. By letting ε tend to 0, and by continuity of F, the preceding results lead to

$$F(x) = \frac{1}{2\pi\mathrm{i}}\,\text{pv}\int_{-\infty}^{+\infty} \frac{F(x')}{x'-x}\,\mathrm{d}x' + \frac{1}{2}F(x),$$

hence

$$F(x) = \frac{1}{\mathrm{i}\pi}\,\text{pv}\int_{-\infty}^{+\infty} \frac{F(x')}{x'-x}\,\mathrm{d}x'.$$

This equation concerning F becomes much more interesting by taking successively its real and imaginary parts; the following theorem then follows:

THEOREM 8.2 (Kramers-Kronig relations) *Let $F : \mathbb{C} \to \mathbb{C}$ be a meromorphic function on $\mathbb{C}$, holomorphic in the upper half-plane and going to 0 sufficiently fast at infinity in this upper half-plane. Then we have*

$$\boxed{\mathrm{Re}\big[F(x)\big] = \frac{1}{\pi}\,\text{pv}\int_{-\infty}^{+\infty} \frac{\mathrm{Im}\big[F(x')\big]}{x'-x}\,\mathrm{d}x'}$$

and
$$\boxed{\operatorname{Im}\big[F(x)\big] = -\frac{1}{\pi}\,\mathrm{pv}\int_{-\infty}^{+\infty} \frac{\operatorname{Re}\big[F(x')\big]}{x'-x}\,\mathrm{d}x'}$$

These are called **dispersion relations** and are very useful in optics (see, e.g., the book by Born and Wolf [14, Chapter 10]) and in statistical physics. Physicists call these formulas the **Kramers-Kronig relations**,[1] while mathematicians say that $\operatorname{Re}(F)$ is the **Hilbert transform** of $\operatorname{Im}(F)$.

Remark 8.3 If F is meromorphic on $\mathbb{C}$ and holomorphic in the lower half-plane (all its poles have positive imaginary parts), then it satisfies relations which are dual to those just proved (also called Kramers-Kronig relations, which does not simplify matters):

$$\operatorname{Re}\big[F(x)\big] = -\frac{1}{\pi}\,\mathrm{pv}\int_{-\infty}^{+\infty} \frac{\operatorname{Im}\big[F(x')\big]}{x'-x}\,\mathrm{d}x' \quad\text{and}\quad \operatorname{Im}\big[F(x)\big] = \frac{1}{\pi}\,\mathrm{pv}\int_{-\infty}^{+\infty} \frac{\operatorname{Re}\big[F(x')\big]}{x'-x}\,\mathrm{d}x'.$$

We will see, in Chapter 13, that the Fourier transform of causal functions $t \longmapsto f(t)$ (those that vanish for negative values of the variable t), when it exists, satisfies the Kramers-Kronig relations of Theorem 8.2.

Remark 8.4 What happens if F is holomorphic on both the upper and the lower half-plane? Since it is assumed that F has no pole on the real axis, it is then an entire function. The assumption that the integral on a circle tends to zero as the radius gets large then leads (by the mean value property) to the vanishing of the function F, which is the only way to reconcile the previous formulas with those of Theorem 8.2.

Remark 8.5 In electromagnetism, the electric induction $\boldsymbol{D}$ and the electric field $\boldsymbol{E}$ are linked, for a monochromatic wave, by

$$\boldsymbol{D}(\boldsymbol{x},\omega) = \varepsilon(\omega)\,\boldsymbol{E}(\boldsymbol{x},\omega),$$

where $\varepsilon(\omega)$ is the dielectric constant of the material, depending on the pulsation ω of the waves. This relation can be rewritten (via a Fourier tranform) in an integral relation between $\boldsymbol{D}(\boldsymbol{x},t)$ and $\boldsymbol{E}(\boldsymbol{x},t)$:

$$\boldsymbol{D}(\boldsymbol{x},t) = \boldsymbol{E}(\boldsymbol{x},t) + \int_{-\infty}^{+\infty} G(\tau)\,\boldsymbol{E}(\boldsymbol{x},t-\tau)\,\mathrm{d}\tau,$$

where $G(\tau)$ is the Fourier transform of $\varepsilon(\omega)-1$. In fact, since the electric field is the physical field and the field $\boldsymbol{D}$ is derived from it, the previous relation must be causal, that is, $G(\tau)=0$ for any $\tau<0$. One of the main consequences is that the function $\omega \longmapsto \varepsilon(\omega)$ is *analytic* in the lower half-plane of the complex plane (see Chapter 12 on the Laplace transform). From the Kramers-Kronig relations, interesting information concerning the function $\omega \longmapsto \varepsilon(\omega)$ can be deduced. Thus, if the **plasma frequency** is defined by ${\omega_{\mathrm{p}}}^2 = \lim_{\omega\to\infty} \omega^2[1-\varepsilon(\omega)]$, we obtain the **sum rule**

$${\omega_{\mathrm{p}}}^2 = \frac{2}{\pi}\int_0^{+\infty} \omega\,\operatorname{Im}\big[\varepsilon(\omega)\big]\,\mathrm{d}\omega.$$

The reader is invited to read, for instance, the book by Jackson [50, Chapter 7.10] for a more detailed presentation of this sum rule.

---

[1] Dispersion relations first appeared in physics in the study of the dielectric constant of materials by R. de L. KRONIG in 1926 and, independently, in the theory of scattering of light by atoms by H. A. KRAMERS in 1927.

♦ **Exercise 8.1** Using the Kramers-Kronig relations, show that

$$\int_0^{+\infty} \frac{\sin x}{x}\,\mathrm{d}x = \frac{\pi}{2}.$$

(*Solution page 245*)

8.1.e A few equations in the sense of distributions

Recall (Theorem 7.28 on page 191) that the solutions of the equation $x \cdot T(x) = 0$ are the multiples of δ.

PROPOSITION 8.6 *We have* $x \cdot \operatorname{pv}\frac{1}{x} = \mathbf{1}$ *(where* $\mathbf{1}$ *is the constant function* $x \mapsto 1$*).*

Proof. Indeed, for any test function $\varphi \in \mathscr{D}(\mathbb{R})$, we have by definition of the product of a distribution ($\operatorname{pv}\frac{1}{x}$) by a $\mathscr{C}^\infty$ function:

$$\left\langle x \operatorname{pv}\frac{1}{x}, \varphi \right\rangle = \left\langle \operatorname{pv}\frac{1}{x}, x\varphi \right\rangle = \lim_{\varepsilon\to 0^+} \int_{|x|>\varepsilon} \frac{x\,\varphi(x)}{x}\,\mathrm{d}x = \int_{-\infty}^{+\infty} \varphi(x)\,\mathrm{d}x = \langle \mathbf{1}, \varphi \rangle .$$

THEOREM 8.7 (Solutions of $x \cdot T(x) = 1$) *The solutions, in the space of distributions, of the equation*

$$x \cdot T(x) = 1,$$

are the distributions given by $T(x) = \operatorname{pv}\frac{1}{x} + \alpha\delta$, *with* $\alpha \in \mathbb{C}$.

Proof. If T satisfies $x \cdot T(x) = 1$, then $S = T - \operatorname{pv}(1/x)$ satisfies $x \cdot S(x) = 0$ and thus, according to Theorem 7.28 on page 191, S is a multiple of δ.

Theorem 7.28 generalizes as follows:

PROPOSITION 8.8 *Let* $n \in \mathbb{N}$. *The solutions of equation* $x^n \cdot T(x) = 0$ *are the linear combinations of* $\delta, \delta', \ldots, \delta^{n-1}$.

Let f be a function continuous on $\mathbb{R}$, with isolated zeros. For each such zero x_i of f, we say that x_i is of multiplicity m if there exists a continuous function g, which does not vanish in a neighborhood of x_i, such that $f(x) = x^m\, g(x)$ for $x \in \mathbb{R}$.

If f has only isolated zeros with finite multiplicity, then one can treat the equation $f(x) \cdot T(x) = 0$ locally around each zero in the same manner as $x^m\, T(x) = 0$.

THEOREM 8.9 (Solutions of $f(x) \cdot T(x) = 0$) *Let* f *be a continuous function on* $\mathbb{R}$. *Denote by* $Z(f) = \{x_i\ i \in E\}$ *the set of its zeros, and assume these zeros are isolated and with finite multiplicities* m_i.

The distributions solutions to the equation $f(x) \cdot T(x) = 0$ *are given by*

$$T(x) = \sum_{i\in I} \sum_{k=0}^{m_i-1} \alpha_{i,k}\, \delta^{(k)}(x - x_i),$$

where $\alpha_{i,k}$ *are arbitrary complex numbers.*

Example 8.10 The equation $\sin x \cdot T(x) = 0$ has for solutions the distributions of the form

$$\sum_{n\in\mathbb{Z}} \alpha_n\, \delta(x - n\pi).$$

The equation $(\cos x - 1) \cdot T(x) = 0$ has for solutions the distributions of the form

$$\sum_{n\in\mathbb{Z}} \big(\alpha_n\, \delta(x - n\pi) + \beta_n\, \delta'(x - n\pi)\big).$$

THEOREM 8.11 (Solutions of $(x^2 - a^2) \cdot T(x) = 1$) *Let $a > 0$ be given. The distribution solutions of the equation $(x^2 - a^2) \cdot T(x) = 1$ are those given by*

$$\operatorname{pv} \frac{1}{x^2 - a^2} + \alpha\, \delta(x - a) + \beta\, \delta(x + a),$$

where α and β are complex numbers, and where we denote

$$\operatorname{pv} \frac{1}{x^2 - a^2} = \frac{1}{2a} \left[\operatorname{pv} \frac{1}{x + a} - \operatorname{pv} \frac{1}{x - a} \right].$$

8.2 Topology in $\mathscr{D}'$

8.2.a Weak convergence in $\mathscr{D}'$

DEFINITION 8.12 A sequence of distributions $(T_k)_{k\in\mathbb{N}}$ **converges weakly** (or simply "converges") in $\mathscr{D}'$ if, for any test function $\varphi \in \mathscr{D}$, the sequence of complex numbers $\langle T_k, \varphi \rangle$ converges in $\mathbb{C}$. We then denote by $\langle T, \varphi \rangle$ this limit, which defines a linear function T on $\mathscr{D}$, called the **weak limit** of $(T_k)_{k\in\mathbb{N}}$ in $\mathscr{D}'$.

THEOREM 8.13 *If $(T_k)_{k\in\mathbb{N}}$ converges weakly in $\mathscr{D}'$ to a functional T, then T is continuous and is therefore a distribution. The weak convergence in $\mathscr{D}'$ is denoted by $T_k \xrightarrow{\mathscr{D}'} T$.*

♦ **Exercise 8.2** For $n \in \mathbb{N}$, let T_n be the regular distribution associated to the locally integrable function $x \mapsto \sin(nx)$. Show that the sequence $(T_n)_{n\in\mathbb{N}}$ converges weakly to 0 in $\mathscr{D}'$.

THEOREM 8.14 (Continuity of differentiation) *The operation of differentiation in $\mathscr{D}'(\mathbb{R})$ is a continuous linear operation: if $T_k \xrightarrow{\mathscr{D}'} T$, then for any $m \in \mathbb{N}$, we have $T_k^{(m)} \xrightarrow{\mathscr{D}'} T^{(m)}$.*

Obviously this extends to partial derivatives of distributions in $\mathbb{R}^n$.

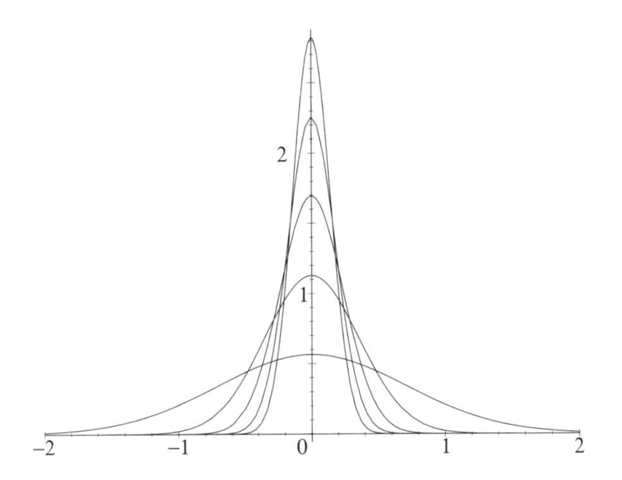

Fig. 8.3 – The sequence of gaussian functions $(g_n)_{n\in\mathbb{N}}$, for $n = 1, \ldots, 5$.

This theorem is extremely powerful and justifies the interchange of limits and derivatives, which is of course impossible in the sense of functions.[2]

8.2.b Sequences of functions converging to δ

DEFINITION 8.15 **A Dirac sequence (of functions)** is any sequence of functions $f_k : \mathbb{R} \to \mathbb{R}$ locally integrable, such that

① there exists a real number $A > 0$ such that, for any $x \in \mathbb{R}$ and any $k \in \mathbb{N}$, we have

$$\big(\,|x| \leqslant A\big) \Longrightarrow \big(f_k(x) \geqslant 0\big),$$

② for any $a \in \mathbb{R}$, $a > 0$,

$$\int_{|x|\leqslant a} f_k(x)\,\mathrm{d}x \xrightarrow[k\to\infty]{} 1 \qquad \text{and} \qquad f_k(x) \xrightarrow{\text{CV.U.}} 0 \quad \text{on } a < |x| < \frac{1}{a}.$$

Example 8.16 An example of a Dirac sequence is the sequence $(f_n)_{n\in\mathbb{N}}$ defined by $f_n(x) = n\Pi(nx)$ for any $x \in \mathbb{R}$ and any $n \in \mathbb{N}$. (Check this on a drawing.)

Example 8.17 The sequence of functions $(g_n)_{n\in\mathbb{N}}$ defined by

$$g_n(x) \stackrel{\text{def}}{=} \frac{n}{\sqrt{\pi}}\mathrm{e}^{-n^2x^2} \qquad \text{for } x \in \mathbb{R}$$

---

[2] Consider, for instance, the sequence of functions $(f_n)_{n\in\mathbb{N}}$ defined by

$$f_n(t) = \sum_{k=1}^{n} \frac{1}{2k}\cos 2kt \qquad \text{for any } t \in \mathbb{R}.$$

The functions f_n are all differentiable, but their limit f is a "sawtooth" function, which is not differentiable at the points of the form $n\pi$ with $n \in \mathbb{Z}$.

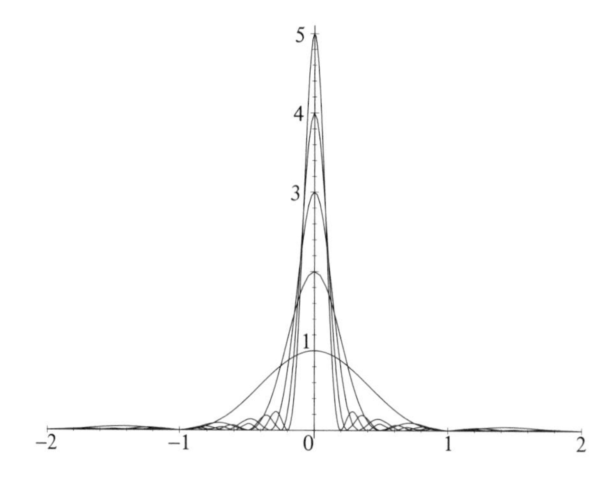

Fig. 8.4 – The sequence of functions $\psi_n(x) = \dfrac{\sin^2(\pi n x)}{n\pi^2 x^2}$ for $n = 1$ to 5.

is a Dirac sequence (see Figure 8.3 on the page before).

THEOREM 8.18 *Any Dirac sequence converges weakly to the Dirac distribution δ in $\mathscr{D}'(\mathbb{R})$.*

PROOF. See Appendix D.

Example 8.19 Similarly, the sequence of functions $(\psi_n)_{n\in\mathbb{N}}$ defined by

$$\psi_n(x) = \frac{\sin^2(\pi n x)}{n\pi^2 x^2} \qquad \text{for } x \in \mathbb{R}$$

is a Dirac sequence (see Figure 8.4) and thus converges also to δ.

Remark 8.20 Without the positivity assumption for x close to 0, there is nothing to prevent contributions proportional to δ' to arise in the limit! Thus, consider the sequence of functions $(k_n)_{n\in\mathbb{N}}$ given by the graph

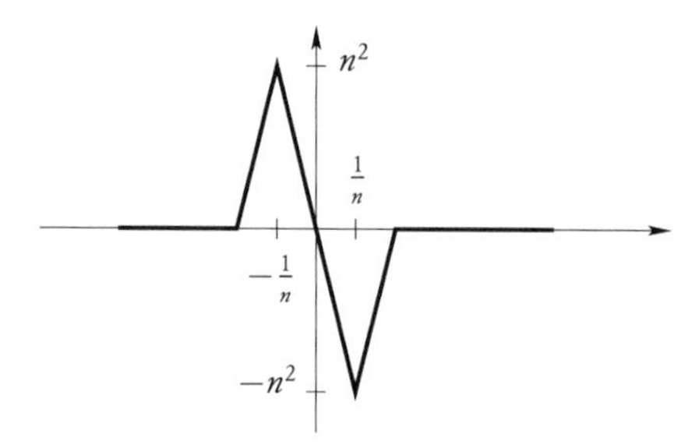

It is easy to show that $k_n \xrightarrow{\mathscr{D}'} \delta'$. Let $(f_n)_{n\in\mathbb{N}}$ be a Dirac sequence; then the sequence $(f_n + k_n)_{n\in\mathbb{N}}$, which still satisfies ②, but not ①, converges to $\delta + \delta'$ and not to δ.

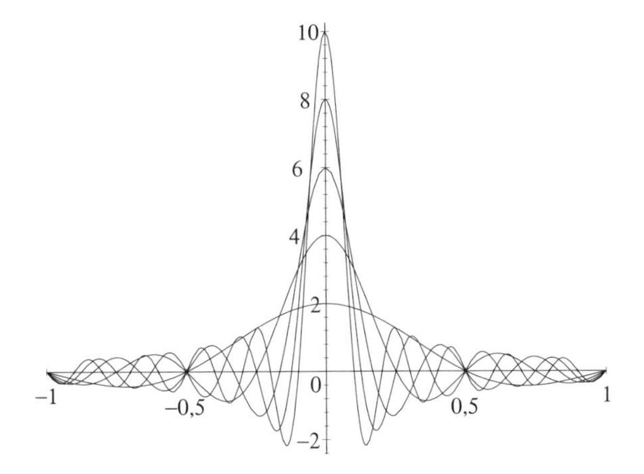

Fig. 8.5 – The functions $b_n : x \longmapsto \dfrac{\sin 2\pi n x}{\pi x}$, for $n = 1, \ldots, 5$.

Consider now the sequence of functions

$$b_n(x) = \frac{\sin 2\pi n x}{\pi x}.$$

All the functions b_n share a common "envelope" $x \mapsto 1/\pi x$ and the sequence of real numbers $(b_n(x))_{n\in\mathbb{N}}$ does not converge if $x \notin \mathbb{Z}$ (see Figure 8.5). We do have $b_n(0) \to \infty$, but, for instance,

$$b_n\left(\frac{1}{4}\right) = \begin{cases} (-1)^k/2\pi & \text{if } n = 2k+1, \\ 0 & \text{if } n = 2k. \end{cases}$$

In particular, this sequence $(b_n)_{n\in\mathbb{N}}$ is not a Dirac sequence. Yet, despite this we have the following result:

PROPOSITION 8.21 *The sequence $(b_n)_{n\in\mathbb{N}}$ converges to δ in the sense of distributions.*

How to explain this phenomenon?

Proving the convergence of $(b_n)_{n\in\mathbb{N}}$ to δ is, because of the lack of positivity, harder than for a Dirac sequence. In the case that concerns us, the sequence of functions $(b_n)_{n\in\mathbb{N}}$ exhibits ever faster oscillations when we let n tend to infinity (see Figure 8.5), and, after getting rid of the singularity at 0, we can hope to be able to invoke the Riemann-Lebesgue lemma (see Theorem 9.44). Moreover, it is easy to compute that $\int b_n = 1$ for all $n \in \mathbb{N}^*$ (for example, by the method of residues).

Proof of Proposition 8.21. Let $\varphi \in \mathscr{D}$ be a test function and let $[-M, M]$ be an interval containing the support of φ.

We can always write $\varphi(x) = \varphi(0) + x\,\psi(x)$, where ψ if of $\mathscr{C}^\infty$ class. Then we have

$$\langle b_n, \varphi\rangle = \varphi(0)\int_{-M}^{M} \frac{\sin 2\pi n x}{\pi x}\,\mathrm{d}x + \int_{-M}^{M} \psi(x)\sin(2\pi n x)\,\mathrm{d}x,$$

and the first integral, *via* the substitution $y = nx$, tends to the improper integral

$$\int_{-\infty}^{+\infty} \frac{\sin \pi x}{\pi x}\,\mathrm{d}x = 1,$$

whereas the second integral tends to 0, by the Riemann-Lebesgue lemma.

Finally, the convergence to δ is highly useful because of the following theorem:

THEOREM 8.22 *Let $(f_n)_{n\in\mathbb{N}}$ be a Dirac sequence of functions. If g is a bounded continuous function on $\mathbb{R}$, then $f_n * g$ exists for all $n \in \mathbb{N}$ and the sequence $(f_n * g)_{n\in\mathbb{N}}$ converges uniformly to g.*

If g is not continuous but the limits on the left or the right of g exist at the point x, then $f_n * g(x)$ tends to $\frac{1}{2}\left[g(x^-) + g(x^+)\right]$.

8.2.c Convergence in $\mathscr{D}'$ and convergence in the sense of functions

The previous examples show that a sequence $(f_n)_{n\in\mathbb{N}}$ of regular distributions may very well converge in $\mathscr{D}'$ to a regular distribution f without converging pointwise almost everywhere (and even more without converging uniformly) in the sense of functions. In particular, the Riemann-Lebesgue lemma is equivalent to the following proposition:

PROPOSITION 8.23 *The sequence of regular distributions $(f_n)_{n\in\mathbb{N}}$ defined by the functions $f_n : x \longmapsto \sin(nx)$ converges to 0 in $\mathscr{D}'$, but it converges pointwise only for $x \in \pi\mathbb{Z}$.*

8.2.d Regularization of a distribution

What happens when we take the convolution of a function f of $\mathscr{C}^1$ class with a function g of $\mathscr{C}^2$ class? Assuming that this convolution exists and denoting $h = f * g$, the function h will be infinitely differentiable *in the sense of distributions*, but how many times will it be differentiable in the sense of functions?

Compute the derivative. It is possible to write $h' = f' * g$ and, since f' and g are functions, h' is well-defined in the sense of functions. Continue. It is not possible to differentiate f more than that, but g is still available, and so $h'' = f' * g'$ is still a function, as is the third derivative $h''' = f' * g''$. But going farther is not possible.

Similarly, the convolution of a function of $\mathscr{C}^k$ class with a function of $\mathscr{C}^\ell$ class will produce a function of $\mathscr{C}^{k+\ell}$ class.

These results generalize to distributions. Of course, it is not reasonable to speak of "$\mathscr{C}^\infty$ class" (since distributions are always infinitely differentiable anyway). But "regularity" for a distribution consists precisely in being what is called a "regular distribution," namely, a distribution associated to a locally integrable function.

An interesting result (which is, in fact, quite intuitive) is that the convolution by a $\mathscr{C}^\infty$ function "smoothes" the object which is subject to the convolution: the result is a function which is $\mathscr{C}^\infty$. This is the case also if we take the convolution of a distribution with a $\mathscr{C}^\infty$ function: the result is a regular distribution associated to a function, and in certain cases one can show that the result is also $\mathscr{C}^\infty$.

THEOREM 8.24 (Regularization of a distribution) *Let $T \in \mathscr{D}'$ be a distribution and $\psi \in \mathscr{C}^\infty(\mathbb{R})$, and assume that $T * \psi$ exists. Then $T * \psi$ is a regular distribution associated to a function f, which is given by*

$$f(x) = T * \psi(x) = \big\langle T(t), \psi(x-t)\big\rangle.$$

Let $T \in \mathscr{D}'$ and $\varphi \in \mathscr{D}$ (so φ is now not only of $\mathscr{C}^\infty$ class, but also with bounded support). Then $T * \varphi$ exists and is a regular distribution associated to a function g of $\mathscr{C}^\infty$ class, with derivatives given by the formula

$$g^{(k)}(x) = \big\langle T(t), \varphi^{(k)}(x-t)\big\rangle.$$

8.2.e Continuity of convolution

The following theorem shows that the operation of convolution with a fixed distribution is continuous for the weak topology on $\mathscr{D}'$:

THEOREM 8.25 (Continuity of convolution) *Let $T \in \mathscr{D}'$. If $(S_k)_{k\in\mathbb{N}}$ converges weakly in $\mathscr{D}'$ to T, and if moreover all the S_k are supported in a fixed closed subset such that all the $S_k * T$ exist, then the sequence $(S_k * T)$ converges in $\mathscr{D}'$ to $S * T$.*

Consider now a Dirac sequence $(\gamma_n)_{n\in\mathbb{N}}$, with elements belonging to $\mathscr{D}$. It converges weakly to δ in $\mathscr{D}'$ (this is Theorem 8.18); moreover, we know (Theorem 8.24) that if T is a distribution, we have $\gamma_n * T \in \mathscr{C}^\infty(\mathbb{R})$. Since $\gamma_n * T \xrightarrow[n\to\infty]{} T$, we deduce:

THEOREM 8.26 (Density of $\mathscr{D}$ in $\mathscr{D}'$) *Any distribution $T \in \mathscr{D}'$ is the limit in $\mathscr{D}'$ of a sequence of functions in $\mathscr{D}$. One says that $\mathscr{D}$ is* **dense** *in $\mathscr{D}'$.*

In other words, any distribution is the weak limit of a sequence of functions of $\mathscr{C}^\infty$ class with bounded support.

♦ **Exercise 8.3** Find a sequence $(\varphi_n)_{n\in\mathbb{N}}$ in $\mathscr{D}$ which converges weakly to Ш in $\mathscr{D}'$.

(See Exercise 8.11 on page 242)

8.3

Convolution algebras

DEFINITION 8.27 A **convolution algebra** is any vector space of distributions containing the Dirac distribution δ and for which the convolution product of an arbitrary number of factors is defined.

There are two well-known and very important examples.

THEOREM 8.28 ($\mathscr{E}'$, $\mathscr{D}'_+$ and $\mathscr{D}'_-$) *The space $\mathscr{E}'$ of distributions with bounded support is a convolution algebra.*

The space of distributions with support bounded on the left, denoted $\mathscr{D}'_+$ and the space of distributions with support bounded on the right, denoted $\mathscr{D}'_-$, are also convolution algebras.

These algebras, as the name indicates, provide means of solving certain equations by means of *algebraic manipulations*. In particular, many physical problems can be put in the form of a convolution equation of the type

$$A * X = B$$

where A and B are distributions (B is often called the "source") and the unknown X is a distribution.[3] The problem is to find a solution and to determine if it is unique. So we are, quite naturally, led to the search for a distribution denoted A^{*-1}, which is called the **convolution inverse of** A or, sometimes, **Green function** of A, which will be such that

$$A * A^{*-1} = A^{*-1} * A = \delta.$$

Remark 8.29 This inverse, if it exists, is *unique in the algebra under consideration*. Indeed, suppose there exist, in a same algebra, two distributions A' and A'' such that

$$A * A' = A' * A = \delta \quad \text{and} \quad A * A'' = A'' * A = \delta.$$

Then

$$A'' = A'' * \delta = A'' * (A * A') = (A'' * A) * A' = \delta * A' = A'$$

and therefore

$$A'' = A'.$$

[3] For instance, notice that any linear differential equation with constant coefficients

$$\left[a_0 + a_1 \frac{\mathrm{d}}{\mathrm{d}t} + \cdots + a_n \frac{\mathrm{d}^n}{\mathrm{d}t^n}\right] X(t) = B(t)$$

can be reexpressed in the form of a convolution product

$$\left[a_0\delta + a_1\delta' + \cdots + a_n\delta^{(n)}\right] * X(t) = B(t).$$

The convolution equation is then solved by putting

$$X = A^{*-1} * B,$$

and this solution is unique.

To summarize, we can state the following theorem:

THEOREM 8.30 *Let $\mathscr{A}$ be a convolution algebra. The convolution equation*

$$A * X = B$$

admits a solution in $\mathscr{A}$ for arbitrary $B \in \mathscr{A}$ if and only if A has a convolution inverse A^{*-1} in this algebra. In this case, the solution is unique and is given by

$$X = A^{*-1} * B.$$

Remark 8.31 It is important to notice that we have always spoken of the inverse of a distribution *in a given algebra* $\mathscr{A}$. In fact, a distribution A may very well have a certain convolution inverse in one algebra and another inverse in a different algebra. Take, for instance, the case of a harmonic oscillator characterized by a convolution equation of the type

$$(\delta'' + \omega^2\delta) * X(t) = B(t),$$

where $B(t)$ is a distribution characterizing the input signal. The reader is invited to check that

$$(\delta'' + \omega^2\delta) * \frac{1}{\omega} H(t) \sin \omega t = \delta.$$

Since the distribution $\frac{1}{\omega} H(t) \sin \omega t$ is an element of $\mathscr{D}'_+$, it follows that, for any $B \in \mathscr{D}'_+$, that is, for any input which started at a finite instant $t \in \mathbb{R}$ and was zero previously, the equation has a solution in $\mathscr{D}'_+$ which is *unique* and is given by

$$\boxed{X(t) = B(t) * \frac{1}{\omega} H(t) \sin \omega t} \qquad \text{causal solution.}$$

But it is also possible to check that

$$(\delta'' + \omega^2\delta) * \left[-\frac{1}{\omega} H(-t) \sin \omega t\right] = \delta,$$

and therefore that $-\frac{1}{\omega} H(-t) \sin \omega t$ is a convolution inverse of $(\delta'' + \omega^2\delta)$ in $\mathscr{D}'_-$. If the input signal $B(t)$ belongs to $\mathscr{D}'_-$, it is therefore necessary, to get the solution, to carry out the convolution of B by $-\frac{1}{\omega} H(-t) \sin \omega t$:

$$\boxed{X(t) = B(t) * \frac{-1}{\omega} H(-t) \sin \omega t} \qquad \text{anti-causal solution.}$$

We also deduce that this convolution operator does not have an inverse in $\mathscr{E}'$.

One of the great computational problems that can confront a physicist is to find the convolution inverse of an operator (its Green function). For instance, the Green function of the Laplace operator in $\mathbb{R}^3$ is given by the Coulomb potential $1/r$, as we have seen by performing the relevant convolution product. But how, given an operator, can we find its inverse? If the convolution operator is a differential operator (more precisely, a differential operator applied to δ), there exist tools which are particularly adapted to computing its inverse, namely the Fourier and Laplace transforms (see Chapters 10 and 12). Infinitely many technical details for the inversion of the laplacian (with refinements thereof) are given in the reference [24], which is somewhat hard to read, however.

8.4 Solving a differential equation with initial conditions

Most students in physics remember very well that the evolution of a system is given by a differential equation, but forget that the real problem is to solve this equation **for given initial conditions**. It is true that, in the linear case, these initial conditions are often used to *parameterize* families of solutions; however, this is not a general rule.[4]

The **Cauchy problem** (an equation or system of differential equations, with initial conditions) is often very difficult to solve, as every one of you knows. Sometimes, it is very hard to show that a solution exists (without even speaking of computing it!), or to show that it is unique.[5]

In the cases below, we have autonomous linear differential equations with constant coefficients. The theoretical issues are therefore much simpler, and we are just presenting a convenient computational method. Some cases which are more complicated (and more useful in physics) will be presented on page 348.

8.4.a First order equations

Consider a differential equation of the type

$$\dot{u} + \alpha u = 0, \tag{e}$$

where we only wish to know $u(t)$ for positive values of t, and $u : \mathbb{R}^+ \to \mathbb{R}$ is a function satisfying the initial condition

$$u(0) = u_0 \in \mathbb{R}.$$

It is possible to use the tools of the theory of distributions to solve this equation with its initial conditions.

---

[4] The most spectacular counterexamples are given by chaotic systems.

[5] For instance, the "old" problem of an electric charge in an electromagnetic field is only *partly* solved [11]. The difficulty of this system is that it is described by highly nonlinear equations where the charge influences the fields, which in turn influence the charge. The reader interested in these problems of mathematical physics can read [84].

We then slightly change our point of view, by looking no longer for a solution in the sense of functions to this differential equation, but for a solution in the sense of distributions, of the type $U(t) = H(t)\,u(t)$. (This solution will then be zero for negative time, which is of no importance since we are interested in positive time only.)

If u is a solution of (e), which equation will the distribution U satisfy? Since $\dot{U} = \dot{u} + u_0\delta$, we see that U satisfies the equation $\dot{U} + \alpha U = u_0\delta$, which may be put in the form of a convolution

$$[\delta' + \alpha\delta] * U = u_0\delta. \tag{E}$$

There only remains to find the unique solution to this equation (in the sense of distributions) which lies in the convolution algebra $\mathscr{D}'_+$ of distributions with support bounded on the left.

THEOREM 8.32 *The convolution inverse of* $[\delta + \alpha\delta']$ *in* $\mathscr{D}'_+$ *is*

$$G(t) = H(t)\,\mathrm{e}^{-\alpha t}.$$

(The inverse in $\mathscr{D}'_-$ *is* $t \mapsto -H(-t)\,\mathrm{e}^{-\alpha t}$*.)*

Proof. Indeed, we have $G'(t) = -\alpha\,\mathrm{e}^{-\alpha t} + \delta(t)$, which shows that $[\delta' + \alpha\delta] * G = \delta$.

Taking the convolution of (E) with $G(t) = H(t)\,\mathrm{e}^{-\alpha t}$, we find that

$$U = G * (u_0\delta), \qquad \text{or} \qquad U = u_0\,H(t)\,\mathrm{e}^{-\alpha t}.$$

It may seem that we used something like a hammer to crush a very small fly. Quite. But this systematic procedure becomes more interesting when considering a differential equation of higher order, or more complicated initial conditions, for instance on surfaces (see Section 15.4, page 422).

8.4.b The case of the harmonic oscillator

Consider now the case of a **harmonic oscillator**, controlled by the equation

$$\frac{\mathrm{d}^2 u}{\mathrm{d}t^2} + \omega^2 u = 0, \tag{e'}$$

where we seek a function $u(t)$ satisfying this differential equation for $t \geqslant 0$, with given initial conditions $u(0)$ and $u'(0)$. For this purpose, we will look for a distribution $U(t)$, of the type $U(t) = H(t)\,u(t)$, which satisfies the differential equation in the sense of functions. What differential equation in the sense of distributions will it satisfy? This will be, taking the discontinuities into account,

$$\left[\delta'' + \omega^2\delta\right] * U = u(0)\,\delta' + u'(0)\,\delta. \tag{E'}$$

THEOREM 8.33 (Harmonic oscillator) *The convolution inverse of* $[\delta'' + \omega^2\delta]$ *in* $\mathscr{D}'_+$ *is*

$$\mathscr{G}^+(t) = \frac{1}{\omega} H(t) \sin \omega t.$$

The convolution inverse of $[\delta'' + \omega^2\delta]$ *in* $\mathscr{D}'_-$ *(see Remark 8.31 on page 237) is*

$$\mathscr{G}^-(t) = \frac{-1}{\omega} H(-t) \sin \omega t.$$

Remark 8.34 There exist methods to rediscover this result, the proof of which is immediate: such convolution inverses, or Green functions, of course do not come out of thin air. These methods are explained in Section 15.2 on page 409.

Then we find directly, by taking the convolution of equation (E′) with $\mathscr{G}^+$, that

$$\begin{aligned} U(t) &= \frac{1}{\omega} H(t) \sin \omega t * \left[u(0)\,\delta'(t) + u'(0)\,\delta(t)\right] \\ &= H(t) \left[u(0) \cos \omega t + \frac{u'(0)}{\omega} \sin \omega t\right]. \end{aligned}$$

To summarize, the same problem can be seen from two different angles:

1. the first, classical, shows a differential equation with given initial conditions at the point 0;

2. the second shows a differential equations valid in the sense of distributions, without conditions at 0, for which a solution is sought in a given convolution algebra (for instance, that of distributions with support bounded on the left).

The right-hand side of (E′) shows what distribution is necessary to move the system at rest (for $t < 0$) "instantaneously" to the initial state $\big(u(0), u'(0)\big)$.

8.4.c Other equations of physical origin

We will see in the following chapters, devoted to the Fourier and Laplace transforms, how to compute the convolution inverses (also called the **Green functions**) of a differential operator.

Here we merely state an interesting result:

THEOREM 8.35 (Heat equation) *The convolution inverse of the heat operator*

$$\left[c\frac{\partial}{\partial t} - \varkappa\frac{\partial^2}{\partial x^2}\right]$$

in $\mathscr{D}'_+$*, called the* ***heat kernel****, is*

$$G(x,t) = \frac{H(t)}{\sqrt{\varkappa\pi c t}}\, \mathrm{e}^{-cx^2/4\pi t},$$

that is,

$$\left[c\delta'(t) - \varkappa\delta''(x)\right] * G(x,t) = \delta^{(2)}(x,t),$$

or

$$c\frac{\partial G}{\partial t} - \varkappa\frac{\partial^2 G}{\partial x^2} = \delta(x)\,\delta(t).$$

PROOF. This result is proved *ab initio* in Section 15.4, page 422.

EXERCISES

Examples of distributions

♦ **Exercise 8.4 (Cauchy principal value)** Show that the map defined on $\mathscr{D}$ by

$$\operatorname{pv}\frac{1}{x} : \varphi \in \mathscr{D} \longmapsto \lim_{\varepsilon \to 0} \int_{|x|>\varepsilon} \frac{\varphi(x)}{x}\,\mathrm{d}x$$

which defines the **Cauchy principal value** is the derivative, in the sense of distributions, of the regular distribution associated to the locally integrable function $x \mapsto \log|x|$.

♦ **Exercise 8.5 (Beyond principal value)** Show that the map

$$\varphi \longmapsto \lim_{\varepsilon \to 0^+} \left[\int_\varepsilon^\infty \frac{\varphi(x)}{x^2}\,\mathrm{d}x - \frac{\varphi(0)}{\varepsilon} + \varphi'(0)\log\varepsilon\right]$$

defines a distribution, that is, that it is linear and *continuous*.

♦ **Exercise 8.6 (Finite part of $1/x^2$)** We denote by $\operatorname{fp}(1/x^2)$ the opposite of the distribution derivative of the distribution $\operatorname{pv}(1/x)$ ("fp" for **finite part**):

$$\operatorname{fp}\frac{1}{x^2} \stackrel{\text{def}}{=} -\left[\operatorname{pv}\frac{1}{x}\right]'.$$

Show that

$$\left\langle \operatorname{fp}\frac{1}{x^2}, \varphi \right\rangle = -\int_{-\infty}^{+\infty} \log|x|\,\varphi''(x)\,\mathrm{d}x = \lim_{\varepsilon \to 0^+} \int_{|x|>\varepsilon} \frac{\varphi(x) - \varphi(0)}{x^2}\,\mathrm{d}x,$$

and that, if $\mathbf{1}$ denotes as usual the constant function equal to 1, or the associated distribution, we have

$$x^2 \cdot \operatorname{fp}\frac{1}{x^2} = \mathbf{1}.$$

Generalize these results to define distributions related to the (not locally integrable) functions $x \mapsto 1/x^n$ for any $n \in \mathbb{N}$.

♦ **Exercise 8.7** Show that we have, for all $a \in \mathbb{R}$,

$$\delta(ax) = \frac{1}{|a|}\delta(x),$$

whereas, if $a > 0$,

$$H(ax) = H(x).$$

♦ **Exercise 8.8** Prove the relation : $x\delta' = -\delta$. Compute similarly $x\delta''$. Generalize by computing $f(x)\,\delta'$, where f is a (locally integrable) function of $\mathscr{C}^\infty$ class. Is it possible to relax those assumptions?

Composition of a δ distribution and a function

♦ **Exercise 8.9 (Special relativity)** We consider the setting of special relativity, with the momentum vector $\mathbf{p} = (E, \boldsymbol{p}) = (p^0, \boldsymbol{p})$, with a Minkowski metric of signature $(+,-,-,-)$ and the convention $c = 1$. Thus, we have $\mathbf{p}^2 = E^2 - \boldsymbol{p}^2$.

Recall that for a classical particle, we have $\mathbf{p}^2 = m^2$, where m is the mass of the particle. In field theory, one is often led to compute integrals of the type

$$\int \delta(\mathbf{p}^2 - m^2)\, H(p^0)\, f(\mathbf{p})\, \mathrm{d}^4\mathbf{p}.$$

What is the interpretation of the functions δ and H? These integrals have the nice property of being invariant under Lorentz transformations. Show that, if we put

$$E(\boldsymbol{p}) \stackrel{\text{def}}{=} \sqrt{\boldsymbol{p}^2 + m^2},$$

then we have

$$\delta(\mathbf{p}^2 - m^2)\, H(p^0)\, \mathrm{d}^4\mathbf{p} = \frac{\mathrm{d}^3\boldsymbol{p}}{2E(\boldsymbol{p})},$$

which is the usual way of writing that for any sufficiently smooth function (indicate precisely the necessary regularity) we have

$$\int \delta(\mathbf{p}^2 - m^2)\, H(p^0) f(\mathbf{p})\, \mathrm{d}^4\mathbf{p} = \int f\big(E(\boldsymbol{p}), \boldsymbol{p}\big) \frac{\mathrm{d}^3\boldsymbol{p}}{2E(\boldsymbol{p})}.$$

The important point to remember is the following:

THEOREM 8.36 *The differential element $\mathrm{d}^3\boldsymbol{p}/2E$, where $E^2 = \boldsymbol{p}^2 + m^2$, is Lorentz invariant.*

♦ **Exercise 8.10** Let $\mathscr{S}$ denote the sphere centered at 0 with radius R. Can we write $\delta_{\mathscr{S}}$ in the form $\delta(r - R)$, where $r = \sqrt{x^2 + y^2 + z^2}$? Can we write it $\delta(r^2 - R^2)$?

Convergence in $\mathscr{D}$

♦ **Exercise 8.11** Let $n \in \mathbb{N}^*$ and $\alpha \in \mathbb{R}$. We consider the function

$$\begin{aligned} f_n : \mathbb{R} \setminus \pi\mathbb{Z} &\longrightarrow \mathbb{R}, \\ x &\longmapsto \frac{1}{2n} \frac{\sin^2 \pi n x/2}{\sin^2 \pi x/2}. \end{aligned}$$

Show that f_n can be extended by continuity to $\mathbb{R}$. Compute $\int_{-1}^{+1} f_n(t)\,\mathrm{d}t$ and show that the result is independent of n.

Show that the sequence $(f_n)_{n\in\mathbb{N}}$ tends to Ш in the sense of distributions.

♦ **Exercise 8.12** Show that

$$\left(x \mapsto \frac{1}{a}\Pi\left(\frac{x}{a}\right)\right) \xrightarrow{\mathscr{D}'} \delta(x) \qquad [a \to 0^+].$$

♦ **Exercise 8.13** Let α be a real number. Show that, in the sense of distributions, we have

$$\sum_{n\in\mathbb{Z}} \sin\alpha\,|n| = \frac{\sin\alpha}{1-\cos\alpha}.$$

Differentiation in higher dimensions

♦ **Exercise 8.14** We consider a two-dimensional space-time setting. Define

$$\mathscr{E}(x,t) = \begin{cases} 1 & \text{if } ct > |x|, \\ 0 & \text{if } ct < |x|, \end{cases}$$

or, in short-hand notation, $\mathscr{E}(x,t) = H(ct - x)$. Compute $\square\mathscr{E}$ in the sense of distributions, where the d'Alembertian operator $\square$ is defined by

$$\square \stackrel{\text{def}}{=} \frac{1}{c^2}\frac{\partial^2}{\partial t^2} - \frac{\partial^2}{\partial x^2}.$$

♦ **Exercise 8.15** In the plane $\mathbb{R}^2$, we define the function $E(x,t) = H(x)H(t)$, and its associated regular distribution, also denoted E. Compute, in the sense of distributions, $\dfrac{\partial^2}{\partial x\,\partial t}E$ and $\triangle E$.

♦ **Exercise 8.16 (The wages of fear)** We model a truck and its suspension by a spring with strength k carrying a mass m at one of its extremities. The wheels at the bottom part of the spring are forced to move on an irregular road, the height of which with respect to a fixed origin is described by a function of the distance $E(x)$. The horizontal speed of the truck is assumed to be constant equal to v, so that at any time t, the wheels are at height $E(vt)$. Denote by $R(t)$ the height of the truck, compared to its height at rest (it is the *response* of the system).

i) Show that

$$\left[\frac{\mathrm{d}^2}{\mathrm{d}t^2} + \omega^2\right] R(t) = \omega^2 E(vt), \qquad \text{where } \omega^2 = \frac{k}{m}.$$

ii) In order to generalize the problem, we assume that E is a distribution. What is then the equation linking the response $R(t)$ of the truck to a given road profile $E(x) \in \mathscr{D}'_+$, where $\mathscr{D}'_+$ is the space of distributions with support bounded on the left? Comment on the choice of $\mathscr{D}'_+$.

iii) Compute the response of the truck to the profile of a road with the following irregularities:

- a very narrow bump $E(x) = \delta(x)$;
- a flat road $E(x) = \ell\, H(x)$;
- a wavy road $E(x) = \ell\, H(x)\sin(x/\lambda)$.

Compute the amplitude of the movements of the truck. How does it depend on the speed?

iv) In the movie *The Wages of Fear* by Henri-Georges Clouzot (1953, original title *Le salaire de la peur*), a truck full of nitroglycerin must cross a very rough segment of road. One of the characters, who knows the place, explains that the truck can only cross safely at very low speed, or on the contrary at high speed, but not at moderate speed. Explain.

PROBLEM

♦ **Problem 3 (Kirchhoff integrals)** The wave equation in three dimensions, for a *scalar* quantity ψ, is

$$\left[\triangle - \frac{1}{c^2}\frac{\partial^2}{\partial t^2}\right] \psi(\boldsymbol{r},t) = 0,$$

where we denote $\boldsymbol{r} \stackrel{\text{def}}{=} (x,y,z)$. Monochromatic waves with pulsation ω are given by

$$\psi(\boldsymbol{r},t) = \varphi(\boldsymbol{r})\,\mathrm{e}^{-\mathrm{i}\omega t}.$$

1. Show that the complex amplitude φ then satisfies the equation called the Helmoltz equation:

$$\left[\triangle + k^2\right] \varphi(\boldsymbol{r}) = 0, \qquad \text{with} \quad k \stackrel{\text{def}}{=} \frac{\omega}{c}.$$

Let $\mathscr{S}$ be an arbitrary closed surface. We are looking for a distribution $\Phi(\boldsymbol{r})$ such that

$$\Phi(\boldsymbol{r}) = \begin{cases} \varphi(\boldsymbol{r}) & \text{inside } \mathscr{S}, \\ 0 & \text{outside } \mathscr{S}, \end{cases}$$

and which is a solution in the sense of distributions of the Helmoltz equation.

2. Show that differentiating in the sense of distribution leads to

$$\left[(\triangle + k^2)\delta(\boldsymbol{r})\right] * \Phi(\boldsymbol{r}) = \varphi\frac{\mathrm{d}}{\mathrm{d}\boldsymbol{n}}\delta_{\mathscr{S}}(\boldsymbol{r}) + \left(\frac{\mathrm{d}\varphi}{\mathrm{d}\boldsymbol{n}}\right)\delta_{\mathscr{S}}(\boldsymbol{r}),$$

where $\boldsymbol{n}$ is the *interior normal* to $\mathscr{S}$ (the choices of orientation followed here are those in the book *Principles of Optics* by Born and Wolf [14]).

3. Show that a Green function of the operator $[\triangle + k^2]$, that is, its convolution inverse, is given by

$$G(\boldsymbol{r}) = -\frac{1}{4\pi}\frac{\mathrm{e}^{\mathrm{i}kr}}{r}, \qquad \text{where} \quad r \stackrel{\text{def}}{=} \|\boldsymbol{r}\|.$$

4. Deduce that

$$\Phi(\boldsymbol{r}) = \frac{1}{4\pi}\iint_{\mathscr{S}} \left[\varphi\frac{\mathrm{d}}{\mathrm{d}\boldsymbol{n}}\frac{\mathrm{e}^{\mathrm{i}kq}}{q} - \frac{\mathrm{e}^{\mathrm{i}kq}}{q}\left(\frac{\mathrm{d}\varphi}{\mathrm{d}\boldsymbol{n}}\right)\right] \mathrm{d}^2\boldsymbol{x} \qquad \text{with} \quad q = \|\boldsymbol{r} - \boldsymbol{x}\|.$$

5. Interpretation: the complex amplitude inside a closed surface $\mathscr{S}$, with given limit conditions, is the same as that produced by a certain surface density of fictive sources[6] located on $\mathscr{S}$. This result is the basis of what is called the *method of images* in electromagnetism and it is, in the situation considered here, the founding principle of the scalar theory of diffraction in optics.

   A more complete version of Kirchhoff's Theorem, for an arbitrary dependence on time, is presented in the book of Born and Wolf [14, p. 420].

[6] Possibly very difficult to compute explicitly! Note that this computation requires knowing both the amplitude φ *and* its normal derivative with respect to the surface. In some cases only we may hope to exploit this exact formula, where symmetries yield additional information.

SOLUTIONS

♦ **Solution of exercise 7.2 on page 198**

$$\langle \delta'(ax), \varphi(x) \rangle = \frac{1}{|a|} \left\langle \delta'(x), \varphi\left(\frac{x}{a}\right) \right\rangle = -\frac{1}{a\,|a|}\varphi'(0)$$
$$= \frac{1}{a\,|a|} \langle \delta', \varphi \rangle .$$

♦ **Solution of exercise 7.4 on page 211.** Put $h(x) = \int f(t)\,g(x-t)\,\mathrm{d}t$ and make the substitution $y = x - t$, with jacobian $|\mathrm{d}y/\mathrm{d}t| = 1$, which yields

$$h(x) = \int f(x-y)\,g(y)\,\mathrm{d}y = \int g(y)\,f(x-y)\,\mathrm{d}y = [g * f](x).$$

♦ **Solution of exercise 7.5 on page 211.** A simple but boring computation shows that

$$\mathrm{e}^{-|ax|} * \mathrm{e}^{-|bx|} = \frac{2}{a^2 - b^2}\left[a\,\mathrm{e}^{-b|x|} - b\,\mathrm{e}^{-a|x|}\right].$$

♦ **Solution of exercise 8.1 on page 229.** Notice that $\sin x = \operatorname{Im}(\mathrm{e}^{\mathrm{i}x})$ and that $z \mapsto \mathrm{e}^{\mathrm{i}z}$ is holomorphic on $\mathbb{C}$ and decays in the upper half-plane (at least in the sense of Jordan's second lemma). Using then the Kramers-Kronig formula from Theorem 8.2, we obtain

$$\begin{aligned}\int_0^{\infty} \frac{\sin x}{x}\,\mathrm{d}x &= \frac{1}{2}\int_{-\infty}^{+\infty} \frac{\sin x}{x}\,\mathrm{d}x \\ &= \frac{1}{2}\lim_{\varepsilon \to 0^+}\left[\int_{-\infty}^{-\varepsilon} + \int_{\varepsilon}^{+\infty}\right] \frac{\sin x}{x}\,\mathrm{d}x \qquad \text{(continuous function)} \\ &= \frac{1}{2}\,\mathrm{pv}\int_{-\infty}^{+\infty} \frac{\operatorname{Im}\mathrm{e}^{\mathrm{i}x}}{x}\,\mathrm{d}x = \frac{\pi}{2}\operatorname{Re}(\mathrm{e}^{\mathrm{i}0}) = \frac{\pi}{2}.\end{aligned}$$

♦ **Solution of exercise 8.4.** Let $\varphi \in \mathscr{D}$. We then have

$$\begin{aligned}\left\langle \left[\log|x|\right]', \varphi \right\rangle &= -\langle \log|x|, \varphi'(x)\rangle = -\int \log|x|\,\varphi'(x)\,\mathrm{d}x \\ &= \lim_{\varepsilon \to 0+}\left[\int_{-\infty}^{-\varepsilon} + \int_{+\varepsilon}^{+\infty}\right]\big(-\log|x|\,\varphi(x)\big)\,\mathrm{d}x,\end{aligned}$$

since $x \mapsto \log|x|$ is locally integrable (the integral is well defined around 0). Integrating each of the two integrals by parts, we get

$$\begin{aligned}\left\langle \left[\log|x|\right]', \varphi \right\rangle &= \lim_{\varepsilon \to 0^+}\left\{\log(\varepsilon)\,\varphi(\varepsilon) - \log(\varepsilon)\,\varphi(-\varepsilon) + \int_{|x|>\varepsilon} \frac{\varphi(x)}{x}\,\mathrm{d}x\right\} \\ &= \mathrm{pv}\int_{-\infty}^{+\infty} \frac{\varphi(x)}{x}\,\mathrm{d}x,\end{aligned}$$

which shows that $\left[\log|x|\right]' = \mathrm{pv}\,\frac{1}{x}$.

♦ **Solution of exercise 8.5.** First, we show that the stated limit does exist. Let $\varphi \in \mathscr{D}$. Integrating by parts twice, we get

$$\int_{\varepsilon}^{+\infty} \frac{\varphi(x)}{x^2}\,\mathrm{d}x = -\int_{\varepsilon}^{+\infty} \varphi''(x)\log x\,\mathrm{d}x - \varphi'(\varepsilon)\log\varepsilon + \frac{\varphi(\varepsilon)}{\varepsilon},$$

and therefore

$$\int_{\varepsilon}^{+\infty} \frac{\varphi(x)}{x^2}\,\mathrm{d}x - \frac{\varphi(0)}{\varepsilon} + \varphi'(0)\log\varepsilon =$$

$$-\int_{\varepsilon}^{+\infty} \varphi''(x)\log x\,\mathrm{d}x + \big[\varphi'(0) - \varphi'(\varepsilon)\big]\log\varepsilon + \frac{\varphi(\varepsilon) - \varphi(0)}{\varepsilon}.$$

The second term tends to 0, and the third to $\varphi'(0)$ as $[\varepsilon \to 0^+]$. The first term is integrable when $\varepsilon = 0$. Hence the expression given has a limit, which we denote by $\langle T, \varphi\rangle$.

The linearity is obvious, and hence we have to prove the continuity, and it is sufficient to show the continuity at 0 (the zero function of $\mathscr{D}$).

Let $(\varphi_n)_{n\in\mathbb{N}}$ be a sequence of functions in $\mathscr{D}$, converging to 0 in $\mathscr{D}$. All the functions φ_n have support included in a fixed interval $[-A, A]$, where we can assume for convenience and without loss of generality that $A > \mathrm{e}$. Then

$$\langle T, \varphi_n\rangle = -\int_0^A \varphi_n''(x)\log x\,\mathrm{d}x + \varphi_n'(0),$$

and hence

$$\big|\langle T, \varphi_n\rangle\big| \leqslant \|\varphi_n''\|_\infty (A\log A - A) + \|\varphi_n'\|_\infty.$$

Choose $\varepsilon > 0$. There exists an integer N such that for $n \geqslant N$ we have $\|\varphi_n''\|_\infty \leqslant \varepsilon/(A\log A - A)$ and $\|\varphi_n'\|_\infty \leqslant \varepsilon$, so that $|\langle T, \varphi_n\rangle| \leqslant 2\varepsilon$ for any $n \geqslant N$. This proves that $\langle T, \varphi_n\rangle$ tends to 0 as $[n \to \infty]$.

Since we have proved the continuity of T, this linear functional is indeed a distribution.

♦ **Solution of exercise 8.6.** To generalize, define:

$$\operatorname{fp}\frac{1}{x^n} = \frac{(-1)^{n-1}}{(n-1)!}\frac{\mathrm{d}^n}{\mathrm{d}x^n}\log|x| = \frac{(-1)^{n-1}}{(n-1)!}\frac{\mathrm{d}^{n-1}}{\mathrm{d}x^{n-1}}\operatorname{pv}\frac{1}{x}.$$

♦ **Solution of exercise 8.7.** Let $a \neq 0$. Then for any $\varphi \in \mathscr{D}$ we have

$$\big\langle\delta(ax), \varphi\big\rangle = \frac{1}{|a|}\big\langle\delta(x), \varphi(x/a)\big\rangle = \frac{1}{|a|}\varphi(0) = \frac{1}{|a|}\langle\delta, \varphi\rangle.$$

For any $a > 0$, the functions $x \mapsto H(x)$ and $x \mapsto H(ax)$ are equal. Hence the associated regular distributions are also equal. If a proof by substitution really is required, simply compute, for any test function φ

$$\big\langle H(ax), \varphi\big\rangle = \frac{1}{|a|}\big\langle H, \varphi(x/a)\big\rangle = \frac{1}{|a|}\int_0^{+\infty} \varphi(x/a)\,\mathrm{d}x = \int_0^{+\infty} \varphi(x)\,\mathrm{d}x = \langle H, \varphi\rangle.$$

For $a < 0$, the reader can check that $H(ax) = 1 - H(x)$.

♦ **Solution of exercise 8.8.** For any test function $\varphi \in \mathscr{D}$ we have

$$\langle x\delta', \varphi\rangle = \langle\delta', x\varphi\rangle = -[x\varphi]'(0) = -\varphi(0) - 0\cdot\varphi'(0) = -\langle\delta, \varphi\rangle.$$

Similarly, by letting it act on any test function φ, one shows that $x\delta'' = -2\delta'$. Note that this relation can also be recovered directly by differentiating the relation $x\delta' = -\delta$, which yields $\delta' + x\delta'' = -\delta'$ and thus $x\delta'' = -2\delta'$.

If f is a locally integrable function of $\mathscr{C}^\infty$ class, we can define the product $f\delta'$. By the same method as before, we find that

$$f(x)\,\delta'(x) = f(0)\,\delta' - f'(0)\,\delta.$$

The assumptions can be weakened by asking only that f be continuous and differentiable at 0. Without a more complete theory of distributions, the relation above can be taken as a definition for $f(x)\,\delta'(x)$ in this case.

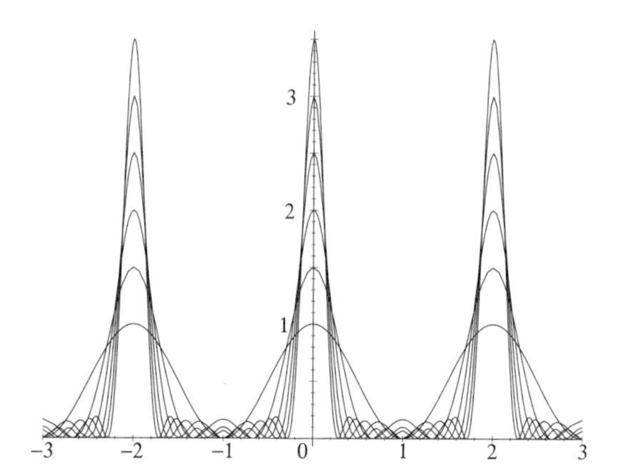

Fig. 8.6 – The sequence of functions $f_n : x \longmapsto \dfrac{1}{2n}\left(\dfrac{\sin \pi n x/2}{\sin \pi x/2}\right)^2$, for $n = 2, \ldots, 7$. The limit in the sense of distributions of this sequence is Ш.

♦ **Solution of exercise 8.10.** We have $\delta_{\mathscr{S}} = \delta(r - R)$. On the other hand,

$$\delta(r^2 - R^2) = \frac{1}{2R}\delta(r - R) = \frac{1}{2R}\delta_{\mathscr{S}}.$$

♦ **Solution of exercise 8.11.** By writing the difference of two consecutive terms and integrating by parts, it follows that $\int_{-1}^{1} f_n(t)\,\mathrm{d}t$ is independent of n. Hence (see the graph in Figure 8.6)

$$\int_{-1}^{1} f_n(t)\,\mathrm{d}t = 1 \qquad \forall n \in \mathbb{N}.$$

Moreover, f_n takes only *positive* values and converges uniformly to 0 on any interval of the type $]+\varepsilon, 1]$, with $\varepsilon > 0$, which shows that

$$\Pi\left(\frac{x}{2}\right) f_n(x) \underset{n\to\infty}{\longrightarrow} \delta,$$

and since $f_n(x) = \sum_{k=-\infty}^{+\infty} \Pi\left(\frac{x-k}{2}\right) f_n(x)$ and each term tends to δ_k, it follows that $f_n \underset{n\to\infty}{\longrightarrow}$ Ш. Of course, it is possible to find simpler sequences of functions that converge weakly to Ш in $\mathscr{D}'$, for instance, the sequence of functions

$$f_n(x) = n\,\Pi\left(\frac{x - E(x)}{n}\right).$$

♦ **Solution of exercise 8.13.** Expand the sine as sum of two complex exponentials, rearrange the terms, and compute the sums of geometric series. The convergence must be justified.

♦ **Solution of exercise 8.15.** We have $E(x,t) = H(x)\,H(t)$ and therefore

$$\frac{\partial E}{\partial x}(x,t) = \delta(x)\,H(t),$$

whence

$$\frac{\partial^2 E}{\partial x \partial t}(x,t) = \delta(x)\,\delta(t).$$

As for the laplacian, it is equal to $\triangle E = \delta'(x)\,H(t) + H(x)\,\delta'(t)$.

♦ **Solution of exercise 8.16.**

i) The force exerted on the truck by the spring is $k \cdot \big(E(vt) - R(t)\big)$, so by Newton's law we can write

$$m\frac{\mathrm{d}^2 R}{\mathrm{d}t}(t) + k \cdot \big(R(t) - E(vt)\big) = 0,$$

hence the required formula.

ii) The previous equation can also be written

$$\left[\delta'' + \omega^2 \delta\right] * R = \omega^2 E(vt),$$

and remains valid if E is a distribution. The choice of $\mathscr{D}'_+$ is justified by the desire to study a *causal* system. More precisely, if the excitation (the road) starts at a finite time in the past, we want the response to be identically zero before that time. The convolution inverse of $\left[\delta'' + \omega^2\delta\right]$ in $\mathscr{D}'_+$ has this property.

iii) In $\mathscr{D}'_+$, the solutions are given by $R(t) = E(t) * \omega H(t) \sin \omega t$.

- In the case of a narrow bump, we have $E(t) = \delta(vt) = \delta(t)/v$ and $R(t) = (\omega/v) H(t) \sin \omega t$.
- For a flat road, we find $R(t) = \ell H(t)[1 - \cos \omega t]$.
- For a wavy road, denoting $\omega' = v/\lambda$ we obtain

$$R(t) = \omega \int_0^t \sin(\omega s) \sin\big(\omega'(t-s)\big)\,\mathrm{d}s = \omega \ell H(t)\,\frac{\omega \sin(\omega' t) - \omega' \sin(\omega t)}{\omega^2 - \omega'^2}.$$

iv) The nitroglycerin will explode if the amplitudes of the vibrations of the truck are too important. The previous formula indicates a resonance at the speed $v = \lambda\omega$. A very slow or very fast speed will avoid this resonance.

♦ **Solution of problem 3.** Using Theorem 7.51 on page 206, the formula asked in Question 2 follows. To show that $(\triangle + k^2)G = \delta$, use the property

$$\triangle(1/r) = -4\pi\,\delta$$

and the formula for differentiating a product: it follows that

$$\Phi(\boldsymbol{r}) = \left[\varphi \frac{\mathrm{d}}{\mathrm{d}\boldsymbol{n}} \delta_{\mathscr{S}}(\boldsymbol{r}) + \left(\frac{\mathrm{d}\varphi}{\mathrm{d}\boldsymbol{n}}\right), \delta_{\mathscr{S}}(\boldsymbol{r})\right] * \frac{(-1)}{4\pi} \frac{\mathrm{e}^{ikr}}{r},$$

and then that

$$\Phi(\boldsymbol{r}) = \left\langle \varphi \frac{\mathrm{d}}{\mathrm{d}\boldsymbol{n}} \delta_{\mathscr{S}}(\boldsymbol{x}) + \left(\frac{\mathrm{d}\varphi}{\mathrm{d}\boldsymbol{n}}\right) \delta_{\mathscr{S}}(\boldsymbol{x}), \frac{(-1)}{4\pi} \frac{\mathrm{e}^{ik\|\boldsymbol{r}-\boldsymbol{x}\|}}{\|\boldsymbol{r} - \boldsymbol{x}\|} \right\rangle.$$

Chapter 9

Hilbert spaces, Fourier series

9.1 Insufficiency of vector spaces

To indicate how vector spaces and the usual notion of basis (which is called an **algebraic basis**) are insufficient, here is a little problem: let $E = \mathbb{R}^{\mathbb{N}}$, the space of sequences of real numbers. Clearly E is an infinite-dimensional vector space. So, the general theory of linear algebra teaches that[1]:

THEOREM 9.1 *Any vector space has at least one algebraic basis.*

The question is then

Can you describe a basis of E?

(Before turning the page, you should try your skill!)

---

[1] Though the theorem that states this does not give a way of constructing such a basis. Even worse, the proof uses the axiom of choice – or rather an equivalent version, called *Zorn's lemma*

LEMME (Zorn) *Let Z be a non-empty ordered set in which any totally ordered subset admits an upper bound in Z. Then Z contains at least one maximal element.*

An elegant statement certainly, but which, for a physicist, looks obviously fishy: not much that is constructive will come of it!

Let's try. One may think that the family

$$\Big\{(1,0,0,0,\dots),\quad (0,1,0,0,\dots),\quad (0,0,1,0,\dots),\quad \dots\Big\} \tag{9.1}$$

of elements of E is a basis. But this is an illusory hope, remembering that a basis is a free and *generating* family:

DEFINITION 9.2 (Algebraic basis) Let I be an arbitrary index set (finite, countable, or uncountable). The **sub-vector space generated** by a family $(x_i)_{i\in I}$ of vectors in a vector space E, denoted $\operatorname{Vect}\{x_i \,;\, i \in I\}$, is the set of (*finite*, by definition) linear combinations of these vectors:

$$\operatorname{Vect}\{x_i \,;\, i\in I\} \stackrel{\text{def}}{=} \Big\{x = \sum_{i\in I'} a_i x_i \,;\, I' \subset I \text{ finite and } a_i \in \mathbb{K}\Big\}.$$

If $\operatorname{Vect}\{x_i \,;\, i \in I\} = E$, the family $(x_i)_{i\in I}$ is called a **generating** family, and the family $(x_i)_{i\in I}$ is **free** if the only (finite) linear combination which is equal to zero is the one where every coefficient is zero.

The family $(x_i)_{i\in I}$ is an **algebraic basis** if it is both free and generating.

This means that any element of E must be expressible as a *finite* linear combination of the basis vectors. But it is clear that $(1,1,1,\dots,1,\dots)$, which is an element of E, is not a finite linear combination of the family (9.1).

Finite sums, schminite sums! We just need to suppress this condition of finiteness and permit infinite sums. But what this means needs to be clarified. More precisely, we must give a meaning to the convergence of a sum

$$\sum_{i=0}^{\infty} \alpha_i e_i,$$

where the α_i are scalars and the e_i are vectors.

To speak of convergence, the simplest solution is to consider the setting of normed vector spaces. Indeed, on the space E of real sequences, there is a natural norm associated to the scalar product

$$\langle u, v\rangle = \sum_{k=0}^{\infty} u_k v_k, \qquad \text{namely,} \qquad \|u\|^2 = \sum_{k=0}^{\infty} |u_k|^2 .$$

However, this will not do, since the norm of the vector $(1,1,1,\dots,1,\dots)$ is infinite. Without additional assumptions, speaking of norms brings other difficulties, and we are back at the beginning. Well, then, one may wish to keep the finiteness requirement for the sums, but to restrict the vector space under consideration by looking only at the space E^0 made of sequences with bounded support (i.e., sequences where all but finitely many elements are zero). This time, the family (9.1) is indeed an algebraic basis of E_0, but another unpleasant phenomenon arises. Consider the sequence

$$x^{(1)} = (1,0,0,0,0,\dots), \qquad x^{(2)} = \left(1,\tfrac{1}{2},0,0,0,\dots\right),$$
$$x^{(3)} = \left(1,\tfrac{1}{2},\tfrac{1}{4},0,0,\dots\right), \qquad x^{(4)} = \left(1,\tfrac{1}{2},\tfrac{1}{4},\tfrac{1}{8},0,\dots\right),$$
$$x^{(n)} = \left(1,\tfrac{1}{2},\dots,\tfrac{1}{2^{n-1}},0,\dots\right), \qquad \dots$$

of elements of E_0.

If we put

$$x = \left(\tfrac{1}{2^k}\right)_{k\in\mathbb{N}^*} = \left(1,\tfrac{1}{2},\tfrac{1}{4},\tfrac{1}{8},\dots,\tfrac{1}{2^{n-1}},\dots\right),$$

it seems "obvious" that the sequence $(x^{(n)})_n$ converges to x, since

$$\left\|x^{(n)} - x\right\|^2 = \sum_{k=n}^{\infty} \frac{1}{4^k} = \frac{1}{3\cdot 4^{n-1}}.$$

Unfortunately, the element x does not belong to E^0 (it does not have bounded support). One says that E^0 is not *complete*.

So we can see that the difficulties pile up when we try to extend the notion of basis to an infinite family of vectors. To stay in a "well-behaved" setting, we are led, following Hilbert and Schmidt, to introduce the notion of a *Hilbert space*. Before doing this, we will review the general techniques linked to the scalar product and to projection onto subspaces of finite dimension.

9.2 Pre-Hilbert spaces

DEFINITION 9.3 Let E be a real or complex vector space. A **complex scalar product** or a **hermitian (scalar) product** on E is any application (form) $(x,y)\mapsto (x|y)$ such that

(a) $(x|y) = \overline{(y|x)}$ for all $x,y\in E$;

(b) $(x|\alpha y + y') = \alpha\,(x|y) + (x|y')$ for any $\alpha\in\mathbb{C}$, and $x,y,y'\in E$;

(c) $(x|x)\geqslant 0$ for all $x\in E$;

(d) $(x|x) = 0$ if and only if $x=0$.

If E is equipped with a hermitian product, it is called a **pre-Hilbert space**.

Property (b) expresses the linearity with respect to the second variable. The scalar product is thus **semilinear** (or **antilinear**) with respect to the first, which means that

$$\forall \alpha, \beta \in \mathbb{C}, \quad \forall x, x', y, \in E \qquad \left(\alpha x + \beta x' \middle| y\right) = \overline{\alpha}\,(x|y) + \overline{\beta}\,\left(x' \middle| y\right).$$

The hermitian product is also called **definite** (property d) and **positive** (property c).

DEFINITION 9.4 (Norm) Let E be a pre-Hilbert space. The **norm** (or **hermitian norm**) of an element $x \in E$ is defined using the scalar product by

$$\|x\| \stackrel{\text{def}}{=} \sqrt{(x|x)}.$$

(Theorem 9.6 on the facing page shows that this is indeed a norm.)

The hermitian product has the following essential property:

THEOREM 9.5 (Cauchy-Schwarz[2] inequality) *Let E be a pre-Hilbert space. Let $x, y \in E$ be two vectors. Then we have*

$$\left|(x|y)\right| \leqslant \|x\| \cdot \|y\|,$$

with equality if and only if x and y are colinear.

PROOF. Put $\theta \stackrel{\text{def}}{=} \frac{(x|y)}{|(x|y)|}$ (assuming that $(x|y) \neq 0$, which is not a restriction since otherwise the inequality is immediate). Then we have $\theta\overline{\theta} = 1$ and, for any real number λ, the inequalities

$$\begin{aligned} 0 \leqslant (\theta x + \lambda y | \theta x + \lambda y) &= \overline{\theta}\theta\,(x|x) + \overline{\theta}\lambda\,(x|y) + \lambda\theta\,(y|x) + \lambda^2\,(y|y) \\ &= \lambda^2\,(y|y) + 2\lambda\,|(x|y)| + (x|x). \end{aligned}$$

If $\|y\|^2 \neq 0$ (otherwise the result is also proved), this last quantity is a polynomial of degree 2 in λ, positive or zero for any value of $\lambda \in \mathbb{R}$; this is only possible if the discriminant is negative, namely, if

$$4\,|(x|y)|^2 - 4\,(y|y)\,(x|x) \leqslant 0,$$

or equivalently if $|(x|y)| \leqslant \sqrt{(x|x)}\sqrt{(y|y)}$, which is the required inequality. Moreover, there is equality if and only if the polynomial has a double root; in that case, taking λ to be equal to this root, we obtain $\left\|\theta x + \lambda y\right\| = 0$, which shows that the vectors x and y are linearly dependent.

Another importance inequality, called the *Minkowski inequality*, shows that the map $x \longmapsto \sqrt{(x|x)}$ is indeed a norm (in the sense defined in Appendix A):

[2] One should not confuse Hermann SCHWARZ (1843–1921) and Laurent SCHWARTZ (1915–2002). For biographical indications, see, respectively, pages 161 and 222.

THEOREM 9.6 (Minkowski inequality) *Let E be a pre-Hilbert space. Then we have*

$$\forall x, y \in E \qquad \|x+y\| \leqslant \|x\| + \|y\|,$$

with equality if and only if x and y are linearly dependent, and their proportionality coefficient is positive or zero.

PROOF. Use the Cauchy-Schwarz inequality in the expansion of $\|x+y\|^2$.

Moreover, we have the following equality:

THEOREM 9.7 (Parallelogram identity) *If E is a pre-Hilbert space, then for any $x, y \in E$ we have*

$$\|x+y\|^2 + \|x-y\|^2 = 2\|x\|^2 + 2\|y\|^2.$$

This is called the the **parallelogram identity** (or law) because of the following analogy with the euclidean norm in the plane: the sum of the squares of the lengths of the diagonals of a parallelogram is equal to the sum of the squares of the lengths of the four sides.

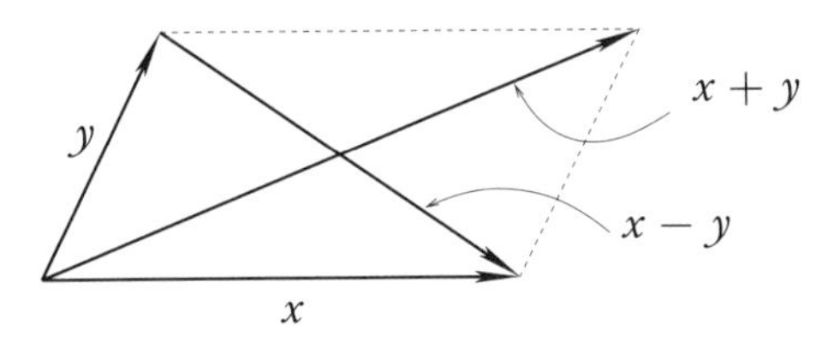

DEFINITION 9.8 (Orthogonality) Let E be a pre-Hilbert space.

i) Two vectors x and y in E are **orthogonal** if $(x|y) = 0$. This is denoted $x \perp y$.

ii) Two subsets A and B are **orthogonal** if, for any $a \in A$ and any $b \in B$, we have $a \perp b$. This is denoted $A \perp B$.

iii) If A is an arbitrary subset of E, the **orthogonal of A**, denoted $A^\perp$, is the vector space of elements in E which are orthogonal to all elements of A:

$$A^\perp = \{x \in E\ ;\ \forall y \in A,\ x \perp y\}.$$

DEFINITION 9.9 (Orthonormal system) Let E be a pre-Hilbert space and let $(e_i)_{i \in I}$ be an arbitrary family of vectors of E indexed by a set I that may or not be countable. The family $(e_i)_{i \in I}$ is

- an **orthogonal system** if $(e_i|e_j) = 0$ as soon as $i \neq j$;
- an **orthonormal system** if $(e_i|e_j) = \delta_{ij}$ for all i, j.

9.2.a The finite-dimensional case

We assume in this section that E is a pre-Hilbert space of finite dimension (which is also called a **hermitian space** in the complex case and a **euclidean space** in the real case).

Now let $(e_1, \ldots, e_n)$ be a given orthonormal basis of E[3]. If x is any vector in E, it can be decomposed according to this basis: $x = \sum x_i \, e_i$. Computing the scalar product of x with one of the basis vectors e_k, we obtain, using the linearity on the right of the scalar product

$$(e_k|x) = \Big(e_k \Big| \sum_{i=1}^{n} x_i \, e_i\Big) = \sum_{i=1}^{n} x_i \, (e_k|e_i) = \sum_{i=1}^{n} x_i \, \delta_{i,k} = x_k.$$

Hence, the coordinates of a vector in an orthonormal basis are simply given by the scalar products of this vector with each basis vector.

Similarly, if x and y are vectors, their scalar product is

$$(x|y) = \Big(\sum_{i=1}^{n} x_i \, e_i \Big| \sum_{j=1}^{n} y_j \, e_j\Big) = \sum_{i,j=1}^{n} \overline{x_i} \, y_j \, \underbrace{(e_i|e_j)}_{\delta_{i,j}} = \sum_{i=1}^{n} \overline{x_i} \, y_i.$$

THEOREM 9.10 (Computations in orthonormal basis) *Let $(e_1, \ldots, e_n)$ be an orthonormal basis of E. Then, for any $x, y \in E$, we have*

$$x = \sum_{i=1}^{n} (e_i|x) \, e_i, \qquad \textit{i.e.,} \quad x_i = (e_i|x),$$

$$(x|y) = \sum_{i=1}^{n} \overline{x_i} \, y_i = \sum_{i=1}^{n} (x|e_i)(e_i|y), \qquad \|x\|^2 = \sum_{i=1}^{n} |x_i|^2.$$

9.2.b Projection on a finite-dimensional subspace

In any pre-Hilbert space, the techniques of the previous section provide a means to project an arbitrary vector on a subspace V of finite dimension p. For this, let $(e_1, \ldots, e_p)$ be an orthonormal basis of V. Define then

$$x_V \stackrel{\text{def}}{=} \sum_{i=1}^{p} (e_i|x) \, e_i.$$

Then x_V is an element of V. Moreover, $(x - x_V|e_k) = 0$ for $k = 1, \ldots, p$, hence $x - x_V$ is orthogonal to V. Finally, for any $y \in V$, since $y - x_V$ is in V, hence orthogonal to $x - x_V$, we can apply Pythagoras' theorem and get

$$\|x - y\|^2 = \|x - x_V\|^2 + \|y - x_V\|^2 \geqslant \|x - x_V\|^2.$$

[3] To show that such a basis exists is easy: just take an arbitrary basis, then apply the Gram-Schmidt algorithm to make it orthonormal.

The distance from x to V, which is defined by $d(x; V) = \inf_{y \in V} \|x - y\|$, is therefore attained at a unique point, which is none other than x_V.

THEOREM 9.11 (Orthogonal projection theorem) *Let E be a pre-Hilbert space of arbitrary dimension. Let V be a subspace of E of* finite dimension. *For $x \in E$, there exists a unique element x_V in V such that $x - x_V \perp V$, which is given by the* ***orthogonal projection*** *of x on V, defined by*

$$x_V \stackrel{\text{def}}{=} \sum_{i=1}^{p} (e_i|x)\, e_i.$$

for any orthonormal basis $(e_1, \ldots, e_p)$ of V.

Moreover, x_V is the unique vector in V such that $d(x; V) = \|x - x_V\|$.

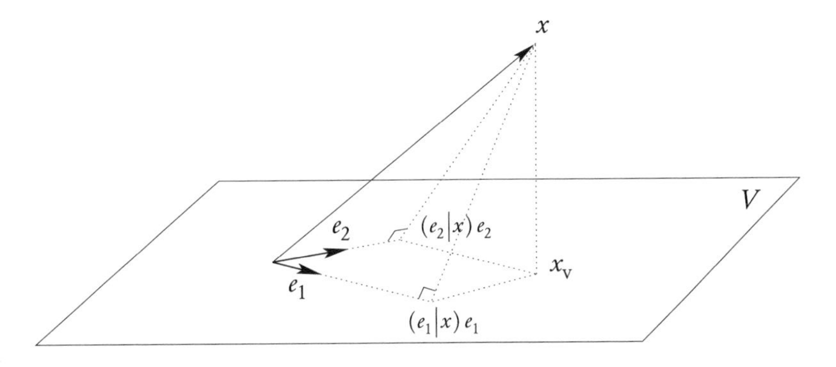

THEOREM 9.12 (Orthogonal complement) *Let V be a finite-dimensional subspace of E.*

i) $V^\perp$ is a complementary subspace of V, called the ***orthogonal complement*** *of V:*

$$E = V \oplus V^\perp.$$

ii) The orthogonal of $V^\perp$ is V: $V = (V^\perp)^\perp$.

Proof

i) It is immediate that $V \cap V^\perp = \{0\}$. Moreover, any vector x is decomposed in the form $x = x_V + (x - x_V)$, where the first vector is an element of V and the second an element of $V^\perp$, which shows that $E = V \oplus V^\perp$.

ii) The inclusion $V \subset (V^\perp)^\perp$ is clear.

Let now $x \in V^{\perp\perp}$. Using the known decomposition $E = V \oplus V^\perp$, we can write $x = y + z$ with $y \in V$ and $z \in V^\perp$.Then z is orthogonal to any vector of V; in particular, $(y|z) = 0$, that is $(x - z|z) = 0$. But x is orthogonal to $V^\perp$ so $(x|z)$. Thus $(z|z) = 0$: it follows that $z = 0$, so $x = y$ and hence $x \in V$ as desired.

Remark 9.13 If V is a subspace of infinite dimension, the previous result may be false. For instance, consider the pre-Hilbert space E of continuous functions on $[0, 1]$, with the natural scalar product $(f|g) = \int_0^1 \overline{f}\, g$, and the subspace V of functions f such that $f(0) = 0$. If g is a function in $V^{\perp}$, then it is orthogonal to $x \mapsto x\, g(x)$ (which belongs to V); therefore $\int_0^1 x\, |g(x)|^2 \, dx = 0$ and $g = 0$. Therefore $V^{\perp} = \{0\}$, although $\{0\}$ is not a complementary subspace of V.

9.2.c Bessel inequality

Let E be a pre-Hilbert space. Consider a countable orthonormal family of E. For a vector $x \in E$, we denote

$$x_p \stackrel{\text{def}}{=} \sum_{n=0}^{p} (e_n|x)\, e_n \qquad \text{for any } p \in \mathbb{N},$$

the orthogonal projection of x on the finite-dimensional subspace $V_p = \operatorname{Vect}\{e_0, \ldots, e_p\}$. Since $x - x_p \perp \operatorname{Vect}\{e_0, \ldots, e_p\}$, Pythagoras' theorem implies that

$$\|x\|^2 = \|x - x_p\|^2 + \|x_p\|^2, \qquad \text{and hence} \quad \|x_p\|^2 = \sum_{n=0}^{p} \big|\, (x|e_n) \,\big|^2 \leqslant \|x\|^2 .$$

Since this inequality holds for all $p \in \mathbb{N}$, we deduce:

THEOREM 9.14 (Bessel inequality) *Let E be a pre-Hilbert space. Let $(e_n)_{n \in \mathbb{N}}$ be a countable orthonormal system and let $x \in E$. Then the series $\sum \big|\, (e_n|x) \,\big|^2$ converges and we have*

$$\sum_{n=0}^{\infty} \big|\, (e_n|x) \,\big|^2 \leqslant \|x\|^2 .$$

9.3 Hilbert spaces

Recall that a normed vector space E is **complete** if any Cauchy sequence in E is convergent in E.

Complete spaces are important in mathematics, because the notion of convergence is much more natural in such a space. These are therefore the spaces that a physicist "requires" intuitively.[4]

[4] Since the formalization of quantum mechanics by János (John) VON NEUMANN (1903–1967) in 1932, many physics courses start with the following axiom: *The space of states of a quantum system is a Hilbert space.*

David HILBERT (1862–1943), born in Königsberg in Germany, obtained his first academic position in this town at the age of 22. He left for Göttingen eleven years later and he ended his career there in 1930. A many-faceted genius, he was interested in the foundations of mathematics (in particular the problem of axiomatization), but also in mechanics, general relativity, and number theory, and invented the notion of completeness, opening the way for the study of what would later be called *Hilbert spaces*. His name remains attached to a list of 23 problems he proposed in 1900 for the attention of mathematicians present and future. Most have been resolved, and some have been shown to be indecidable.[5]

DEFINITION 9.15 Let H be a pre-Hilbert space. If H is complete for its hermitian norm, then it is called a **Hilbert space.**

9.3.a Hilbert basis

The spaces we are going to consider from now on will be of infinite dimension in general. It is harder to work with spaces of infinite dimension since many convenient properties of finite-dimensional vector spaces (for instance, that $E^* \simeq E$) are not true in general. The interest of Hilbert spaces is to keep many of these properties; in particular, the notion of basis can be generalized to infinite-dimensional Hilbert spaces.

DEFINITION 9.16 (Total system) Let H be a Hilbert space. A family $(e_n)_{n\in I}$ of vectors indexed by a set I, countable or not, is a **total system** if

$$\{e_i\,;\ i \in I\}^\perp = \{0\},$$

or, in other words, if

$$\big((x|e_i) = 0,\ \forall i \in I\big) \iff (x = 0).$$

We now wish to know if it is possible to extend completely the notion of orthonormal basis of a finite-dimensional Hilbert space, and especially to recover the formulas of Theorem 9.10 – that is, is it possible to write

$$x = \sum_{n\in\mathbb{N}} (x|e_n)\, e_n \qquad \text{and} \qquad \|x\|^2 = \sum_{n\in\mathbb{N}} \big|\,(x|e_n)\,\big|^2\ ?$$

In what follows, we consider a countable orthonormal system $(e_n)_{n\in\mathbb{N}}$, and define an increasing sequence of subspaces $(V_n)_{n\in\mathbb{N}}$ given by

$$V_p \stackrel{\text{def}}{=} \text{Vect}\{e_0, e_1, \dots, e_p\} \qquad \text{for any } p \in \mathbb{N}.$$

[5] Such is the case of the continuum hypothesis, a result due to P. Cohen in 1963.

Erhard SCHMIDT (1876–1959) studied in his native city of Derpt, in Estonia, then in Berlin (with SCHWARZ) and in Göttingen. There, he defended his thesis (supervised by HILBERT) in 1905. After positions in Zurich, Erlangen, and Breslau, he was hired as professor at the University of Berlin in 1917. He continued the ideas of Hilbert concerning integral equations and defined the notion of a Hilbert space (1905). He also introduced self-adjoint operators in Hilbert space, was interested in functional analysis and integral equations, and, in his studies of systems of infinitely many equations with infinitely many unknowns, introduced "geometric" notation, which led to the geometric interpretation of Hilbert spaces.

Also we introduce the sequence of projections of a vector x on the subspaces V_p.

DEFINITION 9.17 Let $x \in H$. The **partial Fourier series** of x with respect to the orthonormal system $(e_n)_{n\in\mathbb{N}}$ is the sequence of sums $S_p(x)$ given by

$$S_p(x) \stackrel{\text{def}}{=} \sum_{n=0}^{p} (e_n|x)\, e_n, \qquad \text{for any } p \in \mathbb{N}.$$

In other words, $S_p(x)$ is the orthogonal projection of x on V_p. The coefficient $c_n(x) \stackrel{\text{def}}{=} (e_n|x)$ is called the ***n*-th Fourier coefficient of *x***.

Does the partial Fourier series converge to x, which is what we would like? The answer is *no* in general. However, we have the following result:

PROPOSITION 9.18 *For any $x \in H$, the sequence $\big(S_n(x)\big)_{n\in\mathbb{N}}$ is a Cauchy sequence for the hermitian norm, and therefore it converges in H.*

PROOF. From Bessel's inequality it follows that the series $\sum |(e_n|x)|^2 = \sum |c_n|^2$ has bounded partial sums; since it is a series with positive terms, it is therefore convergent, and consequently it is a Cauchy sequence. Hence the quantity

$$\left\|S_p(x) - S_q(x)\right\|^2 = \left\| \sum_{n=p+1}^{q} c_n e_n \right\|^2 = \sum_{n=p+1}^{q} |c_n|^2$$

is the Cauchy remainder of a convergent series of real numbers, and it tends to 0 when $[p, q \to \infty]$; this shows that the sequence $\big(S_n(x)\big)_{n\in\mathbb{N}}$ is a Cauchy sequence in H which is, by assumption, complete. So it converges in H.

This is not sufficient to show that the sequence $\big(S_n(x)\big)_{n\in\mathbb{N}}$ converges *to* x; as far as we know, it may very well converge to an entirely different vector. This is the reason for the introduction of the Hilbert basis.

DEFINITION 9.19 (Hilbert basis) Let H be a Hilbert space. A family $(e_i)_{i\in I}$ is a **Hilbert basis** of H if it is a free family such that $\text{Vect}\{e_i\ ;\ i \in I\}$ is dense[6] in H.

An **orthonormal Hilbert basis** is a Hilbert basis which is also an orthonormal system.

DEFINITION 9.20 (Separable space) A Hilbert space H is **separable** if there exists a countable (or finite) Hilbert basis.[7]

In what follows, all Hilbert spaces considered will be separable.

Take an orthonormal system $(e_n)_{n\in\mathbb{N}}$. What does it mean for this system to "be a Hilbert basis"? As the next theorem shows, it is a very strong property, since it provides a unique decomposition of arbitrary vectors as (infinite) linear combination of the basis vectors (the system is therefore *total*).

THEOREM 9.21 *Let H be a Hilbert space and let $(e_n)_{n\in\mathbb{N}}$ be an orthonormal system. The following statements are equivalent:*

A) $(e_n)_{n\in\mathbb{N}}$ *is a Hilbert basis, that is,* $\text{Vect}\{e_n\ ;\ n \in \mathbb{N}\}$ *is dense in* H;

B) *for any* $x \in H$, *if we put* $x_n \stackrel{\text{def}}{=} \sum_{i=0}^{n} (e_i|x)\, e_i$, *then we have*

$$\lim_{n\to\infty} \|x - x_n\| = 0;$$

C) *for any* $x, y \in H$, *we have* $(x|y) = \sum_{i=0}^{\infty} (x|e_i)(e_i|y)$;

D) *for any* $x \in H$, *we have the **Parseval identity*** $\|x\|^2 = \sum_{i=0}^{\infty} \left|(e_i|x)\right|^2$;

E) *the system* $(e_n)_{n\in\mathbb{N}}$ *is a **total system:** for* $x \in H$, *we have*

$$\big((e_i|x) = 0 \text{ for any } i \in \mathbb{N}\big) \iff (x = 0).$$

PROOF. Denote $T_n = \text{Vect}\{e_0, \dots, e_n\}$ and $T = \bigcup_{n\in\mathbb{N}} T_n$.

A) ⇒ B): by assumption, T is dense in H, hence $\overline{T} = H$. Let $\varepsilon > 0$. There exists $y \in T$ such that $\|x - y\| \leqslant \varepsilon$. Moreover, since $y \in T$, there exists $N \in \mathbb{N}$ such that $y \in T_N$, so that, since the sequence $(T_n)_{n\in\mathbb{N}}$ is increasing for inclusion, $y \in T_n$ for any $n \geqslant N$. In addition the projection of x on T_n achieves the shortest distance (Pythagoras' theorem), hence $\|x - x_n\| \leqslant \|x - y\| \leqslant \varepsilon$ for any $n \geqslant N$.

B) ⇒ C): an easy computation shows that $(x_n|y_n) = \sum_{i=0}^{n} (x|e_i)(e_i|y)$ and letting n go to infinity gives the result.

C) ⇒ D): immediate.

---

[6] In other words, for a Hilbert basis $(e_i)_{i\in I}$, any vector in H can be approached, with arbitrarily high precision, by a finite linear combination of the vectors e_i.

[7] In certain books, countability is part of the definition of a Hilbert basis.

D) $\Rightarrow$ E): immediate.

E) $\Rightarrow$ A): let $x \in H$ and denote $y = \lim_{n\to\infty} x_n$ (from Theorem 9.18, we know that the limit exists). Then $(x - y|e_i) = 0$ for any $i \in \mathbb{N}$, so that $x = y$. In particular, the sequence $(x_n)_{n\in\mathbb{N}}$ of elements of T converges to x, and T is therefore dense in H.

So, an orthonormal Hilbert basis is a total orthonormal system – and is something very close to an algebraic basis, since it is a free family and it is "almost" generating. For an arbitrary vector $x \in H$, the sequence $\big(S_n(x)\big)_{n\in\mathbb{N}}$ of the partial Fourier series of x with respect to the Hilbert basis $(e_n)_{n\in\mathbb{N}}$ converges to x, so that it is possible to write

$$x = \sum_{n\in\mathbb{N}} (e_n|x)\, e_n = \sum_{n\in\mathbb{N}} c_n(x)\, e_n \qquad \text{and} \qquad (x|y) = \sum_{n\in\mathbb{N}} \overline{c_n(x)}\, c_n(y). \tag{9.2}$$

This means that the computational formulas seen in the finite-dimensional case are equally valid here.

THEOREM 9.22 (Computations in a Hilbert basis) *Let $(e_1, \dots, e_n)$ be a Hilbert basis of H. For any $x, y \in H$, we have*

$$x = \sum_{n=0}^{\infty} (e_n|x)\, e_n, \qquad c_n(x) = (e_n|x),$$

$$(x|y) = \sum_{n=0}^{\infty} \overline{c_n(x)}\, c_n(y) = \sum_{n=0}^{\infty} (x|e_n)(e_n|y), \qquad \|x\|^2 = \sum_{n=0}^{\infty} \left|c_n(x)\right|^2.$$

COROLLARY 9.23 *Two elements of a Hilbert space which have the same Fourier coefficients with respect to a fixed orthonormal Hilbert basis are equal.*

PROOF. Let $x, y \in H$ have the same Fourier coefficients with respect to an orthonormal Hilbert basis $(e_n)_{n\in\mathbb{N}}$. We then have $c_n(x - y) = c_n(x) - c_n(y) = 0$ for any $n \in \mathbb{N}$; hence, according to Property E), it follows that $x - y = 0$.

PROPOSITION 9.24 *In a Hilbert space, all Hilbert basis have the same cardinality. In particular, in a separable Hilbert space, any Hilbert basis is countable.*[8]

COROLLARY 9.25 *Every separable Hilbert space is isometric to the space ℓ^2, to the space $L^2(\mathbb{R})$, and to the space $L^2[0, a]$ (see the next pages for definitions).*

Remark 9.26 In quantum mechanics, it is customary to choose a hermitian scalar product which is linear *on the right* and semilinear *on the left*, and we have followed this convention. However, it should be noted that many mathematics references use the opposite convention. It is then necessary to write, for instance, $c_n(x) = (x|e_n)$ instead of $(e_n|x)$.

Following Dirac, the vector e_i is sometimes denoted $|e_i\rangle$ and the linear form $x \mapsto (e_i|x)$ is denoted $\langle e_i|$. The effect of the application of a "bra" $\langle x|$ on a "ket" $|y\rangle$ is conveniently denoted $\langle x|\, y\rangle = (x|y)$. Then, the property

[8] This is important in quantum mechanics. Indeed, since in general the space considered is L^2, which is separable, it follows that the family $\{|x\rangle\}_{x\in\mathbb{R}}$ is not a Hilbert basis.

$$\langle x|y\rangle = \sum_{n\in\mathbb{N}} \overline{c_n(x)}\, c_n(y) = \sum_{n\in\mathbb{N}} (x|e_n)(e_n|y) = \sum_{n\in\mathbb{N}} \langle x|e_n\rangle \langle e_n|y\rangle,$$

which is valid for all $x, y \in H$, can be written symbolically as

$$\mathrm{Id} = \sum_{n\in\mathbb{N}} |e_n\rangle \langle e_n| \qquad \text{and} \qquad \langle x|y\rangle = \langle x|\,\mathrm{Id}\,|y\rangle .$$

Similarly, the formula (9.2) can be interpreted as

$$|x\rangle = \sum_{n\in\mathbb{N}} c_n(x)\, e_n = \sum_{n\in\mathbb{N}} (e_n|x)\, e_n = \underbrace{\sum_{n\in\mathbb{N}} |e_n\rangle \langle e_n|}_{\mathrm{Id}} x\rangle.$$

All of this is detailed in Chapter 14 on page 377.

Two Hilbert spaces will be important in the sequel, the spaces denoted ℓ^2 and L^2; both are extensively used in quantum mechanics as well as in signal theory. The two following sections give the definitions and the essential properties of those two spaces. It will be seen that it is possible to translate from one to the other, and this is precisely the objective of the classical Fourier series expansion.

9.3.b The ℓ^2 space

DEFINITION 9.27 The space of complex-valued sequences $(a_n)_{n\in\mathbb{N}}$ such that the series $\sum |a_n|^2$ converges is called the **ℓ^2 space.**

THEOREM 9.28 *The space ℓ^2, with the hermitian product*

$$(a|b) \stackrel{\text{def}}{=} \sum_{n=0}^{\infty} \overline{a_n}\, b_n,$$

is a Hilbert space.

PROOF. Because of its length and difficulty, the proof is given in Appendix D, page 602.

A Hilbert basis of ℓ^2 is given by the sequence $(e_n)_{n\in\mathbb{N}}$, where

$$e_n \stackrel{\text{def}}{=} (\underbrace{0,\ldots,0}_{n-1 \text{ times}}, 1, 0, 0, \ldots).$$

(In this is recognized the "basis" which we were looking for at the beginning of this chapter!) Since this Hilbert basis is countable, the space ℓ^2 is separable.

Remark 9.29 Because $\mathbb{N}$ and $\mathbb{Z}$ have same cardinality, ℓ^2 can also be defined as the space of families $(a_n)_{n\in\mathbb{Z}}$ which are square integrable.

9.3.c The space $L^2\,[0, a]$

Consider the vector space of functions $f : [0, a] \to \mathbb{C}$, which are square integrable. One wishes to define a scalar product and norm defined by

$$(f, g) \longmapsto \int_0^a \overline{f(x)}\, g(x)\, \mathrm{d}x \qquad \text{and} \qquad \|f\|^2 = \int_0^a |f(x)|^2\, \mathrm{d}x.$$

The problem is that if f and g are equal almost everywhere, even without being equal, then $\|f - g\| = 0$, so this is not really a norm. The way out of this difficulty is to *identify* functions which coincide almost everywhere, namely, to say[9] that if f and g are equal almost everywhere, then they represent the same object.

DEFINITION 9.30 Let $a \in \mathbb{R}^{*+}$. The $\boldsymbol{L^2[0, a]}$ **space** is the space of measurable functions defined on $[0, a]$ with values in $\mathbb{C}$ which are square integrable, defined up to equality almost everywhere, with the hermitian scalar product and the norm given by

$$(f|g)_{L^2} \stackrel{\text{def}}{=} \int_0^a \overline{f(x)}\, g(x)\, \mathrm{d}x, \qquad \text{and} \qquad \|f\|_2 \stackrel{\text{def}}{=} \left(\int_0^a |f(x)|^2\, \mathrm{d}x \right)^{1/2}.$$

When speaking of a "function" in $L^2[0, a]$, the actual meaning implied will be the *equivalence class* of this function *modulo* almost everywhere equality.

THEOREM 9.31 (A Hilbert basis) *The system $(e_n)_{n \in \mathbb{Z}}$ defined by*

$$e_n(x) = \frac{1}{\sqrt{a}}\, \mathrm{e}^{2\pi i n x / a} \qquad \textit{for any } n \in \mathbb{Z} \textit{ and any } x \in [0, a]$$

is an orthonormal Hilbert basis of $L^2\,[0, a]$.

PROOF. The proof, which is somewhat delicate, can be found in most real analysis books, such as [76].

This means that any function in $L^2\,[0, a]$ can be represented as a sum (infinite) of periodic functions, in the sense of *mean-square* convergence of the partial sums toward the original function. This result is the foundation of the theory of Fourier analysis of periodic functions (see Section 9.4).

In other words, a function $f \in L^2\,[0, a]$ may be represented by a sequence $(c_n)_{n \in \mathbb{Z}}$ such that $\sum_{n \in \mathbb{Z}} |c_n|^2 < +\infty$, i.e. an element of ℓ^2.

[9] Mathematicians speak of "taking the quotient by the subspace of functions which are zero almost everywhere."

9.3.d The $L^2(\mathbb{R})$ space

DEFINITION 9.32 The **$L^2(\mathbb{R})$ space** is the space of measurable functions defined on $\mathbb{R}$, with complex values (up to equality almost everywhere) which are square integrable, with the norm

$$\|f\|_2 \stackrel{\text{def}}{=} \sqrt{\int_{-\infty}^{+\infty} |f(x)|^2 \, \mathrm{d}x}.$$

A fundamental result due to Riesz and Fischer, which is not obvious at all, shows that $L^2(\mathbb{R})$ is complete. It is therefore a Hilbert basis, and it can be shown to admit a countable Hilbert basis. Thus the Hilbert spaces ℓ^2, $L^2[0,a]$, and $L^2(\mathbb{R})$ are isometric.

Remark 9.33 Although a Hilbert basis has many advantages, it will also be very interesting to consider on the space $L^2(\mathbb{R})$ a "basis" which is not a Hilbert basis, but rather has elements which are distributions, such as the (noncountable) family $(\delta_x)_{x\in\mathbb{R}}$, which quantum physicists denote $(|x\rangle)_{x\in\mathbb{R}}$, or the family $(\mathrm{e}^{2\pi i p x})_{p\in\mathbb{R}}$, denoted (not without some risk of confusion!) $(|p\rangle)_{p\in\mathbb{R}}$. Instead of representing a function $f \in L^2$ by a discrete sum, it is represented by a "continuous" sum, that is, an integral. Such a point of view lacks rigor if a basis "in the sense of distributions" is not defined (note, for instance, that $x \longmapsto \mathrm{e}^{2\pi i p x}$ is not even in $L^2(\mathbb{R})$!). However, the representation of a function by Fourier transformation is simply (in another language) the decomposition of a function "in the basis $(x \longmapsto \mathrm{e}^{2\pi i p x})_{p\in\mathbb{R}}$."

♦ **Exercise 9.1** Define H' to be the space of functions on $\mathbb{R}$, up to equality almost everywhere, such that $(f|f)' < +\infty$, where

$$(f|g)' \stackrel{\text{def}}{=} \int_{\mathbb{R}} \overline{f(t)}\, g(t)\, \mathrm{e}^{-t^2} \, \mathrm{d}t.$$

Then H' with this norm is a Hilbert space. Show that by orthogonalizing the family $(X^0, \dots, X^n, \dots)$ (using the method due to Erhard SCHMIDT), one obtains a family of polynomials $(H_0, \dots, H_n, \dots)$, called the **Hermite polynomials**.

Show (see, for instance, [20, complement B.v]) that

$$H_n(t) = (-1)^n \, \mathrm{e}^{t^2} \frac{\mathrm{d}^n}{\mathrm{d}t^n} \mathrm{e}^{-t^2}.$$

The family $(\mathscr{H}_0, \dots, \mathscr{H}_n, \dots)$ defined by $\mathscr{H}_n(t) = H_n(t)\,\mathrm{e}^{-t^2/2}$ is then a Hilbert basis of $L^2(\mathbb{R})$ (see, for instance, [47] for a proof). It is used in quantum mechanics, in particular as a Hilbert eigenbasis of the hamiltonian of the harmonic oscillator; it has the interesting property of being a family of eigenvectors for the Fourier transform.

Remark 9.34 Another Hilbert basis of $L^2(\mathbb{R})$ is given by the wavelets of $\mathscr{C}^\infty$ class of Yves MEYER [37, 54, 65], which generalize the Fourier series expansion on $L^2[0,a]$ (see Section 11.5, page 321).

9.4
Fourier series expansion

It is also possible (and sometimes more convenient) to define the space $L^2[0,a]$ as the space of functions $f : \mathbb{R} \to \mathbb{C}$ which are periodic with period a and square integrable on $[0,a]$ (since any function defined on $[0,a]$ may be extended to $\mathbb{R}$ by periodicity[10]).

Recall that
$$e_n(x) = \frac{1}{\sqrt{a}}\,\mathrm{e}^{2\pi i n x/a}$$

9.4.a Fourier coefficients of a function

DEFINITION 9.35 Let $f \in L^2[0,a]$. For $n \in \mathbb{Z}$, the ***n*-th Fourier coefficient** of f is the complex number

$$c_n \stackrel{\text{def}}{=} \frac{1}{\sqrt{a}} \int_0^a f(t)\,\mathrm{e}^{-2\pi i n t/a}\,\mathrm{d}t = (e_n|f).$$

The **sequence of Fourier partial sums** is the sequence $(f_n)_{n\in\mathbb{N}}$ defined by

$$f_n(x) \stackrel{\text{def}}{=} \frac{1}{\sqrt{a}} \sum_{k=-n}^{n} c_k\,\mathrm{e}^{2\pi i k x/a} = \sum_{k=-n}^{n} c_k\,e_k(x).$$

The Fourier coefficients of the conjugate, the transpose, and the translates of f are given by the following proposition:

PROPOSITION 9.36 *Let $f \in L^2[0,a]$, and denote by $\bar{f}$ its conjugate, by $\check{f}$ its transpose and by f_τ, for $\tau \in \mathbb{R}$, its translate $f_\tau(x) = f(x-\tau)$. We have*

i) $\forall n \in \mathbb{Z}, \qquad c_n(\bar{f}) = \overline{c_{-n}(f)}\;;$

ii) $\forall n \in \mathbb{Z}, \qquad c_n(\check{f}) = \overline{c_n(\bar{f})} = c_{-n}(f)\;;$

iii) $\forall n \in \mathbb{Z}, \qquad c_n(f_\tau) = \mathrm{e}^{-2\pi i n a/\tau} c_n(f).$

THEOREM 9.37 *If $f \in L^2[0,a]$ is continuous and is piecewise of $\mathscr{C}^1$ class, then $f' \in L^2[0,a]$ and has Fourier coefficients given by*

$$c_n(f') = \frac{2\pi i n}{a}\, c_n(f) \qquad \forall n \in \mathbb{Z}.$$

PROOF. This is a simple integration by parts.

[10] The value at points 0 and a may not coincide, but "forgetting" this value is of no importance for functions f which are defined up to equality almost everywhere.

9.4.b Mean-square convergence

Let $f \in L^2[0,a]$, so that it can be seen as a function $f : \mathbb{R} \to \mathbb{C}$ which is a-periodic and square integrable on $[0,a]$. Since $L^2[0,a]$ is a Hilbert space and the family $(e_n)_{n\in\mathbb{N}}$ is an orthonormal Hilbert basis of this space (by Theorem 9.31), the partial Fourier sums of f converge to f in mean-square (Theorem 9.21):

THEOREM 9.38 (Mean-square convergence) *Let $f \in L^2[0,a]$. Denote by c_n its Fourier coefficients and by $(f_n)_{n\in\mathbb{N}}$ the sequence of Fourier partial sums of f. Then the sequence $(f_n)_{n\in\mathbb{N}}$ converges in mean-square to f:*

$$\|f - f_n\|_2^2 = \int_0^a |f(t) - f_n(t)|^2 \, \mathrm{d}t \xrightarrow[n\to\infty]{} 0.$$

DEFINITION 9.39 It is customary to write (*abusing notation*, since the equality holds only in the sense of the L^2-norm, and not pointwise) simply that

$$f(t) = \frac{1}{\sqrt{a}} \sum_{n=-\infty}^{+\infty} c_n \, \mathrm{e}^{2\pi i n t/a} = \sum_{n=-\infty}^{+\infty} c_n \, e_n(t) \tag{9.3}$$

and to call this expression the **Fourier series expansion** of f.

Remark 9.40 It is important to fully understand why the identity

$$f = \sum_{n=-\infty}^{+\infty} c_n \, e_n \qquad \textbf{true}$$

does not allow us to write

$$f(t) = \frac{1}{\sqrt{a}} \sum_{n=-\infty}^{+\infty} c_n \, \mathrm{e}^{2\pi i n t/a}. \qquad \textbf{false!}$$

The first equality is proved in the general case of Hilbert spaces. But the Hilbert space which we use is the space $L^2[0,a]$, where functions are only defined up to almost-everywhere equality, and where convergence means convergence in mean-square. The space of a-periodic functions, with the norm defined as the supremum of the absolute values of a function, is not complete. One cannot hope for a pointwise convergence result purely in the setting of Hilbert spaces. (This type of convergence will be considered with other tools in Section 9.4.d on page 267.)

The Parseval identity is then written as follows:

THEOREM 9.41 (Parseval identity) *Let $f \in L^2[0,a]$ and let c_n denote its Fourier coefficients. Then the family $(|c_n|^2)_{n\in\mathbb{Z}}$ is summable and we have*

$$\sum_{n\in\mathbb{Z}} |c_n|^2 = \int_0^a |f(t)|^2 \, \mathrm{d}t = \|f\|_2^2.$$

Moreover, if $f, g \in L^2[0,a]$, then $(\overline{f}g) \in L^1[0,a]$ and we have

$$\sum_{n\in\mathbb{Z}} \overline{c_n(f)} \, c_n(g) = \int_0^a \overline{f(t)} \, g(t) \, \mathrm{d}t = (f|g).$$

In other words, the representation as Fourier series is an isometry between the space $L^2[0,a]$ and the space ℓ^2. Sometimes the following "real" notation is used (with the same abuse of notation previously mentioned):

$$\boxed{f(t) = \frac{a_0}{2\sqrt{a}} + \frac{1}{\sqrt{a}} \sum_{\substack{n=-\infty \\ n\neq 0}}^{+\infty} \left[a_n \cos\left(2\pi n \frac{t}{a}\right) + \frac{1}{\sqrt{a}} b_n \sin\left(2\pi n \frac{t}{a}\right) \right],} \tag{9.4}$$

with
$$a_n \stackrel{\text{def}}{=} \frac{2}{\sqrt{a}} \int_0^a f(t) \cos\left(2\pi n \frac{t}{a}\right) \mathrm{d}t = c_n + c_{-n},$$
$$b_n \stackrel{\text{def}}{=} \frac{2}{\sqrt{a}} \int_0^a f(t) \sin\left(2\pi n \frac{t}{a}\right) \mathrm{d}t = \mathrm{i}(c_n - c_{-n}),$$

and
$$\frac{|a_0|^2}{4a} + \frac{1}{2a} \sum_{\substack{n=-\infty \\ n\neq 0}}^{+\infty} \left(|a_n|^2 + |b_n|^2\right) = \int_0^a |f(t)|^2 \,\mathrm{d}t.$$

Remark 9.42 The Fourier series expansion gives a concrete isometry between the space $L^2[0,a]$ and the space ℓ^2. To each square-integrable function, we can associate a sequence $(c_n)_{n\in\mathbb{Z}} \in \ell^2$, namely, the sequence of its Fourier coefficients. Conversely, given an arbitrary sequence $(c_n)_{n\in\mathbb{Z}} \in \ell^2$, the trigonometric series $\sum c_n e_n$ converges (in the sense of mean-square convergence) to a function which is square integrable (this is the **Riesz-Fischer Theorem**). Frédéric Riesz said that the Fourier series expansion was "a perpetual round-trip ticket between two spaces of infinite dimension" [52].

9.4.c Fourier series of a function $f \in L^1[0,a]$

This business of convergence in L^2 norm is all well and good, but it does not even imply pointwise convergence (and, a fortiori, does not imply uniform convergence[11]). Isn't it possible to do better? Can we write (9.3) or (9.4) rigorously?

We start the discussion of this question by extending the notion of Fourier series to functions which are not necessarily in $L^2[0,a]$, but are in the larger space $L^1[0,a]$ of Lebesgue-integrable functions on $[0,a]$.[12] Indeed, for $f \in$

---

[11] To take an example independent of Fourier series, the sequence $(x \mapsto \sin^n x)_{n\in\mathbb{N}}$ converges to 0 in mean-square, but it does not converge pointwise or uniformly.

[12] Notice that, on the contrary, if $f \in L^2[0,a]$, then $f \in L^1[0,a]$ since, by the Cauchy-Schwarz inequality we have

$$\left(\int_0^a |f(t)|\,\mathrm{d}t\right)^2 = \langle |f|, \mathbf{1} \rangle^2 \leqslant \int_0^a |f(t)|^2\,\mathrm{d}t \cdot \int_0^a 1\,\mathrm{d}t = \|f\|_2^2 \times a.$$

$L^1[0,a]$, it is perfectly possible to define the coefficient

$$c_n \stackrel{\text{def}}{=} \frac{1}{\sqrt{a}} \int_0^a f(t)\, \mathrm{e}^{-2\pi \mathrm{i} n t/a}\, \mathrm{d}t,$$

since a function is Lebesgue-integrable if and only if its modulus is integrable. Hence:

PROPOSITION 9.43 *For any function $f \in L^1[0,a]$, the Fourier coefficients of f are defined.*

It is then possible to construct the sequence $(f_n)_{n\in\mathbb{N}}$ of partial Fourier sums of f. What is it possible to say about this sequence? Since we have left the setting of Hilbert spaces, the general results such as Theorem 9.21 can not be applied. We will subdivide this problem into several questions.

9.4.d Pointwise convergence of the Fourier series

☞ **First question:** is it true that the coefficients c_n tend to 0?

This is not a mathematician's gratuitous question, but rather a crucial point for the practical use of Fourier series: in an approximate computation, it is important to be able to truncate the series and compute only with finitely many terms. A necessary (but not sufficient) condition is therefore that $c_n \to 0$ as $[n \to \pm\infty]$.

The (positive) answer is given by the celebrated Riemann-Lebesgue lemma (a more general version of which will be given in the next chapter, see Theorem 10.12):

THEOREM 9.44 (Riemann-Lebesgue lemma) *Let $f \in L^1[0,a]$ be an integrable function. Then we have*

$$\int_0^a f(t)\, \mathrm{e}^{2\pi \mathrm{i} n t/a}\, \mathrm{d}t \xrightarrow[n\to\pm\infty]{} 0.$$

PROOF. The proof is in two steps.

• Consider first the case where f is of $\mathscr{C}^1$ class; a simple integration by parts implies the result as follows:

$$\int_0^a f(t)\, \mathrm{e}^{2\pi \mathrm{i} n t/a}\, \mathrm{d}t = \frac{\mathrm{i}}{2\pi n} \int_0^a f'(t)\, \mathrm{e}^{2\pi \mathrm{i} n t/a}\, \mathrm{d}t + \frac{\mathrm{i}a}{2\pi n}\big[f(a) - f(0)\big],$$

which, since f and f' are both bounded by assumptions, tends to 0 as n tends to infinity (the integral is bounded).

• Then this is extended to all integrable functions by showing that the space of functions of $\mathscr{C}^1$ class is *dense* in $L^1[0,a]$, for instance, by applying the theorem of regularization by convolution (take convolutions of $f \in L^1[0,a]$ by a gaussian with variance $1/p$ and norm 1, where p is sufficiently large so that $\|f - f * g_p\|_1 \leqslant \varepsilon$). This is technically delicate, but without conceptual difficulty.

Joseph FOURIER (1768–1830) studied at the Royal Military School of Auxerre and, from the age of thirteen, showed an evident interest for mathematics, although he was tempted to become a priest (he entered a seminary at Saint-Benoît-sur-Loire, which he left in 1789). In 1793, he joined a revolutionary committee but, under the Terror, barely escaped the guillotine (the death of Robespierre spared him). At the École Normale Supérieure, Lagrange and Laplace were his teachers. He obtained a position at the École Centrale de Travaux Publics, then was a professor at the École Polytechnique. He participated in Napoléon's Egyptian expedition, was prefect of the Isère, and then studied the theory of heat propagation, which led him to the expansion of periodic functions.

☞ **Second question:** does the sequence of partial sums $(f_n)_{n\in\mathbb{N}}$ converge *pointwise* to f?

In the general case, the answer is unfortunately negative. To obtain a convergence result, additional assumptions of regularity of required on f.

In practice, the result which follows will be the most important.

DEFINITION 9.45 (Regularized function) Let f be a function which admits a limit on the left and on the right at any point. The **regularized** function associated to f is the function

$$f^* \,:\, x \longmapsto \frac{1}{2}\Big[f(x^+) + f(x^-)\Big].$$

THEOREM 9.46 (Dirichlet) *Let $f \in L^1[0,a]$ and let $t_0 \in [0,a]$. If both the limits on the left and on the right $f(t_0^-)$ and $f(t_0^+)$ and the limits on the left and on the right $f'(t_0^-)$ and $f'(t_0^+)$ exist, then the sequence of partial sums $\big(f_n(t_0)\big)_{n\in\mathbb{N}}$ converges to the regularized value of f at t_0, that is, we have*

$$\lim_{n\to\infty} f_n(t_0) = \frac{1}{2}\Big[f(t_0^+) + f(t_0^-)\Big] = f^*(t_0).$$

Notice that not only the function itself, but also its derivative, must have a certain regularity.

It is possible to weaken the assumptions of this theorem, and to ask only that f be of bounded variation [37]. In practice, Dirichlet's theorem is sufficient. Sidebar 4 (page 276) gives an illustration of this theorem.

9.4.e Uniform convergence of the Fourier series

☞ **Third question:** is there *uniform* convergence of the sequence of partial sums $(f_n)_{n\in\mathbb{N}}$ to f?

A positive answer holds in the case of continuous functions which are piecewise of $\mathscr{C}^1$ class

THEOREM 9.47 *Let* $f : \mathbb{R} \to \mathbb{C}$ *be a* continuous *periodic function, such that the derivative of* f *exists except at finitely many points, and moreover* f' *is* piecewise continuous. *Then we have:*

i) the sequence of partial sums $(f_n)_{n\in\mathbb{N}}$ *converges uniformly to* f *on* $\mathbb{R}$;

ii) the Fourier series of f' *is obtained by differentiating termwise the Fourier series of* f, *that is, we have* $c_n(f') = 2i\pi n\, c_n(f)$ *for any* $n \in \mathbb{Z}$;

iii) $\sum_{n=-\infty}^{+\infty} \left|c_n(f)\right| < +\infty$.

If f is not continuous, uniform convergence may not be true (even if the Fourier series converges pointwise); this is the Gibbs phenomenon, described in more detail below.

☞ **Fourth question:** what link is there between the regularity of f and the speed of decay of the Fourier coefficients to 0?

(a) We know already that for $f \in L^1[0,a]$, we have $c_n \to 0$ (Riemann-Lebesgue lemma).

(b) Moreover, if $f \in L^2[0,a]$, then $\sum_{n=-\infty}^{+\infty} |c_n|^2 < +\infty$ (by Hilbert space theory).

(c) Theorem 9.47 shows that if $f \in \mathscr{C}^1[0,a]$, then $\sum_{n=-\infty}^{+\infty} |c_n| < +\infty$, which is an even stronger result.

(d) It is also possible to show that if f is of $\mathscr{C}^{k-1}$ class on $\mathbb{R}$ and piecewise of $\mathscr{C}^k$ class, then $c_n(f^{(k)}) = (2\pi i n/a)^k c_n(f)$ and therefore $c_n = o(1/n^k)$. Moreover, the following equivalence holds:

$$f \in \mathscr{C}^\infty[0,a] \Longleftrightarrow c_n = o(1/n^k) \quad \text{for any } k \in \mathbb{N}.$$

Remark 9.48 It is not sufficient for this last result that f be of $\mathscr{C}^k$ class on $[0,a]$: it is indeed required that the graph of f in the neighborhood of 0 "glues" nicely with that in the neighborhood of a.

Remark 9.49 It has been proved (the Carleson-Hunt theorem) that the Fourier series of a function f which is in L^p for some $p > 1$ converges almost everywhere to f. It is also known (due to Kolmogorov) that the Fourier series of an integrable function may diverge *everywhere*.

The settting of L^2 functions is the only "natural" one for the study of Fourier series. To look for simple (or uniform) convergence is not entirely natural, then. This is why other expansions, such as those provided by wavelets, are sometimes required.

Remark 9.50 Although many physics courses introduce Fourier analysis as a wonderful instrument for the study of sound (music particularly), it should be noted that the expansion in harmonics of a sound with given pitch (from which Ohm postulated that the "timber" of the sound could be obtained) is not sufficient, at least if only the amplitudes $|c_n|$ of the harmonics are considered. The phases of the various harmonics play an important part in the timber of an instrument. A sound recording played in reverse does not give a recognizable sound, although the power spectrum is identical [25].

9.4.f The Gibbs phenomenon

On the figures of Sidebar 4 on page 276, we can see that the partial Fourier sums, although converging pointwise to the square wave function, do not converge uniformly. Indeed, there always exists a "crest" close to the discontinuity. It is possible to show that

- this crest get always narrower;
- it gets closer to the discontinuity;
- its height is constant (roughly 8% of the jump of the function).

An explanation of this phenomenon will be given in the chapter concerning the Fourier transform, page 311.

EXERCISES

♦ **Exercise 9.2** Let $f, g : \mathbb{R} \to \mathbb{C}$ be two continuous functions which are 2π-periodic. Denote by h the **convolution product of f and g**, defined by

$$h(x) = [f * g](x) \stackrel{\text{def}}{=} \frac{1}{2\pi} \int_0^{2\pi} f(x-t)\, g(t)\, dt.$$

i) Show that h is a 2π-periodic continuous function. Compute the Fourier coefficients of h in terms of those of f and g.

ii) Fix the second argument g. Determine the eigenvalues and eigenvectors of the linear application $f \mapsto f * g$.

♦ **Exercise 9.3** Show that

$$|\sin x| = \frac{8}{\pi} \sum_{n=1}^{\infty} \frac{\sin^2 nx}{4n^2 - 1}$$

for all $x \in \mathbb{R}$.

♦ **Exercise 9.4 (Fejér sums)** The goal of this exercise is to indicate a way to improve the representation of a function by its Fourier series when the conditions of Dirichlet's theorem are not valid.

Let $f : \mathbb{R} \to \mathbb{C}$ be 2π-periodic and integrable on $[0, 2\pi]$. Define

$$S_n f(x) = \sum_{k=-n}^{n} c_k\, e^{ikx}, \qquad \text{where} \quad c_k = \frac{1}{2\pi} \int_{-\pi}^{\pi} f(x)\, e^{-ikx}\, dx.$$

The **Fejér sum** of order n of f is the function

$$\sigma_n f = \frac{1}{n}(S_0 f + \cdots + S_{n-1} f),$$

that is, the Cesàro mean of the n first Fourier partial sums.

i) Show that

$$S_n f(x) = \int_{-\pi}^{\pi} f(x-u)\, D_n(u)\, \mathrm{d}u \qquad \text{and} \qquad \sigma_n f(x) = \int_{-\pi}^{\pi} f(x-u)\, \Phi_n(u)\, \mathrm{d}u,$$

where

$$D_n(u) = \frac{1}{2\pi} \frac{\sin\left(n+\frac{1}{2}\right)u}{\sin u/2} \qquad \text{and} \qquad \Phi_n(u) = \frac{1}{2n\pi}\left(\frac{\sin nu/2}{\sin u/2}\right)^2,$$

which are respectively called the **Dirichlet kernel** and the **Fejér kernel**.

ii) Show that the sequence $(\Phi_n)_{n\in\mathbb{N}}$, restricted to $[-\pi, \pi]$, is a Dirac sequence.

iii) Deduce from this **Fejér's theorem:**[13] *if f is continuous, the sequence $(\sigma_n f)_{n\in\mathbb{N}}$ converges uniformly to f; if f merely admits right and left limits at any point, then the sequence $(\sigma_n f)_{n\in\mathbb{N}}$ converges to the regularized function associated to f.*

Remark: the positivity of the Fejér kernel explains the weaker assumptions of Fejér's Theorem, compared with Theorem 9.47 (where assumptions concerning the derivative are required).

♦ **Exercise 9.5 (Poisson summation formula)** Let f be a function of $\mathscr{C}^2$ class on $\mathbb{R}$, such that $\lim_{\pm\infty} x^2 f(x) = \lim_{\pm\infty} x^2 f'(x) = 0$.

Show that the **Poisson summation formula**

$$\sum_{n=-\infty}^{+\infty} f(n) = \sum_{n=-\infty}^{+\infty} \int_{-\infty}^{+\infty} \mathrm{e}^{2\pi i n x} f(x)\, \mathrm{d}x$$

holds.

HINT: Define $F(x) = \sum_{n\in\mathbb{Z}} f(x+n)$ and show that F is defined, 1-periodic, and of $\mathscr{C}^1$ class. Compute the Fourier coefficients of F and compare them with the values stated.

This identity will be proved in a more general situation, using the Fourier transforms of distributions (Theorem 11.22 on page 309).

PROBLEM

♦ **Problem 4 (Isoperimetric inequality)**

① Let $u : \mathbb{R} \to \mathbb{C}$ be a function of $\mathscr{C}^1$ class which is periodic of period L. Show that

$$\int_0^L |u(x)|^2\, \mathrm{d}x \leqslant \frac{L^2}{4\pi^2} \int_0^L |u'(x)|^2\, \mathrm{d}x + \frac{1}{L}\left|\int_0^L u(x)\, \mathrm{d}x\right|^2.$$

Determine the cases where equality holds.

---

[13] Lipót FEJÉR (1880–1959), Hungarian mathematician, was professor at the University of Budapest. He worked in particular on Fourier series, a subject with a bad reputation at the time because it did not conform to the standards of rigor that Cauchy and Weierstrass had imposed. For an interesting historical survey of Fejér's theorem and other works of this period, see [54], which also explores the theory of wavelets, one of the modern developments of harmonic analysis.

② Let γ be a simple closed path of $\mathscr{C}^1$ class in the complex plane, L its length, and $\mathscr{A}$ the area of the region it encloses. Show, using the previous question, that

$$L^2 \geqslant 4\pi \mathscr{A},$$

with equality if and only if γ is a circle.

HINT: Use the complex Green formula (which states that

$$\int_{\partial \mathscr{S}} \mathscr{F}(z, \bar{z})\, \mathrm{d}z = 2\mathrm{i} \iint_{\mathscr{S}} \frac{\partial \mathscr{F}}{\partial \bar{z}}\, \mathrm{d}\mathscr{S}$$

for any function $\mathscr{F}$ continuous with respect to z and $\bar{z}$ and admitting continous partial derivatives with respect to x and y) to show that if γ is parameterized by the map $f : [0, L] \to \mathbb{C}$, with $f = u + \mathrm{i}v$ and $|f'|^2 = u'^2 + v'^2 = 1$, then the area is equal to

$$\mathscr{A} = \int_0^L u(s)\, v'(s)\, \mathrm{d}s.$$

SOLUTIONS

◆ **Solution of exercise 9.2.** Using Fubini's theorem, we compute

$$\begin{aligned} c_n(h) &= \frac{1}{4\pi^2} \int_0^{2\pi} \left(\int_0^{2\pi} f(x-t)\, g(t)\, \mathrm{d}t \right) \mathrm{e}^{-\mathrm{i}nx}\, \mathrm{d}x \\ &= \frac{1}{4\pi^2} \int_0^{2\pi} \left(\int_0^{2\pi} f(x-t)\, \mathrm{e}^{-\mathrm{i}n(x-t)}\, \mathrm{d}x \right) g(t)\, \mathrm{e}^{-\mathrm{i}nt}\, \mathrm{d}t \\ &= c_n(f)\, c_n(g). \end{aligned}$$

For an eigenvector f, the coefficients $c_n(f)$ must therefore satisfy

$$c_n(f)\, c_n(g) = \lambda c_n(f)$$

for any $n \in \mathbb{N}$.

The set of the eigenvalues of $f \mapsto f * g$ is exactly the set of the $c_n(g)$'s. Let λ be an eigenvalue and let $I \subset \mathbb{N}$ be the set of indices n such that $c_n(g) = \lambda$; then the eigenfunctions are the nonzero functions of the kind $\sum_{n \in I} \beta_n\, \mathrm{e}^{\mathrm{i}nx}$.

◆ **Solution of exercise 9.3.** The function $x \mapsto |\sin x|$ is even and 2π-périodic; computing its Fourier coefficients (only those involving cosines being nonzero) yields

$$|\sin x| = \frac{2}{\pi} + \sum_{n=1}^{\infty} \frac{2}{\pi} \left(\frac{1}{2n+1} - \frac{1}{2n-1} \right) \cos 2nx$$

after a few lines of calculations. Expanding $\cos 2\theta = 1 - 2\sin^2\theta$ and noticing that

$$\sum_{n=1}^{\infty} \left(\frac{1}{2n+1} - \frac{1}{2n-1} \right) = -1,$$

it follows that

$$|\sin x| = \frac{8}{\pi} \sum_{n=1}^{\infty} \frac{\sin^2 nx}{(2n+1)(2n-1)} = \frac{2}{\pi} \sum_{n=1}^{\infty} \frac{\sin^2 nx}{4n^2 - 1}.$$

Note that Dirichlet's theorem ensures the pointwise convergence of the series (to the value $|\sin x|$) at any point $x \in \mathbb{R}$. This convergence is moreover uniform.

♦ **Solution of exercise 9.4**

i) It suffices to write

$$D_n(u) = \frac{1}{2\pi} \sum_{k=-n}^{n} \mathrm{e}^{-iku}$$

and to compute this geometric sum. Similarly, we have

$$\Phi_n(u) = \frac{1}{n} \sum_{k=0}^{n-1} D_k(u) = \frac{1}{n} \sum_{k=0}^{n-1} \frac{1}{2\pi} \frac{\operatorname{Im} \exp \mathrm{i}\left(n + \frac{1}{2}\right)u}{\sin u/2},$$

and evaluating the geometric sum gives the stated formula.

ii) A direct calculation shows that $\int_{-\pi}^{\pi} D_n = 1$ and so $\int_{-\pi}^{\pi} \Phi_n = 1$.

Let $\delta > 0$ be given. Then for any $u \in [\delta, \pi]$, we have $\sin u/2 \geqslant \sin \delta/2$, hence

$$0 \leqslant \Phi_n(u) \leqslant \frac{1}{2n\pi} \frac{1}{\sin^2 \delta/2} \qquad \forall u \in [-\pi, -\delta] \cup [\delta, \pi],$$

and this upper bound tends to 0 as $[n \to \infty]$; this shows that $[-\pi, -\delta] \cup [\delta, \pi]$ converges uniformly to 0 and proves therefore that $(\Phi_n)_{n\in\mathbb{N}}$ (restricted to $[-\pi, \pi]$) is a Dirac sequence.

iii) The sequence $(\Phi_n)_{n\in\mathbb{N}^*}$ converges weakly to δ in $\mathscr{D}'$, and so the sequence $(\Phi_n * f)_n$ converges uniformly to f by Theorem 8.22 whenever f is continuous; if f has limits on the right and left, it converges pointwise to the associated regularized function.

The use of Fejér sums resolves the Gibbs phenomenon close to the discontinuities. The convergence is not uniform, yet it is much more regular, as shows by the figure below, which represents the Fourier partial sums and the Fejér partial sums of order 25 for the square wave:

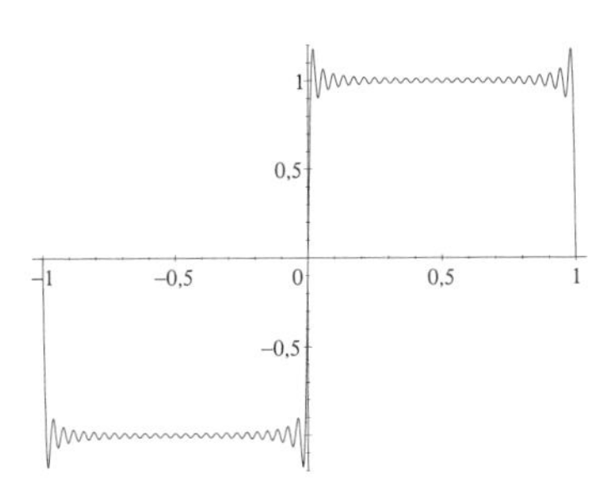

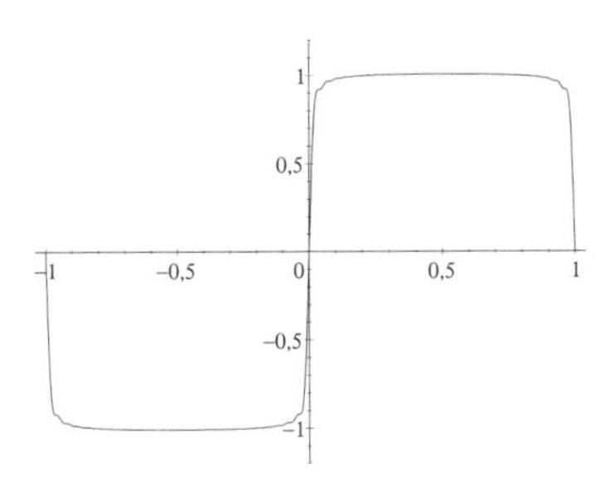

One can show [24] that for $f \in L^1[0,2\pi]$, the sequence $(\sigma_n f)_{n\in\mathbb{N}}$ converges to f in L^1, that is, we have

$$\lim_{n\to\infty} \int_0^{2\pi} \left|\sigma_n f(t) - f(t)\right| \mathrm{d}t = 0.$$

♦ **Solution of exercise 9.5.** We compute

$$\begin{aligned} u_p &= \int_{-\infty}^{+\infty} f(x)\,\mathrm{e}^{2\pi\mathrm{i}px}\,\mathrm{d}x = \sum_{n\in\mathbb{Z}} \int_n^{n+1} f(x)\,\mathrm{e}^{-2\pi\mathrm{i}px}\,\mathrm{d}x \\ &= \sum_{n\in\mathbb{Z}} \int_0^1 f(x+n)\,\mathrm{e}^{-2\pi\mathrm{i}p(x+n)}\,\mathrm{d}x = \sum_{n\in\mathbb{Z}} \int_0^1 f(x+n)\,\mathrm{e}^{-2\pi\mathrm{i}px}\,\mathrm{d}x \\ &= \int_0^1 \sum_{n\in\mathbb{Z}} f(x+n)\,\mathrm{e}^{-2\pi\mathrm{i}px}\,\mathrm{d}x = \int_0^1 F(x)\,\mathrm{e}^{-2\pi\mathrm{i}px}\,\mathrm{d}x = c_p(F), \end{aligned}$$

where $F(x) = \sum_{n\in\mathbb{Z}} f(x+p)$, which is welldefined because of the assumption $f(x) = \underset{x\to\pm\infty}{\mathrm{o}}(1/x^2)$, and where the interversion of the sum and integral are justified by the normal convergence of the series defining F on any compact set.

Moreover, one notes that F is of $\mathscr{C}^1$ class (because the series converges normally on any compact set) and it can be differentiated termwise because of the condition $f'(x) = \mathrm{o}(1/x^2)$, which implies the normal convergence of the series of derivatives.

Since F is of $\mathscr{C}^1$ class and obviously is 1-periodic, the Fourier series converges normally. In particular,

$$F(0) = \sum_{n\in\mathbb{Z}} f(n) = \sum_{p\in\mathbb{Z}} c_p(F) = \sum_{p\in\mathbb{Z}} u_p = \sum_{p\in\mathbb{Z}} \int_{-\infty}^{+\infty} f(x)\,\mathrm{e}^{2\pi\mathrm{i}px}\,\mathrm{d}x.$$

♦ **Solution of problem 4.** Denote by c_n the Fourier coefficients of the function u.

① By Parseval's identity, we have

$$\frac{1}{L}\int_0^L \left|u(x)\right|^2 \mathrm{d}x = \sum_{n=-\infty}^{+\infty} \left|c_n\right|^2 = \sum_{n=-\infty}^{+\infty} \left|\frac{1}{L}\int_0^L u(x)\mathrm{e}^{-2\pi\mathrm{i}nx/L}\,\mathrm{d}x\right|^2$$

and hence

$$\begin{aligned} \int_0^L \left|u(x)\right|^2 \mathrm{d}x &= \frac{1}{L}\left|\int_0^L u(x)\,\mathrm{d}x\right|^2 + \sum_{\substack{n=-\infty \\ n\neq 0}}^{+\infty} \frac{1}{L}\left|\int_0^L u(x)\,\mathrm{e}^{-2\pi\mathrm{i}nx/L}\,\mathrm{d}x\right|^2 \\ &= \frac{1}{L}\left|\int_0^L u(x)\,\mathrm{d}x\right|^2 + \frac{L^2}{4\pi^2}\sum_{\substack{n=-\infty \\ n\neq 0}}^{+\infty} \frac{1}{n^2 L}\left|\int_0^L u'(x)\,\mathrm{e}^{-2\pi\mathrm{i}nx/L}\,\mathrm{d}x\right|^2 \end{aligned}$$

by integration by parts. But, in the last sum, we have $n^2 \geqslant 1$, hence the inequality

$$\begin{aligned} \sum_{\substack{n=-\infty \\ n\neq 0}}^{+\infty} \frac{1}{n^2 L}\left|\int_0^L u'(x)\,\mathrm{e}^{-2\pi\mathrm{i}nx/L}\,\mathrm{d}x\right|^2 &\leqslant \sum_{n=-\infty}^{+\infty} \frac{1}{L}\left|\int_0^L u'(x)\,\mathrm{e}^{-2\pi\mathrm{i}nx/L}\,\mathrm{d}x\right|^2 \\ &\leqslant \int_0^L \left|u'(x)\right|^2 \mathrm{d}x. \end{aligned}$$

This proves the formula stated.

Since all the terms of the last sum are non-negative, equality can hold only if all the terms other than $n=1$ or $n=-1$ are zero, which means only if $u'(x)$ is a linear combination of $\mathrm{e}^{2\pi\mathrm{i}x/L}$ and $\mathrm{e}^{-2\pi\mathrm{i}x/L}$.

② The path being parameterized by the functions $u(s)$ and $v(s)$, the integration element dz can be written

$$\mathrm{d}z = \bigl(u'(s) + \mathrm{i}\,v'(s)\bigr)\,\mathrm{d}s.$$

Take the function $\mathscr{F}(z,\bar{z}) = \bar{z}$ to apply the Green formula. The area of the surface $\mathscr{S}$ enclosed by the path $\gamma = \partial\mathscr{S}$ is

$$\begin{aligned} 2\mathrm{i}\,\mathscr{A} = 2\mathrm{i}\iint_{\mathscr{S}} \frac{\partial \mathscr{F}}{\partial \bar{z}}\,\mathrm{d}\mathscr{S} &= \int_{\gamma} \bar{z}\,\mathrm{d}z = \int_0^L (u - \mathrm{i}\,v)(u' + \mathrm{i}\,v')\,\mathrm{d}s \\ &= \int_0^L (uu' + vv')\,\mathrm{d}s - \mathrm{i}\int_0^L vu'\,\mathrm{d}s + \mathrm{i}\int_0^L uv'\,\mathrm{d}s. \end{aligned}$$

The integrand in the first term is the derivative of $\frac{1}{2}(u^2 + v^2)$; since $u(0) = u(L)$ and $v(0) = v(L)$, this first integral is therefore zero.

The second integral can be integrated by parts, which yields

$$\int_0^L vu'\,\mathrm{d}s = -\int_0^L uv'\,\mathrm{d}s$$

(the boundary term vanishes) and proves that

$$\mathscr{A} = \int_0^L u(s)\,v'(s)\,\mathrm{d}s.$$

For any $a \neq 0$, we have $(au - v'/a)^2 \geqslant 0$, hence the inequality $a^2u^2 + v'^2/a^2 \geqslant 2uv'$ and therefore

$$\mathscr{A} \leqslant \frac{a^2}{2}\int_0^L u^2\,\mathrm{d}s + \frac{1}{2a^2}\int_0^L v'^2\,\mathrm{d}s.$$

Now use the inequality proved in ①, making the assumption that

$$\int_0^L u\,\mathrm{d}s = 0$$

(translating the path on the real axis if need be, which does not change the area); it follows that

$$\mathscr{A} \leqslant \frac{1}{2}\left\{\frac{a^2L^2}{4\pi^2}\int_0^L u'(s)^2\,\mathrm{d}s + \frac{1}{a^2}\int_0^L v'(s)^2\,\mathrm{d}s\right\}.$$

It suffices now to fix a so that the coefficients in front of the integrals are equal, which means $a^2 = 2\pi/L$; with this value, we get

$$\mathscr{A} \leqslant \frac{L}{4\pi}\int_0^1 (u'^2 + v'^2)\,\mathrm{d}s = \frac{L^2}{4\pi}.$$

Equality holds if $au = v'/a$, which means $v' = 2\pi u/L$. Together with the relation $u'^2 + v'^2 = 1$, this implies $u(s) = \frac{L}{2\pi}\cos(2\pi s/L + \varphi)$ and $v(s) = \frac{L}{2\pi}\sin(2\pi s/L + \varphi)$, with $\varphi \in \mathbb{R}$, which is, indeed, the equation of a circle with circumference L.

Note that this result, which is rather intuitive, is by no means easy to prove!

Sidebar 4 (Pointwise convergence and Gibbs phenomenon)

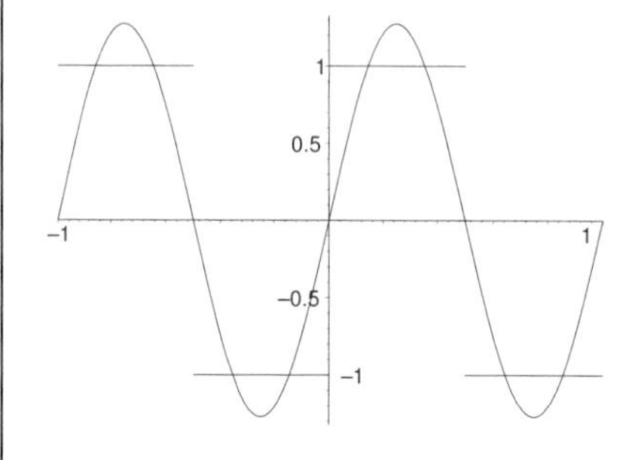

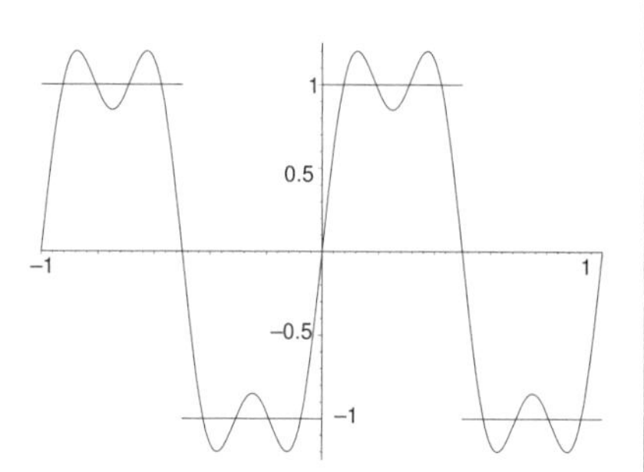

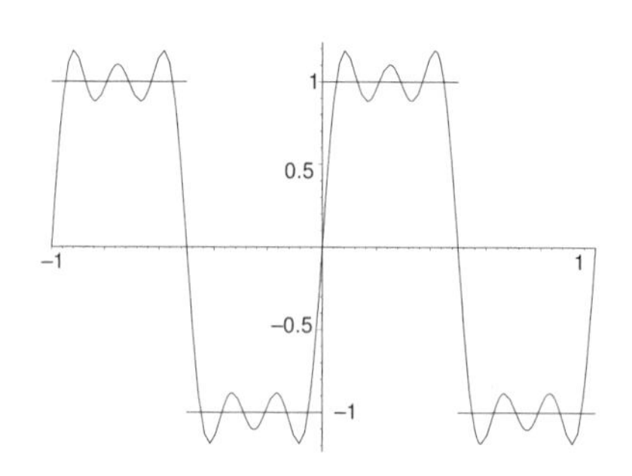

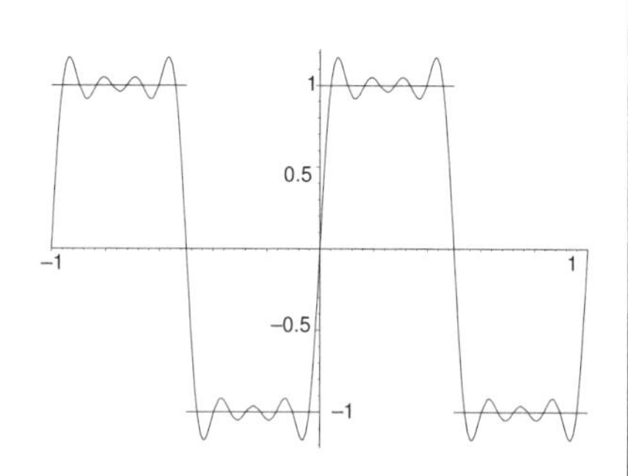

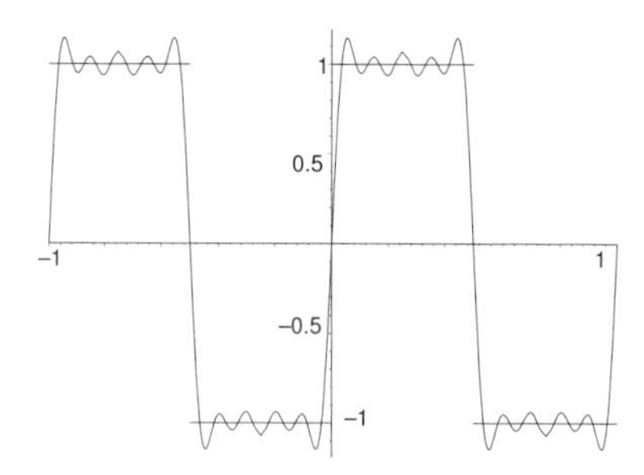

The partial Fourier sums for the square wave, with $1, \ldots, 5$ *terms:*

$$S_n(t) = \frac{4}{\pi} \sum_{k=0}^{n-1} \frac{1}{2k+1} \sin(2k+1)\pi t.$$

The pointwise convergence at any point of continuity is visible, but, close to the discontinuities, it is clear that the convergence is numerically very bad: there always remains a "crest," getting closer to the discontinuity, with constant amplitude. This is the ***Gibbs Phenomenon,*** *first explained by Josiah* GIBBS *[39, 40].*

Chapter 10

Fourier transform of functions

This chapter introduces an integral transform called the "Fourier transform," which generalizes the Fourier analysis of periodic functions to the case of functions defined on the whole real axis $\mathbb{R}$.

We start with the definition of the Fourier transform for functions which are Lebesgue-integrable (elements of L^1). One problem of the Fourier transform thus defined is that it does not leave the space L^1 stable. It will then be extended to functions which are square integrable (elements of L^2), which have a physical interpretation in terms of energy. The space L^2 being stable, any square integrable function has a Fourier transform which is also square integrable.

10.1 Fourier transform of a function in L^1

In this section, we start by defining the Fourier transform of an integrable function and deriving its main properties.

Although the Fourier transform is a generalization of the notion of a Fourier series, it is not the L^2 or Hilbert space setting which is the simplest. Therefore we start with functions in $L^1(\mathbb{R})$,[1] before generalizing to square integrable functions.

[1] The reason is that, in contrast with functions defined on a finite interval, $L^2(\mathbb{R})$ is not contained in $L^1(\mathbb{R})$.

10.1.a Definition

DEFINITION 10.1 Let f be a function, real- or complex-valued, depending on one *real* variable. The **Fourier transform** (or **spectrum**) of f, *if it exists*, is the complex-valued function defined for the real variable ν by

$$\widetilde{f}(\nu) \stackrel{\text{def}}{=} \int_{-\infty}^{\infty} f(x)\,\mathrm{e}^{-2\pi\mathrm{i}\nu x}\,\mathrm{d}x \qquad \text{for all } \nu \in \mathbb{R}. \tag{10.1}$$

This is denoted symbolically

$$\widetilde{f} = \mathscr{F}\left[f\right] \qquad \text{or} \qquad \widetilde{f}(\nu) = \mathscr{F}\left[f(x)\right].$$

Similarly, the **conjugate Fourier transform** of a function f is given (when it exists) by

$$\overline{\mathscr{F}}\left[f\right](\nu) \stackrel{\text{def}}{=} \int_{-\infty}^{\infty} f(x)\,\mathrm{e}^{2\pi\mathrm{i}\nu x}\,\mathrm{d}x.$$

Remark 10.2 The integral that appears does not always exist. For instance, if we consider $f : x \mapsto x^2$, the integral

$$\int_{-\infty}^{+\infty} x^2\,\mathrm{e}^{-2\pi\mathrm{i}\nu x}\,\mathrm{d}x$$

does not exist, for any value of ν, and f does not have a Fourier transform (in the sense of functions; it will be seen in the next chapter how to define its Fourier transform in the sense of distributions).

It is also possible to have to deal with non-integrable functions for which the integral defining $\widetilde{f}(\nu)$ converges as an improper integral. As an exemple, consider $g : x \mapsto (\sin x)/x$. Then g is not integrable but, on the other hand, the limit

$$\lim_{M\to+\infty} \int_{-M}^{M} \frac{\sin x}{x}\,\mathrm{e}^{-2\pi\mathrm{i}\nu x}\,\mathrm{d}x$$

exists for all values of ν. More details about this (not a complete description of what can happen! this is very difficult to obtain and state) will be given in Section 10.3, which discusses the extension of the Fourier transform to the space of square integrable functions.

At least if f is integrable on $\mathbb{R}$, it is clear that $\widetilde{f}$ is defined on $\mathbb{R}$. Indeed, recall that for a function $f : \mathbb{R} \to \mathbb{R}$, the following equivalence holds:

$$f \text{ is Lebesgue-integrable} \iff |f| \text{ is Lebesgue-integrable},$$

which shows that if f is integrable, then so is $x \mapsto f(x)\,\mathrm{e}^{-2\pi\mathrm{i}\nu x}$ for any real value of ν.

THEOREM 10.3 (Fourier transform of an integrable function) *Let f be a function which is Lebesgue-integrable on $\mathbb{R}$. Then the Fourier transform $\nu \mapsto \widetilde{f}(\nu)$ of f is defined for all $\nu \in \mathbb{R}$.*

Remark 10.4 A number of different conventions regarding the Fourier transform are in use in various fields of physics and mathematics; in the table on page 612, the most common ones are presented with the corresponding formula for the inverse Fourier transform.

10.1.b Examples

Example 10.5 Consider the **"rectangle" function** (already considered in Chapter 7):

$$\Pi(x) \stackrel{\text{def}}{=} \begin{cases} 0 & \text{for } |x| > \frac{1}{2}, \\ 1 & \text{for } |x| < \frac{1}{2}. \end{cases}$$

Then its Fourier transform is a sine cardinal (or cardinal sine):

$$\widetilde{\Pi}(\nu) = \int_{-\frac{1}{2}}^{\frac{1}{2}} \mathrm{e}^{-2\pi \mathrm{i} \nu t} \, \mathrm{d}t = \operatorname{sinc}(\pi\nu) \stackrel{\text{def}}{=} \begin{cases} \dfrac{\sin \pi\nu}{\pi\nu} & \text{if } \nu \neq 0, \\ 1 & \text{if } \nu = 0. \end{cases}$$

Similarly, the Fourier transform of the characteristic function $\chi_{[a,b]}$ of an interval is

$$\widetilde{\chi}_{[a,b]}(\nu) = \mathscr{F}\left[\chi_{[a,b]}\right](\nu) = \begin{cases} \dfrac{\sin\left(\pi(b-a)\nu\right)}{\pi\nu} \, \mathrm{e}^{-\mathrm{i}\pi(a+b)\nu} & \text{if } \nu \neq 0, \\ b-a & \text{if } \nu = 0. \end{cases}$$

Example 10.6 Consider the **gaussian** $x \mapsto \mathrm{e}^{-\pi x^2}$. Then one can show (see Exercice 10.1 on page 295 or, using the method of residues, Exercise 4.13 on page 126) that

$$\mathscr{F}\left[\mathrm{e}^{-\pi x^2}\right] = \mathrm{e}^{-\pi\nu^2}.$$

Example 10.7 The **lorentzian** function with parameter $a > 0$ is defined by

$$f(x) = \frac{2a}{a^2 + 4\pi^2 x^2}.$$

Then, by the method of residues, one can show that

$$\mathscr{F}\left[\frac{2a}{a^2 + 4\pi^2 x^2}\right] = \mathrm{e}^{-a|\nu|}.$$

The table on page 614 lists some of the most useful Fourier transforms.

10.1.c The L^1 space

Let $\mathcal{L}^1(\mathbb{R})$ denote temporarily the space of functions on $\mathbb{R}$ which are Lebesgue-integrable. Can it be normed by putting

$$\|f\| \stackrel{\text{def}}{=} \int \left|f(t)\right| \mathrm{d}t \; ?$$

The answer is "no." The reason is similar to what happened in the definition of the space L^2 in the previous chapter: if f and g are two functions in $\mathcal{L}^1$ which are equal almost everywhere,[2] then $\|f - g\| = 0$, although $f \neq g$; this means that the map $f \mapsto \|f\|$ is not a norm. To resolve this difficulty,

[2] For instance, one can take $f = \chi_{\mathbb{Q}}$, the characteristic function of the set $\mathbb{Q}$, also called the **Dirichlet function**, and g the zero function.

those functions which are equal almost everywhere are *identified*. The set of all functions which are equal to f almost everywhere is the **equivalence class of f**. The set of equivalence classes of integrable functions is denoted $L^1(\mathbb{R})$ or more simply L^1. Thus, when speaking of a "function" $f \in L^1$, what is meant is in fact any function in the same equivalence class. For instance, the zero function and the Dirichlet function belong to the same equivalence class, denoted 0. In order to not complicate matters, there are both said to be "equal to the zero function."

In the space L^1, the map $f \mapsto \int |f|$ is indeed a norm.

DEFINITION 10.8 The space $\boldsymbol{L^1(\mathbb{R})}$ or $\boldsymbol{L^1}$ is the vector space of integrable functions, defined "up to equality almost everywhere," with the norm

$$\|f\|_1 \stackrel{\text{def}}{=} \int |f(t)| \, \mathrm{d}t.$$

Consider two functions f and g which coincide almost everywhere. Then $\int f = \int g$. Moreover, if φ is any other function, it follows that $f\varphi$ is equal to $g\varphi$ almost everywhere. So applying this result to the case of $\varphi : x \mapsto \mathrm{e}^{-2\pi i \nu x}$, it follows that *two functions which are equal almost everywhere have the same Fourier transform.*

Hence, we can consider that the Fourier transform is defined on the space $L^1(\mathbb{R})$ of integrable functions defined up to equality almost everywhere.

10.1.d Elementary properties

THEOREM 10.9 (Properties of $\tilde{f}$) *Let $f \in L^1(\mathbb{R})$ be an integrable function and denote by $\tilde{f}$ its Fourier transform. Then $\tilde{f}$ is a continuous function on $\mathbb{R}$. Moreover, it is bounded, and in fact $\|\tilde{f}\|_\infty \leqslant \|f\|_1$.*

DEFINITION 10.10 The space $\boldsymbol{L^\infty(\mathbb{R})}$ is the vector space of bounded measurable functions on $\mathbb{R}$, with the sup norm defined by

$$\|f\|_\infty \stackrel{\text{def}}{=} \sup_{x \in \mathbb{R}} |f(x)|.$$

(It is possible to define functions of L^∞ up to equality almost everywhere. The definition of the norm must then be modified accordingly. This refinement is not essential here).

THEOREM 10.11 (Continuity of the Fourier transform in L^1) *The Fourier transform, defined on $L^1(\mathbb{R})$ and taking values in $L^\infty(\mathbb{R})$, is a* continuous linear *operator, that is*

*i) **(linearity)** for any $f, g \in L^1(\mathbb{R})$ and any $\lambda, \mu \in \mathbb{C}$, we have*

$$\mathscr{F}[\lambda f + \mu g] = \lambda \mathscr{F}[f] + \mu \mathscr{F}[g] \;;$$

ii) **(continuity)** *if a sequence $(f_n)_{n\in\mathbb{N}}$ of integrable functions converges to 0 in the sense of the L^1 norm, then the sequence $(\widetilde{f_n})_{n\in\mathbb{N}}$ tends to 0 in the sense of the sup norm:*

$$\left(\int \left|f_n(x)\right| \mathrm{d}x \xrightarrow[n\to\infty]{} 0\right) \Longrightarrow \left(\sup_{\nu\in\mathbb{R}} \left|\widetilde{f_n}(\nu)\right| \xrightarrow[n\to\infty]{} 0\right).$$

PROOF OF THE TWO PRECEDING THEOREMS. To show the continuity of the function $\widetilde{f}$, we apply the theorem of continuity of an integral depending on a parameter (Theorem 3.11 on page 77), using the domination relation

$$\forall x \in \mathbb{R} \qquad \left|f(x)\,\mathrm{e}^{-2\pi\mathrm{i}\nu x}\right| \leqslant \left|f(x)\right|.$$

Moreover, for any $\nu \in \mathbb{R}$, we have

$$\left|\widetilde{f}(\nu)\right| = \left|\int f(x)\,\mathrm{e}^{-2\pi\mathrm{i}\nu x}\,\mathrm{d}x\right| \leqslant \int \left|f(x)\right| \mathrm{d}x = \|f\|_1,$$

which shows that

$$\|\widetilde{f}\|_\infty = \sup_{\nu\in\mathbb{R}} \left|\widetilde{f}(\nu)\right| \leqslant \|f\|_1.$$

Since the Fourier transform is obviously linear, this inequality is sufficient to establish its continuity (see Theorem A.51 on page 583 concerning continuity of a linear map).

The continuity of the Fourier transform is an important fact. It means that if a sequence $(f_n)_{n\in\mathbb{N}}$ of integrable functions converges to $f \in L^1$, in the sense of the L^1 norm, namely,

$$\int \left|f_n(t) - f(t)\right| \mathrm{d}t \xrightarrow[n\to\infty]{} 0,$$

then there will be convergence in the sense of the L^∞ norm, that is, *uniform convergence* of the sequence $(\widetilde{f_n})_{n\in\mathbb{N}}$:

$$\left(f_n \xrightarrow{L^1} f\right) \Longrightarrow \left(\widetilde{f_n} \xrightarrow{\text{CV.U.}} \widetilde{f} \text{ on } \mathbb{R}\right).$$

Example 10.5 on page 279 shows that the Fourier transform of a function in L^1 is not necessarily integrable itself. So the "Fourier transform" operator does not preserve the space L^1, which is somewhat inconvenient, as one often prefers an honest endomorphism of a vector space to an arbitrary linear map. (It will be seen in Section 10.3 that it is possible to extend the Fourier transform to the space L^2 and that L^2 is then *stable* by the Fourier transform thus defined.)

On the other hand, one can see that each of the Fourier transforms previously computed has the property that it tends to 0 at infinity. This property is confirmed in general by the following theorem, which we will not prove (see, e.g., [76]). It is the analogue of the Riemann-Lebesgue Lemma 9.44 for Fourier series.

THEOREM 10.12 (Riemann-Lebesgue lemma) *Let $f \in L^1(\mathbb{R})$ be an integrable function and denote by $\widetilde{f}$ its Fourier transform. Then the continuous function $\widetilde{f}$ tends to 0 at infinity:*

$$\lim_{\nu \to \pm\infty} \left| \widetilde{f}(\nu) \right| = 0.$$

10.1.e Inversion

The question we now ask is: can one inverse the Fourier transform? The answer is *partially* "yes"; moreover, the inverse transform is none other than the conjugate Fourier transform.

We just saw that, even if $f \in L^1(\mathbb{R})$, its Fourier transform $\widetilde{f}$ is not itself always integrable, and so the conjugate Fourier transform of $\widetilde{f}$ may not be defined. However, in the special case where $\widetilde{f} \in L^1$, we have the following fundamental result.

THEOREM 10.13 (Inversion in $L^1(\mathbb{R})$) *Let $f \in L^1(\mathbb{R})$ be an integrable function such that the Fourier transform $\widetilde{f}$ of f is itself integrable. Then we have*

$$\overline{\mathscr{F}}\left[\widetilde{f}\right](x) = f(x) \qquad \textit{for almost all } x;$$

and also

$$\overline{\mathscr{F}}\left[\widetilde{f}\right](x) = f(x) \qquad \textit{at any point } x \textit{ where } f \textit{ is continuous.}$$

In particular, if f is also continuous,[3] *we have $\overline{\mathscr{F}}\left[\widetilde{f}\right] = f$.*

PROOF. We prove the inversion formula only at points of continuity. The proof starts with the following lemma.

LEMMA 10.14 *Let f and g be two integrable functions. Then $\widetilde{f} \cdot g$ and $f \cdot \widetilde{g}$ are integrable and we have*

$$\int f(t)\widetilde{g}(t)\,\mathrm{d}t = \int \widetilde{f}(t)g(t)\,\mathrm{d}t.$$

PROOF OF THE LEMMA. Since $\widetilde{f}$ is bounded and g is integrable, the product $\widetilde{f} \cdot g$ is also integrable and similarly for $f \cdot \widetilde{g}$. Hence by Fubini's theorem, we have

$$\begin{aligned}
\int f(t)\widetilde{g}(t)\,\mathrm{d}t &= \int f(t)\left(\int g(s)\,\mathrm{e}^{-2\pi i t s}\,\mathrm{d}s\right)\mathrm{d}t \\
&= \int g(s)\left(\int f(t)\,\mathrm{e}^{-2\pi i s t}\,\mathrm{d}t\right)\mathrm{d}s = \int g(s)\widetilde{f}(s)\,\mathrm{d}s,
\end{aligned}$$

which proves the lemma.

---

[3] Then we are not looking at it simply up to "equality almost everywhere."

Now introduce the sequence of functions $(h_n)_{n\in\mathbb{N}}$ defined by

$$h_n(x) = \exp(-\pi^2 x^2/n^2).$$

It is a sequence of gaussians which are more and more "spread out," and converges pointwise to the constant function $\mathbf{1}$. Since $\widetilde{f}$ is integrable and

$$\left|\widetilde{f}(x)\, h_n(x)\, e^{2\pi i x t}\right| \leqslant \left|\widetilde{f}(x)\right|$$

for any reals $x, t \in \mathbb{R}$ and any n, Lebesgue's dominated convergence theorem proves that

$$\int \widetilde{f}(x)\, h_n(x)\, e^{2\pi i x t}\, dx \xrightarrow[n\to\infty]{} \int \widetilde{f}(x)\, e^{2\pi i x t}\, dx = \overline{\mathscr{F}}\left[\widetilde{f}\right](t) \qquad (10.2)$$

for any $t \in \mathbb{R}$. The Fourier transform of $x \mapsto h_n(x)\, e^{2\pi i x t}$ is, as shown by an immediate calculation, equal to $x \mapsto \widetilde{h}_n(x-t)$, and the previous lemma yields

$$\int \widetilde{f}(x)\, h_n(x)\, e^{2\pi i x t}\, dx = \int f(x)\, \widetilde{h}_n(x-t)\, dx. \qquad (10.3)$$

But, as seen in Example 8.17 on page 231, the sequence of Fourier transforms of h_n is

$$\widetilde{h}_n : x \longmapsto \frac{n}{\sqrt{\pi}}\, e^{-n^2 x^2},$$

which is a *Dirac sequence*. It follows that, if t is a point of continuity for f, we have

$$\int f(x)\, \widetilde{h}_n(x-t)\, dx \xrightarrow[n\to\infty]{} f(t). \qquad (10.4)$$

Putting equations (10.2), (10.3), and (10.4) together, we obtain, for any point where f is continuous, that

$$\int \widetilde{f}(x)\, e^{2\pi i x t}\, dx = \overline{\mathscr{F}}\left[\widetilde{f}\right](t) = f(t),$$

which is the second result stated.

COROLLARY 10.15 *If f is integrable and is not equal almost everywhere to a continuous function, then its Fourier transform is not integrable.*

Proof. If the Fourier transform $\widetilde{f}$ is integrable, then f coincides almost everywhere with the inverse Fourier transform of $\widetilde{f}$, which is continuous.

Remark 10.16 With the conventions we have chosen for the Fourier transform, we have, when this makes sense, $\mathscr{F}^{-1} = \overline{\mathscr{F}}$. One could use indifferently $\mathscr{F}^{-1}$ or $\overline{\mathscr{F}}$. The reader should be aware that, in other fields, the convention used for the definition of the Fourier transform is different from the one we use (for instance, in quantum mechanics, in optics, or when dealing with functions of more than one variable), and the notation is not equivalent then. A table summarizes the main conventions used and the corresponding inversion formulas (page 612).

Remark 10.17 Theorem 10.13 may be proved more simply using Lemma 10.14 and the fact that the Fourier transform of the constant function $\mathbf{1}$ is the Dirac δ. But this property will only be established in the next chapter. For this reason we use instead a sequence of functions $(h_n)_{n\in\mathbb{N}}$ converging to $\mathbf{1}$, while the sequence of their Fourier transforms converges to δ.

To apply the inversion formula of Theorem 10.13, one needs some information concerning not only f, but also $\widetilde{f}$ (which must be integrable); this may be inconvenient. In the next proposition, information relative to f suffices to obtain the inversion formula.

PROPOSITION 10.18 *Let f be a function of $\mathscr{C}^2$ class such that f, f', and f'' are all integrable. Then $\widetilde{f}$ is also integrable; the Fourier inversion formula holds, and we have*

$$f = \overline{\mathscr{F}}\,[\widetilde{f}\,].$$

Example 10.19 Consider again the example of the lorentzian function. We have

$$f(x) = \frac{2}{1+4\pi^2x^2}.$$

Then

$$f'(x) = \frac{-16\,\pi^2 x}{(1+4\pi^2x^2)^2} \qquad \text{and} \qquad f''(x) = \frac{192\,\pi^4x^2 - 16\,\pi^2x^2}{(1+4\pi^2x^2)^3},$$

which are both integrable. We deduce from this that $\overline{\mathscr{F}}\left[\mathrm{e}^{-|\nu|}\right] = 2/(1+4\pi^2x^2)$. Moreover, since all the functions considered are real-valued, it follows by taking the complex conjugate and exchanging the variables x and ν that also

$$\mathscr{F}\left[\mathrm{e}^{-|x|}\right] = \frac{2}{1+4\pi^2\nu^2}.$$

Hence the inversion formula is also a tool that can be used to compute new Fourier transforms easily!

10.1.f Extension of the inversion formula

We can now discuss the case where f has a Fourier transform $\widetilde{f}$ which is not integrable, that is, such that

$$\int \widetilde{f}(\nu)\,\mathrm{e}^{2\pi\mathrm{i}\nu x}\,\mathrm{d}x$$

is not defined in the sense of Lebesgue. In some situations, the limit

$$\lim_{R\to+\infty} \int_{-R}^{+R} \widetilde{f}(\nu)\,\mathrm{e}^{2\pi\mathrm{i}\nu x}\,\mathrm{d}\nu$$

may still be defined. (It is not quite an improper integral, because the bounds of integration are symmetrical.) For example, the Fourier transform of the "rectangle" function is a cardinal sine. The latter is not Lebesgue-integrable, but the integral

$$I(x,R) \stackrel{\text{def}}{=} \int_{-R}^{R} \frac{\sin \pi\nu}{\pi\nu}\,\mathrm{e}^{2\pi\mathrm{i}\nu x}\,\mathrm{d}\nu$$

does have a limit as $[R \to \infty]$. In such cases, the following theorem may be used.

THEOREM 10.20 *Let $f(x)$ be an integrable function of class $\mathscr{C}^1$ except at finitely many points of discontinuity, such that $f' \in L^1$. Then for all x we have*

$$\lim_{R\to+\infty} \int_{-R}^{+R} \widetilde{f}(\nu)\, e^{2\pi i \nu x}\, d\nu = \frac{1}{2}\Big[f(x^+) + f(x^-) \Big].$$

By abuse of notation, this limit will still be denoted $\overline{\mathscr{F}}\big[\widetilde{f}\,\big](x)$.

In other words, the inversion formula, in the sense of a limit of integrals (called the **principal value**), gives the *regularized* function associated to f. This should be compared to Dirichlet's theorem on page 268.

In particular, if f is continuous and f and f' are integrable, the inversion formula of Theorem 10.20 gives back the original function f.

Example 10.21 We saw that the Fourier transform of the "rectangle" function is $\operatorname{sinc}(\pi\nu)$. Since the "rectangle" function has only two discontinuities and since its derivative (in the sense of functions) is always zero and therefore integrable, it follows that

$$\lim_{R\to+\infty} \int_{-R}^{+R} \operatorname{sinc}(\pi\nu)\, e^{2\pi i \nu x}\, d\nu = \overline{\mathscr{F}}\big[\operatorname{sinc}(\pi\nu)\big](x) = \begin{cases} 0 & \text{if } |x| > \frac{1}{2}, \\ 1 & \text{if } |x| < \frac{1}{2}, \\ \frac{1}{2} & \text{if } |x| = \frac{1}{2}. \end{cases}$$

10.2

Properties of the Fourier transform

10.2.a Transpose and translates

THEOREM 10.22 *Let f be an integrable function and $a \in \mathbb{R}$ a real number. We have*

$$\mathscr{F}\Big[\overline{f(x)}\Big] = \overline{\widetilde{f}(-\nu)}, \qquad \mathscr{F}\big[f(-x)\big] = \overline{\mathscr{F}}\big[f(x)\big] = \widetilde{f}(-\nu),$$

$$\mathscr{F}\big[f(x-a)\big] = e^{-2i\pi\nu a}\widetilde{f}(\nu), \quad \mathscr{F}\big[f(x)\, e^{2\pi i \nu_0 x}\big] = \widetilde{f}(\nu-\nu_0).$$

These results should be compared to the formulas for similar operations on the Fourier coefficients of the periodic function; see Theorem 9.36 on page 264.

COROLLARY 10.23 *In the following table, the properties of f indicated in the first column are transformed into the properties of $\widetilde{f}$ in the second column, and conversely:*

$f(x)$	$\widetilde{f}(\nu)$
even	*even*
odd	*odd*
real-valued	*hermitian*
purely imaginary	*antihermitian*

*Here a function g is a **hermitian** function if $g(-x) = \overline{g(x)}$ and an **antihermitian** function if $g(-x) = -\overline{g(x)}$.*

10.2.b Dilation

If a dilation is performed (also called a change of scale), that is, if we put

$$g(x) = f(ax) \qquad \forall x \in \mathbb{R}, \qquad \text{for some} \quad a \in \mathbb{R}^*,$$

the Fourier transform of g can be computed easily:

$$\widetilde{g}(\nu) = \int f(ax)\,\mathrm{e}^{-2\pi\mathrm{i}\nu x}\,\mathrm{d}x = \int f(y)\,\mathrm{e}^{-2\pi\mathrm{i}\nu y/a}\,\frac{\mathrm{d}y}{|a|} = \frac{1}{|a|}\widetilde{f}\left(\frac{\nu}{a}\right),$$

which gives the following theorem.

THEOREM 10.24 *Let $a \in \mathbb{R}^*$ and $f \in L^1$. Then we have*

$$\mathscr{F}\left[f(ax)\right] = \frac{1}{|a|}\widetilde{f}\left(\frac{\nu}{a}\right).$$

In other words, a dilation in the real world corresponds to a compression in the Fourier world, and conversely.

10.2.c Derivation

There is a very important relation between the Fourier transform and the operation of differentiation:

THEOREM 10.25 (Fourier transform of the derivative) *Let $f \in L^1$ be a function which decays sufficiently fast at infinity so that $x \mapsto x^k f(x)$ belongs also to L^1 for $k = 0, \ldots, n$. Then $\widetilde{f}$ can be differentiated n times and we have*

$$\mathscr{F}\left[(-2\mathrm{i}\pi x)^k f(x)\right] = \widetilde{f}^{(k)}(\nu) \qquad \textit{for } k = 1, \ldots, n.$$

Conversely, if $f \in L^1$ and if f is of $\mathscr{C}^n$ class and, moreover, its successive derivatives $f^{(k)}$ are integrable for $k = 1, \ldots, n$, then we have

$$\mathscr{F}\left[f^{(m)}(x)\right] = (2\mathrm{i}\pi\nu)^m \widetilde{f}(\nu) \qquad \textit{for } k = 1, \ldots, n.$$

In particular, it will be useful to remember that

$$\mathscr{F}\left[f'(x)\right] = 2\mathrm{i}\pi\nu\tilde{f}(\nu) \qquad \textit{and} \qquad \mathscr{F}\left[-2\mathrm{i}\pi x\, f(x)\right] = \frac{\mathrm{d}}{\mathrm{d}\nu}\tilde{f}(\nu).$$

PROOF. For any fixed $x \in \mathbb{R}$, the function $\nu \mapsto f(x)\,\mathrm{e}^{-2\pi\mathrm{i}\nu x}$ is of $\mathscr{C}^\infty$ class, and its k-th derivative is bounded in modulus by $|(2\pi x)^k f(x)|$, which is integrable by assumption. Applying the theorem of differentiation under the integral sign, we obtain

$$\tilde{f}'(\nu) = \frac{\mathrm{d}}{\mathrm{d}\nu}\tilde{f}(\nu) = \int \frac{\mathrm{d}}{\mathrm{d}\nu}\left[f(x)\,\mathrm{e}^{-2\pi\mathrm{i}\nu x}\right]\mathrm{d}x = \int (-2\pi\mathrm{i}x)\, f(x)\,\mathrm{e}^{-2\pi\mathrm{i}\nu x}\,\mathrm{d}x,$$

and then, by an immediate induction, the first formula stated.

Now for the second part, recall that for any integrable function φ, we have

$$\int \varphi(x)\,\mathrm{d}x = \lim_{R\to+\infty}\int_{-R}^{R}\varphi(x)\,\mathrm{d}x.$$

Since f' is assumed to be integrable, an integration by parts yields

$$\begin{aligned}\mathscr{F}\left[f'\right](\nu) &= \lim_{R\to+\infty}\int_{-R}^{R}\underbrace{f'(x)}_{u'}\,\underbrace{\mathrm{e}^{-2\pi\mathrm{i}\nu x}}_{v}\,\mathrm{d}x \\ &= \lim_{R\to+\infty}\left\{\left[f(x)\,\mathrm{e}^{-2\pi\mathrm{i}\nu x}\right]_{-R}^{R} + \int_{-R}^{R}(2\pi\mathrm{i}\nu)\, f(x)\,\mathrm{e}^{-2\pi\mathrm{i}\nu x}\,\mathrm{d}x\right\}.\end{aligned}$$

As f *and* f' are integrable, f tends to zero at infinity (see Exercise 10.7 on page 296). The previous formula then shows, by letting R tend to infinity, that

$$\int f'(x)\,\mathrm{e}^{-2\pi\mathrm{i}\nu x}\,\mathrm{d}x = \int (2\pi\mathrm{i}\nu)\, f(x)\,\mathrm{e}^{-2\pi\mathrm{i}\nu x}\,\mathrm{d}x,$$

which establishes the second formula for $k = 1$. An induction on k concludes the proof.

Noticing that if f is integrable and has bounded support then $x \mapsto x^k f(x)$ is also integrable for any $k \geqslant 0$, we get the following corollary:

COROLLARY 10.25.1 *If $f \in L^1$ is a function with bounded support, then its Fourier transform is of $\mathscr{C}^\infty$ class.*

The formula of Theorem 10.25 on the preceding page also implies a very important result:

COROLLARY 10.25.2 *Let f and g be measurable functions, and let $p, q \in \mathbb{N}$. Assume that the derivatives of f, of order 1 to p, exist and are integrable, and that $x \mapsto x^k\, g(x)$ is also integrable for $k = 0, \ldots, q$. Then, for any $\nu \in \mathbb{R}$, we have*

$$\left|\tilde{f}(\nu)\right| \leqslant |2\pi\nu|^{-p}\int \left|f^{(p)}(x)\right|\mathrm{d}x$$

and

$$\left|\tilde{g}^{(q)}(\nu)\right| \leqslant \int_{-\infty}^{\infty}|2\pi x|^q\,\left|g(x)\right|\mathrm{d}x,$$

These inequalities give very strong bounds: if f is five times differentiable, for instance, and $f^{(5)}$ is integrable, then $\widetilde{f}$ decays at least as fast as $1/\nu^5$.

So there exists a link between the regularity of f and the rate of decay of $\widetilde{f}$ at infinity, and similarly between the rate of decay of f and the regularity of $\widetilde{f}$.

10.2.d Rapidly decaying functions

DEFINITION 10.26 A **rapidly decaying function** is any function f such that

$$\lim_{x\to\pm\infty} \left|x^k f(x)\right| = 0 \qquad \text{for any } k \in \mathbb{N}.$$

In other words, a rapidly decaying function is one that decays at infinity faster than any power of x.

Example 10.27 The functions $x \mapsto \mathrm{e}^{-x^2}$ and $x \mapsto x^5(\log x)\,\mathrm{e}^{-|x|}$ are rapidly decaying.

This is a useful notion, for instance, because if f is rapidly decaying and is locally integrable, then $x^k f(x)$ is integrable for all $k \in \mathbb{N}$. From Theorem 10.25, we deduce:

THEOREM 10.28 *If f is an integrable rapidly decaying function, then its Fourier transform is of $\mathscr{C}^\infty$ class.*

Still using Theorem 10.25, one can show (exercise!) that

THEOREM 10.29 *Let f be an integrable function of $\mathscr{C}^\infty$ class. If $f^{(k)}$ is integrable for all $k \in \mathbb{N}$, then $\widetilde{f}$ is a rapidly decaying function.*

10.3 Fourier transform of a function in L^2

The are some functions which are not necessarily integrable, but whose square is. Such is the "sine cardinal" function:

$$x \longmapsto \operatorname{sinc}(x) \stackrel{\text{def}}{=} \frac{\sin x}{x}$$

(extended to $\mathbb{R}$ by continuity by putting $\operatorname{sinc}(0) = 1$), which belongs to L^2 but not to L^1.

It turns out that in physics, square integrable functions are of paramount importance and occur frequently:

- in quantum mechanics, the wave function ψ of a particle is a square integrable function such that $\int |\psi(x)|^2 \,\mathrm{d}x = 1$;
- in optics, the square of the modulus of the light amplitude represents a density of energy; the total energy $\int |A(x)|^2 \,\mathrm{d}x$ is finite;
- similarly in electricity, or more generally in signal theory, $\int |f(t)|^2 \,\mathrm{d}t$ is the total energy of a temporal signal $t \mapsto f(t)$.

It is therefore advisable to extend the definition of the Fourier transform to the class of square integrable functions. (It will be seen in Chapter 13, Section 13.6, how to extend this further to functions with infinite total energy but finite *power*.) For this purpose, we start by introducing the Schwartz space, a technical device to reach our goal.[4]

10.3.a The space $\mathscr{S}$

DEFINITION 10.30 The **Schwartz space**, denoted $\mathscr{S}$, is the space of functions of $\mathscr{C}^\infty$ class which are rapidly decaying along with all their derivatives.

Example 10.31 Let $f \in \mathscr{D}$. Then $\widetilde{f}$ is defined and $\widetilde{f} \in \mathscr{S}$.

Indeed, f has bounded support, so, according to Corollary 10.25.1, $\widetilde{f}$ is infinitely differentiable. Moreover, f can be differentiated p times and $f^{(p)}$ is integrable for any $p \in \mathbb{N}$, so that Corollary 10.25.2 implies that $\widetilde{f}$ decays at least as fast as $1/\nu^p$ at infinity, for any $p \in \mathbb{N}$. Moreover (still by Theorem 10.25), the derivatives of $\widetilde{f}$ are also Fourier transforms of functions in $\mathscr{D}$, so the same reasoning implies that $\widetilde{f} \in \mathscr{S}$.

The most important result concerning $\mathscr{S}$ is the following:

THEOREM 10.32 *The Fourier transform defines a continuous linear operator from $\mathscr{S}$ into itself. In other words, for $f \in \mathscr{S}$, we have $\widetilde{f} \in \mathscr{S}$, and if a sequence $(f_n)_{n\in\mathbb{N}}$ tends to 0 in $\mathscr{S}$ then $(\widetilde{f_n})_{n\in\mathbb{N}}$ also tends to 0 in $\mathscr{S}$.*

PROOF THAT $\mathscr{S}$ IS STABLE. We first show that if $f \in \mathscr{S}$, we have $\widetilde{f} \in \mathscr{S}$.

The function f is rapidly decaying, so its Fourier transform $\widetilde{f}$ is of $\mathscr{C}^\infty$ class. For all $k \in \mathbb{N}$, the function $f^{(k)}$ is also rapidly decaying (and integrable). By Theorem 10.25 on page 286, we deduce that $\widetilde{f}$ is rapidly decaying.

There only remains to show that the derivatives of $\widetilde{f}$ are also rapidly decaying. But by Theorem 10.25, the derivatives of $\widetilde{f}$ are Fourier transforms of functions of the type $x \mapsto (-2\mathrm{i}\pi x)^k f(x)$, and it is clear that by definition of $\mathscr{S}$, these are also in $\mathscr{S}$. So applying the previous argument to those functions shows that $\widetilde{f}^{(k)}$ is rapidly decaying for any $k \in \mathbb{N}$.

To speak of continuity of the Fourier transform $\mathscr{F}[\cdot]$ on $\mathscr{S}$, we must make precise what the notion of convergence is in $\mathscr{S}$.

---

[4] The Schwartz space will also be useful to define the Fourier transform in the sense of distributions.

DEFINITION 10.33 A sequence $(f_n)_{n\in\mathbb{N}}$ of elements of $\mathscr{S}$ **converges in $\mathscr{S}$ to 0** if, for any $p, q \in \mathbb{N}$, we have

$$\lim_{n\to\infty} \sup_{x\in\mathbb{R}} \left| x^p f_n^{(q)}(x) \right| = 0.$$

This notion of convergence in $\mathscr{S}$ is very strong; in particular, it implies the uniform convergence on R of $(f_n)_{n\in\mathbb{N}}$ together with that of all its derivatives $\left(f_n^{(k)}\right)_{n\in\mathbb{N}}$.

To end this discussion of the most important properties of the Fourier transform on the space $\mathscr{S}$, it is natural to ask what happens for the inversion formula. We know that if $f \in \mathscr{S}$, its Fourier transform $\widetilde{f}$ is in $\mathscr{S}$ and, since $\widetilde{f}$ is continuous and integrable, we will have (by Theorem 10.13) the relation $\overline{\mathscr{F}}\,[\mathscr{F}\,[f]] = f$. In fact even more is true:

THEOREM 10.34 (Inversion in $\mathscr{S}$) *The Fourier transform is a bicontinuous linear isomorphism (i.e., bijective and continuous along with its inverse) from $\mathscr{S}$ into $\mathscr{S}$. Its inverse is given by $\mathscr{F}^{-1}[\cdot] = \overline{\mathscr{F}}\,[\cdot]$.*

10.3.b The Fourier transform in L^2

To define the Fourier transform on the space L^2 of square integrable functions, we will use the results of the previous section concerning the Fourier transform on $\mathscr{S}$, together with the following lemma, which is given without proof.

LEMMA 10.35 *The space $\mathscr{S}$ is a dense subspace of the space $L^2(\mathbb{R})$.*

The next step is to show that, for two functions f and g in $\mathscr{S}$, we have

$$\int_{-\infty}^{\infty} \overline{f(x)}\, g(x)\, \mathrm{d}x = \int_{-\infty}^{\infty} \overline{\widetilde{f}(\nu)}\, \widetilde{g}(\nu)\, \mathrm{d}\nu.$$

This is an easy consequence of Fubini's theorem and some manipulations with complex conjugation.[5] We can guess that a Hilbert space structure is in the offing!

Then we use the properties

- $\mathscr{S}$ is dense in L^2,
- L^2 is complete,

to show that the Fourier transform defined on $\mathscr{S}$ can be extended to a continuous linear operator on L^2. This step, essentially technical, is explained in Sidebar 5 on page 294.

We end up with an operator "Fourier transform," defined for any function $f \in L^2(\mathbb{R})$ which satisfies the identities of the next theorem:

---

[5] See, for instance, the proof of Lemma 10.14.

THEOREM 10.36 (Parseval-Plancherel) *The Fourier transform $\mathscr{F}[\cdot]$ is an isometry on L^2: for any two square integrable functions f and g, the Fourier transforms $\widetilde{f}$ and $\widetilde{g}$ also in L^2 and we have*

$$\int_{-\infty}^{\infty} \overline{f(x)}\, g(x)\, \mathrm{d}x = \int_{-\infty}^{\infty} \overline{\widetilde{f}(\nu)}\, \widetilde{g}(\nu)\, \mathrm{d}\nu.$$

In particular, taking $f = g$, we have

$$\int_{-\infty}^{\infty} |f(x)|^2\, \mathrm{d}x = \int_{-\infty}^{\infty} |\widetilde{f}(\nu)|^2\, \mathrm{d}\nu$$

or, in other words,

$$\|f\|_2 = \|\widetilde{f}\|_2.$$

THEOREM 10.37 (Inversion in L^2) *For any $f \in L^2$, we have*

$$\overline{\mathscr{F}}\big[\mathscr{F}[f]\big] = \mathscr{F}\big[\overline{\mathscr{F}}[f]\big] = f \quad \text{almost everywhere.}$$

The inverse Fourier transform is thus defined by

$$\mathscr{F}^{-1}[\cdot] = \overline{\mathscr{F}}[\cdot].$$

It is natural to wonder if there is a relation between the Fourier transform thus defined on $L^2(\mathbb{R})$ (after much wrangling and sleight of hand) and the original Fourier transform on $L^1(\mathbb{R})$.

PROPOSITION 10.38 *The Fourier transform on L^1 and that on L^2 coincide on the subspace $L^1 \cap L^2$.*

In other words, if f is square integrable on the one hand, but is also integrable itself on the other hand, then its Fourier transform defined in the section is indeed given by

$$\mathscr{F}[f] : \nu \longmapsto \int f(x)\, \mathrm{e}^{-2\pi \mathrm{i} \nu x}\, \mathrm{d}x.$$

But then, how does one express the Fourier transform of a square integrable function which is *not* integrable? One can show that, *in practice*, its Fourier transform in L^2 can be defined by the following limit:

$$\widetilde{f}(\nu) = \lim_{R \to +\infty} \int_{-R}^{R} f(x)\, \mathrm{e}^{-2\pi \mathrm{i} \nu x}\, \mathrm{d}x \qquad \text{for almost all } \nu \in \mathbb{R}$$

(which is again a special case of improper integral, defined in the sense of a principal value).

10.4

Fourier transform and convolution

10.4.a Convolution formula

Recall first the definition of the convolution of two functions:

DEFINITION 10.39 Let f and g be two locally integrable functions. Their **convolution**, when it exists, is the function denoted $h = f * g$ such that

$$h(x) \stackrel{\text{def}}{=} \int_{-\infty}^{+\infty} f(t)\, g(x-t)\, \mathrm{d}t.$$

Fubini's theorem (page 79) provides the proof of the following fundamental theorem, also known as the *Faltung theorem* (Faltung being the German word for "convolution"):

THEOREM 10.40 *Let f and g be two functions such that their Fourier transforms exist and the convolution $f * g$ is defined and is integrable. Then we have*

$$\mathscr{F}[f*g] = \mathscr{F}[f] \cdot \mathscr{F}[g], \qquad \widetilde{f*g}(\nu) = \widetilde{f}(\nu) \cdot \widetilde{g}(\nu).$$

Similarly, when those expressions are defined, we have

$$\mathscr{F}[f \cdot g] = \mathscr{F}[f] * \mathscr{F}[g], \qquad \widetilde{f \cdot g}(\nu) = \widetilde{f} * \widetilde{g}(\nu).$$

PROOF. We consider the special case where f and g are integrable. Then the Fourier transforms of f and g are defined and $f * g$ is indeed integrable. Define $\varphi(x,t) = f(t)\, g(x-t)\, \mathrm{e}^{-2\pi i \nu x}$. Then

$$\mathscr{F}[f*g](\nu) = \int \left(\int f(t)\, g(x-t)\, \mathrm{d}t \right) \mathrm{e}^{-2\pi i \nu x}\, \mathrm{d}x = \int \left(\int \varphi(x,t)\, \mathrm{d}t \right) \mathrm{d}x.$$

Since φ is integrable on $\mathbb{R}^2$ by Fubini's theorem (and the integrability of f and g), we obtain

$$\begin{aligned} \mathscr{F}[f*g](\nu) &= \int \left(\int \varphi(x,t)\, \mathrm{d}x \right) \mathrm{d}t \\ &= \int \left(\int f(t)\, g(x-t)\, \mathrm{e}^{-2\pi i \nu x}\, \mathrm{d}x \right) \mathrm{d}t \\ &= \int f(t) \left(\int g(x-t)\, \mathrm{e}^{-2\pi i \nu (x-t)}\, \mathrm{d}x \right) \mathrm{e}^{-2\pi i \nu t}\, \mathrm{d}t \\ &= \int f(t)\, \widetilde{g}(\nu)\, \mathrm{e}^{-2\pi i \nu t}\, \mathrm{d}t = \widetilde{f}(\nu)\, \widetilde{g}(\nu). \end{aligned}$$

The second formula is in general more delicate than the first; if we assume that $\widetilde{f}$ and $\widetilde{g}$ are both integrable, it follows from the first together with the Fourier inversion formula.

Example 10.41 We want to compute without a problem the Fourier transform of the function Λ defined by

$$\Lambda(x) = \begin{cases} 1+x & \text{for } -1 \leqslant x \leqslant 0, \\ 1-x & \text{for } 0 \leqslant x \leqslant 1, \\ 0 & \text{for } |x| \geqslant 1. \end{cases}$$

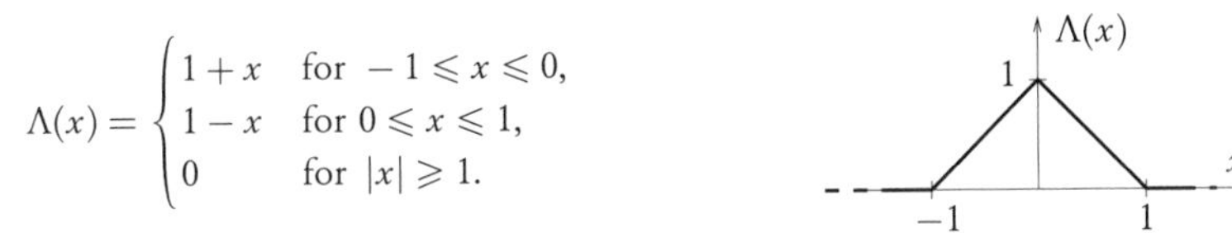

First we show that $\Lambda(x) = \big[\Pi * \Pi\big](x)$ and then we deduce that

$$\mathscr{F}\big[\Lambda(x)\big] = \left(\frac{\sin \pi\nu}{\pi\nu}\right)^2.$$

Note that the Parseval-Plancheral formula gives

$$\int \Lambda(x)\,\mathrm{d}x = \int \Pi(x)\,\mathrm{d}x \times \int \Pi(x)\,\mathrm{d}x = 1,$$

which is the expected result.

10.4.b Cases of the convolution formula

One can show, in particular, that the following cases of the convolution formula are valid:

i) if $f, g \in L^1$, then $\widetilde{f * g}(\nu) = \widetilde{f} \cdot \widetilde{g}(\nu)$ for any $\nu \in \mathbb{R}$;

ii) if $f, g \in L^1$ and their Fourier transforms are also in L^1, then $\widetilde{f \cdot g}(\nu) = \widetilde{f} * \widetilde{g}(\nu)$ for any $\nu \in \mathbb{R}$;

iii) if f and g are in L^2, one can take the inverse Fourier transform of the first formula and write $f * g(t) = \overline{\mathscr{F}}\left[\widetilde{f} \cdot \widetilde{g}\right](t)$ for all $t \in \mathbb{R}$, which is a good way to compute $f * g$ (see Exercise 10.2 on page 295); moreover, the second formula is valid: $\widetilde{f \cdot g}(\nu) = \widetilde{f} * \widetilde{g}(\nu)$ for all $\nu \in \mathbb{R}$;

iv) if $f \in L^1$ and $g \in L^2$, then $f * g(t) = \overline{\mathscr{F}}\left[\widetilde{f} \cdot \widetilde{g}\right](t)$ for almost all $t \in \mathbb{R}$.

Sidebar 5 (Extension of a continuous linear operator) *Let E be a normed vector space and V a* dense *subspace of E. If F is a complete normed vector space and $\varphi : V \to F$ is a continuous linear map, then*

a) there exists a unique continuous linear extension of φ, say $\widehat{\varphi} : E \to F$ (i.e., $\widehat{\varphi}$ satisfies $\widehat{\varphi}(x) = \varphi(x)$ for any $x \in V$);

b) moreover, the norm of $\widehat{\varphi}$ and the norm of φ are equal.

Let $x \in E$. We want to find the vector which will be called $\widehat{\varphi}(x)$. Using the fact that V is dense in E, there exists a sequence $(x_n)_{n\in\mathbb{N}}$ with values in V which converges to x in E. To show that the sequence $(\varphi(x_n))_{n\in\mathbb{N}}$ converges, we start by noting that $(x_n)_{n\in\mathbb{N}}$ is a Cauchy sequence. Since φ is linear and continuous, we have, denoting by $|||\varphi|||$ the norm of φ (see Appendix A, page 583), the inequality

$$\left\|\varphi(x_p) - \varphi(x_q)\right\| \leqslant |||\varphi||| \left\|x_p - x_q\right\|$$

for any $p, q \in \mathbb{N}$. The sequence $\big(\varphi(x_n)\big)_{n\in\mathbb{N}}$ is therefore also a Cauchy sequence, and since it takes values in the complete space F, it is convergent. There only remains to define

$$\widehat{\varphi}(x) = \lim_{n\to\infty} \varphi(x_n).$$

It is easy to check that $\varphi(x)$ does not depend on the chosen sequence $(x_n)_{n\in\mathbb{N}}$ converging to x: indeed, if $(y_n)_{n\in\mathbb{N}}$ also converges to x, we have

$$\lim_{n\to\infty} \varphi(x_n) - \lim_{n\to\infty} \varphi(y_n) = \lim_{n\to\infty} \varphi(x_n - y_n) = 0$$

by continuity of φ.

The map $\widehat{\varphi}$ thus constructed is linear and satisfies

$$\left\|\widehat{\varphi}(x)\right\| = \left\|\lim_{n\to\infty} \varphi(x_n)\right\| = \lim_{n\to\infty} \left\|\varphi(x_n)\right\| \leqslant |||\varphi||| \lim_{n\to\infty} \left\|x_n\right\| = |||\varphi||| \cdot \left\|x\right\|,$$

showing that $|||\widehat{\varphi}||| \leqslant |||\varphi|||$.

To prove the converse inequality, note that for $y \in V$, we have $\widehat{\varphi}(y) = \varphi(y)$ and therefore

$$|||\widehat{\varphi}||| = \sup_{x\in E\setminus\{0\}} \frac{\left\|\widehat{\varphi}(x)\right\|}{\left\|x\right\|} \geqslant \sup_{x\in V\setminus\{0\}} \frac{\left\|\widehat{\varphi}(x)\right\|}{\left\|x\right\|} = |||\varphi|||.$$

Hence the equality $|||\widehat{\varphi}||| = |||\varphi|||$ holds as stated.

EXERCISES

♦ **Exercise 10.1 (Fourier transform of the gaussian)** Using the link between the Fourier transform and derivation, find a differential equation satisfied by the Fourier transform of the function $x \longmapsto \exp(-\pi x^2)$. Solve this differential equation and deduce from this the Fourier transform of this function. Recall that

$$\int_{-\infty}^{\infty} \mathrm{e}^{-\pi x^2}\,\mathrm{d}x = 1.$$

Deduce from this the Fourier transform of the centered gaussian with standard deviation σ:

$$f_\sigma(x) = \frac{1}{\sigma\sqrt{2\pi}} \exp\left(-\frac{x^2}{2\sigma^2}\right).$$

♦ **Exercise 10.2** Using the convolution theorem for functions $f, g \in L^2$ and the preceding exercise, compute the convolution product of a centered gaussian f_σ with standard deviation σ and a centered gaussian $f_{\sigma'}$ with standard deviation σ'.

♦ **Exercise 10.3** Let $\alpha \in \mathbb{C}$ such that $\operatorname{Re}(\alpha) > 0$, and let $n \in \mathbb{N}^*$. Compute the Fourier transform of the function

$$x \longmapsto f(x) = \frac{1}{(2\pi \mathrm{i} x - \alpha)^{n+1}}.$$

In particular, show that $\widetilde{f}(\nu)$ is zero for $\nu < 0$.

♦ **Exercise 10.4** Compute the following integrals, using the known properties of convolution and the Parseval-Plancheral formula, and by finding a relation with the functions Π and $\Lambda = \Pi * \Pi$:

$$\int_{-\infty}^{+\infty} \frac{\sin \pi x}{\pi x}\,\mathrm{d}x, \qquad \int_{-\infty}^{+\infty} \left(\frac{\sin \pi x}{\pi x}\right)^2 \mathrm{d}x, \qquad \int_{-\infty}^{+\infty} \left(\frac{\sin \pi x}{\pi x}\right)^3 \mathrm{d}x.$$

♦ **Exercise 10.5 (sine and cosine Fourier transforms)** Let $f : \mathbb{R} \to \mathbb{R}$ be an *odd* integrable function. Define

$$\widetilde{f}_s(\nu) \stackrel{\text{def}}{=} 2 \int_0^{+\infty} f(t) \sin(2\pi\nu t)\,\mathrm{d}t.$$

Show that

$$f(t) = 2 \int_0^{+\infty} \widetilde{f}_s(\nu) \sin(2\pi\nu t)\,\mathrm{d}\nu.$$

The function $\widetilde{f}_s$ thus defined is called the **sine Fourier transform** of f, and f is the **inverse sine Fourier transform** of $\widetilde{f}_s$. These transforms can be extended to arbitrary (integrable) functions, the values of which are only of interest for $t > 0$ or $\nu > 0$.

Similarly, show that if f is even and if the **cosine Fourier transform** of f is defined by

$$\widetilde{f}_c(\nu) \stackrel{\text{def}}{=} 2 \int_0^{+\infty} f(t) \cos(2\pi\nu t)\,\mathrm{d}t,$$

then we have

$$f(t) = 2 \int_0^{+\infty} \widetilde{f}_c(\nu) \cos(2\pi\nu t)\,\mathrm{d}\nu.$$

Compute the cosine Fourier transform of $x \longmapsto \mathrm{e}^{-\alpha x}$, where α is a strictly positive real number.

Deduce then the identity

$$\forall t > 0 \quad \forall \alpha > 0 \qquad \int_0^{+\infty} \frac{\cos 2\pi \nu t}{\nu^2 + \alpha^2}\, \mathrm{d}\nu = \frac{\pi}{2\alpha} \mathrm{e}^{-2\pi\alpha t}.$$

♦ **Exercise 10.6** Given functions $a : \mathbb{R} \to \mathbb{R}$ and $b : \mathbb{R} \to \mathbb{R}$, and assuming their Fourier transforms are defined, solve the integral equation

$$y(t) = \int_{-\infty}^{+\infty} y(s)\, a(t-s)\, \mathrm{d}s + b(t).$$

♦ **Exercise 10.7 (proof of Theorem 10.25)** Let $f \in L^1(\mathbb{R})$ be a function which is differentiable with a derivative which is Lebesgue-integrable: $f' \in L^1(\mathbb{R})$. Show that $\lim\limits_{x \to \pm\infty} f(x) = 0$.

SOLUTIONS

♦ **Solution of exercise 10.1.** Define $f(x) = \mathrm{e}^{-\pi x^2}$, which is an integrable function, and let $\widetilde{f}$ be its Fourier transform. From

$$\forall x \in \mathbb{R} \qquad f'(x) + 2\pi x f(x) = 0,$$

we deduce, evaluating the Fourier transform,

$$\forall \nu \in \mathbb{R} \qquad 2\pi\nu \widetilde{f}(\nu) + \frac{\mathrm{d}}{\mathrm{d}\nu}\widetilde{f}(\nu) = 0,$$

which has solutions given by $\widetilde{f}(\nu) = A\, \mathrm{e}^{-\pi\nu^2}$ for some constant A to be determined.

Since $\widetilde{f}(0) = \int_{-\infty}^{+\infty} f(x)\, \mathrm{d}x = 1$, it follows that $A = 1$.

By a simple change of variable, it also follows that

$$\frac{1}{\sigma\sqrt{2\pi}} \exp\left(-\frac{x^2}{2\sigma^2}\right) \xrightarrow{\text{F.T.}} \mathrm{e}^{-2\pi^2\sigma^2\nu^2},$$

a result which will be very useful in probability theory.

Compare this with the method given in Exercise 4.9 on page 125, which uses the Cauchy formula in the complex plane.

♦ **Solution of exercise 10.2.** Using the convolution formula, we get $\mathscr{F}\left[f_\sigma * f_{\sigma'}\right] = \widetilde{f_\sigma} \cdot \widetilde{f_{\sigma'}}$, and since f_σ and $f_{\sigma'}$ are both square integrable, their Fourier transforms are also square integrable. Using the Cauchy-Schwarz inequality, it follows that $\widetilde{f_\sigma} \cdot \widetilde{f_{\sigma'}}$ is integrable; there its conjugate Fourier transform is defined and

$$\begin{aligned} f_\sigma * f_{\sigma'}(t) &= \overline{\mathscr{F}}\left[\widetilde{f_\sigma} \cdot \widetilde{f_{\sigma'}}\right](t) = \overline{\mathscr{F}}\left[\mathrm{e}^{-2\pi^2\sigma^2\nu^2} \cdot \mathrm{e}^{-2\pi^2\sigma'^2\nu^2}\right](t) \\ &= \overline{\mathscr{F}}\left[\mathrm{e}^{-2\pi(\sigma^2+\sigma'^2)\nu^2}\right](t) = f_{\sqrt{\sigma^2+\sigma'^2}}(t). \end{aligned}$$

The squares of the standard deviations (i.e., the variances) add up when taking the convolution of gaussians.

♦ **Solution of exercise 10.3.** We first notice that f is integrable: $f \in L^1$. So its Fourier transform is defined by the integral

$$\widetilde{f}(\nu) = \int \frac{\mathrm{e}^{-2\pi \mathrm{i} \nu x}}{(2\pi \mathrm{i} x - \alpha)^{n+1}}\, \mathrm{d}x.$$

We compute this using the method of residues. The only pole of the integrand in $\mathbb{C}$ is at $x = \mathrm{i}\alpha/2\pi$.

• If $\nu < 0$, we close the contour in the lower half-plane to apply Jordan's second lemma, and no residue appears; therefore the function $\widetilde{f}$ vanishes identically for $\nu < 0$.

• For $\nu > 0$, we close the contour in the upper half-plane and, by applying the residue formula, we get $\widetilde{f}(\nu) = (-1)^{n+1}\nu^n \mathrm{e}^{\alpha\nu}/n!$.

To conclude, the Fourier transform of f can be written

$$\widetilde{f}(\nu) = (-1)^{n+1} \frac{\nu^n}{n!} \mathrm{e}^{\alpha\nu} H(\nu).$$

♦ **Solution of exercise 10.5.** An immediate calculation shows that if f is an odd function, we have $\widetilde{f}_s = (\mathrm{i}/2)\widetilde{f}$. Moreover, $\widetilde{f}_s$ is odd; hence the Fourier inversion formula yields

$$f(t) = \int_{-\infty}^{+\infty} \widetilde{f}(\nu)\,\mathrm{e}^{2\pi\mathrm{i}\nu t}\,\mathrm{d}\nu = \int_{-\infty}^{+\infty} \widetilde{f}_s(\nu)\sin(2\pi\nu t)\,\mathrm{d}\nu = 2\int_{0}^{+\infty} \widetilde{f}_s(\nu)\sin(2\pi\nu t)\,\mathrm{d}\nu.$$

By definition, the cosine Fourier transform of $x \longmapsto \mathrm{e}^{-\alpha x}$ is given by

$$2\int_{0}^{+\infty} \mathrm{e}^{-\alpha t}\cos(2\pi\nu t)\,\mathrm{d}t = \frac{2\alpha}{\alpha^2 + 4\pi^2\nu^2}.$$

The inverse cosine Fourier transform leads to $x \longmapsto \mathrm{e}^{-\alpha|x|}$, which is the original input function, after it has been made *even*. The formula follows.

♦ **Solution of exercise 10.6.** Assume that the Fourier transform of y also exists. Denote by Y, A, and B the Fourier transforms of y, a, and b, respectively. Then, by the convolution theorem, we have

$$Y(\nu) = Y(\nu)A(\nu) + B(\nu) \qquad \text{and hence} \qquad Y(\nu) = \frac{B(\nu)}{1 - A(\nu)}.$$

So, under the assumption that it exists, the solution is given by the inverse Fourier transform of the function $\nu \longmapsto Y(\nu)$ thus defined.

♦ **Solution of exercise 10.7.** Since f' is integrable, we can write

$$\int_{0}^{+\infty} f'(t)\,\mathrm{d}t = \lim_{x\to+\infty} \int_{0}^{x} f'(t)\,\mathrm{d}t.$$

Moreover, for any $x \in \mathbb{R}$ we have

$$\int_{0}^{x} f'(t)\,\mathrm{d}t = f(x) - f(0),$$

which shows that f admits a limit at $+\infty$, and similarly at $-\infty$. If those limits were nonzero, f would certainly not be integrable on $\mathbb{R}$.

Chapter 11

Fourier transform of distributions

11.1 Definition and properties

The next objective is to define the Fourier transform for distributions. This is interesting for at least two reasons:

1. it makes it possible to define the Fourier transform of distributions such as δ or Ш;

2. possibly, it provides an extension of the Fourier transform to a larger class of *functions*; in particular, functions which are neither in $L^1(\mathbb{R})$, nor in $L^2(\mathbb{R})$, but which are of constant use in physics, such as the Heaviside function.

In order to define the Fourier transform of a distribution, we begin, as is our custom, by restricting to the special case of a regular distribution. Consider therefore a locally integrable function. But then, unfortunately, we realize that "being locally integrable" is not, for a function, a sufficient condition for the Fourier transform to be defined.

Restricting even more, consider a function $f \in L^1$. Being integrable, it is also locally integrable and has an associated regular distribution, also denoted f. Its Fourier transform $\widetilde{f}$ is continuous, hence also locally integrable, and defines therefore a distribution $\widetilde{f}$, which acts on a test function φ by the formula

$$\langle \widetilde{f}, \varphi \rangle = \int \widetilde{f}(t)\, \varphi(t)\, \mathrm{d}t = \int \left(\int f(x)\, \mathrm{e}^{-2\pi\mathrm{i}xt}\, \mathrm{d}x \right) \varphi(t)\, \mathrm{d}t,$$

which is the same, using Fubini's theorem to exchange the order of integration (this is allowed because f is integrable and φ is also), as

$$\langle \widetilde{f}, \varphi \rangle = \int f(x) \left(\int \mathrm{e}^{-2\pi\mathrm{i}xt} \varphi(t)\, \mathrm{d}t \right) \mathrm{d}x = \int f(x)\, \widetilde{\varphi}(x)\, \mathrm{d}x = \langle f, \widetilde{\varphi} \rangle .$$

This computation justifies the following definition:

DEFINITION 11.1 The **Fourier transform of a distribution** T is defined by its action on compactly supported infinitely differentiable test functions φ, which is given by

$$\langle \mathscr{F}[T], \varphi \rangle \stackrel{\text{def}}{=} \langle T, \mathscr{F}[\varphi] \rangle.$$

Sometimes the notation $\widetilde{T}$ is used instead of $\mathscr{F}[T]$:

$$\langle \widetilde{T}, \varphi \rangle = \langle T, \widetilde{\varphi} \rangle.$$

Remark 11.2 It is known that if φ has compact support and is nonzero, $\mathscr{F}[\varphi]$ does not have compact support; this means that the quantity $\langle T, \widetilde{\varphi} \rangle$ is not always defined. Hence not all distribution have an associated Fourier transform. For this reason, the notion of *tempered distribution* is introduced.

11.1.a Tempered distributions

DEFINITION 11.3 ($\mathscr{S}'$) Recall that a **rapidly decaying function** is a function f such that $\lim\limits_{x \to \pm\infty} \left| x^k f(x) \right| = 0$ for all $k \in \mathbb{N}$, and that the Schwartz space (of functions which are infinitely differentiable and rapidly decaying along with all their derivatives) is denoted $\mathscr{S}$.

The topological dual of $\mathscr{S}$ (i.e., the space of continuous linear forms on $\mathscr{S}$) is denoted $\mathscr{S}'$. A **tempered distribution** is an element of $\mathscr{S}'$.

Hence, to say that T is a tempered distribution implies that T is continuous, that is, that

$$\text{if} \quad \varphi_n \xrightarrow[n\to\infty]{} 0 \quad \text{in } \mathscr{S}, \qquad \text{then} \quad \langle T, \varphi_n \rangle \xrightarrow[n\to\infty]{} 0 \quad \text{in } \mathbb{C}.$$

PROPOSITION 11.4 *Any tempered distribution is a "classical" distribution. In other words, we have* $\mathscr{S}' \subset \mathscr{D}'$.

PROOF. Note first that $\mathscr{D} \subset \mathscr{S}$. Let now $T \in \mathscr{S}'$ be a tempered distribution; it acts on any test function $\varphi \in \mathscr{S}$ and hence, *a fortiori*, on any function $\varphi \in \mathscr{D}$, so T can be restricted to a linear functional on $\mathscr{D}$.

Moreover, if $(\varphi_n)_{n\in\mathbb{N}}$ converges to 0 in $\mathscr{D}$, it converges also to 0 in $\mathscr{S}$ by the definition of those two convergences, so $\langle T, \varphi_n \rangle$ tends to 0 in $\mathbb{C}$. This shows that T is continuous, and hence that $T \in \mathscr{D}'$.

Now let us look fc. examples of tempered distributions.

Let f be a locally integrable function, which increases at most like a power function $|x|^k$, where k is some fixed integer, i.e., we have $f(x) = O(|x|^k)$ in the neighborhood of $\pm\infty$. Then for any function φ which decays rapidly, the integral $\int f\varphi$ exists. The map

$$\varphi \in \mathscr{S} \longmapsto \int_{-\infty}^{+\infty} f(x)\,\varphi(x)\,\mathrm{d}x$$

is also linear, of course, and continuous – which the reader is invited to check. Therefore, it defines a tempered distribution, also denoted f.

DEFINITION 11.5 (Slowly increasing function) A function $f : \mathbb{R} \to \mathbb{C}$ is **slowly increasing** if it increases at most like a power of $|x|$ at infinity[1].

PROPOSITION 11.6 *Any locally integrable slowly increasing function defines a regular tempered distribution.*

♦ **Exercise 11.1** Let T be a tempered distribution. Show that for any $k \in \mathbb{N}$, $x^k T$ and $T^{(k)}$ are also tempered distributions.

♦ **Exercise 11.2** Show that the distribution $\exp(x^4)\,\text{Ш}(x)$ is not tempered.

(Solution page 326)

11.1.b Fourier transform of tempered distributions

Let T be a tempered distribution. For $\varphi \in \mathscr{S}$, that is, for a $\mathscr{C}^\infty$ function which is rapidly decaying, and with all its derivatives also rapidly decaying, the Fourier transform $\mathscr{F}[\varphi]$ is also in the space $\mathscr{S}$ (Theorem 10.32). Hence the quantity $\langle T, \mathscr{F}[\varphi] \rangle$ is defined and, consequently, it is possible to define the Fourier transform of T.

THEOREM 11.7 *Any tempered distribution has a Fourier transform in the sense of distributions, which is also tempered.*

Remark 11.8 Does the Fourier transform of tempered distributions defined in this manner coincide with the Fourier transform of functions, in the case of a regular distribution?

---

[1] A slowly increasing function, multiplied by a rapidly decaying function, is therefore integrable; the terminology is coherent!

Yes indeed. Let f be an integrable function, and denote this time by $T(f)$ the associated regular distribution (to make the distinction clearly). The Fourier transform of g (in the sense of functions) is a bounded function $\widetilde{f}$, which therefore defines a regular tempered distribution. The associated distribution, denoted $T(\widetilde{f})$, satisfies therefore for any $\varphi \in \mathscr{S}$

$$\langle T(\widetilde{f}), \varphi \rangle = \int \widetilde{f}(x)\,\varphi(x)\,\mathrm{d}x = \int f(x)\,\widetilde{\varphi}(x)\,\mathrm{d}x = \langle T(f), \widetilde{\varphi} \rangle = \langle \mathscr{F}\,[T(f)], \varphi \rangle.$$

This implies that $T(\widetilde{f})$ is indeed equal to the Fourier transform, in the sense of distributions, of $T(f)$.

Similar reasoning holds for $f \in L^2(\mathbb{R})$ (Exercise!) and shows that the Fourier transform defined on $\mathscr{S}'$ coincides with that defined on $L^2(\mathbb{R})$.

The properties of the Fourier transform with respect to scaling, translation, and differentiation remain valid in the setting of tempered distributions:

THEOREM 11.9 *Let T be a tempered distribution. Then*

$$\mathscr{F}\,[T'] = (2\pi\mathrm{i}\nu)\,\mathscr{F}\,[T], \qquad \mathscr{F}\,[T^{(n)}] = (2\pi\mathrm{i}\nu)^n\,\mathscr{F}\,[T],$$
$$\mathscr{F}\,[T(x-a)] = \mathrm{e}^{-2\pi\mathrm{i}\nu a}\,\mathscr{F}\,[T], \qquad \mathscr{F}\,[\mathrm{e}^{2\pi\mathrm{i}ax}\,T] = \widetilde{T}(\nu - a).$$

PROOF. Take the first property, for instance. Before starting, remember that the letters x and ν, usually used to denote the variable for a function and its Fourier transform, respectively, are in fact interchangeable. So, it is perfectly legitimate to write $\langle \mathscr{F}\,[T](\nu), \varphi(\nu) \rangle$ as well as $\langle \mathscr{F}\,[T](x), \varphi(x) \rangle$.

This being said, let T be a tempered distribution and $\varphi \in \mathscr{S}$. Then we have

$$\begin{aligned}\langle \mathscr{F}\,[T'], \varphi \rangle &= \langle T', \mathscr{F}\,[\varphi] \rangle \\ &= -\langle T, \mathscr{F}\,[-2\mathrm{i}\pi x\,\varphi(x)] \rangle \\ &= -\langle \mathscr{F}\,[T], -2\pi\mathrm{i}x\,\varphi(x) \rangle = \langle 2\pi\mathrm{i}x\mathscr{F}\,[T](x), \varphi(x) \rangle,\end{aligned}$$

which is the stated formula (with the variable x replacing ν).

The other statements are proved in an identical manner and are left as warm-up exercises.

Finally, here is the theorem linking convolution and Fourier transform.

THEOREM 11.10 *Let S be a tempered distribution and T a distribution with bounded support. Then the convolution product $S * T$ makes sense and is a tempered distribution, and therefore it has a Fourier transform which is given by*

$$\boxed{\mathscr{F}\,[S * T] = \widetilde{S} \cdot \widetilde{T}.}$$

If S is a regular distribution associated to a function of $\mathscr{C}^\infty$ class, and T a distribution such that the product $S \cdot T$ exists and is a tempered distribution, then its Fourier transform is given by

$$\boxed{\mathscr{F}\,[S \cdot T] = \widetilde{S} * \widetilde{T}.}$$

Remark 11.11 The preceding formulas need some comment. Indeed, the product of two distributions is generally not defined (Jean-François COLOMBEAU worked on the possibility of defining such a product, but his results [23] are far from the scope of this book). However, it is defined when one of the two distributions is regular and associated with a function of class $\mathscr{C}^\infty$.

In the first formula ($\mathscr{F}[S * T] = \widetilde{S} \cdot \widetilde{T}$), the distribution $\widetilde{T}$ is really a regular distribution, associated with the function f, of class $\mathscr{C}^\infty$, defined by

$$f(\nu) = \left\langle T(t), \mathrm{e}^{-2\pi\mathrm{i}\nu t} \right\rangle$$

(the right-hand side is well defined for all ν because T has bounded support).

11.1.c Examples

The first important example of the Fourier transform of a tempered distribution is that of a constant function.

THEOREM 11.12 ($\mathscr{F}[1] = \delta$) *The Fourier transform of the constant function* $\mathbf{1}$ *is the Dirac distribution.*

$$\boxed{\mathbf{1} \xrightarrow{\text{F.T.}} \delta(\nu).}$$

PROOF. Let $\varphi \in \mathscr{S}$ be a Schwartz function. We have, by definition and the Fourier inversion formula,

$$\left\langle \widetilde{\mathbf{1}}, \varphi \right\rangle \stackrel{\text{def}}{=} \left\langle \mathbf{1}, \widetilde{\varphi} \right\rangle = \int \widetilde{\varphi}(\nu)\, \mathrm{d}\nu.$$

Furthermore, as $\widetilde{\varphi}$ is in $\mathscr{S}$, the inversion formula for the Fourier transform yields

$$\int \widetilde{\varphi}(\nu)\, \mathrm{d}\nu = \overline{\mathscr{F}}\left[\widetilde{\varphi}\right](0) = \varphi(0) = \langle \delta, \varphi \rangle .$$

This theorem indicates that for a constant function, the "signal" contains a single frequency, namely the zero frequency $\nu = 0$. Similarly, we have:

THEOREM 11.13 *The Fourier transforms of the trigonometric functions are given by*

$$\mathscr{F}\left[\mathrm{e}^{2\pi\mathrm{i}\nu_0 x}\right] = \delta(\nu - \nu_0),$$

$$\mathscr{F}\left[\cos(2\pi\nu_0 x)\right] = \frac{1}{2}\Big[\delta(\nu - \nu_0) + \delta(\nu + \nu_0)\Big],$$

$$\mathscr{F}\left[\sin(2\pi\nu_0 x)\right] = \frac{1}{2\mathrm{i}}\Big[\delta(\nu - \nu_0) - \delta(\nu + \nu_0)\Big]$$

for all $\nu_0 \in \mathbb{R}$.

This result is quite intuitive: for the functions considered, only the frequency ν_0 occurs, but it may appear with a sign ± 1 (see Section 13.4, page 365).

THEOREM 11.14 ($\mathscr{F}[\delta] = 1$) *The Fourier transform of the Dirac distribution is the constant function* $\mathbf{1} : x \mapsto 1$.

$$\boxed{\delta(x) \xrightarrow{\text{F.T.}} \mathbf{1}.}$$

The Dirac distribution is so "spiked" that all frequencies are represented.

PROOF. Let $\varphi \in \mathscr{S}$ be a Schwartz function. We have

$$\langle \widetilde{\delta}, \varphi \rangle \stackrel{\text{def}}{=} \langle \delta, \widetilde{\varphi} \rangle = \widetilde{\varphi}(0).$$

But, from the definition of $\widetilde{\varphi}$, it follows that

$$\widetilde{\varphi}(0) = \int \varphi(x) \underbrace{\mathrm{e}^{-2\pi \mathrm{i} 0 \cdot x}}_{\mathbf{1}(x)} \, \mathrm{d}x = \langle \mathbf{1}, \varphi \rangle .$$

Using Theorem 11.9 for differentiation and translation, we deduce:

THEOREM 11.15 *Let $m \in \mathbb{N}$ and $a \in \mathbb{R}$. Then we have*

$$\mathscr{F}\left[\delta'\right] = 2\pi \mathrm{i}\nu, \qquad \mathscr{F}\left[\delta^{(m)}\right] = (2\pi \mathrm{i}\nu)^m,$$
$$\mathscr{F}\left[\delta(x-a)\right] = \mathrm{e}^{2\pi \mathrm{i}\nu a}.$$

We now compute the Fourier transform of the Heaviside distribution H (which is tempered). From $H' = \delta$, we deduce, using the differentiation property of the Fourier transform, that $2\pi \mathrm{i}\nu \cdot \widetilde{H}(\nu) = 1$. But, according to Theorems 8.6 on page 229 and 7.28 on page 191, the distribution solutions of the equation $\nu \cdot T(\nu) = 1$ are of the form

$$\alpha\delta(\nu) + \frac{1}{2\pi \mathrm{i}} \operatorname{pv} \frac{1}{\nu}, \qquad \text{with} \quad \alpha \in \mathbb{C}.$$

To determine the constant α, note that $H - \frac{1}{2}$ is a real-valued function which is odd, so that its Fourier transform, namely, $\widetilde{H} - \frac{1}{2}\delta$, must also be odd. Since the δ distribution is even, it follows that $\alpha = \frac{1}{2}$. Denoting by "sgn" the sign function, which is given by $\operatorname{sgn}(x) = 2H(x) - 1$, we further obtain:

THEOREM 11.16 (Fourier transform of H and $\operatorname{pv} \frac{1}{x}$) *The Fourier transform of the Heaviside distribution is*

$$\mathscr{F}\left[H(x)\right] = \frac{1}{2\pi \mathrm{i}} \operatorname{pv}\left(\frac{1}{\nu}\right) + \frac{1}{2}\,\delta.$$

Similarly,

$$\mathscr{F}\left[H(-x)\right] = \frac{\mathrm{i}}{2\pi} \operatorname{pv}\left(\frac{1}{\nu}\right) + \frac{1}{2}\,\delta.$$

Moreover, the distribution $\operatorname{pv} \frac{1}{x}$ *satisfies*

$$\mathscr{F}\left[\operatorname{pv} \frac{1}{x}\right] = -\mathrm{i}\pi \operatorname{sgn} \nu, \qquad \mathscr{F}\left[\operatorname{sgn} x\right] = \frac{1}{\mathrm{i}\pi} \operatorname{pv} \frac{1}{\nu}.$$

11.1.d Higher-dimensional Fourier transforms

Consider now distributions defined on $\mathbb{R}^n$. For a function $f : \mathbb{R}^n \to \mathbb{R}$ (with suitable conditions), its Fourier transform is defined on $\mathbb{R}^n$ by

$$\tilde{f}(\boldsymbol{\nu}) \stackrel{\text{def}}{=} \int \cdots \int_{\mathbb{R}^n} f(\boldsymbol{x})\, \mathrm{e}^{-2\pi \mathrm{i} \boldsymbol{\nu}\cdot \boldsymbol{x}}\, \mathrm{d}^n \boldsymbol{x}$$

for n-uples $\boldsymbol{x}$ and $\boldsymbol{\nu}$

The vector $\boldsymbol{\nu}$ is a "frequency vector." It is often advantageous to replace it by the "wave vector" $\boldsymbol{k} = 2\pi\, \boldsymbol{\nu}$. Then the conventions defining the Fourier transform are

$$F(\boldsymbol{k}) \stackrel{\text{def}}{=} \int \cdots \int_{\mathbb{R}^n} f(\boldsymbol{x})\, \mathrm{e}^{-\mathrm{i}\boldsymbol{k}\cdot \boldsymbol{x}}\, \mathrm{d}^n \boldsymbol{x}, \quad f(\boldsymbol{x}) = \frac{1}{(2\pi)^n} \int \cdots \int_{\mathbb{R}^n} F(\boldsymbol{k})\, \mathrm{e}^{\mathrm{i}\boldsymbol{k}\cdot \boldsymbol{x}}\, \mathrm{d}^n \boldsymbol{k}.$$

Those definitions provide a way of defining the Fourier transform of a tempered distribution on $\mathbb{R}^n$, imitating the process in Definition 11.1. The differentiation theorem is easily generalized, and has the following consequence, for instance:

THEOREM 11.17 (Fourier transform of the laplacian) *Let $T(\boldsymbol{x})$ be a distribution in $\mathbb{R}^3$, which admits a Fourier transform. Then the Fourier transform of its laplacian $\triangle T(\boldsymbol{x})$ is $-4\pi\boldsymbol{\nu}^2\, \tilde{T}(\boldsymbol{\nu})$.*

Consider now the following partial differential equation (coming from particle physicis, plasma physics, or ionic solutions, and explained in Problem 5 on page 325):

$$\left(\triangle - m^2\right) f(\boldsymbol{x}) = -4\pi\, \delta(\boldsymbol{x}).$$

Taking the Fourier transform of this relation, it follows that

$$(-4\pi\boldsymbol{\nu}^2 - m^2)\, \tilde{f}(\boldsymbol{\nu}) = -4\pi.$$

This is now an *algebraic* equation, the solution of which is obvious[2]:

$$\tilde{f}(\boldsymbol{\nu}) = \frac{4\pi}{m^2 + 4\pi\boldsymbol{\nu}^2}. \tag{11.1}$$

[2] But is that really true? Not entirely! Since we are looking for distribution solutions, we must remember that there are many distribution solutions to the equation

$$(m^2 + 4\pi\boldsymbol{\nu}^2) \cdot T(\boldsymbol{\nu}) = 0:$$

any distribution which is proportional to $\delta(m^2 + 4\pi\boldsymbol{\nu}^2)$ will do. But, if the function $m^2 + 4\pi\boldsymbol{\nu}^2$ does not vanish on $\mathbb{R}^3$, on the other hand it does vanish for complex values of $\boldsymbol{\nu}$. The fact that $\tilde{f}(\boldsymbol{\nu})$ takes nonzero values for complex values of $\boldsymbol{\nu}$ is more easily interpreted in terms of a (bilateral) Laplace transform than in terms of a Fourier transform. For instance, a solution is given by $f(x, y, z) = \mathrm{e}^{mx}$, which does indeed satisfy $(\Delta - m^2) f(\boldsymbol{x}) = 0$, but this f has no Fourier transform. The solution (11.1) is, in fact, the only one which vanishes at infinity. It is therefore the one we will keep.

There only remains to compute the inverse Fourier transform of this function to obtain a solution to the equation we started with. For this, the method of residues is still the most efficient. This computation is explained in detail in the physics problem on page 325. In addition, the whole of Chapter 15 is devoted to the same subject in various physical cases. The result is

$$f(\boldsymbol{r}) = \frac{\mathrm{e}^{-m\|\boldsymbol{r}\|}}{\|\boldsymbol{r}\|}.$$

Letting m to 0, we deduce[3] the following result, of great use in electromagnetism:

THEOREM 11.18 (Fourier transform of the Coulomb potential) *In $\mathbb{R}^3$, the Fourier transform of the Coulomb potential $\boldsymbol{r} \mapsto 1/\|\boldsymbol{r}\|$ is given by*

$$F(\boldsymbol{k}) = \iiint_{\mathbb{R}^3} \frac{1}{\|\boldsymbol{x}\|} \mathrm{e}^{-\mathrm{i}\boldsymbol{k}\cdot\boldsymbol{x}} \, \mathrm{d}^3\boldsymbol{x} = \frac{4\pi}{\boldsymbol{k}^2}.$$

Remark 11.19 The function $\boldsymbol{k} \mapsto 1/\|\boldsymbol{k}\|^2$ is locally integrable (in $\mathbb{R}^3$), and the right-hand side is therefore a regular distribution.

11.1.e Inversion formula

The inversion formula, known for functions that belong to L^1, has a generalization for tempered distributions. First, let us define the conjugate Fourier transform of a tempered distribution $T \in \mathscr{S}'$ by

$$\forall \varphi \in \mathscr{S} \qquad \left\langle \overline{\mathscr{F}}[T], \varphi \right\rangle = \left\langle T, \overline{\mathscr{F}}[\varphi] \right\rangle = \left\langle T(\nu), \widetilde{\varphi}(-\nu) \right\rangle.$$

Then $\overline{\mathscr{F}}$ is the inverse of $\mathscr{F}$, that is to say,

THEOREM 11.20 (Inversion formula in $\mathscr{S}'$) *Let $T \in \mathscr{S}'$ be a tempered distribution. Its Fourier transform $U = \mathscr{F}[T]$ is also a tempered distribution and $T = \overline{\mathscr{F}}[U]$.*

---

[3] With no rigorous justification. Here is how to proceed to check this. Start by regularizing the Coulomb potential by putting, for $R > 0$,

$$f_R(\boldsymbol{r}) = \begin{cases} 1/r & \text{if } r \leqslant R, \\ 1/R & \text{if } r \geqslant R. \end{cases}$$

This regularized Coulomb potential converges to $f : \boldsymbol{r} \mapsto 1/r$ in the sens of distributions (the verification of this is immediate). Since the Fourier transform is a continuous operator, the Fourier transform of f_R tends to $\widetilde{f}$ as R tends to infinity. After a few lines of computation (isolating the constant part $1/R$), the Fourier transform of f_R is computed explicitly and gives

$$f_R(\boldsymbol{k}) = \frac{(2\pi)^3}{R}\,\delta(\boldsymbol{k}) + \frac{4\pi}{k^2} - \frac{4\pi}{R\,k^3}\sin(kR).$$

The first and third terms tend to 0 in the sense of distributions as $[R \to +\infty]$, as the reader can easily show.

11.2

The Dirac comb

11.2.a Definition and properties

Recall that the "Dirac comb" is the distribution defined by

$$\text{Ш}(x) = \sum_{n=-\infty}^{+\infty} \delta(x-n).$$

We can then show –and this is a fundamental result – that

THEOREM 11.21 *The Dirac comb is a tempered distribution; hence it has a Fourier transform in the sense of distributions, which is also equal to the Dirac comb:*

$$\boxed{\text{Ш}(x) \xrightarrow{\text{F.T.}} \text{Ш}(\nu).}$$

This formula can be stated differently; since Ш is defined as a sum of Dirac distribiutions, we have

$$\mathscr{F}\left[\text{Ш}\right](\nu) = \mathscr{F}\left[\sum_{n=-\infty}^{+\infty} \delta_n\right](\nu) = \sum_{n=-\infty}^{+\infty} \mathrm{e}^{-2\pi i n\nu}.$$

Renaming "x" the free variable "ν" (which is perfectly permissible since the names of the variables are only a matter of choice!) and replacing n by $-n$, as we can, we obtain therefore:

$$\boxed{\sum_{n=-\infty}^{+\infty} \mathrm{e}^{2\pi i n x} = \sum_{n=-\infty}^{+\infty} \delta(x-n).} \tag{11.2}$$

This formula can be interpreted also as the **Fourier series expansion of Ш**. Notice that, in contrast with the case of "usual" periodic functions, the Fourier coefficients of this periodic distribution do not converge to 0 (they are all equal to 1).

PROOF OF THEOREM 11.21. The theorem is equivalent to the formula (11.2), and we will prove the latter.

Let $f(t) = \left[t^2\Pi(t)\right] * \text{Ш}(t)$. This is indeed a function, and its graph is

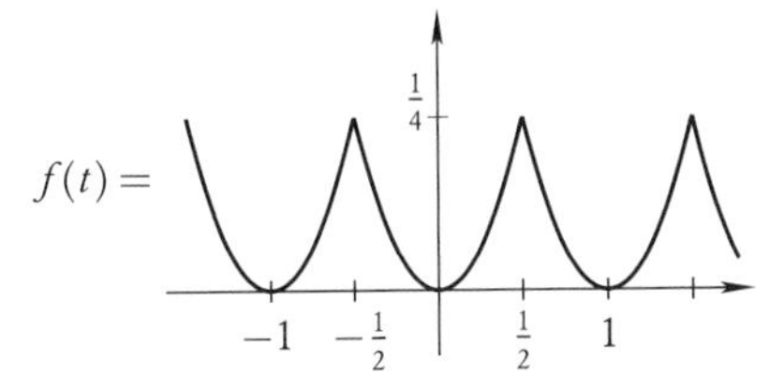

As a periodic function, f can be expanded in Fourier series. Since f is even, only the cosine coefficients a_n occur in this expansion, and after integrating by parts twice, we get

$$\begin{aligned}
a_n &= 4\int_0^{1/2} x^2 \cos(2\pi n x)\,\mathrm{d}x \\
&= -4\int_0^{1/2} \frac{x}{\pi n}\sin(2\pi n x)\,\mathrm{d}x + \left[\frac{x^2\sin(2\pi n x)}{2\pi n}\right]_0^{1/2} \quad \text{(the boundary terms vanish)} \\
&= -\frac{4}{\pi n}\int_0^{1/2} \frac{\cos(2\pi n x)}{2\pi n}\,\mathrm{d}x + \left[\frac{8x}{(2\pi n)^2}\cos(2\pi n x)\right]_0^{1/2}, \quad \text{(and the integral also)}
\end{aligned}$$

that is,

$$a_n = \frac{(-1)^n}{\pi^2 n^2}.$$

Moreover, differentiating twice the function f, in the sense of distributions, we obtain first

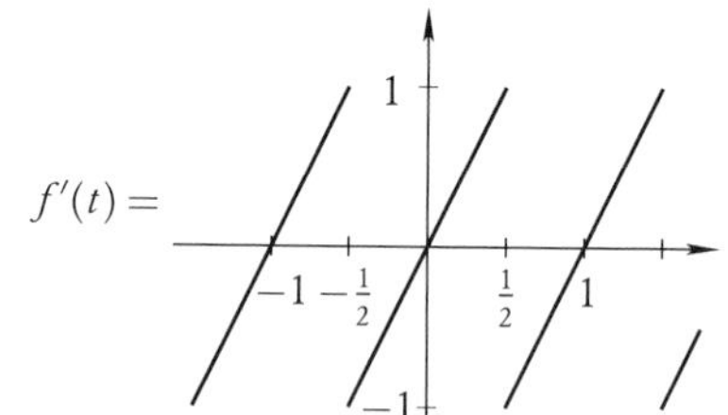

and then

$$f''(t) = 2 - 2\sum_{n=-\infty}^{+\infty} \delta\left(x - n - \tfrac{1}{2}\right) = 2 - 2\,\text{Ш}\left(x - \tfrac{1}{2}\right).$$

Hence, differentiating twice the series $f(t) = \sum_{n=0}^{\infty} a_n \cos(2\pi n t)$ yields

$$2 - 2\text{Ш}(x - \tfrac{1}{2}) = -4\sum_{n=1}^{+\infty}(-1)^n \mathrm{e}^{2\pi \mathrm{i} n x}$$

$$\text{Ш}(x - \tfrac{1}{2}) = 1 + 2\sum_{n=1}^{+\infty}(-1)^n \mathrm{e}^{2\pi \mathrm{i} n x}$$

and therefore

$$\text{Ш}(x) = 1 + 2\sum_{n=1}^{+\infty}(-1)^n \mathrm{e}^{2\pi \mathrm{i} n x}\mathrm{e}^{\mathrm{i}\pi n} = 1 + 2\sum_{n=1}^{+\infty} \mathrm{e}^{2\pi \mathrm{i} n x} = \sum_{n=-\infty}^{+\infty} \mathrm{e}^{2\pi \mathrm{i} n x}$$

(since Ш is even), which was to be proved.

11.2.b Fourier transform of a periodic function

The case of a periodic function seems at first sight to require some care, since a 2π-periodic function f (for instance), only belongs to $L^1(\mathbb{R})$ or $L^2(\mathbb{R})$ if is almost everywhere zero. However, any function in $L^2[0, a]$ defines a tempered distribution, and therefore has a Fourier transform.

So take $f \in L^2[0,1]$, which can be expanded in a Fourier series

$$f(x) = \sum_{n=-\infty}^{+\infty} c_n \, e^{2\pi i n x},$$

which converges to f in quadratic mean. By linearity, we easily compute the Fourier transform of such a series,

$$\widetilde{f}(\nu) = \sum_{n=-\infty}^{+\infty} c_n \, \delta(\nu - n). \tag{11.3}$$

Hence the Fourier coefficients of a function in $L^2[0,1]$ occur as the "weights" of the Dirac distributions which appear in its Fourier transform. This is what is observed when a *periodic* signal (for instance, a note which is held on a musical instrument) is observed on a modern oscilloscope which computes (via a *fast Fourier transform*) the spectrum of the signal.

11.2.c Poisson summation formula

Let f be a continuous function and T an arbitrary nonzero real number. Assume that the following convolution exists:

$$f(x) * \frac{1}{T} Ш\left(\frac{x}{T}\right) = f(x) * \sum_{n=-\infty}^{+\infty} \delta(x - nT) = \sum_{n=-\infty}^{+\infty} f(x - nT)$$

(it suffices, for instance, that f decays faster than $1/|x|$ at infinity). The Fourier transform of $f(x) * \frac{1}{T} Ш(x/T)$ is $\widetilde{f}(\nu) \cdot \widetilde{Ш}(T\nu)$, or equivalently

$$\sum_{n \in \mathbb{Z}} \widetilde{f}\left(\frac{n}{T}\right) \delta(T\nu - n),$$

the inverse Fourier transform of which is equal to

$$\frac{1}{T} \sum_{n \in \mathbb{Z}} \widetilde{f}\left(\frac{n}{T}\right) e^{2i\pi n x/T}.$$

THEOREM 11.22 (Poisson summation formula) *Let f be a continuous function which has a Fourier transform. When the series make sense, we have*

$$\sum_{n=-\infty}^{+\infty} f(x - nT) = \frac{1}{T} \sum_{n=-\infty}^{+\infty} \widetilde{f}\left(\frac{n}{T}\right) e^{2i\pi n x/T}.$$

In the special case $x = 0$ and $T = 1$, the Poisson summation formula is simply:

$$\boxed{\sum_{n=-\infty}^{+\infty} f(n) = \sum_{n=-\infty}^{+\infty} \widetilde{f}(n).}$$

11.2.d Application to the computation of series

Take, for instance, the problem of computing the sum of the series[4]

$$S \stackrel{\text{def}}{=} \sum_{n=-\infty}^{\infty} \frac{1}{a^2+4\pi^2n^2}\,\frac{1}{b^2+4\pi^2n^2},$$

with $a, b \in \mathbb{R}^{*+}$ and $a \neq b$. First, note that if we define

$$S(\tau) \stackrel{\text{def}}{=} \sum_{n=-\infty}^{\infty} \frac{1}{a^2+4\pi^2n^2}\,\frac{1}{b^2+4\pi^2n^2}\,\mathrm{e}^{2\pi i n\tau} \qquad \text{for all } \tau \in \mathbb{R},$$

then we have $S = S(0)$ and we can write

$$S(\tau) = \sum_{n=-\infty}^{\infty} \widetilde{f}(n)\,\mathrm{e}^{2\pi i n\tau},$$

with

$$\widetilde{f}(\nu) = \frac{1}{4\pi^2\nu^2+a^2}\cdot\frac{1}{4\pi^2\nu^2+b^2} = \widetilde{g}(\nu)\cdot\widetilde{h}(\nu).$$

To exploit the Poisson summation formula, it is necessary to compute the inverse Fourier transform of f and hence those of g and h. We know that

$$\mathrm{e}^{-a|x|} \xrightarrow{\text{F.T.}} \frac{2a}{a^2+4\pi^2\nu^2},$$

and therefore we deduce that

$$g(x) = \frac{\mathrm{e}^{-a|x|}}{2a} \qquad \text{and} \qquad h(x) = \frac{\mathrm{e}^{-b|x|}}{2b}.$$

From this we can compute f: indeed, from $\widetilde{f} = \widetilde{g}\cdot\widetilde{h}$ it follows that $f = g * h$; this convolution product is only an elementary (tedious) computation (see Exercise 7.5 on page 211) and yields

$$f(x) = \frac{1}{4ab}\,\mathrm{e}^{-a|x|} * \mathrm{e}^{-b|x|} = \frac{1}{2}\frac{1}{b^2-a^2}\left[\frac{\mathrm{e}^{-a|x|}}{a} - \frac{\mathrm{e}^{-b|x|}}{b}\right].$$

Hence we have

$$\begin{aligned} S = S(0) &= \sum_{n=-\infty}^{+\infty} \widetilde{f}(n) = \sum_{n=-\infty}^{\infty} f(n) \\ &= \frac{1}{b^2-a^2}\frac{1}{2a}\sum_{n=-\infty}^{\infty} \mathrm{e}^{-a|n|} - \frac{1}{b^2-a^2}\frac{1}{2b}\sum_{n=-\infty}^{\infty} \mathrm{e}^{-b|n|}. \end{aligned}$$

Each series is the sum of two geometric series (one for positive n and one for negative n), and we derive after some more work that

$$\sum_{n=-\infty}^{\infty} \mathrm{e}^{-a|n|} = \frac{1+\mathrm{e}^{-a}}{1-\mathrm{e}^{-a}} \qquad \text{and} \qquad \sum_{n=-\infty}^{\infty} \mathrm{e}^{-b|n|} = \frac{1+\mathrm{e}^{-b}}{1-\mathrm{e}^{-b}}.$$

Hence, after rearranging the terms somewhat, we get

$$\begin{aligned} S &= \frac{1}{b^2-a^2}\left(\frac{1}{a}\frac{1}{1-\mathrm{e}^{-a}} - \frac{1}{b}\frac{1}{1-\mathrm{e}^{-b}}\right) - \frac{1}{2ab}\frac{1}{b+a} \\ &= \frac{1}{2(b^2-a^2)}\left(\frac{\coth a}{a} - \frac{\coth b}{b}\right). \end{aligned}$$

---

[4] I did not make it up for the mere pleasure of computing a series; it appeared during a calculation in a finite-temperature quantum field theory problem.

Josiah Willard GIBBS (1839–1903), American physicist, was professor of mathematical physics at Yale University. Gibbs revolutionized the study of thermodynamics in 1873 by a geometric approach and then, in 1876, by an article concerning the equilibrium properties of mixtures. He had the idea of using diagrams with temperature-entropy coordinates, where the work during a cyclic transformation is given by the area of the cycle. It took a long time for chemists to understand the true breadth of this paper of 1876, which was written in a mathematical spirit. Gibbs also worked in pure mathematics, in particular in vector analysis. Finally, his works in statistical mechanics helped provide its mathematical basis.

11.3 The Gibbs phenomenon

Although the Gibbs phenomenon may be explained purely with the tools of Fourier series,[5] it is easier to take advantage of the Fourier transform of distributions to clarify things.

Let f be a function, with Fourier transform $\widetilde{f}(\nu)$ and denote

$$f_\xi(x) = \int_{-\xi}^{\xi} \widetilde{f}(\nu)\, \mathrm{e}^{2\pi \mathrm{i} \nu x}\, \mathrm{d}\nu$$

for any $\xi > 0$. To obtain f_ξ, the process is therefore:

- decompose f in its spectral components $\widetilde{f}(\nu)$;
- remove the higher frequencies (those with $|\nu| > \xi$);
- reconstruct a function by summing the low-frequencies part of the spectrum, "forgetting" about the higher frequencies (which characterize the finer details of the function).

So we see that the function f_ξ is a "fuzzy" version of the original function f. It can be expressed as a convolution, since

$$f_\xi = \mathscr{F}^{-1}\left[\Pi(\nu/2\xi) \cdot \widetilde{f}(\nu)\right] = f(x) * \frac{\sin 2\pi\xi\, x}{\pi x}.$$

Now recall that, in the sense of distributions, we have

$$\lim_{\xi \to +\infty} \frac{\sin 2\pi\xi\, x}{\pi x} = \delta(x),$$

[5] Using, for instance, the Dirichlet kernel, as in [8].

in accordance with the result stated on page 233, which proves that, by forgetting frequencies at a higher and higher level, we will finally recover the original signal:

$$\lim_{\xi \to +\infty} f_\xi = f \qquad \textit{(in the sense of distributions).}$$

In the case where f has a discontinuity at 0, we can simply write $f = g + \alpha H$, where α is the "jump" at the discontinuity and g is now a continuous functions. Now let us concentrate on what happens to H when put through the filtering process described above. We compute

$$H_\xi(x) = H(x) * \frac{\sin 2\pi\xi\, x}{\pi x} = \int_{-\infty}^{x} \frac{\sin 2\pi\xi\, y}{\pi y}\, \mathrm{d}y = \frac{1}{2} + \frac{1}{\pi}\, \mathrm{Si}(2\pi\xi\, x), \quad \text{(II.4)}$$

where Si is the **sine integral**, defined (see [2,42]) by the formula

$$\mathrm{Si}(x) \stackrel{\text{def}}{=} \int_0^x \frac{\sin t}{t}\, \mathrm{d}t.$$

If we plot the graph of $H_1(x) = \frac{1}{2} + \frac{1}{\pi}\,\mathrm{Si}(2\pi x)$, we can see oscillations, with amplitude around 8% of the unit height, on each side of the discontinuity.[6]

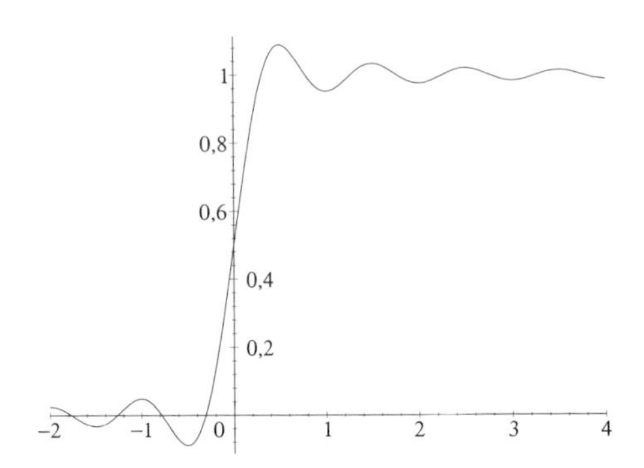

When the cutoff frequency is increased, formula (II.4) shows that the size of the oscillations remains constant, but that they are concentrated in the neighborhood of the point of discontinuity (here, we have $\xi = 2, 3, 4, 5$ and the graph has been enlarged in the horizontal direction to make it more readable):

[6] Notice in passing that a low-pass filter in frequencies is not physically possible, since it is not causal.

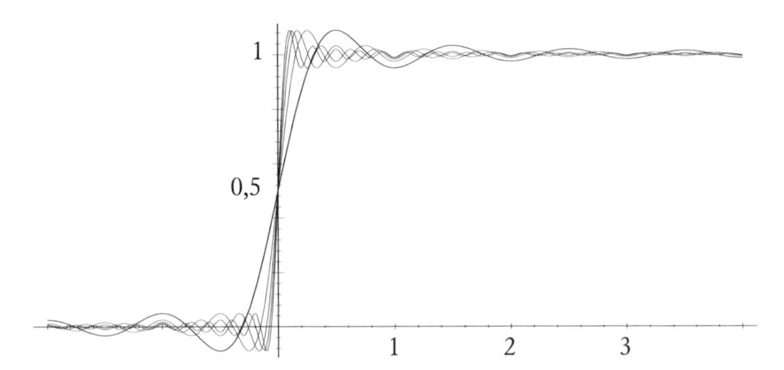

What happens now if we consider a function f which is 1-periodic, piecewise continuous, and piecewise of $\mathscr{C}^1$ class? Then f has a Fourier series expansion

$$f(x) = \sum_{n\in\mathbb{Z}} c_n \, e^{2\pi i n x},$$

which corresponds to a Fourier transform

$$\widetilde{f}(\nu) = \sum_{n\in\mathbb{Z}} c_n \, \delta(\nu - n).$$

Then the function f_ξ, obtained by removing from the spectrum the frequencies $|\nu| \leqslant \xi$, is none other than

$$f_\xi(x) = \sum_{|n|\leqslant\xi} c_n \, e^{2\pi i n x},$$

that is, the partial sum the Fourier series of f of order $E[\xi]$ (integral part of ξ).

Hence the sequence of partial sums of the Fourier series will exhibit the same aspect of oscillations with constant amplitude, concentrated closer and closer to the discontinuities.[7]

It should be noted that this phenomenon renders Fourier series rather unreliable in numerical computations, but that it can be avoided by using, instead of the partial sums of the Fourier series, the Cesàro means of those series, which are called **Fejér sums** (see Exercise 9.4 on page 270).

[7] In the preceding example with the Heaviside function, the oscillations get narrower in a continuous manner. Here, since the function, being 1-periodic, has really infinitely many discontinuities, there appears a superposition of oscillating figures which has the property, rather unintuitive, of remaining "fixed" on the intervals $\xi \in [n, n+1[$ and of changing rapidly when going through the points $\xi = n$.

11.4
Application to physical optics

11.4.a Link between diaphragm and diffraction figure

We are going to investigate the phenomenon of light diffraction through a diaphragm. For this purpose, consider an experiment that gives the possibility of observing the **diffraction at infinity** (i.e., the diffration corresponding to the **Fraunhofer approximation**; see [14] for a detailed study), with a light source S considered as being *purely monochromatic*.[8]

To simplify matters, we consider a system which is invariant under translation along the $\mathscr{O}y$ axis. The diaphragm, placed at the focal point of a lens (with focal length F on the figure), is characterized by a transmittance function $f(x)$ which expresses the ratio between the light amplitude just in front and behind the diaphragm. Typically, this transmittance function is a characteristic function. For instance, a slit of length ℓ (infinitely long) is modeled by a transmittance function equal to

$$f_{\text{slit}}(x) = \Pi\left(\frac{x}{\ell}\right).$$

However, it may happen that one wishes to model an infinitely narrow slit. In order to still be able to observe a figure, it is required to *increase* the intensity of the light arriving in inverse proportion to the width of the slit. This intensity coefficient is incorporated to the the transmittance, which gives

$$f_{\text{slit}} = \lim_{\ell \to 0} \frac{1}{\ell}\Pi\left(\frac{x}{\ell}\right) = \delta(x).$$

With this convention, the "transmittance function" thus becomes a "transmittance distribution."

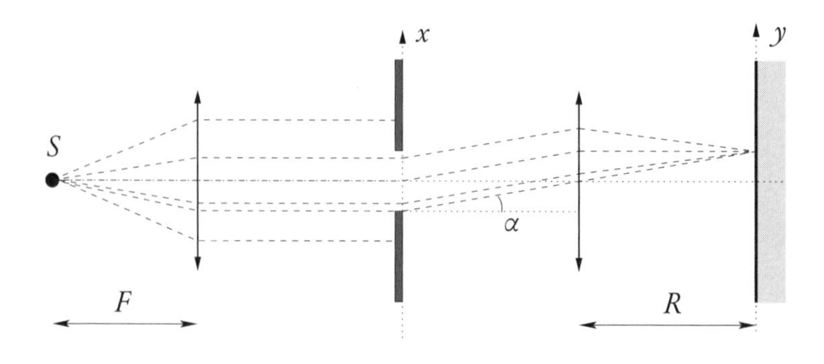

[8] Nonmonochromatic effects will the studied in Section 13.7 on page 371

The amplitude which is observed in the direction α is, with the Fraunhofer approximation and linearizing $\sin\alpha \approx \alpha$, equal to

$$\mathrm{C^{\underline{nt}}} \cdot \int f(x)\, \mathrm{e}^{-2\pi \mathrm{i} \alpha x/\lambda}\, \mathrm{d}x,$$

up to a multiplicative constant. On a screen at the focal point of a lens with focal length R, the amplitude is therefore equal to

$$A(y) = \mathrm{C^{\underline{nt}}} \cdot \int f(x)\, \mathrm{e}^{-2\pi \mathrm{i} yx/R\lambda}\, \mathrm{d}x.$$

Changing the scale for the screen and putting $\theta = y/R\lambda$, which has the dimension of the inverse of length, we obtain therefore

$$A(\theta) = \mathrm{C^{\underline{nt}}} \cdot \int f(x)\, \mathrm{e}^{-2\pi \mathrm{i} \theta x}\, \mathrm{d}x = \widetilde{f}(\theta).$$

Hence, up to a multiplicative constant (which we will ignore in our computations from now on), *the amplitude of the diffraction figure at infinity of a diaphragm is given by the Fourier transform of its transmittance distribution.*

Remark 11.23 From the mathematical point of view, interference and diffraction phenomena are strictly equivalent.

11.4.b Diaphragm made of infinitely many infinitely narrow slits

Consider a diaphragm made of infinitely many slits, placed at a fixed distance a from each other, with coordinates equal to na with n ranging over all integers in $\mathbb{Z}$. The diaphragm is modeled by a "Dirac comb" distribution

$$f_1(x) = \sum_{n\in\mathbb{Z}} \delta(x - na) = \sum_{n\in\mathbb{Z}} a\,\delta\left(\frac{x}{a} - n\right) = a\,\text{Ш}\left(\frac{x}{a}\right).$$

The diffraction figure produced by the diffraction apparatus thus described is then (in terms of amplitude) given by the Fourier transform of

$$f_1(x) = a\,\text{Ш}\left(\frac{x}{a}\right) \xrightarrow{\text{F.T.}} A_1(\theta) = a^2\,\text{Ш}(a\theta).$$

In other words, infinitely many infinitely narrow slits produce a diffraction figure consisting of infinitely many infinitely narrow bands. The step between bands is $1/a$.

Thus, using the properties of the Fourier transform, we notice that *the larger the spacing between the slits, the smaller the spacing between the steps of the diffraction figure.*

Of course, the situation described here is never seen in practice, as there can only be finitely many slits, and they are never infinitely narrow.

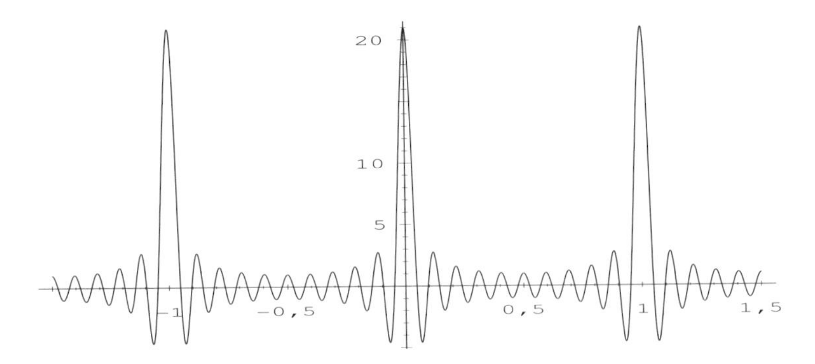

Fig. 11.1 – Amplitude of the diffraction figure given by $2N + 1 = 21$ infinitely narrow slits.

11.4.c Finite number of infinitely narrow slits

Consider now a diaphragm made of $2N + 1$ slits, separated by a fixed distance a. The coordinates of the slits are then $-Na, \dots, -a, 0, a, \dots, Na$. The transmittance of the diaphragm is then modeled by

$$f_2(x) = \sum_{n=-N}^{N} \delta(x - na).$$

The diffraction figure is the Fourier transform of f_2, which may be computed by summing directly the Fourier transforms of each Dirac component:

$$A_2(\theta) = \sum_{n=-N}^{N} \mathrm{e}^{-2\pi \mathrm{i} na\theta},$$

or, summing the partial geometric series,

$$\begin{aligned} A_2(\theta) &= \mathrm{e}^{2\pi\mathrm{i}a\theta N}\left[\frac{1-\mathrm{e}^{-2\pi\mathrm{i}a\theta(N+1)}}{1-\mathrm{e}^{2\pi\mathrm{i}a\theta}}\right] \\ &= \frac{\mathrm{e}^{2\pi\mathrm{i}a\theta\left(N+\frac{1}{2}\right)} - \mathrm{e}^{-2\pi\mathrm{i}a\theta\left(N+\frac{1}{2}\right)}}{\mathrm{e}^{\mathrm{i}\pi a\theta} - \mathrm{e}^{-\mathrm{i}\pi a\theta}} \\ &= \frac{\sin\left((2N+1)\pi a\theta\right)}{\sin \pi a\theta}. \end{aligned} \tag{11.5}$$

This function can be checked to be continuous (by extending by continuity). It is represented in Figure 11.1; the corresponding intensity is represented in Figure 11.2. The shape of this curve is, however, not easy to "see" obviously by just looking at formula (11.5).

The amplitude A_2 may also be obtained using the convolution formulas. Indeed, the transmittance and the diffraction figure for infinitely many slits

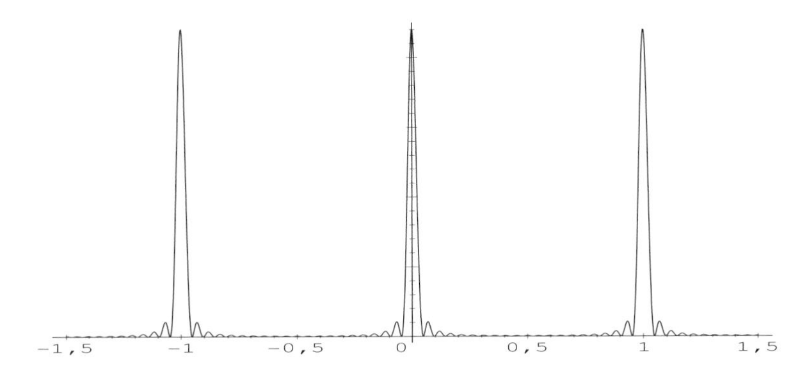

Fig. 11.2 – Intensity of the diffraction figure produced by $2N+1=21$ infinitely narrow slits. This is the square of the function represented on Figure 11.1.

are linked by

$$a\,\text{Ш}\left(\frac{x}{a}\right) \xrightarrow{\text{F.T.}} a^2\,\text{Ш}(a\theta),$$

and hence the $(2N+1)$ slits are modelized, for instance, with the distribution

$$f_1(x) = a\,\text{Ш}\left(\frac{x}{a}\right)\cdot \Pi\left(\frac{x}{(2N+1)a}\right).$$

(We took care that the boundaries of the "rectangle" function do not coincide with any of the Dirac spikes of the comb, which would have led to an ambiguity.)

The Fourier transform of a rectangle function with width $(2N+1)a$ is

$$\Pi\left(\frac{x}{(2N+1)a}\right) \xrightarrow{\text{F.T.}} (2N+1)\,a\,\frac{\sin\big((2N+1)\pi a\theta\big)}{(2N+1)\pi a\theta},$$

and therefore

$$f_2(x) \xrightarrow{\text{F.T.}} A_2(\theta)$$

with

$$\begin{aligned} A_2(\theta) &= \Big[a^2\,\text{Ш}(a\theta)\Big] * \left[(2N+1)\,a\,\frac{\sin\big((2N+1)\pi a\theta\big)}{(2N+1)\pi a\theta}\right] \\ &= (2N+1)a^2\left[\sum_{n=-\infty}^{+\infty}\delta\left(\theta-\frac{n}{a}\right)\right] * \left[\frac{\sin\big((2N+1)\pi a\theta\big)}{(2N+1)\pi a\theta}\right], \end{aligned}$$

that is,

$$f_2(x) \xrightarrow{\text{F.T.}} A_2(\theta) = (2N+1)a^2 \sum_{n=-\infty}^{+\infty} \frac{\sin\big((2N+1)\pi(a\theta-n)\big)}{(2N+1)\pi(a\theta-n)}.$$

So the diffraction figure is given by the superposition of infinitely many sine cardinal functions centered at k/a with k ranging over $\mathbb{Z}$. But the graph of the sine cardinal function has the following shape:

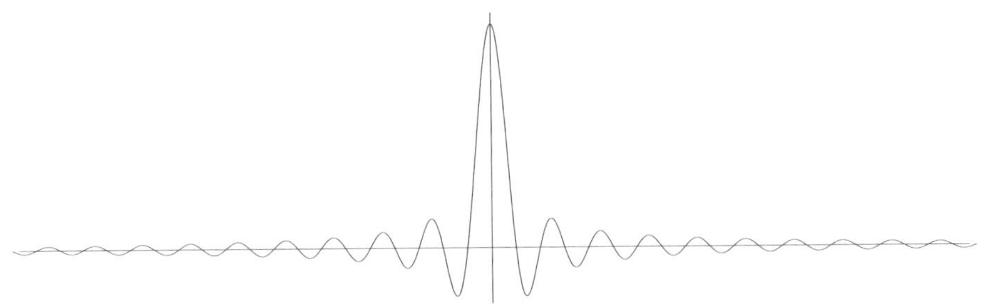

namely, a fairly narrow spike surrounded by small oscillations. One can then easily imagine that the whole diffraction figure is made of a series of spikes, spaced at a distance $1/a$ from each other; this is what has been observed in Figure 11.1.

Notice that the width of the spikes is proportional to $1/(2N+1)$, so *the more slits there are in the diaphragm, the narrower the diffraction bands are.* The number of slits may be read off directly from the figure, since it corresponds to the number of local extrema between two main spikes (the latter being included in the counting). When, as in the example given here, there is an *odd* number of slits, the oscillations from neighboring sine cardinals add up; this implies in particular that the small diffraction bands between the main spikes are fairly visible. On the other hand, for an *even* number of slits, the oscillations compensate for each other; the secondary bands are then much less obvious, as can be seen in Figure 11.3, to be compared with Figure 11.1.

Note in passing that we obtained two different formulas for A_2, which are of course equivalent; this equivalence is, however, difficult to prove using classical analytic means.

11.4.d Finitely many slits with finite width

If we consider now a diaphragm made up of $(2N+1)$ slits of width ℓ (with $\ell < a$ so that the slits are physically separated), its transmittance distribution can be expressed in the form

$$\begin{aligned} f_3(x) &= \sum_{n=-N}^{N} \Pi\left(\frac{x-na}{\ell}\right) \\ &= \left[\sum_{n=-N}^{N} \delta(x-na)\right] * \Pi\left(\frac{x}{\ell}\right) \end{aligned}$$

that is,

$$f_3(x) = \left[a Ш\left(\frac{x}{a}\right) \cdot \Pi\left(\frac{x}{(2N+1)a}\right)\right] * \Pi\left(\frac{x}{\ell}\right).$$

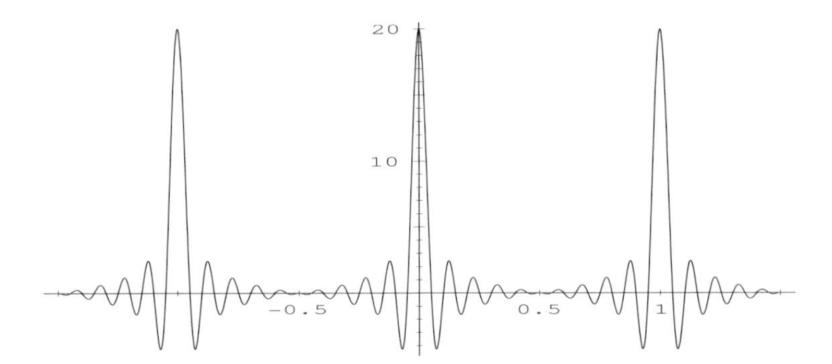

Fig. 11.3 – Diffraction figure of a system with $2N$ infinitely narrow slits (here $2N = 20$). Note that the secondary extremums are much more attenuated than on Figure 11.1, where $2N + 1$ slits occurred.

(Despite the appearance of the last line, this is a regular distribution.) The transmittance f_3 is therefore simply the convolution of the transmittance distribution f_2 with a rectangle function with width ℓ. The diffraction figure is therefore given by the product of the Fourier transforms of these two distributions. The Fourier transform of $\Pi(x/\ell)$ is a sine cardinal function with width $1/\ell$. In general, in such systems, the width of the slits is very small compared to the spacing, that is, we have $\ell \ll a$. This implies that the width of this sine cardinal function $(1/\ell)$ is much larger than the spacing between the bands. The amplitude of the diffraction figure

$$A_3(\theta) = A_2(\theta) \cdot \frac{\sin \pi\ell\theta}{\pi\ell\theta}$$

is shown in Figure 11.4. One sees clearly there how the intensity of the spikes decreases. Since the ratio a/ℓ is equal to 4 in this example, the fourth band from the center is located exactly at the first minimum of the sine cardinal, and hence vanishes completely.

Remark 11.24 In this last example, three different lengths appear in the problem: the width ℓ of the slits, the spacing a between slits, and the total length $(2N + 1)a$ of the system. These three lengths are related by

$$\ell \ll a \ll (2N + 1)a,$$

and reappear, in the Fourier world, in the form of the characteristic width $1/\ell$ of the disappearance of the bands (the largest characteristic length in the diffraction figure), the distance $1/a$ between the principal bands of diffraction, and the typical width $1/(2N + 1)a$ of the fine bands (the shortest), with

$$\frac{1}{\ell} \gg \frac{1}{a} \gg \frac{1}{(2N + 1)a}.$$

Simply looking at the diffraction figure is therefore sufficient in principle to determine the relative scales of the system of slits in use.

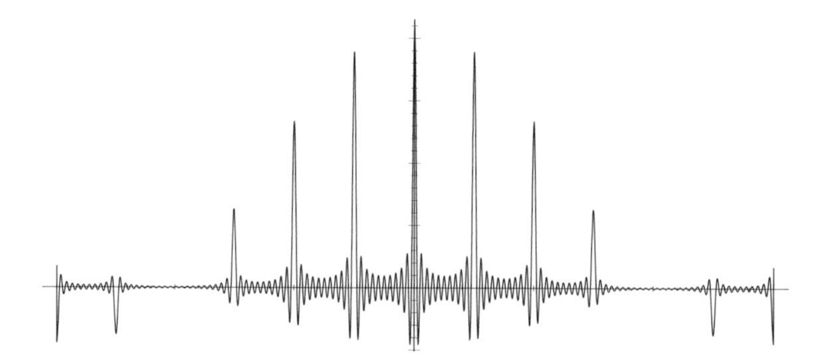

Fig. 11.4 – Diffraction figure with 21 slits, with width 4 times smaller than their spacing: $\ell = a/4$.

11.4.e Circular lens

Consider now the diffraction by a circular lens with diameter D. It is necessary to use here a two-dimensional Fourier transform. The transmittance function of the lens is

$$f(x,y) = \begin{cases} 0 & \text{if } \sqrt{x^2+y^2} > \frac{D}{2}, \\ 1 & \text{if } \sqrt{x^2+y^2} < \frac{D}{2}. \end{cases}$$

Since this function is radially symmetric, it is easier to change variables, putting $r \overset{\text{def}}{=} \sqrt{x^2+y^2}$ and writing $f(x,y) = F(r) = H(D/2 - r)$.

The amplitude $A(\theta, \psi)$ of the diffraction figure is then given by

$$A(\theta,\psi) = \iint f(x,y)\, \mathrm{e}^{-2\pi\mathrm{i}(\theta x+\psi y)}\, \mathrm{d}x\, \mathrm{d}y,$$

which is also, substituting $(x,y) \mapsto (r,\varphi)$ and defining ρ by $\rho = \sqrt{\theta^2+\psi^2}$, equal to

$$A(\rho) = 2\pi \int_0^{D/2} r\, F(r) \int_{-\pi}^{\pi} \mathrm{e}^{-2\pi\mathrm{i}\rho r\cos\varphi}\, \mathrm{d}\varphi\, \mathrm{d}r.$$

DEFINITION 11.25 The **Bessel function of order 0** is defined by

$$\forall z \in \mathbb{C} \qquad J_0(z) \overset{\text{def}}{=} \frac{1}{2\pi} \int_{-\pi}^{\pi} \mathrm{e}^{-\mathrm{i}z\cos\varphi}\, \mathrm{d}\varphi$$

The **Bessel function of order 1** is defined by

$$\forall z \in \mathbb{C} \qquad J_1(z) \overset{\text{def}}{=} \frac{1}{\mathrm{i}\pi} \int_{-\pi}^{\pi} \mathrm{e}^{-\mathrm{i}z\cos\varphi} \cos\varphi\, \mathrm{d}\varphi$$

These functions satisfy $\int_0^x y\, J_0(y)\, \mathrm{d}y = x\, J_1(x)$.

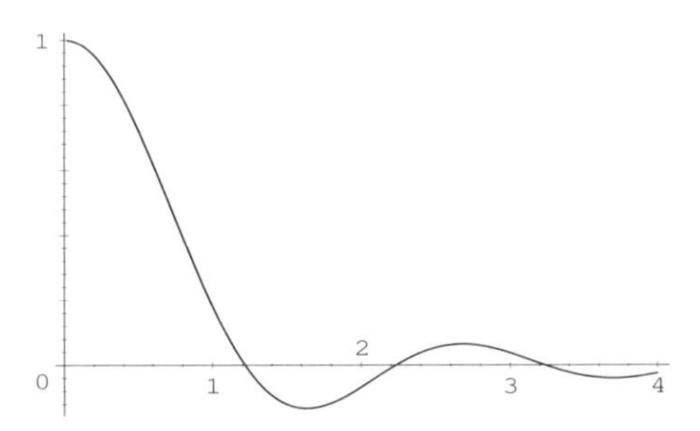

Fig. 11.5 – The function $x \mapsto 2\,J_1(\pi x)/x$. The first zero is located at a point $x \approx 1.220$, the second at $x \approx 2.233$. The distance between successive zeros decreases.

The amplitude $A(\rho)$ can therefore be expressed as

$$A(\rho) = 2\pi \int_0^{D/2} r\,J_0(2\pi\rho r)\,\mathrm{d}r = \frac{1}{2\pi\rho^2}\int_0^{\pi D\rho} y\,J_0(y)\,\mathrm{d}y = \frac{D}{2\rho}\,J_1(\pi D\rho)$$

$$= S\cdot\left[\frac{2\,J_1(\pi D\rho)}{\pi D\rho}\right], \qquad \text{with } S = \frac{\pi D^2}{4} = \text{surface of the lens.}$$

The graph of the function $x \mapsto 2J_1(\pi x)/\pi x$ is represented in Figure 11.5. This function has a smallest positive zero at $x \approx 1.22$, which corresponds to an angle equal to $1.22\,\lambda/D$; this is the origin of the numerical factor in the **Rayleigh criterion** for the resolution of an optical instrument. The diffraction figure has the shape presented in Figure 11.6.

11.5
Limitations of Fourier analysis and wavelets

Fourier analysis can be used to extract from a signal $t \mapsto f(t)$ its composing frequencies. However, it is not suitable for the analysis of a musical or vocal signal, for instance.

The reason is quite simple. Assume that we are trying to determine the various frequencies that occur in the performance of a Schubert *lied*. Then we have to integrate the sound signal from $t = -\infty$ to $t = +\infty$, which is already somewhat difficult (the recording must have a beginning and an ending if we don't want to spend a fortune in magnetic tape). If, moreover, knowing $\widetilde{f}$ we wish to reconstruct f, we must again integrate over $]-\infty, +\infty[$. Such an integration is usually numerical, and apart from the approximate knowledge of $\widetilde{f}$, the integration range must be *limited* to a finite interval, say $[-\nu_0, \nu_0]$. But, even if $\widetilde{f}$ decays rapidly, forgetting the tail of $\widetilde{f}$ may spectacularly affect the reconstitution of the original signal at these instants when the latter changes very suddenly, which is the case during the "attack"

Fig. 11.6 – The diffraction figure of a circular lens. Other larger rings appear in practice, but their intensity is much smaller.

of a note, or when consonants are sung or spoken. Thus, to recreate the attack of a note (a crucial moment to recognize the timber of an instrument), it is necessary to compute the function $\widetilde{f}(\nu)$ with high precision for large values of ν, and hence to know precisely the attack of *all the notes* in the partition. When the original signal is recreated, their musical characteristics and their temporal localization may be altered.

This is why some researchers (for instance, the geophysicist Jean Morlet, Alex Grossman in Marseilles, Yves Meyer at the École Polytechnique, Pierre Gilles Lemarié-Rieusset in Orsay and then Évry, and many others) have developed a different approach, based on a time-frequency analysis.

We know that the modulus of the Fourier spectrum $|\widetilde{f}(\nu)|$ of a signal provides excellent information on the frequency aspects, but little information on the temporal aspect.[9] However, to encode a musical signal, the customary way is to proceed as follows:

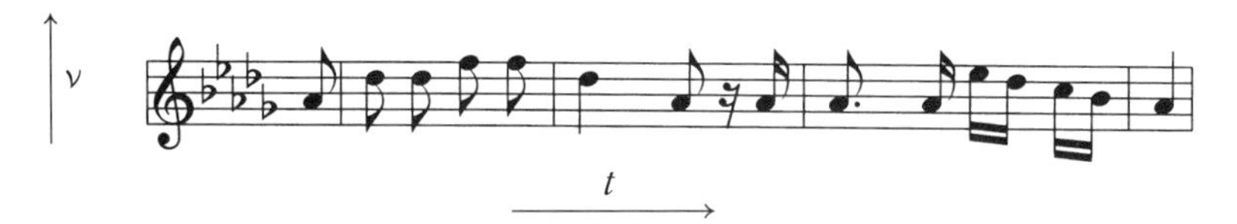

On this staff, we can read:

- vertically, the frequency information (the height of the notes);
- horizontally, the timing information (the rhythm, or even indications concerning attack).

[9] A given signal $f(t)$ and the same translated in time $f(t-a)$ have the same power spectrum, since only the phase changes between the two spectra.

The musical signal is therefore encoded in matters of both time and frequency.

Yves Meyer has constructed a **wavelet** ψ, that is, a function of $\mathscr{C}^\infty$ class, all moments of which are zero, that produces an orthonormal basis $(\psi_{np})_{n,p}$ of L^2, where

$$\psi_{np}(t) \stackrel{\text{def}}{=} 2^{n/2}\,\psi(2^n t - p) \qquad \forall n, p \in \mathbb{Z}.$$

The larger $n \in \mathbb{Z}$ is, the narrower the wavelet ψ_{np} is; when $n \to -\infty$, on the other hand, the wavelet is extremely "wide." Moreover, the parameter p characterizes the average position of the wavelet. The decomposition in the wavelet basis is given by

$$f = \sum_{n=-\infty}^{+\infty} \sum_{p=-\infty}^{+\infty} \left(f \middle| \psi_{np}\right) \psi_{np} \qquad \forall f \in L^2(\mathbb{R}),$$

in the sense of convergence in L^2. Similarly, a formula analogous to the Parseval identity holds:

$$\sum_{(n,p)\in\mathbb{Z}^2} \left| \left(f \middle| \psi_{np}\right) \right|^2 = \int_{-\infty}^{+\infty} |f(t)|^2 \, \mathrm{d}t.$$

Whereas the "basis" for Fourier expansion was a complex exponential, which is extremely well localized in terms of frequency[10] but not localized at all in time, for wavelets, the time localization is more precise when n is large, and the frequency localization is wider when n is large. Thus, both time and frequency information are available. This makes it possible to describe the attack of a note at time $t = t_0$ by concentrating on the wavelet coefficients localized around t_0; these, even at the highest frequencies, may be computed without any knowledge of the signal at other instants.

Research around wavelets is still developing constantly. The interested reader is invited to consult the book by Kahane and Lemarié-Rieusset [54] for a detailed study, or Gasquet [37] for a very clear short introduction, or Mallat [65] for applications to the analysis of signals. The book by Stein and Shakarchi [85] should also be of much interest.

[10] A single frequency occurs.

EXERCISES

♦ **Exercise 11.3** Compute $\Pi(x/a) * \sin x$ using the Fourier transform.

♦ **Exercise 11.4** Compute the Fourier transform, in the sense of distributions, of $x \longmapsto |x|$.

♦ **Exercise 11.5** Compute the Fourier transform of $H(x) \cdot x^n$ for any integer n. (One may use the distributions $\mathrm{fp}(1/x^k)$, introduced in Exercise 8.6 on page 241.)

♦ **Exercise 11.6** Compute the sum

$$S \stackrel{\text{def}}{=} \sum_{n \in \mathbb{N}} \frac{1}{a^2 + 4\pi^2 n^2} \frac{1}{b^2 - 4\pi^2 n^2},$$

with $a, b \in \mathbb{R}^{+*}$.

The result of Exercise 8.13 on page 243 may be freely used.

♦ **Exercise 11.7** Recall that the **Bessel function of order 0** is defined by

$$J_0(x) \stackrel{\text{def}}{=} \frac{1}{2\pi} \int_{-\pi}^{\pi} \mathrm{e}^{-ix\cos(\theta-\alpha)} \,\mathrm{d}\theta \qquad \text{for any } x \in \mathbb{R}.$$

i) Show that this definition is independent of α. Deduce from this that J_0 takes real values.

ii) In the plane $\mathbb{R}^2$, let $f(x,y)$ be a *radial* function or distribution, that is, such that there exists $\varphi : \mathbb{R}^+ \to \mathbb{C}$ (or a distribution with support in $\mathbb{R}^+$) with $f(x,y) = \varphi(r)$, where $r \stackrel{\text{def}}{=} \sqrt{x^2+y^2}$. Assume moreover that f has a Fourier transform. Show that its Fourier transform is also radial: $\widetilde{f}(u,v) = \psi(\rho)$ with $\rho \stackrel{\text{def}}{=} \sqrt{u^2+v^2}$.

   Prove the formula giving ψ as a function of φ, using the Bessel function J_0.

   The function ψ is called the **Hankel transform** of φ, denoted $\varphi(r) \xrightarrow{\text{H. T.}} \psi(\rho)$.

iii) Show that for continuous functions φ and ψ, if $\varphi \xrightarrow{\text{H. T.}} \psi$, then also $\psi \xrightarrow{\text{H. T.}} \varphi$ (i.e., the Hankel transform is its own inverse).

iv) Show that

$$\frac{\mathrm{d}^2\varphi}{\mathrm{d}r^2} + \frac{1}{r}\frac{\mathrm{d}\varphi}{\mathrm{d}r} \xrightarrow{\text{H. T.}} -4\pi^2\rho^2\,\psi(\rho).$$

v) Show that if $\varphi_1 \xrightarrow{\text{H. T.}} \psi_1$ and $\varphi_2 \xrightarrow{\text{H. T.}} \psi_2$, then we have the relation

$$\int_0^\infty r\,\overline{\varphi_1(r)}\,\varphi_2(r)\,\mathrm{d}r = \int_0^\infty \rho\,\overline{\psi_1(\rho)}\,\psi_2(\rho)\,\mathrm{d}\rho.$$

Physical optics

♦ **Exercise 11.8** Describe the diffraction figure formed by a system made of infinitely many regularly spaced slits of finite width ℓ (the spacing between slits is a).

♦ **Exercise 11.9** Compute, in two dimensions, the diffraction figure of a diaphragm made of two circular holes of diameter ℓ, placed at a distance $a > 2\ell$.

PROBLEM

♦ Problem 5 (Debye electrostatic screen)

- **Presentation of the problem**

We consider a plasma with multiple components, or in other words, a system of particles of different types $\alpha \in I$, where I is a finite set, each type being characterized by a density ρ_α, a mass m_α, and a charge e_α. (Examples of such systems are an ionic solution, the stellar matter, the solar wind or solar crown.) It is assumed that the system is initially *neutral*, that is, that

$$\sum_\alpha e_\alpha \rho_\alpha = 0.$$

The system is considered in nonrelativistic terms, and it is assumed that the interactions between particles are restricted to Coulomb interactions. The particles are also assumed to be *classical* (not ruled by quantum mechanics) and *point-like*. The system may be described by the following hamiltonian:

$$\mathscr{H} = \sum_i \frac{\boldsymbol{p}_i^2}{2m_i} + \frac{1}{2} \sum_{\substack{i,j \\ i \neq j}} \frac{e_i\, e_j}{r_{ij}} \qquad \text{with} \quad r_{ij} \stackrel{\text{def}}{=} \left\| \boldsymbol{r}_j - \boldsymbol{r}_i \right\|.$$

The indices i and j ranges successively over all particles in the system.

Now an *exterior charge* q_0, located at $\boldsymbol{r} = 0$, is added to the system. We wish to know the repartition of the other charges where *thermodynamic equilibrium* is reached.

- **Solution of the problem**

i) Let $\rho_\alpha(\boldsymbol{r})$ denote the distribution of density of particles of type α, at equilibrium, in the presence of q_0. Express the distribution $q(\boldsymbol{r})$ in terms of the $\rho_\alpha(\boldsymbol{r})$.

ii) Denoting by $\varphi(\boldsymbol{r})$ the electrostatic potential, write the two equations relating $\varphi(\boldsymbol{r})$ and $\rho_\alpha(\boldsymbol{r})$ – the Poisson equation for electrostatics and the Boltzmann equation for thermodynamics.

iii) Linearize the Boltzmann equation and deduce an equation for $\varphi(\boldsymbol{r})$.

iv) Passing to Fourier transforms, compute the algebraic expression of $\widetilde{\varphi}(\boldsymbol{k})$, the Fourier transform of $\varphi(\boldsymbol{r})$.

v) Using complex integration and the method of residues, find the formula for the potential $\varphi(\boldsymbol{r})$, which is called the **Debye potential** [26].

vi) Show that, at short distances, this potential is indeed of Coulomb type.

vii) Compute the total charge of the system and comment.

SOLUTIONS

♦ **Solution of exercise 11.2 on page 301.** It must be shown that $T(x) = \exp(x^4)\,Ш(x)$ cannot act on certain functions of the Schwartz space. It is easy to check that the gaussian $g : x \mapsto \exp(-x^2)$ decays rapidly as do all its derivatives, and therefore belongs to $\mathscr{S}$, whereas the evaluation of $\langle T, g\rangle$ leads to an infinite result.

Strictly speaking, we must exclude also the possibility that there exists *another* definition of $\langle T, \varphi\rangle$, for φ a Schwartz function, which makes it a tempered distribution and coincides with $\int T(x)\,\varphi(x)\,\mathrm{d}x$ for φ with bounded support; but that is easy by taking a sequence φ_n of test functions with compact support that converges in $\mathscr{S}$ to $g(x)$. Then on the one hand

$$\langle T, \varphi_n\rangle \to \langle T, g\rangle$$

by the assumed continuity of T as tempered distribution, and on the other hand

$$\langle T, \varphi_n\rangle = \int T(x)\,\varphi_n(x)\,\mathrm{d}x \to +\infty,$$

a contradiction.

♦ **Solution of exercise 11.3.** The Fourier transform of the function given is

$$\frac{\sin \pi\nu a}{\pi\nu} \cdot \frac{1}{2\mathrm{i}}\left[\delta\left(\nu - \frac{1}{2\pi}\right) - \left(\nu + \frac{1}{2\pi}\right)\right] = 2\sin\frac{a}{2}\cdot\frac{1}{2\mathrm{i}}\left[\delta\left(\nu - \frac{1}{2\pi}\right) - \left(\nu + \frac{1}{2\pi}\right)\right]$$

(using the formula $f(x)\,\delta(x) = f(0)\,\delta(x)$), which, in inverse Fourier transform, gives

$$2\sin(a/2)\sin x.$$

Of course a direct computation leads to the same result.

♦ **Solution of exercise 11.4.** Notice first that $f : x \mapsto |x|$ is the product of $x \mapsto x$ and $x \mapsto \operatorname{sgn} x$. The Fourier transform of $x \mapsto x$ is, by the differentiation theorem, equal to $-\mathrm{i}\delta'/2\pi$. Moreover, we know that $\operatorname{sgn} \xrightarrow{\text{F.T.}} \operatorname{pv} 1/\mathrm{i}\pi\nu$. Using the convolution theorem yields

$$x \operatorname{sgn} x \xrightarrow{\text{F.T.}} \frac{\mathrm{i}\delta'}{2\pi} * \frac{1}{\mathrm{i}\pi}\operatorname{pv}\frac{1}{\nu} = \frac{1}{2\pi^2}\frac{\mathrm{d}}{\mathrm{d}\nu}\operatorname{pv}\frac{1}{\nu} = -\frac{1}{2\pi^2}\operatorname{fp}\frac{1}{\nu^2}.$$

♦ **Solution of exercise 11.5.** The Fourier transforme of the constant function $\mathbf{1}$ is δ. The formula $\mathscr{F}\left[(-2\mathrm{i}\pi x)^n f(x)\right] = f^{(n)}(x)$ gives

$$x^n \xrightarrow{\text{F.T.}} \frac{\delta^{(n)}}{(-2\mathrm{i}\pi)^n}. \qquad \text{and} \qquad H(x) \xrightarrow{\text{F.T.}} \frac{\delta}{2} + \frac{1}{2\mathrm{i}\pi}\operatorname{pv}\left(\frac{1}{\nu}\right).$$

The convolution formula then yields

$$H(x)\,x^n \xrightarrow{\text{F.T.}} \frac{\delta^{(n)}}{2(-2\mathrm{i}\pi)^n} - \frac{1}{(-2\mathrm{i}\pi)^{n+1}}\frac{\mathrm{d}^n}{\mathrm{d}\nu^n}\operatorname{pv}\left(\frac{1}{\nu}\right).$$

Finally, the distribution $\operatorname{fp}(1/x^{n+1})$ is linked to the n-th derivative of $\operatorname{pv}(1/\nu)$ by a multiplicative constant:

$$\frac{\mathrm{d}^n}{\mathrm{d}\nu^n}\operatorname{pv}\left(\frac{1}{\nu}\right) = (-1)^n\, n!\operatorname{fp}\left(\frac{1}{\nu^{n+1}}\right),$$

from which we derive

$$H(x)\,x^n \xrightarrow{\text{F.T.}} \frac{\delta^{(n)}}{2(-2\mathrm{i}\pi)^n} + \frac{n!}{(2\mathrm{i}\pi)^{n+1}}\operatorname{fp}\left(\frac{1}{\nu^{n+1}}\right).$$

♦ **Solution of exercise 11.6.** Using the technique described in the text on page 310, we first compute that

$$e^{-a|x|} * \sin b\,|x| = \frac{2b}{a^2+b^2} e^{-a|x|} + \frac{2a}{a^2+b^2} \sin b\,|x|,$$

and using the formula of exercise 8.13 on page 243 we get

$$S = \frac{1}{2a(a^2+b^2)} \coth a + \frac{1}{2b(a^2+b^2)} \cdot \frac{\sin b}{1-\cos b}.$$

♦ **Solution of exercise 11.7**

i) A simple change of variable and the periodicity of the cosine function imply that the integral does not depend on $\alpha \in \mathbb{R}$. Moreover, putting $\alpha = \pi/2$, a sine appears, and as this is an odd function, it follows that the integral is real.

ii) Passing in polar coordinates $(x, y) \mapsto (r, \theta)$, we get

$$\begin{aligned}\widetilde{f}(u,v) = \psi(\rho) &= \iint_{\mathbb{R}^2} e^{2i\pi(ux+vy)}\,dx\,dy \\ &= \int_0^{+\infty} \varphi(r) \left(\int_{-\pi}^{\pi} e^{2i\pi\rho r \cos\theta}\,d\theta \right) r\,dr,\end{aligned}$$

or

$$\psi(\rho) = 2\pi \int_0^{+\infty} r\,\varphi(r)\, J_0(2\pi\rho r)\,dr.$$

This will be denoted, as indicated, by $\varphi(r) \xrightarrow{\text{H. T.}} \psi(\rho)$.

iii) This is simply the translation of the Fourier inversion formula in this context.

iv) In the formula

$$\frac{d^2\varphi}{dr^2} + \frac{1}{r}\frac{d\varphi}{dr}$$

we recognize the expression for the laplacian of f in polar coordinates, since there is no dependency on θ. The Fourier transform of $\triangle f$ is equal to $-4\pi^2\rho^2\widetilde{f}(u,v)$ with $\rho = (u,v)$ and $\rho^2 = \rho^2$, hence the formula follows.

v) The Parseval-Plancherel identity

$$\int f_1(x,y)\,\overline{f_2(x,y)}\,dx\,dy = \int \widetilde{f_1}(u,v)\,\overline{\widetilde{f_2}(u,v)}\,du\,dv,$$

gives, in polar coordinates, the required identity:

$$2\pi \int_0^{\infty} r\,\varphi_1(r)\,\overline{\varphi_2(r)}\,dr = 2\pi \int_0^{\infty} \rho\,\psi_1(\rho)\,\overline{\psi_2(\rho)}\,d\rho.$$

♦ **Solution of exercise 11.9.** As seen on page 320, the amplitude of the diffraction figure is given *radially* by the Hankel transform of the *radial* transmittance function, which is given here by

$$f(x,y) = f(r) = \begin{cases} 0 & \text{if } r > \ell, \\ 1 & \text{if } r < \ell, \end{cases}$$

where $r = \sqrt{x^2+y^2}$.

The amplitude is therefore

$$A(\theta,\varphi) = A(\rho) = \frac{\ell}{\rho}\, J_1(2\pi\rho\ell) \qquad \text{with} \qquad \rho = \sqrt{\theta^2+\varphi^2}.$$

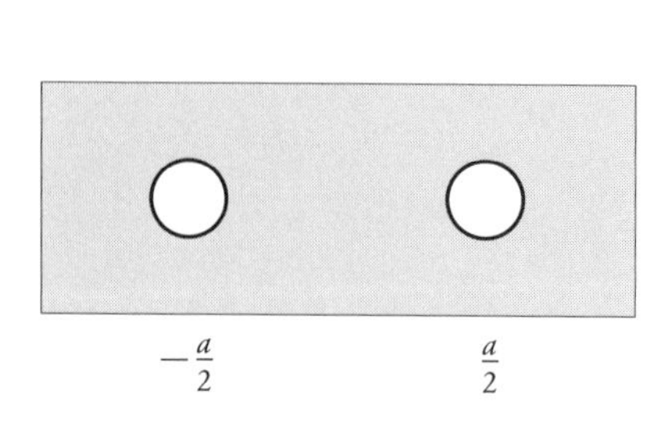

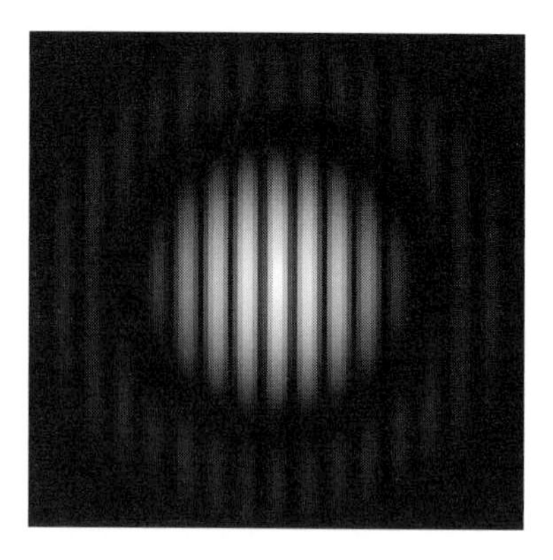

Fig. 11.7 – The diffraction figure created by two circular holes. The spacing between the two lenses is here taken to be 4 times their diameter.

The transmittance function of two holes is

$$g(x,y) = f(x,y) * \left[\delta\left(x - \frac{a}{2}\right) + \delta\left(x + \frac{a}{2}\right)\right].$$

The diffraction figure due to the two holes is obtained by multiplying the diffraction amplitude due to *a single* hole by the Fourier transform of the distribution $\delta(x - a/2) + \delta(x + a/2)$, which is equal to $2\cos(\pi\theta a)$.

We obtain then the result sketched in Figure 11.7.

♦ **Solution of problem 5.** The total density of charge is

$$q(\boldsymbol{r}) = \sum_\alpha e_\alpha \rho_\alpha(\boldsymbol{r}) + q_0 \delta(\boldsymbol{r}).$$

Denote by $\varphi(\boldsymbol{r})$ the effective electrostatic potential created by the distribution of charge $q(\boldsymbol{r})$ (including therefore both the exterior charge and the induced polarization cloud). The quantities $\varphi(\boldsymbol{r})$ and $q(\boldsymbol{r})$ are related by two equations. The first, an electrostatic relation, is the **Poisson equation**, which is

$$\triangle\varphi(\boldsymbol{r}) = -4\pi\, q(\boldsymbol{r})\,; \tag{$*$}$$

in Gauss units. The second, a thermodynamic relation, is the **Boltzmann equation**

$$\rho_\alpha(\boldsymbol{r}) = \rho_\alpha\, \mathrm{e}^{-\beta e_\alpha \varphi(\boldsymbol{r})}, \tag{$**$}$$

where the constant value of ρ_α is justifed by the fact that, for $r \to \infty$, we have $\varphi(\boldsymbol{r}) \to 0$ in the suitable gauge and $\rho_\alpha(\boldsymbol{r}) \to \rho_\alpha$.

In the limit of weak couplings (or low densities), we can linearize equation ($**$) and obtain

$$\rho_\alpha(\boldsymbol{r}) = \rho_\alpha(1 - \beta e_\alpha \varphi(\boldsymbol{r})),$$

which, inserted into ($*$), gives

$$\boxed{(\triangle - \varkappa^2)\,\varphi(\boldsymbol{r}) = -4\pi q_0 \delta(\boldsymbol{r}) \qquad \text{with} \quad \varkappa^2 \stackrel{\text{def}}{=} 4\pi\beta \sum_\alpha \rho_\alpha e_\alpha^2,}$$

linearized Poisson-Boltzmann

taking into account that the system is neutral ($\sum \rho_\alpha e_\alpha = 0$).

We will now try to solve this equation. First, assuming that $\varphi(\boldsymbol{r})$ has a Fourier transform, we obtain the conjugate equation

$$(-\boldsymbol{k}^2 - \varkappa^2)\,\widetilde{\varphi}(\boldsymbol{k}) = -4\pi q_0$$

with the conventions

$$\widetilde{\varphi}(\boldsymbol{k}) \stackrel{\text{def}}{=} \int \mathrm{e}^{-\mathrm{i}\boldsymbol{k}\cdot\boldsymbol{r}}\,\varphi(\boldsymbol{r})\,\mathrm{d}^3\boldsymbol{r} \qquad \text{and} \qquad \varphi(\boldsymbol{r}) = \frac{1}{(2\pi)^3}\int \mathrm{e}^{\mathrm{i}\boldsymbol{k}\cdot\boldsymbol{r}}\,\widetilde{\varphi}(\boldsymbol{k})\,\mathrm{d}^3\boldsymbol{k}.$$

We deduce that[11]

$$\widetilde{\varphi}(\boldsymbol{k}) = \frac{4\pi q_0}{\boldsymbol{k}^2 + \varkappa^2}.$$

There only remains to compute the inverse Fourier transform

$$\varphi(\boldsymbol{r}) = \int \frac{4\pi q_0}{\boldsymbol{k}^2 + \varkappa^2}\,\mathrm{e}^{\mathrm{i}\boldsymbol{k}\cdot\boldsymbol{r}}\,\frac{\mathrm{d}^3\boldsymbol{r}}{(2\pi)^3}.$$

For this purpose, we use spherical coordinates on the space of vectors $\boldsymbol{k}$, with the polar axis oriented in the direction of $\boldsymbol{r}$, and with the notation $\boldsymbol{k}\cdot\boldsymbol{r} = kr\cos\theta$, where $k > 0$ and $r > 0$. Then

$$\varphi(\boldsymbol{r}) = \int_0^{2\pi}\mathrm{d}\varphi\int_0^{\pi}\mathrm{d}\theta\int_0^{+\infty}\frac{4\pi q_0}{k^2+\varkappa^2}k^2\sin\theta\,\mathrm{e}^{\mathrm{i}kr\cos\theta}\frac{\mathrm{d}k}{(2\pi)^3},$$

or

$$\begin{aligned}\varphi(\boldsymbol{r}) &= \int_{-1}^{1}\mathrm{d}(\cos\theta)\int_0^{+\infty}\frac{8\pi^2 q_0}{k^2+\varkappa^2}k^2\,\mathrm{e}^{\mathrm{i}kr\cos\theta}\frac{\mathrm{d}k}{(2\pi)^3}\\ &= \int_0^{+\infty}\frac{q_0}{\pi}\frac{k^2}{k^2+\varkappa^2}\frac{1}{\mathrm{i}kr}\left(\mathrm{e}^{\mathrm{i}kr} - \mathrm{e}^{-\mathrm{i}kr}\right)\mathrm{d}k = \frac{2q_0}{\pi r}\int_0^{+\infty}\frac{k\sin kr}{k^2+\varkappa^2}\,\mathrm{d}k.\end{aligned}$$

We must still evaluate this integral relative to the variable k. For this, we put

$$\mathscr{I}(r) \stackrel{\text{def}}{=} \int_0^{+\infty}\frac{k\sin kr}{k^2+\varkappa^2}\,\mathrm{d}k = \frac{1}{2}\operatorname{Im}\left(\int_{-\infty}^{+\infty}\frac{k\,\mathrm{e}^{\mathrm{i}kr}}{k^2+\varkappa^2}\,\mathrm{d}k\right),$$

and we use the method of residues by extending the variable k to the complex plane. The poles of the meromorphic function $f(z) \stackrel{\text{def}}{=} z\,\mathrm{e}^{\mathrm{i}zr}/(z^2+\varkappa^2)$ are located at $z_1 = \mathrm{i}\varkappa$ and $z_2 = -\mathrm{i}\varkappa$. Since $r > 0$, we can integrate on the contour described in the following figure:

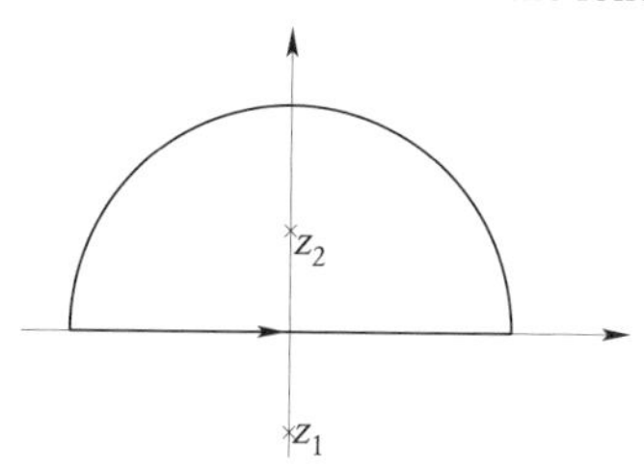

and, after applying Jordan's second lemma 4.85, as justified by the fact that $k/(k^2+\varkappa^2)$ tends to 0 when $[k \to \infty]$, and after taking the limit where the radius in the contour goes to infinity, we derive

$$\int_{-\infty}^{+\infty}\frac{k\,\mathrm{e}^{\mathrm{i}kr}}{k^2+\varkappa^2}\,\mathrm{d}k = 2\mathrm{i}\pi\operatorname{Res}(f;z_1) = 2\mathrm{i}\pi\,\frac{\mathrm{i}k\,\mathrm{e}^{-\varkappa r}}{2\mathrm{i}\varkappa} = \mathrm{i}\pi\,\mathrm{e}^{-\varkappa r},$$

[11] Being careful of the fact that, in the general case, the equation $xT(x) = a$ has solutions given $T(x) = a\,\mathrm{pv}(1/x) + b\delta$, with $b \in \mathbb{C}$. Here, $\widetilde{\varphi}(\boldsymbol{k})$ is defined up to the addition of the Fourier transform of an harmonic functions.

hence
$$\mathscr{I} = \frac{\pi}{2}\,\mathrm{e}^{-\varkappa r}$$

and
$$\boxed{\varphi(\boldsymbol{r}) = \frac{q_0}{r}\,\mathrm{e}^{-\varkappa r}}$$
Debye potential.

Using the Boltzmann equation, each of the functions $\rho_\alpha(\boldsymbol{r})$ can be computed now. Note that the Debye electrostatic screen occurs at a length scale of $\varkappa^{-1}$, which tends to infinity in the limit of weak couplings (small densities). For distances $r \ll \varkappa$, the potential is $\varphi(\boldsymbol{r}) \approx q_0/r$, and we recover the Coulomb potential.

The total charge present around q_0 is

$$\begin{aligned}
Q &= \iiint \sum_\alpha e_\alpha \rho_\alpha(\boldsymbol{r})\,\mathrm{d}^3\boldsymbol{r} \\
&= \iiint \sum_\alpha e_\alpha \rho_\alpha\big(1 - \beta e_\alpha\,\varphi(\boldsymbol{r})\big)\,\mathrm{d}^3\boldsymbol{r} \\
&= -\iiint \sum_\alpha \rho_\alpha \beta e_\alpha^2\,\varphi(\boldsymbol{r})\,\mathrm{d}^3\boldsymbol{r} && \text{by initial neutrality,} \\
&= -q_0 \iiint \sum_\alpha \beta \rho_\alpha e_\alpha^2\,\frac{\mathrm{e}^{-\varkappa r}}{r}\,\mathrm{d}^3\boldsymbol{r} = -q_0.
\end{aligned}$$

Thus is exactly compensates the additional charge q_0. In other words, the system has managed to become neutral again.[12] The Debye screen is said to be a **total screen**.

[12] This may seem surprising; whichever charge is added to the system, it remains neutral. This is a strange property of an infinite system, which may bring in charges from infinity, and make them disappear there. This type of paradox also happens in the famous "infinite hotels," all rooms of which are occupied, but in which, nevertheless, a new customer can always be accommodated: it suffices that the person in Room 1 move to Room 2, the person in Room 2 move to Room 3, and more generally, the person in Room n move to Room $n+1$. All previous occupants have a new room, and Room 1 becomes available for the new customer.

Chapter 12

The Laplace transform

The Laplace transform is an integral transformation, which is a kind of generalization of the Fourier transform. It has a double interest:

1. it avoids using distributions in some cases where a function does not have a Fourier transform in the sense of functions;
2. it can be used to solve problems described by differential equations taking initial conditions into account, that is, it can attack the Cauchy problem associated to the system. An example will be given to the dynamics of the electromagnetic field without sources.

12.1 Definition and integrability

In this chapter, we are interested in an integral transformation operating on functions f which *vanish* for negative values of the variable: $f(t) = 0$ for all $t < 0$ (no continuity at 0 is imposed). An example is the function $t \mapsto H(t) \cos t$.

In the literature about the Laplace transform, the factor $H(t)$ is frequently omitted; we will follow this custom, except where some ambiguity may result, and we will therefore speak of the function $t \mapsto \cos t$, tacitly assuming that this definition is restricted to *positive* values of t.

DEFINITION 12.1 A **causal function** is a function $t \mapsto f(t)$ defined on $\mathbb{R}$ which is zero for negative values of its argument:

$$f(t) = 0 \qquad \text{for any } t < 0.$$

12.1.a Definition

DEFINITION 12.2 Let $f(t)$ be a real- or complex-valued locally integrable function defined for real values of t. The **(unilateral) Laplace transform of $f(t)$** is the complex function of the complex variable p denoted $\widehat{f}(p)$ (or $F(p)$) which is defined by

$$\boxed{\widehat{f}(p) \stackrel{\text{def}}{=} \int_0^{+\infty} f(t)\,\mathrm{e}^{-pt}\,\mathrm{d}t.}$$

Whereas the function $\widehat{f}$ itself is called the Laplace tranform, the operation $f \mapsto \widehat{f}$ should be called the Laplace transformation.

DEFINITION 12.3 Let f be a locally integrable function on $\mathbb{R}$. The **bilateral Laplace transform of f** is the function

$$p \longmapsto \int_{-\infty}^{+\infty} f(t)\,\mathrm{e}^{-pt}\,\mathrm{d}t.$$

Its properties are of the same kind as those of the unilateral transform, with variants that the reader may easily derive by herself.

The bilateral Laplace transform of $H(t)\,f(t)$ is simply the Laplace transform of the function $f(t)$. In the remainder of this chapter, only the unilateral transform will be considered.

Remark 12.4 If f is a function *zero for $t < 0$* which has a Fourier transform, then the relation

$$\widehat{f}(\mathrm{i}\omega) = \widetilde{f}\left(\frac{\omega}{2\pi}\right) \qquad \text{with } \omega \in \mathbb{R}$$

holds. The Laplace transform is an extension of the notion of Fourier transform. More precisely, $\widehat{f}(x + \mathrm{i}\omega)$ is, up to the change of variable $\omega = 2\pi\nu$, the Fourier transform of $t \mapsto f(t)\,\mathrm{e}^{-xt}$ evaluated at $\omega/2\pi$.

Sometimes the following definition is found in the literature:

DEFINITION 12.5 (Original, image) An **original** is any function of a real variable which is locally integrable, with real or complex values, such that

i) f is causal ($f(t) = 0$ for all $t < 0$);

ii) $|f(t)|$ does not increase faster than any exponential function, that is, there exist constants $M > 0$ and $s \in \mathbb{R}$ such that

$$|f(t)| \leqslant M\,\mathrm{e}^{st} \qquad \text{for all } t \in \mathbb{R}.$$

The Laplace transform of an original is the **image** of the original.

Pierre Simon, marquis de LAPLACE (1749–1827), befriended d'Alembert (p. 415) at eighteen, found a position as professor of mathematics at the Paris military school, and then became a professor at the École Polytechnique. He was passionate about astronomy and developed many concepts and useful tools for the study of mechanics, in particular in the field of differential equations. He studied cosmology and cosmogony (the primitive nebula of Laplace), rediscovered the forgotten works of Bayes (p. 518) and was the first to compute the integral of the gaussian. When Napoléon asked why he did not mention God in his *Traité de mécanique céleste*, his answer was: "Je n'ai pas eu besoin de cette hypothèse" (I did not need this assumption).

The following notation are used: if $f(t)$ is an original, its Laplace transform is denoted $\widehat{f}(p)$ or $F(p)$. The symbol $\sqsupset$ is also in widespread use, used as follows: $f(t) \sqsupset F(p)$. (In certain russian books, the notation $f(t) \risingdotseq F(p)$ also appears.)

12.1.b Integrability

In the remainder of this chapter, we write $p = x + \mathrm{i}\omega$.

Let f be a locally integrable function. We are looking for the domain where its Laplace transform $\widehat{f}(p)$ is defined. Note first that integrability of $t \longmapsto f(t)\,\mathrm{e}^{-pt}$ is equivalent to integrability of $t \longmapsto f(t)\,\mathrm{e}^{-xt}$.

Moreover, if this function is integrable for some x_0, then it is also integrable for any $x > x_0$, since

$$\left|f(t)\,\mathrm{e}^{-xt}\right| = \left|f(t)\,\mathrm{e}^{-x_0 t}\right| \mathrm{e}^{(x_0-x)t} \leqslant \left|f(t)\,\mathrm{e}^{-x_0 t}\right|.$$

It follows that the set of complex numbers p, where $t \longmapsto f(t)\,\mathrm{e}^{-pt}$ is integrable, either is empty or is a (right) half-plane in the complex plane, or indeed is the entire complex plane $\mathbb{C}$.

DEFINITION 12.6 The **convergence abscissa** of the original function f is the lower bound of all real numbers x for which the function above is integrable:

$$\alpha \stackrel{\text{def}}{=} \inf\left\{x \in \mathbb{R}\,;\ t \longmapsto \left|f(t)\right|\mathrm{e}^{-xt} \text{ is integrable}\right\}.$$

From the previous reasoning, we deduce the following result:

PROPOSITION 12.7 *Let $f(t)$ be an original and let α be its convergence abscissa. Denote as before $p = x + \mathrm{i}\omega$. Then*

i) for $x < \alpha$, the function $t \longmapsto f(t)\,\mathrm{e}^{-pt}$, is not integrable;

ii) for $x > \alpha$, this function is integrable;

iii) for $x = \alpha$, it may, or may not, be Lebesgue-integrable. If it is not Lebesgue-integrable, the improper integral

$$\int_{-\infty}^{+\infty} f(t)\,\mathrm{e}^{-pt}\,\mathrm{d}t = \lim_{\substack{R\to+\infty\\ R'\to-\infty}} \int_{R'}^{R} f(t)\,\mathrm{e}^{-pt}\,\mathrm{d}t$$

may still converge for certain values of $p = \alpha + \mathrm{i}\omega$, which provides an extension of the Laplace transform of f to those values of p.

Example 12.8 Consider the Heaviside function. Its Laplace transform is given by

$$\widehat{H}(p) = \int_0^{+\infty} H(t)\,\mathrm{e}^{-pt}\,\mathrm{d}t = \frac{1}{p}\left[\mathrm{e}^{-pt}\right]_0^{+\infty},$$

so that the convergence abscissa of H is equal to 0, and for all $p \in \mathbb{C}$ such that $\mathrm{Re}(p) > 0$, we have $\widehat{H}(p) = 1/p$. The domain where $\widehat{H}$ is defined is thus an *open* half-plane.

Notice that, if the function $\widehat{H}$ is extended to the imaginary axis (except the origin) by defining, for any nonzero ω, $H(\mathrm{i}\omega) = 1/\mathrm{i}\omega$, we obtain a function which is "close" to the Fourier transform of H, the latter being equal to

$$\widetilde{H}(\nu) = \mathrm{pv}\,\frac{1}{2\mathrm{i}\pi\nu} + \frac{\delta}{2},$$

which is the same as $\mathrm{pv}(1/2\mathrm{i}\pi\nu)$ on $\mathbb{R}^*$. This resemblance will be explained in Section 12.4.e, page 345.

Example 12.9 Let $f(t) = 1/(1+t^2)$. Then we have

$$F(p) = \int_0^{+\infty} \frac{\mathrm{e}^{-pt}}{1+t^2}\,\mathrm{d}t,$$

which is defined for all $p \in \mathbb{C}$ such that $\mathrm{Re}(p) \geqslant 0$. The domain of definition of F is therefore a *closed* half-plane.

Example 12.10 Let $a \in \mathbb{C}$ be an arbitrary complex number, and consider the Laplace transform of the function $t \mapsto H(t)\,\mathrm{e}^{at}$. We have

$$H(t)\,\mathrm{e}^{at} \sqsupset \int_0^{+\infty} \mathrm{e}^{at}\,\mathrm{e}^{-pt}\,\mathrm{d}t = \int_0^{+\infty} \mathrm{e}^{-(p-a)t}\,\mathrm{d}t,$$

so that the convergence abscissa is $\alpha = \mathrm{Re}(a)$, and for all $p \in \mathbb{C}$ such that $\mathrm{Re}(p) > \mathrm{Re}(a)$, the Laplace transform of $H(t)\,\mathrm{e}^{at}$ is $1/(p-a)$.

THEOREM 12.11 *In the open half-plane on the right of the abcissa of convergence, the Laplace transform $\widehat{f}$ is* holomorphic *and hence also analytic.*

PROOF. Let $x \in \mathbb{R}$ with $x > \alpha$. Then, for any $m \in \mathbb{N}$, the function $t \mapsto t^m f(t)\mathrm{e}^{-pt}$ is integrable: since $x > \alpha$, the real number $y = (x+\alpha)/2$ satisfies $\alpha < y < x$; moreover, $t \mapsto f(t)\,\mathrm{e}^{-yt}$ is integrable, and we may write

$$t^m \left|f(t)\right| \mathrm{e}^{-xt} = \left|f(t)\right| \mathrm{e}^{-yt} \cdot t^m \mathrm{e}^{-(x-a)t/2},$$

where the first factor is integrable while the second is bounded. The derivative of F can be computed using the theorem of differentiation under the integral sign, and is given by

$$\frac{\mathrm{d}F}{\mathrm{d}p} = F'(p) = \int_0^{\infty} (-t)\,f(t)\,\mathrm{e}^{-pt}\,\mathrm{d}t.$$

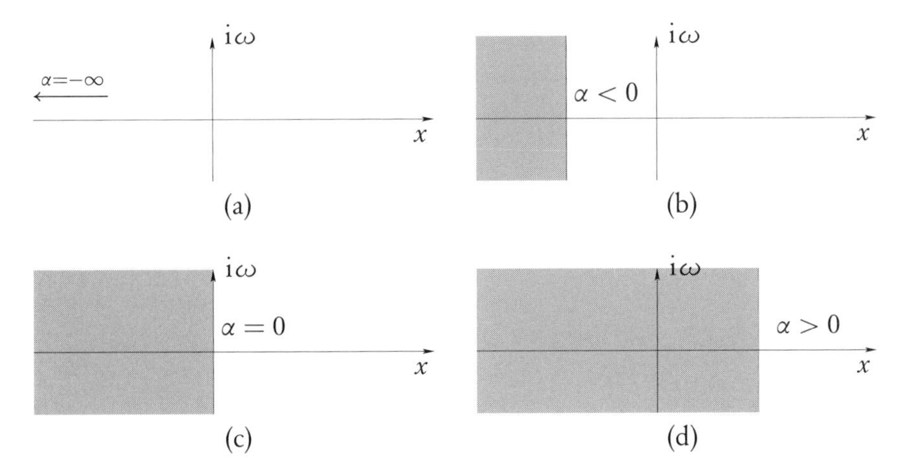

Fig. 12.1 – Convergence abcissas for unilateral Laplace transforms. (a) Function with bounded support on the right. (b) Function rapidly decaying on the right. (c) Tempered function. (d) Rapidly increasing function. In the non-gray open set, the Laplace transform is holomorphic.

Since F is differentiable in the complex sense at any point of the open half-plane $\{p \in \mathbb{C}\,;\ \operatorname{Re} p > \alpha\}$, it is an analytic function.

Notice that the convergence abcissa for $t \mapsto tf(t)$ is the same as that for f. Also, an obvious induction shows that

$$F^{(n)}(p) = \int_0^{\infty} (-t^n) f(t) \, e^{-pt} \, dt.$$

We will admit the following result, which is summarized in Figure 12.1:

PROPOSITION 12.12 *Let f be an original.*

i) If the function $f(t)$ has bounded support on the right, then $\alpha = -\infty$.

ii) If $f(t)$ decays rapidly on the right, then $-\infty \leqslant \alpha < 0$ and $\widehat{f}(p)$ is holomorphic in a half-plane containing the imaginary axis (in particular, the Fourier transform of f exists).

iii) If $f(t)$ is tempered (in the sense of distributions) and does not tend to 0 at infinity, then $\alpha = 0$; its Fourier transform does not necessarily exist in the sense of functions, but it exists in the sens of distributions.

iv) Finally, if $f(t)$ increases rapidly at infinity, then $0 < \alpha \leqslant +\infty$; then $\widehat{f}$ (if it exists) is holomorphic on a half-plane not containing the imaginary axis and the Fourier transform of f does not exist.

Example 12.13 The convergence abcissa for the "rectangle" function is $-\infty$ and we have

$$\widehat{\Pi}(p) = \int_0^{1/2} e^{-pt} \, dt = \frac{1 - e^{-p/2}}{p},$$

which is analytic on $\mathbb{C}$; the singularity at 0 is indeed an artificial one, since $\widehat{\Pi}$ is bounded in a neighborhood of 0; $\widehat{\Pi}$ has the power series expansion

$$\widehat{\Pi}(p) = \sum_{n=0}^{\infty} \left(\frac{(-1)^{n+1}}{(n+1)!\, 2^{n+1}} \right) p^n.$$

Any nonzero polynomial function or rational function has convergence abcissa equal to 0.

♦ **Exercise 12.1** Show that the function $g : t \mapsto \mathrm{e}^{t^2}$ does not have a Laplace transform (that is, $\alpha = +\infty$).

12.1.c Properties of the Laplace transform

PROPOSITION 12.14 *Let $f(t)$ be an original and $F(p)$ its image by the Laplace transform. For $x > \alpha$, the function $F(x + \mathrm{i}\omega)$ tends to 0 as $\omega \to \pm\infty$.*

Proof. This is an easy consequence of the Riemann-Lebesgue lemma (see page 282):

$$\lim_{\omega \to \pm\infty} \int_0^{+\infty} \left[f(t)\, \mathrm{e}^{-xt} \right] \mathrm{e}^{-\mathrm{i}\omega t}\, \mathrm{d}t = 0.$$

We know that, on the right of the abscissa of convergence, the Laplace transform is holomorphic. What happens elsewhere?

Take, for instance, $f(t) = \mathrm{e}^{3t}$. The convergence abscissa is $\alpha = 3$. Its Laplace transform is $F(p) = 1/(p-3)$ which is holomorphic everywhere except at $p = 3$. More precisely, F is defined in the convergence half-plane $\{z \in \mathbb{C}\ ;\ \mathrm{Re}(z) > 3\}$, but it can be *analytically continued* to the plane with the single point $\{3\}$ removed, namely, to $\mathbb{C} \setminus \{3\}$. It often happens that the Laplace transform of an original is a *meromorphic* function.[1] Thus we will show that

$$\cos t \sqsupset \frac{p}{p^2+1};$$

this image function is meromorphic and has poles at $p = \pm\mathrm{i}$. It is represented in Figure 12.2, together with its analytic continuation to $\mathbb{C} \setminus \{-\mathrm{i}, \mathrm{i}\}$.

12.2 Inversion

It is possible to find an inversion formula for the Laplace transform using our knowledge of the Fourier inversion formula. Indeed, if $F(p)$ is the Laplace transform of the function $f(t)$, we have, for any $x > \alpha$,

$$F(x + 2\mathrm{i}\pi\nu) = \int_{-\infty}^{+\infty} H(t)\, f(t)\, \mathrm{e}^{-xt} \mathrm{e}^{-2\mathrm{i}\pi\nu t}\, \mathrm{d}t,$$

[1] This is not always the case. It may have branch points, as in the case of the Laplace transform of $t \mapsto H(t)/(t+1)$.

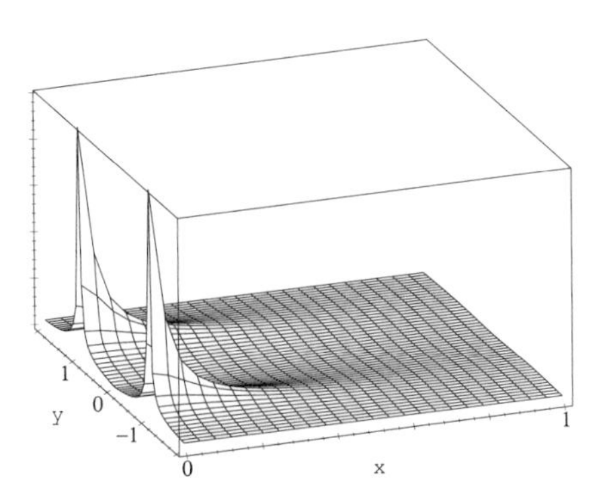

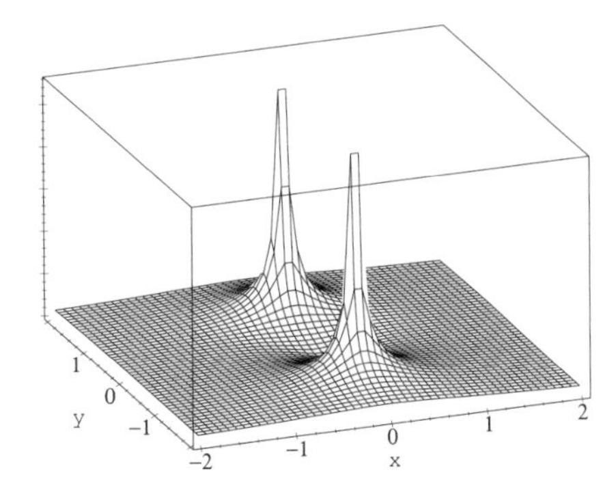

Fig. 12.2 – Representation of the Laplace transform of the function $x \mapsto \cos x$. The z-axis is the modulus of the function $p \mapsto p/(p^2+1)$. This function may be *analytically* continued to the complex plane without i and −i.

which shows that, for fixed x, the function $\nu \mapsto F(x + 2i\pi\nu)$ is the Fourier transform of $H(t)\,f(t)\,e^{-xt}$. The Fourier inversion formula leads, when t is a point where $H(t)\,f(t)$ is continuous, to

$$H(t)\,e^{-xt} f(t) = \int_{-\infty}^{+\infty} F(x + 2i\pi\nu)\,e^{2i\pi\nu t}\,d\nu$$

or

$$H(t)\,f(t) = \int_{-\infty}^{+\infty} F(x + 2i\pi\nu)\,e^{xt} e^{2i\pi\nu t}\,d\nu = \frac{1}{2i\pi}\int_{D_x} F(p)\,e^{pt}\,dp,$$

where we have denoted $p = x + 2i\pi\nu$ and D_x is the vertical line

$$D_x \stackrel{\text{def}}{=} \{x + i\omega\ ;\ \omega \in \mathbb{R}\} \qquad \textbf{(Bromwich contour)}.$$

Remark 12.15 Note that the resulting formula gives the original as a function of the image, for an *arbitrary* $x > \alpha$.

The result of the integral is indeed independent of the integration contour chosen since, F being analytic on the convergence half-plane, Cauchy's theorem guarantees that the difference between the integral on D_x and on $D_{x'}$ vanishes (using the fact that $\lim_{\omega\to\pm\infty} F(p) = 0$, as given by Proposition 12.14).

The integration is performed by cutting the line D_x and by closing it, on the left, by a circle of arbitrarily large radius, containing all the singularities of F. (Sometimes it is necessary to use a more subtle contour; see, for instance [31, §7.5].)

THEOREM 12.16 (Inversion of the Laplace transform) *Let f be an original and F its Laplace transform. Let α be the convergence abcissa. For any point where f is continuous, we have*

$$\boxed{f(t) = \frac{1}{2\mathrm{i}\pi} \int_{x_0-\mathrm{i}\infty}^{x_0+\mathrm{i}\infty} F(p)\,\mathrm{e}^{pt}\,\mathrm{d}p} \qquad \textit{for any } x_0 > \alpha.$$

Remark 12.17 If we start with an arbitrary function, this formula leads us back to $H(t)f(t)$. Can you show this using the fact that F is analytic in the half-plane $x > a$ and decays on vertical lines?

12.3

Elementary properties and examples of Laplace transforms

Just as in the case of the Fourier transform, there are relations linking the Laplace transform and operations like translation, convolution, differentiation, and integration.

12.3.a Translation

There is a translation theorem similar to what holds for the Fourier transform, but which requires some care here: since we are looking at the *unilateral* Laplace transform, we will write explicitly the function $H(t)f(t)$ instead of $f(t)$. The, writing $p = x + \mathrm{i}\omega$ as usual and denoting by α the convergence abcissa for f, we have

$$\begin{array}{lll} \text{if} & H(t)f(t) \sqsupset F(p) & \text{for } \alpha < x, \\ \text{then} & H(t)f(t)\,\mathrm{e}^{-a\tau} \sqsupset F(p+a) & \text{for } \alpha - \mathrm{Re}(a) < x. \end{array}$$

Similarly

$$\begin{array}{lll} \text{if} & H(t)f(t) \sqsupset F(p) & \text{for } \alpha < x, \\ \text{then} & H(t-\tau)f(t-\tau) \sqsupset F(p)\,\mathrm{e}^{-\tau p} & \text{for } \alpha - \tau < x. \end{array}$$

Be careful to consider $H(t-\tau)f(t-\tau)$, that is, the function which vanishes for $t < \tau$. There is, a priori, no relation between the Laplace transform of the function $H(t)f(t-\tau)$ and that of $H(t)f(t)$.

12.3.b Convolution

For two causal functions f and g, we define the *convolution product*, if it exists, in the same manner as was done in 7.64 on page 211, and the definition simplifies here to

$$f * g(t) = \int_0^{+\infty} f(s)\, g(t-s)\, \mathrm{d}s = \int_0^t f(s)\, g(t-s)\, \mathrm{d}s.$$

In particular, the quantity $f * g(t)$ depends only on the values of f and g between the times 0 and t. This convolution, although formally identical to what occurred with Fourier transforms, still has a causal aspect. To remember this causality, we sometimes write explicitly $Hf * Hg$ instead of $f * g$.

As for the Fourier transform, there is a convolution theorem, the proof of which is left as an exercise:

THEOREM 12.18 *Let f and g be originals with convergence abcissas respectively equal to α and α'.*

$$\begin{array}{llll} \textit{If} & f(t) \sqsupset F(p) & \textit{for } \alpha < x, \\ \textit{and} & g(t) \sqsupset G(p) & \textit{for } \alpha' < x, \\ \textit{then} & [Hf * Hg](t) \sqsupset F(p)\, G(p) & \textit{for } \max(\alpha, \alpha') < x. \end{array}$$

Similarly, we have:

$$\begin{array}{llll} \textit{if} & f(t) \sqsupset F(p) & & \textit{for } \alpha < x, \\ \textit{and} & g(t) \sqsupset G(p) & & \textit{for } \alpha' < x, \\ \textit{then} & f(t)\, g(t) \sqsupset \dfrac{1}{2\mathrm{i}\pi} \displaystyle\int_{x_0 - \mathrm{i}\infty}^{x_0 + \mathrm{i}\infty} F(s)\, G(p-s)\, \mathrm{d}s & & \textit{for } \begin{cases} \alpha < x_0, \\ \alpha + \alpha' < x_0. \end{cases} \end{array}$$

12.3.c Differentiation and integration

Let us now try to find the Laplace transform of the derivative of a function *in the sense of functions*. Take therefore an original function f and denote by F its Laplace transform:

$$H(t)\, f(t) \sqsupset F(p).$$

We assume that f is differentiable in the sense of functions; then we have

$$H(t)\, f'(t) \sqsupset \int_0^{+\infty} f'(t)\, \mathrm{e}^{-pt}\, \mathrm{d}t = \int_0^{+\infty} p\, f(t)\, \mathrm{e}^{-pt}\, \mathrm{d}t + \Big[f(t)\, \mathrm{e}^{-pt} \Big]_0^{+\infty}.$$

One shows (see Exercise 12.6) that $\lim_{t \to +\infty} f(t)\, \mathrm{e}^{-pt} = 0$ so that the boundary term is equal to $-f(0)$ or, more precisely, to $-\lim_{t \to 0^+} f(t)$, which is also denoted $-f(0^+)$.

THEOREM 12.19 (Laplace transform and differentiation) *Let f be an original which is continuous, differentiable, and such that its derivative is also an original. Then, if $f(t) \sqsupset F(p)$, the Laplace transform of the derivative of f* in the sense of functions *is given by*

$$H(t)\,\frac{\mathrm{d}}{\mathrm{d}t}\,f(t) \sqsupset p\,F(p) - f(0^+).$$

Similarly, if f can be differentiated n times and if its n-th derivative if an original, we have

$$H(t)\,\frac{\mathrm{d}^n}{\mathrm{d}t^n}\,f(t) \sqsupset p^n\,F(p) - p^{n-1}\,f(0^+) - p^{n-2}\,f'(0^+) - \cdots - f^{(n-1)}(0^+).$$

Example 12.20 Let $f(t) = 1 + \mathrm{e}^{-t}$, which is an original with convergence abscissa $\alpha = 0$, and with Laplace transform equal to

$$1 + \mathrm{e}^{-t} \sqsupset \frac{1}{p} + \frac{1}{p+1}.$$

The function f is differentiable, with derivative $g : t \mapsto -\mathrm{e}^{-t}$, which is an original with convergence abscissa $\alpha' = -1$. Then we have

$$p\,F(p) - f(0^+) = 1 + \frac{p}{p+1} - 2 = -\frac{1}{p+1} = G(p).$$

It is very important to remark that the derivative is considered in the sense of functions and the derivative at 0 must not involve a Dirac distribution δ. This is why the term $H(t)$ was explicitly included *outside* the derivative sign. We will see later what happens if the derivative is considered in the sense of distributions.

There is also a converse result:

THEOREM 12.21 *Let $f(t)$ be an original with convergence abscissa α. Then $t^n f(t)$ is an original with the same convergence abscissa. If $f(t) \sqsupset F(p)$, then we have*

$$(-t)^n f(t) \sqsupset \frac{\mathrm{d}^n}{\mathrm{d}p^n}\,F(p).$$

Similarly, there is the following integration result:

THEOREM 12.22 (Laplace transform and integration) *Let f be an original, with convergence abscissa equal to α.*

$$\text{If} \qquad f(t) \sqsupset F(p) \qquad \text{for } \alpha < x,$$

$$\text{then} \qquad \int_0^t f(u)\,\mathrm{d}u \sqsupset \frac{F(p)}{p} \qquad \text{for } \max(\alpha, 0) < x.$$

Remark 12.23 This may be deduced from the convolution theorem since

$$\int_0^t f(u)\,\mathrm{d}u = [f * H](t).$$

Conversely, we have

THEOREM 12.24 *Let f be an original.*

$$\text{If} \quad f(t) \sqsupset F(p) \qquad \text{for } \alpha < x,$$

$$\text{then} \quad \frac{f(t)}{t} \sqsupset \int_p^{\infty} F(z)\,\mathrm{d}z \quad \text{for } \sup(\alpha, 0) < x.$$

In this last relation, the path of integration joins the point p at infinity, in any direction in the half-plane of convergence. The result of the integration is independent of the chosen path since F is holomorphic in this half-plane. It is still necessary that $F(p)$ decay faster than $1/p$ at infinity (this is true for nice functions, but will not hold in general for the Laplace transform of distributions).

12.3.d Examples

We have already shown that

$$H(t) \sqsupset \frac{1}{p} \quad \text{for } 0 < \operatorname{Re}(p).$$

Using the integration and translation theorems, we obtain

$$H(t)\,\mathrm{e}^{at} \sqsupset \frac{1}{p-a} \qquad \text{for } \operatorname{Re}(a) < \operatorname{Re}(p),$$

$$H(t)\,\mathrm{e}^{at}\frac{t^{n-1}}{(n-1)!} \sqsupset \frac{1}{(p-a)^n} \qquad \text{for } \operatorname{Re}(a) < \operatorname{Re}(p).$$

Using a simple fraction expansion, this provides a way to find the original for the Laplace transform of any rational function. Note (although we do not prove it) that the formula remains valid if we take a value of n which is not an integer; then the factor $(n-1)!$ in the formula must be replaced by $\Gamma(n)$, where Γ is the Euler function (see page 154).

In particular, we have:

$$H(t)\,\mathrm{e}^{\mathrm{i}\omega t} \sqsupset \frac{1}{p-\mathrm{i}\omega} \quad \text{and} \quad H(t)\,\mathrm{e}^{-\mathrm{i}\omega t} \sqsupset \frac{1}{p+\mathrm{i}\omega}, \qquad 0 < \operatorname{Re}(p).$$

By linearity of the Laplace transform, we deduce that

$$H(t)\cos(\omega t) \sqsupset \frac{p}{p^2+\omega^2} \qquad H(t)\sin(\omega t) \sqsupset \frac{\omega}{p^2+\omega^2}, \qquad 0 < \operatorname{Re}(p).$$

Similarly,

$$H(t)\cosh(\omega t) \sqsupset \frac{p}{p^2-\omega^2} \qquad H(t)\sinh(\omega t) \sqsupset \frac{\omega}{p^2-\omega^2} \qquad 0 < \operatorname{Re}(p).$$

Remark 12.25 Be careful not to use the linearity to state something like:

$$H(t)\cos\omega t = \operatorname{Re}\big(H(t)\,e^{i\omega t}\big) \sqsupset \operatorname{Re}\left(\frac{1}{p - i\omega}\right),$$

since taking the *real* part of some expression is **certainly not** a $\mathbb{C}$-linear operation.

12.4 Laplace transform of distributions

12.4.a Definition

Since we are only discussing the unilateral Laplace transform (or, what amounts to the same thing, the Laplace transform of *causal* functions, zero for $t < 0$), we will only generalize the Laplace transformation to distributions with support bounded on the left, that is, distributions in the space $\mathscr{D}'_+$. Recall that there is a convolution algebra structure defined on this space – namely, the product $T * S$ exists for all $T, S \in \mathscr{D}'_+$.

DEFINITION 12.26 Let T be a distribution with support bounded on the left. If there exists an $\alpha \in \mathbb{R}$ such that, for all $x > \alpha$, the distribution $e^{-xt}\,T(t)$ is tempered, then the **Laplace transform** of T is defined by the formula

$$T(t) \sqsupset \widehat{T}(p) \stackrel{\text{def}}{=} \big\langle T(t), e^{-pt} \big\rangle.$$

$\widehat{T}$ is a *function*. The lower bound of the possible values of α is called the **convergence abscissa**.

The fundamental example is the following:

PROPOSITION 12.27 (Laplace transform of δ) *The Dirac distribution being tempered and with support bounded on the left, the Laplace transform of δ exists and is equal to the constant function* 1, *with convergence abscissa* $\alpha = -\infty$.

PROOF. We have $e^{-\alpha t}\delta = \delta$ for all $\alpha \in \mathbb{R}$, and δ is tempered. This proves that $\alpha = -\infty$; furthermore, we have $\big\langle \delta(t), e^{-pt} \big\rangle = e^{-0} = 1$.

12.4.b Properties

Let T be a distribution with support bounded on the left with Laplace transform $\widehat{T}$. Then the function $\widehat{T}$ is holomorphic at any point $p \in \mathbb{C}$ such that $\operatorname{Re}(p) > \alpha$.

In addition, the following result may be shown:

PROPOSITION 12.28 (Sufficent condition for a Laplace transform) *Any function $F(p)$ which is holomorphic and bounded in modulus by a polynomial function of p in a half-plane* $\text{Re}(p) > \alpha$ *is the Laplace transform of a distribution in* $\mathscr{D}'_+$.

The following important result then holds:

THEOREM 12.29 (Laplace transform and convolution) *In the convolution algebra $\mathscr{D}'_+$, the fundamental relation between convolution of distributions and product of Laplace transforms holds:*

$$T * U \sqsupset \widehat{T}(p) \cdot \widehat{U}(p).$$

We may then state a differentiation theorem for distributions:

THEOREM 12.30 (Laplace transform and differentiation) *Let $T \in \mathscr{D}'_+$ be a distribution with Laplace transform $F(p)$, and let T' denote its derivative in the sense of distributions. Then T' has a Laplace transform given by*

$$T'(t) \sqsupset p\,\widehat{T}(p).$$

Remark 12.31 This result is compatible with Theorem 12.19: let f be a continuous function on $\mathbb{R}$ (hence locally integrable); the function $t \mapsto H(t) f(t)$ is a priori discontinuous at 0. The derivative of $H(t) f(t)$ is, in the sense of distributions, equal to $H(t)\{f'(t)\} + f(0)\,\delta(t)$. Applying the differentiation theorem 12.19 *for functions* says that

$$H(t)\{f'(t)\} \sqsupset pF(p) - f(0).$$

Since in addition we have $f(0)\,\delta(t) \sqsupset f(0)$, it follows that

$$\left[H(t) f(t)\right]' \sqsupset pF(p),$$

which is the statement in the differentiation theorem 12.30 in the sense of distributions.

Also, the following results remain valid for the Laplace transform of distributions:

THEOREM 12.32 *Let $T(t) \in \mathscr{D}'_+$ have Laplace transform $\widehat{T}(p)$ with convergence abscissa α. For $\tau, a \in \mathbb{R}$, we have*

$$\begin{aligned}
T(t-\tau) &\sqsupset \widehat{T}(p)\,\mathrm{e}^{-p\tau} && \textit{for } \alpha < x, \\
T(t)\,\mathrm{e}^{-a\tau} &\sqsupset \widehat{T}(p+a) && \textit{for } \alpha - \text{Re}(a) < x, \\
H * T &\sqsupset \frac{\widehat{T}(p)}{p} && \textit{for } \max(\alpha, 0) < x.
\end{aligned}$$

Remark 12.33 The last equation is the translation of the integration theorem, since $H * f$ is a *primitive* of the distribution f.

12.4.c Examples

Using the differentiation and translation theorems for the Laplace transform, we find:

THEOREM 12.34 *The Dirac δ distribution δ and its derivatives have convergence abscissas $\alpha = -\infty$ and*

$$\delta(t) \sqsupset 1, \qquad \delta'(t) \sqsupset p,$$
$$\delta^{(n)}(t) \sqsupset p^n, \qquad \delta(t-\tau) \sqsupset \mathrm{e}^{-\tau p}.$$

Putting

$$\text{Ш}^+(t) \stackrel{\text{def}}{=} \sum_{n=0}^{\infty} \delta(t-n),$$

it follows by linearity (and continuity) that

$$\text{Ш}^+(t) \sqsupset \frac{1}{1-\mathrm{e}^{-p}}.$$

The convergence abscissa is $\alpha = 0$.

It is possible to write symbolically $\text{Ш}^+(t) = H(t)\,\text{Ш}(t)$, although in principle this is an ambiguous notation: indeed, H is not continuous at 0, so $H(t)\,\delta(t)$ is not well defined. *By abuse of notation*, we will consider that $H(t)\,\delta(t) = \delta(t)$.

♦ **Exercise 12.2** Show that

$$\cos(\pi t)\,\text{Ш}^+(t) \sqsupset \frac{1}{1-\mathrm{e}^{-p}} \quad \text{and} \quad \mathrm{e}^t\,\text{Ш}^+(t) \sqsupset \frac{1}{1-\mathrm{e}^{1-p}}.$$

12.4.d The z-transform

Let f be a function of $\mathscr{C}^\infty$ class. We then have

$$H(t)\,f(t)\,\text{Ш}(t) = \sum_{n=0}^{\infty} f(n)\,\delta(t-n) \sqsupset \sum_{n=0}^{\infty} f(n)\,\mathrm{e}^{-np}.$$

Putting $z = \mathrm{e}^{-p}$, the Laplace transform thus obtained may also be written $\sum_{n=0}^{\infty} f(n)\,z^n$.

DEFINITION 12.35 The **z-transform** of a function $t \mapsto f(t)$ is the power series

$$z \longmapsto \sum_{n=0}^{\infty} f(n)\,z^n.$$

The z-transform of a function is used in signal theory, when it is necessary to work with sampling (we do not know $f(t)$ for all real t, but only for integer values).

12.4.e Relation between Laplace and Fourier transforms

Consider a distribution $f \in \mathscr{D}'_{+}$, with Laplace transform $F(p)$ and convergence abscissa α.

• **If $\alpha > 0$**, then $f(t)$ is not a tempered distribution; it does not a priori have a well-defined Fourier transform.

• **If $\alpha < 0$**, then $f(t)$ has a Fourier transform; it is linked to the Laplace transform by the relation

$$\widetilde{f}(\nu) = F(2\mathrm{i}\pi\nu).$$

• **If $\alpha = 0$**, which is the most interesting case, it is possible to show that $F(p)$ is a *meromorphic* function which has poles only on the imaginary axis (remember that although F is *not* defined on the left half-plane of the complex plane, it may possibly have an analytic continuation to a larger domain). It is not possible to consider directly the function $F(2\mathrm{i}\pi\nu)$, which is not locally integrable and consequently does not correctly define a distribution – the problems occur of course at the poles. Now expand $F(p)$ as the sum of a holomorphic function and rational functions with poles on the imaginary axis:

$$F(p) = G(p) + \sum_{n\in I} \frac{\lambda_n}{(p - \mathrm{i}\omega_n)^{m_n}},$$

with G holomorphic for $\mathrm{Re}(p) \geqslant 0$. The integers m_n are the orders of the poles $\mathrm{i}\omega_n$. By means of the inverse Laplace transform, we derive

$$f(t) = g(t) + H(t) \sum_{n\in I} \lambda_n \,\mathrm{e}^{\mathrm{i}\omega_n t} \frac{t^{m_n - 1}}{(m_n - 1)!}.$$

The Fourier transform of this function may be computed, as long as g has a Fourier transform. Denoting $\omega_n = 2\mathrm{i}\pi\nu_n$, and using the "finite part" distribution (see Exercises 8.6 on page 241 and 11.5 on page 324), we obtain:

$$\widetilde{f}(\nu) = \widetilde{g}(\nu) + \sum_{n\in I} \left\{ \frac{\lambda_n \delta^{(m_n-1)}(\nu - \nu_n)}{2(-2\mathrm{i}\pi)^{m_n-1}(m_n - 1)!} + \operatorname{fp} \frac{\lambda_n}{\left(2\mathrm{i}\pi(\nu - \nu_n)\right)^{m_n}} \right\},$$

that is,

$$\widetilde{f}(\nu) = \operatorname{fp} F(2\mathrm{i}\pi\nu) + \sum_{n\in I} \frac{\lambda_n}{2(-2\mathrm{i}\pi)^{m_n-1}\,(m_n - 1)!}\,\delta^{(m_n-1)}(\nu - \nu_n),$$

or also

$$\widetilde{f}\left(\frac{\omega}{2\pi}\right) = \operatorname{fp} F(\mathrm{i}\omega) + \sum_{n\in I} \frac{(\mathrm{i})^{m_n-1}\,\lambda_n}{2(m_n - 1)!}\,\delta^{(m_n-1)}(\omega - \omega_n), \qquad (12.1)$$

where fp is the finite part (for multiple poles), or the Cauchy principal value (for simple poles).

♦ **Exercise 12.3** Compute this way the Fourier transform of the Heaviside distribution and of $H(x)x^n$ (see Exercise 11.5 on page 324).

12.5

Physical applications, the Cauchy problem

The Laplace transform may be used to solve problems linked to the evolution of a system, when the initial condition is known. Suppose we have a linear system of differential equations describing the evolution of a physical system – for which the equations of motion are supposed to be known. We are looking for the solutions $t \longmapsto f(t)$, where the value of the function and of a certain number of its derivatives at $t = 0$ are imposed. The **Cauchy problem** is the data consisting of the differential equation (or the system of differential equations) together with the initial conditions.

12.5.a Importance of the Cauchy problem

The physically essential notion – which is also philosophically fascinating – of *determinism* is founded on the more mathematical concept of the Cauchy problem. A system is deterministic if the knowledge, at a given time $t = 0$, of the various parameters characterizing this system (typically, the positions and velocities of the particles that it is made of, in the case of a classical system), determines theoretically the future (or past) evolution of the system without ambiguity.

In other words, the differential equations that describe the dynamics of the system must have a solution, for a given initial condition, and this solution must be unique.

A whole branch of mathematical physics deals with the proof of existence and uniqueness theorems for solutions of physical systems (often simplified). But it should not be forgotten, taking as excuse that this kind of work is highly technical (and thus often arid-looking), that this is the key to the precise notion of determinism.[2]

[2] Many quarrels between scientists have been based on this concept. It is to be seen already in Newton's equations and appears, in particular, later with Laplace's intuition of the deterministic character of classical mechanics (the famous "I did not need this assumption"). These disagreements continued into the last century: Einstein, with his "God does not play dice," and the ever-lively discussions between partisans of various interpretations of quantum mechanics. For a theatrical approach to this subject, the reader is invited to go see (or to read) the play *Arcadia* by Tom Stoppard [88].

12.5.b A simple example

The Laplace transform can be used to solve a differential equation, taking initial conditions into account, in a more natural manner than with the Fourier transform.

The best is to start with a simple example. Suppose we want to solve the following Cauchy problem:

$$(C) \qquad \begin{cases} \ddot{x} + x = 2\cos t, \\ x(0) = 0 \quad \dot{x}(0) = -1. \end{cases}$$

We assume that the function x has a Laplace transform: $x(t) \sqsupset X(p)$. Then the second derivative of x has the following Laplace transform, which incorporates the initial conditions:

$$\ddot{x}(t) \sqsupset p^2 X(p) - p\,x(0) - \dot{x}(0) = p^2 X(p) + 1,$$

while

$$\cos t \sqsupset \frac{p}{p^2+1}.$$

The differential equation (with the given initial conditions) can therefore be written simply in the form

$$p^2 X(p) + 1 + X(p) = \frac{2p}{p^2+1},$$

which is solved, in the sense of functions[3], by putting

$$X(p) = \frac{2p}{(p^2+1)^2} - \frac{1}{p^2+1}.$$

There only remains to find the original of $X(p)$. The second term on the right-hand side of the preceeding relation is the image of the sine function: $\sin t \sqsupset 1/(p^2+1)$. Moreover, using Theorem 12.21, we obtain

$$t \sin t \sqsupset -\frac{\mathrm{d}}{\mathrm{d}p}\left(\frac{1}{p^2+1}\right) = \frac{2p}{(p^2+1)^2},$$

which gives the solution $X(p) \sqsubset t\sin t - \sin t = (t-1)\sin t$.

It is easy to check that $x(t) = (t-1)\sin t$ is indeed a solution of the stated Cauchy problem.

The reader is referred to the table on page 614 for a list of the principal Laplace transforms.

---

[3] One can see the advantage of working with Laplace transforms, which are functions, instead of Fourier transforms, which are distributions.

12.5.c Dynamics of the electromagnetic field without sources

The example treated here is somewhat more involved.

Consider the *free* evolution of the electromagnetic field in the space $\mathbb{R}^3$, in the absence of any source. For reasons of simplicity, it is easier to study the evolution of the potential four-vector $\mathsf{A}(\boldsymbol{r},t)$ in Lorentz gauge instead of the evolution of the electric and magnetic fields. The initial conditions at $t=0$ are given by the data of the field $\mathsf{A}(\boldsymbol{r},0)=\mathsf{A}_0(\boldsymbol{r})$ and of its first derivative with respect to time, which we denote by $\dot{\mathsf{A}}_0(\boldsymbol{r})$.

The equations which determine the field are therefore

$$\begin{cases} \Box\mathsf{A}(\boldsymbol{r},t)=0 & \text{for all } \boldsymbol{r}\in\mathbb{R}^3 \text{ and } t>0,\\ \mathsf{A}(\boldsymbol{r},0)=\mathsf{A}_0(\boldsymbol{r}) & \text{for all } \boldsymbol{r}\in\mathbb{R}^3,\\ \dot{\mathsf{A}}(\boldsymbol{r},0)=\dot{\mathsf{A}}_0(\boldsymbol{r}) & \text{for all } \boldsymbol{r}\in\mathbb{R}^3. \end{cases}$$

We denote by $\mathscr{A}(\boldsymbol{r},p)$ the Laplace transform of $\mathsf{A}(\boldsymbol{r},t)$, and by $\widetilde{\mathsf{A}}(\boldsymbol{k},t)$ its Fourier transform with respect to the space variable. We also denote by $\widetilde{\mathscr{A}}(\boldsymbol{k},p)$ the double transform (Laplace transform for the variable t and Fourier transform for the variable $\boldsymbol{r}$).

We have, by the differentiation theorem 12.19 :

$$\begin{aligned} \mathsf{A}(\boldsymbol{r},t) &\sqsupset \mathscr{A}(\boldsymbol{r},p),\\ \frac{\partial}{\partial t}\mathsf{A}(\boldsymbol{r},t) &\sqsupset p\mathscr{A}(\boldsymbol{r},p)-\mathsf{A}(\boldsymbol{r},0),\\ \frac{\partial^2}{\partial t^2}\mathsf{A}(\boldsymbol{r},t) &\sqsupset p^2\mathscr{A}(\boldsymbol{r},p)-p\mathsf{A}(\boldsymbol{r},0)-\dot{\mathsf{A}}(\boldsymbol{r},0), \end{aligned}$$

and the evolution of the potential is therefore described, in the Laplace world, by the equation

$$\left(\frac{p^2}{c^2}-\triangle\right)\mathscr{A}(\boldsymbol{r},p)=\frac{p}{c^2}\mathsf{A}_0(\boldsymbol{r})+\frac{1}{c}\dot{\mathsf{A}}_0(\boldsymbol{r}).$$

Take then the Fourier transform for the space variable, which gives

$$\left(\frac{p^2}{c^2}+\boldsymbol{k}^2\right)\widetilde{\mathscr{A}}(\boldsymbol{k},p)=\frac{p}{c^2}\widetilde{\mathsf{A}}_0(\boldsymbol{k})+\frac{1}{c}\tilde{\dot{\mathsf{A}}}_0(\boldsymbol{k}),$$

or

$$\widetilde{\mathscr{A}}(\boldsymbol{k},p)=\frac{p}{p^2+\boldsymbol{k}^2c^2}\widetilde{\mathsf{A}}_0(\boldsymbol{k})+\frac{1}{p^2+\boldsymbol{k}^2c^2}\tilde{\dot{\mathsf{A}}}_0(\boldsymbol{k}),$$

from which we can now take the inverse Laplace transform to derive, for any $t\geqslant 0$, that

$$\widetilde{\mathsf{A}}(\boldsymbol{k},t)=\widetilde{\mathsf{A}}_0(\boldsymbol{k})\,H(t)\cos(kct)+\tilde{\dot{\mathsf{A}}}_0(\boldsymbol{k})\,\frac{H(t)\sin(kct)}{kc}.$$

Finally we need only compute the inverse Fourier of $t\mapsto(\sin kct)/kc$ and $t\mapsto\cos(kct)$.

THEOREM 12.36 *The inverse Fourier transforms of those functions are given by*

$$\mathscr{F}^{-1}\left[\frac{\sin(kct)}{kc}\right] = \frac{1}{(2\pi)^3}\int \frac{\sin(kct)}{kc}\,\mathrm{e}^{\mathrm{i}\boldsymbol{k}\cdot\boldsymbol{r}}\,\mathrm{d}^3\boldsymbol{k}$$
$$= \frac{1}{4\pi c r}\Big[\delta(r-ct)-\delta(r+ct)\Big]$$
$$= \begin{cases} \frac{1}{4\pi c r}\delta(r-ct) & \text{if } t>0,\\ 0 & \text{if } t=0,\end{cases}$$

$$\text{and}\quad \mathscr{F}^{-1}\left[\cos(kct)\right] = \begin{cases} \frac{1}{4\pi r}\delta'(r-ct) & \text{if } t>0,\\ \delta(\boldsymbol{r}) & \text{if } t=0.\end{cases}$$

PROOF. See Appendix D, page 603. The derivative $\delta'(v-ct)$ is assumed to be a *radial* derivative.

The expression for the evolution of the field is then obtained by this last inverse Fourier transform using the the convolution theorem. Denoting by $\partial\mathcal{B}(\boldsymbol{r}\,;\,ct)$ the sphere centered at the point $\boldsymbol{r}$ with radius ct (i.e., the *boundary* of the corresponding ball), we derive for $t>0$ the formula

$$\boxed{\mathsf{A}(\boldsymbol{r},t) = -\frac{1}{4\pi}\iint_{\partial\mathcal{B}(\boldsymbol{r};ct)} \frac{\partial}{\partial\boldsymbol{n}}\frac{\mathsf{A}_0(\boldsymbol{r}')}{|\boldsymbol{r}-\boldsymbol{r}'|}\,\mathrm{d}^2\boldsymbol{r}' + \frac{1}{4\pi c^2 t}\iint_{\partial\mathcal{B}(\boldsymbol{r};ct)} \dot{\mathsf{A}}_0(\boldsymbol{r}')\,\mathrm{d}^2\boldsymbol{r}'.}$$

where $\partial/\partial\boldsymbol{n}$ is the exterior normal derivative (in the direction the vector exterior to the surface).

Notice that the value of the field at point $\boldsymbol{r}$ and time t is expressed in terms of the initial values of the field and its derivative at points $\boldsymbol{r}'$ such that $|\boldsymbol{r}'-\boldsymbol{r}| = ct$ (this is what appears in the integration domain $\partial\mathcal{B}(\boldsymbol{r}\,;\,ct)$).

Note also that if we let t tend to 0 in this formula, the second term tends to 0, whereas the first one tends to $\mathsf{A}_0(\boldsymbol{r})$ if A_0 is sufficiently regular (it is enough that it be of $\mathscr{C}^1$ class here).

♦ **Exercise 12.4** As an exercise, the reader may show that the previous formulas may be recovered using the Kirchhoff integral presented in Problem 3 on page 244.

Start by showing that, if $\mathscr{S}$ is an arbitrary (smooth) surface, and if $\psi(\boldsymbol{r},t)$ is a scalar field such that $\Box\psi = 0$, then by decomposing ψ in monochromatic waves via a Fourier transform with respect to t, and using the results of Problem 3, we get

$$\psi(\boldsymbol{r},t) = \frac{1}{4\pi}\iint_{\mathscr{S}} \left\{ -[\psi]\frac{\partial}{\partial\boldsymbol{n}}\frac{1}{q} + \frac{1}{cq}\cdot\frac{\partial q}{\partial\boldsymbol{n}}\left[\frac{\partial\psi}{\partial t}\right] + \frac{1}{q}\left[\frac{\partial\psi}{\partial\boldsymbol{n}}\right]\right\}\mathrm{d}^2\mathscr{S},$$

where q is the distance from $\boldsymbol{r}$ to the point of surface which is the integration variable, and $\partial/\partial\boldsymbol{n}$ is the exterior normal derivative. The quantities between curly brackets are evaluated at the delayed time $t-q/c$.

To conclude, take $\mathscr{S} = \partial\mathcal{B}(\boldsymbol{r}\,;\,ct)$ and show that the last formula recovers the framed equation above.

We can also express the potential evolving from $t = 0$ in presence of a current four-vector $\mathsf{j}(\boldsymbol{r}, t)$ which is assumed to be known at any time, namely,

$$\mathsf{A}(\boldsymbol{r}, t) = \iiint_{\mathcal{B}(\boldsymbol{r};ct)} \frac{\mathsf{j}\left(\boldsymbol{r}, t - \frac{1}{c}|\boldsymbol{r} - \boldsymbol{r}'|\right)}{|\boldsymbol{r} - \boldsymbol{r}'|} \, \mathrm{d}^3\boldsymbol{r}'$$
$$- \frac{1}{4\pi} \iint_{\partial\mathcal{B}(\boldsymbol{r};ct)} \frac{\partial}{\partial \boldsymbol{n}} \frac{\mathsf{A}_0(\boldsymbol{r}')}{|\boldsymbol{r} - \boldsymbol{r}'|} \, \mathrm{d}^2\boldsymbol{r}' + \frac{1}{4\pi c^2 t} \iint_{\partial\mathcal{B}(\boldsymbol{r};ct)} \dot{\mathsf{A}}_0(\boldsymbol{r}') \, \mathrm{d}^2\boldsymbol{r}'.$$

In this last expression, the first term, produced by the sources, is given by a volume integral; it is therefore clear that the current $\mathsf{j}(\boldsymbol{r}, t)$ may be a distribution instead of a function.

Note that to be physically pertinent, we must impose that the inital conditions (the potential at $t = 0$ and its first derivative) must be compatible with the initial conditions concerning the charge and current distribution.

EXERCISES

♦ **Exercise 12.5** Let $f(t)$ be an original which is periodic with period a. Show that its Laplace transform may be written

$$F(p) = \frac{1}{1 - e^{-pa}} \int_0^a f(t)\, e^{-pt}\, dt.$$

♦ **Exercise 12.6** We prove here an intermediate result used in the proof of Theorem 12.19.

i) Let $h : \mathbb{R} \to \mathbb{R}$ be a differentiable function such that h and h' are integrable. Show that $\lim_{x \to +\infty} h(x) = 0$.

ii) Deduce that if f is an original with convergence abscissa α, such that f is differentiable, and if f' is also an original with convergence abscissa α', then for any complex number p such that $\text{Re}(p) > \max(\alpha, \alpha')$, we have

$$\lim_{t \to +\infty} f(t)\, e^{-pt} = 0.$$

♦ **Exercise 12.7** Let J_0 be the Bessel function of order 0. Show that

$$\forall t \geqslant 0, \qquad \int_0^t J_0(s)\, J_0(t-s)\, ds = \sin t.$$

(See the table on page 614.)

♦ **Exercise 12.8** Prove the following results, and draw the original functions in each case (k is a given real number):

a) $$H(t-k) \sqsupset \frac{1}{p}\, e^{-kp},$$

b) $$(t-k)\, H(t-k) \sqsupset \frac{1}{p^2}\, e^{-kp},$$

c) $$H(t) - H(t-k) \sqsupset \frac{1 - e^{-kp}}{p},$$

d) $$\sum_{n=0}^{\infty} H(t-nk) \sqsupset \frac{1}{p(1 - e^{-kp})},$$

e) $$H(t) + 2 \sum_{n=1}^{\infty} (-1)^n H(t-2nk) \sqsupset \frac{\tanh(kp)}{p},$$

f) $$\sum_{n=0}^{\infty} (-1)^n H(t-nk) \sqsupset \frac{1}{p(1 + e^{-kp})}$$

g) $$t\, H(t) + 2 \sum_{n=1}^{\infty} (-1)^n (t-2nk)\, H(t-2nk) \sqsupset \frac{1}{p^2} \tanh(kp).$$

These results are particularly useful in electronic and signal theory.

♦ **Exercise 12.9** Consider a mechanical system given by a dampened harmonic oscillator, described by the equation $\ddot{x} - 2\dot{x} + x = f(t)$, where $f(t)$ is an outside perturbation. We assume that the system is initally at rest, that is, $x(0) = \dot{x}(0) = 0$, and that the outside perturbation has the form

$$f(t) = \sum_{n=0}^{5} (-1)^n H(t - n\tau) \quad \text{with } \tau > 0.$$

Draw the graph of f, then find the solution $t \mapsto x(t)$ to the given Cauchy problem.

SOLUTIONS

♦ **Solution of exercise 12.5.** Put $\varphi \stackrel{\text{def}}{=} f\big|_{[0,a]}$. Then we have $f = \varphi * H(t)\,Ш(t/a)$. Apply then the convolution formula, remembering that $H(t)\,Ш(t) \sqsupset 1/(1-\mathrm{e}^{-p})$.

♦ **Solution of exercise 12.6**

i) Since f' is integrable, the function f has a limit at $+\infty$ (indeed, for any real number x we have x: $f(x) = f(0) + \int_0^x f'(t)\,\mathrm{d}t$ and this last expression, by definition, has a limit as x tends to $+\infty$).
ii) By definition of the convergence abscissa, for $\mathrm{Re}(p) > \alpha$, we know that $t \mapsto f(t)\,\mathrm{e}^{-pt}$ is integrable. Moreover, its derivative is then $t \mapsto f'(t)\,\mathrm{e}^{-pt} - p\,f(t)\,\mathrm{e}^{-pt}$, which is also integrable if $\mathrm{Re}(p)$ is larger than α and α'. Apply then the previous result to the function $h : t \mapsto f(t)\,\mathrm{e}^{-pt}$.

♦ **Solution of exercise 12.7.** The Laplace transform of J_0 is given by

$$H(t)\,J_0(t) \sqsupset \frac{1}{\sqrt{p^2+1}}.$$

We want to find the convolution $H(t)\,J_0(t)$ with itself, which has image

$$H(t)\,J_0(t) * H(t)\,J_0(t) \sqsupset \frac{1}{\sqrt{p^2+1}} \cdot \frac{1}{\sqrt{p^2+1}} = \frac{1}{p^2+1},$$

the original of which is $H(t)\sin t$.

Note that the formula stated is still valid for $t \leqslant 0$ by the parity of J_0.

♦ **Solution of exercise 12.8.** The graphs are

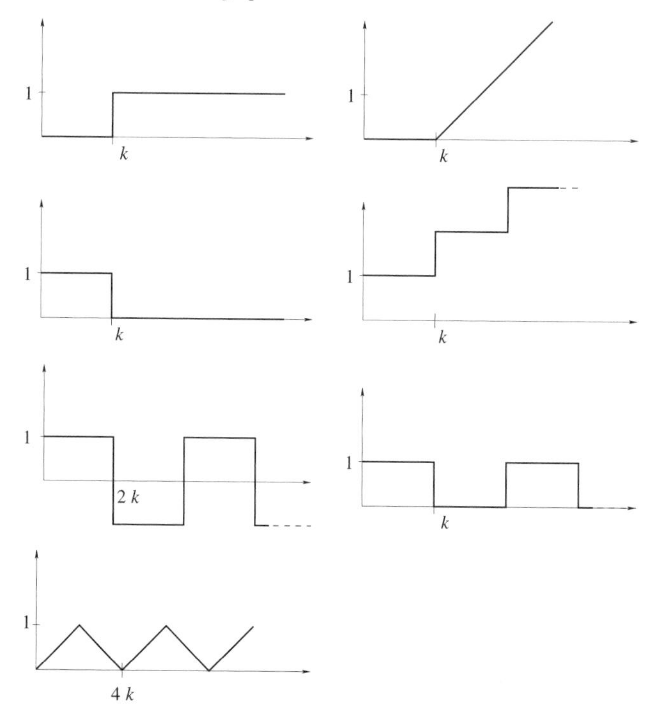

The formulas are obtained by simple application of the various properties of the Laplace transform and summations.

♦ **Solution of exercise 12.9.**

$$x(t) = \sum_{n=0}^{5} (-1)^n \big(1 + e^{t-n\tau}(t - n\tau - 1)\big) H(t - n\tau).$$

Chapter 13

Physical applications of the Fourier transform

13.1 Justification of sinusoidal regime analysis

When studying a *linear system*[1] it is customary to reason as follows: consider a situation where the signal is purely *monochromatic*, and where the response to an arbitrary signal is the sum of responses to the various frequencies that make up the signal. The sinusoidal analysis is then justified by stating that

If the system has for input a sinusoidal signal with given frequency ω, then the output signal is necessarily sinusoidal with the same frequency ω.

For definiteness, let us look at the very pedestrian example of an electrical circuit of RLC type (Resistors, Inductors, Capacitors). Assume that the input signal is the circuit voltage $t \mapsto u(t)$, and the output signal is the resistor voltage $t \mapsto v(t)$ (i.e., up to the factor R, the intensity of current in the circuit).

[1] For instance, a mechanical system with springs and masses, or an electric circuit built out of linear components such as resistors, capacitors or inductors, or in the study of the free electromagnetic field, ruled by the Maxwell equations.

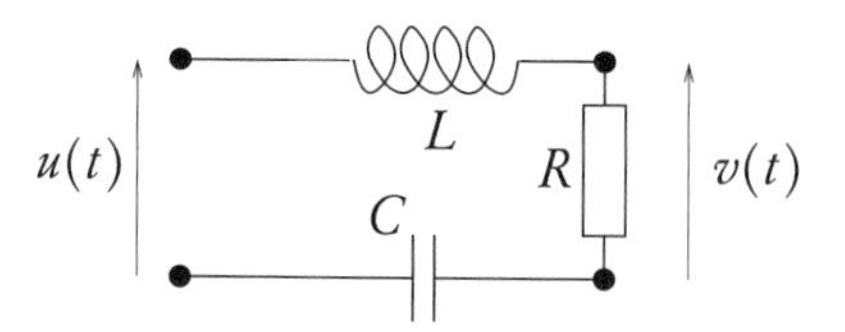

Writing down the differential equation linking u and v, we obtain the identity (valid in the sense of distributions):

$$u = \left[\frac{L}{R}\delta' + \frac{1}{RC}H + \delta\right] * v = D * v,$$

which may be inverted (using the whole apparatus of Fourier analysis!) to yield an equation of the type

$$v(t) = T(t) * u(t),$$

where T is the impulse response of the system, that is, $T = D^{*-1}$.

Let us restrict our attention to the case where the input signal is sinusoidal, with pulsation ω (or frequency ν related to ω by $\omega = 2\pi\nu$). The output signal is assumed to be sinusoidal, with same pulsation ω. The equations are much simplified because, formally, we have

$$\text{“}\mathrm{d}/\mathrm{d}t = \mathrm{i}\omega\text{”}$$

(a differential operator is replaced by a simple algebraic operation). Define the **transfer function** $Z(\omega)$ as the ratio between the output signal $v(t) = v_0\,\mathrm{e}^{\mathrm{i}\omega t}$ and the input signal $u(t) = u_0\,\mathrm{e}^{\mathrm{i}\omega t}$:

$$Z(\omega) = \frac{v(t)}{u(t)} = \frac{v_0}{u_0}.$$

When the input signal is a superposition of sinusoidal signals (which is, generally speaking, always the case), the linearity of the system may be used to find the output signal: it suffices to multiply *each sinusoidal component* of the input signal by the value of the transfer function for the given pulsation and to sum (or integrate) over the whole set of frequencies:

$$\text{if}\quad u(t) = \sum_n a_n\,\mathrm{e}^{\mathrm{i}\omega_n t}, \qquad \text{then}\quad v(t) = \sum_n a_n\,Z(\omega_n)\,\mathrm{e}^{\mathrm{i}\omega_n t},$$

$$\text{and if}\quad u(t) = \int a(\omega)\,\mathrm{e}^{\mathrm{i}\omega t}\,\mathrm{d}\omega, \quad \text{then}\quad v(t) = \int a(\omega)\,Z(\omega)\,\mathrm{e}^{\mathrm{i}\omega t}\,\mathrm{d}\omega.$$

The only subtle point in this argument is to justify that the output signal is monochromatic if the input signal is. This step is provided by the following theorem:

THEOREM 13.1 *Any exponential signal (either real $t \mapsto \mathrm{e}^{-pt}$ or complex $t \mapsto \mathrm{e}^{2\pi\mathrm{i}\nu t}$) is an* eigenfunction *of a convolution operator.*

Indeed, assume that the input signal u and the output signal v are related by a convolution equation $v(t) = T(t) * u(t)$, where T is either a function or a tempered distribution. If u is of the form $u(t) = \mathrm{e}^{2\pi i \nu t}$, then $v(t) = T(t) * \mathrm{e}^{2\pi i \nu t}$ is infinitely differentiable (because the exponential is) and Theorem 8.24 yields

$$v(t) = \left\langle T(\theta), \mathrm{e}^{2\pi i \nu (t-\theta)} \right\rangle = \left\langle T(\theta), \mathrm{e}^{-2\pi i \nu \theta} \right\rangle \mathrm{e}^{2\pi i \nu t},$$

which is indeed a complex exponential with the same frequency as the input signal. The transfer function is, for the frequency ν, equal to $Z(\nu) = \left\langle T(\theta), \mathrm{e}^{-2\pi i \nu \theta} \right\rangle$ or, in other words, it is the value of the Fourier transform of the impulse response T:

$$Z = \widetilde{T}.$$

The quantity $Z(\nu)$ may also be seen as the *eigenvalue* associated to the *eigenvector* $t \longmapsto \mathrm{e}^{2\pi i \nu t}$ of the convolution operator T.

The same reasoning applies for a real-valued exponential. If the input signal is of the type $u(t) = \mathrm{e}^{pt}$ (where p is a real number), the output signal can be written

$$v(t) = \left\langle T(\theta), \mathrm{e}^{p(t-\theta)} \right\rangle = \left\langle T(\theta), \mathrm{e}^{-p\theta} \right\rangle \mathrm{e}^{pt},$$

still according to Theorem 8.24. The "transfer function" $p \longmapsto \left\langle T(\theta), \mathrm{e}^{-p\theta} \right\rangle$ is then of course the *Laplace transform* of the impulse response T (see Chapter 12).

Remark 13.2 (Case of a resonance) This discussion must be slightly nuanced: if $Z(\nu)$ is infinite, which means the Fourier transform of the distribution T is not regular at ν, one cannot conclude that the output signal will be sinusoidal.

A simple example is given by the equation of the driven harmonic oscillator:

$$v'' + \omega^2 v = \mathrm{e}^{2\pi i \nu t} = u(t).$$

This equation can be written $u = [\delta'' + \omega^2 \delta] * v$, the solution of which, in $\mathscr{D}'_+$, is (see Theorem 8.33 on page 239)

$$v = T * u \qquad \text{with } T(t) = \frac{1}{\omega} H(t) \sin \omega t.$$

The Fourier transform T (computed more easily by taking the Laplace transform evaluated at $2\pi i \nu$, and applying the formula (12.1), page 345) is then[2]

$$Z(\nu) = \widetilde{T}(\nu) = \operatorname{fp} \frac{1}{\omega^2 - 4\pi^2 \nu^2} + \frac{1}{4i} \delta\left(\nu - \frac{\omega}{2\pi}\right) - \frac{1}{4i} \delta\left(\nu + \frac{\omega}{2\pi}\right).$$

For the values $\nu = \omega/2\pi$ and $\nu = -\omega/2\pi$, there is divergence.

[2] The transfer function $Z(\nu)$ is computed, *ab initio*, at the beginning of Chapter 15. This will explain why imaginary Dirac distributions appear, in relation with the choice of a solution of the equation $u = [\delta'' + \omega^2 \delta] * v$; for a solution in $\mathscr{D}'_-$, the Dirac spikes would have the opposite sign. A solution which does not involve any Dirac distribution exists also, but it does not belong to any convolution algebra.

In fact, this reflects the fact that the solutions of the differential equation are not sinusoidal; instead, they are of the form

$$v(t) = \frac{1}{2i\omega}\, t\, e^{it},$$

at the resonance frequency. More details are given in Chapter 15.

13.2 Fourier transform of vector fields: longitudinal and transverse fields

The Fourier transform can be extended to an n-dimensional space.

DEFINITION 13.3 The **Fourier transform** of a function $\boldsymbol{x} \longmapsto f(\boldsymbol{x})$ with variable $\boldsymbol{x}$ in $\mathbb{R}^n$ is given by

$$\tilde{f}(\boldsymbol{\nu}) = \iint \cdots \int_{\mathbb{R}^n} f(\boldsymbol{x})\, e^{-2\pi i \boldsymbol{\nu}\cdot\boldsymbol{x}}\, d^n\boldsymbol{x}.$$

All the properties of the Fourier transform extend easily to this case, *mutatis mutandis*; for instance, some care is required for scaling:

$$\mathscr{F}\big[f(a\boldsymbol{x})\big] = \frac{1}{|a|^n}\tilde{f}\left(\frac{\boldsymbol{\nu}}{a}\right).$$

One may also define, *coordinate-wise*, the three-dimensional Fourier transform of a *vector field* $\boldsymbol{x} \longmapsto \boldsymbol{A}(\boldsymbol{x})$. Using this, it is possible to define *transverse fields* and *longitudinal fields*.

Let $\boldsymbol{x} \longmapsto \boldsymbol{A}(\boldsymbol{x})$ be a vector field, defined on $\mathbb{R}^3$ and with values in $\mathbb{R}^3$. This field may be decomposed into three components $\boldsymbol{A} = (A_x, A_y, A_z)$, each of which is a scalar field, that is, a function from $\mathbb{R}^3$ to $\mathbb{R}$. Denote by $\mathscr{A}_x$ the Fourier transform of A_x, $\mathscr{A}_y$ that of A_y, and $\mathscr{A}_z$ that of A_z. Putting together these three scalar fields into a vector field $\boldsymbol{k} \longmapsto \boldsymbol{\mathscr{A}}(\boldsymbol{k})$, we obtain the Fourier transform of the original vector field:

$$\boldsymbol{\mathscr{A}}(\boldsymbol{k}) = \iiint_{\mathbb{R}^3} \boldsymbol{A}(\boldsymbol{x})\, e^{-2\pi i \boldsymbol{k}\cdot\boldsymbol{x}}\, d^3\boldsymbol{x}.$$

DEFINITION 13.4 Let $\boldsymbol{A}(\boldsymbol{x})$ be a vector field $\boldsymbol{x} \in \mathbb{R}^n$ with values in $\mathbb{R}^n$. Denote by $\boldsymbol{k} \longmapsto \boldsymbol{\mathscr{A}}(\boldsymbol{k})$ its Fourier transform. For any $\boldsymbol{k} \in \mathbb{R}^n$, decompose the vector $\boldsymbol{\mathscr{A}}(\boldsymbol{k})$ in two components:

- the first is parallel with $\boldsymbol{k}$ and is denoted $\boldsymbol{\mathscr{A}}_{/\!/}(\boldsymbol{k})$; it is given by the formula

$$\boldsymbol{\mathscr{A}}_{/\!/}(\boldsymbol{k}) \stackrel{\text{def}}{=} \frac{\boldsymbol{\mathscr{A}}(\boldsymbol{k})\cdot\boldsymbol{k}}{\|\boldsymbol{k}\|^2}\,\boldsymbol{k}\ ;$$

- the second is orthogonal to $\boldsymbol{k}$ and is denoted $\mathscr{A}_{\perp}(\boldsymbol{k})$; it is given by

$$\mathscr{A}_{\perp}(\boldsymbol{k}) \stackrel{\text{def}}{=} \mathscr{A}(\boldsymbol{k}) - \mathscr{A}_{/\!/}(\boldsymbol{k}).$$

The **longitudinal component** of the field A is given by the inverse Fourier transform of the field $\boldsymbol{k} \mapsto \mathscr{A}_{/\!/}(\boldsymbol{k})$:

$$A_{/\!/}(\boldsymbol{x}) \stackrel{\text{def}}{=} \mathscr{F}^{-1}\left[\boldsymbol{k}\frac{\mathscr{A}(\boldsymbol{k})\cdot\boldsymbol{k}}{|\boldsymbol{k}|^2}\right].$$

Its **transverse component** is then

$$A_{\perp}(\boldsymbol{x}) \stackrel{\text{def}}{=} \mathscr{F}^{-1}\left[\mathscr{A}_{\perp}(\boldsymbol{k})\right] = A(\boldsymbol{x}) - A_{/\!/}(\boldsymbol{x}).$$

DEFINITION 13.5 A vector field is **transverse** if its longitudinal component is zero, and is **longitudinal** if its transverse component is zero.

From the differentiation theorem, we can derive the well-known result:

PROPOSITION 13.6 *A transverse field satisfies the relation*

$$\nabla\cdot A(\boldsymbol{x}) = 0 \qquad \text{at any } \boldsymbol{x}\in\mathbb{R}^3,$$

that is, its divergence is identically zero.

It is important to notice that the property of "being transverse" or "being longitudinal" is by no means a *local* property of the vector field (it is not sufficient to know the value of the field at a given point, or in the neighborhood of a point), but a property that depends on the value of the field at all points in space (see Exercise 13.1 on page 375).

Remark 13.7 A monochromatic field, with wave vector $\boldsymbol{k}$, is transverse if it is perpendicular to $\boldsymbol{k}$ at any $\boldsymbol{k}$; it is easy to have an intuitive idea of this. A general transverse field is then a sum of monochromatic transverse fields.

13.3 Heisenberg uncertainty relations

In nonrelativistic quantum mechanics, the state of a spinless particle is represented by a complex-valued function ψ which is square integrable and of $\mathscr{C}^2$ class. More precisely, for any $t\in\mathbb{R}$, the function $\psi(\cdot,t) : \boldsymbol{r}\mapsto\psi(\boldsymbol{r},t)$ is in $L^2(\mathbb{R}^3)$. This function represents the probability amplitude of the particle at the given point, that is, the probability density associated with the presence of the particle at point $\boldsymbol{r}$ in space and at time t is $|\psi(\boldsymbol{r},t)|^2$. The function $\psi(\cdot,t)$ is normalized (in the sense of the L^2 norm) to have norm 1:

$$\text{for any } t \in \mathbb{R}, \qquad \|\psi(\cdot,t)\|_2^2 = \iiint_{\mathbb{R}^3} |\psi(\boldsymbol{r},t)|^2 \, \mathrm{d}^3\boldsymbol{r} = 1.$$

DEFINITION 13.8 The **wave function in position representation** is the square-integrable function thus defined. The vector space $\mathscr{H}$ of the wave functions, with the usual scalar product, is a *Hilbert space.*[3]

DEFINITION 13.9 The **position operator** along the $\mathcal{O}x$-axis is denoted X, and similarly Y and Z are the position operators along the $\mathcal{O}y$- and $\mathcal{O}z$-axes, where $\boldsymbol{R} \stackrel{\text{def}}{=} (X,Y,Z)$. The operator X is defined by its action on wave functions given by

$$X\psi = \varphi \qquad \text{with } \varphi(\cdot,t) : \boldsymbol{r} \longmapsto x\,\psi(\boldsymbol{r},t),$$

where we write generically $\boldsymbol{r} = (x,y,z)$.

Be careful not to make the (classical!) mistake of stating that ψ and $X\psi$ are proportional, with proportionality constant equal to x!

It is customary to denote $\left|\psi(t)\right\rangle$ the function $\psi(\cdot,t)$, considered as a vector in the vector space of wave functions.[4] The scalar product of two vectors is then given by

$$\left\langle \psi(t) \middle| \varphi(t) \right\rangle \stackrel{\text{def}}{=} \iiint \overline{\psi(\boldsymbol{r},t)}\,\varphi(\boldsymbol{r},t)\,\mathrm{d}^3\boldsymbol{r},$$

(recall the chosen convention for the hermitian product in quantum mechanics: it is *linear on the right* and *semilinear on the left*), and the position operator acts as follows:

$$X\left|\psi(t)\right\rangle = \left|\varphi(t)\right\rangle \qquad \text{with } \varphi(\boldsymbol{r},t) = x\,\psi(\boldsymbol{r},t).$$

With each vector $|\psi\rangle$ is associated a linear form on the space $\mathscr{H}$ by means of the scalar product, which is denoted $\langle\psi| \in \mathscr{H}^*$. Thus, the notation $\langle\varphi|\psi\rangle$ can be interpreted in two (equivalent) ways: as the scalar product of φ with ψ (in the sense of the product in L^2), or as the action of the linear form $\langle\psi|$ on the vector $|\varphi\rangle$ (see Chapter 14 for a thorough discussion of this notation).

[3] Only "physical" wave functions belong to this space. Some useful functions, such as monochromatic waves $\boldsymbol{r} \mapsto \mathrm{e}^{i\boldsymbol{k}\cdot\boldsymbol{r}}$ do not; they are the so-called "generalized kets," which are discussed in Chapter 14, in particular, in Section 14.2.d.

[4] In fact $\left|\psi(t)\right\rangle$ is a vector in an abstract Hilbert space H. However, there exists an isometry between H and $\mathscr{H} = L^2(\mathbb{R}^3)$, that is, the vector $\left|\psi(t)\right\rangle$ can be *represented* by the function $\boldsymbol{r} \mapsto \psi(\boldsymbol{r},t)$: this is the **position representation**. But, since the Fourier transform is itself an isometry of $L^2(\mathbb{R}^3)$ into itself, the vector $\left|\psi(t)\right\rangle$ may also be represented by the function $\boldsymbol{p} \mapsto \widetilde{\psi}(\boldsymbol{p},t)$, which is the Fourier transform of the previous function. This is called the **momentum representation** of the wave vector.

DEFINITION 13.10 The **average position on the x-axis** of a particle characterized by the wave function $\psi(\boldsymbol{r},t)$ is the quantity

$$\langle x\rangle(t) \stackrel{\text{def}}{=} \langle\psi(t)|X\psi(t)\rangle = \langle\psi(t)|X|\psi(t)\rangle = \iiint x\,|\psi(\boldsymbol{r},t)|^2\,\mathrm{d}^3\boldsymbol{r}.$$

In addition, to every wave function $\psi(\boldsymbol{r},t)$ is associated its Fourier transform with respect to the space variable $\boldsymbol{r}$, namely,

$$\Psi(\boldsymbol{p},t) \stackrel{\text{def}}{=} \iiint \psi(\boldsymbol{r},t)\,\mathrm{e}^{-\mathrm{i}\boldsymbol{p}\cdot\boldsymbol{r}/\hbar}\,\frac{\mathrm{d}^3\boldsymbol{r}}{\sqrt{(2\pi\hbar)^3}}.$$

This function represents the probability amplitude for finding the particle with momentum $\boldsymbol{p}$. Of course, this function is also normalized to have norm 1, since for every $t\in\mathbb{R}$ we have

$$\langle\Psi(t)|\Psi(t)\rangle = \iiint |\Psi(\boldsymbol{p},t)|^2\,\mathrm{d}^3\boldsymbol{p} = \iiint |\psi(\boldsymbol{r},t)|^2\,\mathrm{d}^3\boldsymbol{r} = 1$$

by the Parseval-Plancherel identity.

DEFINITION 13.11 The **wave function in momentum representation** is the Fourier transform $\Psi(\boldsymbol{p},t)$ of the wave function $\psi(\boldsymbol{r},t)$ of a particle.

DEFINITION 13.12 The **momentum operator** $P \stackrel{\text{def}}{=} -\mathrm{i}\hbar\nabla = (P_x,P_y,P_z)$ is the linear operator acting on differentiable functions by

$$P_x\psi = \varphi \qquad \left(\text{or}\quad P_x|\psi\rangle = |\varphi\rangle\right) \qquad \text{with } \varphi(\cdot,t):\boldsymbol{r}\longmapsto -\mathrm{i}\hbar\,\frac{\partial\psi}{\partial x}(\boldsymbol{r},t).$$

The **average momentum** along the x-direction of a particle characterized by the wave function $\psi(\boldsymbol{r},t)$ is

$$\langle p_x\rangle(t) \stackrel{\text{def}}{=} \langle\psi(t)|P_x\psi(t)\rangle = -\mathrm{i}\hbar\iiint \overline{\psi(\boldsymbol{r},t)}\,\frac{\partial\psi}{\partial x}(\boldsymbol{r},t)\,\mathrm{d}^3\boldsymbol{r}.$$

From the known properties of the Fourier transform, we see that it is possible to extend the definition of the operator P_x on wave functions "in momentum representation" by

$$P_x\Psi \stackrel{\text{def}}{=} \mathscr{F}\left[P_x\mathscr{F}^{-1}[\Psi]\right],$$

which gives explicitly, by the differentiation theorem,

$$P_x\Psi = \Phi \qquad \text{with} \qquad \Phi(\cdot,t):\boldsymbol{p}\longmapsto p_x\Psi(\boldsymbol{p},t).$$

DEFINITION 13.13 The **position uncertainty** of a particle characterized by the wave function $\psi(\boldsymbol{r},t)$ is the quantity Δx defined by the following equivalent expressions:

$$(\Delta x)^2 \stackrel{\text{def}}{=} \left\langle (x - \langle x \rangle)^2 \right\rangle = \langle x^2 \rangle - \langle x \rangle^2 = \langle X\psi | X\psi \rangle - \langle \psi | X\psi \rangle^2$$
$$= \iiint \left(x - \langle x \rangle\right)^2 |\psi(\boldsymbol{r},t)|^2 \, \mathrm{d}^3\boldsymbol{r} = \|X\psi - \langle x \rangle \psi\|^2 .$$

DEFINITION 13.14 The **momentum uncertainty** of a particle characterized by the wave function $\psi(\boldsymbol{r},t)$ is the quantity Δp_x defined by the following equivalent expressions:

$$(\Delta p_x)^2 = \langle p_x^2 \rangle - \langle p_x \rangle^2 = \langle P\psi | P\psi \rangle - \langle \psi | P\psi \rangle^2$$
$$= \iiint \hbar^2 \left| \frac{\partial \psi}{\partial x}(\boldsymbol{r},t) \right|^2 \mathrm{d}^3\boldsymbol{r} - \langle p_x \rangle^2$$
$$= \iiint p_x^2 \left| \Psi(\boldsymbol{p},t) \right|^2 \mathrm{d}^3\boldsymbol{p} - \langle p_x \rangle^2 = \|P_x\psi - \langle p_x \rangle \psi\|^2$$

Note that $\langle p_x^2 \rangle$ may also be expressed as the average value of the operator $P_x^2 = -\hbar^2 \frac{\partial^2}{\partial x^2}$ in the state $|\psi\rangle$, since an easy integration by parts reveals that

$$\langle P_x^2 \rangle \stackrel{\text{def}}{=} \langle \psi | P_x^2 \psi \rangle = -\hbar^2 \iiint \overline{\psi(\boldsymbol{r},t)} \, \frac{\partial^2 \psi}{\partial x^2}(\boldsymbol{r},t) \, \mathrm{d}^3\boldsymbol{r}$$
$$= \iiint \hbar^2 \left| \frac{\partial \psi}{\partial x}(\boldsymbol{r},t) \right|^2 \mathrm{d}^3\boldsymbol{r}.$$

To perform this integration by parts, it is necessary that the derivative of ψ decay fast enough, which is the case when $\langle \Delta p_x \rangle$ exists.

Remark 13.15 The average values and the uncertainties should be compared with the expectation value and standard deviation of a random variable (see Chapter 20).

We are going to show now that there exists a relation between Δx and Δp_x. To start with, we have the following lemma:

LEMMA 13.16 *Let ψ be a wave function.*

i) If the wave function is translated by the vector $\boldsymbol{a} \in \mathbb{R}^3$ by putting

$$\psi_{\boldsymbol{a}}(\boldsymbol{r},t) \stackrel{\text{def}}{=} \psi(\boldsymbol{r} - \boldsymbol{a}, t),$$

then neither Δx nor Δp_x is changed.

ii) Similarly, if a translation by the quantity $\boldsymbol{k} \in \mathbb{R}^3$ is performed in momentum space, or, what amounts to the same thing, if the wave function in position representation is multiplied by a phase factor by defining

$$\psi_{\boldsymbol{k}}(\boldsymbol{r}, t) \stackrel{\text{def}}{=} \psi(\boldsymbol{r}, t)\, \mathrm{e}^{\mathrm{i}\boldsymbol{k}\cdot\boldsymbol{r}/\hbar},$$

then neither Δx nor Δp_x is changed.

iii) *Consequently, it is possible to define a wave function $\hat{\psi}(\boldsymbol{r}, t)$ which is centered in position and momentum (i.e., such that the average position and the average momentum are zero) and has the same uncertainties as the original function $\psi(\boldsymbol{r}, t)$ with respect to position and momentum.*

Proof. The average position of the translated wave function $\psi_{\boldsymbol{a}}$ is

$$\langle x\rangle_{\boldsymbol{a}} = \iiint x\left|\psi_{\boldsymbol{a}}(\boldsymbol{r}, t)\right|^2 \mathrm{d}^3\boldsymbol{r} = \iiint (x - a_x)\left|\psi(\boldsymbol{r}, t)\right|^2 \mathrm{d}^3\boldsymbol{r} = \langle x\rangle - a_x.$$

The position uncertainty is then

$$\begin{aligned}
(\Delta x)^2_{\boldsymbol{a}} = \langle x^2\rangle_{\boldsymbol{a}} - \langle x\rangle^2_{\boldsymbol{a}} &= \iiint x^2\left|\psi(\boldsymbol{r} - \boldsymbol{a}, t)\right|^2 \mathrm{d}^3\boldsymbol{r} - \big(\langle x\rangle - a_x\big)^2 \\
&= \iiint (x - a_x)^2\left|\psi(\boldsymbol{r}, t)\right|^2 \mathrm{d}^3\boldsymbol{r} - \big(\langle x\rangle^2 - 2a_x\langle x\rangle + a_x^2\big) \\
&= \langle x^2\rangle - 2a_x\langle x\rangle + a_x^2 - \big(\langle x\rangle^2 - 2a_x\langle x\rangle + a_x^2\big) \\
&= \langle x^2\rangle - \langle x\rangle^2 = (\Delta x)^2.
\end{aligned}$$

The momentum uncertainty of $|\psi\rangle$ is of course unchanged.

The proof is identical for the translates in momentum space.

Finally, the centered wave function is, for a given t, equal to

$$\hat{\psi}(\boldsymbol{r}, t) \stackrel{\text{def}}{=} \psi(\boldsymbol{r} - \langle\boldsymbol{r}\rangle, t)\, \mathrm{e}^{\mathrm{i}\langle\boldsymbol{p}\rangle\cdot\boldsymbol{r}/\hbar}.$$

THEOREM 13.17 (Heisenberg uncertainty relations) *Let ψ be a wave function for which the average position and average momentum, as well as the position and momentum uncertainties, are defined. Then we have*

$$\Delta x \cdot \Delta p_x \geqslant \frac{\hbar}{2} \tag{13.1}$$

(and similarly for y and z).

Moreover, equality holds only if ψ is a gaussian function.

Proof. Translating the wave function as in Lemma 13.16, we may assume that $\langle x\rangle = 0$ and $\langle p_x\rangle = 0$. Moreover, to simplify notation, we omit the dependency on t. We have

$$(\Delta x)^2 = \iiint x^2\left|\psi(\boldsymbol{r})\right|^2 \mathrm{d}^3\boldsymbol{r} \qquad \text{and} \qquad (\Delta p_x)^2 = \hbar^2 \iiint \left|\frac{\partial\psi}{\partial x}(\boldsymbol{r})\right|^2 \mathrm{d}^3\boldsymbol{r}.$$

The Cauchy-Schwarz inequality for functions in L^2 states that

$$\left|\iiint \overline{f(\boldsymbol{r})}\, g(\boldsymbol{r})\, \mathrm{d}^3\boldsymbol{r}\right|^2 \leqslant \iiint \left|f(\boldsymbol{r})\right|^2 \mathrm{d}^3\boldsymbol{r} \cdot \iiint \left|g(\boldsymbol{r})\right|^2 \mathrm{d}^3\boldsymbol{r}.$$

Applying this inequality to $f(\boldsymbol{r}) = x\,\psi(\boldsymbol{r})$ and $g(\boldsymbol{r}) = \frac{\partial\psi}{\partial x}(\boldsymbol{r})$, we obtain

$$\begin{aligned}
\left|\iiint \overline{x\,\psi(\boldsymbol{r})}\, \frac{\partial\psi}{\partial x}(\boldsymbol{r})\, \mathrm{d}^3\boldsymbol{r}\right|^2 &\leqslant \iiint x^2\left|\psi(\boldsymbol{r})\right|^2 \mathrm{d}^3\boldsymbol{r} \cdot \iiint \left|\frac{\partial\psi}{\partial x}(\boldsymbol{r})\right|^2 \mathrm{d}^3\boldsymbol{r} \\
&\leqslant \frac{1}{\hbar^2}(\Delta x)^2\,(\Delta p_x)^2.
\end{aligned} \tag{$*$}$$

But, using an integration by parts, since x is real and the boundary terms cancel (see Remark 13.19 concerning this point), we can write

$$\iiint \overline{x\,\psi(\boldsymbol{r})}\,\frac{\partial \psi}{\partial x}(\boldsymbol{r})\,\mathrm{d}^3\boldsymbol{r} = -\iiint |\psi(\boldsymbol{r})|^2\,\mathrm{d}^3\boldsymbol{r} - \iiint x\,\psi(\boldsymbol{r})\,\overline{\frac{\partial \psi}{\partial x}(\boldsymbol{r})}\,\mathrm{d}^3\boldsymbol{r},$$

or
$$2\,\mathrm{Re}\left(\iiint \overline{x\,\psi(\boldsymbol{r})}\,\frac{\partial \psi}{\partial x}(\boldsymbol{r})\,\mathrm{d}^3\boldsymbol{r}\right) = -\iiint |\psi(\boldsymbol{r})|^2\,\mathrm{d}^3\boldsymbol{r} = -1.$$

Since $|z| \geqslant |\mathrm{Re}(z)|$ for all $z \in \mathbb{C}$, the left-hand side of $(*)$ satisfies

$$\left|\iiint \overline{x\,\psi(\boldsymbol{r})}\,\frac{\partial \psi}{\partial x}(\boldsymbol{r})\,\mathrm{d}^3\boldsymbol{r}\right|^2 \geqslant \left|\mathrm{Re}\iiint \overline{x\,\psi(\boldsymbol{r})}\,\frac{\partial \psi}{\partial x}(\boldsymbol{r})\,\mathrm{d}^3\boldsymbol{r}\right|^2 = \frac{1}{4}.$$

This immediately implies (13.1).

The Cauchy-Schwarz inequality is an equality only if the two functions considered are proportional; but it is easy to see that $X\psi$ and $\partial\psi/\partial x$ can only be proportional if ψ is a gaussian as a function of x (this is a simple differential equation to solve).

If the usual definition of the Fourier transform is kept (with the factor 2π and without $\hbar$), the following equivalent theorem is obtained:

THEOREM 13.18 (Uncertainty relation) *Let $f \in L^2(\mathbb{R})$ be a function such that the second moment $\int x^2\,|f(x)|^2\,\mathrm{d}x$ and $\int |f'(x)|^2\,\mathrm{d}x$ both exist. Define*

$$\langle x^2\rangle \stackrel{\text{def}}{=} \frac{1}{\|f\|_2^2}\int x^2 |f(x)|^2\,\mathrm{d}x,$$

$$\langle \nu^2\rangle \stackrel{\text{def}}{=} \frac{1}{\|f\|_2^2}\int \nu^2 |\widetilde{f}(\nu)|^2\,\mathrm{d}\nu = \frac{4\pi^2}{\|f\|_2^2}\int |f'(x)|^2\,\mathrm{d}x.$$

Then we have $\langle x^2\rangle \cdot \langle \nu^2\rangle \geqslant \frac{1}{16\pi^2}$ with equality if and only if f is a gaussian.

Remark 13.19 A point that was left slightly fuzzy above requires explanation: the cancellation of the boundary terms during the integration by parts. Denote by $\mathscr{H}$ the space of square integrable functions ψ, such that $X\psi$ and $P_x\psi$ are also square integrable. This space, with the norm N defined by

$$\begin{aligned} N(\psi)^2 &= \|\psi\|_2^2 + \|X\psi\|_2^2 + \|P\psi\|_2^2 \\ &= \iiint |\psi(\boldsymbol{r})|^2\,\mathrm{d}^3\boldsymbol{r} + \iiint x^2|\psi(\boldsymbol{r})|^2\,\mathrm{d}^3\boldsymbol{r} + \iiint p_x^2|\Psi(\boldsymbol{p})|^2\,\mathrm{d}^3\boldsymbol{p}, \end{aligned}$$

is a Hilbert space.

If $\psi \in \mathscr{H}$ is of $\mathscr{C}^\infty$ class with bounded support with respect to the variable x, the proof above (and in particular the integration by parts with the boundary terms vanishing) is correct.

A classical result of analysis (proved using convolution with a Dirac sequence) is that the space of $\mathscr{C}^\infty$ functions with bounded support is *dense* in $\mathscr{H}$. By continuity, the preceding result then holds for all $\psi \in \mathscr{H}$.

Remark 13.20 Another approach of the uncertainty relations is proposed in Exercise 14.6 on page 404.

13.4
Analytic signals

It is customary, in electricity or optics, for instance, to work with complex sinusoidal signals of the type $e^{i\omega t}$, for which only the real part has physical meaning. This step from a real signal $f_{(R)}(t) = \cos(\omega t)$ to a complex signal $f(t) = e^{i\omega t}$ may be generalized in the case of a nonmonochromatic signal.

First of all, notice that the real signal may be recovered in two different ways: as the real part of the associated complex signal, or as the average of the signal and its complex conjugate.

We can generalize this. We consider a real signal $f_{(R)}(t)$ and search for the associated complex signal. In the case $f_{(R)}(t) = \cos(2\pi\nu_0 t)$ (with $\nu_0 > 0$), the spectrum of $f_{(R)}$ contains both frequencies ν_0 and $-\nu_0$. However, this information is redundant, since the intensity for the frequency $-\nu_0$ is the same as that intensity for the frequency ν_0, since $f_{(R)}$ is real-valued. The complex signal $f(t) = e^{2\pi i\nu t}$ associated to the real signal $f_{(R)}$ is obtained as follows: take the spectrum of $f_{(R)}$, keep only the positive frequencies, multiply by two, then reconstitute the signal.

In a general way, for a real-valued signal $f_{(R)}(t)$ for which the Fourier transform $\widetilde{f}_{(R)}(\nu)$ is defined, its spectrum is hermitian, that is, we have

$$\widetilde{f}_{(R)}(\nu) = \overline{\widetilde{f}_{(R)}(-\nu)} \qquad \text{for all } \nu \in \mathbb{R};$$

thus, "half" the spectrum $\widetilde{f}_{(R)}$ suffices to recover $f_{(R)}$.

If the spectrum $\widetilde{f}_{(R)}(\nu)$ does not have a singularity at the origin, or at least if is sufficiently regular, we can consider the truncated spectrum

$$2H(\nu) \cdot \widetilde{f}_{(R)}(\nu).$$

The inverse Fourier transform of this truncated spectrum will give the analytic signal associated to the function $f_{(R)}$. Since the inverse Fourier transform of $2H(\nu)$ is

$$\mathscr{F}^{-1}\big[2H(\nu)\big] = \mathscr{F}^{-1}\left[1 + \operatorname{sgn}\nu\right] = \delta(t) + \operatorname{pv}\frac{i}{\pi t}$$

(see Theorem 11.16 on page 304), we deduce that the inverse Fourier transform of $2H(\nu) \cdot \widetilde{f}_{(R)}(\nu)$ is

$$\mathscr{F}^{-1}\left[2H(\nu) \cdot \widetilde{f}_{(R)}(\nu)\right] = \left(\delta + \operatorname{pv}\frac{i}{\pi t}\right) * f_{(R)} = f_{(R)}(t) - \frac{i}{\pi}\operatorname{pv}\int \frac{f_{(R)}(t')}{t'-t}\,dt'.$$

We make the following definition:

DEFINITION 13.21 An **analytic signal** is a function with causal Fourier transform, that is, a function f such that the Fourier transform vanishes for negative frequencies.

DEFINITION 13.22 Let $f_{(R)}$ be a real signal with sufficiently regular Fourier transform.[5] The **imaginary signal** associated to $f_{(R)}$ is the opposite of its Hilbert transform:

$$f_{(I)}(t) \stackrel{\text{def}}{=} -\frac{1}{\pi} \,\text{pv} \int \frac{f_{(R)}(t')}{t'-t} \, \mathrm{d}t'.$$

DEFINITION 13.23 The **analytic signal** associated to the real signal $f_{(R)}$ is the function

$$t \longmapsto f(t) \stackrel{\text{def}}{=} f_{(R)}(t) + \mathrm{i} f_{(I)}(t) = f_{(R)}(t) - \frac{\mathrm{i}}{\pi} \,\text{pv} \int \frac{f(t')}{t'-t} \, \mathrm{d}t',$$

the spectrum of which is $\widetilde{f}(\nu) = 2H(\nu) \cdot \widetilde{f}_{(R)}(\nu)$.

The analytic signal associated with $f_{(R)}$ is analytic in the sense of Definition 13.21: its spectrum is identically zero for frequencies $\nu < 0$. Since $\widetilde{f}(\nu)$ is causal, it has a Laplace transform $F(p)$ defined at least on the right-hand half-plane of the complex plane. Since the signal f is the inverse Fourier transform of $\widetilde{f}$, it is given by

$$f(x) = F(-2\pi \mathrm{i} x).$$

Hence there is an *analytic continuation* of f to the upper half-plane,[6] which can be obtained using the Laplace transform of the spectrum:

$$f(z) = F(-2\pi \mathrm{i} z).$$

This analytic continuation is holomorphic in this *upper half-plane*, and therefore satisfies the Kramers-Kronig relations – which are neither more nor less than the relation defining $f_{(I)}$: we've made a complete loop!

The fact that f may be continued to an analytic *function on the upper half-place justifies also the terminology "analytic signal."*

All this may be summarized as follows:

THEOREM 13.24 *Let $f : \mathbb{R} \to \mathbb{C}$ be a function which has a Fourier transform. If the spectrum of f is causal, then f may be analytically continued to the complex upper half-plane.*

Conversely, if f may be analytically continued to this half-plane, and if the integral of f along a half-circle in the upper half-plane tends to 0 as the radius tends to infinity, then the spectrum of f is causal.

(See [79] for more details.)

[5] In the sense that the product $H(\nu)\widetilde{f}_{(R)}(\nu)$ exists; this means that $\widetilde{f}_{(R)}$ cannot involve a singular distribution at the origin 0.

[6] The upper half-plane because of the similitude $x \mapsto -2\pi \mathrm{i} x$: the image of the upper half-plane is the right-hand half-plane.

An example of the use of analytic signals is given by a light signal, characterized by its real amplitude $A_{(R)}(t)$. We have

$$A_{(R)}(t) = \int_{-\infty}^{+\infty} \mathscr{A}_{(R)}(\nu)\, e^{2\pi i \nu t}\, d\nu,$$

where $\mathscr{A}_{(R)}(\nu)$ is the Fourier transform of $A_{(R)}(t)$. However, since $A_{(R)}(t)$ is real, it is also possible to write

$$A_{(R)}(t) = \int_{0}^{+\infty} a(\nu) \cos\left[\varphi(\nu) + 2\pi i \nu t\right] d\nu,$$

where for $\nu > 0$ we denote $2\mathscr{A}_{(R)}(t)(\nu) = a(\nu)\, e^{i\varphi(\nu)}$ with a and φ real.

The imaginary signal associated to $A_{(R)}$ is of course given by

$$A_{(I)}(t) = \int_{0}^{+\infty} a(\nu) \sin\left[\varphi(\nu) + 2\pi i \nu t\right] d\nu,$$

and the analytic signal is

$$A(t) = \int_{0}^{+\infty} a(\nu)\, e^{i\varphi(\nu)} e^{2\pi i \nu t}\, d\nu.$$

The quantity $a(\nu) = 2\left|\mathscr{A}_{(R)}(\nu)\right|$ has the following interpretation: its square $W(\nu) = a^2(\nu)$ is the **power spectrum**, or *spectral density* of the complex signal A (see Definition 13.26 on the next page). This complex spectral density contains the energy carried by the real signal, and an equal amount of energy carried by the imaginary signal. The **total light intensity** carried by the real light signal is then equal to

$$\mathscr{E} = \frac{1}{2} \int_{0}^{+\infty} W(\nu)\, d\nu = \int_{-\infty}^{+\infty} |\mathscr{A}_{(R)}(\nu)|^2\, d\nu,$$

namely, half of the total complex intensity (the other half is of course carried by the imaginary signal).

13.5
Autocorrelation of a finite energy function

13.5.a Definition

Let $f \in L^2(\mathbb{R})$ be a square integrable function. We know it has a Fourier transform $\widetilde{f}(\nu)$.

DEFINITION 13.25 A **finite energy signal** is any function $f \in L^2(\mathbb{R})$. The **total energy** of a finite energy signal is

$$\mathscr{E} \stackrel{\text{def}}{=} \int |f(t)|^2 \, dt.$$

The Parseval identity states that

$$\int |f(t)|^2 \, dt = \int \left|\widetilde{f}(\nu)\right|^2 d\nu,$$

so we can interpret $\left|\widetilde{f}(\nu)\right|^2$ as an energy density per frequency interval $d\nu$.

DEFINITION 13.26 The **spectral density** of a finite energy signal f is the function $\nu \longmapsto \left|\widetilde{f}(\nu)\right|^2$.

In optics and quantum mechanics, for instance, the autocorrelation function of f is often of interest:

DEFINITION 13.27 Let f be a square integrable function. The **autocorrelation function** of f is the function $\Gamma = f * \check{\bar{f}}$ given by $\Gamma(x) = f(x) * \overline{f(-x)}$, that is,

$$\Gamma(x) \stackrel{\text{def}}{=} \int f(t)\,\overline{f(t-x)}\, dt \qquad \text{for any } x \in \mathbb{R}.$$

13.5.b Properties

Since the autocorrelation function is the convolution of f with the conjugate of its transpose $\Gamma(x) = f(x) * \overline{f(-x)}$, it is possible to use the known properties of the Fourier transform to derive some interesting properties. In particular,

THEOREM 13.28 *The Fourier transform of the autocorrelation function of f is equal to the spectral density of f:*

$$\mathscr{F}\left[\Gamma(t)\right] = \mathscr{F}\left[f(x) * \overline{f(-x)}\right] = \left|\widetilde{f}(\nu)\right|^2.$$

PROPOSITION 13.29 *The autocorrelation function of any finite energy function f is hermitian: $\Gamma(-x) = \overline{\Gamma(x)}$ for all $x \in \mathbb{R}$.*

Moreover, if f is real-valued, then Γ is an even, real-valued function.

In addition, since $\left|e^{2\pi i\nu x}\right| \leqslant 1$, we have the inequality

$$\left|\Gamma(x)\right| = \left|\int \left|\widetilde{f}(\nu)\right|^2 e^{2\pi i\nu x}\,d\nu\right| \leqslant \int \left|\widetilde{f}(\nu)\right|^2 d\nu = \Gamma(0),$$

from which we deduce:

THEOREM 13.30 *The modulus of the autocorrelation function of a function f is maximal at the origin: $|\Gamma(x)| \leqslant \Gamma(0)$ for any $x \in \mathbb{R}$. $\Gamma(0)$ is real, positive, and equal to the energy of f: $\Gamma(0) = \int |f|^2$.*

DEFINITION 13.31 The **reduced autocorrelation function**, also called the **self-coherence function** of f, is given by

$$\gamma(x) \stackrel{\text{def}}{=} \frac{\Gamma(x)}{\Gamma(0)}.$$

This function $\gamma(x)$ is thus always of modulus between 0 and 1.

13.5.c Intercorrelation

DEFINITION 13.32 The **intercorrelation function** of two square integrable functions f and g is the function

$$\Gamma_{fg}(x) \stackrel{\text{def}}{=} \int f(t)\,\overline{g(x-t)}\,dt.$$

From the Cauchy-Schwarz inequality, we derive

$$\left|\Gamma_{fg}(x)\right|^2 \leqslant \Gamma_f(0)\,\Gamma_g(0).$$

This brings us to the definition of the *coherence function*:

DEFINITION 13.33 The **intercoherence function**, or **coherence function**, of f and g is the reduced intercorrelation function

$$\gamma_{fg}(x) \stackrel{\text{def}}{=} \frac{\Gamma_{fg}(x)}{\sqrt{\Gamma_f(0)\,\Gamma_g(0)}}.$$

The modulus of this coherence function is therefore also always between 0 and 1.

13.6 Finite power functions

13.6.a Definitions

Up to now, the functions which have been considered are mostly square integrable, that is, functions in $L^2(\mathbb{R})$. In many physical situations (in optics, for instance), the integral $\int |f|^2$ represents an energy. Sometimes, however, the objects of interest are not functions with finite energy but those with finite *power*, such as $f(t) = \cos \omega t$.

DEFINITION 13.34 A function $f : \mathbb{R} \to \mathbb{C}$ is a **finite power function** if the limit

$$\lim_{T\to+\infty} \frac{1}{2T} \int_{-T}^{T} |f(t)|^2 \, dt$$

exists and is finite. This limit is then the **average power** carried by the signal f.

The spectral density may also be redefined:

DEFINITION 13.35 Let f be a finite power signal. The **spectral density**, or **power spectrum**, of f is the function defined by

$$W(\nu) \stackrel{\text{def}}{=} \lim_{T\to+\infty} \frac{1}{2T} \left| \int_{-T}^{T} f(t)\, e^{-2\pi i \nu t} \, dt \right|^2 .$$

An analogue of the Parseval formula holds for a finite power signal:

THEOREM 13.36 *Let $f(t)$ be a finite power signal and $W(\nu)$ its spectral density. Then we have*

$$\lim_{T\to+\infty} \frac{1}{2T} \int_{-T}^{T} |f(t)|^2 \, dt = \int_{-\infty}^{+\infty} W(\nu) \, d\nu.$$

13.6.b Autocorrelation

For a finite power function, the previous definition 13.27 of the autocorrelation is not valid anymore, and is replaced by the following:

DEFINITION 13.37 The **autocorrelation function** of a finite power signal f is given by

$$\Gamma(x) \stackrel{\text{def}}{=} \lim_{T\to+\infty} \frac{1}{2T} \int_{-T}^{T} f(t)\, \overline{f(t-x)} \, dt.$$

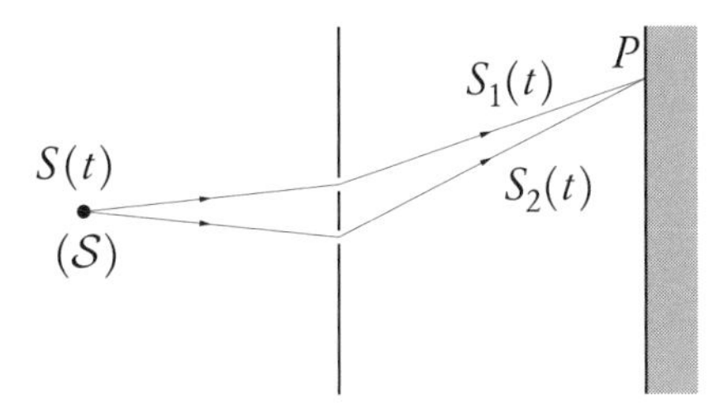

Fig. 13.1 – Light interference experiment with Young slits.

The analogue of Theorem 13.28 is

THEOREM 13.38 *The Fourier transform of the autocorrelation function of a finite power signal is equal to the spectral density:*

$$\mathscr{F}\left[\Gamma(t)\right] = W(\nu).$$

Example 13.39 Let $f(t) = \mathrm{e}^{2\pi \mathrm{i}\nu_0 t}$. The autocorrelation function is $\Gamma(x) = \mathrm{e}^{2\pi \mathrm{i}\nu_0 x}$ and the spectral density is $W(\nu) = \delta(\nu - \nu_0)$. Using the definition for a finite energy function, we would have found instead that $\widetilde{f}(\nu) = \delta(\nu - \nu_0)$ and the spectral density would have to be something like $\delta^2(\nu - \nu_0)$. But, as explained in Chapter 7, the square of a Dirac distribution does not make sense.

13.7 Application to optics: the Wiener-Khintchine theorem

We consider an optical interference experiment with Young slits which are lighted by a point-like light source (see Figure 13.1).

We will show that, in the case where the source is not quite monochromatic but has a certain spectral width $\Delta\nu$, the interference figure will get blurry at points of the screen corresponding to differences in the optical path between the two light rays which are too large. The threshold difference between the optical paths such that the diffraction rays remain visible is called the **coherence length**, and we will show that it is given by $\ell = c/\Delta\nu$.

Consider a "nonmonochromatic" source. We assume that the source $\mathcal{S}$ of the figure emits a real signal $t \mapsto S_{(\mathrm{R})}(t)$, and that an analytic complex signal $t \mapsto S(t)$ is associated with it. Hence, we have $S_{(\mathrm{R})}(t) = \mathrm{Re}\{S(t)\}$. The Fourier transform of $S_{(\mathrm{R})}(t)$ is denoted $\mathscr{S}_{(\mathrm{R})}(\nu)$.

Moreover, we assume that the signal emitted has finite power and is characterized by a spectral power density $W(\nu)$ for $\nu > 0$. The signal is **non-**

monochromatic if the spectral density $W(\nu)$ is not a Dirac distribution.

In the proposed experiment, we look for the light intensity arriving at the point P, given by the time average of the square of the signal $S_P(t)$ arriving at the point P:

$$\mathscr{I}_P = \left\langle \left| S_P^2(t) \right| \right\rangle,$$

where we put

$$\left\langle \left| S_P^2(t) \right| \right\rangle \stackrel{\text{def}}{=} \lim_{T\to+\infty} \frac{1}{2T} \int_{-T}^{T} \left| S_P(t) \right|^2 \mathrm{d}t.$$

The signal reaching P is the sum of the two signals coming from the two slits. The signal coming from one slit is thus the same as the signal coming from the other, *but* with a time shift $\tau = \Delta L/c$, where ΔL is the path difference between the two rays (and hence depends on the point P):

$$S_P(t) = S_1(t) + S_2(t) = S_1(t) + S_1(t-\tau).$$

(The function $S_1(t)$ is directly related to the signal $S(t)$ emitted by the source.)

From this we derive

$$\begin{aligned} \mathscr{I}_P = \left\langle S_P \cdot \overline{S}_P \right\rangle &= \left\langle (S_1 + S_2) \cdot (\overline{S_1} + \overline{S_2}) \right\rangle \\ &= \left\langle S_1 \cdot \overline{S_1} \right\rangle + \left\langle S_2 \cdot \overline{S_2} \right\rangle + 2\mathrm{Re} \left\langle S_1 \cdot \overline{S_2} \right\rangle \\ &= 2I + 2\mathrm{Re} \left\langle S_1 \cdot \overline{S_2} \right\rangle, \end{aligned}$$

denoting by I the average value $\left\langle S_1 \cdot \overline{S_1} \right\rangle$, which is the light intensity of the signal that passes by a single slit without interference. All information concerning the interference phenomenon is given by the term $2\mathrm{Re} \left\langle S_1 \cdot \overline{S_2} \right\rangle$, which is easy to compute:

$$\left\langle S_1 \cdot \overline{S_2} \right\rangle = \left\langle S_1(t) \cdot \overline{S_1(t-\tau)} \right\rangle = \Gamma(\tau),$$

where $\Gamma(\tau)$ is the autocorrelation function of the complex signal S_1, also called here the **autocoherence function**.[7] (Recall that τ is a function of the observation point P.)

So finally the light intensity at the point P is given by

$$\mathscr{I}_P = 2I + 2\,\mathrm{Re}\big(\Gamma(\tau)\big) = 2I\Big[1 + \mathrm{Re}\big(\gamma(\tau)\big)\Big]$$

with

$$\gamma(\tau) \stackrel{\text{def}}{=} \frac{\Gamma(\tau)}{\Gamma(0)} = \frac{\Gamma(\tau)}{I}.$$

[7] In some optics books, the normalized autocorrelation function $\gamma(\tau) = \Gamma(\tau)/\Gamma(0)$ is also called the autocoherence function.

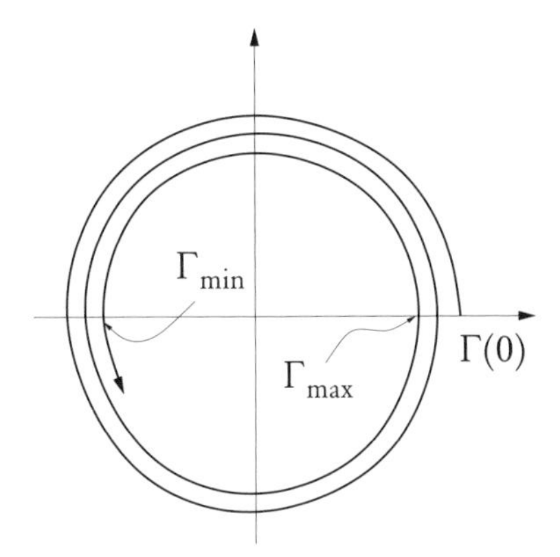

Fig. 13.2 – The autocorrelation function for an almost monochromatic source, varying with the parameter τ in the complex plane. The value at the origin $\Gamma(0)$ is real and positive. The modulus of $\Gamma(t)$ varies *slowly* compared to its phase.

We may then estimate the visibility factor of the diffraction figure, using the Rayleigh criterion [14]:

$$\mathscr{V} = \frac{I_{max} - I_{min}}{I_{max} + I_{min}}. \tag{13.2}$$

Here, I_{max} represents a *local maximum* of the light intensity, not at the point P, but *in a close neighborhood* of P; that is, the observation point will be moved slightly and hence the value of the parameter τ will change until an intensity which is a local maximum is found, which is denoted I_{max}. Similarly, the local minimum in the vicinity of P is found and is denoted I_{min}.

Estimating this parameter is only possible if the values τ_{max} and τ_{min} corresponding to the maximal and minimal intensities close to a given point are such that the difference $\tau_{max} - \tau_{min}$ is very small compared to the duration of a wave train. If this is the case, the autocorrelation function will have the shape in Figure 13.2, that is, its phase will vary very fast compared to its modulus; in particular, the modulus of Γ will remain almost constant between τ_{max} and τ_{min}, whereas the phase will change by an amount equal to π.[8] For $\tau = 0$, we have of course $\Gamma = I$, and then $|\Gamma(\tau)|$ will decrease with time.

Thus, by moving slightly P, τ varies little and $|\Gamma|$ remains essentially constant. Then $\mathrm{Re}(\Gamma)$ varies between $-|\Gamma|$ and $+|\Gamma|$, and the visibility factor

[8] *This is what the physics says, not the mathematics!* This phenomenon may be easily understood. Suppose τ is made to vary so that it remains *small* during the duration of a wave train. Then, taking a value of τ equal to half a period of the signal (which is almost monochromatic), the signals S_1 and S_2 are simply shifted by π: they are in opposition and $\Gamma(\tau)$ is real, negative, and with absolute value almost equal to $\Gamma(0)$. If we take values of τ much larger, the signals become more and more decorrelated because it may be that S_1 comes from one wave train whereas S_2 comes from another. A simple statistical model of a wave train gives an exponential decay of $|\Gamma(\tau)|$.

for the rays is then simply

$$\mathscr{V}(\tau) = |\gamma(\tau)| = \frac{|\Gamma(\tau)|}{I}.$$

But, according to Theorem 13.38, the autocorrelation function is given by the inverse Fourier transform of the power spectrum:

$$\Gamma(\tau) = \int_0^{+\infty} W(\nu)\, \mathrm{e}^{2\pi i \nu \tau}\, \mathrm{d}\nu.$$

The following theorem follows:

THEOREM 13.40 (Wiener-Khintchine) *The visibility factor of the diffraction figure in the Young interference experiment is equal to the modulus of the normalized Fourier transform of the spectral density of the source evaluated for the value τ corresponding to the time difference between the two rays:*

$$\mathscr{V} = \left| \frac{\int_0^{+\infty} W(\nu)\, \mathrm{e}^{2\pi i \nu \tau}\, \mathrm{d}\nu}{\int_0^{+\infty} W(\nu)\, \mathrm{d}\nu} \right|.$$

So, take the case of an almost monochromatic source – for instance, a spectral lamp, which *should* emit a monochromatic light but, because of the phenomenon of emission of light by wave trains, is characterized by a nonzero spectral width. If we denote by $\Delta\nu$ the spectral width, we know that the Fourier transform of the spectral energy density has typical width $\Delta t = 1/\Delta\nu$; for a time difference $\tau \gg 1/\Delta\nu$, that is, for a path difference $\gg c/\Delta\nu$, the diffraction pattern will *become blurred.* The quantity $\Delta t = 1/\Delta\nu$ is the **coherence time** and the length $\ell = c/\Delta\nu$ is the **coherence length**.

Remark 13.41 A similar theorem, due to P. K. VAN CITTERT and Frederik ZERNIKE (winner of the Nobel prize in 1953), links the visibility factor of a diffraction figure to the Fourier transform of the spacial distribution of a monochromatic extended source. The interested reader may consult [14] or look at Exercise 13.2.

EXERCISES

♦ **Exercise 13.1 (Electromagnetic field in Coulomb gauge)** We consider an electromagnetic field $[\boldsymbol{E}(\boldsymbol{r},t), \boldsymbol{B}(\boldsymbol{r},t)]$ compatible with a charge and current distribution $[\rho(\boldsymbol{r},t), \boldsymbol{j}(\boldsymbol{r},t)]$, that is, such that the fields satisfy the differential equations (Maxwell equations):

$$\operatorname{div} \boldsymbol{E} = -\rho/\varepsilon_0, \qquad \mathbf{curl}\, \boldsymbol{E} = -\frac{\partial \boldsymbol{B}}{\partial t},$$

$$\operatorname{div} \boldsymbol{B} = 0, \qquad \mathbf{curl}\, \boldsymbol{B} = \frac{1}{c^2}\frac{\partial \boldsymbol{E}}{\partial t} + \mu_0\, \boldsymbol{j}.$$

i) Show that the field $\boldsymbol{B}$ is transverse.

ii) Show that the longitudinal component of the electric field is given by the *instantaneous* Coulomb potential associated with the charge distribution. What may be said from the point of view of relativity theory?

iii) Is the separation between transverse and longitudinal components preserved by a Lorentz transformation?

♦ **Exercise 13.2 (van Cittert-Zernike theorem)** Consider a light source which is translation invariant along the $\mathscr{O}y$-axis, characterized by a light intensity $I(x)$ on the $\mathscr{O}x$-axis, such that the light passes through an interferential apparatus as described below. The distance between the source and the slits is f and the distance to the screen is D.

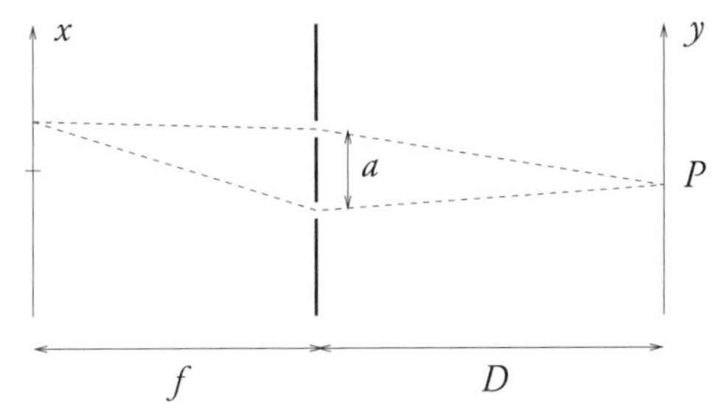

We assume that the source may be modeled by a set a point-like sources, emitting wave trains independently of each other, with the same frequency ν. Since the sources are not coherent with respect to each other, we admit that the intensity observed at a point of the interference figure is the sum of intensities coming from each source.[9] Show:

THEOREM 13.42 (van Cittert-Zernike) *The visibility factor of the source $I(x)$ is equal to the modulus of the Fourier transform of the normalized spatial intensity of the source for the spatial frequency $k = a/\lambda f$, where a is the distance between the interference slits and f is the distance between the source and the system:*

$$|\gamma| = \left|\widetilde{\mathscr{I}}\left(\frac{a}{\lambda f}\right)\right| \qquad \textit{with} \qquad \mathscr{I}(x) \stackrel{\text{def}}{=} \frac{I(x)}{\int_{-\infty}^{+\infty} I(s)\, \mathrm{d}s}.$$

The reader will find in [14] a generalization of this result, due to VAN CITTERT (1934) and ZERNIKE (1938).

---

[9] We saw in Exercise 1.3 on page 43 that it is in fact necessary to have a nonzero spectral width before the sources can be considered incoherent.

SOLUTIONS

♦ **Solution of exercise 13.1.** Applying the Fourier transform, the Maxwell equations become

$$\boldsymbol{k}\cdot\mathscr{E}=\mathrm{i}\widetilde{\rho}/\varepsilon_0, \qquad \mathrm{i}\boldsymbol{k}\wedge\mathscr{E}=\frac{\partial\mathscr{B}}{\partial t},$$

$$\boldsymbol{k}\cdot\mathscr{B}=0, \qquad \mathrm{i}\boldsymbol{k}\wedge\mathscr{B}=\frac{1}{c^2}\frac{\partial\mathscr{E}}{\partial t}+\mu_0\widetilde{\boldsymbol{j}}.$$

The third of these shows that $\boldsymbol{B}$ is transverse.

Moreover, since $\boldsymbol{k}\cdot\mathscr{E}=\boldsymbol{k}\cdot\mathscr{E}_{/\!/}(\boldsymbol{k})=-\mathrm{i}\widetilde{\rho}(\boldsymbol{k})/\varepsilon_0$, we find that

$$\mathscr{E}_{/\!/}(\boldsymbol{k})=\frac{\boldsymbol{k}\cdot\mathscr{E}(\boldsymbol{k})}{\boldsymbol{k}^2}=\mathrm{i}\frac{\boldsymbol{k}}{\boldsymbol{k}^2}\,\widetilde{\rho}(\boldsymbol{k})/\varepsilon_0,$$

and hence, taking the inverse Fourier transform, we have

$$\boldsymbol{E}_{/\!/}(\boldsymbol{r},t)=\frac{1}{4\pi\varepsilon_0}\int\frac{\rho(\boldsymbol{r}')\,[\boldsymbol{r}-\boldsymbol{r}']}{\|\boldsymbol{r}-\boldsymbol{r}'\|^3}\,\mathrm{d}^3\boldsymbol{r}'=\boldsymbol{E}_{\text{Coulomb}}(\boldsymbol{r},t).$$

Thus, the longitudinal field propagates instantaneously through space, which contradicts special relativity. This indicates that the decomposition in transverse and longitudinal components *is not physical.*

In fact, it is not preserved by a Lorentz transformation, since the Coulomb field is not.

♦ **Solution of exercise 13.2.** The intensity at a point P with coordinate y on the screen is given by the integral of intensities produced by $I(x)$, for $x\in\mathbb{R}$. An elementary interference computation shows that, for small angles, we have

$$\begin{aligned}J(x)&=K\cdot\int_{-\infty}^{+\infty}I(x)\left[1+\cos\frac{2\pi a}{\lambda}\left(\frac{x}{f}+\frac{y}{D}\right)\right]\mathrm{d}x\\&=K I_0\left\{1+\int_{-\infty}^{+\infty}\mathscr{I}(x)\cos\frac{2\pi a}{\lambda}\left(\frac{x}{f}+\frac{y}{D}\right)\mathrm{d}x\right\},\end{aligned}$$

with $I_0=\int_{-\infty}^{+\infty}I(s)\,\mathrm{d}s$, and where $\mathscr{I}(x)=I(x)/I_0$ is the normalized intensity. Since I and $\mathscr{I}$ are real-valued functions, we may write

$$\begin{aligned}J(y)&=K I_0\left\{1+\operatorname{Re}\int_{-\infty}^{+\infty}\mathscr{I}(x)\exp\left[-\frac{2\pi\mathrm{i}a}{\lambda}\left(\frac{x}{f}+\frac{y}{D}\right)\right]\mathrm{d}x\right\}\\&=K I_0\left\{1+\operatorname{Re}\left[\widetilde{\mathscr{I}}\left(\frac{a}{\lambda f}\right)\mathrm{e}^{-2\pi\mathrm{i}ay/\lambda D}\right]\right\}.\end{aligned}$$

If we decompose $\widetilde{\mathscr{I}}(a/\lambda f)=|\gamma|\,\mathrm{e}^{\mathrm{i}\alpha}$ in terms of a phase and modulus, we obtain

$$J(y)=K I_0\left\{1+|\gamma|\cos\left(\frac{2\pi a y}{\lambda D}-\alpha\right)\right\}.$$

Now, if we observe the interference figure close to the center and vary y slowly, the intensity J on the screen varies between $K I_0(1-|\gamma|)$ and $K I_0(1+|\gamma|)$, which shows that $|\gamma|$ is the visibility factor of the interference figure, which is what was to be shown.

This result is useful in particular to measure the angular distance between two stars, or the apparent diameter of a star, using two telescopes set up as interferometers.

Chapter 14

Bras, kets, and all that sort of thing

14.1 Reminders about finite dimension

In this section, we consider a vector space E, real or complex, with finite dimension n, equipped with a scalar product $(\cdot|\cdot)$ and with a given orthonormal basis $(e_1, \ldots, e_n)$. Some elementary properties concerning the scalar product, orthonormal bases, and the adjoint of an endomorphism will be recalled; these results should be compared with those holding in infinite dimensions, which are quite different.

14.1.a Scalar product and representation theorem

THEOREM 14.1 (Representation theorem) *The space E is isomorphic to its dual space E^*. More precisely, for any linear form φ on E, there exists a unique $a \in E$ such that*

$$\forall x \in E \qquad \varphi(x) = (a|x).$$

PROOF

Existence. Using the orthonormal basis $(e_1, \ldots, e_n)$, define $a = \sum_{j=1}^{n} \varphi(e_i)\, e_i$.

For any $x = \sum_{i=1}^{n} x_i\, e_i \in E$, we have $\varphi(x) = \sum_{i=1}^{n} x_i\, \varphi(e_i)$, hence $\varphi(x) = (a|x)$.

Uniqueness. Assume a and $a' \in E$ satisfy

$$(a|x) = (a'|x) \qquad \forall x \in E.$$

Then $(a - a'|x) = 0$ for any $x \in E$, and in particular $(a - a'|a - a') = 0$, which implies that $(a - a') = 0$ and so $a = a'$.

14.1.b Adjoint

DEFINITION 14.2 (Adjoint) Let u be an endomorphism of E. The **adjoint** of u, if it exists, is the unique $v \in \mathscr{L}(E)$ such that, for any $x, y \in E$, we have $\big(v(x)\big|y\big) = \big(x\big|u(y)\big)$. It is denoted u^*, and so by definition we have

$$\forall x, y \in E \qquad \big(u^*(x)\big|y\big) = \big(x\big|u(y)\big).$$

PROOF OF UNIQUENESS. Let v, w both be adjoints of u. Then we have

$$\forall x, y \in E \qquad \big(x\big|u(y)\big) = \big(v(x)\big|y\big) = \big(w(x)\big|y\big),$$

which shows that $v(x) - w(x) \in E^\perp = \{0\}$ for any $x \in E$ and hence $v = w$.

THEOREM 14.3 (Existence of the adjoint) *In a euclidean or hermitian space E (hence of finite dimension), any endomorphism admits an adjoint. Moreover, if $u \in \mathscr{L}(E)$ and if $\mathcal{B}$ is an orthonormal basis of E, the matrix representing u^* in the basis $\mathcal{B}$ is the conjugate transpose of that of u:*

$$\operatorname{mat}_{\mathcal{B}}(u^*) = {}^t\overline{\operatorname{mat}_{\mathcal{B}}(u)}.$$

PROOF. For any $x \in E$, the map $\varphi_x : y \longmapsto \big(x\big|u(y)\big)$ is a linear form; according to Theorem 14.1, there exists a unique vector, which we denote $u^*(x)$, such that φ_x is the map $y \longmapsto \big(u^*(x)\big|y\big)$.

It is easy to see that the map $x \longmapsto u^*(x)$ is linear. So it is the adjoint of u.

Another viewpoint, less elegant but more constructive, is the following.

Choose an orthonormal basis $\mathcal{B}$. Notice first that if $x, y \in E$ are represented by matrices X and Y in this basis, we have $(x|y) = {}^t\overline{X}Y$. Notice also that if ${}^t\overline{X}AY = {}^t\overline{X}BY$ for any $X, Y \in \mathfrak{M}_{n1}(\mathbb{C})$, then in fact $A = B$.

Let now $u, v \in \mathscr{L}(E)$, and define $U = \operatorname{mat}_{\mathcal{B}}(u)$ and $V = \operatorname{mat}_{\mathcal{B}}(v)$. Thanks to the preceding remark,

$$\begin{aligned} v \text{ is an adjoint of } u &\iff \forall x, y \in E \quad \big(v(x)\big|y\big) = \big(x\big|u(y)\big) \\ &\iff \forall X, Y \in \mathfrak{M}_{n1}(\mathbb{R}) \quad {}^t(\overline{VX})Y = {}^t\overline{X}(UY) \\ &\iff {}^t\overline{V} = U. \end{aligned}$$

Hence it suffices to define u^* as the endomorphism represented by the matrix ${}^t\overline{U}$ in the basis $\mathcal{B}$.

COROLLARY 14.4 (Properties of the adjoint) *Let $u, v \in \mathscr{L}(E)$ and let $\lambda \in \mathbb{C}$:*

$$(\lambda u + v)^* = \overline{\lambda} u^* + g^*, \qquad \mathrm{Id}_E^* = \mathrm{Id}_E, \qquad (u \circ v)^* = v^* \circ u^*,$$

$$(u^*)^* = u, \qquad \operatorname{rank}(u^*) = \operatorname{rank}(u), \qquad \operatorname{tr}(u^*) = \overline{\operatorname{tr}(u)},$$

$$\det u^* = \overline{\det u}, \qquad \lambda \in \operatorname{Sp}(u) \iff \overline{\lambda} \in \operatorname{Sp}(u^*).$$

Moreover, if u is invertible, then u^* is invertible and $(u^*)^{-1} = (u^{-1})^*$.

14.1.c Symmetric and hermitian endomorphisms

DEFINITION 14.5 (Hermitian/symmetric endomorphisms) Let $u \in \mathscr{L}(E)$ be an endomorphism of E. Then u is called **hermitian**, **symmetric**, or **self-adjoint** if

$$\forall x, y \in E \qquad \big(u(x)\big|y\big) = \big(x\big|u(y)\big),$$

or equivalently if $u^* = u$. In other words, u is hermitian if and only if the matrix representing u in an *orthonormal* basis satisfies $A = {}^t\overline{A}$ (**hermitian matrix**).

In the euclidean case (for a real vector space), the terminology used is **symmetric** or **self-adjoint**, and amounts to saying that the matrix representing u in an *orthonormal* basis is symmetric.

THEOREM 14.6 (Spectral theorem) *Let u be a **normal** endomorphism of E, that is, such that $u\,u^* = u^*\,u$. Then there exists an orthonormal basis $\mathscr{E} = (e_1, \ldots, e_n)$ such that the matrix representing u in $\mathscr{E}$ is diagonal.*

Similarly, if $A \in \mathfrak{M}_n(\mathbb{C})$ commutes with ${}^t\overline{A}$, then there exist a unitary matrix U and a diagonal matrix D such that $A = U\,D\,{}^t\overline{U}$.

Since a self-adjoint endomorphism obviously commutes with its adjoint, the following corollary follows:

COROLLARY 14.7 (Reduction of a self-adjoint endomorphism) *Let u be a self-adjoint endomorphism (i.e., hermitian in the complex case, or symmetric in the real case). Then u can be diagonalized in an orthonormal basis.*

Also the reader should be able to prove the following result as an exercise:

THEOREM 14.8 (Simultaneous reduction) *If u are v two commuting self-adjoint endomorphisms of E, then they can be diagonalized simultaneously in the same orthonormal basis.*

14.2 Kets and bras

In the remainder of this chapter, $\mathcal{H}$ is a separable Hilbert space (i.e., admitting a countable Hilbert basis).

14.2.a Kets $|\psi\rangle \in H$

DEFINITION 14.9 (Kets) Following Dirac, vectors in the space $\mathcal{H}$ will be generically denoted $|\psi\rangle$. They are also called the **kets** of the space $\mathcal{H}$.

A function with values in $\mathcal{H}$ will be denoted $t \longmapsto \big|\psi(t)\big\rangle$.

An important Hilbert space is the space $L^2(\mathbb{R})$. A function $\psi \in L^2(\mathbb{R})$ will be denoted $|\psi\rangle$. If this function is furthermore a function of time t, the function $x \mapsto \psi(x,t)$ will be denoted $\big|\psi(t)\big\rangle$.

Remark 14.10 (Spaces of quantum mechanics) In the setting of quantum mechanics, it is usual to work at the same time with *two* or even *three* Hilbert spaces:

- an "abstract" Hilbert space $\mathcal{H}$, the vectors of which are denoted $|\psi\rangle$;
- the space L^2, in which the vectors are denoted ψ (meaning a function $x \mapsto \psi(x)$); this space will be denoted $L^2(\mathbb{R}, \mathrm{d}x)$ for added precision;
- the space L^2 again, but in which the vectors are denoted $\Psi : p \mapsto \Psi(p)$; this space will be denoted $L^2(\mathbb{R}, \mathrm{d}p)$.

The square integrable functions ψ correspond to Schrödinger's formalism of wave mechanics, in **position representation** in the case of the space $L^2(\mathbb{R}, \mathrm{d}x)$, and in **momentum representation** in the case of the space $L^2(\mathbb{R}, \mathrm{d}p)$. The function $p \mapsto \Psi(p)$ is the Fourier transform of the function $x \mapsto \psi(x)$. We know that the Fourier transform gives an isometry between $L^2(\mathbb{R}, \mathrm{d}x)$ and $L^2(\mathbb{R}, \mathrm{d}p)$.

When working at the same time with the abstract space $\mathcal{H}$ and the space $L^2(\mathbb{R})$, the distinction can be be made explicit, for instance, by writing φ a function in $L^2(\mathbb{R})$ and $|\varphi\rangle$ the corresponding abstract vector. However, we will not make this distinction here.

In the case of a particle confined in an interval $[0,a]$ (because of an infinite potential well), the space used is $L^2\,[0,a]$ for the position representation and ℓ^2 for the momentum representation; going from one to the other is then done by means of Fourier series.

Moreover, one should notice that all that is done here in the space $L^2(\mathbb{R})$ may be easily generalized to $L^2(\mathbb{R}^3)$ or $L^2(\mathbb{R}^n)$.

We will denote $\langle\cdot,\cdot\rangle_{\mathcal{H}}$ the scalar product in $\mathcal{H}$. The norm of a vector $|\psi\rangle$ is therefore

$$\big\|\,|\psi\rangle\,\big\|^2_{\mathcal{H}} \stackrel{\text{def}}{=} \big\langle\,|\psi\rangle\,,|\psi\rangle\,\big\rangle_{\mathcal{H}}.$$

14.2.b Bras $\langle\psi| \in H'$

DEFINITION 14.11 (Topological dual space) The **(topological) dual space** of $\mathcal{H}$, denoted $\mathcal{H}'$, is the space of continuous linear forms on $\mathcal{H}$. Often, we will simply write "dual" for "topological dual."

DEFINITION 14.12 (Bras) Elements of $\mathcal{H}'$ are called **bras** and are denoted generically $\langle\varphi|$. The result of the application of a bra $\langle\varphi|$ on a ket $|\psi\rangle$ is denoted $\langle\varphi|\,\psi\rangle$, which is therefore a **bracket**.[1]

Example 14.13 Consider the Hilbert space $L^2(\mathbb{R})$. The map

$$\omega : \left|\begin{array}{l} L^2(\mathbb{R}) \longrightarrow \mathbb{C} \\[2mm] \psi \longmapsto \displaystyle\int_{-\infty}^{+\infty} \psi(x)\,\mathrm{e}^{-x^2}\,\mathrm{d}x \end{array}\right.$$

[1] Dirac did not simply make a pun when inventing the terminology "bra" and "ket": this convenient notation became quickly indispensable for most physicists. Note that Dirac also coined the words *fermion* and *boson*, and introduced the terminology of c-number, the δ-function, and the useful notation $\hbar = h/2\pi$.

is linear. Defining $g : x \mapsto e^{-x^2}$, we have $\omega(\psi) = \langle g, \psi \rangle_{\mathcal{H}}$ for any $\psi \in L^2$; by the Cauchy-Schwarz inequality, we have

$$\forall \psi \in L^2 \qquad \left|\omega(\psi)\right| = \left| \langle g, \psi \rangle_{\mathcal{H}} \right| \leqslant \|g\|_2 \ \|\psi\|_2 ,$$

and hence the linear form ω is continuous and is an element of the topological dual of $L^2(\mathbb{R})$.

More generally, if we choose a ket $\varphi \in \mathcal{H}$, a linear form

$$\omega_\varphi : |\psi\rangle \longmapsto \big\langle\, |\varphi\rangle \,, |\psi\rangle \,\big\rangle_{\mathcal{H}}$$

is associated to φ using the scalar product. The Cauchy-Schwarz inequality shows that this linear form ω_φ is continuous (see Theorem A.51 on page 583). It corresponds therefore to a bra, which is denoted $\langle \varphi|$:

$$\langle \varphi|\psi\rangle \stackrel{\text{def}}{=} \big\langle\, |\varphi\rangle \,, |\psi\rangle \,\big\rangle_{\mathcal{H}}.$$

In other words:

To each ket $|\varphi\rangle \in \mathcal{H}$ is associated a bra $\langle\varphi| \in \mathcal{H}'$.

The converse is also true, as proved by the Riesz representation theorem:

THEOREM 14.14 (Riesz representation theorem) *The map $|\varphi\rangle \mapsto \langle\varphi|$ is a semilinear isomorphism of the Hilbert space $\mathcal{H}$ with its topological dual $\mathcal{H}'$. In other words, for any continuous linear form $\langle\varphi|$, there exists a unique vector $|\varphi\rangle$ such that*

$$\forall\, |\psi\rangle \in \mathcal{H} \qquad \langle \varphi|\psi\rangle = \big\langle\, |\varphi\rangle \,, |\psi\rangle \,\big\rangle_{\mathcal{H}}.$$

Hence there is a one-to-one correspondance between bras and kets:

To each bra $\langle\varphi| \in \mathcal{H}'$ one can associate a ket $|\varphi\rangle \in \mathcal{H}$.

Because of this, it could be conceivable to work only with $\mathcal{H}$ and to forget about $\mathcal{H}'$, stating that $\langle\varphi|\psi\rangle$ is nothing more nor less than a notation for $\big\langle\, |\varphi\rangle \,, |\psi\rangle \,\big\rangle_{\mathcal{H}}$. However, this result will not remain true when we consider "generalized bras."

When we define on $\mathcal{H}'$ the operator norm

$$\big\|\, \langle\varphi| \,\big\|_{\mathcal{H}'} \stackrel{\text{def}}{=} \sup_{\||\psi\rangle\|_{\mathcal{H}}=1} \big|\, \langle\varphi|\psi\rangle \,\big|,$$

the isomorphism above,

$$\begin{aligned} \mathcal{H} &\longrightarrow \mathcal{H}', \\ |\varphi\rangle &\longmapsto \langle\varphi|, \end{aligned}$$

becomes a semilinear isometry:

$$\big\|\, \langle\varphi| \,\big\|_{\mathcal{H}'} = \big\|\, |\varphi\rangle \,\big\|_{\mathcal{H}}.$$

In particular it is permissible to write $\|\varphi\|$ without ambiguity whether φ corresponds to a bra or a ket.

Remark 14.15 The semilinearity of the map above means that $\langle \lambda\varphi| = \overline{\lambda}\,\langle\varphi|$ for any $\lambda \in \mathbb{C}$ and $\varphi \in \mathcal{H}$. This is a result of the convention chosen; if we associate to a ket $|\varphi\rangle$ the linear map

$$|\psi\rangle \longmapsto \langle\, |\psi\rangle\,, |\varphi\rangle\,\rangle_{\mathcal{H}},$$

the result map $\mathcal{H} \to \mathcal{H}'$ is a *linear* isomorphism.

Counterexample 14.16 Consider again the Hilbert space $\mathcal{H} = L^2(\mathbb{R})$. Let $x_0 \in \mathbb{R}$. Then the map $\delta_{x_0} : \psi \mapsto \psi(x_0)$ is a linear form on the subspace D of continuous functions[2] in $\mathcal{H}$.

However, it does not take much effort to discover that it is *not* representable as a scalar product with a square integrable function. What's the problem? Certainly, it must be the case that δ_{x_0} is continuous?

Well, in fact, it is not! It is continuous on the space D with its topology (of uniform convergence), but not on L^2 with the hermitian topology (based on integrals, so defined "up to equality almost everywhere").

To see this, it suffices to define a sequence of functions $\psi_n : x \mapsto \mathrm{e}^{-nx^2/2}$. Then, with $x_0 = 0$ and $\delta = \delta_0$ (which is no loss of generality), we have

$$\|\psi_n\|^2_{\mathcal{H}} = \int_{-\infty}^{+\infty} \mathrm{e}^{-nx^2}\,\mathrm{d}x = \sqrt{\frac{\pi}{n}} \xrightarrow[n\to\infty]{} 0,$$

whereas $\delta(\psi_n) = 1$ for all $n \in \mathbb{N}$. So there does not exist a constant $\alpha \in \mathbb{R}^+$ such that $\left|\delta(\psi)\right| \leqslant \alpha\,\|\psi\|_{\mathcal{H}}$ for all $\psi \in \mathcal{H}$, which means that δ is not continuous.

14.2.c Generalized bras

We will denote also with the generic notation $\langle\psi|$ the **generalized bras**, namely linear forms which are not necessarily in $\mathcal{H}'$. Such a form may not be continuous, or it may not be defined on the whole of $\mathcal{H}$ (as seen in the case of δ, in counterexample 14.16); still it will be asked that it be defined at least on a *dense* subspace of $\mathcal{H}$.

One of the most important example is the following:

DEFINITION 14.17 (The bra $\langle x_0|$) Let $x_0 \in \mathbb{R}$. We denote by $\langle x_0|$ the *discontinuous* linear form defined on the dense subspace of continuous functions by

$$\langle x_0|\,\psi\rangle = \psi(x_0).$$

The bra $\langle x_0|$ is thus another name for the Dirac distribution $\delta(x - x_0)$.

Consider now a broader approach to the notion of generalized bra. For this, let T be a tempered distribution. Since $\mathscr{S}$ is a dense subspace of $L^2(\mathbb{R})$, the dual $\mathscr{S}'$ of $\mathscr{S}$ contains the dual of $L^2(\mathbb{R})$, which is $L^2(\mathbb{R})$ itself because of the Riesz representation theorem.[3] Hence we have the following situation:

$$\mathscr{S} \subset L^2(\mathbb{R}) \subset \mathscr{S}'.$$

---

[2] Since functions in L^2 are defined up to equality almost everywhere, δ_{x_0} is not well defined on the whole of $\mathcal{H}$. We therefore restrict it to the *dense* subspace of continuous functions.

[3] In fact, the dual of $L^2(\mathbb{R})$ is isomorphic to $L^2(\mathbb{R})$, but via this isomorphism one can identify any continuous linear form on $L^2(\mathbb{R})$ to a function in $L^2(\mathbb{R})$.

This is called a **Gelfand triple**. It is tempting to define the bra $\langle T|$ by $\langle T|\varphi\rangle = \langle T, \varphi\rangle$; but this last expression is *linear* in T, in the sense that $\langle \alpha T, \varphi\rangle = \alpha\,\langle T, \varphi\rangle$, whereas the hermitian product is *semilinear* on the left. To deal with this minor inconvenience, the following definition is therefore used:

DEFINITION 14.18 (Generalized bra) Let T be a tempered distribution. The **generalized bra** $\langle T|$ is defined on the Schwartz space $\mathscr{S}$ by its action on any $|\varphi\rangle \in \mathscr{S}$ given by

$$\langle T|\varphi\rangle \stackrel{\text{def}}{=} \overline{\langle T, \overline{\varphi}\rangle}.$$

Remark 14.19 It is immediate to check that, if T_φ is a regular distribution, associated to a function $\varphi \in \mathscr{S}$, then we have

$$\forall \psi \in \mathcal{H} \qquad \langle T_\varphi|\psi\rangle = \int_{-\infty}^{+\infty} \overline{\varphi(x)}\,\psi(x)\,\mathrm{d}x,$$

which is equal to the hermitian scalar product $\langle \varphi|\psi\rangle$ or $(\varphi|\psi)$ in $L^2(\mathbb{R})$.

If T is a "real" distribution (such as δ), we do, however, have $\langle T|\varphi\rangle = (T|\varphi)$.

14.2.d Generalized kets

In the case where $\mathcal{H} = L^2(\mathbb{R})$ or $L^2[a,b]$, it is possible to associate a generalized ket to *some* generalized bras. If $\langle\psi|$ is a linear form defined by a regular distribution

$$\langle\psi|\varphi\rangle = \int_{-\infty}^{+\infty} \overline{\psi}\,\varphi,$$

where ψ is a locally integrable function, we define the associated **generalized ket** $|\psi\rangle$, namely, simply the function ψ. This does not necessarily belong to L^2 and is therefore not necessarily an element of $\mathcal{H}$. All that is asked is that the integral be defined.

DEFINITION 14.20 (Ket $|p\rangle$ and bra $\langle p|$) Let p be a real number. The function

$$\omega_p : x \longmapsto \frac{1}{\sqrt{2\pi\hbar}}\,\mathrm{e}^{\mathrm{i}px/\hbar}$$

is not square integrable. It still defines a noncontinuous linear form on the dense space $L^1 \cap L^2$:

$$\Omega_p : \left|\begin{array}{l} L^1(\mathbb{R}) \cap L^2(\mathbb{R}) \longrightarrow \mathbb{C} \\ \displaystyle \varphi \longmapsto \int_{-\infty}^{+\infty} \overline{\omega_p(x)}\,\varphi(x)\,\mathrm{d}x = \widetilde{\varphi}(p), \end{array}\right.$$

where $\widetilde{\varphi}$ is the Fourier transform of φ, with the conventions usual in quantum mechanics. This generalized bra is denoted $\langle p| = \langle\Omega_p|$. The ket that corresponds to the function ω_p is denoted $|p\rangle = |\omega_p\rangle \notin \mathcal{H}$.

Moreover, the notion of distribution can be seen as a generalization of that of function. Hence, the generalized bra $\langle p|$ not only acts on (a subspace of) $L^2(\mathbb{R})$, associating to a function its Fourier transform evaluated at p, but can be extended to tempered distributions. If $|T\rangle$ is a ket formally associated to a tempered distribution $T \in \mathscr{S}'$, then we have $\langle p|\, T\rangle = \widetilde{T}(\omega_p)$.

THEOREM 14.21 *The family of kets $|p\rangle$ is orthonormal, in the sense that for all $p, p' \in \mathbb{R}$, we have $\langle p|p'\rangle = \delta(p - p')$.*

PROOF. Apply the definition $\langle p|\, p'\rangle = \widetilde{\omega_p}(\omega_{p'})$, and remember that the Fourier transform of $\omega_{p'}$ in the sense of distributions is $\delta_{p'}$.

The boundary between "bra" and "ket" is not clear. A distribution is a linear form, thus technically a generalized "bra." But as a generalization of a function, it can – formally – be denoted as a ket. Thus we define the $|x_0\rangle$ associated to the "generalized function" $\delta(x - x_0)$. In particular, for any function $\varphi \in \mathscr{S}$, we have

$$\langle \varphi|\, x_0\rangle = \overline{\langle x_0|\, \varphi\rangle} = \overline{\varphi(x_0)}.$$

Also, the kets $|x\rangle$ satisfy the orthonormality relation:

$$\langle x|\, x_0\rangle = \delta(x - x_0) = \delta_x(x_0),$$

valid in the sense of tempered distributions – that is, which makes sense only when applied to a function in the Schwartz space.

14.2.e $\mathrm{Id} = \sum_n |\varphi_n\rangle \langle \varphi_n|$

Let $(\varphi_n)_{n\in\mathbb{N}}$ be an orthonormal Hilbert basis of $\mathcal{H}$. We know that this property is characterized by the fact that

$$\forall x \in \mathcal{H} \qquad x = \sum_{n=0}^{\infty} (\varphi_n|x)\, \varphi_n.$$

Using Dirac's notation, this equation is written as follows:

$$\forall\, |\psi\rangle \in \mathcal{H} \qquad |\psi\rangle = \sum_{n=0}^{\infty} \langle \varphi_n|\, \psi\rangle\, |\varphi_n\rangle,$$

or also, putting the scalar $\langle \varphi_n|\, \psi\rangle$ *after* the vector $|\varphi_n\rangle$

$$\forall\, |\psi\rangle \in \mathcal{H} \qquad |\psi\rangle = \sum_{n=0}^{\infty} |\varphi_n\rangle\, \langle \varphi_n|\, \psi\rangle\,.$$

Or still, formally, the operator $\sum_{n=0}^{\infty} |\varphi_n\rangle \langle \varphi_n|$ is the identity operator. This is what physicists call a **closure relation**.

For easy remembering, this will be stated as

THEOREM 14.22 (Closure relation) *A family $(\varphi_n)_{n\in\mathbb{N}}$ is a Hilbert basis of $\mathcal{H}$ if and only if*

$$\mathrm{Id} = \sum_{n=0}^{\infty} |\varphi_n\rangle \langle\varphi_n| .$$

This point of view allows us to recover easily another characterization of a Hilbert basis, namely the Parseval identity. Indeed, for any $|\psi\rangle \in \mathcal{H}$, we have

$$\begin{aligned}\|\psi\|^2 = \langle\psi|\,\psi\rangle = \langle\psi| \left(\sum_{n=0}^{\infty} |\varphi_n\rangle \langle\varphi_n|\right) |\psi\rangle &= \sum_{n=0}^{\infty} \langle\psi|\,\varphi_n\rangle \langle\varphi_n|\,\psi\rangle \\ &= \sum_{n=0}^{\infty} \left| \langle\varphi_n|\,\psi\rangle \right|^2,\end{aligned}$$

which can be compared with the formula D) in Theorem 9.21 on page 259.

14.2.f Generalized basis

We now extend the notion of basis of the space $L^2(\mathbb{R})$, to include distributions, seen as generalizations of functions. Let $(T_\lambda)_{\lambda\in\Lambda}$ be a family of tempered distributions, indexed by a discrete or continuous set[4] Λ. To generalize the idea of a Hilbert basis, the first idea would be to impose the condition

$$\mathrm{Id} = \int_\Lambda |T_\lambda\rangle \langle T_\lambda| \,\mathrm{d}\lambda, \qquad \text{that is,} \qquad |\varphi\rangle = \int_\Lambda |T_\lambda\rangle \langle T_\lambda|\varphi\rangle \,\mathrm{d}\lambda,$$

where the notation $\int \mathrm{d}\lambda$ may represent an integral, a sum, or a mixture of both. Still, there would be the problem of defining correctly the integral on the right: the integrand is not a scalar but a vector in an infinite-dimensional space.

It is easier to consider another characterization of a Hilbert basis: the Parseval formula, which only involves scalar quantities.

DEFINITION 14.23 (Generalized basis) The family $(T_\lambda)_{\lambda\in\Lambda}$ is a **generalized basis**, or a **total orthonormal system**, if and only if we have

$$\forall\, |\varphi\rangle \in \mathscr{S} \qquad \|\varphi\|_2^2 = \langle\varphi|\,\varphi\rangle = \int_\Lambda \left| \langle T_\lambda|\,\varphi\rangle \right|^2 \mathrm{d}\lambda.$$

The family $(T_\lambda)_{\lambda\in\Lambda}$ is a **total system** (or **a complete system**) if, for all $\varphi \in \mathscr{S}$, we have

$$\big(\langle T_\lambda, \varphi\rangle = 0 \text{ for any } \lambda \in \Lambda\big) \iff (\varphi = 0).$$

---

[4] The choice of index is not without meaning. A generalized basis can be useful to "diagonalize" a self-adjoint operator which has, in addition to eigenvalues, a *continuous spectrum*.

As in the case of the usual Hilbert basis (see Theorem 9.21), a generalized basis is a total system. However, the normalization of such a family (the elements of which are not a priori in L^2) is delicate.

Note that a generalized basis is *not* in general a Hilbert basis of L^2, for instance, because any Hilbert basis is countable.

THEOREM 14.24 (The basis $\{\langle x| \ ; \ x \in \mathbb{R}\}$) *Denote by $\langle x| = \langle \delta_x|$ the generalized bra associated to the distribution δ_x. Then the family of distributions $\{ \langle x| \ ; \ x \in \mathbb{R}\}$ is a generalized basis of $L^2(\mathbb{R})$.*

PROOF. Indeed, for any $\varphi \in \mathscr{S}$, by the very definition of the L^2 norm, we have $\|\varphi\|_2^2 = \int_{-\infty}^{+\infty} |\varphi(x)|^2 \,\mathrm{d}x = \int_{-\infty}^{+\infty} |\langle x|\varphi\rangle|^2 \,\mathrm{d}x$.

THEOREM 14.25 (The basis $\{\langle p| \ ; \ p \in \mathbb{R}\}$) *Denote by $\langle p| = \langle \Omega_p|$ the generalized bra defined by the regular distributions Ω_p associated to the function*

$$\omega_p : x \longmapsto \frac{e^{i\pi x/\hbar}}{\sqrt{2\pi\hbar}}.$$

Then the family $\{ \langle p| \ ; \ p \in \mathbb{R}\}$ is a generalized basis of $L^2(\mathbb{R})$.

PROOF. Let $\varphi \in \mathscr{S}$. Then $\langle p|\varphi\rangle = \widetilde{\varphi}(p)$; the Parseval-Plancherel identity (Theorem 10.36), which is a translation of the fact that the Fourier transform is an isometry of $L^2(\mathbb{R})$, can be written as

$$\|\varphi\|_2^2 = \|\widetilde{\varphi}\|_2^2 = \int_{-\infty}^{+\infty} |\widetilde{\varphi}(p)|^2 \,\mathrm{d}p = \int_{-\infty}^{+\infty} |\langle p|\varphi\rangle|^2 \,\mathrm{d}p.$$

Finally, it can be shown that the closure relation is still valid for two generalized bases $\{ \langle x| \ ; \ x \in \mathbb{R}\}$ and $\{ \langle p| \ ; \ p \in \mathbb{R}\}$. Indeed, if $\varphi, \psi \in L^2(\mathbb{R})$, the relation $\langle\varphi|\psi\rangle = \int \langle\varphi|p\rangle \langle p|\psi\rangle \,\mathrm{d}p$ holds: this is (once more) the Parseval-Plancherel identity! Similarly, we have

$$\langle\varphi|\psi\rangle = \int_{-\infty}^{+\infty} \overline{\varphi(x)}\, \psi(x) \,\mathrm{d}x = \int_{-\infty}^{+\infty} \overline{\langle x|\varphi\rangle}\, \langle x|\psi\rangle \,\mathrm{d}x$$
$$= \int_{-\infty}^{+\infty} \langle\varphi|x\rangle \langle x|\psi\rangle \,\mathrm{d}x.$$

It will be convenient to state this as follows:

THEOREM 14.26 (Generalized closure relation)

$$\mathrm{Id} = \int_{-\infty}^{+\infty} |x\rangle \langle x| \,\mathrm{d}x \qquad \textit{and} \qquad \mathrm{Id} = \int_{-\infty}^{+\infty} |p\rangle \langle p| \,\mathrm{d}p.$$

Remark 14.27 This shows that it is possible to write, formally, identities like

$$|\varphi\rangle = \int_{-\infty}^{+\infty} |x\rangle \langle x|\varphi\rangle \,\mathrm{d}x, \qquad \text{that is,} \qquad \varphi(\cdot) = \int_{-\infty}^{+\infty} \delta_x(\cdot)\, \varphi(x) \,\mathrm{d}x.$$

This formula has a sense if evaluated at a point y; indeed, this yields the correct formula

$$\varphi(y) = \int_{-\infty}^{+\infty} \varphi(x)\, \delta_x(y)\, \mathrm{d}y.$$

The great strength of Dirac's notation is to lead, very intuitively, to many correct formulas. For instance, writing

$$|\varphi\rangle = \int_{-\infty}^{+\infty} |p\rangle\, \langle p|\, \varphi\rangle\, \mathrm{d}p$$

amounts to saying that, for any $x \in \mathbb{R}$, we have

$$\varphi(x) = \langle x|\, \varphi\rangle = \int_{-\infty}^{+\infty} \langle x|\, p\rangle\, \langle p|\, \varphi\rangle\, \mathrm{d}p = \frac{1}{\sqrt{2\pi\hbar}} \int_{-\infty}^{+\infty} \mathrm{e}^{\mathrm{i}px/\hbar}\, \widetilde{\varphi}(p)\, \mathrm{d}p,$$

which is none other than the inverse Fourier transform formula!

14.3 Linear operators

We will only describe here the simplest properties of linear operators acting on Hilbert spaces. This is a very complex and subtle theory, which requires by itself a full volume. The interested reader can read, for instance, the books [4, 95] or [71], which deals with quantum mechanics in Hilbert spaces. We will define bounded operators, closed operators, eigenvalues and eigenvectors, and adjoints of operators.

14.3.a Operators

DEFINITION 14.28 (Linear operator) A **linear operator** on a Hilbert space $\mathcal{H}$ is a pair (A, D_A), where

- D_A is a (not necessarily closed) subspace of $\mathcal{H}$, called the **domain of** A;
- A is a linear map defined on D_A and with values in $\mathcal{H}$.

The image of D_A by A is called the **image of** A, and denoted $\mathbf{Im}(A)$. Finally the image of an element $\varphi \in \mathcal{H}$ by A will be denoted either $A(\varphi)$ or, more often, simply $A\varphi$.

Remark 14.29 To say that two operators A **and** B **are equal** means, in consequence, that:

- their domains coincide: $D_A = D_B$;
- they act identically on their domains: $A\varphi = B\varphi$ for any $\varphi \in D_A$.

Very often, the action is only used as a shorthand for the whole operator; for instance, the position operator in quantum mechanics, which acts on a wave function ψ by sending it to the function $x \mapsto x\,\psi(x)$. It may be dangerous, however, to leave the domain unspecified; a number of paradoxes arising from such sloppiness in notation are pointed out in a paper by F. Gieres [41].

DEFINITION 14.30 (Extension) Let A and B be two linear operators on $\mathcal{H}$. If $D_A \subset D_B$ and if $A\varphi = B\varphi$ for any $\varphi \in D_A$, then we say that **B is an extension A** or that **A is the restriction of B to D_A**.

Remark 14.31 There is often a large choice available for the domain on which to define a linear operator. For instance, the maximal domain on which the position operator, denoted X, is defined is the subspace

$$D_1 = \{\psi \in L^2(\mathbb{R}) \,;\, x \mapsto x\,\psi(x) \in L^2(\mathbb{R})\}.$$

However, this space may not be the most convenient to use. It may be simpler to restrict X to the Schwartz space $D_2 = \mathscr{S}$ of $\mathscr{C}^\infty$ class functions with all derivatives rapidly decaying. This space is dense in $\mathcal{H} = L^2(\mathbb{R})$, and the operator X is defined on $\mathscr{S}$. Moreover, this space has the advantage of being stable with respect to X.

We must insist: the operator (X, D_1) is not the same as the operator (X, D_2), even if they are both usually denoted by the same name "operator X." The first is an extension of the second (to a strictly larger space).

An important example of an operator is the momentum operator of quantum mechanics. Consider the space $\mathcal{H} = L^2[0,a]$ of square integrable functions on the interval $[0,a]$. The wave functions of a particle confined to this interval all satisfy $\psi(0) = \psi(a) = 0$ (this is a physical constraint). We wish to define an operator (the momentum operator P) by

$$P\psi = -\mathrm{i}\hbar\psi'.$$

If $\psi \in \mathcal{H}$, its derivative ψ' must be taken in the sense of distributions. But we wish to define an operator P with image contained in $\mathcal{H}$. Hence this distribution must be a regular distribution,[5] which is associated to a square integrable function. This justifies the following definition:

DEFINITION 14.32 (Momentum operator P in $L^2[0,a]$) Let $\mathcal{H} = L^2[0,a]$. The **momentum operator P** is defined, on its domain

$$D_P = \{\psi \in \mathcal{H} \,;\, \psi' \in \mathcal{H} \quad \text{and} \quad \psi(0) = \psi(a) = 0\},$$

by

$$\forall \psi \in D_P \qquad P\psi = -\mathrm{i}\hbar\,\psi'.$$

Remark 14.33 (Boundary conditions for the operator P) One may wonder what the boundary conditions $\psi(0) = \psi(a)$ really mean; as a matter of fact, ψ, like every function of the L^2 space, is defined only up to equality almost everywhere, and therefore, it would seem that a boundary condition is meaningless. But the fact that $\psi' \in L^2[a,b]$ implies that $\psi' \in L^1[a,b]$ and, in particular, that ψ is continuous.[6] This explains why the boundary conditions make sense.

The momentum operator for a particle that can range over $\mathbb{R}$ is defined similarly:

---

[5] This does not imply that the function ψ is differentiable; but it must be continuous and its derivative exists almost everywhere.

[6] Even absolutly continuous, but this is not of concern here.

DEFINITION 14.34 (Momentum operator P in $L^2(\mathbb{R})$) Let $\mathcal{H} = L^2(\mathbb{R})$. The **momentum operator P** is defined, on its domain

$$D_P = \{\psi \in \mathcal{H}\,;\ \psi' \in \mathcal{H}\},$$

by

$$P\psi = -\mathrm{i}\hbar\,\psi' \qquad \forall \psi \in D_P.$$

The limit conditions $\psi(-\infty) = \psi(+\infty) = 0$ are automatically satisfied when $\psi, \psi' \in \mathcal{H}$.

14.3.b Adjoint

DEFINITION 14.35 (Adjoint) Let (A, D_A) be a linear operator on $\mathcal{H}$. The **domain of $A^\dagger$**, denoted **$D_{A^\dagger}$**, is the set of vectors $\varphi \in \mathcal{H}$ for which there exists a *unique* vector $\omega \in \mathcal{H}$ satisfying

$$\forall \psi \in D_A \qquad \langle \omega | \psi \rangle = \langle \varphi | A\psi \rangle.$$

If $D_{A^\dagger}$ is not empty, the **operator adjoint to A**, denoted $A^\dagger$, is defined by $A^\dagger\varphi = \omega$ for any $\varphi \in D_{A^\dagger}$, with the preceding notation. In other words,

$$\forall \varphi \in D_{A^\dagger} \quad \forall \psi \in D_A \qquad \langle A^\dagger\varphi | \psi \rangle = \langle \varphi | A\psi \rangle.$$

PROPOSITION 14.36 (Existence of the adjoint) *A has an adjoint if and only if D_A is dense in $\mathcal{H}$.*

PROOF. Assume first that D_A is dense. In the case $\varphi = 0$, the vector $\omega = 0$ satisfies the equation $\langle \omega | \psi \rangle = \langle \varphi | A\psi \rangle$ for all $\psi \in D_A$. So we must show that this vector is unique. More generally, for a given arbitrary φ, let ω and ω' satisfying the required property. Then $\langle \omega - \omega' | \psi \rangle = 0$ for all $\psi \in D_A$. By density, we may find a sequence $(\psi_n)_{n\in\mathbb{N}}$, with values in D_A, which converges to $\omega - \omega'$. Taking the limit in the scalar product (which is a continuous map), we obtain $\|\omega - \omega'\|^2 = \langle \omega - \omega' | \omega - \omega' \rangle = 0$ and hence $\omega = \omega'$.

Conversely, assume that D_A is not dense. Then the orthogonal of D_A in not equal to $\{0\}$ (it is a corollary of Theorem 9.21). Choose any $\chi \neq 0$ in this orthogonal space. Then $\langle \chi | \psi \rangle = 0$ for any $\psi \in D_A$. Let $\varphi \in \mathcal{H}$. If ω is a candidate vector such that the relation $\langle \omega | \psi \rangle = \langle \varphi | A\psi \rangle$ holds for all $\psi \in D_A$, it is obvious that the vector $\omega + \chi$ also satisfies this property. So there is no uniqueness and therefore $D_{A^\dagger} = \varnothing$.

Remark 14.37 If D_A is dense, it may still be the case that $D_{A^\dagger}$ is equal to $\{0\}$.

In the rest of this chapter, we only consider linear operators with dense *domain in $\mathcal{H}$. So the adjoint is always defined.*

PROPOSITION 14.38 (Properties of the adjoint) *Let A and B be linear operators.*

i) $A^\dagger$ is a linear operator;

ii) $(\alpha A)^\dagger = \overline{\alpha} A^\dagger$ for all $\alpha \in \mathbb{C}$;

iii) if $A \subset B$, then $B^\dagger \subset A^\dagger$;

iv) if $D_{A^\dagger}$ is dense, then $A^\dagger$ has un adjoint, denoted $A^{\dagger\dagger}$, which is an extension of A: $A \subset A^{\dagger\dagger}$.

In Section 14.4, we will determine the adjoints of some important operators.

14.3.c Bounded operators, closed operators, closable operators

In the terminology of Hilbert spaces, a "bounded" operator is simply a continuous operator. This is justified by the following definition and proposition:

DEFINITION 14.39 An operator A is **bounded** if and only if it is bounded on the unit ball of $\mathcal{H}$, or in other words, if the set

$$\{\, \|A\varphi\| \ ; \ \varphi \in D_A \text{ and } \|\varphi\| \leqslant 1\}$$

is bounded. If A is bounded, we denote

$$|\!|\!|A|\!|\!| = \sup_{\substack{\varphi \in D_A \\ \|\varphi\| \leqslant 1}} \|A\varphi\| = \sup_{\substack{\varphi \in D_A \\ \varphi \neq 0}} \frac{\|A\varphi\|}{\|\varphi\|}.$$

Just as in any other complete normed vector space, the following holds (it is just a reminder of Theorem A.51 on page 583 and of sidebar 5 on page 294):

PROPOSITION 14.40 *A is bounded if and only if A is continuous.*

If A is bounded, it extends uniquely to a continuous linear operator defined on the whole of $\mathcal{H}$.

A property weaker than boundedness but which is also important is that of being a closed operator.

DEFINITION 14.41 (Closed operator, graph) Let A be a linear operator. The **graph of A** is the subspace

$$\Gamma(A) = \{(\varphi, A\varphi) \ ; \ \varphi \in D_A\}$$

of $\mathcal{H} \times \mathcal{H}$.

The operator A is **closed** if and only if its graph is closed,[7] that is, if for any sequence $(\varphi_n)_{n\in\mathbb{N}}$ with values in D_A which converges to a limit φ, *if* moreover the sequence $(A\varphi_n)_n$ converges to a vector ψ, then we have $\varphi \in D_A$ and $A\varphi = \psi$.

---

[7] For the topology defined on $\mathcal{H} \times \mathcal{H}$ by the natural scalar product

$$\langle(\varphi,\psi)|(\varphi',\psi')\rangle = \langle\varphi|\varphi'\rangle + \langle\psi|\psi'\rangle.$$

The reader will find the following elementary properties easy to check:

PROPOSITION 14.42 *Any bounded operator is closed.*

THEOREM 14.43 *For any operator A, its adjoint $A^\dagger$ is closed.*

DEFINITION 14.44 (Closable operator, closure) An operator A is **closable** if there exists a closed extension of A; the smallest such closed extension is called the **closure of A** and is denoted $\overline{A}$.

PROPOSITION 14.45 *A is closable if and only if the domain of $A^\dagger$ is dense. In this case, we have $\overline{A} = (A^\dagger)^\dagger$ and $\overline{A}^\dagger = A^\dagger$.*

This notion of closure of an operator may seem very abstract. We will see below how to construct it in certain special cases important in physics.

14.3.d Discrete and continuous spectra

DEFINITION 14.46 Let (A, D_A) be an operator. An **eigenvalue** of A is a complex number λ such that there exists a nonzero vector $\varphi \in D_A$ satisfying the relation

$$A\varphi = \lambda\,\varphi.$$

Such a vector φ is an **eigenvector**. The set of eigenvectors of A is the **discrete spectrum** of A, and is denoted $\sigma_d(A)$.

Example 14.47 (Discrete spectrum of the momentum operator P) Consider the **momentum operator** P on the Hilbert space $\mathcal{H} = L^2(\mathbb{R})$. It is defined on $D_P = \{\varphi \in \mathcal{H}\,;\ \varphi' \in \mathcal{H}\}$ by

$$P\psi = -\mathrm{i}\hbar\psi'.$$

Let $\lambda \in \mathbb{C}$. We try to solve, in $\mathcal{H}$, the equation

$$P\varphi = \lambda\varphi.$$

This implies $\varphi' = (\mathrm{i}\lambda/\hbar)\,\varphi$, and hence $\varphi(x) = \alpha\,\mathrm{e}^{\mathrm{i}\lambda x/\hbar}$, where α is a constant. However, functions of this type are square integrable only if $\alpha = 0$; since the zero function is not an eigenvector, it follows that P has no eigenvalue: $\sigma_d(P) = \varnothing$.

Example 14.48 (Discrete spectrum of the position operator X) Consider now the **position operator** X on the Hilbert space $\mathcal{H} = L^2(\mathbb{R})$. It is defined on $D_X = \{\varphi \in \mathcal{H}\,;\ x \mapsto x\,\varphi(x) \in \mathcal{H}\}$ by $X\varphi = \psi$, where $\psi : x \mapsto x\,\varphi(x)$.

Let $\lambda \in \mathbb{C}$. As before we try to solve, in $\mathcal{H}$, the equation

$$X\varphi = \lambda\varphi.$$

It follows that for almost all $x \in \mathbb{R}$ we have $(x - \lambda)\,\varphi(x) = 0$. Hence the function φ is zero almost everywhere, in other words,[8] $\varphi = 0$. Hence X has no eigenvalue: $\sigma_d(X) = \varnothing$.

---

[8] Since, we recall, the space L^2 is constructed by identifying together functions which are equal almost everywhere. Any function which is zero almost everywhere is therefore identified with the zero function.

DEFINITION 14.49 Let (A, D_A) be an operator. A **generalized eigenvalue** of A is a complex number λ such that

- λ is not an eigenvalue of A;
- there exists a sequence $(\varphi_n)_{n\in\mathbb{N}}$ of vectors of norm 1 satisfying
$$\lim_{n\to\infty} \|A\varphi_n - \lambda\varphi_n\| = 0.$$

The set of generalized eigenvalues of A is called the **pure continuous spectrum** of A, and is denoted $\sigma_c(A)$. It is disjoint from the discrete spectrum: $\sigma_d(A) \cap \sigma_c(A) = \varnothing$.

Example 14.50 (Continuous spectrum of the operator X) We saw that the operator X of the Hilbert space $L^2(\mathbb{R})$ has no eigenvalue. On the other hand, a function φ with graph which is a very narrow "spike" around a real number λ should satisfy "$X\varphi \approx \lambda\varphi$." Hence we will check that any real number is a generalized eigenvalue, the idea being to construct such a sequence of spikier and spikier functions. For reasons of normalization, this is not quite the same as a Dirac sequence.

Let $x_0 \in \mathbb{R}$. We consider the sequence $(\psi_n)_{n\in\mathbb{N}}$ of functions defined by $\psi_n(x) = \psi_n^0(x - x_0)$ and $\psi_n^0(x) = \sqrt{n}\,\Pi(nx)$; recall that the "rectangle" function Π satisfies $\Pi(x) = 1$ for $|x| \leqslant 1$ and $\Pi(x) = 0$ otherwise.

It is easy to check that the functions ψ_n are of norm 1. For all $n \in \mathbb{N}$, we have

$$\|(X - x_0)\psi_n\| = \int_{-\infty}^{+\infty} x^2\, \varphi_n(x)^2 \,\mathrm{d}x = \int_{-1/2n}^{1/2n} nx^2 \,\mathrm{d}x = \frac{1}{12\, n^2}.$$

Hence x_0 is a generalized eigenvalue of the operator X.

One can also show that the continuous spectrum is a subset of $\mathbb{R}$, and therefore conclude that the continuous spectrum of X is the whole real line: $\sigma_c(X) = \mathbb{R}$.

Remark 14.51 In the previous example, the sequence $(\psi_n)_{n\in\mathbb{N}}$ converges neither in $L^2(\mathbb{R})$ nor in the sense of distributions (because the normalization is chosen for the purpose of the L^2 norm). However, the sequence $(\sqrt{n}\,\psi_n)$ is a sequence that converges, in the sense of distributions, to $T = \delta(x - x_0)$, which satisfies the relation

$$x\,\delta(x - x_0) = x_0\,\delta(x - x_0),$$

in other words $XT = x_0 T$ in the sense of distributions. This is the (generalized) eigenvector associated to the generalized eigenvalue!

Defining properly what is a (tempered) distribution which is an eigenvector T_λ of an operator A requires a proper definition of what is meant by $A\,T_\lambda$. This is in fact a delicate problem, which we will only solve for a certain type of operators (see Definition 14.68).

Example 14.52 (Continuous spectrum of the operator P) Consider the operator P on the space $L^2(\mathbb{R})$. We have already shown that it has no eigenvalue. On the other hand, we can now show that any $p \in \mathbb{R}$ is a generalized eigenvalue. The idea is, once more, to find a function which "looks like" the function ω_p (which suffers itself from the essential defect of not being square integrable). For this purpose, we will simply average the functions ω_λ, for values of λ close to p. Such an average is, up to unimportant factors, the same as the Fourier transform of a function similar to the rectangle function: hence it is a function that decays like $1/x$ at infinity, and is therefore square integrable.

To be more precise, define

$$\omega_p^{(\varepsilon)}(x) \stackrel{\text{def}}{=} \frac{1}{\sqrt{2\varepsilon}} \int_{p-\varepsilon}^{p+\varepsilon} \omega_\lambda(x)\,\mathrm{d}\lambda = \sqrt{\frac{\hbar}{\pi\varepsilon}}\,\frac{\sin(\varepsilon x/\hbar)}{\mathrm{i}x}$$

for $x \in \mathbb{R}$ and $\varepsilon > 0$. It is easily checked that this function is of norm 1 in $L^2(\mathbb{R})$, and that

$$\left\|(P-p)\,\omega_p^{(\varepsilon)}\right\|^2 = \mathrm{C}^{\underline{\mathrm{nt}}}\cdot\varepsilon^2,$$

so that taking $\varepsilon = 1/n$ with n going to infinity shows that p is indeed a generalized eigenvalue.

Later on (see Theorem 14.82) we will see that the same operator defined on $L^2[0,a]$ has no generalized eigenvalue.

To finish this section, we remark that, in addition to the continuous spectrum and discrete spectrum, there exists a third type.

DEFINITION 14.53 Let A be an operator with dense domain D_A. The **residual spectrum** of A, denoted $\boldsymbol{\sigma}_{\mathrm{r}}(\boldsymbol{A})$, is the set of complex numbers λ such that

$$\lambda \notin \sigma_{\mathrm{d}}(A) \qquad \text{and} \qquad \overline{\lambda} \in \sigma_{\mathrm{d}}(A^\dagger).$$

The **spectrum of A** is the union of the three types of spectra:

$$\sigma(A) = \sigma_{\mathrm{d}}(A) \cup \sigma_{\mathrm{c}}(A) \cup \sigma_{\mathrm{r}}(A).$$

The **resolvant set** of A is the set of complex numbers which are not in the spectrum of A: $\rho(A) = \mathbb{C}\setminus\sigma(A)$. It is possible to prove that for any $\lambda \in \rho(A)$, the operator $A - \lambda\,\mathrm{Id}$ has a continuous inverse.

Since the residual spectrum of a self-adjoint operator is empty, we will not mention it anymore. The interested reader can look at Exercise 14.3 on page 403 for another definition of the various spectra, and may consult a general reference in functional analysis, such as [71, 94].

14.4 Hermitian operators; self-adjoint operators

The goal of this short section is to prove the following result, which is a generalization of Theorem 14.7: *any self-adjoint operator admits an orthogonal basis of eigenvectors*, where the basis may be a generalized basis (i.e., its elements may be distributions). This fact is absolutely fundamental for quantum mechanics, where an **observable** is defined, according to various textbooks, as being a hermitian operator which has an orthogonal basis of eigenvectors or, better, as a self-adjoint operator. In contrast with the case of finite dimension, the notions of hermitian operator and of self-adjoint operator no longer coincide, and this causes some difficulties.

Finally, we assume in this section that $\mathcal{H} = L^2(\mathbb{R})$.

Example 14.54 (A classical problem of quantum mechanics) Consider a particle in an infinite potential well, confined to the interval $[0,a]$. The operator P satisfies $\langle\varphi|\,P\psi\rangle = \langle P\varphi|\,\psi\rangle$ for all functions $\varphi, \psi \in D_P$. However, P has no eigenvalue, and no generalized eigenvalue. Hence P is not an observable.

14.4.a Definitions

DEFINITION 14.55 Let A be a linear operator, with dense domain D_A in $\mathcal{H}$. Then A is **hermitian** (or **symmetric**) if and only if we have

$$\forall \varphi, \psi \in D_A \qquad \langle A\varphi | \psi \rangle = \langle \varphi | A\psi \rangle .$$

THEOREM 14.56 (Characterizations of a hermitian endomorphism) *Let A be an operator on $\mathcal{H}$. Then the following statements are equivalent:*

i) A is hermitian;

ii) $D_A \subset D_{A^\dagger}$ and $A^\dagger \varphi = A\varphi$ for any $\varphi \in D_A$;

iii) $A \subset A^\dagger$.

PROOF. First of all, notice that iii) is just a rephrasing of ii).
Assume that A is hermitian. Then, for any $\varphi \in D_A$, the element $\omega = A\varphi$ satisfies

$$\langle \omega | \psi \rangle = \langle A\varphi | \psi \rangle = \langle \varphi | A\psi \rangle ,$$

which proves that $\varphi \in D_{A^\dagger}$ and $A^\dagger \varphi = A\varphi$.
The converse is immediate.

DEFINITION 14.57 Let A be an operator, with D_A dense in $\mathcal{H}$. Then A is **self-adjoint** if and only if $A = A^\dagger$, that is, if

$$D_A^\dagger = D_A \qquad \text{and} \qquad \forall \varphi \in D_A \quad A^\dagger \varphi = A\varphi .$$

In a finite-dimensional space, it is equivalent for an operator to be hermitian or self-adjoint. Here, however, there is a crucial distinction: the domain of the adjoint of a hermitian operator $A^\dagger$ may well be strictly larger than that of A. In other words, $A^\dagger$ may be acting in the same way as A on the domain of A, but it may also act on a larger space. Disregarding this distinction may lead to unwelcome surprises (an apparent paradox of quantum mechanics is explained in Exercise 14.5). It is often easy to check that an operator is hermitian, but much more delicate to see whether it is, or is not, self-adjoint.

If A is defined on the whole of $\mathcal{H}$, then things simplify enormously; since $A \subset A^\dagger$, it follows that the domain of $A^\dagger$ is also equal to $\mathcal{H}$ and so if A is hermitian, it is in fact self-adjoint. The situation is even simpler than that:

THEOREM 14.58 (Hellinger-Toeplitz) *Let A be a hermitian operator defined on $\mathcal{H}$. Then A is self-adjoint and bounded.*

The adjoint of any operator is closed, and therefore we have

THEOREM 14.59 *Any self-adjoint operator is closed.*

This result can be made more precise:

PROPOSITION 14.60 *Let A be a hermitian operator. Then $A \subset A^\dagger$; hence $D_{A^\dagger}$ is dense and A is closable. Moreover, $A \subset \overline{A} \subset A^\dagger$, and A is self-adjoint if and only if $A = \overline{A} = A^\dagger$.*

If A is hermitian and closed, then A is self-adjoint if and only if $A^\dagger$ is hermitian.

A last result may be useful.

PROPOSITION 14.61 *Let A be a hermitian operator. The* ***defect indices*** *of A are the quantities, either integers or $+\infty$, defined by*

$$n_+ = \dim \operatorname{Ker}(A^\dagger - \mathrm{i}) \qquad \text{and} \qquad n_- = \dim \operatorname{Ker}(A^\dagger + \mathrm{i}).$$

The following statements are equivalent:

i) A is self-adjoint;

ii) A is closed and $n_+ = n_- = 0$;

iii) $\operatorname{Im}(A \pm \mathrm{i}) = \mathcal{H}$.

PROPOSITION 14.62 *The operator X on the space $L^2[0,a]$ is self-adjoint and bounded.*

The operator X on the space $L^2(\mathbb{R})$ is self-adjoint and unbounded.

Proof

• Consider the operator X on $\mathcal{H} = L^2[0,a]$. Its domain is the whole of $\mathcal{H}$. Moreover, for any $\varphi, \psi \in \mathcal{H}$, we have

$$\langle X\varphi|\,\psi\rangle = \int_{-\infty}^{+\infty} \overline{x\,\varphi(x)}\,\psi(x)\,\mathrm{d}x = \int_{-\infty}^{+\infty} \varphi(x)\,\overline{x\,\psi(x)}\,\mathrm{d}x = \langle \varphi|\,X\psi\rangle,$$

which proves that $X^\dagger = X$ and that X is self-adjoint. It is bounded because

$$\|X\varphi\|^2 = \int_0^a \left|x\,\varphi(x)\right|^2 \mathrm{d}x \leqslant a^2 \int_0^a \left|\varphi(x)\right|^2 \mathrm{d}x = a^2\,\|\varphi\|^2,$$

from which we derive in fact that $|||X||| \leqslant a$. (As an exercise, the reader is invited to prove that $|||X||| = a$.)

• Consider now the operator X on the space $L^2(\mathbb{R})$. The domain of X is

$$D_X = \left\{\varphi \in \mathcal{H}\,;\; x \mapsto x\,\varphi(x) \in \mathcal{H}\right\}.$$

By the same computation as before, it follows easily that X is hermitian. Let's prove that it is even self-adjoint.

Let $\psi \in D_{X^\dagger}$ and write $\psi^* = X^\dagger\psi$. Then, for any $\varphi \in D_X$, we have $\langle X\varphi|\,\psi\rangle = \langle\varphi|\,\psi^*\rangle$, by definition of $X^\dagger$. Expressing the two scalar products as integrals, this gives, for any $\varphi \in D_X$,

$$\int_{-\infty}^{+\infty} \overline{\varphi(x)}\left[x\,\psi(x) - \psi^*(x)\right]\mathrm{d}x = 0.$$

In particular, taking $\varphi(x) = x\,\psi(x) - \psi^*(x)$ for $x \in [-M, M]$ and $\varphi(x) = 0$ for $|x| > M$, we have $\varphi \in D_X$ and

$$\int_{-M}^{M} \left|x\,\psi(x) - \psi^*(x)\right|^2 \mathrm{d}x = 0,$$

hence $\psi^*(x) = x\,\psi(x)$ almost everywhere on $[-M, M]$, and hence almost everywhere on $\mathbb{R}$ since M was arbitrarily large. It follows that $x \longmapsto x\,\psi(x) \in L^2(\mathbb{R})$, which means $\psi \in D_X$, and in addition $\psi^* = X\psi$.

Thus, we have shown that $D_{X^\dagger} \subset D_X$ and that $X^\dagger$ is the restriction of X, that is, $X^\dagger \subset X$. Since we know that $X \subset X^\dagger$, we have $X = X^\dagger$.

Remark 14.63 The operator X on $L^2\,[0, +\infty[$ is also self-adjoint, the previous proof being valid without change.

PROPOSITION 14.64 *The operator P on $L^2\,[0, a]$ is hermitian (symmetric), but not self-adjoint.*

PROOF. A function $\varphi \in L^2\,[0, a]$ belongs to the domain of P if and only if it satisfies the following two conditions:

$$\varphi \text{ and } \varphi' \text{ are in } L^2\,[0, a], \tag{$*$}$$

$$\varphi(0) = \varphi(a) = 0. \tag{$**$}$$

(See Remark 14.33 on page 388 for the meaning of these boundary conditions.) The subspace D_P is dense. Moreover, for any $\varphi, \psi \in D_P$, we have

$$\begin{aligned}\langle P\varphi|\,\psi\rangle &= \int_0^a \overline{-\mathrm{i}\hbar\,\varphi'(x)}\,\psi(x)\,\mathrm{d}x \\ &= -\mathrm{i}\hbar \int_0^a \overline{\varphi(x)}\,\psi'(x)\,\mathrm{d}x + \mathrm{i}\hbar\left[\overline{\varphi(a)}\,\psi(a) - \overline{\psi(0)}\,\psi(0)\right] = \langle\varphi|\,P\psi\rangle,\end{aligned}$$

since the boundary terms cancel out from condition ($**$). This establishes that P is symmetric.

However the domain of $P^\dagger$ is *strictly larger* than that of P. Indeed, any function ψ which satisfies *only* the condition ($*$) is such that $\langle P\varphi|\,\psi\rangle = \langle\varphi|\,P\psi\rangle$ for any $\varphi \in D_P$. Using techniques similar to that in the proof of Theorem 14.62, it is possible to prove the converse inclusion, namely, that

$$D_{P^\dagger} = \left\{\varphi\ ;\ \varphi, \varphi' \in L^2\,[0, a]\right\}.$$

On $D_{P^\dagger}$, the operator $P^\dagger$ is defined by $P^\dagger\psi = -\mathrm{i}\hbar\psi$. So the adjoint of P acts in the same manner as P on the domain of P, but it has a larger domain.

14.4.b Eigenvectors

The first important result concerning hermitian operators is in perfect analogy with one in finite dimensions: eigenvectors corresponding to distinct eigenvalues are orthogonal.

THEOREM 14.65 *Let A be a hermitian operator. Then:*

i) For any $\varphi \in D_A$, $\langle\varphi|A\varphi\rangle$ is real.

ii) The eigenvalues of A are real.

iii) Eigenvectors of A associated to distinct eigenvalues are orthogonal.

iv) If $(\varphi_n)_{n\in\mathbb{N}}$ is an orthonormal Hilbert basis where each φ_n is an eigenvector of A, then the discrete spectrum of A is exactly $\{\lambda_n\ ;\ n \in \mathbb{N}\}$; in other words, there is no other eigenvalue of A.

Remark 14.66 Concerning point iv), there is nothing a priori that ensures that such a basis of eigenvectors exists.

PROPOSITION 14.67 *The residual spectrum of a self-adjoint operator A is empty.*

PROOF. Let $\lambda \in \sigma_{\mathrm{r}}(A)$. Then $\overline{\lambda} \in \sigma_{\mathrm{d}}(A^{\dagger}) = \sigma_{\mathrm{d}}(A)$ since $A = A^{\dagger}$; but the eigenvalues of A are real, hence $\lambda = \overline{\lambda}$ and therefore $\lambda \in \sigma_{\mathrm{d}}(A)$, which contradicts the definition of the residual spectrum.

14.4.c Generalized eigenvectors

If A is an operator and λ is a generalized eigenvalue, we are going to try to solve the equation $AT = \lambda T$ in the sense of (tempered) distributions. The first thing to do is to define AT. We have already seen the method: start by looking at what happens in the case of a regular distribution T_{ψ} associated with a function $\overline{\psi} \in \mathscr{S}$, that is, defined by

$$\left\langle T_{\psi}, \varphi \right\rangle = \langle \psi | \varphi \rangle = \int_{-\infty}^{+\infty} \overline{\psi(x)}\, \varphi(x)\, \mathrm{d}x.$$

If A is a *symmetric* operator, we have

$$\forall \varphi \in \mathscr{S} \qquad \langle A\psi | \varphi \rangle = \langle \psi | A\varphi \rangle,$$

hence

$$\forall \varphi \in \mathscr{S} \qquad \langle AT_{\psi} | \varphi \rangle = \left\langle T_{\psi} \middle| A\varphi \right\rangle.$$

This property can be extended, as long as $A\varphi$ is defined for all $\varphi \in \mathscr{S}$ and is also in $\mathscr{S}$:

DEFINITION 14.68 (Image of a distribution by a symmetric operator) Let A be a *symmetric* operator, such that

- $\mathscr{S} \subset D_A$;
- $\mathscr{S}$ is stable under A;
- A is continuous on $\mathscr{S}$ (for its topology).

Then **A is of class $\mathscr{S}$**. For any tempered distribution $T \in \mathscr{S}'$, the tempered distribution AT is defined by

$$\forall \varphi \in \mathscr{S} \qquad \langle AT | \varphi \rangle = \langle T | A\varphi \rangle .$$

♦ **Exercise 14.1** Prove that AT is indeed a tempered distribution.

DEFINITION 14.69 (Generalized eigenvector) Let A be a symmetric operator of class $\mathscr{S}$. A nonzero tempered distribution T is an **eigendistribution** or a **generalized eigenvector**, associated to the eigenvalue λ, if and only if

$$AT = \lambda T,$$

in other words if $\langle T | A\varphi \rangle = \lambda \langle T | \varphi \rangle$ for any $\varphi \in \mathscr{S}$.

The main result is the following:

THEOREM 14.70 (Spectral theorem) *For any self-adjoint operator of a Hilbert space $\mathcal{H}$ which is of class $\mathscr{S}$, there exists an orthonormal basis of eigendistributions which are associated to real eigenvalues or generalized eigenvalues.*

Remark 14.71 There exists a much more general version of this theorem, which applies to any self-adjoint operator. This generalized spectral theorem states that there always exists a **spectral family**, but this notion is beyond the scope of this book. It is discussed in many books, for instance [71, §III.5, 6], [95, p. 424], and [38, vol. 4, ch.1, §4].

PROPOSITION 14.72 *The momentum operator P on the space $L^2(\mathbb{R})$ is of class $\mathscr{S}$. The family $\{ \langle p| \; ; \; p \in \mathbb{R} \}$ is a generalized orthonormal basis of generalized eigenvectors for P.*

PROOF. We already know, by Theorem 14.25, that this family is a generalized basis. The orthonormality is the statement that

$$\forall p, p' \in \mathbb{R} \qquad \langle p | p' \rangle = \delta(p - p').$$

We now check that, denoting by Ω_p the regular distribution associated to the function $\omega_p(x) = (2\pi\hbar)^{1/2}\, e^{i\pi x/\hbar}$, we have $P\Omega_p = p\Omega_p$.

Indeed, let $\varphi \in \mathscr{S}$. Then we have

$$\begin{aligned} \langle P\Omega_p | \varphi \rangle &= \langle \Omega_p | P\varphi \rangle && \text{by Definition 14.68} \\ &= \langle \Omega_p | -i\hbar\, \varphi' \rangle = -i\hbar\, \widetilde{\varphi'}(p) \\ &= p\, \widetilde{\varphi}(p) && \text{since } \mathscr{F}[\varphi'](p) = (ip/\hbar)\widetilde{\varphi}(p) \\ &= p\, \langle \Omega_p | \varphi \rangle = \langle p\, \Omega_p | \varphi \rangle && \text{since } p \text{ is real.} \end{aligned}$$

Remark 14.73 The distribution Ω_p is, up to normalization, the limit of the $\omega_p^{(\varepsilon)}$ defined in Example 14.52.

PROPOSITION 14.74 *The operator X on $L^2(\mathbb{R})$ is of class $\mathscr{S}$. Moreover, the family $\{ \langle x| \; ; \; x \in \mathbb{R} \}$ is a generalized orthonormal basis of generalized eigenvectors for X.*

PROOF. It is only necessary to check that $X\,\delta(x - x_0) = x_0\,\delta(x - x_0)$ in the sense of Definition 14.68. For any $\varphi \in \mathscr{S}$, we have indeed

$$\langle X\delta(x - x_0) | \varphi \rangle = \langle \delta(x - x_0) | X\varphi \rangle = (X\varphi)(x_0) = x_0\,\varphi(x_0) = x_0\,\langle \delta(x - x_0) | \varphi \rangle.$$

14.4.d "Matrix" representation

Let A be an operator on $\mathcal{H}$, with *dense* domain as usual. Let $(\varphi_n)_{n\in\mathbb{N}}$ be an orthonormal basis of $\mathcal{H}$, all elements of which are also in D_A. The associated **matrix elements** are defined by

$$a_{i,k} \stackrel{\text{def}}{=} \langle \varphi_i | A\,\varphi_k \rangle .$$

The question is: is it true or not that the knowledge of the infinite "matrix" $M = (a_{ik})_{i,k}$ is sufficient to reconstruct entirely the operator A?

DEFINITION 14.75 The operator A **has a matrix representation associated with $(\varphi_n)_n$** if and only if

$$\forall \psi = \sum_{k=0}^{\infty} x_k \, \varphi_k \qquad A\psi = \sum_{i=0}^{\infty} \left(\sum_{k=0}^{\infty} a_{ik} \, x_k \right) \varphi_i . \tag{14.1}$$

Assume that A has a matrix representation. Using Dirac's notation, we have $x_k = \langle \varphi_k | \psi \rangle$; still following Dirac, define[9]

$$\langle \varphi_i | A | \varphi_k \rangle \stackrel{\text{def}}{=} \langle \varphi_i | A \varphi_k \rangle$$

and derive

$$A |\psi\rangle = \sum_{i=0}^{\infty} \sum_{k=0}^{\infty} |\varphi_i\rangle \langle \varphi_i | A | \varphi_k \rangle \langle \varphi_k | \psi \rangle ,$$

which is again the closure relation. Symbolically, this is written:

$$A = \sum_{i,k} a_{ik} \, |\varphi_i\rangle \langle \varphi_k| .$$

The difficulty is that these formulas are not always correct, since the right-hand side of (14.1) is not always correctly defined for all $\psi \in D_A$. It is easy to see that, if A is bounded, then by continuity everything works out. More generally:

THEOREM 14.76 (Matrix representation) *If A be a self-adjoint operator, then it has a matrix representation associated to any orthonormal basis of D_A.*

This property is also true if A is bounded, or if it is hermitian and closed.[10]

As an example, the operators of orthogonal projection on a closed subspace are self-adjoint operators which admit a matrix representation:

THEOREM 14.77 *Let $(\omega_k)_{k\in\mathbb{N}}$ be an orthonormal family, and let M be the closure of* $\text{Vect}\{\omega_k \ ; \ k \in \mathbb{N}\}$. *Then M is a closed subspace. Let $M^\perp$ be its orthogonal complement. Any vector $\psi \in \mathcal{H}$ can be decomposed in a unique way as*

$$\psi = \psi_M + \chi \qquad \textit{with } \psi_M \in M \textit{ and } \chi \in M^\perp.$$

---

[9] This is where the trap lies for those who manipulate the notation blindly. The definition $\langle \varphi | A \stackrel{\text{def}}{=} \langle A^\dagger \varphi |$, which is often found, permits two readings of the expression $\langle \varphi | A | \psi \rangle$:

- as $\langle A^\dagger \varphi | \psi \rangle$ – which imposes that $|\varphi\rangle \in D_{A^\dagger}$ and $|\psi\rangle \in \mathcal{H}$;
- as $\langle \varphi | A\psi \rangle$ – which imposes that $|\varphi\rangle \in \mathcal{H}$ and $|\psi\rangle \in D_A$.

Those are *not quite* equivalent – always because of the issue of the exact domain of an operator. Note that, for a self-adjoint operator, there is no ambiguity; for a hermitian operator, mistakes can arise because of this [41].

[10] For instance, if $\sum_{i=0}^{\infty} |a_{ik}|^2 < +\infty$ for all $k \in \mathbb{N}$, the operator $A : \varphi_k \mapsto \sum_{i=0}^{\infty} a_{ik} \, \varphi_i$ defines, by density of $\text{Vect}\{\varphi_i \ ; \ i \in \mathbb{N}\}$, a closed operator, which is hermitian if $a_{ki} = \overline{a_{ik}}$.

The operator Π *which associates* ψ_M *to* ψ *is the* ***orthogonal projector on*** $\boldsymbol{M}$. *It satisfies* $\Pi^2 = \Pi$ *and* $\Pi^\dagger = \Pi$. *In matrix representation it is given by*

$$\Pi = \sum_{k=0}^{\infty} |\omega_k\rangle \langle\omega_k| .$$

Let A be a self-adjoint operator, and assume that there is an orthonormal Hilbert basis $(\varphi_n)_{n\in\mathbb{N}}$ of eigenvectors of A, with corresponding eigenvalues $(\lambda_n)_{n\in\mathbb{N}}$ (recall that this is by no means always the case). Then

$$A = \sum_{n=0}^{\infty} \lambda_n \, |\varphi_n\rangle \langle\varphi_n| ,$$

that is,

$$\forall\psi \in D_A \qquad A\,|\varphi\rangle = \sum_{n=0}^{\infty} \lambda_n \, \langle\varphi_n|\,\psi\rangle \, |\varphi_n\rangle ,$$

and

$$\psi \in D_A \quad \text{if and only if} \quad \sum \lambda_n^2 |\, \langle\varphi_n|\,\psi\rangle \,|^2 \text{ converges.}$$

With this the *formal calculus* of the operator A can be defined:

DEFINITION 14.78 (Function of an operator) Let A be a self-adjoint operator, $(\varphi_n)_{n\in\mathbb{N}}$ an orthonormal Hilbert basis of eigenvectors of A, associated to the eigenvalues $(\lambda_n)_{n\in\mathbb{N}}$. Let $F : \mathbb{R} \to \mathbb{R}$ be a measurable function. The operator $F(A)$ is defined by

$$F(A) = \sum_{n=0}^{\infty} F(\lambda) \, |\varphi_n\rangle \langle\varphi_n| ,$$

with domain

$$D_{F(A)} = \Big\{ \psi \in \mathcal{H} \, ; \, \sum_{n=0}^{\infty} F(\lambda)^2 \, |\langle\varphi_n|\,\psi\rangle|^2 < +\infty \Big\} .$$

All the φ_n are in $D_{F(A)}$, hence $D_{F(A)}$ is also dense. In fact we have [95, §5.8]:

THEOREM 14.79 *The operator* $F(A)$ *is self-adjoint.*

Since the identity operator is obviously self-adjoint and defined on $\mathcal{H}$, for any orthonormal Hilbert basis $(\varphi_n)_{n\in\mathbb{N}}$ and any sequence $(\beta_n)_{n\in\mathbb{N}}$ of real numbers, the operator

$$B \stackrel{\text{def}}{=} \sum_{n=0}^{\infty} \beta_n \, |\varphi_n\rangle \langle\varphi_n|$$

is a self-adjoint operator.

This technique can provide the extension of an operator which is merely hermitian, but admits an orthonormal basis of eigenvectors, to a self-adjoint operator – defined therefore on a larger domain, and which is the only one for which many computations make sense.

For instance, consider the operator H_0 defined on $D_{H_0} = \mathscr{D}(\mathbb{R}^3)$ (space of $\mathscr{C}^\infty$ functions with bounded suppoert), defined by

$$\forall \psi \in \mathscr{D} \qquad H_0\psi(x) = \left(-\frac{\hbar^2}{2m}\,\triangle + V(x)\right)\,\psi(x),$$

where V is a given function. In many cases, H_0 is **essentially self-adjoint,** that is, the closure $\overline{H_0}$ is self-adjoint; this closure of H_0 is then given by $\overline{H_0} = H_0^\dagger$. The self-adjoint extension $H = \overline{H_0}$, is the **hamiltonian operator of the system**. It is this operator which is observable. In practice, it is obtained by finding an orthonormal basis $(\varphi_n)_{n\in\mathbb{N}}$ for H_0 and applying the method above.

Example 14.80 (Hamiltonian for the harmonic oscillator) Consider a harmonic oscillator with mass m and pulsation ω. We first define a *formal hamiltonian* H_0 on $D_{H_0} = \mathscr{S}$ by

$$H_0\,|\psi\rangle = \frac{P^2}{2m}\,|\psi\rangle + \frac{m\omega^2}{2}\,X^2\,|\psi\rangle\,.$$

This operator is hermitian, and has a Hilbert basis $(\varphi_n)_{n\in\mathbb{N}}$ such that

$$H_0\,|\varphi_n\rangle = E_n\,|\varphi_n\rangle \qquad \text{with } E_n = \left(n + \frac{1}{2}\right)\hbar\omega.$$

The functions φ_n are of the type $\exp(-m\omega^2x^2/2\hbar)\,h_n(x)$, where h_n is a Hermite polynomial of degree n [20, Ch. V §C.2]. The operator H_0 is not self-adjoint.

Let us now define the operator H by

$$H = \sum_{n=0}^{\infty} E_n\,|\varphi_n\rangle\,\langle\varphi_n|\,, \qquad \text{i.e.,} \qquad H\,|\psi\rangle = \sum_{n=0}^{\infty} E_n\,\langle\varphi_n|\,\psi\rangle\,|\varphi_n\rangle$$

for any function $|\psi\rangle$ such as the series $\sum E_n^2\,\big|\,\langle\varphi_n|\,\psi\rangle\,\big|^2$ converges.

The operator H is self-adjoint; it is an extension of H_0, which is called the **hamiltonian** of the harmonic oscillator.

An example of this is given in Exercise 14.5.

14.4.e Summary of properties of the operators P and X

PROPOSITION 14.81 (Operator P on $L^2(\mathbb{R})$) *The operator P on $L^2(\mathbb{R})$ is self-adjoint (and hence closed). It is defined on*

$$D_P = \{\varphi \in \mathcal{H}\,;\ \varphi' \in \mathcal{H}\}.$$

The discrete spectrum of P is empty; its continuous spectrum is equal to $\mathbb{R}$. The family $\big\{\langle p|\ ;\ p \in \mathbb{R}\big\}$ is a generalized basis of eigenvectors such that

$$\langle p\,|\,p'\rangle = \delta(p - p'), \qquad \textit{(orthonormal basis)}$$

$$\mathrm{Id} = \int_{-\infty}^{+\infty} |p\rangle\,\langle p|\;\mathrm{d}p. \qquad \textit{(closure relation)}$$

Proof. See Example 14.52 for the continuous spectrum and Theorem 14.72 for the basis of eigenvectors.

PROPOSITION 14.82 (Operator P on $L^2[0,a]$) *The operator P on $L^2[0,a]$ is hermitian (symmetric), but not self-adjoint. Its domain is*

$$D_P = \{\varphi \in \mathcal{H}\,;\ \varphi' \in \mathcal{H} \text{ and } \varphi(0) = \varphi(a) = 0\}.$$

P is closed (and therefore admits matrix representations).

The discrete spectrum of P is empty; its continuous spectrum is also empty and its residual spectrum is equal to $\mathbb{C}$.

PROOF. The first property has already been proved in Proposition 14.64.

Before trying to prove the absence of any eigenvalue, even generalized, we can try to reason "physically." A vector such that $P\varphi$ is close to $\lambda\varphi$ is characterized by a very small "uncertainty" Δp with respect to the momentum, which, according to Heisenberg's uncertainty principle is only possible if Δx is very large. But, here, Δx is bounded by a.

Such an argument is not simply heuristic: it can be justified by noting that P and X (on $L^2(\mathbb{R})$) are unitarily equivalent via the Fourier transform, namely, we have $P = \mathscr{F} X \mathscr{F}^{-1}$. Let now p_0 be any real number; we will show that p_0 *is not a generalized eigenvalue*. Let φ be a function with norm 1. The Fourier transform $\widetilde{\psi}$ of $(P - p_0)\varphi = -\mathrm{i}\hbar\varphi' - p_0\varphi$ is $p' \mapsto (p - p_0)\,\widetilde{\varphi}(p)$. The Parseval-Plancherel identify shows that

$$\left\|(P - p_0)\varphi\right\|^2 = \left\|(p - p_0)\,\widetilde{\varphi}(p)\right\|^2.$$

The norm squared $\left\|\widetilde{\psi}\right\|^2$ is simply the average value, in the state corresponding to φ, of the quantity $(p - p_0)^2$, which was denoted $\left\langle (p - p_0)^2 \right\rangle$ in the section on uncertainty relations. It can be checked that the quantity $\left\langle (p - p_0)^2 \right\rangle$ is smallest for $p_0 = \langle p \rangle$, hence

$$\left\|\widetilde{\psi}(p)\right\|^2 = \left\langle (p - p_0)^2 \right\rangle \geqslant (\Delta p)^2.$$

By the uncertainty relations (Theorem 13.17), we know that $\Delta p \geqslant \hbar/2\Delta x \geqslant \hbar/2a$, which finally implies

$$\left\|\widetilde{\psi}\right\|^2 \geqslant \frac{\hbar^2}{4a}.$$

Consequently, it is not possible to make $\left\|p \mapsto (p - p_0)\,\widetilde{\varphi}(p)\right\|^2 = \left\|(P - p_0)\varphi\right\|^2$ as small as possible, so that p_0 is not a generalized eigenvalue.

One easily checks that the domain of $P^\dagger$ is $D_{P^\dagger} = \{\varphi \in \mathcal{H}\,;\ \varphi' \in \mathcal{H}\}$, and that any complex λ is an eigenvalue of $P^\dagger$, with associated eigenvector $\varphi_\lambda : x \mapsto \mathrm{e}^{\mathrm{i}\lambda x/\hbar}$. This proves that $\sigma_{\mathrm{r}}(P) = \mathbb{C}$.

PROPOSITION 14.83 (Operator X on $L^2(\mathbb{R})$) *The operator X on $L^2(\mathbb{R})$ is self-adjoint; its domain is*

$$D_X = \{\varphi \in L^2(\mathbb{R})\,;\ x \mapsto x\,\varphi(x) \in L^2(\mathbb{R})\}.$$

The discrete spectrum of X is empty; its continuous spectrum is $\mathbb{R}$. The family

$$\{\,\langle x|\ ;\ x \in \mathbb{R}\}$$

is a generalized basis of generalized eigenvectors such that

$$\langle x | x' \rangle = \delta(x - x'), \qquad \textit{(orthonormal basis)}$$

$$\mathrm{Id} = \int_{-\infty}^{+\infty} |x\rangle\,\langle x|\,\mathrm{d}x. \qquad \textit{(closure relation)}$$

Proof. See Theorem 14.62 to show that X is self-adjoint. See Remark 14.50 for the spectrum.

PROPOSITION 14.84 (Operator X on $L^2[0,a]$) *The operator X on $L^2[0,a]$ is self-adjoint and bounded. Its domain is $D_X = L^2[0,a]$. The discrete spectrum of X is empty. Its continuous spectrum is $[0,a]$. The family*

$$\{ \langle x| \; ; \; x \in [0,a] \}$$

is a generalized basis of eigenvectors of X on $L^2[0,a]$.

Proof. See Theorem 14.62 to show that X is self-adjoint. The computation of the spectrum is left as an exercise; the proof in Remark 14.50 can be used as a clue.

EXERCISES

♦ **Exercise 14.2** Check that any eigenvector is also an eigendistribution.

♦ **Exercise 14.3** Let A be a linerar operator defined on the subspace $D_A = W$. For any complex number λ, the operator $A_\lambda = A - \lambda \, \mathrm{Id}$ is also defined on W. One and only one of the following four cases applies for λ (see, e.g., [94]):

- A_λ is not injective; then λ is an **eigenvalue**, or equivalently **λ belongs to the discrete spectrum**.
- A_λ is injective, $\mathrm{Im}(A_\lambda)$ is dense but the inverse[11] of A_λ is not continuous; then λ is a **generalized eigenvalue** and **λ belongs to the continous spectrum**.
- A_λ is injective but $\mathrm{Im}(A_\lambda)$ is not dense; then λ belongs to the **residual spectrum**;
- A_λ is injective, $\mathrm{Im}(A_\lambda)$ is dense and A_λ^{-1} is continuous (and hence may be extended by continuity to an operator defined on the whole of $\mathcal{H}$); then λ belongs to the **resolvant set** of A.

Show that those definitions agree with those in Definitions 14.46 and 14.49.

Note that in some books (for instance [4]), the definitions somewhat different, and the discrete and continuous spectrum, for instance, are not necessarily disjoint. (This may actually be desirable to account for natural situations, such as some hyperbolic billiards arising in "arithmetic quantum chaos," where there is a seemingly chaotic discrete spectrum embedded in a very regular continuous spectrum; see, e.g., [78].)

♦ **Exercise 14.4 (Continuous spectrum of X)** Consider the Hilbert space $L^2[a,b]$. Show that X has no eigenvalue, but that any $\lambda \in [a,b]$ is a generalized eigenvalue, that is, $\sigma_{\mathrm{d}}(X) = \varnothing$ and $\sigma_{\mathrm{c}}(X) = [a,b]$.

♦ **Exercise 14.5 (A "paradox" of quantum mechanics, following F. Gieres [41])** Consider a particle, in an infinite potential well, constrained to remain in the interval $[-a,a]$. Physical considerations impose that the wave functions satisfy $\varphi(-a) = \varphi(a) = 0$. The **hamiltonian** is the operator on $L^2[-a,a]$ defined by $H\psi = -\hbar^2\varphi''/2m$. Assume H is a self-adjoint operator.

---

[11] The target space of A_λ can be restricted to $\mathrm{Im}(A_\lambda)$; then A_λ is a bijection from W to $\mathrm{Im}(A_\lambda)$.

i) Solve the eigenvalue equation $H\psi = E\psi$, associated to the boundary conditions $\psi(\pm a) = 0$; deduce from this the spectrum $(E_n)_{n\in\mathbb{N}^*}$ and an orthonormal basis of eigenfunctions $(\varphi_n)_{n\in\mathbb{N}^*}$.

ii) Consider the wave function given by $\psi(x) = \mu(a^2 - x^2)$, where μ is a normalizing factor so that $\langle\psi|\psi\rangle = 1$. Show that $\langle H\psi|H\psi\rangle > 0$. Isn't it the case however that $H^2\psi = 0$ and hence $\langle\psi|H^2\psi\rangle = 0$? How should the average value of the energy square in the state ψ be computed? Comment.

♦ **Exercise 14.6 (Uncertainty principle)** Let A and B be two self-adjoint operators on $\mathcal{H}$ (i.e., "observables"). Let $\psi \in \mathcal{H}$ be a normed vector (i.e., a "physical state") such that

$$\psi \in D_A \cap D_B, \qquad A\psi \in D_B, \qquad \text{and} \qquad B\psi \in D_A.$$

The **average value of A** and the **uncertainty on A in the state ψ** are defined by

$$a \stackrel{\text{def}}{=} \langle\psi|A\psi\rangle \qquad \text{and} \qquad \Delta a = \|A\psi - a\psi\|.$$

Show that

$$\Delta a \cdot \Delta b \geqslant \tfrac{1}{2}\left|\langle\psi|[A,B]\psi\rangle\right|, \qquad \text{where } [A,B] \stackrel{\text{def}}{=} AB - BA.$$

Check that $[X,P] = \mathrm{i}\hbar\,\mathrm{Id}$ and deduce from this that $\Delta x \cdot \Delta p \geqslant \hbar/2$.

SOLUTIONS

♦ **Solution of exercise 14.5.**

i) Solving the eigenvalue equation leads to the discrete spectrum given by:

$$E_n = \frac{\hbar^2 \pi^2 n^2}{8ma^2} \qquad (n \geqslant 1)$$

associated with the eigenfunctions

$$\varphi_n(x) = \frac{1}{\sqrt{a}}\cos\left(\frac{\sqrt{2mE_n}\,x}{\hbar}\right) = \frac{1}{\sqrt{a}}\cos\left(\frac{n\pi x}{2a}\right) \qquad \text{if } n \text{ is odd,}$$

$$\varphi_n(x) = \frac{1}{\sqrt{a}}\sin\left(\frac{\sqrt{2mE_n}\,x}{\hbar}\right) = \frac{1}{\sqrt{a}}\sin\left(\frac{n\pi x}{2a}\right) \qquad \text{if } n \text{ is even.}$$

It is easy to check that the sequence $(\varphi_n)_{n\in\mathbb{N}^*}$ is a complete orthonormal system of eigenfunctions of H.

ii) First compute that $\mu = \sqrt{15}/4a^{5/2}$. Then another computation shows that

$$\|H\psi\|^2 = \langle H\psi|H\psi\rangle = \frac{15\,\hbar^4}{8m^2a^4}.$$

Moreover, since the fourth derivative of ψ is zero, it is *tempting* to write $H^2\psi = 0$ and, consequently, $\langle\psi|H^2\psi\rangle = 0$; which is quite annoying since H is self-adjoint, and one would expect to have identical results.

The point is that one must be more careful with the definition of the operator H^2. Indeed, the domain of H is

$$D_H = \{\varphi \in \mathcal{H}\,;\ \varphi'' \in \mathcal{H} \text{ and } \varphi(\pm a) = 0\};$$

hence $H\psi \notin D_H$ and therefore it is not permitted to compute $H^2\psi$ as the composition $H(H\psi)$. (It may be checked that the elements of D_{H^2} must satisfy $\varphi''(\pm a) = 0$.) On the other hand, using the orthonormal Hilbert basis $(\varphi_n)_{n\in\mathbb{N}^*}$, we can write

$$H = \sum_{n=1}^{\infty} E_n \, |\varphi_n\rangle \langle\varphi_n|$$

and, using the techniques of Section 14.4.d, Definition 14.78, it is possible to define a new operator, denoted $\overline{H^2}$, by

$$\overline{H^2} \stackrel{\text{def}}{=} \sum_{n=1}^{\infty} E_n^2 \, |\varphi_n\rangle \langle\varphi_n| \,.$$

Notice that since the eigenfunctions φ_n are in the domain of H^2, we have

$$H^2\varphi_n = \overline{H^2}\varphi_n.$$

This new operator is thus an *extension* of the previous one. Its domain is the subspace of functions ψ such that $\sum E_n^4 \big| \langle\varphi_n|\,\psi\rangle \big|^2 < +\infty$. Then one checks that the function ψ, which did not belong to the domain[12] of H^2, is indeed in the domain of $\overline{H^2}$.

This extension is self-adjoint. A (boring) computation with Fourier series shows that

$$\langle\psi|\overline{H}^2\psi\rangle \stackrel{\text{def}}{=} \sum_{n=1}^{\infty} E_n^2 \, \langle\psi|\,\varphi_n\rangle \langle\varphi_n|\,\psi\rangle = \frac{15\,\hbar^4}{8m^2a^4},$$

i.e., that $\langle\psi|\overline{H}^2\psi\rangle = \langle H\psi|\,H\psi\rangle$ as it should be.

♦ **Solution of exercise 14.6.** First of all, notice that a and b are real numbers since A and B are hermitian operators (Theorem 14.65). Also, notice that

$$\langle\psi|(AB - BA)\psi\rangle = \langle\psi|AB\psi\rangle - \langle\psi|BA\psi\rangle = \langle A\psi|\,B\psi\rangle - \langle B\psi|\,A\psi\rangle = 2\mathrm{Im}\,\langle A\psi|\,B\psi\rangle\,.$$

Since a and b are real, we can expand

$$\langle(A-a)\psi|(B-b)\psi\rangle = \langle A\psi|\,B\psi\rangle - ab - b\overline{a} + ab\,\|\psi\|^2 = \langle A\psi|\,B\psi\rangle - ab,$$

and obtain
$$\langle\psi|(AB - BA)\psi\rangle = 2\,\mathrm{Im}\langle(A-a)\psi|(B-b)\psi\rangle.$$

The conclusion now follows using the Cauchy-Schwarz inequality:

$$\begin{aligned}\Delta a \cdot \Delta b \geqslant \Big|\langle(A-a)\psi|(B-b)\psi\rangle\Big| &= \Big|\mathrm{Im}\langle(A-a)\psi|(B-b)\psi\rangle\Big| \\ &\geqslant \big|\mathrm{Im}\,\langle A\psi|\,B\psi\rangle\,\big| = \frac{1}{2}\big|\langle\psi|[A,B]\,\psi\rangle\big|.\end{aligned}$$

[12] This is of course linked to the Fourier series of ψ, which converges uniformly but cannot be differentiated twice termwise.

Chapter 15

Green functions

This chapter, properly speaking, is not a course on Green functions, and it does not really introduce any new object or concept. Rather, through some simple physical examples, it explains how the various techniques discussed previously (Fourier and Laplace transforms, conformal maps, convolution, differentiation in the sense of distributions) can be used to solve easily certain physical problems related to linear differential equations.

The first problem concerns the propagation of electromagnetic waves in the vacuum. There the Green function of the d'Alembertian is recovered from scratch, as well as the *retarded potentials formula* for an arbitrary source distribution.

The second problem is the resolution of the heat equation, using either the Fourier transform, or the Laplace transform.

Finally, it will be seen how Green functions occur naturally in quantum mechanics.

15.1 Generalities about Green functions

Consider a *linear* system with input signal I and output (or response) signal R. This is described by an equation of the type

$$\Phi(R) = I.$$

What we called "input signal" and "output signal" could be many things, such as:

- electrical signals in a circuit (for instance, power as input and response of a component as output);

- charges and currents as input and electromagnetic fields as output;
- heat sources as input and temperature as output;
- forces as input and position or velocity as output.

The operator Φ is *linear* and *continuous*. It may depend on variables such as time or position. In this chapter, we are interested in the case where it is *translation (spacial or temporal) invariant.*[1] Most of the time, Φ is a differential operator (such as a Laplacian, d'Alembertian, etc.).

Since Φ is at the same time continuous, linear, and translation invariant, it is known that it may be expressed as a convolution operator; hence there exists[2] a distribution D such that $I = \Phi(R) = D * R$. The **Green function** of the system is defined to be any distribution G satisfying the equation

$$\Phi(G) = D * G = G * D = \delta,$$

where the Dirac distribution δ "applies" to the spacial or temporal variable or variables of the system. Computing the convolution product of the input signal with the Green function, we get

$$G * I = G * (D * R) = (G * D) * R = \delta * R = R,$$

which shows that knowing the Green function is sufficient in principle to compute R from the knowledge of the input I. However, complications arise when different Green functions exist in different algebras (for instance, in electromagnetism, two distinct Green functions lead, respectively, to retarded and advanced potentials).

Thus, this chapter's goal is to provide some examples of explicit computation techniques available to compute the Green functions, using mainly the Fourier and Laplace transforms. The strategy used is always the same:

i) take the Fourier or Laplace transform of the equation;

ii) solve the resulting equation, which is now algebraic;

iii) take the inverse transform.

This may also be symbolized as follows:

$$D * G = \delta \Longrightarrow \begin{cases} \widetilde{D} \cdot \widetilde{G} = 1 \\ \textit{or} \\ \widehat{D} \cdot \widehat{G} = 1 \end{cases} \Longrightarrow \begin{cases} \widetilde{G} = 1/\widetilde{D} \\ \textit{or} \\ \widehat{G} = 1/\widehat{D} \end{cases} \Longrightarrow \begin{cases} G = \overline{\mathscr{F}}\,[1/\widetilde{D}] \\ \textit{or} \\ G \sqsupset 1/\widehat{D}. \end{cases}$$

Despite the simplicity of this outline, difficulties sometimes obstruct the way; we will see how to solve them.

---

[1] Still, an example which is not translation invariant is treated in Section 6.2.a on page 165.

[2] It suffices to put $D = \Phi(\delta)$.

15.2

A pedagogical example: the harmonic oscillator

We treat here in detail a simple example, which contains the essential difficulties concerning the computation of Green functions, without extraneous complications (temporal and spacial variables together).

A **Green function of the harmonic oscillator** is a function $t \mapsto G(t)$ that is a solution, in the sense of distributions, of the equation

$$G''(t) + \omega_0^2 G(t) = \delta(t), \tag{15.1}$$

where $\omega_0 > 0$ is a fixed real number. This amounts to saying that G is a convolution inverse of the operator $(\delta'' + \omega_0^2 \delta)$, and we have already seen expressions for this in Theorem 8.33 on page 239.

We discuss here two methods that lead to the discovery of such a Green function of the harmonic oscillator (in addition to the "manual" method of variation of the constant,[3] which does not generalize very well).

---

[3] We quickly recall how the differential equation

$$y'' + y = f(t) \tag{E}$$

can be solved, where f (the *exterior excitation*) is a continuous function, and where we put $\omega_0 = 1$. This is done in two steps:

- **Resolution of the homogeneous equation.** We know that the space of solutions of the associated homogeneous equation (namely $y'' + y = 0$) is a two-dimensional vector space, a basis of which is, for instance, $(u, v) = (\cos, \sin)$.
- **Search for a particular solution.** We use the method of variation of the constants, generalized for order 2 equations, as follows: look for a solution of the type $y_0 = \lambda u + \mu v$, where λ and μ are functions to be determined. Denoting by $\mathscr{W}_{u,v}$ the **Wronskian matrix** associated to the basis (u, v), which is defined by

$$\mathscr{W}_{u,v} = \begin{pmatrix} u & v \\ u' & v' \end{pmatrix},$$

those functions are given by

$$\begin{pmatrix} \lambda' \\ \mu' \end{pmatrix} = \mathscr{W}_{u,v}^{-1} \cdot \begin{pmatrix} 0 \\ f \end{pmatrix} = \begin{pmatrix} \cos t & \sin t \\ -\sin t & \cos t \end{pmatrix}^{-1} \cdot \begin{pmatrix} 0 \\ f(t) \end{pmatrix},$$

that is,

$$\lambda' = -\sin(t) f(t) \qquad \text{and} \qquad \mu' = \cos(t) f(t),$$

so that (up to a constant, which can be absorbed in the solution of the homogeneous equation) we have

$$\lambda(t) = -\int_0^t \sin(s) f(s)\, ds \qquad \text{and} \qquad \mu(t) = \int_0^t \cos(s) f(s)\, ds.$$

Using the trigonometric formula $\sin(t - s) = \sin t \cos s - \cos t \sin s$, we deduce

$$y_0(t) = \int_0^t f(s) \sin(t - s)\, ds.$$

15.2.a Using the Laplace transform

Denote by $\mathscr{G}(p)$ the Laplace transform of $G(t)$. In the Laplace world, equation (15.1) becomes

$$(p^2+\omega_0^2)\cdot\mathscr{G}(p)=1.$$

Since $\mathscr{G}$ is a function, and not a distribution, we therefore have

$$\mathscr{G}(p)=\frac{1}{p^2+\omega_0^2}$$

and hence, using for instance a table of Laplace transforms (see page 613), we have

$$\boxed{G(t)=\frac{1}{\omega_0}H(t)\sin(\omega_0 t).}$$

Remark 15.1 On the one hand, this is quite nice, since the computation was very quick. On the other hand, it may be a little disappointing: we found a *single* expression for a Green function. More precisely, we found the unique inverse image of $\delta''+\delta$ in $\mathscr{D}'_+$, and not the one in $\mathscr{D}'_-$ (see Theorem 8.33 on page 239). This is quite normal: by definition of the unilateral Laplace transform, only *causal* input functions are considered (hence they are in $\mathscr{D}^+$).

15.2.b Using the Fourier transform

Denote by $\widetilde{G}(\nu)$ the Fourier transform of $G(t)$. Equation (15.1) becomes

$$(-4\pi\nu^2+\omega_0^2)\,\widetilde{G}(\nu)=1 \tag{15.2}$$

in the Fourier world. A first trial for a solution, rather naive, leads to

$$\widetilde{G}(\nu)=\frac{1}{\omega_0^2-4\pi\nu^2}. \qquad \text{(naive solution)}$$

Then we should take the inverse Fourier transform of $\widetilde{G}(\nu)$.

But here a serious difficulty arises: the function $\widetilde{G}$ as defined above has two real poles at $\nu=\pm\omega_0/2\pi$, and hence is not integrable (nor is it square integrable). So the inverse Fourier transform is not defined.

In fact, the reasoning was much too fast. The solutions of equation (15.2) in the sense of distributions are those distributions of the form

$$\widetilde{G}(\nu)=\operatorname{pv}\frac{1}{\omega_0^2-4\pi^2\nu^2}+\alpha\,\delta\left(\nu-\frac{\omega_0}{2\pi}\right)+\beta\,\delta\left(\nu+\frac{\omega_0}{2\pi}\right), \tag{15.3}$$

Thus, the general solution of the equation are given by

$$y(t)=\int_0^t f(s)\sin(t-s)\,\mathrm{d}s+\underbrace{\alpha\cos t+\beta\sin t}_{\text{free part}}.$$

We recognize, for $t>0$, the convolution of the function $G: t\mapsto H(t)\sin t$ and the truncated excitation function $H(t)f(t)$. The function G is the causal Green function associated to the equation of the harmonic oscillator (see Theorem 8.33 on page 239).

according to Theorem 8.11 on page 230, where we denote

$$\mathrm{pv}\,\frac{1}{\omega_0^2 - 4\pi^2\nu^2} = \frac{1}{2\omega}\left[\mathrm{pv}\,\frac{1}{2\pi\nu + \omega_0} - \mathrm{pv}\,\frac{1}{2\pi\nu - \omega_0}\right].$$

The last two terms of equation (15.3) are the Fourier transforms of the functions $t \mapsto \alpha\,\mathrm{e}^{\mathrm{i}\omega_0 t}$ and $t \mapsto \beta\,\mathrm{e}^{-\mathrm{i}\omega_0 t}$, in which we recognize free solutions (those that satisfy $y'' + y = 0$). We could leave them aside, but it is more useful to find interesting valus of α and β.

Indeed, if we take $\alpha = \beta = 0$, this amounts to use as integration contour in the inverse Fourier transform the following truncated real line (where $\nu_0 = \omega_0/2\pi$):

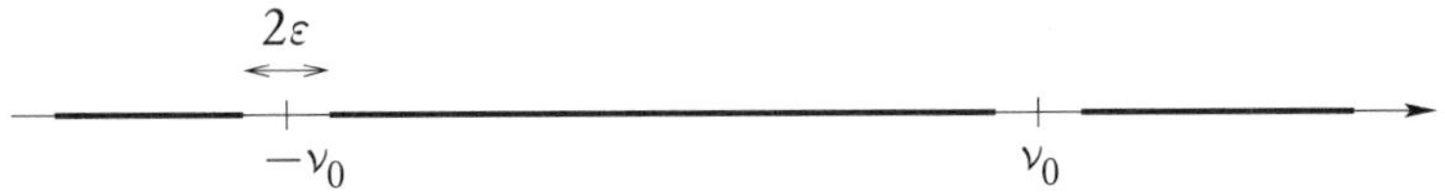

2ε

$-\nu_0$ ν_0

in the limit $\varepsilon \to 0$.

As seen in Section 8.1.c on page 225, we may add to ν_0 a positive or negative imaginary part $\pm\mathrm{i}\varepsilon$, which amounts to closing the contour by small half-circles either above or under the poles. Replacing each occurence of ν_0 by $\nu_0 + \mathrm{i}\varepsilon$, or in other words, putting

$$\widetilde{G}(\nu) = \frac{1}{4\pi\,\omega_0}\left[\frac{1}{\nu + \nu_0 + \mathrm{i}\varepsilon} - \frac{1}{\nu - \nu_0 - \mathrm{i}\varepsilon}\right],$$

leads to a complex-valued function G, which is a priori uninteresting. We will rather define

$$\widetilde{G}^{(\mathrm{ret})}(\nu) \stackrel{\mathrm{def}}{=} \lim_{\varepsilon \to 0^+} \frac{1}{4\pi\,\omega_0}\left[\frac{1}{\nu + \nu_0 - \mathrm{i}\varepsilon} - \frac{1}{\nu - \nu_0 - \mathrm{i}\varepsilon}\right],$$

or, equivalently, put $\beta = -\alpha = 1/4\omega_0$:

$$\widetilde{G}^{(\mathrm{ret})}(\nu) = \frac{1}{4\pi\,\omega_0}\left[\mathrm{pv}\,\frac{1}{\nu + \nu_0} + \mathrm{i}\pi\,\delta(\nu + \nu_0) - \mathrm{pv}\,\frac{1}{\nu - \nu_0} - \mathrm{i}\pi\,\delta(\nu - \nu_0)\right].$$

In the inverse Fourier transform formula

$$G^{(\mathrm{ret})}(t) = \int_\gamma \widetilde{G}^{(\mathrm{ret})}(\nu)\,\mathrm{e}^{2\mathrm{i}\pi\nu t}\,\mathrm{d}\nu,$$

this corresponds to the integration contour γ sketched below:

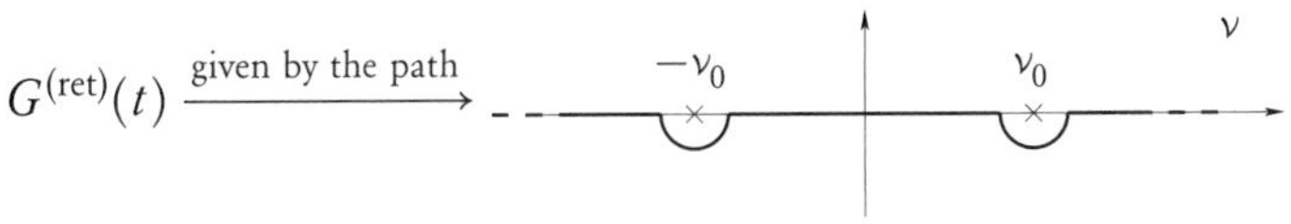

$G^{(\mathrm{ret})}(t)$ given by the path $-\nu_0$ ν_0 ν

Since the inverse Fourier transform of $\frac{1}{2i\pi}\,\text{pv}\left(\frac{1}{\nu}\right)+\frac{\delta}{2}$ is $H(t)$, it follows that

$$\frac{1}{2i\pi}\,\text{pv}\left(\frac{1}{\nu+\nu_0}\right)+\frac{1}{2}\,\delta(\nu+\nu_0)\xrightarrow{\text{T.F.}^{-1}} H(t)\,e^{-i\omega_0 t}$$

and

$$\frac{1}{2i\pi}\,\text{pv}\left(\frac{1}{\nu-\nu_0}\right)+\frac{1}{2}\,\delta(\nu-\nu_0)\xrightarrow{\text{T.F.}^{-1}} H(t)\,e^{i\omega_0 t}.$$

Hence the inverse Fourier transform of $\widetilde{G}^{(\text{ret})}$ is

$$G^{(\text{ret})}(t)=\frac{1}{2\omega_0}\,H(t)\left[i\,e^{-i\omega_0 t}-ie^{i\omega_0 t}\right]=\frac{1}{\omega_0}\,H(t)\,\sin(\omega_0 t).$$

Remark 15.2 It is no more difficult, if the formula is forgotten and no table of Fourier transforms is handy, to perform the integration along the contour indicated using the residue theorem. The sign of t must then be watched carefully, since it dictates whether one should close the integration contour *from above* or *from below* in the complex plane.

In the case discussed here, for $t<0$, the imaginary part of ν must be negative in order that the exponential function $e^{2i\pi\nu t}$ decay fast enough to apply Jordan's second lemma. In this case, the integrand has no pole inside the contour, and the residue theorem leads to the conclusion that $G^{(\text{ret})}(t)=0$. If, on the other hand, we have $t>0$, then we integrate from above, and the residue at each pole must be computed; this is left as an exercise for the reader, who will be able to check that this leads to the same result as previously stated.

In the following, the integration will be executed systematically using the method of residues.

Similarly, we will define

$$\widetilde{G}^{(\text{ad})}(\nu)\stackrel{\text{def}}{=}\lim_{\varepsilon\to 0^+}\frac{1}{4\pi\,\omega_0}\left[\frac{1}{\nu+\nu_0+i\varepsilon}-\frac{1}{\nu-\nu_0+i\varepsilon}\right],$$

or, equivalently

$$\widetilde{G}^{(\text{ad})}(\nu)=\frac{1}{4\pi\,\omega_0}\left[\text{pv}\,\frac{1}{\nu+\nu_0}-i\pi\,\delta(\nu+\nu_0)-\text{pv}\,\frac{1}{\nu-\nu_0}+i\pi\,\delta(\nu-\nu_0)\right],$$

which now corresponds to the integration contour

$G^{(\text{ad})}(t)$ given by the path → (contour along the real axis passing above the poles at $-\nu_0$ and ν_0, in the ν plane)

As before we compute the inverse Fourier transform of $\widetilde{G}^{(\text{ad})}$, and get

$$\boxed{G^{(\text{ret})}(t)=\frac{1}{\omega_0}\,H(t)\sin(\omega_0 t)}$$

and

$$\boxed{G^{(\text{ad})}(t)=\frac{-1}{\omega_0}\,H(-t)\sin(\omega_0 t).}$$

If we chose to take $\alpha = \beta = 0$, we obtain the average of the two preceding Green functions, namely,

$$\widetilde{G}^{(\mathrm{s})} = \frac{1}{2}\left[\widetilde{G}^{(\mathrm{ret})} + \widetilde{G}^{(\mathrm{ad})}\right]$$

(where "s" stands for "symmetric"). Looking separately at $t > 0$ and $t < 0$, this has a more compact expression:

$$\boxed{G^{(\mathrm{s})}(t) = \frac{1}{2}\sin\left(\omega_0\,|t|\right).}$$

So we have three distinct possibilities for a Green function of the harmonic oscillator – the reader is invited to check by hand that all three do satisfy the equation $G'' + \omega_0^2 G = \delta$. The function $G^{(\mathrm{ret})}$ has support bounded on the left – it is the *causal*, or *retarded*, Green function. $G^{(\mathrm{ad})}$ has support bounded on the right, and is the *advanced* Green function. As for $G^{(\mathrm{s})}$, its support is unbounded and belongs to no convolution algebra (in particular, its convolution product with itself is not defined), which explains why it did not come out in earlier chapters, although it is no less interesting than the others.

The retarded Green function is the most useful. Indeed, if a harmonic oscillator which is initally at rest (for $t = t_0$), with pulsation ω_0, is submitted to an excitation $f(t)$ which is zero for $t < t_0$ (the function f is then an element of the convolution algebra $\mathscr{D}'_+$), the length y of the oscillator satisfies

$$\forall t \in \mathbb{R} \qquad y''(t) + y(t) = f(t),$$

the solution of which, taking initial conditions into account, is

$$y = G^{(\mathrm{ret})} * f,$$

that is,

$$y(t) = \int_{t_0}^{t} f(s)\, G^{(\mathrm{ret})}(t-s)\,\mathrm{d}s.$$

Remark 15.3 Other complex-valued Green functions can also be defined. Those are of limited interest as far as classical mechanics is concerned, by are very useful in quantum field theory (see the Feynman propagator on page 431).

15.3
Electromagnetism and the d'Alembertian operator

DEFINITION 15.4 The **d'Alembertian operator** or simply **d'Alembertian** is the differential operator

$$\square \stackrel{\text{def}}{=} \triangle - \frac{1}{c^2}\frac{\partial^2}{\partial t^2}.$$

(Certain authors use a different sign convention.)

15.3.a Computation of the advanced and retarded Green functions

We assume known the charge distribution $\rho(\boldsymbol{r},t)$ and the current distribution $\boldsymbol{j}(\boldsymbol{r},t)$, and we want to find the electromagnetic fields produced by those distributions, up to a free solution. The easiest method is to start by looking for the potentials $\varphi(\boldsymbol{r},t)$ and $\boldsymbol{A}(\boldsymbol{r},t)$ which, in *Lorentz gauge*, are related to the sources by the following second order partial differential equation:

$$-\square \begin{pmatrix} \varphi \\ \boldsymbol{A} \end{pmatrix} = \begin{pmatrix} \rho/\varepsilon_0 \\ \mu_0\,\boldsymbol{j} \end{pmatrix}. \tag{15.4}$$

Hence we start by finding the Green function associated to the d'Alembertian, that is, a the distribution $G(\boldsymbol{r},t)$ satisfying the same differential equation with a source $\delta(\boldsymbol{x})\,\delta(t)$ which is a Dirac distribution, in both space and time aspects:

$$-\square G(\boldsymbol{r},t) = \delta(\boldsymbol{r})\,\delta(t) \tag{15.5}$$

(**Green function of the d'Alembertian**).

Put $\mathsf{A}(\boldsymbol{r},t) \stackrel{\text{def}}{=} \big[\varphi(\boldsymbol{r},t), \boldsymbol{A}(\boldsymbol{r},t)\big]$ and $\mathsf{j}(\boldsymbol{r},t) \stackrel{\text{def}}{=} \big[\rho(\boldsymbol{r},t), \boldsymbol{j}(\boldsymbol{r},t)\big]$. The general solution of (15.4) is given by

$$\mathsf{A}(\boldsymbol{r},t) = [G * \mathsf{j}](\boldsymbol{r},t) = \iiint_{\mathbb{R}^3}\int_{-\infty}^{+\infty} G(\boldsymbol{r}-\boldsymbol{r}', t-t')\,\mathsf{j}(\boldsymbol{r}',t')\,\mathrm{d}t\,\mathrm{d}^3\boldsymbol{r}$$

(up to a free solution).

Passing to the Fourier transform in equation (15.3.a) with the convention[4]

$$\mathscr{G}(\boldsymbol{k},z) \stackrel{\text{def}}{=} \iiiint G(\boldsymbol{r},t)\,\mathrm{e}^{-\mathrm{i}(\boldsymbol{k}\cdot\boldsymbol{r}-zt)}\,\mathrm{d}^3\boldsymbol{r}\,\mathrm{d}t \qquad \text{for all } \boldsymbol{k}\in\mathbb{R}^3 \text{ and } z\in\mathbb{R},$$

[4] Notice the sign in the exponential; this convention is customary in physics.

An illegitimate son abandoned in front of the Saint-Jean-le-Rond Church, the young Jean LE ROND D'ALEMBERT (1717–1783) studied in Jesuit schools and directed his attention to mathematics. Fond of success in the salons, he was interested in Newton's theories, then much in vogue. He collaborated with Diderot to edit and write the *Encyclopédie*; in particular, writing the *Discours préliminaire* and many scientific articles. He studied complex analysis, the equation of vibrating strings, and probability theory, and gave the first essentially complete proof of the fundamental theorem of algebra (any polynomial with real coeffecients can be factorized into polynomials of degree 1 or 2), which was completely proved by Gauss.

we obtain the relation[5]

$$-\left(\frac{z^2}{c^2} - \boldsymbol{k}^2\right) \mathscr{G}(\boldsymbol{k}, z) = 1.$$

A possible solution to this equation is[6]

$$\mathscr{G}(\boldsymbol{k}, z) = \frac{c^2}{c^2\boldsymbol{k}^2 - z^2},$$

which leads, after performing the inverse Fourier transform, to

$$G(\boldsymbol{r}, t) = \frac{1}{(2\pi)^4} \iiiint \frac{c^2}{c^2\boldsymbol{k}^2 - z^2} \mathrm{e}^{\mathrm{i}(\boldsymbol{k}\cdot\boldsymbol{r} - zt)}\, \mathrm{d}^3\boldsymbol{k}\, \mathrm{d}z. \tag{15.6}$$

However, this integral is not well defined, because the integrand has a pole. To see what happens, we start by simplifying the expression by means of an integration in spherical coordinates where the angular components can be easily dealt with. So we use spherical coordinates on the $\boldsymbol{k}$-space, putting $\mathrm{d}^3\boldsymbol{k} = k^2 \sin\theta\, \mathrm{d}k\, \mathrm{d}\theta\, \mathrm{d}\varphi$, with the polar axis in the direction of $\boldsymbol{r}$. We have

[5] To avoid sign mistakes or missing 2π factors, etc, use the following method: start from the relation $-\Box G = \delta$, multiply by $\mathrm{e}^{-\mathrm{i}(\boldsymbol{k}\cdot\boldsymbol{r} - zt)}$, then integrate. Then a further integration by parts leads to the equation above.

[6] We know that there exist distributions ("free solutions") satisfying

$$\left(\frac{z^2}{c^2} - \boldsymbol{k}^2\right) T(\boldsymbol{k}, z) = 0,$$

as, for instance, $T(\boldsymbol{k},z) = \delta^3(\boldsymbol{k})\,\delta(z)$ or $T(\boldsymbol{k},z) = \psi(\boldsymbol{k},z)\,\delta(z^2 - \boldsymbol{k}^2c^2)$. Adding these to the general solution changes in particular the behavior (decay) at infinity; in general, the solution will decay at infinity if its Fourier transform is sufficiently "regular."

therefore

$$G(\boldsymbol{r},t) = \frac{1}{(2\pi)^3} \iiiint \frac{c^2}{c^2\boldsymbol{k}^2 - z^2} \mathrm{e}^{\mathrm{i}(kr\cos\theta - zt)}\, k^2 \sin\theta \,\mathrm{d}k\,\mathrm{d}\theta\,\mathrm{d}\varphi\,\mathrm{d}z$$

$$= \frac{c^2}{4\pi^3 r} \int_0^{+\infty} \sin(kr) \left(\int_{-\infty}^{+\infty} \frac{k}{c^2k^2 - z^2} \mathrm{e}^{-\mathrm{i}zt}\,\mathrm{d}z \right) \mathrm{d}k.$$

For the evaluation of the integral in parentheses, there is the problem of the pole at $z = \pm ck$. To solve this difficulty, we move to the complex plane (that is, we consider that $z \in \mathbb{C}$), and *avoid* the poles by deforming the contour of integration. Different choices are possible, leading to different Green functions.

Remark 15.5 It is important to realize that, with the method followed up to now, there is an ambiguity in (15.6). The ways of deforming the contour lead to *radically different Green functions*,[7] since they are defined by different integrals. These functions differ by a free solution (such that $\Box f = 0$). The solution is the same as in the preceding section concerning the harmonic oscillator: take the *principal value* of the fraction at issue, or this principal value with an added Dirac distribution $\pm\mathrm{i}\delta$ at each pole.

We will make two choices, leading to functions which (for reasons that will be clear at the end of the computation) will be named $G^{(\mathrm{ad})}(\boldsymbol{r},t)$ and $G^{(\mathrm{ret})}(\boldsymbol{r},t)$:

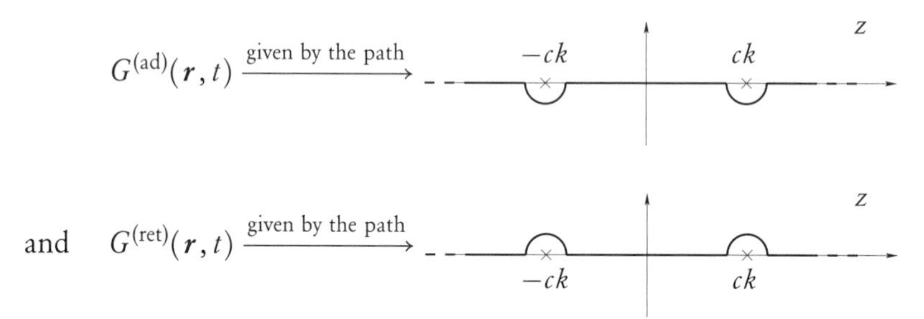

First, we compute the **retarded Green function**. For negative values of the time t, the complex exponential $\mathrm{e}^{-\mathrm{i}zt}$ only decays at infinity if the imaginary part of z is positive; hence, to be able to apply the second Jordan lemma, we must close the contour in the *upper* half-plane (see page 120):

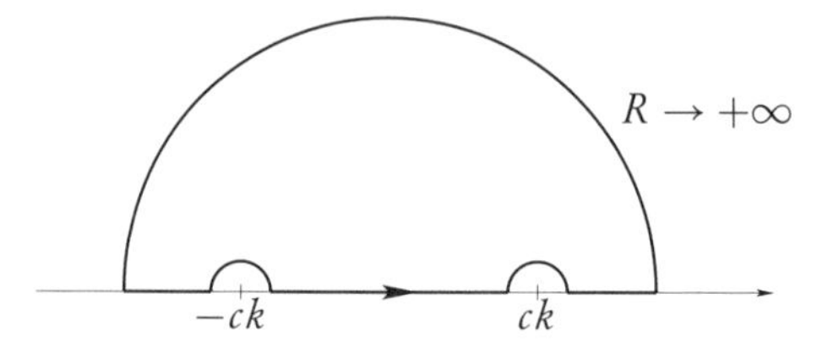

[7] Radically different for a physicist. Some may be causal while some others are not, for instance.

Since the poles are located outside this contour, the integral is equal to zero:

$$\boxed{G^{(\text{ret})}(\boldsymbol{r}, t) \equiv 0 \qquad \text{for } t < 0.}$$

For positive values of t, we have to close the contour in the lower half-plane, and both poles are then inside the contour. Since the path is taken with the *clockwise* (negative) orientation, we find that for $t > 0$ we have

$$\begin{aligned}\int_{(\text{ret})} \frac{\mathrm{e}^{-\mathrm{i}zt}}{c^2k^2 - z^2}\,\mathrm{d}z &= -2\mathrm{i}\pi \operatorname{Res}\left(\frac{\mathrm{e}^{-\mathrm{i}zt}}{c^2k^2 - z^2}\ ;\ z = \pm ck\right) \\ &= \frac{-\mathrm{i}\pi}{ck}\left[\mathrm{e}^{\mathrm{i}ckt} - \mathrm{e}^{-\mathrm{i}ckt}\right],\end{aligned}$$

and hence

$$\begin{aligned}G^{(\text{ret})}(\boldsymbol{r}, t) &= \frac{c^2}{4\pi^3 r}\int_0^{+\infty} \sin(kr)\frac{-\mathrm{i}\pi k}{ck}\left[\mathrm{e}^{\mathrm{i}ckt} - \mathrm{e}^{-\mathrm{i}ckt}\right]\mathrm{d}k \\ &= \frac{c}{8\pi^2 r}\int_0^{+\infty}\left[\mathrm{e}^{\mathrm{i}k(r-ct)} - \mathrm{e}^{-\mathrm{i}k(r-ct)} - \mathrm{e}^{\mathrm{i}k(r+ct)} + \mathrm{e}^{-\mathrm{i}k(r+ct)}\right]\mathrm{d}k \\ &= \frac{c}{8\pi^2 r}\int_{-\infty}^{+\infty}\left[\mathrm{e}^{\mathrm{i}k(r-ct)} - \mathrm{e}^{\mathrm{i}k(r+ct)}\right]\mathrm{d}k \\ &= \frac{c}{4\pi r}\left[\delta(r - ct) - \delta(r + ct)\right],\end{aligned}$$

since $\int \mathrm{e}^{\mathrm{i}kx}\,\mathrm{d}k = 2\pi\,\delta(x)$ (see the table page 612). Because t is now *strictly positive* and $r \geqslant 0$, the distribution $\delta(r + ct)$ is identically zero on the whole space. In conclusion, we have obtained

$$\boxed{G^{(\text{ret})}(\boldsymbol{r}, t) = \frac{c}{4\pi r}\delta(r - ct) \qquad \text{for } t > 0.}$$

Note that this formula is in fact also valid for $t < 0$, since the remaining Dirac distribution is then identically zero.

Similarly, the **advanced Green function** is zero for positive times and is given for all t by

$$\boxed{\begin{aligned}G^{(\text{ad})}(\boldsymbol{r}, t) &= \frac{c}{4\pi r}\delta(r + ct) && \text{for } t < 0, \\ G^{(\text{ad})}(\boldsymbol{r}, t) &\equiv 0 && \text{for } t > 0.\end{aligned}}$$

Physical interpretation

The retarded Green function is represented by a spherical "shell," emitted at $t = 0$ and with increasing radius $r = ct$. The fact that this *pulse* is so localized explains why a light flash (from a camera, for instance) is only seen for a time equal to the length of time it was emitted[8] but with a delay proportional to the distance. Moreover, the amplitude of the Green function varies like $1/r$, which is expected for a potential.

15.3.b Retarded potentials

Coming back to equation (15.4) for potentials in Lorentz gauge, we obtain using the retarded Green function for the d'Alembertian that

$$\varphi^{(\text{ret})}(\boldsymbol{r},t) = \left[G^{(\text{ret})} * \frac{\rho}{\varepsilon_0}\right](\boldsymbol{r},t), \qquad \boldsymbol{A}^{(\text{ret})}(\boldsymbol{r},t) = \left[G^{(\text{ret})} * \mu_0\, \boldsymbol{j}\right](\boldsymbol{r},t),$$

or

$$\varphi^{(\text{ret})}(\boldsymbol{r},t) = \frac{c}{4\pi\varepsilon_0} \iiint_{\mathbb{R}^3} \left(\int_{-\infty}^{+\infty} \frac{\rho(\boldsymbol{r}',t')}{\|\boldsymbol{r}-\boldsymbol{r}'\|}\, \delta\left(\|\boldsymbol{r}-\boldsymbol{r}'\| - c(t-t')\right) \mathrm{d}t' \right) \mathrm{d}^3\boldsymbol{r}'.$$

We can integrate[9] the time variable t'. Since $c\,\delta(r-ct) = \delta(t-r/c)$, only the value $t' = t - \|\boldsymbol{r}-\boldsymbol{r}'\|/c$ need be kept, which gives

$$\varphi^{(\text{ret})}(\boldsymbol{r},t) = \frac{1}{4\pi\varepsilon_0} \iiint_{\mathbb{R}^3} \frac{\rho\left(\boldsymbol{r}', t - \|\boldsymbol{r}-\boldsymbol{r}'\|/c\right)}{\|\boldsymbol{r}-\boldsymbol{r}'\|}\, \mathrm{d}^3\boldsymbol{r}'.$$

Each "integration element" therefore gives the contribution of the charge element situated at the point $\boldsymbol{r}'$ at the "retarded time" $\tau = t - \|\boldsymbol{r}-\boldsymbol{r}'\|/c$.

Similarly, we find that

$$\boldsymbol{A}^{(\text{ret})}(\boldsymbol{r},t) = \frac{\mu_0}{4\pi} \iiint_{\mathbb{R}^3} \frac{\boldsymbol{j}\left(\boldsymbol{r}', t - \|\boldsymbol{r}-\boldsymbol{r}'\|/c\right)}{\|\boldsymbol{r}-\boldsymbol{r}'\|}\, \mathrm{d}^3\boldsymbol{r}'.$$

Remark 15.6 Only a special solution has been described here. Obviously, the potential in space is the sum of this special solution and a *free* solution, such that $\square \mathbf{A} = 0$. If boundary conditions are imposed, the free solution may be quite difficult to find. In two dimensions, at least, techniques of conformal transformation may be helpful.

Lower-dimensional cases

We consider the analogue of the previous problem in one and two dimensions, starting with the latter. First, it must be made clear what is meant by "two-dimensional analogue." Here, it will be *the study of a system described by a potential field satisfying the d'Alembert equation in dimension* 2. Note that the ensuing theory of electromagnetism is extremely different from the theory of "standard" electromagnetism in dimension 3. Especially notable is that the

[8] Disregarding the phenomenon of retinal persistence.

[9] Note that clearly we should *never* write this integral in this manner, since distributions are involved. We forgo absolute rigor in order to show a computation similar to those which are exposed, often very quickly, in books of (classical and quantum) field theory.

Coulomb potential is not in $1/r$ but rather logarithmic,[10] and the magnetic field is not a vector but a simple scalar.[11]

What about the Green function of the d'Alembertian? It is certainly possible to argue using a method identical with the preceding one, but it is simpler to remark that it is possible to pass from the Green function in dimension 3 to the Green function in dimension 2 by integrating over one of the space variables:

THEOREM 15.7 *The Green functions of the d'Alembertian in two and three dimensions are related by the relation*

$$G_{(2)}(x, y, t) = \int_{-\infty}^{+\infty} G_{(3)}(x, y, z', t)\, \mathrm{d}z',$$

which may be seen as the convolution of $G_{(3)}$ with a linear Dirac source:

$$G_{(2)}(x, y, t) = G_{(3)}(x', y', z', t') * \delta(x' - x)\, \delta(y' - y)\, \delta(t' - t)\, \mathbf{1}(z').$$

Proof. Recall that $\Box G_{(3)}(x, y, z, t) = \delta(x)\, \delta(y)\, \delta(z)\, \delta(t)$. Passing to the Fourier transform (and putting $c = 1$ in the proof), we obtain the relation

$$(z^2 - k_x^2 - k_y^2 - k_z^2)\, \mathscr{G}(k_x, k_y, k_z, z) = 1, \tag{15.7}$$

which holds for all k_x, k_y, k_z and z. We now put $T(x, y, t) \stackrel{\text{def}}{=} \int_{-\infty}^{+\infty} G_{(3)}(x, y, z', t)\, \mathrm{d}z'$ and compute $\Box_{(2)} T$. Passing again to the Fourier transform, we obtain with $\mathscr{T} = \mathscr{F}[T]$ that

$$(z^2 - k_x^2 - k_y^2)\, \mathscr{T}(k_x, k_y, z) = (z^2 - k_x^2 - k_y^2)\, \mathscr{G}_{(3)}(k_x, k_y, 0, z),$$

which is equal to 1 according to equation (15.7) for $k_z = 0$. Hence we get

$$(z^2 - k_x^2 - k_y^2)\, \mathscr{T}(k_x, k_y, z) = 1,$$

or

$$\Box_{(2)} T(x, y, t) = \delta(x)\, \delta(y)\, \delta(t),$$

which is the equation defining the Green function of the d'Alembertian in two dimensions, and therefore $G_{(2)} = T$ (up to a free solution).

We have therefore obtained

$$G_{(2)}^{(\text{ret})}(x, y, t) = \frac{c}{4\pi} \int_{-\infty}^{+\infty} \frac{1}{\sqrt{r^2 + z'^2}}\, \delta\left(\sqrt{r^2 + z'^2} - ct\right) \mathrm{d}z',$$

where

$$r \stackrel{\text{def}}{=} \sqrt{x^2 + y^2}.$$

---

[10] Since $f(\boldsymbol{r}) = \log \|\boldsymbol{r}\|$ does satisfy $\triangle f = -2\pi\delta$ in two dimensions, see equation (7.11) page 208.

[11] See Chapter 17, page 483, concerning differential forms for an explanation of this seemingly bizarre phenomenon.

Once more, this function is identically zero for negative values of t. For positive values of t, we use the rule (explained in Theorem 7.45, page 199)

$$\delta\big(f(z)\big) = \sum_{y\in Z} \frac{1}{\big|f'(y)\big|}\delta(z-y) \qquad \text{with } Z \stackrel{\text{def}}{=} \big\{y\,;\, f(y)=0\big\},$$

which is valid if the zeros of f are simple and isolated (i.e., if $f'(y)$ does not vanish at the same time as $f(y)$). Applying this formula gives in particular

$$\int_{-\infty}^{+\infty} \delta\big(f(z')\big)\, g(z')\,\mathrm{d}z' = \sum_{y\in Z} \frac{g(y)}{\big|f'(y)\big|}.$$

In the case considered we have $f(z') = \sqrt{r^2 + z'^2} - ct$ and

$$f'(z') = \frac{z'}{\sqrt{r^2+z'^2}} \qquad Z = \{z_1, z_2\} \quad \text{with} \begin{cases} z_1 = \sqrt{c^2t^2 - r^2}, \\ z_2 = -z_1. \end{cases}$$

After calculations, we find

$$G_{(2)}^{(\text{ret})}(x,y,t) = \begin{cases} \dfrac{c}{2\pi\sqrt{c^2t^2 - (x^2+y^2)}} & \text{if } ct > (x^2+y^2), \\ 0 & \text{otherwise.} \end{cases} \tag{15.8}$$

A cylindrical outgoing wave can be seen which, in contrast to what happens in three dimensions, is not a *pulse*. In figure 15.1, we show the radial function $r \longmapsto G_{(2)}^{(\text{ret})}(r,t)$ for two values t_1 and t_2 with $t_2 > t_1$.

It should be remarked that, if we were living in a flat world,[12] our visual perception would have been entirely different. Instead of lighting a given point (the eye of the subject) for a very short lapse of time, a light flash (for instance, coming from a camera) would light it for an *infinite* amount of time, although with an intensity rapidly decreasing[13] with time (roughly like $1/\sqrt{t}$, as shown by formula (15.8)).

Remark 15.8 A number of electromagnetism books comment (or "prove") the retarded potentials formula by noting that it is perfectly intuitive, "the effect of a charge propagating at the speed of light, etc." Unfortunately, this proof based on intuition is utterly wrong, since it completely disregards the phenomenon of *nonlocality* of the Green function of the d'Alembertian in dimension 2. It is true that the Green function in three dimensions is of *pulse* type, but it is dangerous to found intuition on this fact!

The reader can check (using whichever method she prefers) that the Green function of the d'Alembertian in one dimension is given by

$$G_{(1)}(x,t) = \begin{cases} c/2 & \text{if } ct > x, \\ 0 & \text{if } ct < x. \end{cases}$$

---

[12] As was envisioned by the pastor and academic Edwin ABBOTT (1839–1926) in his extraordinary novel *Flatland* [1].

[13] Physiologically at least.

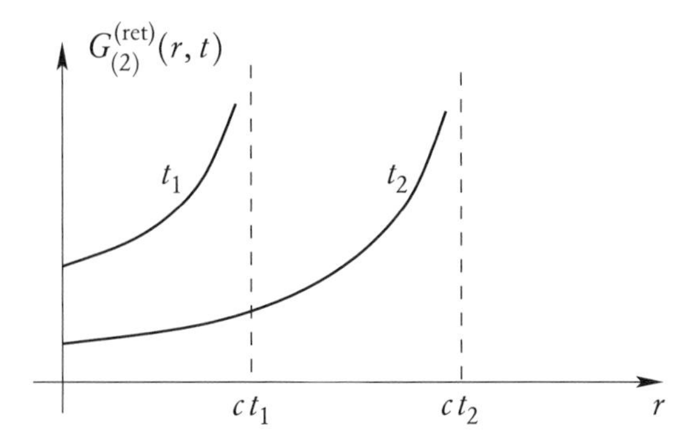

Fig. 15.1 – The retarded Green function of the d'Alembertian in two dimensions, for increasing values $t_2 > t_1$ of the variable t. In the abscissa, $r = \sqrt{x^2 + y^2}$.

15.3.c Covariant expression of advanced and retarded Green functions

The advanced and retarded Green functions may be expressed in a covariant manner. Denoting $\mathsf{x} = (ct, \boldsymbol{x}) = (x_0, \boldsymbol{x})$ and $\mathsf{x}^2 = x_0^2 - \|\boldsymbol{x}\|^2 = x_0^2 - r^2$, we have (see, for instance, Exercise 8.9, page 242):

$$\delta(\mathsf{x}^2) = \delta(c^2t^2 - r^2) = \frac{1}{2r}\big[\delta(r - ct) + \delta(r + ct)\big],$$

which proves that

$$G^{(\text{ret})}(\boldsymbol{x}, t) = \frac{c}{2\pi} H(t)\,\delta(\mathsf{x}^2) \quad \text{and} \quad G^{(\text{ad})}(\boldsymbol{x}, t) = \frac{c}{2\pi} H(-t)\,\delta(\mathsf{x}^2). \quad (15.9)$$

Although the function $H(t)$ is not covariant, it should be noticed that $H(t)\,\delta(\mathsf{x}^2)$ *is* covariant (it is invariant under Lorentz transformations).[14] This is a fundamental property when trying to write the equations of electromagnetism or field theory in a covariant way [49].

15.3.d Radiation

What is the physical interpretation of the difference $G^{(\text{ad})} - G^{(\text{ret})}$?

We know that the vector potential is given, quite generally, by

$$\mathsf{A}(\mathsf{r}) = \text{free solution} + \iiiint G(\mathsf{r} - \mathsf{r}')\,\mathsf{j}(\mathsf{r}')\,\mathrm{d}^4\mathsf{r}',$$

but there are *two* possible choices (at least) for the Green function, since either the retarded or the advanced one may be used (or any linear combination

[14] In fact, this expression is somewhat ambiguous since, for $\mathsf{x} = 0$, the Dirac distribution is singular and yet it is multiplied by a discontinuous function. The same ambiguity was, however, also present in the first expression, since it was not specified what G was at $t = 0$...

$\alpha G^{(\text{ad})} + (1-\alpha)G^{(\text{ret})}$). The incoming and outgoing free solutions,[15] denoted $\mathbf{A}^{(\text{in})}$ and $\mathbf{A}^{(\text{out})}$, respectively, are defined by the relations

$$\forall \mathbf{r} \in \mathbb{R}^4 \qquad \mathbf{A}(\mathbf{r}) = \mathbf{A}^{(\text{in})}(\mathbf{r}) + \iiiint G^{(\text{ret})}(\mathbf{r}-\mathbf{r}')\,\mathbf{j}(\mathbf{r}')\,\mathrm{d}^4\mathbf{r}', \tag{15.10}$$

$$\forall \mathbf{r} \in \mathbb{R}^4 \qquad \mathbf{A}(\mathbf{r}) = \mathbf{A}^{(\text{out})}(\mathbf{r}) + \iiiint G^{(\text{ad})}(\mathbf{r}-\mathbf{r}')\,\mathbf{j}(\mathbf{r}')\,\mathrm{d}^4\mathbf{r}'. \tag{15.11}$$

If the sources $\mathbf{j}$ are localized in space and time, then letting $t \to -\infty$ in (15.10), the integral in the right-hand side vanishes, since the Green function is retarded. The field $\mathbf{A}^{(\text{in})}$ can thus be interpreted as an *incoming* field, coming from $t = -\infty$ and evolving freely in time. One may also say that it is the field from the past, asymptotically free, which we would like to see evolve since $t = -\infty$ without perturbing it.

Similarly, the field $\mathbf{A}^{(\text{out})}$ is the field which escapes, asymptotically freely for $t \to \infty$, that is, it is the field which, if left to evolve freely, becomes (arbitrarily far in the future) equal to the total field for $t \to +\infty$.

The difference between those two fields is therefore what is called the field "radiated" by the current distribution $\mathbf{j}$ between $t = -\infty$ and $t = +\infty$. It is equal to

$$\mathbf{A}^{(\text{ray})}(\mathbf{r}) = \mathbf{A}^{(\text{out})}(\mathbf{r}) - \mathbf{A}^{(\text{in})}(\mathbf{r}) = \iiiint G(\mathbf{r}-\mathbf{r}')\,\mathbf{j}(\mathbf{r}-\mathbf{r}')\,\mathrm{d}^4\mathbf{r}',$$

where
$$G(\mathbf{r}) \stackrel{\text{def}}{=} G^{(\text{ret})}(\mathbf{r}) - G^{(\text{ad})}(\mathbf{r}).$$

Hence the field $\mathbf{A}^{(\text{ray})}(\mathbf{r})$ gives, at any point and any time, the total field radiated by the distribution of charges. It is in particular interesting to notice that, contrary to what is often stated, the unfortunate advanced Green function $G^{(\text{ad})}$ is not entirely lacking in physical sense.

15.4 The heat equation

DEFINITION 15.9 The **heat operator** in dimension d is the differential operator

$$\mathcal{D} \stackrel{\text{def}}{=} c\frac{\partial}{\partial t} - \mu\,\triangle_d \qquad \text{with } \triangle_d \stackrel{\text{def}}{=} \frac{\partial^2}{\partial x_1^2} + \cdots + \frac{\partial^2}{\partial x_d^2},$$

where $c > 0$ and $\mu > 0$ are positive constants.

---

[15] Recall that what is meant by "free" is a solution of the homogeneous Maxwell equations (without source).

15.4.a One-dimensional case

The Green function of the problem

We wish to solve the heat equation in one dimension, which is the equation

$$\left[c\frac{\partial}{\partial t} - \mu\frac{\partial^2}{\partial x^2}\right] T(x,t) = \rho(x,t), \tag{15.12}$$

where $T(x,t)$ is the temperature depending on $x \in \mathbb{R}$ and $t > 0$, and where ρ is a heat source. The constant $c > 0$ represents a calorific capacity and $\mu > 0$ is a thermal diffusion coefficient.

If no limit condition is imposed, the problem is translation invariant. We will start by solving the analogue problem when the heat source is elementary and modeled by a Dirac distribution. This means looking for the solution of the equation

$$\left[c\frac{\partial}{\partial t} - \mu\frac{\partial^2}{\partial x^2}\right] G(x,t) = \delta(x,t),$$

and G will be the **Green function of the heat equation** which is also called the **heat kernel**.[16]

Taking Fourier transforms, we have the correspondences

$$G(x,t) \longmapsto \mathscr{G}(k,\nu) = \iint G(x,t)\,\mathrm{e}^{-2\mathrm{i}\pi(\nu t + kx)}\,\mathrm{d}t\,\mathrm{d}\nu,$$

$$\frac{\partial}{\partial t} \longmapsto 2\mathrm{i}\pi\nu \qquad \text{and} \qquad \frac{\partial^2}{\partial x^2} \longmapsto -4\pi^2 k^2.$$

Hence the equation satisfied by $\mathscr{G}$ is

$$\left(4\pi^2\mu k^2 + 2\mathrm{i}\pi\nu c\right)\mathscr{G}(k,\nu) = 1,$$

so that[17]

$$\mathscr{G}(k,\nu) = \frac{1}{\left(4\pi^2\mu k^2 + 2\mathrm{i}\pi\nu c\right)}.$$

There only remains to take the inverse Fourier transforms. We start with the variable ν. We have

$$\widetilde{G}(k,t) = \int \frac{\mathrm{e}^{2\mathrm{i}\pi\nu t}}{\left(4\pi^2\mu k^2 + 2\mathrm{i}\pi\nu c\right)}\,\mathrm{d}\nu.$$

---

[16] A *kernel* is a function K that can be used to express the solution f of a functional problem, depending on a function g, in the form

$$f(x) = \int K(x\,;\,x')\,g(x')\,\mathrm{d}x'$$

(for instance, the Poisson, Dirichlet, and Fejér kernels already encountered). The Green functions are therefore also called kernels.

[17] *Up to* addition of a distribution T satisfying $\left(4\pi^2\mu k^2 + 2\mathrm{i}\pi\nu c\right) T(k,\nu) = 0$.

If ν is considered as a complex variable, the integrand has a unique pole located at $\nu_0 = 2\mathrm{i}\pi\mu k^2/c$. Moreover, to evaluate this integral with the method of residues, we must close the contour of integration of ν in the *lower* half-plane if $t < 0$ and in the *upper* half-plane if $t > 0$. The pole being situated in the upper half-plane, it follows already that

$$\boxed{G(x,t) \equiv 0 \qquad \text{for } t < 0.}$$

This is a rather good thing, since it shows that the heat kernel is **causal**.

Remark 15.10 It is the presence of unique purely imaginary pole, with strictly positive imaginary part, which accounts for the fact that there is no "advanced Green function" here. In other words, this is due to the fact that we have a partial derivative of order 1 with respect to t and the equation has real coefficients.

In the case of quantum mechanics, the Schrödinger equation is formally identical to the heat equation, but it has complex coefficients. The pole is then real, and both advanced and retarded Green functions can be introduced.

For *positive* values of the time variable t, we close the contour in the upper half-plane, apply Jordan's second lemma, and thus derive

$$\begin{aligned}\widetilde{G}(k,t) &= 2\mathrm{i}\pi \operatorname{Res}\left(\frac{\mathrm{e}^{2\mathrm{i}\pi\nu t}}{(4\pi^2\mu k^2 + 2\mathrm{i}\pi\nu c)}\,;\ \nu = \frac{2\mathrm{i}\pi\mu k^2}{c}\right)\\ &= \frac{\mathrm{e}^{-4\pi^2 k^2 \mu t/c}}{c} \qquad\qquad (\text{for } t > 0).\end{aligned}$$

Remembering that the Fourier transform of a gaussian is a gaussian, and that the formula for scaling leads to

$$\mathscr{F}^{-1}\left[\mathrm{e}^{-\pi a^2 k^2}\right] = \frac{1}{|a|}\mathrm{e}^{-\pi x^2/a^2}$$

(which we apply with $a = 2\sqrt{\mu\pi t/c}$); the inverse Fourier transform in the space variable is easily performed to obtain

$$\boxed{G(x,t) = \frac{1}{2\sqrt{\mu\pi c t}} \exp\left(\frac{-cx^2}{4\mu t}\right) \qquad \text{for } t > 0.} \tag{15.13}$$

Notice that, whereas temperature is zero everywhere for $t < 0$, the heat source, although it is localized at $x = 0$ and $t = 0$, leads to strictly positive temperatures everywhere for all $t > 0$. In other words, *heat has propagated at infinite speed.* (Since this is certainly not literally the case, the heat equation is not completely correct. However, given that the phenomena of thermal diffusion are very slow, it may be considered that the microscopic processes are indeed infinitely fast, and the heat equaton is perfectly satisfactory for most thermal phenomena.)

Remark 15.11 The "typical distance traveled by heat" is $\Delta x^2 = (4\mu/c)\Delta t$. This is also what Einstein had postulated to explain the phenomenon of diffusion. Indeed, a particle under diffusion in a gas or liquid does not travel in a straight line, but rather follows a very roundabout path, known as **Brownian motion**, which is characterized in particular by $\Delta x^2 \propto \Delta t$. The diffusion equation is in fact formally identical with the heat equation.

The gaussian we obtained in (15.13), with variance $\Delta x^2 \propto \Delta t$, can also be interpreted mathematically as a consequence of a probabilistic result, the **central limit theorem**, discussed in Chapter 21.

The problem with an arbitraty heat source (15.12) is now easily solved, since by linearity we can write

$$\mathcal{T}(x,t) = [G * \rho](x,t) = \int_{-\infty}^{\infty} \int_{0}^{+\infty} G(x',t')\,\rho(x-x',t-t')\,\mathrm{d}t\,\mathrm{d}x. \quad (15.14)$$

Note, however, that this general solution again satisfies $T(x,0) = 0$ for all $x \in \mathbb{R}$. We must still add to it a solution satisfying the initial conditions at $t = 0$ and the *free* equation of heat propagation.

Initial conditions

We now want to take into account the initial conditions in the problem of heat propagation. For this, we impose that the temperature at time $t = 0$ be given by $T(x,0) = T_0(x)$ for all $x \in \mathbb{R}$, where T_0 is an arbitrary function.

We need a solution to the *free* problem (i.e., without heat source), which means such that

$$\mathscr{T}(x,t) = H(t)\,u(x,t).$$

The discontinuity of the solution at $t = 0$ provides a way to incorporate the initial condition $\mathscr{T}(x,0) = T_0(x)$. Indeed, in the sense of distributions we have

$$\frac{\partial \mathscr{T}}{\partial t} = \left\{\frac{\partial u}{\partial t}\right\} + T_0(x)\,\delta(t).$$

The distribution $\mathscr{T}(x,t)$ then satisfies the equation

$$\left[c\frac{\partial}{\partial t} + \mu\frac{\partial^2}{\partial x^2}\right] \mathscr{T}(x,t) = c\,T_0(x)\,\delta(t),$$

the solution of which is given by

$$\mathscr{T}(x,t) = c\,T_0(x)\,\delta(t) * G(x,t), \quad \text{or} \quad \mathscr{T}(x,t) = c\int_{-\infty}^{+\infty} G(x-x',t)\,T_0(x)\,\mathrm{d}x.$$

Hence the solution of the problem that incorporates both initial condition and heat source is given by

$$\boxed{T(x,t) = \mathcal{T}(x,t) + \mathscr{T}(x,t) = \Big[\rho(x,t) + c\,T_0(x)\,\delta(t)\Big] * G(x,t).}$$

However, it is easier to take the initial conditions into account when using the *Laplace transform*, as in the next example.

15.4.b Three-dimensional case

We are now interested in the three-dimensional case, which we will treat using the formalism of the Laplace transform.[18]

So we consider a function (or, if need be, a distribution) that satisfies the free heat equation with a given boundary condition:

$$c\frac{\partial T}{\partial t} - \mu \triangle T = 0 \qquad \text{and} \qquad T(\boldsymbol{r},0) = T_0(\boldsymbol{r}).$$

We perform a Laplace transform with respect to the time variable, denoted $T(\boldsymbol{r},t) \sqsupset \mathscr{T}(\boldsymbol{r},p)$. We then have

$$(cp - \mu\triangle)\mathscr{T}(\boldsymbol{r},p) = T_0(\boldsymbol{r}).$$

Notice that, this time, the initial conditions $T_0(\boldsymbol{r})$ are automatically involved in the Laplace transform, since the value of the temperature at $t = 0$ appears in the differentiation theorem.

Taking the Fourier transform of this equation, we obtain

$$\widetilde{\mathscr{T}}(\boldsymbol{k},p) = c\,\widetilde{T}_0(\boldsymbol{k}) \cdot \widetilde{\mathscr{G}}(\boldsymbol{k},p),$$

were we have denoted

$$\widetilde{\mathscr{G}}(\boldsymbol{k},p) = \frac{1}{cp + 4\pi^2\mu\,\boldsymbol{k}^2} = \frac{1}{c} \cdot \frac{1}{p + 4\pi^2\mu\boldsymbol{k}^2/c}. \qquad (15.15)$$

A formal solution is then obtained by taking successively the inverse Laplace transform:

$$\widetilde{T}(\boldsymbol{k},t) = c\,\widetilde{T}_0(\boldsymbol{k}) \cdot \widetilde{G}(\boldsymbol{k},t),$$

and then the inverse Fourier transform:

$$T(\boldsymbol{r},t) = c\left[T_0 * G\right](\boldsymbol{r},t) = \iiint c\,T_0(\boldsymbol{r}')\,G(\boldsymbol{r}-\boldsymbol{r}',t)\,\mathrm{d}^3\boldsymbol{r}.$$

In order to obtain the function $G(\boldsymbol{r},t)$, we come back to the expression (15.15) and use the table of Laplace transforms on page 614:

$$\widetilde{G}(\boldsymbol{k},t) = \frac{H(t)}{c}\,\mathrm{e}^{-4\pi^2\mu\,\boldsymbol{k}^2 t/c}$$

whence (the inverse Fourier transform of a gaussian in three dimensions)

$$G(\boldsymbol{r},t) = \frac{\sqrt{c}}{8(\pi\mu t)^{3/2}}\,\mathrm{e}^{-c r^2/4\mu t}. \qquad (15.16)$$

[18] The reader should not, however, get the impression that this is vastly preferable in three dimensions: we could just as well use the method of the one-dimensional case, or we could have treated the latter with the Laplace transform.

Remark 15.12 We may add a right-hand side to the heat equation: $\mathcal{D}T(\boldsymbol{r},t) = \rho(\boldsymbol{r},t)$, where ρ is the distribution describing the heat source; it allows us to derive a more general solution (where the convolution is both on time and space variables)

$$T(\boldsymbol{r},t) = \Big[\big(\rho + c\,T_0(\boldsymbol{r})\,\delta(t)\big) * G\Big](\boldsymbol{r},t).$$

Remark 15.13 Here also the typical distance of heat propagation is $\Delta x^2 \propto \Delta t$. This result (still a consequence of the central limit theorem!) is independent of the dimension of the system.

15.5 Quantum mechanics

This time, we consider the evolution equation for a wave function:

DEFINITION 15.14 The **Schrödinger operator** is the differential operator

$$\mathrm{i}\hbar\frac{\partial}{\partial t} + \frac{\hbar^2}{2m}\,\triangle\, - V(\boldsymbol{r}),$$

where m is the mass of the particle and $V(\boldsymbol{r})$ is a potential independent of time. If $V \equiv 0$, this is the **free** evolution equation.

In fact, only the *free* case $V(\boldsymbol{r}) \equiv 0$ will be treated here. Notice that this operator is formally similar to the heat operator with $\mu = \mathrm{i}\hbar/2mc$. However, its properties are very different, since in particular the heat equation is irreversible, whereas the Schrödinger equation is perfectly reversible. This is due to the fact that the Schrödinger equation is applied to *complex-valued* functions.[19]

The Green function of the Schrödinger operator is computed in exactly the same manner as in Section 15.4[20] and leads to

$$\mathscr{G}(\boldsymbol{p},t) = \mathrm{e}^{-\mathrm{i}\,p^2\hbar t/2m} = \mathrm{e}^{-\mathrm{i}H\,t/\hbar} \qquad \text{for } t > 0,$$

[19] In the Schrödinger equation, it is therefore possible to perform the transformation $t \mapsto -t$ and *simultaneously* to map $\psi \mapsto \overline{\psi}$, so that the equation is still satisfied. Note that ψ and $\overline{\psi}$ have the same physical interpretation (since the square of the modulus is the probability density of the particle). Finally, notice that the change of ψ into $\overline{\psi}$ simply amounts to a change of convention and to changing i into $-$i, which cannot affect physics. The operator $\mathrm{i}\hbar\partial/\partial t$ is indeed invariant when simultaneously $t \mapsto -t$ and $\mathrm{i} \mapsto -\mathrm{i}$.

[20] This calculation involves a slight additional difficulty in that one must extract the square root of a quantity which is this time imaginary. Hence a determination of the function $z \mapsto \sqrt{z}$ must be specified, involving a cut in the plane. Once this is done, there is no further difference. In the text, we have chosen a cut along the positive imaginary axis.

(where we write $H = p^2/2m$ in momentum representation), taking $\boldsymbol{p}$ as the conjugate variable to $\boldsymbol{x}$ and using the standard convention in quantum mechanics:

$$\mathscr{G}(\boldsymbol{p},t) = \iiint G(\boldsymbol{x},t)\,\mathrm{e}^{\mathrm{i}\boldsymbol{p}\cdot\boldsymbol{x}/\hbar}\,\mathrm{d}^3\boldsymbol{x}.$$

Similarly, with the inverse Fourier transform, we obtain the equivalent of (15.16):

$$\boxed{G(\boldsymbol{r},t) = \left(\frac{m}{2\mathrm{i}\pi\hbar t}\right)^{3/2} \mathrm{e}^{\mathrm{i}m|\boldsymbol{r}|^2/2\hbar t} \qquad \text{for } t > 0.} \tag{15.17}$$

If we want to solve the free Schrödinger equation

$$\mathrm{i}\hbar\frac{\partial\psi}{\partial t} + \frac{\hbar^2}{2m}\,\triangle\,\psi = 0$$

with initial conditions $\psi(\boldsymbol{r},0) = \psi_0(\boldsymbol{r})$, we get[21] the relation for the propagation of the wave function:

$$\boxed{\psi(\boldsymbol{r},t) = \int G(\boldsymbol{r}-\boldsymbol{r}',t)\,\psi_0(\boldsymbol{r}')\,\mathrm{d}^3\boldsymbol{r}' \qquad \text{for } t > 0.} \tag{15.18}$$

The function G is called the **Schrödinger kernel**. Equation (15.18) is therefore perfectly equivalent to the Schrödinger equation written in the form

$$\mathrm{i}\hbar\frac{\partial\psi}{\partial t} = H\psi \qquad \text{with } H \stackrel{\text{def}}{=} -\frac{\hbar^2}{2m}\,\triangle\,. \tag{15.19}$$

Either form may be used as a postulate for the purpose of the formalization of quantum mechanics.

It should be noted that the general form of the Green function (15.17) may be obtained through the very rich physical tool of path integration. The reader is invited to look at the book of Feynman and Hibbs [34] to learn more about this.

Note also that, using Dirac's notation, we can write the evolution equation of the particle as

$$\mathrm{i}\hbar\frac{\mathrm{d}}{\mathrm{d}t}\big|\psi(t)\big\rangle = H\big|\psi(t)\big\rangle,$$

(denoting by $|\psi\rangle$ the wave vector of the particle), which can be solved – formally – by putting

$$\big|\psi(t)\big\rangle = \mathrm{e}^{-\mathrm{i}Ht/\hbar}\,|\psi_0\rangle\,.$$

[21] Once more, those initial conditions are taken into accound by adding a right-hand side given by $\psi_0(\boldsymbol{r})\,\delta(t)$.

Using the "closure relation" $\mathbf{1} = \int |\boldsymbol{r}'\rangle\langle\boldsymbol{r}'|\, \mathrm{d}^3\boldsymbol{r}'$ (see Theorem 14.26) we obtain for all t that

$$\psi(\boldsymbol{r},t) = \langle \boldsymbol{r} | \psi(t)\rangle = \langle \boldsymbol{r} | \mathrm{e}^{-\mathrm{i}Ht/\hbar} \int |\boldsymbol{r}'\rangle \langle \boldsymbol{r}' | \psi_0\rangle \, \mathrm{d}^3\boldsymbol{r}'$$
$$= \int \langle \boldsymbol{r} | \mathrm{e}^{-\mathrm{i}Ht/\hbar} | \boldsymbol{r}'\rangle \, \psi_0(\boldsymbol{r}')\, \mathrm{d}^3\boldsymbol{r}'.$$

This is recognizable as the same integral as in (15.18). Since this relation should be kept only for positive time, we are led to write

$$\boxed{G(\boldsymbol{r}-\boldsymbol{r}',t) = H(t)\, \langle \boldsymbol{r} | \mathrm{e}^{-\mathrm{i}Ht/\hbar} | \boldsymbol{r}'\rangle.}$$

In terms of Fourier transforms, we do indeed get

$$\mathscr{G}(\boldsymbol{p},t) = \langle \boldsymbol{p} | \mathrm{e}^{-\mathrm{i}Ht/\hbar} | \boldsymbol{p}\rangle = \mathrm{e}^{-\mathrm{i}p^2\hbar t/2m},$$

which shows that the two expressions are consistent (and illustrates the simplicity of the Dirac notation).

15.6 Klein-Gordon equation

In this section, the space and time variables will be expressed in a covariant manner, that is, we will put $\mathsf{x} = (ct, \boldsymbol{x}) = (x_0, \boldsymbol{x})$ and $\mathsf{p} = (p_0, \boldsymbol{p})$. We choose a Minkowski metric with signature $(+,-,-,-)$, which means that we put

$$\mathsf{p}^2 = p_0^2 - \boldsymbol{p}^2, \qquad \mathsf{x}^2 = x_0^2 - \boldsymbol{x}^2, \qquad \text{and} \qquad \Box = \frac{\partial^2}{\partial t^2} - \triangle.$$

Remark 15.15 Note that the sign of the d'Alembertian is opposite to that used in Section 15.3; here we conform with relativistic customs, where usually $\Box = \partial_\mu \partial^\mu$ with a metric $(+,-,-,-)$.

We want to solve an equation of the type

$$(\Box + m^2)\, \varphi(\mathsf{x}) = j(\mathsf{x}),$$

which is called the **Klein-Gordon equation** for the field φ (it originates in field theory). In this equation, m represents the mass of a particle, j is an arbitrary source and φ is the field which is the unknown. As usual, we take the constants $\hbar$ and c to be equal to 1.[22]

[22] Otherwise we should write $(\hbar^2 c^2 \Box + m^2 c^2)\, \varphi = j$.

To harmonize notation with those which are customary in field theory, the Fourier transform and its inverse are denoted as follows:

$$\widetilde{f}(\mathsf{p}) = \int f(\mathsf{x})\,\mathrm{e}^{\mathrm{i}\mathsf{p}\cdot\mathsf{x}}\,\mathrm{d}^4\mathsf{x} \quad \text{and} \quad f(\mathsf{x}) = \frac{1}{(2\pi)^4}\int \widetilde{f}(\mathsf{p})\,\mathrm{e}^{-\mathrm{i}\mathsf{p}\cdot\mathsf{x}}\,\mathrm{d}^4\mathsf{p}.$$

We start by trying to solve the equation where the source term is replaced by a Dirac distribution; the Green function of this equation

$$(\Box + m^2)\,G(\mathsf{x}) = \delta^4(\mathsf{x})$$

is called a **propagator**. Taking the Fourier transform, we get

$$(-\mathsf{p}^2 + m^2)\,\mathscr{G}(\mathsf{p}) = 1,$$

or

$$\mathscr{G}(\mathsf{p}) = \frac{-1}{\mathsf{p}^2 - m^2} = \frac{-1}{p_0^2 - \boldsymbol{p}^2 - m^2}.$$

As in the case of electromagnetism, there will arise a problem with this denominator when performing the inverse Fourier transform. Indeed, we will try to write an expression like

$$G(\mathsf{x}) = -\frac{1}{(2\pi)^4}\iiiint \frac{1}{p_0^2 - \boldsymbol{p}^2 - m^2}\,\mathrm{e}^{-\mathrm{i}\mathsf{p}\cdot\mathsf{x}}\,\mathrm{d}^3\boldsymbol{p}\,\mathrm{d}p_0.$$

During the process of integrating over the variable p_0, two poles at $p_0 = \pm\sqrt{\boldsymbol{p}^2 + m^2}$ occur. They must be bypassed by seeing p_0 as a variable in the complex plane and integrating along a deformed contour (see Section 15.2).

For instance, we may go around both poles in small half-circles in the lower half-plane, which provides the advanced Green function:

$$G^{(\mathrm{ad})}(t, \boldsymbol{x}) \xrightarrow{\text{given by the path}}$$

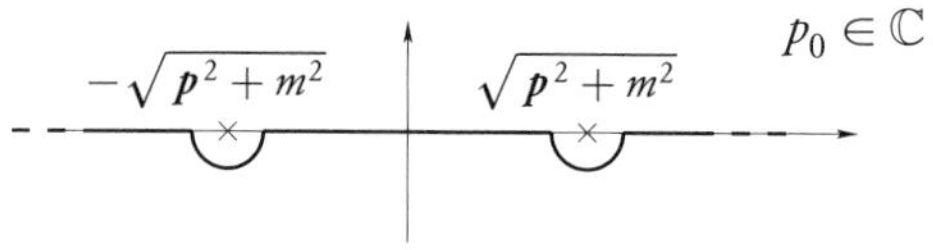

$p_0 \in \mathbb{C}$ $\quad -\sqrt{\boldsymbol{p}^2 + m^2} \quad \sqrt{\boldsymbol{p}^2 + m^2}$

This also amounts to adding a *negative* imaginary part to p_0, if integrating first over $\boldsymbol{p}$ is preferred.

Similarly, the retarded Green function is defined by going around both poles in the upper half-plane, or equivalently, by adding a small *positive* imaginary part to p_0:

$$G^{(\mathrm{ret})}(t, \boldsymbol{x}) \xrightarrow{\text{given by the path}}$$

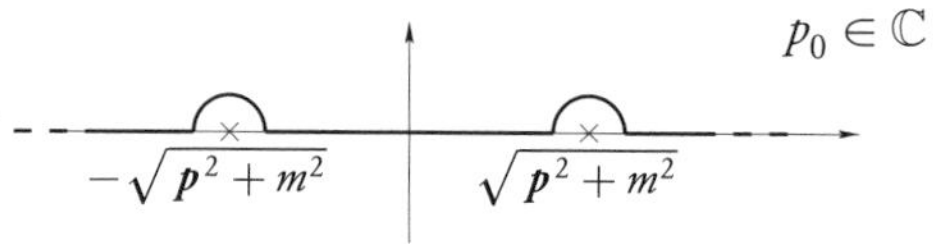

$p_0 \in \mathbb{C}$ $\quad -\sqrt{\boldsymbol{p}^2 + m^2} \quad \sqrt{\boldsymbol{p}^2 + m^2}$

Hence we put

$$\mathscr{G}^{(\mathrm{ad})}(\mathsf{p}) \stackrel{\text{def}}{=} \frac{-1}{(p_0 - \mathrm{i}\varepsilon)^2 - \boldsymbol{p}^2 - m^2}$$

and

$$\mathscr{G}^{(\text{ret})}(\mathsf{p}) \stackrel{\text{def}}{=} \frac{-1}{(p_0 + \mathrm{i}\varepsilon)^2 - \boldsymbol{p}^2 - m^2}.$$

Taking inverse Fourier transforms, we obtain

$$G^{(\text{ad})/(\text{ret})}(\mathsf{x}) = -\frac{1}{(2\pi)^4} \iiiint \frac{\mathrm{e}^{-\mathrm{i}\mathsf{p}\cdot\mathsf{x}}}{(p_0 \mp \mathrm{i}\varepsilon)^2 - \boldsymbol{p}^2 - m^2}\, \mathrm{d}^4\mathsf{p}.$$

By the residue theorem, the function $G^{(\text{ret})}$ is zero for $t < 0$ and $G^{(\text{ad})}$ is zero for $t > 0$. Moreover, these Green functions being defined by Lorentz invariant objects, it follows that $G^{(\text{ret})}$ is zero outside the future light cone (it is said to be **causal** in the relativistic sense) and $G^{(\text{ret})}$ outside of the past light cone. Also, note that both functions are real-valued (despite the presence of the term $\mathrm{i}\varepsilon$) and that $G^{(\text{ret})}(-\mathsf{x}) = G^{(\text{ad})}(\mathsf{x})$.

Whereas, in the case $m^2 = 0$ of electromagnetism,[23] we had a simple explicit formula for the Green functions, we would here get Bessel functions after integrating on the variable $\boldsymbol{p}$. The full calculation is not particularly enlightening, and we leave it aside. Note, however, that the Green functions are no longer supported on the light cone as with (15.9) (the case $m^2 = 0$); there is also propagation of signals at speeds less than the speed of light. This is of course consistent with having a nonzero mass!

To close this very brief section (referring the reader to [49] for a wider survey of the properties of the Green fnctions of the Klein-Gordon equation), we mention that it is possible to introduce a very different Green function, following Stueckelberg and Feynman. Let

$$\mathscr{G}_{\mathrm{F}}(\mathsf{p}) = \frac{-1}{\mathsf{p}^2 - m^2 + \mathrm{i}\varepsilon}.$$

From the formula

$$x^2 - \alpha^2 + \mathrm{i}\varepsilon \approx \left(x + \alpha - \frac{\mathrm{i}\varepsilon}{2\alpha}\right)\left(x - \alpha + \frac{\mathrm{i}\varepsilon}{2\alpha}\right),$$

which is valid to first order in ε, we deduce that changing m^2 into $m^2 - \mathrm{i}\varepsilon$ amounts to changing the poles

$$p_0^{(1)} = -\sqrt{\boldsymbol{p}^2 + m^2} \qquad \text{and} \qquad p_0^{(2)} = \sqrt{\boldsymbol{p}^2 + m^2}$$

into, respectively,

$$-\sqrt{\boldsymbol{p}^2 + m^2} + \mathrm{i}\varepsilon' \qquad \text{and} \qquad \sqrt{\boldsymbol{p}^2 + m^2} - \mathrm{i}\varepsilon'$$

with $\varepsilon' = \varepsilon/2\sqrt{\boldsymbol{p}^2 + m^2}$, or to take the following integration contour:

[23] Which is indeed a zero-mass field theory.

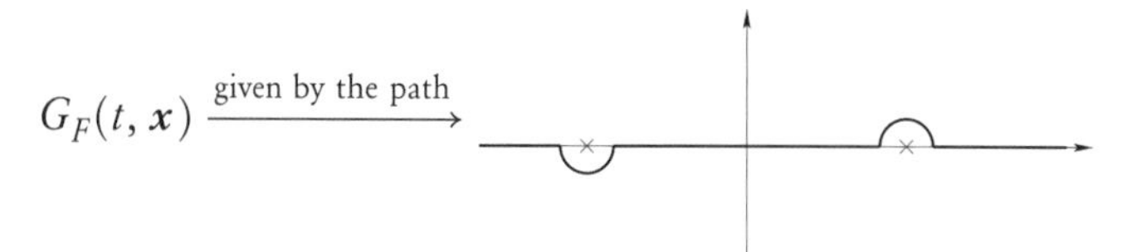

This last Green function is called the **Feynman propagator**[24]; it is complex-valued and does satisfy $(-\Box + m^2)\, G_F(\mathsf{x}) = \delta(\mathsf{x})$, as well as the relation $G_F(\mathsf{x}) = G_F(-\mathsf{x})$. Moreover, it can be shown (see [49]) that this propagator is not zero outside the light cone (i.e., in particular that $G_F(0, \boldsymbol{r})$ is not identically zero for values of $\boldsymbol{r} \neq \boldsymbol{0}$, contrary to what happens with the "advanced" and "retarded" functions).

Remark 15.16 Why does this function not appear in "classical" physics? Simply because only $G^{(\text{ad})}$ and $G^{(\text{ret})}$ are *real-valued* and have a physical meaning that is easy to grasp. The Feynman propagator G_F is complex-valued; however, it makes sense in quantum mechanics, where functions are *intrinsically* complex-valued.

EXERCISES

♦ **Exercise 15.1** Show that the Green function of the d'Alembertian is of *pulse* type in odd dimensions at least equal to 3, but that otherwise it is not localized.

♦ **Exercise 15.2 (Damped harmonic oscillator)** A damped harmonic oscillator is ruled by the following ordinary differential equation:

$$y'' + 2a\, y' + \omega^2\, y = f,$$

where a is some positive parameter. We are looking for the Green function associated to the equation $y'' + 2a\, y' + \omega^2\, y = \delta$.

i) What is the equation which is satisfied by the Fourier transform of the Green function $\widetilde{G}$?

ii) Compute the Green function in $\mathscr{D}'_+$. One may represent graphically the poles of the meromorphic function that appears during the calculations.

iii) Find the solution of the equation with an excitation f (some assumptions on f may be required).

---

[24] Richard Phillips FEYNMAN (1918–1988), American physicist, was one of the inventors of quantum electrodynamics (the quantum relativistic theory of electromagnetism), which is considered by many as "one of the most beautiful current physical theories." With a penetrating mind and a physical sense particularly acute and clear, he simplified the perturbative methods and invented the diagrams that were named after him, which are extremely useful both for making computations *and* for understanding them. His physics lecture notes are indispensable reading and his autobiography is funny and illuminating in many respects.

Chapter 16

Tensors

In this chapter, we will give few details concerning the *practical handling* of tensors, or the usual calculation tricks, juggling with indices, contractions, recognition of tensors, etc. Many excellent books [5,27,35,57,62,92] discuss those rules (and often do much more than that!)

Here, the focus will be on the *mathematical construction* of tensors, in order to complement the "computational" viewpoint of the physicist. This chapter is therefore not directed to physicists trying to learn how to tensor, but rather to those who, having started to manipulate tensors, would like to understand what is hidden behind.

Also, since this is a purely introductory chapter, we only discuss tensors in a flat space. There will be no mention of parallel transport, fiber bundles, connections, or Christoffel symbols. Readers wishing to learn more should consult any of the classical textbooks of differential geometry (see Bibliography, page 617).

16.1 Tensors in affine space

Let $\mathbb{K}$ denote either the real field $\mathbb{R}$ or the complex field $\mathbb{C}$. We identify $E = \mathbb{K}^n$, a vector space of dimension n, with an affine space $\mathscr{E}$ of the same dimension, We consider E with a basis $\mathcal{B} \stackrel{\text{def}}{=} (\mathbf{e}_\mu)_\mu = (\mathbf{e}_1, \dots, \mathbf{e}_n)$.

16.1.a Vectors

Let $\mathbf{u}$ be a vector in $\mathscr{E}$, that is, an element of $\mathbb{K}^n$. There exists a unique expression of $\mathbf{u}$ as linear combination of the basis vectors of $\mathcal{B}$:

DEFINITION 16.1 (Contravariant coordinates of a vector) The **contravariant coordinates** of a vector $\mathbf{u}$ are the unique elements $(u^\mu)_\mu$ in $\mathbb{K}$ such that

$$\mathbf{u} = \sum_{\mu=1}^{n} u^\mu \, \mathbf{e}_\mu = u^1 \mathbf{e}_1 + \cdots + u^n \mathbf{e}_n.$$

The contravariant coordinates are therefore the "usual coordinates" of linear algebra.

Remark 16.2 Obviously, for a given vector $\mathbf{u}$, the μ-th coordinate depends not only on the vector $\mathbf{e}_\mu$, but also on *all other vectors of the chosen basis.*

Remark 16.3 From the point of view of a physicist, what quantities may be modeled by a vector? If working in $\mathbb{R}^3$, a vector is not simply "a triplet of real numbers"; rather, it possesses an additional property, related to changes of reference frame: if an observer performs a rotation, the basis vectors turn also, and all triplets that are called "vectors" are transformed in a uniform manner. A triplet "(temperature, pressure, density)," for instance, does not transform in the same way, since each component is invariant under rotations (they are **scalars**).

In what manner are the coordinates of a vector transformed during a change of basis? Let $\mathcal{B}' = (\mathbf{e}'_\mu)_\mu$ denote another basis of $\mathscr{E}$. We write L the matrix for the change of basis from $\mathcal{B}$ to $\mathcal{B}'$, that is, the matrix where the μ-th column represents the vector $\mathbf{e}'_\mu$, as expressed in coordinates in the basis $\mathcal{B}$ (see Appendix C). Writing this matrix $L = (L^\nu{}_\mu)_{\nu\mu}$, where ν is the line index and μ the column index, it follows that for all $\mu \in [\![1, n]\!]$ we have

$$\boxed{\mathbf{e}'_\mu = \sum_{\nu=1}^{n} L^\nu{}_\mu \, \mathbf{e}_\nu.} \tag{16.1}$$

The left index in $L^\nu{}_\mu$ is the row index, and the right index is the column index. For reasons which will become clearer later on, the first index is indicated in superscript and the second in subscript. Symbolically, we write[1]

$$\mathcal{B}' = L\,\mathcal{B}.$$

Let $\Lambda \stackrel{\text{def}}{=} L^{-1}$ be the inverse matrix L, i.e., the unique matrix with coefficients $(\Lambda^\mu{}_\nu)_{\mu\nu}$ that satisfies

$$\sum_{\nu=1}^{n} L^\mu{}_\nu \, \Lambda^\nu{}_\rho = \sum_{\nu=1}^{n} \Lambda^\mu{}_\nu \, L^\nu{}_\rho = \delta^\mu_\rho$$

for all $\mu, \nu \in [\![1, n]\!]$. The vectors of $\mathcal{B}$ can be expressed as functions of those of $\mathcal{B}'$ by means of the matrix Λ:

---

[1] Note that this is *not* a matrix relation. However, one can formally build a line vector with the $\mathbf{e}_\mu$'s and write

$$(\mathbf{e}'_1, \ldots, \mathbf{e}'_n) = (\mathbf{e}_1, \ldots, \mathbf{e}_n) \cdot L.$$

$$\mathcal{B} = \Lambda\,\mathcal{B}' \qquad \text{or} \qquad \mathbf{e}_\mu = \sum_{\nu=1}^{n} \Lambda^\nu{}_\mu\, \mathbf{e}'_\nu.$$

By writing the decomposition of a vector in the two bases, we can find the transformation law for the coordinates of a vector:

$$\begin{aligned}
\mathbf{u} &= \sum_\mu u^\mu\, \mathbf{e}_\mu &&= \sum_\mu u'^\mu\, \mathbf{e}'_\mu, \\
&= \sum_{\mu,\nu} u^\mu\, (\Lambda^\nu{}_\mu\, \mathbf{e}'_\nu) &&= \sum_{\mu,\nu} u'^\mu\, (L^\nu{}_\mu\, \mathbf{e}_\nu), && \text{(use (16.1))} \\
&= \sum_{\mu,\nu} u^\nu\, (\Lambda^\mu{}_\nu\, \mathbf{e}'_\mu) &&= \sum_{\mu,\nu} u'^\nu\, (L^\mu{}_\nu\, \mathbf{e}_\mu), && \text{(change names)}
\end{aligned}$$

hence, by uniqueness of the decomposition in the basis $\mathcal{B}$ and $\mathcal{B}'$, respectively, we have

$$\boxed{u'^\mu = \sum_{\nu=1}^{n} \Lambda^\mu{}_\nu\, u^\nu} \qquad \text{and} \qquad u^\mu = \sum_{\nu=1}^{n} L^\mu{}_\nu\, u'^\nu. \tag{16.2}$$

Remark 16.4 These last formulas are also a **criterion for being a vector**: an n-uple, the components of which are transformed according to (16.2) during a change of basis (for example, when an observer performs a rotation) will be a vector. In what follows, saying that "we want to prove that a certain quantity is a vector" will simply mean that we wish to establish the relations (16.2).

16.1.b Einstein convention

Very quickly, the formulas for changes of coordinates become unreadable because of the increasing accumulation of indices of summation signs. This is why Einstein[2] suggested *omitting* the summation sign, which is implied each time a Greek letter appears *one time, and one time only, as a subscript* and *one time, and one time only, as a superscript*. Thus, relations (16.1) and (16.2) are expressed as follows using the Einstein convention:

$$\boxed{\mathbf{e}'_\mu = L^\nu{}_\mu\, \mathbf{e}_\nu} \qquad \text{and} \qquad \mathbf{e}_\mu = \Lambda^\nu{}_\mu\, \mathbf{e}'_\nu, \tag{16.1'}$$

$$\boxed{u'^\mu = \Lambda^\mu{}_\nu\, u^\nu} \qquad \text{and} \qquad u^\mu = L^\mu{}_\nu\, u'^\nu. \tag{16.2'}$$

Remark 16.5 One should be aware also that many books of relativity theory use *the same letter* Λ to denote both matrices Λ and L. Two solutions are used to avoid conflicts:

- In the book of S. Weinberg [92], for instance, they are distinguished by the position of indices. Thus, $\Lambda_\mu{}^\nu$ represents what we denoted $L^\nu{}_\mu$, whereas $\Lambda^\mu{}_\nu$ is the same as ours.

[2] Albert EINSTEIN (1879–1955), German – then Swiss, then American when, being Jewish, he fled Nazi Germany – physicist and mathematician, student of Hermann Minkowski, father of special relativity (1905) then general relativity (around 1915), used and popularized the tensor calculus, in particular by the introduction of riemannian geometry in physics.

- In the book of Ch. Misner, K. Thorne, and J. A. Wheeler [67], for instance, a primed superscript indicates that Λ is being understood, and a primed subscript that L is understood.

Each notation has its advantages; the one we chose is closer to "classical" linear algebra. Here is a short summary of various conventions in use, where $\bullet$ is used to indicate any tensor object:

this book	$L^{\mu}{}_{\nu}$	$\Lambda^{\mu}{}_{\nu}$	$\bullet_{\mu} = L^{\nu}{}_{\mu}\, \bullet'_{\nu}$	$\bullet'^{\mu} = \Lambda^{\mu}{}_{\nu}\, \bullet^{\nu}$
Weinberg	$\Lambda_{\nu}{}^{\mu}$	$\Lambda^{\mu}{}_{\nu}$	$\bullet_{\mu} = \Lambda_{\mu}{}^{\nu}\, \bullet'_{\nu}$	$\bullet'^{\mu} = \Lambda^{\mu}{}_{\nu}\, \bullet^{\nu}$
Wheeler	$\Lambda^{\mu}{}_{\nu'}$	$\Lambda^{\mu'}{}_{\nu}$	$\bullet_{\mu} = \Lambda^{\nu}{}_{\mu'}\, \bullet'_{\nu}$	$\bullet'^{\mu} = \Lambda^{\mu'}{}_{\nu}\, \bullet^{\nu}$

16.1.c Linear forms

We recall the following definitions:

DEFINITION 16.6 A **linear form** on $\mathscr{E}$ in any linear map from $\mathscr{E}$ to $\mathbb{K}$.

DEFINITION 16.7 The **dual** of $\mathscr{E}$, denoted $\mathscr{E}^*$, is the vector space of linear forms on $\mathscr{E}$. For $\boldsymbol{\omega} \in \mathscr{E}^*$, we denote either $\boldsymbol{\omega}(\mathbf{v})$ or $\langle \boldsymbol{\omega}, \mathbf{v} \rangle$ the effect of evaluating the linear form $\boldsymbol{\omega}$ on a vector $\mathbf{v} \in \mathscr{E}$. Similarly, we sometimes write $\langle \boldsymbol{\omega}, \cdot \rangle$ instead of $\boldsymbol{\omega}$.

Example 16.8 The **coordinate linear forms**

$$\begin{aligned} \mathrm{d}x^i : \qquad \mathbb{R}^n &\longrightarrow \mathbb{R}, \\ (h^1, \dots, h^n) &\longmapsto h^i, \end{aligned}$$

defined for $i = 1, \dots, n$, are linear forms.

For a fixed vector $\mathbf{u} \in \mathscr{E}$, the map

$$\begin{aligned} \psi : \mathscr{E}^* &\longrightarrow \mathbb{K}, \\ \boldsymbol{\omega} &\longmapsto \langle \boldsymbol{\omega}, \mathbf{u} \rangle \end{aligned}$$

is $\mathbb{K}$-linear, hence it is an element in $\mathscr{E}^{**}$. This application can only be identically zero if $\mathbf{u} = 0$. This shows that by associating it to $\mathbf{u}$, we define an injective map from $\mathscr{E}$ into $\mathscr{E}^{**}$. In general, this is not a surjection. However, it is so in two important cases: in Hilbert spaces (when considering only continuous linear forms and linear forms defined by scalar products), and when $\mathscr{E}$ is a finite-dimensional vector space:

THEOREM 16.9 (Bidual) *Let $\mathscr{E}$ be a finite-dimensional $\mathbb{K}$-vector space. Then we have $\mathscr{E} \simeq \mathscr{E}^*$,* $\dim \mathscr{E}^* = \dim \mathscr{E}$*, and moreover $\mathscr{E}$ and $\mathscr{E}^{**}$ are canonically isomorphic.*

THEOREM 16.10 (Dual basis) *Let $(\mathbf{e}_\mu)_\mu$ be a basis of $\mathscr{E}$. The family $(\boldsymbol{\alpha}^\mu)_\mu$ of elements of $\mathscr{E}^*$ is defined by its action on any basis vector $(\mathbf{e}_\mu)_\mu$ given by*

$$\forall \mu, \nu \in [\![1, n]\!] \qquad \langle \boldsymbol{\alpha}^\mu, \mathbf{e}_\nu \rangle = \delta^\mu_\nu = \begin{cases} 1 & \text{if } \mu = \nu, \\ 0 & \text{if } \mu \neq \nu. \end{cases}$$

Then the family $(\boldsymbol{\alpha}^\mu)_\mu$ *is a basis of* $\mathscr{E}^*$, *called the* ***dual basis of*** $(\mathbf{e}_\mu)_{\mu\in I}$.

The linear forms $(\boldsymbol{\alpha}^\mu)_\mu$ *are also called* ***coordinate forms,*** *and often denoted* $\boldsymbol{\alpha}^\mu = \mathrm{d}x^\mu$; *to each vector* $\mathbf{u} = u^\nu\,\mathbf{e}_\nu$, *the linear form* $\mathrm{d}x^\mu$ *associates its coordinate* u^μ:

$$\mathrm{d}x^\mu : \mathbf{u} = u^\nu\,\mathbf{e}_\nu \longmapsto u^\mu.$$

Those two definitions are of course consistent.

PROOF OF THEOREMS 16.9 AND 16.10. Let $\boldsymbol{\omega} \in \mathscr{E}^*$ be a linear form. For all $\mu \in [\![1,n]\!]$, denote $\omega_\mu = \boldsymbol{\omega}(\mathbf{e}_\mu)$. Then we have

$$\boldsymbol{\omega}(\mathbf{u}) = \boldsymbol{\omega}(u^\mu\,\mathbf{e}_\mu) = u^\mu\,\boldsymbol{\omega}(\mathbf{e}_\mu) = \omega_\mu\,\boldsymbol{\alpha}^\mu(\mathbf{u}),$$

for all $\mathbf{u} \in \mathscr{E}$, showing that $\boldsymbol{\omega} = \omega_\mu\,\boldsymbol{\alpha}^\mu$. Hence the family $(\boldsymbol{\alpha}^\mu)_\mu$ is generating. It is also free: if $\lambda_\nu\boldsymbol{\alpha}^\nu = 0$, applying this form to the basis vector $\mathbf{e}_\mu$ yields $\lambda_\mu = 0$. So it is a basis of $\mathscr{E}^*$ which consequently has the same dimension as $\mathscr{E}$.

Since any linear form can be expressed uniquely in terms of the vectors of the dual basis $(\boldsymbol{\alpha}^\mu)_\mu$, we make the following definition:

DEFINITION 16.11 (Covariant coordinates of a linear form) Let $\boldsymbol{\omega}$ be a linear form, that is, an element in $\mathscr{E}^*$. The **covariant coordinates** of the linear form $\boldsymbol{\omega}$ in the basis $(\boldsymbol{\alpha}^\mu)_\mu$ are the coefficients ω_μ appearing in its expression in this basis:

$$\boldsymbol{\omega} = \omega_\mu\,\boldsymbol{\alpha}^\mu$$

(with the Einstein convention applied).

Hence the covariant coordinates of a linear form, just like the contravariant coordinates of a vector, are its usual linear algebra coordinates.

They are computed using the following result:

THEOREM 16.12 (Expression of vectors and linear forms) *Let* $(\mathbf{e}_\mu)_\mu$ *be a basis of* E *and let* $(\boldsymbol{\alpha}^\mu)_\mu$ *be the dual basis of* $(\mathbf{e}_\mu)_\mu$. *Then, for any vector* $\mathbf{u} \in \mathscr{E}$ *and any linear form* $\boldsymbol{\omega} \in \mathscr{E}^*$, *we have the decompositions*

$$\boxed{\begin{aligned} &\boldsymbol{\omega} = \langle \boldsymbol{\omega}, \mathbf{e}_\mu\rangle\,\boldsymbol{\alpha}^\mu = \omega_\mu\,\boldsymbol{\alpha}^\mu && \text{with } \omega_\mu = \langle \boldsymbol{\omega}, \mathbf{e}_\mu\rangle \\ &\mathbf{u} = \langle \boldsymbol{\alpha}^\mu, \mathbf{u}\rangle\,\mathbf{e}_\mu = u^\mu\,\mathbf{e}_\mu && \text{with } u^\mu = \langle \boldsymbol{\alpha}^\mu, \mathbf{u}\rangle. \end{aligned}}$$

PROOF. Indeed, we have

$$\langle \boldsymbol{\alpha}^\mu, \mathbf{u}\rangle = \langle \boldsymbol{\alpha}^\mu, u^\nu\,\mathbf{e}_\nu\rangle = u^\nu\,\langle \boldsymbol{\alpha}^\mu, \mathbf{e}_\nu\rangle = u^\nu\,\delta^\mu_\nu = u^\mu.$$

Similarly

$$\langle \boldsymbol{\omega}, \mathbf{e}_\mu\rangle = \langle \omega_\nu\,\boldsymbol{\alpha}^\nu, \mathbf{e}_\mu\rangle = \omega_\nu\delta^\nu_\mu = \omega_\mu.$$

If the basis of E is changed, it is clear that the dual basis will also change. Therefore, let $\mathbf{e}'_\mu = L^\nu{}_\mu\,\mathbf{e}_\nu$, where L is an invertible matrix. We have then:

PROPOSITION 16.13 *If the basis of $\mathscr{E}$ is changed according to the matrix L, the dual basis is changed according to the inverse matrix $\Lambda = L^{-1}$, that is, if we denote $(\boldsymbol{\alpha}'^{\mu})$ the dual basis of $(\mathbf{e}'_{\mu})_{\mu} = L \cdot (\mathbf{e}_{\mu})_{\mu}$, we have*

$$\boxed{\mathbf{e}'_{\mu} = L^{\nu}{}_{\mu}\, \mathbf{e}_{\nu}} \qquad \textit{and} \qquad \boxed{\boldsymbol{\alpha}'^{\mu} = \Lambda^{\mu}{}_{\nu}\, \boldsymbol{\alpha}^{\nu}}$$

for all μ.

PROOF. We have indeed

$$\left\langle \Lambda^{\mu}{}_{\nu}\boldsymbol{\alpha}^{\nu}, \mathbf{e}'_{\rho} \right\rangle = \left\langle \Lambda^{\mu}{}_{\nu}\boldsymbol{\alpha}^{\nu}, L^{\sigma}{}_{\rho}\, \mathbf{e}_{\sigma} \right\rangle = \Lambda^{\mu}{}_{\nu}\, \delta^{\nu}_{\sigma}\, L^{\sigma}{}_{\rho} = \Lambda^{\mu}{}_{\nu}\, L^{\nu}{}_{\rho} = \delta^{\mu}_{\rho},$$

which proves that the family $(\Lambda^{\mu}{}_{\nu}\boldsymbol{\alpha}^{\nu})_{\mu}$ is the dual basis of $(\mathbf{e}'_{\mu})_{\mu}$.

It is easy now to find the change of coordinates formula for a linear form $\boldsymbol{\omega} = \omega_{\mu}\boldsymbol{\alpha}^{\mu}$, using the same technique as for vectors. We leave the two-line computation as an exercise to the reader, and state the result:

PROPOSITION 16.14 *The covariant coordinates of a linear form are changed in the same manner as the basis vectors, namely,*

$$\boxed{\omega'_{\mu} = L^{\nu}{}_{\mu}\, \omega_{\nu}.}$$

Remark 16.15 The fact that the coordinates of a linear form are transformed *in the same manner as* the vectors of the basis is the reason they are called *co*variant. Conversely, the coordinates of a vector are changed in the *inverse manner* compared to the basis vectors (i.e., with the inverse matrix Λ) and this explains why the coordinates are called *contra*variant coordinates.

Remark 16.16 Some quantities that may look like vectors actually transform like linear forms. Thus, for a differentiable function $f : \mathscr{E} \to \mathbb{R}$, the quantity ∇f transforms as a linear form. The notation for the gradient is in fact a slightly awkward. It is mathematically much more efficient to consider rather the linear form, depending on the point $\mathbf{x}$, given by

$$\mathrm{d}f_{\mathbf{x}} = \frac{\partial f}{\partial x^{\mu}}(\mathbf{x})\, \mathrm{d}x^{\mu}$$

(this is an example of a *differential form*, see the next chapter). This linear form associates the number $\mathrm{d}f_{\mathbf{x}}(\mathbf{h})$, which is equal to the scalar product $\nabla f \cdot \mathbf{h}$, to any vector $\mathbf{h} \in \mathscr{E}$. Hence, to define the vector ∇f from the linear form $\mathrm{d}f|_{\mathbf{x}}$, we must introduce this notion of scalar product (what is called a *metric*). We emphasize this again: because a metric needs to be introduced to define it, the gradient is *not* a vector.

16.1.d Linear maps

Let $\Phi : \mathscr{E} \to \mathscr{E}$ be a linear map. We denote $(\varphi^{\mu}{}_{\nu})_{\mu,\nu}$ its matrix in the basis $\mathcal{B}$. Note that *the row index is indicated by a superscript, and the column index is indicated by a subscript.* The coefficients are given by

$$\varphi^{\mu}{}_{\nu} \stackrel{\text{def}}{=} \left\langle \boldsymbol{\alpha}^{\mu}, \Phi(\mathbf{e}_{\nu}) \right\rangle,$$

and for any vector $\mathbf{x}$, we have $\Phi(\mathbf{x}) = \varphi^{\mu}{}_{\nu}\,\boldsymbol{\alpha}^{\nu}(\mathbf{x})\,\mathbf{e}_{\mu}$. Hence, under a change of basis, the coordinates of of Φ (i.e., its matrix coefficients) are transformed as follows:

$$\varphi'^{\mu}{}_{\nu} = \left\langle \boldsymbol{\alpha}'^{\mu}, \Phi(\mathbf{e}'_{\mu}) \right\rangle = \left\langle \Lambda^{\mu}{}_{\rho}\,\boldsymbol{\alpha}^{\rho}, \Phi(L^{\sigma}{}_{\nu}\,\mathbf{e}_{\sigma}) \right\rangle = \Lambda^{\mu}{}_{\rho}\,L^{\sigma}{}_{\nu}\,\varphi^{\rho}{}_{\sigma}.$$

PROPOSITION 16.17 (Change of coordinates) *The coefficients of the matrix of a linear map are transformed, with respect to the superscript, as the contravariant coordinates of a vector, and with respect to the subscript, as the covariant coordinates of a linear form.*

Thus, if a map is representated by the matrix φ in the basis $\mathcal{B}$ and by φ' in the basis $\mathcal{B}'$, we have

$$\boxed{\varphi'^{\mu}{}_{\nu} = \Lambda^{\mu}{}_{\rho}\,L^{\sigma}{}_{\nu}\,\varphi^{\rho}{}_{\sigma}.}$$

This formula should be compared with the formula (c.2), on page 596.

♦ **Exercise 16.1** Show that the formulas

$$\Phi(\mathbf{x}) = \varphi'^{\mu}{}_{\nu}\,\boldsymbol{\alpha}'^{\nu}(\mathbf{x})\,\mathbf{e}'_{\mu} \qquad \text{and} \qquad \Phi(\mathbf{x}) = \varphi^{\mu}{}_{\nu}\,\boldsymbol{\alpha}^{\nu}(\mathbf{x})\,\mathbf{e}_{\mu}$$

are consistent. *(Solution page 462).*

16.1.e Lorentz transformations

As an illustration, we give here the matrices L and Λ corresponding to the Lorentz "boost": the frame of $(\mathbf{e}'_{\mu})$ has, relative to the frame of $(\mathbf{e}_{\mu})$, a velocity $\boldsymbol{v} = v\,\mathbf{e}_{x}$. Denoting $\beta = v/c$ and $\gamma = (1-\beta^2)^{-1/2}$, we have

$$\Lambda = (\Lambda^{\mu}{}_{\nu}) = \begin{pmatrix} \gamma & -\beta\gamma & 0 & 0 \\ -\beta\gamma & \gamma & 0 & 0 \\ 0 & 0 & 1 & 0 \\ 0 & 0 & 0 & 1 \end{pmatrix} \qquad L = (L^{\nu}{}_{\mu}) = \begin{pmatrix} \gamma & \beta\gamma & 0 & 0 \\ \beta\gamma & \gamma & 0 & 0 \\ 0 & 0 & 1 & 0 \\ 0 & 0 & 0 & 1 \end{pmatrix}.$$

16.2 Tensor product of vector spaces: tensors

16.2.a Existence of the tensor product of two vector spaces

Note: This section may be skipped on a first reading; it contains the statement and proof of the existence theorem for the tensor space of two vector spaces.

THEOREM 16.18 *Let E and F be two finite-dimensional vector spaces. There exists a vector space $E \otimes F$ such that, for any vector space G, the space of **linear** maps from $E \otimes F$ to G is isomorphic to the space of **bilinear** maps from $E \times F$ to G, that is,*

$$\mathscr{L}(E \otimes F, G) \simeq \mathcal{B}il(E \times F, G).$$

More precisely, there exists a bilinear map

$$\varphi : E \times F \longrightarrow E \otimes F$$

such that, for any vector space G and any bilinear map f from $E \times F$ to G, there exists a unique linear map f^* from $E \otimes F$ into G such that $f = f^* \circ \varphi$, which is summarized by the following diagram:

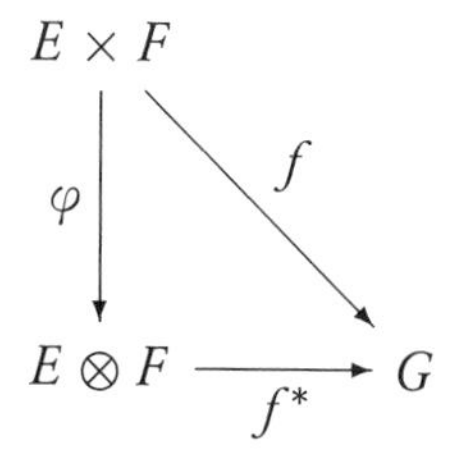

*The space $E \otimes F$ is called the **tensor product** of E and F. It is only unique up to (unique) isomorphism.*

*The bilinear map φ from $E \times F$ to $E \otimes F$ is denoted "$\otimes$" and also called the **tensor product**. It is such that if $(\mathbf{e}_i)_i$ and $(\mathbf{f}_j)_j$ are basis of E and F, respectively, the family $(\mathbf{e}_i \otimes \mathbf{f}_j)_{i,j}$ is a basis of $E \otimes F$.*

All this means that the tensor product "linearizes what was bilinear," and also that "to any bilinear map on $E \times F$, one can associate a unique linear map on the space $E \otimes F$."

PROOF. See Appendix D, page 604.

It is important to remember that the tensor product of two vector spaces is not defined uniquely, but always up to isomorphism. So there isn't a *single* tensor product, but rather infinitely many, which are all isomorphic.

♦ **Exercise 16.2** Let $n, p \in \mathbb{N}$. For $A = (a_{ij}) \in \mathfrak{M}_n(\mathbb{C})$ and $B \in \mathfrak{M}_p(\mathbb{C})$, denote by $A \otimes B$ the matrix of size np which is defined by blocks as follows:

$$A \otimes B = \begin{pmatrix} a_{11}B & \cdots & a_{1n}B \\ \vdots & & \vdots \\ a_{n1}B & \cdots & a_{nn}B \end{pmatrix}.$$

Show that the map $(A, B) \mapsto A \otimes B$ gives an isomorphism

$$\mathfrak{M}_n(\mathbb{C}) \otimes \mathfrak{M}_p(\mathbb{C}) \simeq \mathfrak{M}_{(np)}(\mathbb{C}).$$

In particular, deduce that $(E_{ij} \otimes E_{kl})_{i,j,k,l}$ is a basis of $\mathfrak{M}_n(\mathbb{C}) \otimes \mathfrak{M}_p(\mathbb{C})$. Show that there exist matrices in $\mathfrak{M}_{(np)}(\mathbb{C})$ which are not of the form $A \otimes B$ with $A \in \mathfrak{M}_n(\mathbb{C})$ and $B \in \mathfrak{M}_p(\mathbb{C})$.

(Solution page 462).

THEOREM 16.19 *If E and F are finite-dimensional vector spaces, we have*

$$\dim E \otimes F = \dim E \times \dim F.$$

PROPOSITION 16.20 *In the particular case $G = \mathbb{K}$, Theorem 16.18 gives*

$$\mathscr{L}(E \otimes F, \mathbb{K}) \simeq \mathcal{B}il(E \times F, \mathbb{K}).$$

For E and F finite-dimensional vector spaces, we have

$$E^* \otimes F^* \simeq \mathcal{B}il(E \times F, \mathbb{K}),$$

that is, the tensor product $E^* \otimes F^*$ is isomorphic to the space of bilinear forms on $E \times F$ (see Section 16.2.b).

16.2.b Tensor product of linear forms: tensors of type $\binom{0}{2}$

The abstract construction in the previous section is made more concrete when considering the tensor product of $\mathscr{E}^*$ and $\mathscr{F}^*$, where $\mathscr{E}$ and $\mathscr{F}$ are two finite-dimensional vector spaces.

DEFINITION 16.21 Let $\boldsymbol{\omega} \in \mathscr{E}^*$ and $\boldsymbol{\sigma} \in \mathscr{F}^*$ be two linear forms on $\mathscr{E}$ and $\mathscr{F}$, respectively. The **tensor product of $\boldsymbol{\omega}$ and $\boldsymbol{\sigma}$**, denoted $\boldsymbol{\omega} \otimes \boldsymbol{\sigma}$, is the bilinear form on $\mathscr{E} \times \mathscr{F}$ defined by

$$\begin{aligned} \boldsymbol{\omega} \otimes \boldsymbol{\sigma} : \mathscr{E} \times \mathscr{F} &\longrightarrow \mathbb{K}, \\ (\mathbf{u}, \mathbf{v}) &\longmapsto \boldsymbol{\omega}(\mathbf{u})\, \boldsymbol{\sigma}(\mathbf{v}). \end{aligned}$$

If $(\boldsymbol{\alpha}^{\mu})_{\mu}$ and $(\boldsymbol{\beta}^{\nu})_{\nu}$ are bases of $\mathscr{E}^*$ and $\mathscr{F}^*$, respectively, the family

$$(\boldsymbol{\alpha}^{\mu} \otimes \boldsymbol{\beta}^{\nu})_{\mu\nu}$$

is a basis of the vector space of bilinear forms on $\mathscr{E} \times \mathscr{F}$. In other words, any element of $\mathcal{B}il(\mathscr{E} \times \mathscr{F}, \mathbb{K})$ is a linear combination of the tensor products $\boldsymbol{\alpha}^{\mu} \otimes \boldsymbol{\beta}^{\nu}$. The space of bilinear forms on $\mathscr{E} \times \mathscr{F}$ is a realization of the tensor product $\mathscr{E}^* \otimes \mathscr{F}^*$ defined in the previous section. This is denoted

$$\boxed{\mathcal{B}il(\mathscr{E} \times \mathscr{F}, \mathbb{K}) \simeq \mathscr{E}^* \otimes \mathscr{F}^*,}$$

and in general, with a slight abuse of notation, we identify those two spaces, meaning that we do not hesitate to write an equal sign between the two.

Remark 16.22 Although it is perfectly possible to deal with the general case where $\mathscr{E}$ are $\mathscr{F}$ are distinct spaces, we will henceforth assume that $\mathscr{E} = \mathscr{F}$, which is the most common situation. Generalizing the formulas below causes no problem.

DEFINITION 16.23 A **tensor of type $\binom{0}{2}$** or **$\binom{0}{2}$-tensor** is any element in $\mathscr{E}^* \otimes \mathscr{E}^*$, in other words any bilinear form on $\mathscr{E} \times \mathscr{E}$.

Coordinates of a $\binom{0}{2}$-tensor

THEOREM 16.24 *Let $\mathfrak{T} \in \mathscr{E}^* \otimes \mathscr{E}^*$ be a $\binom{0}{2}$-tensor. It can be expressed in terms of the basis $(\boldsymbol{\alpha}^\mu \otimes \boldsymbol{\alpha}^\nu)_{\mu\nu}$ of $\mathscr{E}^* \otimes \mathscr{E}^*$ by*

$$\mathfrak{T} = T_{\mu\nu}\, \boldsymbol{\alpha}^\mu \otimes \boldsymbol{\alpha}^\nu,$$

with

$$T_{\mu\nu} \stackrel{\text{def}}{=} \mathfrak{T}(\mathbf{e}_\mu, \mathbf{e}_\nu).$$

Change of coordinates of a $\binom{0}{2}$-tensor

The coordinates of a $\binom{0}{2}$-tensor are transformed according to the following theorem during a change of basis:

THEOREM 16.25 *Let $\mathfrak{T} \in \mathscr{E}^* \otimes \mathscr{E}^*$ be a tensor given by $\mathfrak{T} = T_{\mu\nu}\, \boldsymbol{\alpha}^\mu \otimes \boldsymbol{\alpha}^\nu$. During a basis change, the coordinates $T_{\mu\nu}$ are transformed according to the rule*

$$\boxed{T'_{\mu\nu} = L^\rho{}_\mu L^\sigma{}_\nu\, T_{\rho\sigma}.}$$

In other words, each *co*variant index of $\mathfrak{T}$ is transformed using the matrix L.

Another viewpoint for a $\binom{0}{2}$-tensor

Since a $\binom{0}{2}$-tensor is (via the isomorphism in Theorem 16.20) nothing but a bilinear form on $\mathscr{E} \times \mathscr{E}$, it is a "machine that takes two vectors in $\mathscr{E}$ and gives back a number":

$$\mathfrak{T}(\text{vector}, \text{vector}) = \text{number}.$$

Let's continue with this mechanical analogy. If we only put *a single* vector in the machine, what happens? Then the machine expects another vector (which we did not give) in order to produce a number. But a machine that expects a vector to output a number is simply a linear form. Thus, once our tensor is "fed" with *one* vector, it has been transformed into a linear form. Precisely, if $\mathbf{u} \in \mathscr{E}$, the map

$$\mathfrak{T}(\mathbf{u}, \cdot) : \mathbf{v} \longmapsto \mathfrak{T}(\mathbf{u}, \mathbf{v})$$

is indeed a linear form on $\mathscr{E}$.

This means that the tensor $\mathfrak{T} \in \mathscr{E}^* \otimes \mathscr{E}^*$ may also be seen as a machine that turns a vector into a linear form:

$$\begin{aligned} \mathfrak{T}(\cdot, \cdot) : \mathscr{E} &\longrightarrow \mathscr{E}^*, \\ \mathbf{u} &\longmapsto \mathfrak{T}(\mathbf{u}, \cdot) \end{aligned}$$

which does indeed take a vector (denoted $\mathbf{u}$) as input and "turns" it to the linear form $\mathfrak{T}(\mathbf{u}, \cdot)$.

In mathematical terminology, showing that a bilinear form on $\mathscr{E} \times \mathscr{E}$ (a $\binom{0}{2}$-tensor), is "the same thing as" a linear map from $\mathscr{E}$ (the space of vectors) to $\mathscr{E}^*$ (the space of linear forms) is denoted

$$\mathcal{B}il(\mathscr{E} \times \mathscr{E}, \mathbb{R}) \simeq \mathscr{E}^* \otimes \mathscr{E}^* \simeq \mathscr{L}(\mathscr{E}, \mathscr{E}^*).$$

This property is clearly visible in the notation used for the coordinates of the objects involved: if we write $\mathfrak{T} = T_{\mu\nu}\, \boldsymbol{\alpha}^\mu \otimes \boldsymbol{\alpha}^\nu$, then we have

$$\begin{aligned}\mathfrak{T}(\mathbf{u}, \cdot) &= T_{\mu\nu}\, \boldsymbol{\alpha}^\mu \otimes \boldsymbol{\alpha}^\nu(\mathbf{u}, \cdot) \\ &= T_{\mu\nu}\, \langle \boldsymbol{\alpha}^\mu, \mathbf{u} \rangle\, \boldsymbol{\alpha}^\nu \qquad (= T_{\mu\nu}\, \langle \boldsymbol{\alpha}^\mu, \mathbf{u} \rangle\, \langle \boldsymbol{\alpha}^\nu, \cdot \rangle) \\ &= T_{\mu\nu}\, u^\mu\, \boldsymbol{\alpha}^\nu,\end{aligned}$$

which is a linear combination of the linear forms $\boldsymbol{\alpha}^\nu$. The reader can check carefully that those notations precisely describe the isomorphisms given above, and that all this works *with no effort whatsoever.*

A particularly important instance of bilinear forms is the *metric,* and it deserves its own section (see Section 16.3).

16.2.c Tensor product of vectors: tensors of type $\binom{2}{0}$

DEFINITION 16.26 Let $\mathbf{u} \in \mathscr{E}$ and $\mathbf{v} \in \mathscr{E}$ be two vectors. The **tensor product of u and v**, denoted $\mathbf{u} \otimes \mathbf{v}$, is the bilinear form on $\mathscr{E}^* \times \mathscr{E}^*$ defined by

$$\begin{aligned}\mathbf{u} \otimes \mathbf{v} : \mathscr{E}^* \times \mathscr{E}^* &\longrightarrow \mathbb{K}, \\ (\boldsymbol{\omega}, \boldsymbol{\sigma}) &\longmapsto \boldsymbol{\omega}(\mathbf{u})\, \boldsymbol{\sigma}(\mathbf{v}).\end{aligned}$$

Any element in $\mathcal{B}il(\mathscr{E}^* \times \mathscr{E}^*, \mathbb{K})$ is a linear combination of the family $(\mathbf{e}_\mu \otimes \mathbf{e}_\nu)_{\mu\nu}$.

We denote $\mathscr{E} \otimes \mathscr{E} = \mathcal{B}il(\mathscr{E}^* \times \mathscr{E}^*, \mathbb{K})$, and any element of $\mathscr{E} \otimes \mathscr{E}$ is called a **tensor of type** $\binom{2}{0}$ or a $\binom{2}{0}$**-tensor**.

Coordinates of a $\binom{2}{0}$-tensor

The following theorem describes the coordinates of a $\binom{2}{0}$-tensor:

THEOREM 16.27 *Let $\mathfrak{T} \in \mathscr{E} \otimes \mathscr{E}$ be a tensor. Then $\mathfrak{T}$ may be expressed in terms of the basis $(\mathbf{e}_\mu \otimes \mathbf{e}_\nu)_{\mu\nu}$ of $\mathscr{E} \otimes \mathscr{E}$ as*

$$\mathfrak{T} = T^{\mu\nu}\, \mathbf{e}_\mu \otimes \mathbf{e}_\nu,$$

with

$$T^{\mu\nu} \stackrel{\text{def}}{=} \mathfrak{T}(\boldsymbol{\alpha}^\mu, \boldsymbol{\alpha}^\nu).$$

Change of coordinates of a $\binom{2}{0}$-tensor

THEOREM 16.28 *Let $\mathfrak{T} \in \mathscr{E} \otimes \mathscr{E}$ be a tensor given by $\mathfrak{T} = T^{\mu\nu}\,\mathbf{e}_\mu \otimes \mathbf{e}_\nu$. During a basis change, the coordinates $T^{\mu\nu}$ are transformed according to the rule:*

$$\boxed{T'^{\mu\nu} = \Lambda^\mu{}_\rho \Lambda^\nu{}_\sigma\, T^{\rho\sigma}.}$$

In other words, each *contra*variant index of $\mathfrak{T}$ is transformed using the matrix Λ.

16.2.d Tensor product of a vector and a linear form: linear maps or $\binom{1}{1}$-tensors

A special case of tensor product is the case of $\mathscr{E} \otimes \mathscr{E}^*$, since the previous results allow us to write down isomorphisms

$$\mathscr{E} \otimes \mathscr{E}^* \simeq \mathcal{B}il(\mathscr{E}^* \times \mathscr{E}, \mathbb{K}) \simeq \mathscr{L}(\mathscr{E}^*, \mathscr{E}^*) \simeq \mathscr{L}(\mathscr{E}, \mathscr{E}).$$

Therefore, it should be possible to see the tensor product of a vector and a linear form as a linear map from $\mathscr{E}$ into itself!

DEFINITION 16.29 Let $\mathbf{u} \in \mathscr{E}$ and $\boldsymbol{\omega} \in \mathscr{E}^*$. The **tensor product of the vector u and the form $\boldsymbol{\omega}$** is the bilinear form on $\mathscr{E}^* \times \mathscr{E}$ defined by

$$\begin{aligned} \mathbf{u} \otimes \boldsymbol{\omega} : \mathscr{E}^* \times \mathscr{E} &\longrightarrow \mathbb{K}, \\ (\boldsymbol{\sigma}, \mathbf{v}) &\longmapsto \langle \boldsymbol{\sigma}, \mathbf{u} \rangle \langle \boldsymbol{\omega}, \mathbf{v} \rangle . \end{aligned}$$

Any bilinear form on $\mathscr{E}^* \times \mathscr{E}$ may be expressed as linear combination of the family $(\mathbf{e}_\mu \otimes \boldsymbol{\alpha}^\nu)_{\mu\nu}$. We write $\mathscr{E} \otimes \mathscr{E}^* = \mathcal{B}il(\mathscr{E}^* \times \mathscr{E}, \mathbb{K})$ and any element of $\mathscr{E} \otimes \mathscr{E}^*$ is called a **tensor of type $\binom{1}{1}$** or **$\binom{1}{1}$-tensor.**

Now we can spell out concretely the statement above, namely that:

$$\mathscr{E} \otimes \mathscr{E}^* \simeq \mathscr{L}(\mathscr{E}, \mathscr{E}),$$

which gives an identification of a $\binom{1}{1}$-tensor and an endomorphism of $\mathscr{E}$.

Let $\mathfrak{T} = T^\mu_\nu\, \mathbf{e}_\mu \otimes \boldsymbol{\alpha}^\nu$ be such a tensor; then we can construct an associated linear map from $\mathscr{E}$ to $\mathscr{E}$ by

$$\mathbf{v} \longmapsto T^\mu_\nu\, \boldsymbol{\alpha}^\nu(\mathbf{v})\, \mathbf{e}_\mu.$$

Conversely, a linear map may be seen as a $\binom{1}{1}$-tensor:

PROPOSITION 16.30 *Let $\Phi : \mathscr{E} \to \mathscr{E}$ be a linear map. Denote*

$$\varphi^\mu_\nu \stackrel{\text{def}}{=} \left\langle \boldsymbol{\alpha}^\mu, \Phi(\mathbf{e}_\nu) \right\rangle$$

A student of Jacobi and then of Dirichlet in Berlin, Leopold KRONECKER (1823–1891) worked in finance for ten years starting when he was 21. Thus enriched, he retired from business and dedicated himself to mathematics. His interests ranged from Galois theory (e.g., giving a simple proof that the general equation of degree at least 5 cannot be solved with radicals), to elliptic functions, to polynomial algebra. "God created integers, all the rest is the work of Man" is his most famous quote. His constructive and finitist viewpoints, opposed to those of Cantor, for instance, made him pass for a reactionary. Ironically, with the advent of computers and algorithmic questions, those ideas can now be seen as among the most modern and lively.

its matrix coefficients in the basis $\mathcal{B} = (\mathbf{e}_\mu)_\mu$. If $(\boldsymbol{\alpha}^\mu)_\mu$ is the dual basis to $\mathcal{B}$, then the map Φ can also be seen as a bilinear form on $\mathscr{E}^* \times \mathscr{E}$ given by

$$\Phi = \varphi^\mu{}_\nu\, \mathbf{e}_\mu \otimes \boldsymbol{\alpha}^\nu.$$

PROOF. Indeed, we can show that $\varphi^\mu{}_\nu\, \mathbf{e}_\mu \otimes \boldsymbol{\alpha}^\nu$ is equal to Φ by checking that it agrees with Φ when acting on a basis vector $\mathbf{e}_\lambda$. Since

$$\begin{aligned}\varphi^\mu{}_\nu \mathbf{e}_\mu \otimes \boldsymbol{\alpha}^\nu(\mathbf{e}_\lambda) &= \langle \boldsymbol{\alpha}^\mu, \Phi(\mathbf{e}_\nu)\rangle\, \mathbf{e}_\mu \otimes \boldsymbol{\alpha}^\nu(\mathbf{e}_\lambda) = \langle \boldsymbol{\alpha}^\mu, \Phi(\mathbf{e}_\nu)\rangle\, \mathbf{e}_\mu\, \langle \boldsymbol{\alpha}^\nu, \mathbf{e}_\lambda\rangle \\ &= \langle \boldsymbol{\alpha}^\mu, \Phi(\mathbf{e}_\nu)\rangle\, \mathbf{e}_\mu\, \delta^\nu_\lambda = \langle \boldsymbol{\alpha}^\mu, \Phi(\mathbf{e}_\lambda)\rangle\, \mathbf{e}_\mu = \Phi(\mathbf{e}_\lambda),\end{aligned}$$

this is the case, and by linearity it follows that $\Phi = \varphi^\mu{}_\nu\, \mathbf{e}_\mu \otimes \boldsymbol{\alpha}^\nu$.

DEFINITION 16.31 A **tensor of type $\binom{1}{1}$** or **$\binom{1}{1}$-tensor** is any element of $\mathscr{E} \otimes \mathscr{E}^*$ or, equivalently, any linear map from $\mathscr{E}$ to itself.

♦ **Exercise 16.3** With the notation above, show that $\mathscr{E} \otimes \mathscr{E}^* \simeq \mathscr{L}(\mathscr{E}^*, \mathscr{E}^*)$.

Notice, morevoer, that the coordinates of the identity map are the same in any basis:

THEOREM 16.32 *The coordinates of the identity map in any basis $(\mathbf{e}_\mu)_\mu$ are given by the **Kronecker symbol***

$$\delta^\mu_\nu \stackrel{\text{def}}{=} \begin{cases} 1 & \text{if } \mu = \nu, \\ 0 & \text{if } \mu \neq \nu. \end{cases}$$

Thus it is possible to write $\mathrm{Id}^\mu_\nu = \delta^\mu_\nu$ and $\mathrm{Id} = \mathbf{e}_\mu \otimes \boldsymbol{\alpha}^\mu = \mathbf{e}_\mu \otimes \mathrm{d}x^\mu$, independently of the chosen basis.

Remark 16.33 The formula "$\mathrm{Id} = \mathbf{e}_\mu \otimes \boldsymbol{\alpha}^\mu = \mathbf{e}_\mu \otimes \mathrm{d}x^\mu$" should be compared with the closure relation given in Remark 9.26, page 260, namely $\mathrm{Id} = \sum |e_n\rangle \langle e_n|$. The tensor product is implicit there, and the basis $(\langle e_n|)$ is the dual basis of $(|e_n\rangle)$ if the latter is orthonormal.

Change of coordinates of a $\binom{1}{1}$-tensor

The coordinates of a $\binom{1}{1}$-tensor are neither more nor less than the coefficients of the matrix of the associated linear map. Therefore, Proposition 16.17 describes the law that governs the change of coordinates of a $\binom{1}{1}$-tensor.

THEOREM 16.34 *Let $\mathfrak{T} \in \mathscr{E} \otimes \mathscr{E}^*$ be a tensor given by $\mathfrak{T} = T^{\mu}_{\ \nu}\, \mathbf{e}_{\mu} \otimes \boldsymbol{\alpha}^{\nu}$. During a basis change, the coordinates $T^{\mu}_{\ \nu}$ are transformed according to the rule*

$$\boxed{T'^{\mu}_{\ \nu} = \Lambda^{\mu}_{\ \rho}\, L^{\sigma}_{\ \nu}\, T^{\rho}_{\ \sigma}.}$$

Remark 16.35 In other words, each *contra*variant index of $\mathfrak{T}$ is transformed using the matrix Λ and each *co*variant index is transformed using the matrix L. Compare this formula with the change of basis formula (equation (c.2), page 596) in linear algebra.

16.2.e Tensors of type $\binom{p}{q}$

Of course, it is possible to continue the tensor process an arbitrary number of times, either with $\mathscr{E}$ as a factor or with $\mathscr{E}^*$. One has to be careful with the order of the factors, since the tensor product is not commutative, strictly speaking. However, there are canonical isomorphisms $\mathscr{E} \otimes \mathscr{E}^* \simeq \mathscr{E}^* \otimes \mathscr{E}$ (i.e., isomorphisms which are independent of the choice of a basis) which justify some simplification, except that one must be quite careful when performing "index gymnastics," as described in the next section. It suffices therefore, when dealing with higher-order tensors, to consider those of the type

$$\overbrace{\mathscr{E} \otimes \mathscr{E} \otimes \cdots \otimes \mathscr{E}}^{p \text{ times}} \otimes \overbrace{\mathscr{E}^* \otimes \mathscr{E}^* \otimes \cdots \otimes \mathscr{E}^*}^{q \text{ times}},$$

where the order of the factors is irrelevant.

DEFINITION 16.36 We denote $\mathscr{E}^{\otimes p} \stackrel{\text{def}}{=} \mathscr{E} \otimes \mathscr{E} \otimes \cdots \otimes \mathscr{E}$, where the tensor product is iterated p times, and similarly $\mathscr{E}^{*\otimes q} = \mathscr{E}^* \otimes \ldots \otimes \mathscr{E}^*$ (q times).

DEFINITION 16.37 A **tensor of type $\binom{p}{q}$** or **$\binom{p}{q}$-tensor,** is an element of the space $\mathscr{E}^{\otimes p} \otimes \mathscr{E}^{*\otimes q}$; in other words, it is a multilinear form on the space $\mathscr{E}^* \times \cdots \times \mathscr{E}^* \times \mathscr{E} \times \cdots \times \mathscr{E}$.

Coordinates of a $\binom{p}{q}$-tensor

THEOREM 16.38 *Let $\mathfrak{T}$ be a $\binom{p}{q}$-tensor. Then $\mathfrak{T}$ may be expressed in terms of the basis $\left(\mathbf{e}_{\mu_1} \otimes \cdots \otimes \mathbf{e}_{\mu_p} \otimes \boldsymbol{\alpha}^{\nu_1} \otimes \cdots \otimes \boldsymbol{\alpha}^{\nu_q}\right)_{\substack{\mu_1 \cdots \mu_p \\ \nu_1 \cdots \nu_q}}$ of the space $\mathscr{E}^{\otimes p} \otimes \mathscr{E}^{*\otimes q}$:*

$$\mathfrak{T} = T^{\mu_1 \cdots \mu_p}_{\nu_1 \cdots \nu_q}\, \mathbf{e}_{\mu_1} \otimes \cdots \otimes \mathbf{e}_{\mu_p} \otimes \boldsymbol{\alpha}^{\nu_1} \otimes \cdots \otimes \boldsymbol{\alpha}^{\nu_q}$$

with $$T^{\mu_1\cdots\mu_p}_{\nu_1\cdots\nu_q} \stackrel{\text{def}}{=} \mathfrak{T}(\alpha^{\mu_1},\ldots,\alpha^{\mu_p},\mathbf{e}_{\nu_1},\ldots,\mathbf{e}_{\nu_q}).$$

Change of coordinates of a $\binom{p}{q}$-tensor

THEOREM 16.39 *Let* $\mathfrak{T} = T^{\mu_1\cdots\mu_p}_{\nu_1\cdots\nu_q}\,\mathbf{e}_{\mu_1}\otimes\cdots\otimes\mathbf{e}_{\mu_p}\otimes\alpha^{\nu_1}\otimes\cdots\otimes\alpha^{\nu_q}$ *be a* $\binom{p}{q}$-*tensor. During a basis change* $\mathcal{B}' = L\mathcal{B}$ *the coordinates* $T^{\mu_1\cdots\mu_p}_{\nu_1\cdots\nu_q}$ *are transformed according to the rule*

$$T'^{\mu_1\cdots\mu_p}_{\nu_1\cdots\nu_q} = \Lambda^{\mu_1}_{\ \rho_1}\cdots\Lambda^{\mu_p}_{\ \rho_p}\,L^{\sigma_1}_{\ \nu_1}\cdots L^{\sigma_q}_{\ \nu_q}\,T^{\rho_1\cdots\rho_p}_{\sigma_1\cdots\sigma_q}.$$

For this reason, a $\binom{p}{q}$-tensor is said to be ***p*** **times covariant and** ***q*** **times contravariant.**

16.3 The metric, *or*: how to raise and lower indices

16.3.a Metric and pseudo-metric

DEFINITION 16.40 A **metric** (or **scalar product**) on a finite-dimensional *real* vector space $\mathscr{E}$ ($\mathbb{K} = \mathbb{R}$) is a *symmetric, positive definite* bilinear form; in other words it is a bilinear map

$$\begin{aligned} \mathfrak{g}:\quad \mathscr{E}^2 &\longrightarrow \mathbb{R},\\ (\mathbf{u},\mathbf{v}) &\longmapsto \mathfrak{g}(\mathbf{u},\mathbf{v}), \end{aligned}$$

such that $\mathfrak{g}(\mathbf{u},\mathbf{v}) = \mathfrak{g}(\mathbf{v},\mathbf{u})$ for any $\mathbf{u},\mathbf{v}\in\mathscr{E}$, and such that $\mathfrak{g}(\mathbf{u},\mathbf{u}) > 0$ for any nonzero vector $\mathbf{u}$.

A **pseudo-metric** on a finite-dimensional *real* vector space $\mathscr{E}$ is a *symmetric definite* bilinear form (i.e., the only vector $\mathbf{u}$ for which we have $\mathfrak{g}(\mathbf{u},\mathbf{v}) = 0$ for all $\mathbf{v}\in\mathscr{E}$ is the zero vector); however, $\mathfrak{g}(\mathbf{u},\mathbf{u})$ itself may be of arbitrary sign (and even zero, for a so-called light-like vector in special relativity).

Remark 16.41 A metric (resp. pseudo-metric) $\mathfrak{g}$ can be seen, with the isomorphisms described in the previous sections, as a $\binom{0}{2}$-tensor.

DEFINITION 16.42 (Coefficients $g_{\mu\nu}$) Let $\mathfrak{g}$ be a metric on a real vector space $\mathscr{E}$ of dimension n. Let $\mathcal{B} = (\mathbf{e}_1,\ldots,\mathbf{e}_n)$ be a basis of $\mathscr{E}$ and $(\alpha^1,\ldots,\alpha^n)$ its dual basis. The coordinates of $\mathfrak{g}$ in the basis $(\alpha^\mu\otimes\alpha^\nu)_{\mu\nu}$ of $\mathscr{E}^*\otimes\mathscr{E}^*$ are

Hermann MINKOWSKI (1864–1909), German mathematician, was born in Lithuania and spent his childhood in Russia. Since tsarist laws prevented Jews from obtaining a decent education, his parents emigrated to Germany. One of his brothers became a great physician, while he himself turned to mathematics. He taught in Bonn, in Königsberg, and then at the Zürich polytechnic school, where Albert EINSTEIN was his student. His mathematical works concern number theory and mathematic physics. He gave a geometric interpretation of Einstein's special relativity, which is the most natural framework, and his formalism, generalizing the notion of scalar product and distance in space-time is still used today.

denoted $g_{\mu\nu}$. In other words, we have

$$\boxed{\mathfrak{g} = g_{\mu\nu}\,\mathrm{d}x^{\mu} \otimes \mathrm{d}x^{\nu}.}$$

Since $\mathfrak{g}$ is symmetric, it follows that $g_{\mu\nu} = g_{\nu\mu}$ for all $\mu, \nu \in [\![1, n]\!]$.

Example 16.43 The most important example of a metric is the **euclidean metric** which, in the canonical basis $(\mathbf{e}_{\mu})_{\mu}$ of $\mathscr{E}$, is given by coordinates

$$g_{\mu\nu} = \begin{cases} 1 & \text{if } \mu = \nu, \\ 0 & \text{if } \mu \neq \nu. \end{cases}$$

DEFINITION 16.44 A basis $\mathcal{B} = (\mathbf{e}_1, \dots, \mathbf{e}_n)$ is **orthonormal** if

$$\mathfrak{g}(\mathbf{e}_{\mu}, \mathbf{e}_{\nu}) = \delta_{\mu\nu} \qquad \text{for any } \mu, \nu.$$

There can only be an orthonormal basis if the space is carrying a "true" metric, and not a pseudo-metric (since the coordinates above describe the euclidean metric in $\mathcal{B}$).

DEFINITION 16.45 (Minkowski pseudo-metric) The **Minkowski space** is the space $\mathbb{R}^4$ with the pseudo-metric defined by

$$g_{\mu\nu} = \eta_{\mu\nu} \stackrel{\text{def}}{=} \begin{cases} 1 & \text{if } \mu = \nu = 0, \\ -1 & \text{if } \mu = \nu = 1, 2, \text{ or } 3, \\ 0 & \text{if } \mu \neq \nu. \end{cases} \tag{16.3}$$

The space $\mathbb{R}^4$ with this pseudo-metric is denote $\mathcal{M}^{1,3}$. This metric was introduced by H. Minkowski as the natural setting for the description of space-time in einsteinian relativity.

DEFINITION 16.46 (Scalar product) The **scalar product** of the vectors $\mathbf{u}$ and $\mathbf{v}$ in $\mathscr{E}$ is the quantity

$$\mathbf{u} \cdot \mathbf{v} \stackrel{\text{def}}{=} \mathrm{g}(\mathbf{u}, \mathbf{v}).$$

If g is a pseudo-metric, we will also speak of **pseudo-scalar product**. It should be noticed however that *most books of relativity theory do not make an explicit distinction*; it is also customary to speak of **minkowskian scalar product**, or more simply of **minkowskian product**.

PROPOSITION 16.47 *The coordinates of the metric* g *are given by the scalar products (defined by* g*) of the basis vectors*

$$\boxed{g_{\mu\nu} = \mathbf{e}_\mu \cdot \mathbf{e}_\nu.}$$

PROOF. Indeed, we have $\mathbf{e}_\mu \cdot \mathbf{e}_\nu = \mathrm{g}(\mathbf{e}_\mu, \mathbf{e}_\nu) = g_{\rho\sigma}\, \boldsymbol{\alpha}^\rho \otimes \boldsymbol{\alpha}^\sigma (\mathbf{e}_\mu, \mathbf{e}_\nu)$

$$= g_{\rho\sigma} \left\langle \boldsymbol{\alpha}^\rho, \mathbf{e}_\mu \right\rangle \left\langle \boldsymbol{\alpha}^\sigma, \mathbf{e}_\nu \right\rangle = g_{\rho\sigma}\, \delta^\rho_\mu\, \delta^\sigma_\nu = g_{\mu\nu}.$$

16.3.b Natural duality by means of the metric

Once a basis in $\mathscr{E}$ is chosen, we can exhibit an isomorphism between $\mathscr{E}$ and $\mathscr{E}^*$, which associates to a given basis vector $\mathbf{e}_\mu$ the dual vector $\boldsymbol{\alpha}^\mu$, and which, therefore, associates to an arbitrary vector $\mathbf{v} = v^\mu\, \mathbf{e}_\mu$ the linear form

$$\omega = \sum_\mu v^\mu\, \boldsymbol{\alpha}^\mu.$$

However, this isomorphism is not really interesting, because it depends on the basis that has been chosen[3] (the terminology used is that the isomorphism is *noncanonical*.)

Using the metric tensor g, it is possible to introduce a duality of somewhat different nature (a **metric duality**) between $\mathscr{E}$ and $\mathscr{E}^*$ which *will be canonical* (i.e., it does not depend on the choice of a basis[4]).

Indeed, if $\mathbf{u} \in \mathscr{E}$ is a vector, we can introduce a linear form $\widetilde{\mathbf{u}} \in \mathscr{E}^*$ on $\mathscr{E}$ using the bilinear form g: it is given by

$$\begin{aligned} \widetilde{\mathbf{u}} : \mathscr{E} &\longrightarrow \mathbb{R}, \\ \mathbf{v} &\longmapsto \mathrm{g}(\mathbf{u}, \mathbf{v}) = \mathbf{u} \cdot \mathbf{v} \end{aligned}$$

for any vector $\mathbf{v}$. We denote this $\widetilde{\mathbf{u}} = \mathrm{g}(\mathbf{u}, \cdot)$.

DEFINITION 16.48 Let $\mathscr{E}$ be a real vector space with a metric g, and let $\mathcal{B} = (\mathbf{e}_\mu)_\mu$ be a basis of $\mathscr{E}$. The **covariant coordinates of a vector** $\mathbf{u} \in \mathscr{E}$ are the coordinates, in the basis $(\boldsymbol{\alpha}^\mu)_\mu$ dual to $\mathcal{B}$, of the linear form $\widetilde{\mathbf{u}} \in \mathscr{E}^*$ defined by $\langle \widetilde{\mathbf{u}}, \mathbf{v} \rangle = \mathbf{u} \cdot \mathbf{v}$ for all $\mathbf{v} \in \mathscr{E}$.

---

[3] This "artifical" aspect is visible even in the way it is expressed: one has to abandon the Einstein conventions and explicitly write a summation sign in $\omega = \sum_\mu v^\mu \boldsymbol{\alpha}^\mu$, since the two indices μ are "at the same level." This is of course not a coincidence.

[4] It depends on the metric, but the metric is usually *given* and is an intrinsic part of the problem considered.

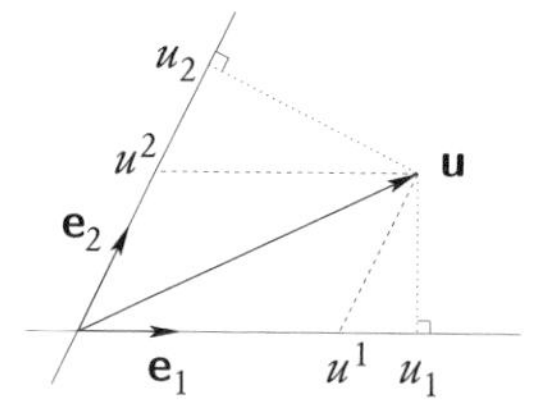

Fig. 16.1 – Contravariant and covariant coordinates of a vector of $\mathbb{R}^2$ with the usual euclidean scalar product. The vectors $\mathbf{e}_1$ and $\mathbf{e}_2$ have norm 1.

Those coordinates are denoted u_μ, so that we have

$$\boxed{\widetilde{\mathbf{u}} = u_\mu\, \boldsymbol{\alpha}^\mu} \qquad \text{whereas} \qquad \mathbf{u} = u^\mu\, \mathbf{e}_\mu.$$

THEOREM 16.49 *Let* $\mathbf{u} \in \mathscr{E}$ *be a vector. Its contravariant and covariant coordinates are given respectively by*

$$\boxed{u^\mu = \langle \boldsymbol{\alpha}^\mu, \mathbf{u} \rangle = \langle \mathrm{d}x^\mu, \mathbf{u} \rangle} \qquad \textit{and} \qquad \boxed{u_\mu = \mathbf{u} \cdot \mathbf{e}_\mu = \mathfrak{g}(\mathbf{u}, \mathbf{e}_\mu).}$$

Proof. The first equality is none other than the definition 16.1 of the contravariant coordinates of a vector. To prove the second equality, remember that, by the very definition of $\widetilde{\mathbf{u}}$, we have $\mathbf{u} \cdot \mathbf{e}_\mu = \langle \widetilde{\mathbf{u}}, \mathbf{e}_\mu \rangle$ and therefore

$$\mathbf{u} \cdot \mathbf{e}_\mu = \langle u_\nu \boldsymbol{\alpha}^\nu, \mathbf{e}_\mu \rangle = u_\nu \langle \boldsymbol{\alpha}^\nu, \mathbf{e}_\mu \rangle = u_\nu \delta^\nu_\mu = u_\mu.$$

It follows that the covariant and contravariant coordinates of a vector are given by totally different means. Fix a vector $\mathbf{u}$ and choose a basis $(\mathbf{e}_\mu)_\mu$. The ν-th contravariant coordinate of $\mathbf{u}$ in $(\mathbf{e}_\mu)$, $u^\nu = \boldsymbol{\alpha}^\nu(\mathbf{u})$, is obtained using the linear form $\boldsymbol{\alpha}^\nu$, which depends not only on $\mathbf{e}_\nu$, but on all vectors of the basis $(\mathbf{e}_\mu)_\mu$. On the other hand, the covariant coordinate u_ν is given by the scalar product of $\mathbf{u}$ with $\mathbf{e}_\nu$, which does not depend on the other basis vectors.

Remark 16.50 We see that the covariant coordinates correspond to *orthogonal projections* on the axis spanned by the basis vector (requiring a metric to make precise the notion of orthogonality), whereas contravariant coordinates correspond to projections on the axis *parallel* to all others (and the metric is not necessary). See figure 16.1.

♦ **Exercise 16.4** Show that, in an orthonormal basis, the covariant and contravariant coordinates are identical.

16.3.c Gymnastics: raising and lowering indices

THEOREM 16.51 *Let* $\mathbf{u} \in \mathscr{E}$ *be a vector with contravariant coordinates* u^μ *and covariant coordinates* u_μ. *Then we have*

$$\boxed{u_\mu = g_{\mu\nu}\, u^\nu.}$$

PROOF. Indeed, we find that $u_\mu = \mathbf{u} \cdot \mathbf{e}_\mu = (u^\nu \, \mathbf{e}_\nu) \cdot \mathbf{e}_\mu = u^\nu (\mathbf{e}_\mu \cdot \mathbf{e}_\nu) = g_{\mu\nu} \, u^\nu$.

Hence the coefficients $(g_{\mu\nu})_{\mu,\nu}$ turn out to be useful to *lower* an index. This does not apply only to vectors, but may be done for tensors of arbitrary type, always using the metric to transform a $\binom{p}{q}$-tensor into a $\binom{p-1}{q+1}$-tensor. However, **beware!** When performing such manipulations, the order of the indices must be carefully taken care of (see Remark 16.60).

THEOREM 16.52 *Let $\boldsymbol{\omega} \in \mathscr{E}^*$ be a linear form. There exists a unique vector $\widehat{\boldsymbol{\omega}} \in \mathscr{E}$ such that*

$$\langle \boldsymbol{\omega}, \mathbf{v} \rangle = \widehat{\boldsymbol{\omega}} \cdot \mathbf{v} \qquad \text{for all } \mathbf{v} \in \mathscr{E}.$$

This result comes from the definiteness of the metric and the fact that $\mathscr{E}$ and $\mathscr{E}^{**}$ are isomorphic in finite dimensions. The explicit construction of $\widehat{\boldsymbol{\omega}}$ is described in Proposition 16.54.

DEFINITION 16.53 The **contravariant coordinates** of the linear $\boldsymbol{\omega}$ are the contravariant coordinates of the vector $\widehat{\boldsymbol{\omega}}$. With a slight abuse of notation, they are denoted ω^μ, so that we have

$$\widehat{\boldsymbol{\omega}} = \omega^\mu \, \mathbf{e}_\mu.$$

PROPOSITION 16.54 *Let $\boldsymbol{\omega} \in \mathscr{E}^*$ be a linear form. The following relations hold between its covariant and contravariant coordinates:*

$$\omega_\mu = \left\langle \boldsymbol{\omega}, \mathbf{e}_\mu \right\rangle \qquad \text{and} \qquad \omega^\mu = \langle \boldsymbol{\alpha}^\mu, \widehat{\boldsymbol{\omega}} \rangle \,. \tag{16.4}$$

We have seen how the metric tensor can be used to lower the contravariant index of a vector, thus obtaining its covariant coordinates. Similarly, it can be used to lower the contravariant index of a linear form:

THEOREM 16.55 *Let $\boldsymbol{\omega}$ be a linear form. Its covariant and contravariant coordinates satisfy the relations*

$$\boxed{\omega_\mu = g_{\mu\nu} \, \omega^\nu.}$$

PROOF. By definition, $\widehat{\boldsymbol{\omega}} = \omega^\mu \mathbf{e}_\mu$, and moreover for any vector $\mathbf{v} \in \mathscr{E}$, we have $\langle \boldsymbol{\omega}, \mathbf{v} \rangle = \widehat{\boldsymbol{\omega}} \cdot \mathbf{v}$. In particular, $\omega_\mu = \left\langle \boldsymbol{\omega}, \mathbf{e}_\mu \right\rangle = \widehat{\boldsymbol{\omega}} \cdot \mathbf{e}_\mu = \omega^\nu \mathbf{e}_\nu \cdot \mathbf{e}_\mu = \omega^\nu g_{\nu\mu}$, and this, together with the relation $g_{\mu\nu} = g_{\nu\mu}$, concludes the proof.

We now pass to the question of *raising indices.*

DEFINITION 16.56 (Coefficients $g^{\mu\nu}$) Let $\mathfrak{g} = g_{\mu\nu} \, \mathrm{d}x^\mu \otimes \mathrm{d}x^\nu$ be the metric tensor. Its dual

$$\widehat{\mathfrak{g}} = g^{\mu\nu} \, \mathbf{e}_\mu \otimes \mathbf{e}_\nu$$

is defined by

$$\widehat{\mathfrak{g}}(\boldsymbol{\omega}, \boldsymbol{\sigma}) \stackrel{\text{def}}{=} \mathfrak{g}(\widehat{\boldsymbol{\omega}}, \widehat{\boldsymbol{\sigma}}) = \widehat{\boldsymbol{\omega}} \cdot \widehat{\boldsymbol{\sigma}} \qquad \forall \boldsymbol{\omega}, \boldsymbol{\sigma} \in \mathscr{E}^*.$$

In particular, the coefficients $g^{\mu\nu}$ satisfy $g^{\mu\nu} = \widehat{\mathfrak{g}}(\boldsymbol{\alpha}^{\mu}, \boldsymbol{\alpha}^{\nu})$ by Theorem 16.27. Moreover, we have $\widehat{\mathfrak{g}} \in \mathscr{E} \otimes \mathscr{E}$; that is, the dual of the metric is a $\binom{2}{0}$-tensor.

The dual first appears in the coordinates of the basis vectors of $\mathscr{E}$ and $\mathscr{E}^*$, which are given by the following proposition:

PROPOSITION 16.57 *The covariant and contravariant coordinates of* $\mathbf{e}_{\mu}$ *and* $\boldsymbol{\alpha}_{\mu}$ *– denoted* $[\mathbf{e}_{\mu}]_{\sigma}$, *with square brackets – are given by*

$$[\mathbf{e}_{\mu}]_{\sigma} = g_{\mu\sigma}, \qquad [\boldsymbol{\alpha}^{\mu}]^{\sigma} = g^{\mu\sigma},$$
$$[\mathbf{e}_{\mu}]^{\sigma} = \delta^{\sigma}_{\mu}, \qquad [\boldsymbol{\alpha}^{\mu}]_{\sigma} = \delta^{\mu}_{\sigma}.$$

PROOF. The first equality comes directly from $v_{\sigma} = \mathbf{v} \cdot \mathbf{e}_{\sigma}$ applied to $\mathbf{v} = \mathbf{e}_{\mu}$. The second equality holds because $(\boldsymbol{\alpha}^{\mu})^{\sigma}$ is the σ-th contravariant coordinate of the vector $\widehat{\boldsymbol{\alpha}^{\mu}}$, and is therefore equal to $\langle \boldsymbol{\alpha}^{\sigma}, \widehat{\boldsymbol{\alpha}^{\mu}} \rangle = \widehat{\boldsymbol{\alpha}^{\sigma}} \cdot \widehat{\boldsymbol{\alpha}^{\mu}} = g^{\mu\nu}$. The third equality is trivial since the σ-th coordinate of $\mathbf{e}_{\mu}$ is $\langle \boldsymbol{\alpha}^{\sigma}, \mathbf{e}_{\mu} \rangle$. Finally, the last equality follows from $(\boldsymbol{\alpha}^{\mu})_{\sigma} = \langle \boldsymbol{\alpha}^{\mu}, \mathbf{e}_{\sigma} \rangle$.

THEOREM 16.58 *The matrices* $g_{\mu\nu}$ *and* $g^{\mu\nu}$ *are inverse to each other, that is, we have*

$$\boxed{g_{\mu\nu}\, g^{\nu\rho} = \delta^{\rho}_{\nu}} \qquad \textit{and} \qquad \boxed{g^{\mu\nu}\, g_{\nu\rho} = \delta^{\mu}_{\rho}.}$$

PROOF. First we prove the second formula. According to Proposition 16.57, we have $g^{\mu\nu} = (\boldsymbol{\alpha}^{\mu})^{\nu}$, hence $g^{\mu\nu} g_{\nu\rho} = (\boldsymbol{\alpha}^{\mu})^{\nu} g_{\nu\rho} = (\boldsymbol{\alpha}^{\mu})_{\rho}$ by Theorem 16.55. Then the desired conclusion follows from Proposition 16.57 again.

The first equality is implied by the second.

From this we can see that the dual tensor $\widehat{\mathfrak{g}} = g^{\mu\nu} \mathbf{e}_{\mu} \otimes \mathbf{e}_{\nu}$ can be used to *raise* indices:

THEOREM 16.59 *Let* $\mathbf{u} \in \mathscr{E}$ *be a vector and let* $\boldsymbol{\omega} \in \mathscr{E}^*$ *be a linear form. Then their covariant and contravariant coordinates satisfy the following relations:*

$$\boxed{u^{\mu} = g^{\mu\nu} u_{\nu}} \qquad \textit{and} \qquad \boxed{\omega^{\mu} = g^{\mu\nu} \omega_{\nu}.}$$

PROOF. This is clear from to the formulas related to lowering indices, since $g_{\mu\nu}$ and $g^{\mu\nu}$ are inverse to each other.

Of course these results are also valid (with identical proofs!) for $\binom{p}{q}$-tensors, which are transformed into $\binom{p+1}{q-1}$-tensors by $\widehat{\mathfrak{g}}$ (i.e., by $g^{\mu\nu}$).

Remark 16.60 This is where the abuse of notation consisting in putting together all covariant indices (and all contravariant indices) becomes dangerous. Indeed, consider, for instance, a tensor with coordinates

$$T^{\mu\ \rho}_{\ \nu},$$

in expanded notation, and suppose now that, to simplify notation, we denote the coordinates simply

$$T^{\mu\rho}_{\nu},$$

using the fact that $\mathscr{E} \otimes \mathscr{E}^* \otimes \mathscr{E}$ is canonically isomorphic to $\mathscr{E}^{\otimes 2} \otimes \mathscr{E}^*$.

Now, suppose we want to lower the index μ by contracting it with the metric tensor. Should the result be denoted

$$T^{\rho}_{\mu\nu} \quad \text{or} \quad T^{\rho}_{\nu\mu}?$$

Only the original "noncondensed" form can give the answer. If the metric tensor is used, it is important to be careful to write all indices properly in the correct position.

16.4

Operations on tensors

Theorem 16.59 may be generalized, noting that the tensors $\mathfrak{g}$ and $\widehat{\mathfrak{g}}$ are used to transform $\binom{p}{q}$-tensors into $\binom{p+1}{q-1}$-tensors or $\binom{p-1}{q+1}$-tensors, respectively.

For instance, suppose we have a tensor denoted $T^{\mu\nu}{}_{\rho}$ in physics. We may lower the index ν using $g_{\mu\nu}$, defining

$$T^{\mu}{}_{\nu\rho} \stackrel{\text{def}}{=} g_{\nu\tau}\, T^{\mu\tau}{}_{\rho}.$$

This is the viewpoint of "index-manipulations." What is the intrinsic mathematical meaning behind this operation?

The tensor $\mathfrak{T}$, written

$$\mathfrak{T} = T^{\mu\nu}{}_{\rho}\, \mathbf{e}_{\mu} \otimes \mathbf{e}_{\nu} \otimes \boldsymbol{\alpha}^{\rho},$$

is the object which associates the real number

$$\mathfrak{T}(\boldsymbol{\omega}, \boldsymbol{\sigma}, \mathbf{v}) = T^{\mu\nu}{}_{\rho}\, \omega_{\mu}\, \sigma_{\nu}\, v^{\rho}$$

to a triple $(\boldsymbol{\omega}, \boldsymbol{\sigma}, \mathbf{v}) \in E^* \otimes E^* \otimes E$. As to the tensor "where one index is lowered," it corresponds intrinsically to

$$\widetilde{\mathfrak{T}} = T^{\mu}{}_{\nu\rho}\, \mathbf{e}_{\mu} \otimes \boldsymbol{\alpha}^{\nu} \otimes \boldsymbol{\alpha}^{\rho} \;:\; (\boldsymbol{\omega}, \mathbf{u}, \mathbf{v}) \longmapsto T^{\mu}{}_{\nu\rho}\, \omega_{\mu}\, u^{\nu}\, v^{\rho}.$$

To define $\widetilde{\mathfrak{T}}$, it suffices simply to be able to pass from $\mathbf{u}$ to $\boldsymbol{\sigma}$. But this is precisely the role of $\mathfrak{g}$, since from $\mathbf{u} \in E$ we may define a linear form $\widetilde{\mathbf{u}} \in E^*$ by the relation

$$\begin{aligned} \widetilde{\mathbf{u}} : E &\longrightarrow \mathbb{R}, \\ \mathbf{w} &\longmapsto \mathfrak{g}(\mathbf{u}, \mathbf{w}). \end{aligned} \tag{16.5}$$

Then we can define $\widetilde{\mathfrak{T}}$ as the "object" which

1. takes as input the triple $(\boldsymbol{\omega}, \mathbf{u}, \mathbf{v})$,

2. transforms it into $(\boldsymbol{\omega}, \widetilde{\mathsf{u}}, \mathsf{v})$, where $\widetilde{\mathsf{u}}$ is defined by the relation (16.5),
3. outputs the real number $\mathfrak{T}(\boldsymbol{\omega}, \widetilde{\mathsf{u}}, \mathsf{v})$:

$$\widetilde{\mathfrak{T}} : E^* \otimes E \otimes E \longrightarrow \mathbb{R},$$
$$(\boldsymbol{\omega}, \mathsf{u}, \mathsf{v}) \longmapsto \mathfrak{T}(\boldsymbol{\omega}, \widetilde{\mathsf{u}}, \mathsf{v}).$$

The same operation may be performed to "raise an index." This time, the tensor $\widehat{\mathfrak{g}}$ is used in order to define a vector, starting with a linear form (still because of the metric duality).

To summarize:

THEOREM 16.61 (Raising and lowering indices) *The metric tensors* $\mathfrak{g}$ *and* $\widehat{\mathfrak{g}}$ *can be used to transform a* $\binom{p}{q}$*-tensor into a* $\binom{p+1}{q-1}$*-tensor or a* $\binom{p-1}{q+1}$*-tensor, respectively.*

There is another fundamental operation defined on tensors, namely, the contraction of indices. Once again, to start with a concrete example, consider a tensor

$$\mathfrak{T} = T^{\mu\nu}{}_{\rho}\, \mathsf{e}_{\mu} \otimes \mathsf{e}_{\nu} \otimes \boldsymbol{\alpha}^{\rho}.$$

We will associate to this $\binom{2}{1}$-tensor a $\binom{1}{0}$-tensor, by defining

$$T^{\mu} \stackrel{\text{def}}{=} T^{\mu\nu}{}_{\nu}.$$

Thus we have performed an **index contraction**. For instance, the trace of a linear operator in a finite-dimensional vector space is given by the contraction of its indices.

Intrinsically, we are looking for a new tensor

$$\widetilde{\mathfrak{T}} = T^{\mu}\, \mathsf{e}_{\mu} \ : \boldsymbol{\omega} \longmapsto T^{\mu} \omega_{\mu}.$$

Notice that, taking an arbitrary basis $(\mathsf{e}_{\nu})_{\nu}$ and its dual basis $(\boldsymbol{\alpha}^{\nu})_{\nu}$, we have

$$\begin{aligned} \mathfrak{T}(\boldsymbol{\omega}, \boldsymbol{\alpha}^{\tau}, \mathsf{e}_{\tau}) = T^{\mu\nu}{}_{\rho}\, \omega_{\mu} [\boldsymbol{\alpha}^{\tau}]_{\nu} [\mathsf{e}_{\tau}]^{\rho} = T^{\mu\nu}{}_{\rho}\, \omega_{\mu}\, \delta^{\tau}_{\nu}\, \delta^{\rho}_{\tau} &= T^{\mu\nu}{}_{\rho}\, \omega_{\mu}\, \delta^{\rho}_{\nu} \\ &= T^{\mu\nu}{}_{\nu}\, \omega_{\mu}. \end{aligned}$$

Hence it suffices to define the tensor $\widetilde{\mathfrak{T}}$ by

$$\widetilde{\mathfrak{T}} : E^* \longrightarrow \mathbb{R},$$
$$\boldsymbol{\omega} \longmapsto \mathfrak{T}(\boldsymbol{\omega}, \boldsymbol{\alpha}^{\tau}, \mathsf{e}_{\tau}),$$

and then the relation $\widetilde{\mathfrak{T}} = T^{\mu}\, \mathsf{e}_{\mu}$ with $T^{\mu} = T^{\mu\nu}{}_{\nu}$ holds indeed (independently of the choice of a basis).

THEOREM 16.62 (Index contraction) *There exists an operation of "index contraction" which, for any choice of one covariant and one contravariant coordinate, transforms a* $\binom{p}{q}$*-tensor into a* $\binom{p-1}{q-1}$*-tensor. This operation does not require the use of the metric, and is defined independently of the choice of a basis.*

16.5 Change of coordinates

One of the most natural question that a physicist may ask is "how does a change of coordinates transform a quantity q?"

First, notice that a change of coordinates may be different things:

- a simple isometry of the ambient space (rotation, translation, symmetry,...);
- a change of galilean reference frame (which is then an isometry in Minkowski space-time with its pseudo-metric);
- a more complex abstract transformation (going from cartesian to polar coordinates, arbitrary coordinates in general relativity,...).

16.5.a Curvilinear coordinates

Consider general coordinates (x, y, z) in $\mathbb{R}^3$, which we rather denote (x^1, x^2, x^3), or even $X = (x^1, \ldots, x^n)$ in order to easily generalize to a space with n dimensions. For instance, this may be euclidean coordinates in $\mathbb{R}^n$.

Suppose given a set of functions which define new coordinates, called **curvilinear coordinates**:

$$U = \begin{cases} u^1 = u^1(x^1, \ldots, x^n), \\ \vdots \qquad \vdots \\ u^n = u^n(x^1, \ldots, x^n). \end{cases}$$

These functions define a $\mathscr{C}^1$ change of coordinates if (and only if) the map

$$\Phi : \mathbb{R}^n \longrightarrow \mathbb{R}^n \qquad (x^1, \ldots, x^n) \longmapsto (u^1, \ldots, u^n)$$

is of $\mathscr{C}^1$-class (its partial derivatives exist and are continuous), it is injective and its image is open, and if Φ^{-1} is of $\mathscr{C}^1$-class on this open subset (this is called a **$\mathscr{C}^1$-diffeomorphism** in mathematics, but physicists simply speak of **coordinate change**). Recall the following result of differential calculus [75]: Φ is a $\mathscr{C}^1$-diffeomorphism if and only if it is of $\mathscr{C}^1$-class and its jacobian is nowhere zero. In other words, its differential $\mathrm{d}\Phi$ is a bijective linear map at every point.

How is a vector such as a velocity vector, for instance, transformed? Its norm, expressed in kilometers per hour (hence, in terms of a first set of coordinates) is not the same expressed in miles per second (in a second set of coordinates). However, the velocity is a physical quantity independent of the manner in which it is measured. Therefore, the intrinsic physical quantity must not be mistaken for its *coordinates* in a given system of measurement.

Unfortunately, physicists often use the same word for those two very different concepts! Or, which amounts to the same thing here, they omit to state whether *active* or *passive* transformations are involved.

In what follows, we consider *passive* transformations: the space is invariant, *as well as the quantities measured*; on the other hand, the coordinate system varies, and hence the coordinates of the quantities measured also change.

16.5.b Basis vectors

Let $\mathbf{Y}$ be a point in $\mathbb{R}$, and let $\big(\mathbf{e}_1(\mathbf{Y}),\ldots,\mathbf{e}_n(\mathbf{Y})\big)$ denote the basis vectors in $\mathbb{R}^n$ attached to the point $\mathbf{Y}$. This family does not depend on $\mathbf{Y}$, since X represents the canonical coordinates on $\mathbb{R}^n$.

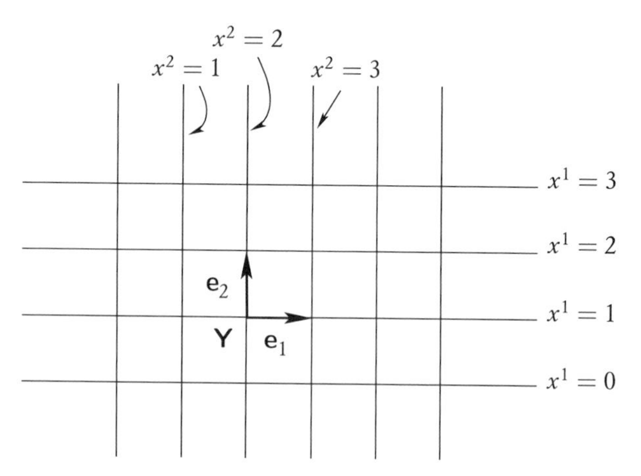

We now want to find the new vectors, attached at the point $\mathbf{Y}' = \Phi(\mathbf{Y})$ and tangent to the "coordinate lines," which represent the vectors $\big(\mathbf{e}_1(\mathbf{Y}),\ldots,\mathbf{e}_n(\mathbf{Y})\big)$ after the change of coordinates. They will then be the new basis vectors. We denote them $\big(\mathbf{e}'_1(\mathbf{Y}'),\ldots,\mathbf{e}'_n(\mathbf{Y}')\big)$.

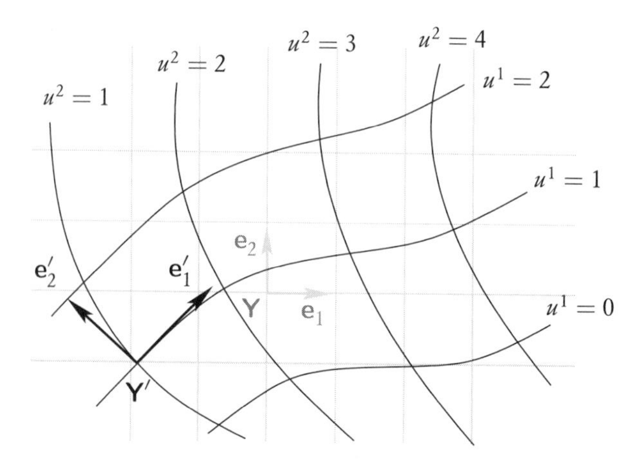

How are they defined? Suppose that we have two nearby points M and M' in space. Then we can write

$$\boldsymbol{MM'} = \delta x^{\mu}_{\mathrm{M}} \, \mathbf{e}_{\mu},$$

where $\delta x^{\mu}_{\mathrm{M}}$ is the variation of the μ-th coordinate x^{μ} when going from M to M'. This same vector, which is an intrinsic quantity, may also be written

$$\boldsymbol{MM'} = \delta u^{\mu}_{\mathrm{M}} \, \mathbf{e}'_{\mu},$$

where $\delta u^{\mu}_{\mathrm{M}}$ is the variation of the μ-th curvilinear coordinate u^{μ} when going from M to M'. But, to first order, we have the relation

$$\delta u^{\mu}_{\mathrm{M}} = \frac{\partial u^{\mu}}{\partial x^{\nu}} \, \delta x^{\nu}_{\mathrm{M}},$$

which, after identifying, gives

$$\delta x^{\mu}_{\mathrm{M}} \, \mathbf{e}_{\mu} = \frac{\partial u^{\mu}}{\partial x^{\nu}} \, \delta x^{\nu}_{\mathrm{M}} \, \mathbf{e}'_{\mu} = \frac{\partial u^{\nu}}{\partial x^{\mu}} \, \delta x^{\mu}_{\mathrm{M}} \, \mathbf{e}'_{\nu}$$

(since μ and ν are *mute* indices), and hence, since the $\delta x^{\mu}_{\mathrm{M}}$ are independent and arbitrary

$$\mathbf{e}_{\mu} = \frac{\partial u^{\nu}}{\partial x^{\mu}} \, \mathbf{e}'_{\nu} \qquad \text{or conversely} \qquad \mathbf{e}'_{\mu} = \frac{\partial x^{\nu}}{\partial u^{\mu}} \, \mathbf{e}_{\nu},$$

or again, expanding the notation,

$$\begin{pmatrix} \mathbf{e}_1 \\ \vdots \\ \mathbf{e}_n \end{pmatrix} = \begin{pmatrix} \dfrac{\partial u^1}{\partial x^1} & \cdots & \dfrac{\partial u^n}{\partial x^1} \\ \vdots & & \vdots \\ \dfrac{\partial u^1}{\partial x^n} & \cdots & \dfrac{\partial u^n}{\partial x^n} \end{pmatrix} \begin{pmatrix} \mathbf{e}'_1 \\ \vdots \\ \mathbf{e}'_n \end{pmatrix},$$

and

$$\begin{pmatrix} \mathbf{e}'_1 \\ \vdots \\ \mathbf{e}'_n \end{pmatrix} = \begin{pmatrix} \dfrac{\partial x^1}{\partial u^1} & \cdots & \dfrac{\partial x^n}{\partial u^1} \\ \vdots & & \vdots \\ \dfrac{\partial x^1}{\partial u^n} & \cdots & \dfrac{\partial x^n}{\partial u^n} \end{pmatrix} \begin{pmatrix} \mathbf{e}_1 \\ \vdots \\ \mathbf{e}_n \end{pmatrix}.$$

Note that the matrices

$$\Lambda^{\nu}{}_{\mu} = \frac{\partial u^{\nu}}{\partial x^{\mu}} \qquad \text{and} \qquad L^{\mu}{}_{\rho} = \frac{\partial x^{\mu}}{\partial u^{\rho}}$$

represent, respectively, the jacobian matrices of Φ and Φ^{-1}. In particular, they are invertible (the differential of Φ is bijective and therefore invertible), and each is in fact the inverse of the other. This may be written as

$$\frac{\partial u^\nu}{\partial x^\mu} \cdot \frac{\partial x^\mu}{\partial u^\rho} = \delta^\nu_\rho.$$

We recover the notation used in an affine space[5]; the only difference is that, now, the matrices L and Λ depend on the point M where the transformation is performed. To summarize:

THEOREM 16.63 (Transformation of basis vectors) *During a change of coordinates $\Phi : (x^\mu)_\mu \mapsto (u^\nu)_\nu$, the basis vectors are transformed according to the following relations:*

$$\boxed{\mathbf{e}'_\mu = \frac{\partial x^\nu}{\partial u^\mu}\,\mathbf{e}_\nu} \qquad \textit{and} \qquad \mathbf{e}_\mu = \frac{\partial u^\nu}{\partial x^\mu}\,\mathbf{e}'_\nu.$$

Note that these formulas are "balanced" from the point of view of indices, obeying the following rule: *a superscripted index of a term in the denominator of a fraction is equivalent to a subscripted index.*

16.5.c Transformation of physical quantities

Now, assume we have a vector which corresponds to some physical quantity (and hence is an intrinsic object, also called a **vector field**). It can be expressed by

$$\mathsf{v} = v^\mu\,\mathbf{e}_\mu = v'^\mu \mathbf{e}'_\mu$$

in each system of coordinates. Applying the rule for the transformation of basis vectors, we derive the rule for the, transformation of the contravariant coordinates of a vector:

THEOREM 16.64 (Transformation of contravariant coordinates)
During a change of coordinates $\Phi : (x^\mu)_\mu \mapsto (u^\nu)_\nu$, the contravariant coordinates v^μ of a vector field are transformed according to the relation

$$\boxed{v'^\mu = \frac{\partial u^\mu}{\partial x^\nu}\,v^\nu} \qquad \textit{and} \qquad v^\mu = \frac{\partial x^\mu}{\partial u^\nu}\,v'^\nu,$$

that is, following the opposite rule as that used for basis vectors.[6]

---

[5] Consider in this manner the formula of Proposition 16.13. We may write $\boldsymbol{\alpha}'^\mu = \Lambda^\mu{}_\nu\,\boldsymbol{\alpha}^\nu$ in the form $\mathrm{d}x'^\mu = \Lambda^\mu{}_\nu\,\mathrm{d}x^\nu$, which gives the well-known formula

$$\mathrm{d}x'^\mu = \frac{\partial x'^\mu}{\partial x^\nu}\,\mathrm{d}x^\nu.$$

[6] One should of course indicate every time at which point the vectors are tangent, which is also where the partial derivatives are evaluated. In this case, the preceding formulas should be

Remark 16.65 Here is a way to memorize those formulas. Let x'^{μ} be the new coordinates. We recover the relation expressing $\mathbf{e}'_{\mu}$ as a function of $\mathbf{e}_{\nu}$, namely,

$$\mathbf{e}'_{\mu} = \frac{\partial x^{\nu}}{\partial x'^{\mu}}\,\mathbf{e}_{\nu},$$

by applying the following rules:

- since the index of the vector on the left-hand side of the equation (which is $\mathbf{e}'_{\mu}$) is subscripted, we write $\mathbf{e}'_{\mu} = \frac{\partial x}{\partial x}\,\mathbf{e}_{\nu}$ and put the "prime" sign at the bottom:

$$\mathbf{e}'_{\mu} = \frac{\partial x}{\partial x'}\,\mathbf{e}_{\nu};$$

- since this index μ is subscripted, it can appear as "x^{μ}" only in the denominator, so we write

$$\mathbf{e}'_{\mu} = \frac{\partial x}{\partial x'^{\mu}}\,\mathbf{e}_{\nu};$$

- finally, we complete the indices with a "mute" ν:

$$\mathbf{e}'_{\mu} = \frac{\partial x^{\nu}}{\partial x'^{\mu}}\,\mathbf{e}_{\nu}$$

(and then one should check that the index is balanced as it should).

Similarly, to recover the formula giving v'^{μ} in terms of the v^{ν}'s,

- we write the formula with a "prime" sign in the numerator, since the index is superscripted:

$$v'^{\mu} = \frac{\partial x'}{\partial x}\,v^{\nu};$$

- then we complete the indices in order that the result be balanced:

$$v'^{\mu} = \frac{\partial x'^{\mu}}{\partial x^{\nu}}\,v^{\nu}.$$

The reader is invited to check that the converse formulas (giving v^{μ} and $\mathbf{e}_{\mu}$ in terms of the v'^{ν}'s and $\mathbf{e}'_{\nu}$'s) may be obtained with similar techniques.

16.5.d Transformation of linear forms

Since linear forms are dual to vectors, the reader will not find the following result very surprising, except for the terminology: what might be called a field of linear forms is called a **differential form**:

---

written more rigorously:

$$v'^{\mu}(\mathbf{Y}') = \frac{\partial u^{\mu}}{\partial x^{\nu}}(\mathbf{Y})\,v^{\nu}(\mathbf{Y}) \qquad \text{and} \qquad v^{\mu}(\mathbf{Y}) = \frac{\partial x^{\mu}}{\partial u^{\nu}}(\mathbf{Y}')\,v^{\nu}(\mathbf{Y}'),$$

which complicates the notation quite a bit.

THEOREM 16.66 (Transformation of linear forms) *Let ω be a differential form. During a change of coordinates $\Phi : (x^\mu)_\mu \mapsto (u^\nu)_\nu$, its covariant coordinates are transformed according to the relations*

$$\boxed{\omega'_\mu = \frac{\partial x^\nu}{\partial u^\mu}\,\omega_\nu} \qquad \text{and} \qquad \omega_\mu = \frac{\partial u^\nu}{\partial x^\mu}\,\omega'_\nu.$$

The basic linear forms are transformed according to

$$\boxed{\boldsymbol{\alpha}'^\mu = \frac{\partial u^\mu}{\partial x^\nu}\,\boldsymbol{\alpha}^\nu} \qquad \text{and} \qquad \boldsymbol{\alpha}^\mu = \frac{\partial x^\mu}{\partial u^\nu}\,\boldsymbol{\alpha}'^\nu.$$

The rule sketched above to memorize those relations remains valid.

16.5.e Transformation of an arbitrary tensor field

Of course, the previous results extend easily to tensors of arbitrary type.

THEOREM 16.67 (Transformation of a tensor) *Let $\mathfrak{T}$ be a tensor field with coordinates*

$$\mathfrak{T}(\mathbf{Y}) = T^{\mu_1\dots,\mu_p}_{\nu_1,\dots,\nu_q}\,\mathbf{e}_{\mu_1}(Y)\otimes\cdots\otimes\mathbf{e}_{\mu_n}(Y)\otimes\boldsymbol{\alpha}^{\nu_1}(\mathbf{Y})\otimes\cdots\otimes\boldsymbol{\alpha}^{\nu_q}(\mathbf{Y}).$$

During a change of coordinates $\Phi : (x^\mu)_\mu \mapsto (u^\nu)_\nu$, these coordinates are transformed according to the rule

$$T'^{\mu_1\dots,\mu_p}_{\nu_1,\dots,\nu_q} = \frac{\partial u^{\mu_1}}{\partial x^{\rho_1}}\cdots\frac{\partial u^{\mu_p}}{\partial x^{\rho_p}}\cdot\frac{\partial x^{\sigma_1}}{\partial u^{\nu_1}}\cdots\frac{\partial x^{\sigma_q}}{\partial u^{\nu_q}}\,T^{\rho_1,\dots,\rho_p}_{\sigma_1,\dots,\sigma_q}.$$

Finally, we note that it is customary, both in relativity theory and in field theory, to introduce the following notation:

$$\partial_\mu \overset{\text{def}}{=} \frac{\partial}{\partial x^\mu} \qquad \text{and} \qquad T_{,\nu} \overset{\text{def}}{=} \partial_\nu T,$$

so that, for instance, the divergence of a vector $\mathfrak{T} = T^\mu \mathbf{e}_\mu$ is written

$$\operatorname{div}\mathfrak{T} = T^\mu{}_{,\mu}.$$

16.5.f Conclusion

Le Calcul Tensoriel sait mieux la physique que le Physicien lui-même.
Tensor calculus knows physics better than the physicist himself does.

Paul LANGEVIN (quoted in [10]).

Physical applications are too numerous to be studied individually. Among others, we may mention:

Special relativity: the natural setting of special relativity was defined by Minkowski. It is a space isomorphic to $\mathbb{R}^4$, with a pseudo-metric with signature $(1, 3)$. Generalizing the scalar product is done very simply. The notion of *covariance*, in other words, of the invariance of a quantity or a the form of an equation, is expressed mathematically in the invariance under change of basis of quantities which are "balanced" in terms of their expressions using indices.

General relativity: Riemannian geometry is unfortunately outside the scope of this book. We simply mention that the notion of Minkowski space is extended to non-euclidean geometry. Tensor calculus (in the context of manifolds) is then, for intrinsic reasons, an indispensable ingredient of the theory.

Fluid mechanics: using tensors, the notion of current can be generalized. The current of a vector is a scalar. The current of a vector may be expressed in matrix form (one index for each component of the vector, one index for the direction of the current). It is then convenient to go farther and to denote as a tensor with three indices the current of a tensor with two indices, and so on.

Electricity, signal theory, optics: tensor calculus may be used as a *mathematical trick* to express general linear transformations. Then the underlying notions such as *linear forms, metric, duality,...* do not carry special physical significance.

Remark 16.68 A reader interested in general tensor calculus is invited to read the books of Delachet [27]; the excellent treatise on modern geometry [30] in three volumes; the bible concerning the mathematics of general relativity (including all non-euclidean geometry) by Wheeler[7] et al. [67]; or finally, one of the best physics books ever, that of Steven Weinberg [92].

On voit [...] se substituer à l'*homo faber* l'*homo mathematicus*.
Par exemple l'outil tensoriel est un merveilleux opérateur de généralité ;
à le manier, l'esprit acquiert des capacités nouvelles de généralisation.

We see [...] homo mathematicus *substituting himself to* homo faber.
For instance, tensor calculus is a wonderful operator in generalities; handling it, the mind develops new capacities for generalization.

Gaston BACHELARD, *Le nouvel esprit scientifique* [10]

[7] Who invented the words "black hole."

SOLUTIONS

♦ **Solution of exercise 16.1 on page 439.** Using the formulas for change of basis, we get

$$\begin{aligned}
\Phi'^{\mu}{}_{\nu}\,\boldsymbol{\alpha}'^{\nu}(\mathsf{x})\,\mathbf{e}'_{\mu} &= \big(\Lambda^{\mu}{}_{\rho}L^{\sigma}{}_{\nu}\Phi^{\rho}{}_{\sigma}\big)\big(\Lambda^{\nu}{}_{\tau}\,\boldsymbol{\alpha}^{\tau}(\mathsf{x})\big)\,\big(L^{\varkappa}{}_{\mu}\,\mathbf{e}_{\varkappa}\big) \\
&= \big(\Lambda^{\mu}{}_{\rho}L^{\varkappa}{}_{\mu}\big)\big(L^{\sigma}{}_{\nu}\Lambda^{\nu}{}_{\tau}\big)\Phi^{\rho}{}_{\sigma}\,\boldsymbol{\alpha}^{\tau}(\mathsf{x})\,\mathbf{e}_{\varkappa} \\
&= \delta^{\varkappa}_{\rho}\,\delta^{\sigma}_{\tau}\,\Phi^{\rho}{}_{\sigma}\,\boldsymbol{\alpha}^{\tau}(\mathsf{x})\,\mathbf{e}_{\varkappa} = \Phi^{\rho}{}_{\sigma}\,\boldsymbol{\alpha}^{\sigma}(\mathsf{x})\,\mathbf{e}_{\rho},
\end{aligned}$$

and this is the required formula.

♦ **Solution of exercise 16.2 on page 440.** The first questions are only a matter of writing things down. In the case $n = p = 2$, we see, for instance, that

$$\begin{pmatrix} 0 & 1 & 0 & -1 \\ 2 & 3 & -2 & -3 \\ 0 & 0 & 0 & 2 \\ 0 & 0 & 4 & 6 \end{pmatrix} = \begin{pmatrix} 1 & -1 \\ 0 & 2 \end{pmatrix} \otimes \begin{pmatrix} 0 & 1 \\ 2 & 3 \end{pmatrix},$$

but

$$\begin{pmatrix} 0 & 0 & 0 & 0 \\ 0 & 1 & 0 & 0 \\ 0 & 0 & 2 & 3 \\ 0 & 0 & 0 & 7 \end{pmatrix}$$

cannot be a tensor product of two matrices.

Chapter 17

Differential forms

In all this chapter, E denotes a real vector space of finite dimension n (hence we have $E \simeq \mathbb{R}^n$, but E does not, a priori, have a euclidean structure, for instance).

17.1 Exterior algebra

17.1.a 1-forms

DEFINITION 17.1 A **linear form** or **exterior 1-form** on E is a linear map from E to $\mathbb{R}$:

$$\omega : E \longrightarrow \mathbb{R}.$$

The **dual of E** is the vector space E^* of linear forms on E.

Let $\mathcal{B} = (\boldsymbol{b}_1, \ldots, \boldsymbol{b}_n)$ be a basis of E. Any vector $\boldsymbol{x} \in E$ can be expressed in a unique way as combination of vectors of the basis $\mathcal{B}$:

$$\boldsymbol{x} = x^1 \boldsymbol{b}_1 + \cdots + x^n \boldsymbol{b}_n.$$

We often denote simply $\boldsymbol{x} = (x^1, \ldots, x^n)$ (identifying E and $\mathbb{R}^n$, the basis of E being assumed given). For any $i \in [\![1, n]\!]$, we denote either b^{*i} or $\mathrm{d}x^i$ the linear form on E which is defined by

$$b^{*i}(\boldsymbol{x}) = \langle b^{*i}, \boldsymbol{x} \rangle = x^i \qquad \text{or} \qquad \boxed{\langle b^{*i}, \boldsymbol{b}_j \rangle = \delta^i_j} \quad \forall i, j \in [\![1, n]\!].$$

THEOREM 17.2 *The family $(b^{*1}, \dots, b^{*n})$ is a basis of E^*.*

PROOF. Let $\varphi \in E^*$. Then, denoting $\alpha_i = \varphi(b_i)$, we have obviously (by linearity) $\varphi = \sum_i \alpha_i b^{*i}$, which shows that the family $(b^{*1}, \dots, b^{*n})$ generates E^*. It is also a free family (indeed, if $\sum \lambda_i b^{*i} = 0$, applying this linear form to the vector $\boldsymbol{b}_j$ yields $\lambda_j = 0$), and hence is a basis.

Any linear form may therefore be written

$$\omega = \omega_1 \, \mathrm{d}x^1 + \cdots + \omega_n \, \mathrm{d}x^n \qquad \text{with } \omega_i = \omega(\boldsymbol{b}_i).$$

17.1.b Exterior 2-forms

DEFINITION 17.3 An **exterior form of degree 2** or a **2-form** is an *antisymmetric* bilinear form on $E \times E$, that is, a bilinear form $\omega^2 : E \times E \to \mathbb{R}$ such that we have

$$\omega^2(\boldsymbol{x}, \boldsymbol{y}) = -\omega^2(\boldsymbol{y}, \boldsymbol{x})$$

for all $\boldsymbol{x}, \boldsymbol{y} \in E$.

Example 17.4 The determinant map in a basis $\mathcal{B} = (\boldsymbol{b}_1, \boldsymbol{b}_2)$ of $\mathbb{R}^2$, namely,

$$\begin{aligned} \det : E \times E &\longrightarrow \mathbb{R}, \\ (\boldsymbol{x}, \boldsymbol{y}) &\longmapsto \det_{\mathcal{B}}(\boldsymbol{x}, \boldsymbol{y}) = \begin{vmatrix} x^1 & y^1 \\ x^2 & y^2 \end{vmatrix}, \end{aligned}$$

is an exterior 2-form on $\mathbb{R}^2$.

DEFINITION 17.5 The vector space of exterior forms of degree 2 on E is denoted $\boldsymbol{\Lambda}^{*2}(E)$.

We now look for a basis of $\Lambda^{*2}(E)$. Clearly, the bilinear form

$$\begin{aligned} \mathrm{d}x^1 \otimes \mathrm{d}x^2 - \mathrm{d}x^2 \otimes \mathrm{d}x^1 : E \times E &\longrightarrow \mathbb{R}, \\ (\boldsymbol{x}, \boldsymbol{y}) &\longmapsto x^1 y^2 - x^2 y^1 \end{aligned}$$

is an element of $\Lambda^{*2}(E)$. Similarly, the bilinear forms $\mathrm{d}x^i \otimes \mathrm{d}x^j - \mathrm{d}x^j \otimes \mathrm{d}x^i$ with $i, j \in [\![1, n]\!]$ are also in this space. A moment of thought suffices to see that any exterior 2-form may be expressed as linear combination of those elements. So we have found a generating family. But clearly, this family is not free, since

- $\mathrm{d}x^i \otimes \mathrm{d}x^i - \mathrm{d}x^i \otimes \mathrm{d}x^i = 0$ for all $i \in [\![1, n]\!]$;
- $\mathrm{d}x^j \otimes \mathrm{d}x^i - \mathrm{d}x^i \otimes \mathrm{d}x^j = -\big(\mathrm{d}x^i \otimes \mathrm{d}x^j - \mathrm{d}x^j \otimes \mathrm{d}x^i\big)$ for all $i, j \in [\![1, n]\!]$.

It is sufficient to keep the elements of this type with $i < j$; there are $n(n-1)/2$ of them.

THEOREM 17.6 *A basis of $\Lambda^{*2}(E)$ is given by the family of the exterior 2-forms*

$$\mathrm{d}x^i \wedge \mathrm{d}x^j \stackrel{\text{def}}{=} \mathrm{d}x^i \otimes \mathrm{d}x^j - \mathrm{d}x^j \otimes \mathrm{d}x^i,$$

where $1 \leqslant i < j \leqslant n$. Hence we have $\dim \Lambda^{*2}(E) = n(n-1)/2$.

Notice that the space $E^{*\otimes 2} \stackrel{\text{def}}{=} E^* \otimes E^*$ admits for basis the family of elements

$$\left\{\mathrm{d}x^i \otimes \mathrm{d}x^j \;;\; i, j \in [\![1, n]\!]\right\},$$

from which it is natural to create two further families:

- the family we have seen of elements of the type $\mathrm{d}x^i \wedge \mathrm{d}x^j$ (with $i < j$);
- the family of elements of the type $\frac{1}{2}(\mathrm{d}x^i \otimes \mathrm{d}x^j + \mathrm{d}x^j \otimes \mathrm{d}x^i)$.

Those two families are bases, respectively, for the space $\Lambda^{*2}(E)$, which is the space of *antisymmetric* $\binom{0}{2}$-tensors, and for the space of *symmetric* $\binom{0}{2}$-tensors.

PROPOSITION 17.7 *Any $\binom{0}{2}$-tensor is the sum of a symmetric and an antisymmetric tensor.*

This may be generalized to $\binom{0}{k}$-tensors.

17.1.c Exterior k-forms

We first recall the following definition:

DEFINITION 17.8 A **permutation** of $[\![1, n]\!]$ is any bijection σ of the set $[\![1, n]\!]$ into itself; in other words, σ permutes the integers from 1 to n. A **transposition** is any permutation which exchanges two elements exactly and leaves the others invariants. The set of permutations of $[\![1, n]\!]$ is denoted $\mathfrak{S}_n$.

The composition of applications defines the **product** of permutations.

Any permutation may be written as a product of finitely many transpositions. The **signature** of a permutation $\sigma \in \mathfrak{S}_n$ is the integer $\varepsilon(\sigma)$ given by: $\varepsilon(\sigma) = 1$ if σ can be written as a product a an *even* number of transpositions, and $\varepsilon(\sigma) = -1$ if σ can be written as the product of an *odd* number of transpositions.[1] For any permutations σ and τ, we have

$$\varepsilon(\sigma\tau) = \varepsilon(\sigma)\,\varepsilon(\tau).$$

DEFINITION 17.9 An **exterior k-form**, or more simply a **k-form**, is any antisymmetric k-multilinear form ω^k on E^k, that is, the map $\omega^k : E^k \to \mathbb{R}$ satisfies the relation

$$\omega^k(\boldsymbol{x}_1, \ldots, \boldsymbol{x}_k) = (-1)^{\varepsilon(\sigma)}\,\omega^k(\boldsymbol{x}_{\sigma(1)}, \ldots, \boldsymbol{x}_{\sigma(k)})$$

for any permutation $\sigma \in \mathfrak{S}_k$.

---

[1] Implicit in this definition is the fact, which we admit, that the parity of the number of transpositions is independent of the choice of a formula expressing σ as a product of transpositions.

Example 17.10 The most famous example is that of an exterior k-form on a space of dimension k, namely, the determinant map expressed using the canonical basis: if $(\boldsymbol{e}_1, \dots, \boldsymbol{e}_k)$ is the canonical basis of $\mathbb{R}^k$, and if $\boldsymbol{\xi}_i \in \mathbb{R}^k$ for $i = 1, \dots, k$, then denoting $\boldsymbol{\xi}_i = \xi_{i1}\boldsymbol{e}_1 + \cdots + \xi_{ik}\boldsymbol{e}_k$, this is a map

$$\begin{aligned} \det : \qquad & (\mathbb{R}^k)^k \longrightarrow \mathbb{R}, \\ & (\boldsymbol{\xi}_1, \dots, \boldsymbol{\xi}_k) \longmapsto \det_{\text{can}}(\boldsymbol{\xi}_1, \dots, \boldsymbol{\xi}_k) = \begin{vmatrix} \xi_{11} & \cdots & \xi_{1k} \\ \vdots & & \vdots \\ \xi_{k1} & \cdots & \xi_{kk} \end{vmatrix}, \end{aligned}$$

which is an exterior k-form.

Example 17.11 For a fixed vector $\boldsymbol{u} \in \mathbb{R}^3$, the map $\omega : (\boldsymbol{x}, \boldsymbol{y}) \mapsto \det(\boldsymbol{x}, \boldsymbol{y}, \boldsymbol{u})$ on $\mathbb{R}^3 \times \mathbb{R}^3$ is an exterior 2-form.

DEFINITION 17.12 The vector space of exterior k-forms on E is denoted $\Lambda^{*k}(E)$.

We will now find a basis of $\Lambda^{*k}(E)$.

To generalize the corresponding result of the preceding section for $\Lambda^{*2}(E)$, we introduce the following definition:

DEFINITION 17.13 Let $k \in \mathbb{N}$ and let $i_1, \dots, i_k \in [\![1, n]\!]$. Then the exterior k-form $\mathrm{d}x^{i_1} \wedge \cdots \wedge \mathrm{d}x^{i_k}$ is defined by the formula

$$\mathrm{d}x^{i_1} \wedge \cdots \wedge \mathrm{d}x^{i_k} \stackrel{\text{def}}{=} \sum_{\sigma \in \mathfrak{S}_k} \varepsilon(\sigma)\, \mathrm{d}x^{\sigma(i_1)} \otimes \cdots \otimes \mathrm{d}x^{\sigma(i_k)}.$$

Example 17.14 Consider the space $\mathbb{R}^4$ and denote by x, y, z, t the coordonnates in this space. Then we have

$$\begin{aligned} \mathrm{d}x \wedge \mathrm{d}y \wedge \mathrm{d}z = \mathrm{d}x \otimes \mathrm{d}y \otimes \mathrm{d}z - \mathrm{d}x \otimes \mathrm{d}z \otimes \mathrm{d}y + \mathrm{d}y \otimes \mathrm{d}z \otimes \mathrm{d}x - \mathrm{d}y \otimes \mathrm{d}x \otimes \mathrm{d}z \\ + \mathrm{d}z \otimes \mathrm{d}x \otimes \mathrm{d}y - \mathrm{d}z \otimes \mathrm{d}y \otimes \mathrm{d}x. \end{aligned}$$

THEOREM 17.15 *The family*

$$\{\mathrm{d}x^{i_1} \wedge \cdots \wedge \mathrm{d}x^{i_k} \ ; \ 1 \leqslant i_1 < \cdots < i_k \leqslant n\}$$

is a basis of $\Lambda^{*k}(\mathbb{R}^n)$, *the dimension of which is equal to the binomial coefficient* $\binom{n}{k}$. *Hence any exterior k-form (i.e., any antisymmetric* $\binom{0}{k}$*-tensor) can be written in the form*

$$\omega^k = \sum_{i_1 < \cdots < i_k} \omega_{i_1, \dots, i_k}\, \mathrm{d}x^{i_1} \wedge \cdots \wedge \mathrm{d}x^{i_k}.$$

In the remainder of this chapter, Einstein's summation conventions will be extended to exterior forms, so we will write

$$\omega = \omega_{i_1, \dots, i_k}\, \mathrm{d}x^{i_1} \wedge \cdots \wedge \mathrm{d}x^{i_k} \quad \text{instead of} \quad \omega = \sum_{i_1 < \cdots < i_k} \omega_{i_1, \dots, i_k}\, \mathrm{d}x^{i_1} \wedge \cdots \wedge \mathrm{d}x^{i_k},$$

that is, when no ambiguity arises, repeated indices (appearing once as superscript and once as subscript) are assumed to be summed in strictly increasing order; so if there are k indices, the sum is over $\binom{n}{k}$ terms instead of n^k terms.

For the special case $k = n$, we deduce the following result:

THEOREM 17.16 *The space $\Lambda^{*n}(\mathbb{R}^n)$ is of dimension* 1.

COROLLARY 17.16.1 *Any exterior n-form on $\mathbb{R}^n$ is a multiple of the form "determinant in the canonical basis."*

DEFINITION 17.17 A **volume form** on $\mathbb{R}^n$ is any nonzero exterior n-form.

Remark 17.18 We will often denote ω^k an exterior k-form, ω^p an exterior p-form, and so on. This supscript notation is only intended to recall the "degree" of the form which is considered. In particular, it should not be mistaken with the exponent in a power expression.

17.1.d Exterior product

There is a product defined on the various spaces of exterior forms, which associates a $(k+\ell)$-exterior form $\omega^k \wedge \omega^\ell$ to a pair (ω^k, ω^ℓ) in $\Lambda^{*k}(E) \times \Lambda^{*\ell}(E)$. We will first give the definition for the product of two 1-forms, then generalize it. Note that the notation will be consistent with the previous introduction of basis elements $\mathrm{d}x^{i_1} \wedge \cdots \wedge \mathrm{d}x^{i_k}$ for $\Lambda^{*k}(E)$ (i.e., those will be products of 1-forms).

Exterior product of two 1-forms

DEFINITION 17.19 Let ω, ω' be exterior 1-forms. The **exterior product** $\omega \wedge \omega'$ of ω and ω' is defined by

$$\omega \wedge \omega' \stackrel{\text{def}}{=} (\omega \otimes \omega' - \omega' \otimes \omega).$$

It is easily checked that $\omega \wedge \omega'$ is an exterior 2-form.

PROPOSITION 17.20 *Let ω and ω' be exterior 1-forms. Then we have*

$$(\omega \wedge \omega')(\boldsymbol{x}, \boldsymbol{y}) = \begin{vmatrix} \omega(\boldsymbol{x}) & \omega'(\boldsymbol{x}) \\ \omega(\boldsymbol{y}) & \omega'(\boldsymbol{y}) \end{vmatrix},$$

and moreover $\omega \wedge \omega' = -\omega' \wedge \omega$. The exterior product is bilinear and antisymmetric.

Similarly, the product of p exterior 1-forms may be defined as follows:

DEFINITION 17.21 Let $\omega_1, \dots, \omega_p$ be exterior 1-forms. Then the exterior product $\omega_1 \wedge \cdots \wedge \omega_p \in \Lambda^{*p}(E)$ is defined as the multilinear map on E^p

such that

$$\omega_1 \wedge \cdots \wedge \omega_p(\boldsymbol{x}_1, \ldots, \boldsymbol{x}_p) \stackrel{\text{def}}{=} \begin{vmatrix} \omega_1(\boldsymbol{x}_1) & \ldots & \omega_p(\boldsymbol{x}_1) \\ \vdots & & \vdots \\ \omega_1(\boldsymbol{x}_p) & \ldots & \omega_p(\boldsymbol{x}_p) \end{vmatrix}.$$

Exterior product of two arbitrary exterior forms

We now introduce the general definition of the exterior product of two exterior forms.

DEFINITION 17.22 (Exterior product) Let $k, \ell \in \mathbb{N}^*$, $\omega^k \in \Lambda^{*k}(E)$ and $\omega^\ell \in \Lambda^{*\ell}(E)$. The **exterior product of $\boldsymbol{\omega}^k$ and $\boldsymbol{\omega}^\ell$** is the exterior $(k+\ell)$-form denoted $\omega^k \wedge \omega^\ell$ such that

$$\omega^k \wedge \omega^\ell(\boldsymbol{x}_1, \ldots, \boldsymbol{x}_{k+\ell}) \stackrel{\text{def}}{=} \sum_{\sigma \in \mathfrak{S}_{k+\ell}} \varepsilon(\sigma)\, \omega^k(\boldsymbol{x}_{\sigma(1)}, \ldots, \boldsymbol{x}_{\sigma(k)}) \cdot \omega^\ell(\boldsymbol{x}_{\sigma(k+1)}, \ldots, \boldsymbol{x}_{\sigma(k+\ell)}).$$

Example 17.23 For a 2-form ω^2 and a 1-form ω^1, we have

$$\omega^2 \wedge \omega^1(a, b, c) = 2\omega^2(a, b)\,\omega^1(c) + 2\omega^2(b, c)\,\omega^1(a) + 2\omega^2(c, a)\,\omega^1(b).$$

The reader is invited, as an exercise, to check the following result:

PROPOSITION 17.24 *The exterior product is*

- *(anti)symmetric:* $\omega^p \wedge \omega^q = (-1)^{pq}\ \omega^q \wedge \omega^p$ *for* $\omega^p \in \Lambda^{*p}(E)$ *and* $\omega^q \in \Lambda^{*q}(E)$;
- *distributive with respect to addition:* $\omega \wedge (\omega' + \omega'') = \omega \wedge \omega' + \omega \wedge \omega''$;
- *associative:* $(\omega \wedge \omega') \wedge \omega'' = \omega \wedge (\omega' \wedge \omega'')$.

Example 17.25 Let ω and ω' be exterior 1-forms. Then using Definition 17.22 we get

$$\omega \wedge \omega'(\boldsymbol{x}, \boldsymbol{y}) = \omega(\boldsymbol{x})\,\omega'(\boldsymbol{y}) - \omega'(\boldsymbol{x})\,\omega(\boldsymbol{y}),$$

i.e. $\omega \wedge \omega' = \omega \otimes \omega' - \omega' \otimes \omega$ as defined in the preceding paragraph.

PROPOSITION 17.26 *Let* $\omega_1, \ldots, \omega_p, \omega_{p+1}, \ldots, \omega_{p+q}$ *be exterior 1-forms. Then we have*

$$(\omega_1 \wedge \cdots \wedge \omega_p) \wedge (\omega_{p+1} \wedge \cdots \wedge \omega_{p+q}) = \omega_1 \wedge \cdots \wedge \omega_p \wedge \omega_{p+1} \wedge \cdots \wedge \omega_{p+q}.$$

PROOF. This formula follows from the definition of the exterior product of 1-forms and the compatibility with the general definition.

DEFINITION 17.27 The **exterior algebra** on $\mathbb{R}^n$ is the vector space

$$\Lambda^*(\mathbb{R}^n) \stackrel{\text{def}}{=} \bigoplus_{k=0}^{n} \Lambda^{*k}(\mathbb{R}^n).$$

The exterior product is extended to the exterior algebra by distributivity.

17.2
Differential forms on a vector space

17.2.a Definition

As before, in all this section, $E = \mathbb{R}^n$ is a real vector space of finite dimension n.

DEFINITION 17.28 Let p be an integer or $+\infty$. A **differential form of degree k on $\mathbb{R}^n$ of $\mathscr{C}^p$-class**, or **differential k-form**, is a map of $\mathscr{C}^p$-class

$$\begin{aligned} \omega : \mathbb{R}^n &\longrightarrow \Lambda^{*k}(\mathbb{R}^n), \\ x &\longmapsto \omega(x). \end{aligned}$$

For any point x in $\mathbb{R}^n$, $\omega(x)$ is therefore an exterior k-form on $\mathbb{R}^n$.

What is the precise meaning of "$\mathscr{C}^p$-class" here? Any differential k-form may be written

$$\omega(x) = \omega_{i_1,\dots,i_k}(x)\,\mathrm{d}x^{i_1} \wedge \cdots \wedge \mathrm{d}x^{i_k},$$

where $x \mapsto \omega_{i_1,\dots,i_k}(x)$ are real-valued functions on $\mathbb{R}^n$. Then $x \mapsto \omega(x)$ is of $\mathscr{C}^p$-class if and only if the $\binom{n}{k}$ functions $x \mapsto \omega_{i_1,\dots,i_k}(x)$ are.

Before going farther, here are some examples of differential k-forms, with $E = \mathbb{R}^n$.

- For $k = 0$: a differential form of degree 0 is none other than a *function* on E, of $\mathscr{C}^p$-class, with values in $\mathbb{R}$.
- For $k = 1$: let ω^1 be a differential 1-form. Then there exists an n-tuple $(f_1, \dots, f_n)$ of functions of $\mathscr{C}^p$-class such that

$$\omega^1 = f_1\,\mathrm{d}x^1 + \cdots + f_n\,\mathrm{d}x^n.$$

- For $k = 2$: let ω^2 be a differential 2-form. There exist $n(n-1)/2$ functions $(f_{ij})_{1\leqslant i<j\leqslant n}$ (of $\mathscr{C}^p$-class) such that

$$\begin{aligned}\omega^2 &= f_{12}\,\mathrm{d}x^1 \wedge \mathrm{d}x^2 + f_{13}\,\mathrm{d}x^1 \wedge \mathrm{d}x^2 + \cdots + f_{1n}\,\mathrm{d}x^1 \wedge \mathrm{d}x^n \\ &\qquad + \cdots + f_{n-1,n}\,\mathrm{d}x^{n-1} \wedge \mathrm{d}x^n \\ &= \sum_{1\leqslant i<j\leqslant n} f_{ij}\,\mathrm{d}x^i \wedge \mathrm{d}x^j = f_{ij}\,\mathrm{d}x^i \wedge \mathrm{d}x^j.\end{aligned}$$

Note again the use of the convention of *partial summation* of repeated indices.

Remark 17.29 **Warning:** in what follows, as in most books, the point x in space where the extorior form is evaluated will be *omitted* from the notation, and we will often write ω instead of $\omega(x)$, when no ambiguity exists. Thus, for a 2-form ω, $\omega(\xi_1, \xi_2)$ will be the real number obtained by applying the exterior 2-form $\omega(x)$ to the vectors ξ_1 and ξ_2 in E.

17.2.b Exterior derivative

DEFINITION 17.30 Let $\omega = \omega_{i_1,\dots,i_k}(x)\,\mathrm{d}x^{i_1} \wedge \cdots \wedge \mathrm{d}x^{i_k}$ be a differential k-form. Since the functions $x \mapsto \omega_{i_1,\dots,i_k}(x)$ are differentiable, the differentials $\mathrm{d}\omega_{i_1,\dots,i_k}(x)$ are exterior 1-forms. The **exterior derivative** of the k-form ω is the differential form of degree $k+1$ given by

$$\mathrm{d}\omega = \mathrm{d}\omega_{i_1,\dots,i_k}(x) \wedge \mathrm{d}x^{i_1} \wedge \cdots \wedge \mathrm{d}x^{i_k}.$$

Example 17.31 Let $\omega = \omega_x \mathrm{d}x + \omega_y \mathrm{d}y + \omega_z \mathrm{d}z$ be a differential 1-form on $\mathbb{R}^3$. Then we have

$$\mathrm{d}\omega_x = \frac{\partial \omega_x}{\partial x}\,\mathrm{d}x + \frac{\partial \omega_x}{\partial y}\,\mathrm{d}y + \frac{\partial \omega_x}{\partial z}\,\mathrm{d}z,$$

and similarly for $\mathrm{d}\omega_y$ and $\mathrm{d}\omega_z$. It follows that

$$\begin{aligned}\mathrm{d}\omega &= \left(\frac{\partial \omega_x}{\partial x}\,\mathrm{d}x + \frac{\partial \omega_x}{\partial y}\,\mathrm{d}y + \frac{\partial \omega_x}{\partial z}\,\mathrm{d}z\right) \wedge \mathrm{d}x + \left(\frac{\partial \omega_y}{\partial x}\,\mathrm{d}x + \frac{\partial \omega_y}{\partial y}\,\mathrm{d}y + \frac{\partial \omega_y}{\partial z}\,\mathrm{d}z\right) \wedge \mathrm{d}y \\ &\qquad + \left(\frac{\partial \omega_z}{\partial x}\,\mathrm{d}x + \frac{\partial \omega_z}{\partial y}\,\mathrm{d}y + \frac{\partial \omega_z}{\partial z}\,\mathrm{d}z\right) \wedge \mathrm{d}z \\ &= \left(\frac{\partial \omega_y}{\partial x} - \frac{\partial \omega_x}{\partial y}\right) \mathrm{d}x \wedge \mathrm{d}y + \left(\frac{\partial \omega_z}{\partial y} - \frac{\partial \omega_y}{\partial z}\right) \mathrm{d}y \wedge \mathrm{d}z + \left(\frac{\partial \omega_x}{\partial z} - \frac{\partial \omega_z}{\partial x}\right) \mathrm{d}z \wedge \mathrm{d}x.\end{aligned}$$

THEOREM 17.32 *Let ω be a differential k-form of class at least $\mathscr{C}^2$. Then* $\mathrm{dd}\omega \equiv 0$, *in other words we have identically* $\boxed{\mathrm{d}^2 \equiv 0.}$

PROOF. We have by definition $\mathrm{d}\omega = \mathrm{d}\omega_{i_1,\dots,i_k}(x) \wedge \mathrm{d}x^{i_1} \wedge \cdots \wedge \mathrm{d}x^{i_k}$, with

$$\mathrm{d}\omega_{i_1,\dots,i_k} = \frac{\partial \omega_{i_1,\dots,i_k}}{\partial x^\alpha}\mathrm{d}x^\alpha.$$

It follows that

$$\mathrm{dd}\omega = \frac{\partial^2 \omega_{i_1,\dots,i_k}}{\partial x^\beta \, \partial x^\alpha} \, \mathrm{d}x^\beta \wedge \mathrm{d}x^\alpha \wedge \mathrm{d}x^{i_1} \wedge \cdots \wedge \mathrm{d}x^{i_k}.$$

But since the form ω is of $\mathscr{C}^2$-class, Schwarz's theorem on the commutativity of second partial derivatives implies that

$$\frac{\partial^2 \omega_{i_1,\dots,i_k}}{\partial x^\beta \, \partial x^\alpha} = \frac{\partial^2 \omega_{i_1,\dots,i_k}}{\partial x^\alpha \, \partial x^\beta} \qquad \text{and} \qquad \mathrm{d}x^\alpha \wedge \mathrm{d}x^\beta = -\mathrm{d}x^\beta \wedge \mathrm{d}x^\alpha,$$

which shows that the preceding expression is identically zero.

17.3 Integration of differential forms

The general theory of integration of differential forms is outside the scope of this book. Hence the results we present are only intended to give an idea of the flavor of this subject. The interested reader can turn for more details to a book of differential geometry such as Cartan [17], Nakahara [68], or Arnold [9]. Before discussing the general case, we mention two special situations: the integration of n-forms on $\mathbb{R}^n$ and of 1-forms on any $\mathbb{R}^n$.

DEFINITION 17.33 (Integral of a differential n-form) Let Ω be a sufficiently regular domain in $\mathbb{R}^n$ (so Ω is an n-dimensional "volume") and let ω be a differential n-form. Then there exists a function f such that we can write

$$\omega(\boldsymbol{x}) = f(\boldsymbol{x}) \, \mathrm{d}x^1 \wedge \cdots \wedge \mathrm{d}x^n,$$

and the **integral of ω on Ω** is defined to be

$$\int_\Omega \omega \stackrel{\text{def}}{=} \int_\Omega f(\boldsymbol{x}) \, \mathrm{d}x^1 \cdots \mathrm{d}x^n.$$

Be careful that this special case of integration of forms of degree n on a space of the same dimension is indeed very special.

The next special case is that of differential 1-forms. Before trying to integrate a 1-form ω on a domain, we must look again at what is the nature of ω. At a point $\boldsymbol{x} \in \mathbb{R}^n$, the value $\omega(\boldsymbol{x})$ of ω is an exterior 1-form, that is, simply a linear form which associates a real number to a vector. Where may we naturally find vectors? In differential geometry, vectors are defined in the same manner that velocity vectors are defined in physics: one chooses a curve $t \mapsto \gamma(t)$ going through $\boldsymbol{x}$ and one looks at what is the "velocity" of the moving point $\gamma(t)$ when it passes at the point $\boldsymbol{x}$.

So, consider a path $\gamma : [0,1] \to \mathbb{R}^n$, $t \mapsto \gamma(t)$. Then for any $t \in [0,1]$, the velocity vector $\gamma'(t) = \mathrm{d}\gamma/\mathrm{d}t$ provides, after being "fed" to $\omega\big(\gamma(t)\big)$ (the

George STOKES (1819–1903), English physicist and mathematician, discovered the concept of *uniform convergence*, but his most important contributions were in physics, in particular in fluid mechanics, where he investigated the differential equations introduced by Navier that now bear their names. The formula linking the circulation of a vector field and the flux of its curl is also due to him.

exterior form given by evaluating ω at the point $\gamma(t)$) a real number, namely, $g(t) = \omega\big(\gamma(t)\big)\big(\gamma'(t)\big)$; this real number may be integrated along the path, and the resulting integral will give the definition of the integral of ω on the path γ:

DEFINITION 17.34 (Integral of a differential 1-form on a path)
Let γ be a path and $\omega = f_i \, \mathrm{d}x^i$ a differential 1-form. The integral of ω on γ is defined as the integral

$$\int_\gamma \omega = \int_\gamma f_i \, \mathrm{d}x^i = \int_0^1 f_i\big(\gamma(t)\big) \frac{\mathrm{d}x^i}{\mathrm{d}t} \, \mathrm{d}t,$$

where the path is parameterized by

$$\begin{aligned} \gamma : [0,1] &\longrightarrow \mathbb{R}^n, \\ t &\longmapsto \gamma(t) = \big(x^1(t), \dots, x^n(t)\big). \end{aligned}$$

Example 17.35 In the plane $\mathbb{R}^2$, with $\omega = f \, \mathrm{d}x + g \, \mathrm{d}y$, we obtain

$$\int_\gamma \omega = \int_\gamma f \, \mathrm{d}x + g \, \mathrm{d}y = \int_\gamma f\big(\gamma(s)\big) \frac{\mathrm{d}x}{\mathrm{d}s} \, \mathrm{d}s + \int_\gamma g\big(\gamma(s)\big) \frac{\mathrm{d}y}{\mathrm{d}s} \, \mathrm{d}s.$$

This recovers the notion already seen (in complex analysis) of contour integration.

In the general case, how do we integrate a differential k-form? Most importantly, on *what domain* should one perform such an integral? Take a 2-form, for instance. At every point of this integration domain, we need *two* vectors. This is easily given if this domain is a smooth surface, for the two vectors can then be a basis of the tangent plane. *A differential 2-form must be integrated along a 2-dimensional surface.*

As an example, we consider a 2-form ω^2 in the space $\mathbb{R}^3$. We denote it

$$\omega = f_x \, \mathrm{d}y \wedge \mathrm{d}z + f_y \, \mathrm{d}z \wedge \mathrm{d}x + f_z \, \mathrm{d}x \wedge \mathrm{d}y.$$

(Note that we have changed conventions somewhat, using $\mathrm{d}z \wedge \mathrm{d}x$ instead of $\mathrm{d}x \wedge \mathrm{d}z$; this only amounts to changing the sign of f_y.) Let $\mathscr{S}$ be an oriented surface. At any point x in $\mathscr{S}$, we can perform the orthogonal projection of the basis vectors of $\mathbb{R}^3$ to the tangent plane of $\mathscr{S}$ at x (this is a linear operation). Hence three vectors u_x, u_y, and u_z are obtained, attached to each point $x \in \mathscr{S}$; they are not linearly independent since they belong to the two-dimensional tangent plane to the surface. Consider then $f_z \, \mathrm{d}x \wedge \mathrm{d}y(u_x, u_y)$, which is a real number. This may be integrated over the whole surface. Note that $\mathrm{d}x \wedge \mathrm{d}y(u_x, u_y) = n \cdot e_z$, and therefore we obtain

$$\begin{aligned} \int_{\mathscr{S}} \omega^2 &= \int_{\mathscr{S}} f_x \, \mathrm{d}y \wedge \mathrm{d}z + f_y \, \mathrm{d}z \wedge \mathrm{d}x + f_z \, \mathrm{d}x \wedge \mathrm{d}y \\ &= \int_{\mathscr{S}} f_x \, \mathrm{d}x + f_y \, \mathrm{d}y + f_z \, \mathrm{d}z = \int_{\mathscr{S}} f \cdot n, \end{aligned}$$

where n is the normal vector to the surface. The first integral involves the *differential form* ω, whereas the integral on the second line is an *integral of functions*.

This is easy to memorize: it suffices to replace $\mathrm{d}x \wedge \mathrm{d}y$ by $\mathrm{d}z$ and, similarly $\mathrm{d}y \wedge \mathrm{d}z$ by $\mathrm{d}x$ and $\mathrm{d}z \wedge \mathrm{d}x$ by $\mathrm{d}y$.

To generalize this result, it is necessary to study the behavior of a differential form during a change of coordinates. This is done in sidebar 6. One shows that a differential k-form may be integrated on a "surface" of dimension k, that is, there appears a duality[2] between

- differential 1-forms and paths (curves);
- differential 2-forms and surfaces;
- differential 3-forms and volumes (of dimension 3) or three-dimensional surfaces in $\mathbb{R}^n$;
- etc.

The following result is then particularly interesting:

THEOREM 17.36 (Stokes) *Let Ω be a smooth $(k+1)$-dimensional domain with k-dimensional boundary $\partial\Omega$ (possibly empty), and let ω be a differential k-form on E. Then we have*

$$\int_{\partial\Omega} \omega = \int_{\Omega} \mathrm{d}\omega.$$

This formula is associated with many names, including Newton, Leibniz, Gauss, Green, Ostrogradski, Stokes, and Poincaré, but it is in general called the "Stokes formula."

Example 17.37 The boundary of a path $\gamma : [0,1] \to \mathbb{R}^n$ is made of only the points $b = \gamma(1)$ and $a = \gamma(0)$. Let us consider the case $n = 1$ and let ω be a differential 0-form, that is, a

[2] The following is meant by "duality": the meeting of a k-form and a k-surface gives a number, just as the meeting of a linear form and a vector gives a number.

function $\omega = f : \mathbb{R} \to \mathbb{R}$. Then the Stokes formula means that we have

$$\int_\gamma \mathrm{d}f = \int_0^1 \frac{\mathrm{d}f}{\mathrm{d}t}(\gamma(t)) \cdot \gamma'(t)\,\mathrm{d}t = \int_a^b f'(s)\,\mathrm{d}s = f(b) - f(a),$$

which is of course known to the youngest children.

17.4 Poincaré's theorem

DEFINITION 17.38 (Exact, closed differential forms) A differential form ω is **closed** if $\mathrm{d}\omega = 0$.

A differential k-form ω is **exact** if there exists a differential $(k-1)$-form A such that $\omega = \mathrm{d}A$.

THEOREM 17.39 *Any exact differential form is closed.*

PROOF. Indeed, if ω is exact, there exists a form A such that $\omega = \mathrm{d}A$, and we then have $\mathrm{d}\omega = \mathrm{dd}A = 0$.

Note that an exact differential 1-form is none other than a **differential** (of a function). Indeed, if ω is an exact differential 1-form, there exists a 0-form, that is, a *function* f, such that $\omega = \mathrm{d}f$.

THEOREM 17.40 (Schwarz) *Let ω be a differential 1-form. If we write $\omega = f_i(x)\,\mathrm{d}x^i$, then the following are equivalent*

A) *ω is closed (i.e., $\mathrm{d}\omega = 0$) ;*

B) *we have* $\dfrac{\partial f_i}{\partial x^j} = \dfrac{\partial f_j}{\partial x^i}$ *for any $i, j \in \{1, 2, 3\}$.*

PROOF. Assume first that ω is closed. Then $\mathrm{d}\omega = 0 = \partial_k f_i\, \mathrm{d}x^k \wedge \mathrm{d}x^i$; since $\mathrm{d}x^k \wedge \mathrm{d}x^i = \mathrm{d}x^k \otimes \mathrm{d}x^i - \mathrm{d}x^i \otimes \mathrm{d}x^k$, this shows that $\partial_k f_i = \partial_i f_k$.

Conversely, $\partial_k f_i = \partial_i f_k$ obviously implies that $\mathrm{d}\omega = 0$.

Knowing whether the converse to Theorem 17.39 holds, that is, whether any closed form is exact, turns out to be of paramount importance. The answer depends in a crucial way on the shape and structure of the domain of definition of the differential form.

DEFINITION 17.41 (Contractible open set) A **contractible open set** in $\mathbb{R}^n$ is an open subset which may be continuously deformed to a point. Hence, Ω is contractible if and only if there exists a *continuous* map $C : \Omega \times [0, 1] \to \Omega$ and a point $p \in \Omega$ such that

Henri POINCARÉ (1854–1912) is one the most universal geniuses in the whole history of mathematics; he studied – and often revolutionized – most areas of the mathematics of his time, and also mechanics, astronomy, and philosophy. He created topology, used this new tool to investigate differential equations, flirted with chaos, studied asymptotic expansions, and proved that the time series used in astronomy are of this type. He also gave a major contribution to the three-body problem.

a) $C(x,0) = x$ for all $x \in \Omega$;

b) $C(x,1) = p$ for all $x \in \Omega$.

Example 17.42 A star-shaped open set is contractible. Here, an open set Ω is **star-shaped** if and only if, for any $x \in \Omega$, the line segment

$$[0,x] \stackrel{\text{def}}{=} \left\{\lambda x \; ; \; \lambda \in [0,1]\right\}$$

is contained is Ω. In particular, $\mathbb{R}^n$ is star-shaped and hence contractible. Here we can choose $p = 0$ and the map C may be defined by

$$\forall x \in \mathbb{R}^n \quad \forall \theta \in [0,1] \qquad C(x,\theta) = (1-\theta)x.$$

Example 17.43 The surface of a torus is not contractible.
$\mathbb{R}^n \setminus \{0\}$ is not contractible (although it is simply connected as soon as $n \geqslant 2$).

THEOREM 17.44 (Poincaré) *Let $\Omega \subset \mathbb{R}^n$ be a contractible open set in $\mathbb{R}$. Then any closed differential form on Ω is exact.*

In one important special case, a more general condition than contractibility is sufficient to ensure that a closed form is exact. Indeed, for differential 1-forms, we have

THEOREM 17.45 *Let $\Omega \subset \mathbb{R}^n$ be a* simply connected *and connected open set. Then any closed differential* 1*-form on Ω is exact.*

Remark 17.46 Be careful that this theorem is only valid for differential 1-forms. In general, it is not sufficient for a closed differential form to be defined on a simply connected set in order for it to be exact. We will see below (see page 480) a simple consequence of this for the field created by a magnetic monopole.

Notice that any contractible subset is simply connected (it suffices to look at the image of a path γ under the "contraction" C to see that it also contracts to a single point), but the converse is not true.

Counterexample 17.47 Consider the differential 1-form

$$\omega = \frac{(x-y)\,\mathrm{d}x + (x+y)\,\mathrm{d}y}{x^2+y^2}$$

in $\mathbb{R}^2$. It is easy to check that $\mathrm{d}\omega = 0$ on $\mathbb{R}^2 \setminus \{(0,0)\}$, using Schwarz's theorem, for instance. However, ω is *not* exact. One may indeed integrate ω to get

$$f(x,y) = \frac{1}{2}\log(x^2+y^2) - \arctan\left(\frac{x}{y}\right) + \mathrm{C}^{\underline{\mathrm{nt}}},$$

but this expression is singular at $y = 0$ and does not admit any continuous continuation on $\mathbb{R}^2 \setminus \{(0,0)\}$. This is possible because $\mathbb{R}^2 \setminus \{(0,0)\}$ is not simply connected.

17.5 Relations with vector calculus: gradient, divergence, curl

17.5.a Differential forms in dimension 3

Consider now the case of differential forms in three-dimensional space $\mathbb{R}^3$. These are

- the 0-forms, that is, functions;
- the 1-forms, which are linear combinations of $\mathrm{d}x$, $\mathrm{d}y$, and $\mathrm{d}z$;
- the 2-forms which are linear combinatons of $\mathrm{d}x \wedge \mathrm{d}y$, $\mathrm{d}y \wedge \mathrm{d}z$, and $\mathrm{d}z \wedge \mathrm{d}x$;
- the forms of degree 3, multiples of the volume form $\mathrm{d}x \wedge \mathrm{d}y \wedge \mathrm{d}z$.

Let ω be a differential 1-form, so that

$$\omega = f_i\,\mathrm{d}x^i = f_x\,\mathrm{d}x + f_y\,\mathrm{d}y + f_z\,\mathrm{d}z.$$

Let now γ be a smooth *closed* path in $\mathbb{R}^3$. Then this path bounds an oriented surface (oriented using the orientation of γ), denoted $\mathscr{S}$. So we have $\gamma = \partial\mathscr{S}$. Since the exterior derivative of ω is given by

$$\mathrm{d}\omega = \left(\frac{\partial f_y}{\partial x} - \frac{\partial f_x}{\partial y}\right)\mathrm{d}x\wedge\mathrm{d}y + \left(\frac{\partial f_z}{\partial y} - \frac{\partial f_y}{\partial z}\right)\mathrm{d}y\wedge\mathrm{d}z + \left(\frac{\partial f_x}{\partial z} - \frac{\partial f_z}{\partial x}\right)\mathrm{d}z\wedge\mathrm{d}x,$$

Stokes's formula gives

$$\int_{\mathscr{S}} \mathrm{d}\omega = \int_{\gamma} \omega,$$

or, putting $\boldsymbol{f} = (f_x, f_y, f_z)$ and using the fact that the components of $\mathrm{d}\omega$ are the components of the curl $\mathbf{curl}\,\boldsymbol{f}$,

$$\iint_{\mathscr{S}} \mathbf{curl}\,\boldsymbol{f}\ \mathrm{d}^2\sigma = \oint_{\gamma} \boldsymbol{f}\cdot \mathrm{d}\boldsymbol{\ell}, \tag{17.1}$$

or in other words still, the flux of the curl of f through a surface is equal to the circulation of f along the boundary of this surface (**Green-Ostrogradski** formula).

Similarly, if ω is a differential 2-form given by

$$\omega \stackrel{\text{def}}{=} f_x\,\mathrm{d}y\wedge\mathrm{d}z + f_y\,\mathrm{d}z\wedge\mathrm{d}x + f_z\,\mathrm{d}x\wedge\mathrm{d}y,$$

its exterior derivative is then

$$\mathrm{d}\omega = \left(\frac{\partial f_x}{\partial x}+\frac{\partial f_y}{\partial y}+\frac{\partial f_z}{\partial z}\right)\mathrm{d}x\wedge\mathrm{d}y\wedge\mathrm{d}z = (\operatorname{div}\boldsymbol{f}\,)\,\mathrm{d}x\wedge\mathrm{d}y\wedge\mathrm{d}z,$$

and if $\mathscr{S}$ is a closed surface bounding a volume $\mathscr{V}$ such that $\mathscr{S} = \partial\mathscr{V}$, we obtain

$$\int_{\mathscr{V}} \omega = \int_{\mathscr{S}} \mathrm{d}\omega, \qquad \text{that is,} \qquad \iint_{\mathscr{S}} \boldsymbol{f}\cdot\mathrm{d}^2\boldsymbol{\sigma} = \iiint_{\mathscr{V}} \operatorname{div}\boldsymbol{f}\ \mathrm{d}^3\boldsymbol{r}.$$

Finally, if $\omega = f$ is a differential 0-form, that is, a function, since the differential $\mathrm{d}f$ satisfies $\mathrm{d}f(\boldsymbol{n}) = \mathbf{grad}\,f\cdot\boldsymbol{n}$, the Stokes formula expresses that

$$\int_{\gamma} \mathbf{grad}\,f\cdot\mathrm{d}\boldsymbol{\ell} = f\big(\gamma(1)\big) - f\big(\gamma(0)\big).$$

Note that the relations just stated between exterior derivatives and the gradient, divergence and curl operators show that the identity $\mathrm{d}^2 \equiv 0$ may simply be expressed, in the language of the man in the street, as

$$\operatorname{div}\mathbf{curl} \equiv 0 \qquad \text{and} \qquad \mathbf{curl}\,\mathbf{grad} \equiv \mathbf{0}.$$

17.5.b Existence of the scalar electrostatic potential

Let $\boldsymbol{f} = (f_1, f_2, f_3)$ be a vector field on $\mathbb{R}^3$, and assume it satisfies the condition

$$\boldsymbol{\nabla}\wedge\boldsymbol{f} = \mathbf{curl}\,\boldsymbol{f} \equiv \mathbf{0} \tag{17.2}$$

at every point of a domain $\Omega \subset \mathbb{R}^3$ which is *simply connected*. Define a differential 1-form on Ω by

$$\omega = f_i\,\mathrm{d}x^i = f_x\,\mathrm{d}x + f_y\,\mathrm{d}y + f_z\,\mathrm{d}z.$$

As we have seen, the exterior derivative of ω is

$$\mathrm{d}\omega = \left(\frac{\partial f_y}{\partial x}-\frac{\partial f_x}{\partial y}\right)\mathrm{d}x\wedge\mathrm{d}y + \left(\frac{\partial f_z}{\partial y}-\frac{\partial f_y}{\partial z}\right)\mathrm{d}y\wedge\mathrm{d}z + \left(\frac{\partial f_x}{\partial z}-\frac{\partial f_z}{\partial x}\right)\mathrm{d}z\wedge\mathrm{d}x,$$

hence condition (17.2) can be expressed more simply by stating that

$$\mathrm{d}\omega \equiv 0. \tag{17.2$'$}$$

Poincaré's theorem (the second version, Theorem 17.45) ensures that there exists a differential 0-form, that is, a function $\varphi : \mathbb{R} \to \mathbb{R}$, such that

$$\omega = \mathrm{d}\varphi = \frac{\partial \varphi}{\partial x}\,\mathrm{d}x + \frac{\partial \varphi}{\partial y}\,\mathrm{d}y + \frac{\partial \varphi}{\partial z}\,\mathrm{d}z.$$

Comparing with the definition $\omega = f_x\,\mathrm{d}x + f_y\,\mathrm{d}y + f_z\,\mathrm{d}z$, it follows that

$$f = \nabla\varphi.$$

This proves the following theorem:

THEOREM 17.48 (Existence of the scalar potential) *Let $\Omega \subset \mathbb{R}^3$ be a simply connected open set in $\mathbb{R}^3$, and let $f : \Omega \to \mathbb{R}^3$ be an irrotational vector field on Ω, that is, such that* $\mathbf{curl}\, f \equiv \mathbf{0}$. *Then there exists a potential $\varphi : \Omega \to \mathbb{R}$ such that $f = \nabla\varphi$.*

Application to electrostatic

Since the electrostatic field satisfies $\nabla \wedge \boldsymbol{E} \equiv \mathbf{0}$ on $\mathbb{R}^3$, both with and without charges[3] it follows that it derives from a potential:

$$\boldsymbol{E} = -\nabla\varphi.$$

It is important to remark that the scalar potential only exists because $\mathbb{R}^3$ is simply connected. If the equation were only satisfied in a non-simply connected subset $\Omega \subset \mathbb{R}^3$, there would be no reason for the potential to exist globally (its local existence being of course always true).

Too many physics books do not hesitate to state things in roughly the following way: "since we know that **curl grad** $\equiv \mathbf{0}$, *it follows that*: if **curl** $\boldsymbol{E} \equiv \mathbf{0}$, then $E = -$ **grad** φ." This, we just saw, is a *mathematical absurdity* and, worse, a dreadful *logical mistake*.[4] Moreover this may discredit physics in the eyes of students who, in general, are perfectly aware that they are being hoodwinked, even if they may not know the necessary tools and terminology to explain the problem.[5] It is perfectly possible to remain rigorous in physics; students should be made aware of this.

Should you, reader, teach physics one day, do not make this mistake!

[3] This is one of the *homogeneous* Maxwell equations, giving the global *structure* of fields in the setting of electrostatics. Hence all magnetic phenomena have been neglected.

[4] This amounts to mixing up the logical statements $(P \Rightarrow Q)$ and $(Q \Rightarrow P)$!

[5] And who, moreover, would be quite incapable of finding a counterexample: *it is true* that E comes from a potential, but for a very different reason.

17.5.c Existence of the vector potential

Let now $f : \Omega \to \mathbb{R}^3$ be a vector field with vanishing divergence:

$$\operatorname{div} f = \nabla \cdot f \equiv 0 \qquad \text{on } \Omega. \tag{17.3}$$

Define a differential 2-form

$$\omega \stackrel{\text{def}}{=} f_1 \, \mathrm{d}x^2 \wedge \mathrm{d}x^3 + f_2 \, \mathrm{d}x^3 \wedge \mathrm{d}x^1 + f_3 \, \mathrm{d}x^1 \wedge \mathrm{d}x^2.$$

Then the exterior derivative of ω is simply

$$\mathrm{d}\omega = \left(\frac{\partial f_1}{\partial x^1} + \frac{\partial f_2}{\partial x^2} + \frac{\partial f_3}{\partial x^3} \right) \mathrm{d}x^1 \wedge \mathrm{d}x^2 \wedge \mathrm{d}x^3,$$

and the condition (17.3) can be expressed again simply as

$$\mathrm{d}\omega = 0.$$

Poincaré's theorem (Theorem 17.44 now) ensures that if Ω is contractible, there exists a differential 1-form $A = A_1 \, \mathrm{d}x^1 + A_2 \, \mathrm{d}x^2 + A_3 \, \mathrm{d}x^3$ such that

$$\begin{aligned} \omega &= \mathrm{d}A \\ &= \left(\frac{\partial A_2}{\partial x^1} - \frac{\partial A_1}{\partial x^2} \right) \mathrm{d}x^1 \wedge \mathrm{d}x^2 + \left(\frac{\partial A_3}{\partial x^2} - \frac{\partial A_2}{\partial x^3} \right) \mathrm{d}x^2 \wedge \mathrm{d}x^3 \\ &\quad + \left(\frac{\partial A_1}{\partial x^3} - \frac{\partial A_3}{\partial x^1} \right) \mathrm{d}x^3 \wedge \mathrm{d}x^1, \end{aligned}$$

which, putting $\boldsymbol{A} = (A_1, A_2, A_3)$ and comparing with the definition of ω, shows that

$$f = \nabla \wedge \boldsymbol{A} = \mathbf{curl}\, \boldsymbol{A}.$$

THEOREM 17.49 (Existence of the vector potential) *Let $\Omega \subset \mathbb{R}^3$ be a contractible open subset and let $f : \Omega \to \mathbb{R}^3$ be a vector field with zero divergence on Ω: $\operatorname{div} f \equiv \mathbf{0}$. Then there exists a vector potential $\boldsymbol{A} : \Omega \to \mathbb{R}^3$ such that $f = \nabla \wedge \boldsymbol{A}$.*

In the space $\mathbb{R}^3$, the magnetic field satisfies the relation $\operatorname{div} \boldsymbol{B} \equiv 0$, and this independently of the distribution of charges. It follows that the magnetic field, on $\mathbb{R}^3$ which is contractible, comes from a vector potential $\boldsymbol{A}$.

Remark 17.50 In the general setting of electromagnetism, one first proves the existence of the vector potential $\boldsymbol{A}$; then, from the equation $\mathbf{curl}\, \boldsymbol{E} = -\partial \boldsymbol{B}/\partial t$, it follows that $\mathbf{curl}(\boldsymbol{E} + \partial \boldsymbol{A}/\partial t) = 0$ and hence the existence of the scalar potential φ such that

$$\boldsymbol{E} = -\nabla \varphi - \frac{\partial \boldsymbol{A}}{\partial t}$$

on any contractible open set.

17.5.d Magnetic monopoles

The magnetic field B satisfies $\operatorname{div} B \equiv 0$, and so it comes from a vector potential; or, more precisely, this is ensured by Poincaré's theorem when considering a contractible set (in particular it is not sufficient to have a simply connected set).

What happens if we consider a *magnetic monopole*, that is, a *point-like* particle which transforms one the Maxwell equations to be

$$\operatorname{div} B(r,t) = q_m \, \delta(r)\ ? \tag{17.4}$$

The relation $\operatorname{div} B \equiv 0$ is satisfied now only on $\mathbb{R}^3 \setminus \{0\}$, *which is not contractible.* It is possible to show then that there *does not exist* a vector potential A defined on $\mathbb{R}^3 \setminus \{0\}$ such that we have $B = \nabla \wedge A$ at every point of $\mathbb{R}^3 \setminus \{0\}$. But since a *local* description of the magnetic field using a vector potential remains possible, it is possible to show that one may defined two potentials A_1 and A_2 defined on a neighborhood of the upper half-space (resp. lower half-space) such that the relations $\nabla \wedge A_i = B$ for $i = 1, 2$ hold on the respective domains where the vectors are defined. For instance, A_1 and A_2 may be defined on the following grayed domains:

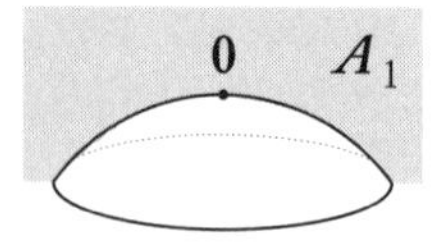

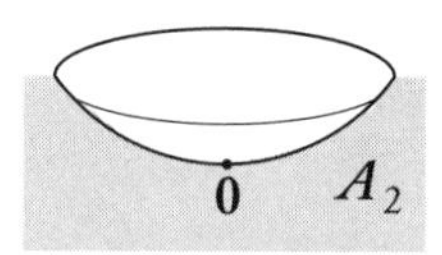

Hence the existence of magnetic monopoles would have as a consequence that the vector potential does not exist. Experiences have been attempted to discover magnetic monopoles (which have amusing theoretical properties; for instance, their existence automatically implies the quantification of the elementary electric charge), but without success.

17.6

Electromagnetism in the language of differential forms

Physicists see the vector potential usually as a *vector*, but it should rather be seen as a *differential form.* Indeed, if A itself does not have an intrinsic meaning, the integral of A along a path has one: for a closed path γ, the integral $\oint_\gamma A \cdot d\ell$ is equal to the flux of the magnetic field through a surface $\mathscr{S}$ bounded by γ (see formula (17.1)).

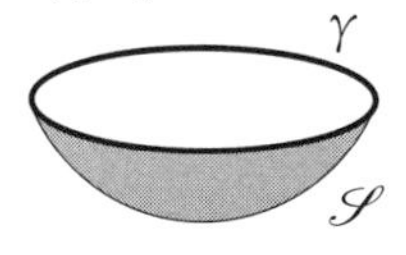

This is why, in special relativity, the components of the potential four-vector (vector potential + scalar potential) are denoted in *contravariant* notation:

$$\mathsf{A} = A_\mu \, \mathrm{d}x^\mu.$$

The **Faraday tensor** (which is a differential 2-form or equivalently a field of $\binom{0}{2}$-tensors) is defined as the exterior derivative of A:

$$F = \mathrm{d}\mathsf{A}, \quad \text{namely,} \quad F = \tfrac{1}{2} F_{\mu\nu} \, \mathrm{d}x^\mu \wedge \mathrm{d}x^\nu = \tfrac{1}{2}\left(\partial_\mu A_\nu - \partial_\nu A_\mu\right) \mathrm{d}x^\mu \wedge \mathrm{d}x^\nu,$$

where *complete* summation on repeated indices is performed. Denoting the potential four-vector $\mathsf{A} = (A_0, \boldsymbol{A})$, $\boldsymbol{E} = \mathbf{grad}\, A_0 - \partial \boldsymbol{A}/\partial t$ and $\boldsymbol{B} = \mathbf{curl}\, \boldsymbol{A}$, the coordinates of F are given by

$$\begin{aligned} F_{01} &= E_x, & F_{12} &= -B_z, \\ F_{02} &= E_y, & F_{13} &= B_y, \\ F_{03} &= E_z, & F_{23} &= -B_x, \end{aligned}$$

that is, we have

$$F = E_x \, \mathrm{d}x \wedge \mathrm{d}t + E_y \, \mathrm{d}y \wedge \mathrm{d}t + E_z \, \mathrm{d}z \wedge \mathrm{d}t - B_z \, \mathrm{d}x \wedge \mathrm{d}y - B_x \, \mathrm{d}y \wedge \mathrm{d}z + B_y \, \mathrm{d}z \wedge \mathrm{d}x.$$

Since $F = \mathrm{d}\mathsf{A}$, its exterior derivative $\mathrm{d}F$ is zero, which means that

$$\begin{aligned} 0 = &\left(\frac{\partial B_x}{\partial x} + \frac{\partial B_y}{\partial y} + \frac{\partial B_z}{\partial z}\right) \mathrm{d}x \wedge \mathrm{d}y \wedge \mathrm{d}z \\ &+ \left(\frac{\partial E_z}{\partial y} - \frac{\partial E_y}{\partial z} + \frac{\partial B_x}{\partial t}\right) \mathrm{d}t \wedge \mathrm{d}y \wedge \mathrm{d}z \\ &+ \left(\frac{\partial E_x}{\partial z} - \frac{\partial E_z}{\partial x} + \frac{\partial B_y}{\partial t}\right) \mathrm{d}t \wedge \mathrm{d}z \wedge \mathrm{d}x \\ &+ \left(\frac{\partial E_y}{\partial x} - \frac{\partial E_x}{\partial y} + \frac{\partial B_z}{\partial t}\right) \mathrm{d}t \wedge \mathrm{d}x \wedge \mathrm{d}y. \end{aligned}$$

If we reorder the terms, we obtain the Maxwell equations without source:

$$\mathrm{d}F = 0 \quad \Longleftrightarrow \quad \begin{cases} \mathbf{curl}\, \boldsymbol{E} = -\dfrac{\partial \boldsymbol{B}}{\partial t} \\ \operatorname{div} \boldsymbol{B} = 0 \end{cases} \quad \textit{Maxwell equations without source.}$$

How is it possible to obtain the Maxwell equations with source? We need to involve *derivatives* of $\boldsymbol{E}$ and $\boldsymbol{B}$, but we cannot get this from $\mathrm{d}F$ (which is identically zero). Another differential form is needed, constructed using $\boldsymbol{E}$ and $\boldsymbol{B}$ and involving derivatives of the first order.

Now notice that the space of exterior 1-forms is a vector space of dimension $\binom{4}{1} = 4$ (generated by $\mathrm{d}t$, $\mathrm{d}x$, $\mathrm{d}y$, and $\mathrm{d}z$), and the the space of exterior

3-forms is of dimension $\binom{4}{3} = 4$ also. Using the *Hodge operator*, an isomorphism between both spaces is constructed:

DEFINITION 17.51 (Hodge operator) The **Hodge operator**, denoted "$\star$", is a linear map defined on the spaces of exterior forms on $\mathbb{R}^n$:

$$\star : \qquad \Lambda^{*k}(\mathbb{R}^n) \longrightarrow \Lambda^{*(n-k)}(\mathbb{R}^n),$$
$$\mathrm{d}x^{i_1} \wedge \cdots \wedge \mathrm{d}x^{i_k} \longmapsto \frac{1}{(n-k)!}\, \varepsilon_{i_1,\ldots,i_k,i_{k+1},\ldots,i_n}\, \mathrm{d}x^{i_{k+1}} \wedge \cdots \wedge \mathrm{d}x^{i_n}$$

with complete summation on repeated indices. For any $k \in [\![0, n]\!]$, the Hodge operator is an isomorphism between the space of k-forms and the space of $(n-k)$-forms.

Example 17.52 *Consider the case* $n = 3$. Then ${}^\star\mathrm{d}x = \frac{1}{(3-1)!}\left[\mathrm{d}y \wedge \mathrm{d}z - \mathrm{d}z \wedge \mathrm{d}y\right] = \mathrm{d}y \wedge \mathrm{d}z$. Similarly

$${}^\star(\mathrm{d}x \wedge \mathrm{d}y) = \frac{1}{(3-2)!}\mathrm{d}z = \mathrm{d}z.$$

The Hodge dual of a k-form is thus a $(3-k)$-form.

In the case $n = 4$, we find for instance that ${}^\star\mathrm{d}x = \mathrm{d}y \wedge \mathrm{d}z \wedge \mathrm{d}t$ and that ${}^\star(\mathrm{d}x \wedge \mathrm{d}y) = \mathrm{d}z \wedge \mathrm{d}t$. A k-form is mapped to a $(4-k)$-form.

From the differential 2-form F, we can thus construct *another* 2-form, namely, ${}^\star F$. It is given by

$${}^\star F = -B_x\,\mathrm{d}x\wedge\mathrm{d}t - B_y\,\mathrm{d}y\wedge\mathrm{d}t - B_z\,\mathrm{d}z\wedge\mathrm{d}t + E_z\,\mathrm{d}x\wedge\mathrm{d}y + E_x\,\mathrm{d}y\wedge\mathrm{d}z + E_y\,\mathrm{d}z\wedge\mathrm{d}x.$$

Let J be the differential 1-form $J = \rho\,\mathrm{d}t + j_x\,\mathrm{d}x + j_y\,\mathrm{d}y + j_z\,\mathrm{d}z$, and consider its dual (image) by the Hodge operator

$${}^\star J = \rho\,\mathrm{d}x \wedge \mathrm{d}y \wedge \mathrm{d}z - j_x\,\mathrm{d}t \wedge \mathrm{d}y \wedge \mathrm{d}z - j_y\,\mathrm{d}t \wedge \mathrm{d}z \wedge \mathrm{d}x - j_z\,\mathrm{d}t \wedge \mathrm{d}x \wedge \mathrm{d}y.$$

The equation

$$\boxed{\mathrm{d}^\star F = {}^\star J} \tag{17.5}$$

is then equivalent to

$$\mathbf{curl}\,\boldsymbol{B} - \frac{\partial \boldsymbol{E}}{\partial t} = -\boldsymbol{j} \qquad \text{div}\,\boldsymbol{E} = \rho$$

Maxwell equations with source.

Before concluding, a last remark about the differential form ${}^\star J$. The equation (17.5) means that this form is *exact* (being the exterior derivative of ${}^\star F$). Hence its exterior derivative is identically zero, that is, we have

$$\mathrm{d}^\star J = \left(\frac{\partial \rho}{\partial t} + \frac{\partial j_x}{\partial x} + \frac{\partial j_y}{\partial y} + \frac{\partial j_z}{\partial z}\right) \mathrm{d}t \wedge \mathrm{d}x \wedge \mathrm{d}y \wedge \mathrm{d}z \equiv 0,$$

which is simply the equation of conservation of charge, which is indeed known to follow from the Maxwell equations.

Note that the components of F and ${}^\star F$ may be written down in matrix form. They are

$$F = (F_{\mu\nu}) = \begin{pmatrix} 0 & E_x & E_y & E_z \\ -E_x & 0 & -B_z & B_y \\ -E_y & B_z & 0 & -B_x \\ -E_z & -B_y & B_x & 0 \end{pmatrix},$$

$${}^\star F = ({}^\star F^{\mu\nu}) = \begin{pmatrix} 0 & -B_x & -B_y & -B_z \\ B_x & 0 & E_z & -E_y \\ B_y & -E_z & 0 & E_x \\ B_z & E_y & -E_x & 0 \end{pmatrix}.$$

The coordinates of F and ${}^\star F$ satisfy also

$${}^\star F^{\alpha\beta} = \frac{1}{2}\varepsilon^{\alpha\beta\mu\nu} F_{\mu\nu},$$

where ε is the totally antisymmetric tensor defined by $\varepsilon^{1234} = 1$, which is called the **Levi-Civita tensor**. It satisfies $\varepsilon^{\alpha\beta\mu\nu} = 1$ if $(\alpha, \beta, \mu, \nu)$ is an even permutation of $(1, 2, 3, 4)$, $\varepsilon^{\alpha\beta\mu\nu} = -1$ if it is an odd permutation, and $\varepsilon^{\alpha\beta\mu\nu} = 0$ if any index is repeated more than once.

Remark 17.53 What are the laws of electromagnetism in two dimensions? If we want to keep the *local* nature of the fundamental laws of the universe, it is their differential expression which should be preserved in order to extend them to such imaginary universes. Thus, one should keep the structure of the Faraday tensor for the electromagnetic field. This may be written here

$$F_{\mu\nu} = \begin{pmatrix} 0 & E_x & E_y \\ -E_x & 0 & B \\ -E_y & B & 0 \end{pmatrix},$$

which shows that the electric field has two components, whereas the magnetic field is a *scalar* field (or rather a *pseudo-scalar*, which changes sign during a change of basis that reverses the orientation).

PROBLEM

♦ **Problem 6 (Proca lagrangian and Cavendish experiment)** We want to change the Maxwell equations in order to describe a "photon with nonzero mass." For this purpose, we start from the lagrangian description of the electromagnetic field, and add a "mass term" to the lagrangian. Then we look for detectable effects of the resulting equations.

I. Recall that a lagrangian density describing the laws[6] of electromagnetism is given by

$$\mathscr{L} \stackrel{\text{def}}{=} -\frac{1}{4} F_{\alpha\beta} F^{\alpha\beta} - J^{\alpha} A_{\alpha} / \varepsilon_0,$$

with units where $\hbar = c = 1$ [49, 50, 57], where the $F_{\alpha\beta}$ are the covariant coordinates of the Faraday tensor, related to the potential four-vector by $F_{\alpha\beta} = \partial_\alpha A_\beta - \partial_\beta A_\alpha$. We have denoted here $\partial_\alpha = \partial / \partial x^\alpha$ and used the Einstein convention for the summation of repeated indices (see Chapter 16 concerning tensors).

1. Show that the equations of movement, given by the Euler-Lagrange equations

$$\frac{\partial \mathscr{L}}{\partial A_\mu} = \partial_\nu \frac{\partial \mathscr{L}}{\partial(\partial_\nu A_\mu)}$$

provide the Maxwell equations with source. Write them in *Lorentz gauge*, in which the condition $\partial \cdot \mathbf{A} = \partial^\mu A_\mu = 0$ is imposed.

2. A mass term of the type

$$\frac{\mu^2}{2} \mathbf{A}^2 = \frac{\mu^2}{2} A_\mu A^\mu,$$

is added to the classical lagrangian to obtain the **Proca lagrangian**

$$\mathscr{L}_{\text{P}} \stackrel{\text{def}}{=} -\frac{1}{4} F_{\alpha\beta} F^{\alpha\beta} - J^{\alpha} A^{\alpha} / \varepsilon_0 + \frac{\mu^2}{2} A_\mu A^\mu.$$

Write the equations of movement associated with this lagrangian. Show that, in Lorentz gauge, they are given by

$$(\square + \mu^2) A_\alpha = J_\alpha / \varepsilon_0.$$

3. Express the relation between the frequency of an electromagnetic wave and its wave number. Show that μ has the dimension of mass (still with $\hbar = c = 1$). Using the de Broglie relations, show that the waves thus described are *massive*.

4. Consider the field created by a *single, immobile, point-like* particle situated at the origin. Show that only the component $A_0 = \varphi$ is nonzero. Show that this scalar potential is of the "Yukawa potential" type and is equal to

$$\varphi(\boldsymbol{r}) = \frac{q}{4\pi\varepsilon_0} \cdot \frac{\mathrm{e}^{-\mu r}}{r}.$$

What is the typical decay distance for this potential?

II. In 1772, CAVENDISH realized an experiment to check the $1/r$ decay of the Coulomb potential [50].

Consider two hollow spheres, perfectly conducting, concentric with respective radii R_1 and R_2, with $R_1 < R_2$. At the beginning of the experiment, both spheres are electrically neutral, and the exterior sphere is given a potential V. Then the spheres are joined (with a thin metallic wire for instance).

---

[6] That is, the Maxwell equations for the electromagnetic field, and the Newton equation with Lorentz force for particles.

1. Show that, if the electrostatic potential created by a point-like charge follows the Coulomb law in $1/r$, the inner sphere *remains electrically neutral.*
2. We assume now that the electromagnetic field is described by the Proca lagrangian and that the electric potential created by a point-like charge follows the Yukawa potential. Show that this potential satisfies the differential equation

$$(-\triangle + \mu^2)\varphi = \rho/\varepsilon_0.$$

3. Show that the electric field has a discontinuity σ/ε_0 when passing through a *surface* density of charge σ.
4. Compute the potential at any point of space after the experiment is ended (both spheres are at the same potential V). Deduce what is the charge of the inner sphere. (One can assume that μ is small compared to $1/R_1$ and $1/R_2$ and simplify the expressions accordingly.)
5. Knowing that Cavendish, in his time, could not measure any charge on the inner sphere, and that he was able to measure charges as small as 10^{-9} C, deduce an upper bound for the mass of a photon. Take $V = 10\,000$ V and $R_1 \approx R_2 = 30$ cm.

   Any comments?

SOLUTION

♦ Solution of problem 6.

I. 1. In Lorentz gauge, one finds $\partial_\mu F^{\mu\nu} = J^\nu$, which is indeed the covariant form of the Maxwell equations.

2. Immediate computation.

3. Notice first that, in analogy with the Klein-Gordon equation, the additional term deserves to be called a "mass term." If we consider a monochromatic wave with frequency ω and wave vector $\boldsymbol{k}$, the equation of movement in vacuum $(\Box + \mu^2)A_\mu = 0$ gives $-\omega^2 + \boldsymbol{k}^2 + \mu^2 = 0$. The de Broglie relations link frequency and energy on the one hand, and momentum and wave vector on the other hand. We obtain then $E^2 = \boldsymbol{p}^2 + \mu^2$, which may be interpreted as the Einstein relation between energy, momentum, and mass, taking μ as the mass. In fact, it is rather $\hbar\mu/c$ which has the dimension of a mass.

4. We have $J_\alpha = (q\delta, \boldsymbol{0})$; the components A_i for $i = 1, 2, 3$ therefore are solutions of the equation $(\Box + \mu^2)J_i \equiv 0$ and are zero up to a free term.

   The component $\varphi = A_0$, on the other hand, satisfies

$$\frac{\mathrm{d}}{\mathrm{d}t}\varphi - \triangle\varphi + \mu^2\varphi = q\,\delta(\boldsymbol{r})/\varepsilon_0,$$

   which is independent of time. Looking for a stationary solution, we write therefore

$$(-\triangle + \mu^2)\varphi = q\,\delta(\boldsymbol{r})/\varepsilon_0,$$

   which (see Problem 5, page 325) can be solved using the Fourier transform and a residue computation, and yields the formula stated. The typical decay distance is therefore μ^{-1}.

II. 1. When the exterior sphere is charged with the potential V, the potential at any point exterior to the two spheres is (using for instance Gauss's theorem on the electric field)

$$\varphi(\boldsymbol{r}) \equiv V\,\frac{R_2}{r} \qquad \text{for all } r > R_2.$$

The inner potential, on the other hand, is uniform and equal to V (same method). In particular, the inner sphere is *already* at the same potential V. The electric wire is also itself entirely at the potential V. Hence there is no electric field, no current is formed, and the inner sphere remains neutral.

2. If electromagnetism follows the Proca lagrangian, then the Poisson law is not valid anymore, and consequently, Gauss's theorem is not either. The potential created by a point-like charge being

$$\Phi(\boldsymbol{r}) = \frac{1}{4\pi\varepsilon_0 r} e^{-\mu r},$$

it satisfies the differential equation $(\mu^2 - \triangle)\Phi = \delta/\varepsilon_0$ (which is indeed the differential equation of Question I.2 in stationary regime); by convolution, the potential created by a distribution of charge ρ therefore satisfies $(\mu^2 - \triangle)\varphi = \rho/\varepsilon_0$ (recall that $\triangle(\Phi * \rho) = (\triangle\Phi) * \rho$).

3. Gauss's theorem is not valid anymore, since we have

$$\operatorname{div} \boldsymbol{E} = -\triangle\varphi = \rho/\varepsilon_0 - \mu^2\varphi.$$

Consider a charged surface element, surrounded by a surface $\mathscr{S}$.

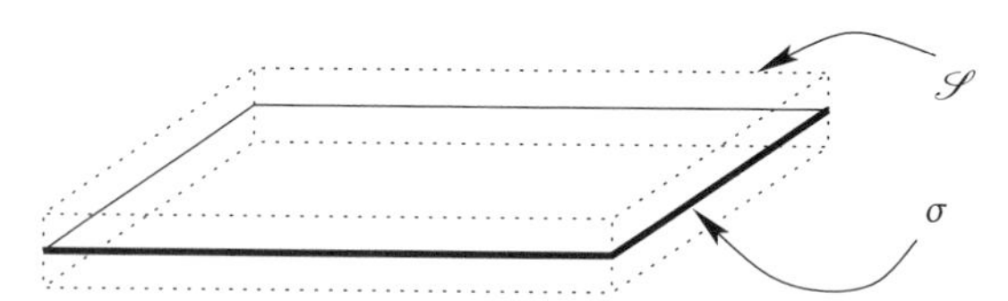

Then we have

$$\iint_{\mathscr{S}} \boldsymbol{E} \cdot \mathrm{d}\boldsymbol{S} = \iiint_{\mathscr{V}} \triangle\varphi \, \mathrm{d}^3\boldsymbol{r} = \iiint_{\mathscr{V}} \left(\frac{\rho}{\varepsilon_0} - \mu^2\varphi\right) \mathrm{d}^3\boldsymbol{r}.$$

But, close to σ, the Yukawa potential is regular (i.e., behaves like the Coulomb potential), and hence $\iiint_{\mathscr{V}} \mu^2\varphi \, \mathrm{d}^3\boldsymbol{r}$ tends to 0 when the volume tends to 0. The distribution ρ is a Dirac surface distribution surfacique. It follows that the discontinuity of the normal component of the electric field is σ/ε_0.

4. Since the potential has radial symmetry, it satisfies the equation

$$\frac{1}{r}\frac{\partial^2}{\partial r^2}(r\varphi) - \mu^2\varphi = 0$$

at every point in space except where $r = R_1$ or $r = R_2$. The general form of the potential is therefore

$$\varphi(r) = \frac{A}{r}\cosh\mu r + \frac{B}{r}\sinh\mu r.$$

Denote by φ, χ, and ψ the functions giving the potential (respectively) inside the inner sphere, between the two spheres, and outside the exterior sphere.

The potential φ is bounded in the neighborhood of the origin, and therefore may be written

$$\varphi(r) = A_1 \frac{\sinh(\mu r)}{r},$$

which, together with the condition of continuity of the potential $\varphi(R_1) = V$, gives

$$\varphi(r) = V\frac{R_1}{r}\frac{\sinh(\mu r)}{\sinh(\mu R_1)}.$$

The potential χ is of the form

$$\chi(r) = \frac{A}{r}\cosh\mu r + \frac{B}{r}\sinh\mu r \qquad \text{for all } R_1 < r < R_2,$$

and moreover satisfies $\chi(R_1) = \chi(R_2) = V$, which leads to the system of equations

$$\begin{cases} A\,\mathrm{e}^{-\mu R_1} + B\,\mathrm{e}^{\mu R_1} = R_1 V, \\ A\,\mathrm{e}^{-\mu R_2} + B\,\mathrm{e}^{\mu R_2} = R_2 V. \end{cases}$$

Solving this linear system, we obtain

$$\chi(r) = \frac{V}{r\sinh(\mu\Delta R)}\left(R_1\sinh\mu(R_2 - r) + R_2\sinh\mu(r - R_1)\right)$$

with $\Delta R = R_2 - R_1$. Then, differentiating, we get the values of the electric field on the left and on the right of R_1, and the difference is

$$\begin{aligned} E(R_1^+) - E(R_1^-) &= \left.\frac{\partial\varphi}{\partial r}\right|_{r=R_1} - \left.\frac{\partial\chi}{\partial r}\right|_{r=R_1} \\ &= \mu V\left(\coth\mu R_1 + \coth\mu\Delta R - \frac{R_2}{R_1\sinh\mu\Delta R}\right) = \frac{\sigma}{\varepsilon_0}. \end{aligned}$$

Thus, for small μ, we have

$$\sigma = \frac{\mu^2 V\varepsilon_0}{6}\frac{R_2}{R_1}(R_1 + R_2).$$

The total charge is therefore

$$Q = 4\pi R_1^2\sigma = \frac{2\pi}{3}\mu^2\, V\, R_1 R_2 \varepsilon_0 (R_1 + R_2).$$

5. With the data in the text, since $Q \leqslant 10^{-9}$ C, we obtain $\mu^{-1} \geqslant 6$ m, or, adding correctly the proper factors $\hbar$ and c,

$$\mu \leqslant 3.\ 10^{-43}\ \text{kg},$$

which is a very remarkable result. (Compare, for instance, with the electron mass $m_e = 9.1 \cdot 10^{-31}$ kg.)

Sidebar 6 (Integration of differential forms) *Let ω be a differential k-form. In $\mathbb{R}^k$, it would be easy to give a meaning to the integral of ω on (a subset of) $\mathbb{R}^k$* (*see Definition 17.33, page 471*).

In order to define the integral of ω on a k-dimensional "surface" $\mathscr{S}$ (also called a k-dimensional manifold), one starts by transforming $\mathscr{S}$ into a manifold Δ, which is indeed a subset of $\mathbb{R}^k$, by means of a change of coordinates. Let f denote the inverse *change of coordinates, that is, the map from $\Delta \subset \mathbb{R}^k$ into $\mathscr{S}$:*

$$f : \Delta \longrightarrow \mathscr{S} \subset \mathbb{R}^n,$$
$$x \longmapsto f(x).$$

A point $x \in \mathbb{R}^k$ in the space $\mathbb{R}^k$ is mapped to $f(x)$. How are the tangent vectors *in $\mathbb{R}^k$ transformed? Since tangent vectors correspond to "differences between two points of $\mathbb{R}^k$," the image by the change of coordinates of a vector $\boldsymbol{v}$ attached to the point x will be $\boldsymbol{v} \mapsto \mathrm{d}f_x.\boldsymbol{v}$, that is, the result of evaluating the differential form $\mathrm{d}f_x$ (given by the differential of f at the point x) at the vector $\boldsymbol{v}$. It is often customary to write $f_* = \mathrm{d}f$ the differential of f (without indicating precisely at which point it is considered).*

From the change of coordinates, a k-differential form on Δ, the "inverse image of ω by f," which is denoted $f^*\omega$, may be defined by putting

$$(f^*\omega): \qquad (\mathbb{R}^k)^k \longrightarrow \mathbb{R},$$
$$(\boldsymbol{v}_1, \ldots, \boldsymbol{v}_k) \longmapsto f^*\omega(\boldsymbol{v}_1, \ldots, \boldsymbol{v}_k) \stackrel{\text{def}}{=} \omega(f_*\boldsymbol{v}_1, \ldots, f_*\boldsymbol{v}_k).$$

Using the definition of the integral of a differential k-form on $\mathbb{R}^k$, it is then possible to define:

$$\int_{\mathscr{S}} \omega \stackrel{\text{def}}{=} \int_{\Delta} f^*\omega.$$

Of course, this definition can be shown to be independent of the choice of f.

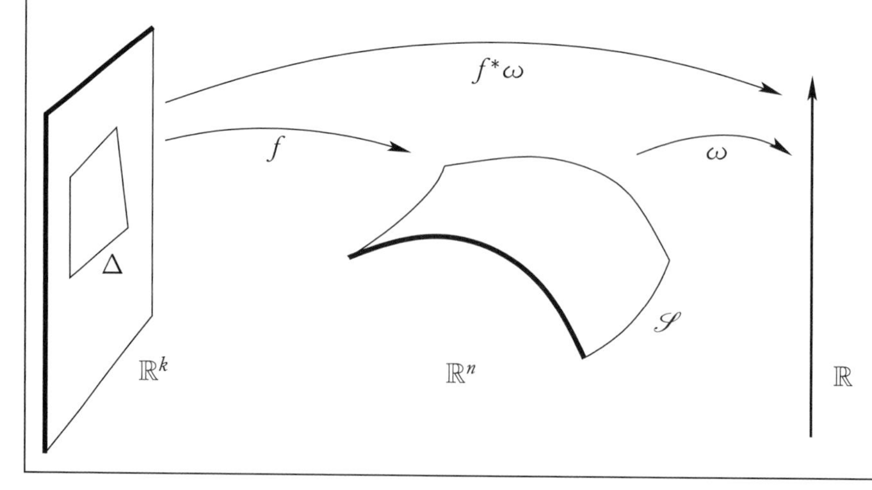

Chapter 18

Groups and group representations

18.1 Groups

DEFINITION 18.1 A **group** $(G, \cdot)$ is a set G with a product law "$\cdot$" defined on $G \times G$, such that

1. "$\cdot$" is associative: $(g \cdot h) \cdot k = g \cdot (h \cdot k)$ for all $g, h, k \in G$;
2. $(G, \cdot)$ has an **identity element** e, that is, an element of G such that $e \cdot g = g \cdot e = g$ for any $g \in G$;
3. any element in G is **invertible**, that is, for any $g \in G$, there exists an element $h \in G$ such that $g \cdot h = h \cdot g = e$. This element is unique; it is denoted g^{-1} and is called the **inverse** of g.

DEFINITION 18.2 A group $(G, \cdot)$ is **abelian** or commutative if the product law is commutative, that is, if $g \cdot h = h \cdot g$ for any $g, h \in G$. An abelian group is often denoted additively: one writes $g + h$ instead of $g \cdot h$.

For commutative groups, the definition can thus be rephrased as follows:

DEFINITION 18.3 An **additive group** is a pair $(G,+)$ such that

1'. "+" is an addition law on G, which is associative and commutative;

2'. $(G,+)$ has an **identity element** 0_G such that

$$\forall g \in G \qquad 0_G + g = g + 0_G = g;$$

3'. any G is **invertible**, that is, for any $g \in G$, there exists an element $h \in G$ such that $g + h = h + g = 0_G$. This element is denoted $-g$ and is called the **opposite** of g.

Example 18.4 The set $\mathrm{GL}_n(\mathbb{R})$ of invertible square matrices with real coefficients of size n, with the law given by the product of matrices, is a group; for $n \geqslant 2$, it is non-abelian.

Example 18.5 The set of symmetries of a molecule, with the law given by composition of maps, is a group.

Example 18.6 The set of rotations of the vector space $\mathbb{R}^2$ (rotations fixing the origin) is a commutative group.

Example 18.7 Consider the oriented space $E = \mathbb{R}^3$ with its euclidean structure. Let G denote the set of rotations of E centered at the origin 0. Then $(G, \circ)$ is a noncommutative group. Indeed, if R_x denotes the rotation with axis $\mathscr{O}x$ and angle $\pi/2$, and if R_y denotes the rotation with axis $\mathscr{O}y$ and angle $\pi/2$, the following transformations of a domino show that $R_x \circ R_y \neq R_y \circ R_x$:

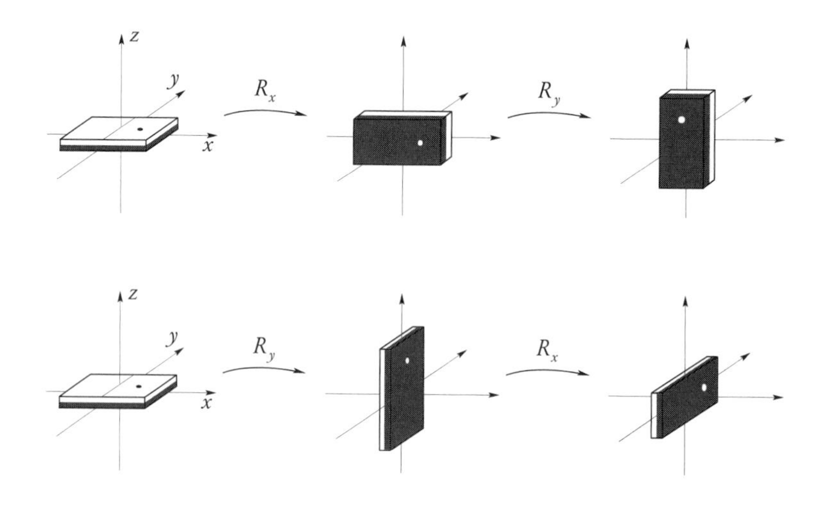

18.2 Linear representations of groups

A group may be seen as an abstract set of "objects," together with a table giving the result of the product for every pair of elements (a "multiplication table"). From this point of view, two groups $(G, \cdot)$ and $(G', \star)$ may well have the same abstract multiplication table: this means there exists a bijective map $\varphi : G \to G'$ which preserves the product laws, that is, such that for any g_1, g_2 in G, the image by φ of the product $g_1 \cdot g_2 \in G$ is the product of the respective images of each argument:

$$\varphi(g_1 \cdot g_2) = \varphi(g_1) \star \varphi(g_2).$$

In such a case, the groups are called **isomorphic**.

DEFINITION 18.8 A map that preserves in this manner the group structure is called an **homomorphism** or a **group morphism**; if it is bijective, it is a **group isomorphism**.

Example 18.9 The set M of matrices

$$M_x = \begin{pmatrix} 1 & x \\ 0 & 1 \end{pmatrix},$$

with the product given by the product of matrices, is a group. Since $M_x \cdot M_y = M_{x+y}$ for any $x, y \in \mathbb{R}$, it follows that the (obviously bijective) map $\varphi : x \mapsto M_x$ is a group isomorphism between $(\mathbb{R}, +)$ and $(M, \cdot)$.

Alternately, one speaks of a **representation** of the group G by G' if there exists a homomorphism $\varphi : G \to G'$ (not necessarily bijective).

Example 18.10 Let $\mathbf{U}$ be the set of complex numbers z such that $|z| = 1$. The map $\varphi : \mathbb{R} \to \mathbf{U}$ defined by $\varphi(\theta) = e^{i\theta}$ is a group morphism between $(\mathbb{R}, +)$ and $(\mathbf{U}, \cdot)$, but it is not an isomorphism. Its **kernel**, defined as the set of elements with image equal to the identity element in $\mathbf{U}$, is given by

$$\operatorname{Ker} \varphi = \{\theta \in \mathbb{R} \,;\, \varphi(\theta) = 1\} = 2\pi\mathbb{Z}.$$

The map $\varphi : \mathbb{R} \to \mathbf{U}$ is a (nonbijective) continuous representation[1] of the additive group $\mathbb{R}$. (Note that the groups $\mathbb{R}$ and $\mathbf{U}$ have very different topological properties; for instance, $\mathbb{R}$ is simply connected, whereas $\mathbf{U}$ is not.)

On the other hand, if we denote by G the group of rotations of $\mathbb{R}^2$ centered at the origin 0, the map $\psi : G \to \mathbf{U}$ which associates $e^{i\theta}$ to a rotation with angle θ is indeed a group isomorphism. Moreover, it is continuous. It follows that G, like U, is not simply connected.

By far the most useful representations are those which map a group $(G, \cdot)$ to the group of automorphisms of a vector space or, equivalently, to the group $\mathrm{GL}_n(\mathbb{K})$ of invertible square matrices of size n:

[1] To speak of continuity, we must have a topology on each group, which here is simply the usual topology of $\mathbb{R}$ and of $\mathbf{U} \subset \mathbb{C}$.

DEFINITION 18.11 Let $\mathbb{K}$ be either $\mathbb{R}$ or $\mathbb{C}$. A **linear representation** of a group G defined over $\mathbb{K}$ is a pair (V, φ), where V is a $\mathbb{K}$-vector space and φ is a homomorphism of G into $\mathscr{GL}(V)$, the group of linear automorphisms of V.

Such a representation is **of finite degree** if V is a finite-dimensional vector space, and the **degree of the representation** is the dimension n of V. In this case, $\mathscr{GL}(V)$ is isomorphic to $\mathrm{GL}_n(\mathbb{K})$, and we have a **matrix representation.** By abuse of notation, one sometimes refers to V only, or to φ only, as a representation of the group G.

DEFINITION 18.12 Let (V, φ) be a linear representation of a G. This representation is **faithful** if φ is injective, that is, if distinct elements of the group have distinct images (this means no information is lost by going through the map φ).

On the other hand, the representation is **trivial** if $\varphi(g) = \mathrm{Id}_V$ for any $g \in G$, i.e., if we have $\varphi(g)(v) = v$ for all $g \in G$ and all $v \in V$ (which means all information in G is lost after applying φ).

Example 18.13 Consider Example 18.9 again. The map $\varphi : \mathbb{R} \to \mathrm{GL}_2(\mathbb{R})$ which associates the matrice M_x to a real number x is a faithful matrix representation of $\mathbb{R}$.

18.3
Vectors and the group SO(3)

In this section, physical space is identified with the vector space $E = \mathbb{R}^3$.

DEFINITION 18.14 The **special orthogonal group** or **group of rotations** of euclidean space $E = \mathbb{R}^3$, denoted $\mathbf{SO}(E)$, is the group of linear maps $\mathscr{R} : \mathbb{R}^3 \to \mathbb{R}^3$ such that

- the scalar product is invariant under $\mathscr{R}$:

$$\forall \boldsymbol{a}, \boldsymbol{b} \in \mathbb{R}^3 \qquad (\boldsymbol{a}|\boldsymbol{b}) = \big(\mathscr{R}(\boldsymbol{a})\big|\mathscr{R}(\boldsymbol{b})\big) \ ;$$

- $\mathscr{R}$ preserves orientation in space: the image of any basis which is positively oriented is also positively oriented.

The rotation group $\mathrm{SO}(E)$ will be identified without further comment with the group $\mathbf{SO(3)}$ of matrices representing rotations in a fixed orthonormal basis.

Denoting by R the matrix representing a rotation $\mathscr{R}$ in the (orthonormal) canonical basis of E, we have

$$(\boldsymbol{a}|\boldsymbol{b}) = (R\boldsymbol{a}|R\boldsymbol{b}) = (\boldsymbol{a}|{}^tRR\boldsymbol{b}) \qquad \text{for all } \boldsymbol{a}, \boldsymbol{b} \in E$$

(using the invariant of the scalar product under R), which shows that R is an invertible matrix and that

$$ {}^tR = R^{-1}. \tag{18.1} $$

A matrix for which (18.1) holds is called an **orthogonal** matrix.

Thus linear maps preserving the scalar product correspond to orthogonal matrices, and they form a group, called the **orthogonal group**, denoted $\mathrm{O}(E)$. The group of orthogonal matrices is itself denoted $\mathrm{O}(3)$. Note that the matrix $-I_3$, although it is orthogonal, is not a rotation matrix; indeed, it reverses orientation, and its determinant is equal to -1.

The group $\mathrm{O}(3)$ is therefore larger than $\mathrm{SO}(3)$. More precisely, we have:

THEOREM 18.15 *The group* $\mathrm{O}(3)$ *is not connected; it is the union of two connected subsets, namely,* $\mathrm{SO}(3)$ *and* $\{-R\ ;\ R \in \mathrm{SO}(3)\}$, *on which the determinant is equal, respectively, to* 1 *and* -1.

Since the identity matrix lies in $\mathrm{SO}(3)$, it is natural to state that $\mathrm{SO}(3)$ is the **connected component of the identity in O(3).**

To summarize, the notation $\mathrm{SO}(3)$ means

$$ \begin{cases} \text{“S” for } \textit{special} \text{ (i.e., with determinant 1),} \\ \text{“O” for } \textit{orthogonal} \text{ (such that } {}^tR = R^{-1}\text{).} \end{cases} $$

Consider then a rotation $\mathscr{R}_{\boldsymbol{\theta}}$, characterized by a vector $\boldsymbol{\theta} \in \mathbb{R}^3$ with norm $\theta \in [0, 2\pi[$ (the angle of the rotation), directed along a unit vector $\boldsymbol{n}$ (the axis of the rotation): $\boldsymbol{\theta} = \theta\, \boldsymbol{n}$.

Denote by $(\mathbf{e}_1, \mathbf{e}_2, \mathbf{e}_3)$ the basis vectors in space and by $(\mathbf{e}'_1, \mathbf{e}'_2, \mathbf{e}'_3)$ their images by a rotation $\mathscr{R}_{\boldsymbol{\theta}}$. With matrix notation (it is easy to come back to tensor notation) we denote by $R(\boldsymbol{\theta})$ the matrix defined by the relations[2]

$$ \mathbf{e}'_i = \sum_{j=1}^{3} \big(R(\boldsymbol{\theta})\big)_{ij}\, \mathbf{e}_j. $$

---

[2] Note : the matrix $R(\boldsymbol{\theta})$ is not the matrix for the change of basis $L^{\mu}{}_{\nu}$ defined Equation (16.1) on page 434, but its transpose. Here, we have, symbolically,

$$ \begin{pmatrix} \mathbf{e}'_1 \\ \mathbf{e}'_2 \\ \mathbf{e}'_3 \end{pmatrix} = R(\boldsymbol{\theta}) \begin{pmatrix} \mathbf{e}_1 \\ \mathbf{e}_2 \\ \mathbf{e}_3 \end{pmatrix}, $$

which means that one can read the coordinates of the new vectors in the old basis in the *rows* of the matrix (rather than in the columns). The matrix linking the coordinates of a (fixed) point in both bases is $R(\boldsymbol{\theta})^{-1} = {}^tR(\boldsymbol{\theta})$:

$$ \begin{pmatrix} x' \\ y' \\ z' \end{pmatrix} = {}^tR(\boldsymbol{\theta}) \begin{pmatrix} x \\ y \\ z \end{pmatrix}. $$

A rotation $\mathscr{R}(\theta\,\mathbf{e}_z)$ with angle θ around the axis $\mathscr{O}z$ is represented by the matrix

$$R_z(\theta) \stackrel{\text{def}}{=} R(\theta\,\mathbf{e}_z) = \begin{pmatrix} \cos\theta & \sin\theta & 0 \\ -\sin\theta & \cos\theta & 0 \\ 0 & 0 & 1 \end{pmatrix}$$

Similarly, the matrices of rotations around the axis $\mathscr{O}x$ and $\mathscr{O}y$ are defined by

$$R_x(\zeta) \stackrel{\text{def}}{=} R(\zeta\,\mathbf{e}_x) = \begin{pmatrix} 1 & 0 & 0 \\ 0 & \cos\zeta & \sin\zeta \\ 0 & -\sin\zeta & \cos\zeta \end{pmatrix}$$

and (be careful with the sign in this last formula!)

$$R_y(\eta) \stackrel{\text{def}}{=} R(\eta\,\mathbf{e}_y) = \begin{pmatrix} \cos\eta & 0 & -\sin\eta \\ 0 & 1 & 0 \\ \sin\eta & 0 & \cos\eta \end{pmatrix}.$$

Remark 18.16 When speaking of space rotations, there are two different points of view, for a physicist. The first point of view (called **active**) is one where the axes of the frame (the observer) are immobile, whereas the physical system is rotated. In the second point of view, called **passive**, the same system is observed by two different observers, the axes of whose respective frames differ from each other by a rotation. In all the rest of the text, *we only consider passive transformations.*

For a rotation $\mathscr{R}(\boldsymbol{\theta})$ parameterized by $\boldsymbol{\theta} = \theta\,\boldsymbol{n}$, we can express the vector $\boldsymbol{\theta}$ in terms of the canonical basis:

$$\boldsymbol{\theta} = \zeta\,\mathbf{e}_1 + \eta\,\mathbf{e}_2 + \theta\,\mathbf{e}_3.$$

How does one deduce the representation for the rotation $\mathscr{R}(\boldsymbol{\theta})$? It is tempting to try the product $R_x(\zeta)\cdot R_y(\eta)\cdot R_z(\theta)$; however, performing the product in this order seems an arbitrary decision, and the example of the domino on page 490 should convince the reader that this is unlikely to be the correct solution. Indeed, in neither the first nor the second line is the final state of the domino the same as its state after a rotation of angle $\pi/\sqrt{2}$ around the first diagonal in the $\mathscr{O}xy$-plane.

A proper solution is in fact quite involved. It is first required to use *infinitesimal transformations*, that is, to consider rotations with a "very small" angle $\delta\theta$. Then it is necessary to explain how non-infinitesimal rotations may be reconstructed from infinitesimal transformations, by means of a process called "exponentiation" (which is of course related to the usual exponential function on $\mathbb{R}$).

Consider then a rotation with axis $\mathscr{O}z$ and "infinitesimal" angle $\delta\theta$: it is given by

$$R(\delta\theta\,\mathbf{e}_z) = \begin{pmatrix} 1 & \delta\theta & 0 \\ -\delta\theta & 1 & 0 \\ 0 & 0 & 1 \end{pmatrix} = I_3 + \delta\theta\,J_z$$

where we put

$$J_z \stackrel{\text{def}}{=} \left.\frac{\mathrm{d}R(\theta\,\mathbf{e}_z)}{\mathrm{d}\theta}\right|_{\theta=0} = \begin{pmatrix} 0 & 1 & 0 \\ -1 & 0 & 0 \\ 0 & 0 & 0 \end{pmatrix},$$

Similarly, define

$$J_x \stackrel{\text{def}}{=} \left.\frac{\mathrm{d}R(\zeta\,\mathbf{e}_x)}{\mathrm{d}\zeta}\right|_{\zeta=0} = \begin{pmatrix} 0 & 0 & 0 \\ 0 & 0 & 1 \\ 0 & -1 & 0 \end{pmatrix}$$

and

$$J_y \stackrel{\text{def}}{=} \left.\frac{\mathrm{d}R(\eta\,\mathbf{e}_y)}{\mathrm{d}\eta}\right|_{\eta=0} = \begin{pmatrix} 0 & 0 & -1 \\ 0 & 0 & 0 \\ 1 & 0 & 0 \end{pmatrix}.$$

DEFINITION 18.17 Denote $J_1 = J_x$, $J_2 = J_y$, and $J_3 = J_z$. The matrices J_i thus defined are the **infinitesimal generators** of rotations. We denote by $\boldsymbol{J}$ the vector $\boldsymbol{J} = (J_1, J_2, J_3)$.

How does one now pass from infinitesimal transformations to "finite" transformations?[3] A first method is to remark that the set of matrices of the form $R(\theta\,\mathbf{e}_z)$ is a representation of the additive group $(\mathbb{R}, +)$, what is called a **one-parameter group,**[4] since we have $R\big((\theta+\theta')\,\mathbf{e}_z\big) = R(\theta\,\mathbf{e}_z)\cdot R(\theta'\,\mathbf{e}_z)$. An explicit computation shows that

$$\left.\frac{\mathrm{d}R(\theta\,\mathbf{e}_z)}{\mathrm{d}\theta}\right|_{\theta=\theta_0} = \begin{pmatrix} -\sin\theta_0 & \cos\theta_0 & 0 \\ -\cos\theta_0 & -\sin\theta_0 & 0 \\ 0 & 0 & 1 \end{pmatrix} = J_z \cdot R(\theta_0\,\mathbf{e}_z),$$

and hence the matrix-valued function $R(\theta\,\mathbf{e}_z)$ is a solution of the ordinary differential equation with constant coefficients

$$M' = J_z \cdot M$$

for which the solutions are given by

$$R(\theta\,\mathbf{e}_z) = R(0)\cdot\exp(\theta J_z) = \exp(\theta J_z) = I_2 + \theta\,J_z + \frac{\theta^2}{2!}\,J_z^2 + \cdots + \frac{\theta^n}{n!}\,J_z^n + \cdots.$$

(This series is absolutely convergent, and therefore convergent.) This is a general fact, and the following theorem holds:

THEOREM 18.18 *Let $\{R(\theta)\}_{\theta\in\mathbb{R}}$ be a one-parameter group of matrices. Then its infinitesimal generator given by $J = R'(0)$ satisfies $R(\theta) = \mathrm{e}^{\theta J}$ for any $\theta \in \mathbb{R}$.*

---

[3] "Finite" in the usual physicist's sense, meaning "not infinitesimal."

[4] I.e., a group with elements of the type g_x, x running over the group of real numbers $\mathbb{R}$, such that $g_y \cdot g_x = g_{y+x}$ and $g_{-x} = g_x^{-1}$ for any $x, y \in \mathbb{R}$.

PROOF. Indeed, we have $R(0) = I$ and

$$\frac{\mathrm{d}R(\theta)}{\mathrm{d}\theta} = \lim_{h\to 0} \frac{R(\theta+h) - R(\theta)}{h} = \lim_{h\to 0} \left(\frac{R(h) - \mathrm{Id}}{h}\right) R(\theta) = J R(\theta),$$

which shows that $R(\theta)$ is the solution of the differential equation with constant coefficient $R' = JR$ such that $R(0) = I$, that is, $R(\theta) = \mathrm{e}^{\theta J}$.

THEOREM 18.19 *Let $\mathscr{R}_{\boldsymbol{\theta}}$ be a rotation characterized by the vector $\boldsymbol{\theta} = \theta \boldsymbol{n}$ (angle θ, axis directed by the vector $\boldsymbol{n}$). Then, denoting as before $\boldsymbol{J} = (J_1, J_2, J_3)$, the matrix representing $R(\boldsymbol{\theta})$ is given by*

$$R(\boldsymbol{\theta}) = \exp(\boldsymbol{\theta} \cdot \boldsymbol{J}) = \exp(\theta_1 J_1 + \theta_2 J_2 + \theta_3 J_3).$$

It is important here to be very careful that the matrices J_i *do not commute with each other,* so that this expression is not a rotation with angle θ_1 around the x-axis, followed by a rotation with angle θ_2 around the y-axis, and a rotation with angle θ_3 around the z-axis:

$$\exp(\theta_1 J_1 + \theta_2 J_2 + \theta_3 J_3) \not\equiv \exp(\theta_1 J_1) \exp(\theta_2 J_2) \exp(\theta_3 J_3).$$

However, using the exponentials, the matrices J_i can provide any rotation matrix. Even better: it turns out that, in practice, it is only necessary to know the *commutation relations* between the matrices J_i. The reader will easily check that they are given by

$$[J_1, J_2] = -J_3, \qquad [J_2, J_3] = -J_1, \qquad \text{and} \qquad [J_3, J_1] = -J_2$$

(recall that $[A, B] = AB - BA$ for any two square matrices A and B), which can be summarized neatly by the single formula

$$\boxed{[J_i, J_j] = -\varepsilon_{ijk}\, J_k,} \tag{18.2}$$

where summation over the repeated index (i.e., over k) is implicit, and where the tensor ε is the totally anti-symmetric Levi-Civita tensor. The commutation relations (18.2) define what is called the **Lie algebra of SO(3)**. The coefficients $-\varepsilon_{ijk}$ are the **structure constants** of the Lie algebra of SO(3).

Remark 18.20 The minus sign in the formula is not important; if we had been dealing with active transformations instead of passive ones, the matrices J_k would have carried a minus sign, and the commutation relations also.

Knowing the Lie algebra of the group (i.e., knowing the structure constants) is enough to recover the local structure of the group. (However, we will see that there exist groups which are "globally" different but have the same Lie algebra, for instance, SO(3) and SU(2), or more simply SO(3) and O(3); but this is not a coincidence, since SU(2) is a covering of SO(3).)

Remark 18.21 Since SO(3) is a group, performing a rotation with parameter $\boldsymbol{\theta}$ followed by a rotation with parameter $\boldsymbol{\theta}'$ is another rotation, with parameter $\boldsymbol{\theta}''$. However, there is no *simple* relation between $\boldsymbol{\theta}$, $\boldsymbol{\theta}'$ and $\boldsymbol{\theta}''$.

DEFINITION 18.22 Objects that are transformed by a matrix of SO(3) during a passive rotation of coordinate axes are called **vectors**. It is sometimes said that SO(3) is the **vector representation** of the group of rotations.

Example 18.23 Consider a rotation $\mathscr{R}$ of $\mathbb{R}^3$. The differential of $\mathscr{R}$ at any point is $\mathrm{d}\mathscr{R} = \mathscr{R}$, which shows that velocity vectors transform according to $\mathscr{R}$. In other words velocity vectors are indeed objects that transform according to a vector representation.

A vector, in the representation-theoretic sense here, is therefore more than simply an element of a vector space: it is an object generating the representation SO(3) of the rotation group [18, 28].

18.4 The group SU(2) and spinors

Consider now the complex vector space $E' = \mathbb{C}^2$, with its canonical Hilbert structure, that is, with the hermitian scalar product

$$(\boldsymbol{x}\,|\,\boldsymbol{y}) = {}^t\overline{\boldsymbol{x}} \cdot \boldsymbol{y} = \overline{x_1}\, y_1 + \overline{x_2}\, y_2, \qquad \text{where } \boldsymbol{x} = \begin{pmatrix} x_1 \\ x_2 \end{pmatrix} \text{ and } \boldsymbol{y} = \begin{pmatrix} y_1 \\ y_2 \end{pmatrix}.$$

DEFINITION 18.24 The **special unitary group** of E', denoted $\mathbf{SU}(\boldsymbol{E'})$, is the group of linear maps from E' to itself which preserve the scalar product and have determinant one.

If M is the matrix representing such a linear map in the canonical basis, we have therefore:

$$(\boldsymbol{x}\,|\,\boldsymbol{y}) = (M\,\boldsymbol{x}\,|\,M\,\boldsymbol{y}) = {}^t(\overline{M\,\boldsymbol{x}}) \cdot M\,\boldsymbol{y} = \left(\boldsymbol{x}\,\middle|\,{}^t\overline{M}\,M\,\boldsymbol{y}\right)$$

for any $\boldsymbol{x}, \boldsymbol{y} \in E'$, which implies that ${}^t\overline{M}\,M = I_2$.

DEFINITION 18.25 The conjugate transpose of a matrix M is denoted

$$M^\dagger \stackrel{\text{def}}{=} {}^t\overline{M}.$$

A matrix such that $M \cdot M^\dagger = I_2$ only also satisfies

$$|\det M|^2 = \det M \cdot \overline{\det {}^t M} = \det(M \cdot M^\dagger) = 1.$$

Such a matrix is a **unitary matrix**. Its determinant is of the form $\mathrm{e}^{\mathrm{i}\alpha}$ for some real number α.

Example 18.26 The matrix $M = \operatorname{diag}(\mathrm{e}^{\mathrm{i}\alpha}, \mathrm{e}^{\mathrm{i}\beta})$ is unitary, but does not belong to SU(2) if $\alpha + \beta \not\equiv 0 \mod 2\pi$.

DEFINITION 18.27 The group of complex 2 by 2 matrices such that

$$M \cdot M^\dagger = I_2 \qquad \text{and} \qquad \det M = 1$$

is denoted **SU(2)**.

The notation SU(2) comes from

$$\begin{cases} \text{“S” for } \textit{special} \text{ (i.e., with determinant 1),} \\ \text{“U” for } \textit{unitary} \text{ (i.e., with } {}^t\overline{M} = M^{-1}). \end{cases}$$

Recall that any rotation matrix could be expressed as the exponential of a linear combination of elementary matrices J_k; a similar phenomenon holds for matrices in SU(2). First, notice that any matrix in SU(2) is of the form

$$U = \begin{pmatrix} \alpha & \beta \\ -\overline{\beta} & \overline{\alpha} \end{pmatrix} \qquad \text{with } |\alpha|^2 + |\beta|^2 = 1.$$

♦ **Exercise 18.1** Prove this property.

If we write $\alpha = a + \mathrm{i}b$ and $\beta = c + \mathrm{i}d$, it follows therefore that $a^2 + b^2 + c^2 + d^2 = 1$. We can write $a = \cos(\theta/2)$ for some θ and then write $(b, c, d) = \sin(\theta/2)\boldsymbol{n}$, where the real vector $\boldsymbol{n} = (n_x, n_y, n_z)$ is of norm one. Then U is given by

$$U(\boldsymbol{\theta}) = \left(\cos\frac{\theta}{2}\right) I_2 + \left(\mathrm{i}\sin\frac{\theta}{2}\right) \boldsymbol{n} \cdot \boldsymbol{\sigma},$$

where

$$\boldsymbol{\sigma} \stackrel{\text{def}}{=} (\sigma_1, \sigma_2, \sigma_3),$$

the matrices σ_i being none other than the **Pauli matrices**, well known to physicists:

$$\sigma_1 = \begin{pmatrix} 0 & 1 \\ 1 & 0 \end{pmatrix} \qquad \sigma_2 = \begin{pmatrix} 0 & -\mathrm{i} \\ \mathrm{i} & 0 \end{pmatrix} \qquad \sigma_3 = \begin{pmatrix} 1 & 0 \\ 0 & -1 \end{pmatrix}.$$

Those matrices have trace zero (they are “traceless” in the physicist’s language) and are **hermitian**, which means that we have

$$\sigma_i^\dagger = \sigma_i \qquad \text{for } i = 1, 2, 3.$$

The Pauli matrices play for SU(2) the same role that the J_k play for SO(3), as we will see. First, notice that the reasoning above can be reversed:

PROPOSITION 18.28 *Let $\boldsymbol{\theta} = \theta\boldsymbol{n}$ be a vector in $\mathbb{R}^3$. The matrix $\boldsymbol{\theta}\cdot\boldsymbol{\sigma}$ is hermitian, and the matrix $U(\boldsymbol{\theta}) = \mathrm{e}^{\mathrm{i}\boldsymbol{\theta}\cdot\boldsymbol{\sigma}/2}$ is unitary. Moreover, we have*

$$U(\boldsymbol{\theta}) = (\cos\theta/2)\, I_2 + (\mathrm{i}\sin\theta/2)\, \boldsymbol{n} \cdot \boldsymbol{\sigma}$$

$$= \begin{pmatrix} \cos\frac{\theta}{2} + \mathrm{i}n_z \sin\frac{\theta}{2} & (\mathrm{i}n_x + n_y)\sin\frac{\theta}{2} \\ (\mathrm{i}n_x - n_y)\sin\frac{\theta}{2} & \cos\frac{\theta}{2} - \mathrm{i}n_z\sin\frac{\theta}{2} \end{pmatrix}.$$

PROOF. A direct computation shows that $U(\boldsymbol{\theta}) \cdot {}^t\overline{U(\boldsymbol{\theta})} = I_2$.

PROPOSITION 18.29 (Properties of the Pauli matrices)
Commutation relations:

$$\left[\frac{\mathrm{i}\sigma_i}{2}, \frac{\mathrm{i}\sigma_j}{2}\right] = -\varepsilon_{ijk}\,\frac{\mathrm{i}\sigma_k}{2};$$

Anticommutation relations:

$$\sigma_i\sigma_j + \sigma_j\sigma_i = 2\,\delta_{ij}\,I_2,$$
$$\sigma_i\sigma_j = \delta_{ij}\,I_2 + \mathrm{i}\varepsilon_{ijk}\sigma_k,$$
$$\operatorname{tr}(\sigma_i) = 0, \qquad \sigma_i^\dagger = \sigma_i.$$

The groups SO(3) and SU(2) are quite similar. Indeed, for any vector $\boldsymbol{\theta} \in \mathbb{R}^3$, we construct two matrices

$$R(\boldsymbol{\theta}) = \exp(\boldsymbol{\theta} \cdot \boldsymbol{J}) \in \mathrm{SO}(3) \qquad \text{and} \qquad U(\boldsymbol{\theta}) = \exp(\mathrm{i}\boldsymbol{\theta} \cdot \boldsymbol{\sigma}/2) \in \mathrm{SU}(2),$$

using the matrices (J_1, J_2, J_3) and $(\sigma_1, \sigma_2, \sigma_3)$, respectively, which satisfy the same commutation relations:

$$[J_i, J_j] = -\varepsilon_{ijk}\,J_k \qquad \text{and} \qquad \left[\frac{\mathrm{i}\sigma_i}{2}, \frac{\mathrm{i}\sigma_j}{2}\right] = -\varepsilon_{ijk}\,\frac{\mathrm{i}\sigma_k}{2}.$$

The structure constants of SO(3) and SU(2) are identical. Since these structure constants are enough to reconstruct (locally) the internal product law of the group, there should exist a strong link between the groups. This is made precise by the following theorem:

THEOREM 18.30 *The group* SO(3) *is a representation of* SU(2)*: there exists a homomorphism*

$$R : \mathrm{SU}(2) \longrightarrow \mathrm{SO}(3),$$
$$U \longmapsto R_U.$$

Moreover, for any $U \in \mathrm{SU}(2)$*, the matrices* U *and* $-U$ *have same image:* $R_U = R_{-U}$.

PROOF. We need a way of constructing a matrix $R_U \in \mathrm{SO}(3)$ for any matrix $U \in \mathrm{SU}(2)$, such that the group structure is preserved. In other words, we need that

$$R_U \cdot R_V = R_{UV} \qquad \text{for all } U, V \in \mathrm{SU}(2).$$

A trick is needed to do this. Denote by

$$\mathscr{M} = \{M \in \mathfrak{M}_2(\mathbb{C})\,;\, M^\dagger = M,\ \operatorname{tr} M = 0\}$$

the vector space of hermitian matrices with trace zero. This may be written

$$\mathscr{M} = \left\{\begin{pmatrix} z & x-\mathrm{i}y \\ x+\mathrm{i}y & -z\end{pmatrix}\ ;\ x, y, z \in \mathbb{R}\right\} = \{\boldsymbol{x}\cdot\boldsymbol{\sigma}\ ;\ \boldsymbol{x} \in \mathbb{R}^3\}.$$

Thus, for any vector $\boldsymbol{x} = (x, y, z) \in \mathbb{R}^3$, we have a unique associated matrix in $\mathscr{M}$, denoted $M(\boldsymbol{x}) = \boldsymbol{x} \cdot \boldsymbol{\sigma}$.

LEMMA 18.31 *For all $x, y \in \mathbb{R}^3$, we have*

$$x \cdot x' = \frac{1}{2} \operatorname{tr} \{M(x) M(x')\} \qquad \text{and} \qquad \det M(x) = -x^2.$$

Conversely, any matrix $M \in \mathcal{M}$ can be written $M = \sum x_i \sigma_i$ with $x_i = \frac{1}{2} \operatorname{tr}(M\sigma_i)$.

PROOF OF THE LEMMA. This is a simple computation.

Now we can describe the construction of the map $U \mapsto R_U$.

Let $U \in SU(2)$. We look for a matrix R_U, that is, a rotation matrix, which we may characterize by its action on $\mathbb{R}^3$. Since $\mathcal{M} \simeq \mathbb{R}^3$, we may start by finding a matrix acting on $\mathcal{M}$. For this purpose, define

$$\begin{aligned} Q(U) : \mathcal{M} &\longrightarrow \mathcal{M}, \\ M &\longmapsto UMU^{-1}. \end{aligned}$$

First notice that obviously we have $Q(-U) = Q(U)$ for all $U \in SU(2)$. It is also simple to check that $M' = UMU^{-1}$ is also in $\mathcal{M}$: indeed, we have

$$M^\dagger = (UMU^{-1})^\dagger = (UMU^\dagger)^\dagger = M$$

and

$$\operatorname{tr} M' = \operatorname{tr}(UMU^{-1}) = \operatorname{tr} M = 0.$$

Hence there exists a unique element $x' \in \mathbb{R}^3$ such that $M' = x' \cdot \sigma$. We then define $x' = R_U(x)$, and in this manner we have defined a map $R_U : \mathbb{R}^3 \to \mathbb{R}^3$. The uniqueness of x' implies that R_U is linear.

LEMMA 18.32 *R_U is a rotation.*

PROOF OF THE LEMMA. We compute the norm of x': we have

$$x^2 = -\det M' = -\det(UMU^{-1}) = -\det M = x^2,$$

and hence R_U is an isometry. It is therefore in O(3). We may compute the coefficients of R_U, remarking that the vector $x = (x_i)_i$ is mapped to $x' = (x'_i)_i$ with

$$\begin{aligned} x'_i = \tfrac{1}{2} \operatorname{tr}(M'\sigma_i) = \tfrac{1}{2} \operatorname{tr}(UMU^{-1}\sigma_i) &= \tfrac{1}{2} \sum_j \operatorname{tr}(Ux_j\sigma_j U^{-1}\sigma_i) \\ &= \tfrac{1}{2} \sum_j \operatorname{tr}(U\sigma_j U^{-1}\sigma_i) x_j, \end{aligned}$$

so

$$(R_U)_{ij} = \frac{1}{2} \operatorname{tr}(\sigma_i U \sigma_j U^{-1}).$$

A direct computation shows finally that we have $\det R_U > 0$; hence R_U is a rotation.

To conclude, we check now that for any $M \in \mathcal{M}$, we have

$$(Q(U) \circ Q(V))(M) = U(VMV^{-1})U^{-1} = (UV)M(UV)^{-1} = Q(UV)(M),$$

showing that $Q(U)Q(V) = Q(UV)$ for any $U, V \in SU(2)$ and therefore that $U \mapsto Q(U)$ is a homomorphism. It also satisfies $Q(-U) = Q(U)$. This implies that these properties are also valid for the map $U \mapsto R_U$ (this may be checked using the coefficients $(R_U)_{ij}$ computed above).

Here is a summary of the course of the proof:

$\mathscr{R}$	$\boldsymbol{\theta} = \theta\,\boldsymbol{n}$	$M(\boldsymbol{\theta}) = \boldsymbol{\theta} \cdot \boldsymbol{\sigma}$	$U = \exp(\mathrm{i}\boldsymbol{\theta} \cdot \boldsymbol{\sigma}/2)$
rotation	vector (θ defined up to 2π)	matrix in $\mathscr{M}$ $\boldsymbol{\theta} = \frac{1}{2}\operatorname{tr}(M\boldsymbol{\sigma})$	matrix in SU(2) (θ defined up to 4π)

In other words, any element of SU(2) may be mapped to an element in SO(3). However, this representation is not faithful: different elements of SU(2) can have the same image in SO(3); indeed this is the case for U and $-U$ which have the same associated representation (this is the only case where this happens). In other words, there is no inverse to this homomorphism (it would be "bi-valued").

In fact, there is no continuous isomorphism between SO(3) and SO(2) because these two spaces are not topologically equivalent:

THEOREM 18.33 SO(3) *is not simply connected, but is* ***doubly connected*** *(i.e., running twice over any closed path gives a path which is homotopic to a point).*

On the other hand SU(2) *is simply connected.*

PROOF (INTUITIVE). We will explain why SO(3) is not simply connected. Any element $R \in \mathrm{SO}(3)$ may be characterized by a unit vector $\boldsymbol{n}$, defining the axis of rotation, and by its angle θ, as before, and R is parameterized again by the vector $\theta\boldsymbol{n}$. This is a continuous parameterization, so the group SO(3) may be represented by a sphere with radius π (any rotation of angle between π and 2π is also a rotation with angle $\theta\,[0,\pi]$ and vector $-\boldsymbol{n}$).

This is an exact description, up to one small detail: the rotations $(\boldsymbol{n},\pi)$ and $(-\boldsymbol{n},\pi)$ are equivalent. The group SO(3) is therefore, topologically, the "same" as the following space: *a sphere of radius* π *on the surface of which diametrically opposite points have been identified.*

Hence the figure below represents a *closed* path in SO(3). It is indeed closed since the north and south poles of the sphere are identified, that is, they represent the same point in SO(3). The reader will easily convince herself that this closed path cannot be continuously deformed into a single point (if we try to move the north pole toward the south pole, the south pole is *obliged* to also move to remain opposite).

On the other hand, suppose we travel along this path *twice*. Then the resulting path may be contracted to a single point, following the sequence of movements below (in the first figure, the path taken twice is also drawn twice for legibility):

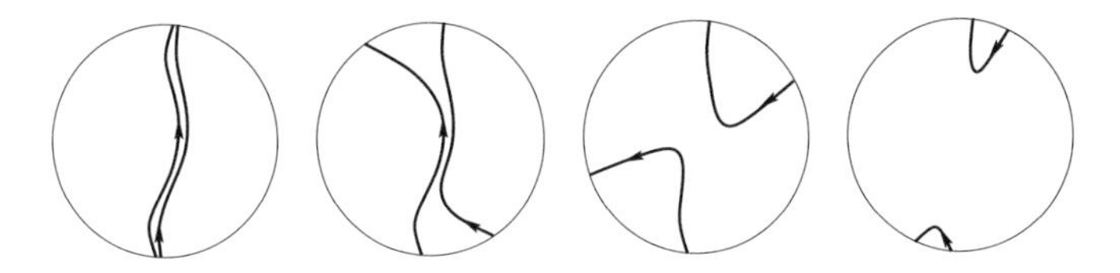

Note, moreover, that a closed path in SU(2) may be represented as a closed path that is repeated twice in SO(3).

A consequence of this theorem is that a rotation of 4π is *always* equivalent with a trivial rotation for a physical object.

What is this useful for?

You may perform the following magic trick. Take any object (a rectangular pencil eraser, for instance), and attach at least two pieces of string to it. Then attach the pieces of string to the walls, floor, or ceiling of the room where you are operating. Then turn the eraser once (a 2π-rotation). The strings become entangled. *Without cutting a string or turning the eraser again, it is impossible to disentangle them and bring them back to their original state.* However, if you turn the eraser once more (another 2π-rotation, so altogether a 4π-rotation), it will now be perfectly possible – in front of an admiring and disbelieving audience – to disentangle the strings without any more turns or cutting (see sidebar 7, page 506).

This shows that rotating the eraser twice (4π-rotation) is equivalent to no rotation at all, whereas one turn (2π-rotation) is not.

But shouldn't a 2π-rotation be, in fact, equivalent with a zero rotation, as is commonly stated and believed? It all depends on the object considered. If, as in this trick, the object is not by itself, the answer is "No." More generally, the mathematical properties of the object will reflect the *path* followed in SO(3) to perform the rotation. In other words, the laws of the transformation must reflect the group structure involved; but such a law is not unique: a physical object may follow the trivial representation of SO(3) (it is then a scalar, such as temperature), or a vector representation (it is then a vector, such as velocity); a *complex* object may follow a representation of SU(2).

This is, for instance, the case of the wave function of the electron.

An electron may be represented (in nonrelativistic quantum mechanics) by a wave function with two components,

$$\Psi(\boldsymbol{r},t) = \begin{pmatrix} \varphi(\boldsymbol{r},t) \\ \psi(\boldsymbol{r},t) \end{pmatrix}.$$

During a rotation, this changes according to a *spinorial transformation*:

$$\Psi(\boldsymbol{r},t) \longmapsto \Psi'(\boldsymbol{r}',t) = \begin{pmatrix} \varphi'(\boldsymbol{r}',t) \\ \psi'(\boldsymbol{r}',t) \end{pmatrix} = \mathrm{e}^{-\mathrm{i}\boldsymbol{\theta}\cdot\boldsymbol{\sigma}/2} \begin{pmatrix} \varphi(\boldsymbol{r},t) \\ \psi(\boldsymbol{r},t) \end{pmatrix}, \tag{18.3}$$

where $\boldsymbol{\theta} = \theta\,\boldsymbol{n}$ is the vector characterizing the rotation. Note that, as seen previously, we have

$$\mathrm{e}^{-\mathrm{i}(\theta+2\pi)\,\boldsymbol{n}\cdot\boldsymbol{\sigma}/2} = -\mathrm{e}^{-\mathrm{i}\theta\,\boldsymbol{n}\cdot\boldsymbol{\sigma}/2},$$

but

$$\mathrm{e}^{-\mathrm{i}(\theta+4\pi)\,\boldsymbol{n}\cdot\boldsymbol{\sigma}/2} = +\mathrm{e}^{-\mathrm{i}\theta\,\boldsymbol{n}\cdot\boldsymbol{\sigma}/2}.$$

Hence, a 4π-rotation is required in order that the wave function remain invariant. How is it, then, that this particular behavior of the electron (it is not invariant under a 2π-rotation) has no consequence in physical observations? After all, electrons exist abundantly in everyday matter. Note that, in fact, after a 2π-rotation, only the *sign of the wave function* has changed; consequently, the physical properties of the particle are unchanged. What changes are properties of interactions of the particles: in particular, the fact that electrons are fermions (and hence obey the Pauli exclusion principle, so that matter is stable) derive from this fact.

An object transforming according to the rule (18.3) is called a **spinor**.

Exercise 18.3 shows that other transformation rules are possible.

Remark 18.34 The fact that SU(2) is like "twice SO(3)" (or, more technically, it is a "double cover" of SO(3)) is used to show that two states of spin exist in the fundamental spinor representation. In special relativity, the group of isometries of space is not SO(3) anymore, but rather the group SO(3, 1), that leaves Minkowski space ($\mathbb{R}^4$ with the Minkowski metric) invariant. One can show that there exists a simply connected double cover of this group, isomorphic to $\mathrm{SL}(2, \mathbb{C})$ and generated by *two* independent representations of SU(2) (one sometimes find the notation $\mathrm{SU}(2) \otimes \mathrm{SU}(2)$ to recall this independance). This is constructed using suitable linear combinations of generators of rotations and Lorentz boosts as infinitesimal generators. This group then affords four degrees of freedom for a particle with spin $\frac{1}{2}$, instead of the two expected; this allows us to predict that not only the particle is thus described, but so is the associated *antiparticle*.

The reader interested by the mathematics underlying the representations of the Dirac equation may begin by looking at the excellent book [68].

18.5
Spin and Riemann sphere

While we are dealing with pretty pictures, I will not resist the temptation to present an application to quantum mechanics of the representation of the complex plane as the Riemann sphere (see page 146); more precisely, this concerns the representation of the spin.

We know that the wave function of an electron (or indeed of any other particle with spin $\frac{1}{2}$) may be represented as the tensor product of a vector $|f\rangle$ corresponding to a space dependency and a vector $|\chi\rangle$ corresponding to the spin:

$$|\Psi\rangle = |f\rangle \otimes |\chi\rangle .$$

This is another way of expressing that the wave function has two components, as in (18.3) (the first, φ, corresponding to the state $|\uparrow\rangle$ or "spin-up," and the second, ψ, corresponding to the state $|\downarrow\rangle$ or "spin-down").

Let us concentrate only on the spin for now. Then the wave function $|\chi\rangle$ is a linear combination of the basis vectors $|\uparrow\rangle$ and $|\downarrow\rangle$. This, however, is related to the *choice* of one particular direction for the measure of the spin, namely, the direction z. We might as well have chosen *another* direction, for instance, the direction x. Any vector $|\chi\rangle$ would then be a linear combination of $|\rightarrow\rangle$ and $|\leftarrow\rangle$, eigenvectors for the measure of spin on the x axis.

How is a wave function (denoted $|\nearrow\rangle$) characterized when it is an eigenvector for the measure of spin in an arbitrarily chosen direction? Since the family $\{\,|\uparrow\rangle, |\downarrow\rangle\,\}$ is a basis of the space of spins, there exist complex numbers $\lambda, \mu \in \mathbb{C}$ such that

$$|\nearrow\rangle = \lambda\,|\uparrow\rangle + \mu\,|\downarrow\rangle.$$

Because the physical meaning of a wave function is not changed when the vector is multiplied by an arbitrary nonzero complex number, it follows that *only the ratio* λ/μ *is necessary to characterize the state* $|\nearrow\rangle$.

Then we have the following very nice result:

THEOREM 18.35 *Let* $z = \overline{\lambda/\mu}$*; then we have* $z \in \overline{\mathbb{C}} = \mathbb{C} \cup \{\infty\}$*. The stereographic projection* z' *of* z *on the Riemann sphere gives the direction characterizing the vector* $|\nearrow\rangle$.

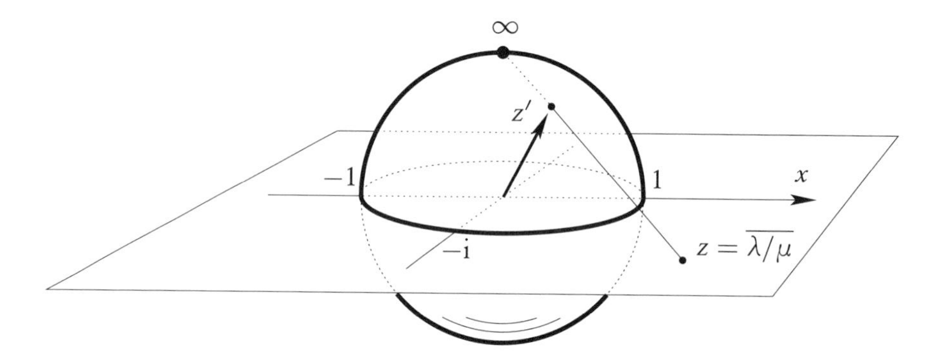

For instance, to obtain the direction of positive x, we should take $z = 1$, which gives (after normalizing appropriately):

$$|\rightarrow\rangle_x = \frac{1}{\sqrt{2}}\Big[\,|\uparrow\rangle + |\downarrow\rangle\,\Big].$$

To obtain the direction $(-Ox)$, we must take $z = -1$:

$$|\leftarrow\rangle_x = \frac{1}{\sqrt{2}}\Big[\,|\uparrow\rangle - |\downarrow\rangle\,\Big].$$

For the direction (Oy), we take $z = \mathrm{i}$:

$$|\rightarrow\rangle_y = \frac{1}{\sqrt{2}}\Big[\,|\uparrow\rangle + \mathrm{i}\,|\downarrow\rangle\,\Big].$$

Similarly, for the direction $(-Oy)$, we take $z = -\mathrm{i}$:

$$|\leftarrow\rangle_y = \frac{1}{\sqrt{2}}\Big[\,|\uparrow\rangle - \mathrm{i}\,|\downarrow\rangle\,\Big].$$

Finally, for direction (Oz), we must of course take $z = \infty$:

$$|\uparrow\rangle = |\uparrow\rangle + 0\,|\downarrow\rangle,$$

whereas the direction $(-Oz)$ corresponds to $z = 0$, so:

$$|\downarrow\rangle = 0\,|\uparrow\rangle + |\downarrow\rangle.$$

EXERCISES

♦ **Exercise 18.2** Let σ_i $(i = 1, \ldots, 3)$ denote the Pauli matrices. For a real number γ, compute $\exp(-\mathrm{i}\gamma\sigma_3/2) \cdot \sigma_k \cdot \exp(\mathrm{i}\gamma\sigma_3/2)$ when $k = 1, 2, 3$. Interpret this result.

♦ **Exercise 18.3 (Rotations acting on functions)** Let $\mathscr{R}$ be a rotation. To a smooth function $\psi(\boldsymbol{r})$, we associate the function $R\psi$ such that $R\psi(\boldsymbol{r}) = \psi(\mathscr{R}^{-1}(\boldsymbol{r}))$ (so that the axes have been transformed by the rotation $\mathscr{R}$). For a rotation $\mathscr{R}_z(\theta)$ around the $\mathscr{O}z$-axis, show that we have $R_z(\delta\theta)\psi(x, y, z) = \psi(x - y\,\delta\theta, y + x\,\delta\theta, z)$ and then that

$$\frac{\mathrm{d}R_z(\theta)}{\mathrm{d}\theta}\,\psi = \mathrm{i}L_z R_z(\theta)\,\psi \qquad \text{with } L_z = -\mathrm{i}\left(x\frac{\partial}{\partial y} + y\frac{\partial}{\partial x}\right).$$

Deduce that $R_z(\theta) = \exp(\mathrm{i}\theta L_z)$ and generalize this to the other axes. Show that

$$[\mathrm{i}L_i, \mathrm{i}L_j] = -\varepsilon_{ijk}\,\mathrm{i}L_k.$$

Sidebar 7 (Double connectedness of $SO(3)$ and magic tricks) *Start with a sphere attached to the ceiling and floor with three strings (or more). At the beginning of the experiment, the sphere is rotated by an angle of 4π (i.e., turned twice completely); hence the strings that attach it become entangled. However, without turning the sphere anymore, and of course without putting knife or scissors to the strings, it is possible to free the strings, following the sequence below, at the end of which all the strings are exactly as they where before the rotation was performed.*

(It is necessary to assume that the strings are somewhat flexible; all this is topological, that is, it requires continuous deformation.)

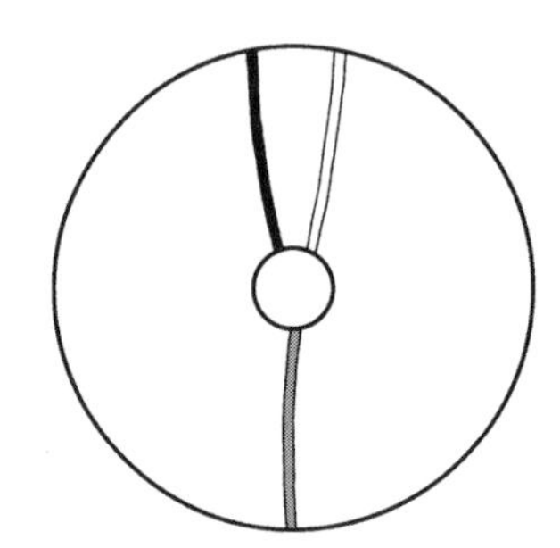

Initial positon.

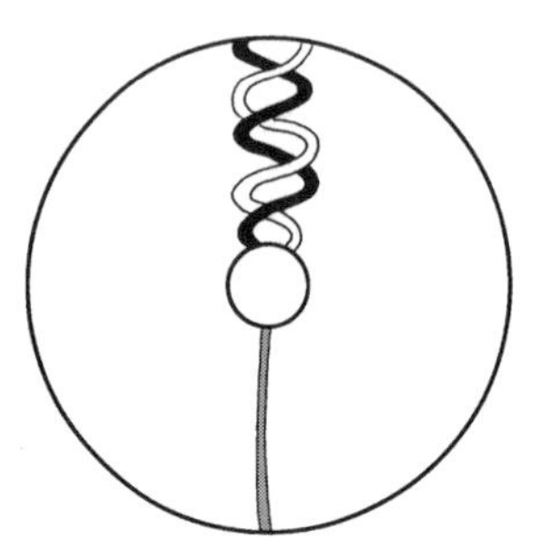

After two rotations of the sphere, the strings are entangled.

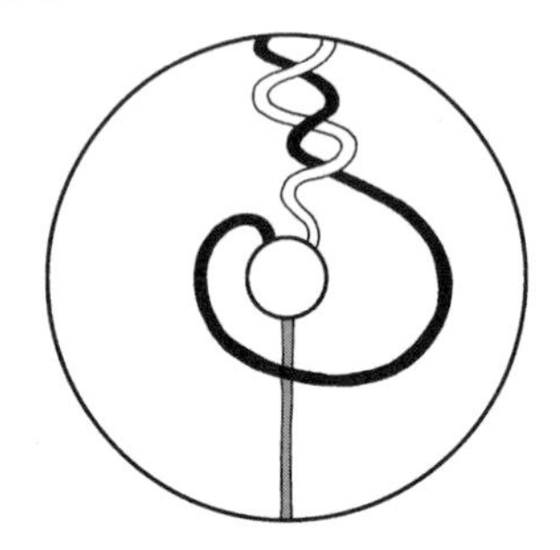

Beginning of the trick.

The white string goes over the rest.

The gray string starts moving.

Sidebar 7 (Cont.)

The gray string goes behind the rest...

... continues...

... and goes back to its initial position.

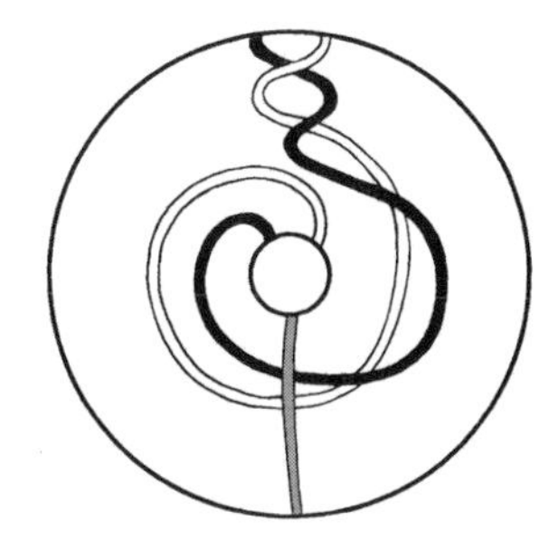

The white string goes behind the black string...

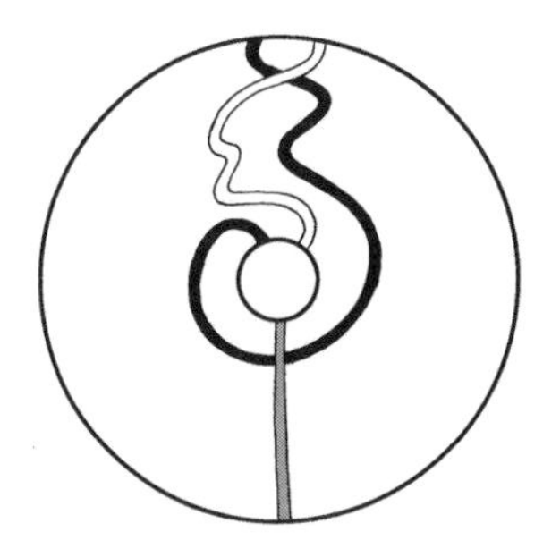

... and into position.

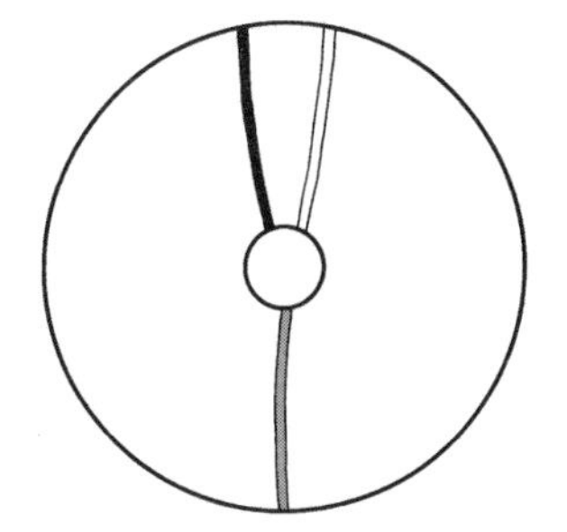

The black string is now free to take its original place.

This magic trick (which may very well be done at home, using a pencil eraser, some tacks and a yard or so of ordinary string) is a consequence of the double connectedness of SO(3). *Another illustration of this fact is "Feynmann's hand trick" [36].*

Chapter 19

Introduction to probability theory

Statistics is that science which proves that 99.99% of human beings have more legs than the average number.

The goal of probability theory is to provide mathematical tools to discuss rigorously the elusive concept of *randomness*.

Randomness means in general that we have incomplete knowledge of some system. Probability theory is interested, for instance, in what can be said of an event when an "experiment" is repeated a large number of times. For instance, this "experiment" may be the throwing of a die, and the event might be "the result of the throw is at most 3."

However, probability theory is not only concerned with "random" events, since it is also at the heart of the formalism of statistical mechanics, which, at the classical level, is based on deterministic equations (the Newton equations, in the Lagrangian or Hamiltonian formalism). It can also be used to *describe* in a simpler (maybe incomplete) manner some very complicated distributions, such as in signal theory or in image processing.

Finally, it should be mentioned that probability theory may be used to *discuss* various events, and – as the case may be – to *mislead, bluff, or hoodwink* the unsuspecting public. Although this is not something that students should do, it is important that they know some of those techniques, not only to be capable of clear scientific judgment (some experimental results are presented as "proved,"[1] whereas rigorous statistical analysis shows them to be no better than purely random), but also to enlighten the daily life of an informed citizen.

19.1

Introduction

Dice are the embodiment of randomness. The Latin name, *alea*, is indeed the source of the French word "aléatoire" for "random." The Arabic word *az-zahr*, which also means "dice," has produced Spanish *azar*, French "hasard" ("randomness"), as well as English "hazard." The way dice fall, through the Latin *cadere* (to fall), has brought the Old French word *chaance* and then the word "chance."

According to a dictionary, chance is *"The unknown and unpredictable element in happenings that seems to have no assignable cause."*

This is what happens with the throw of a die. Even if we hold that the laws that dictate the fall and rebounds of the die are perfectly deterministic, it is a fact that this system is chaotic and thus, in practice, has unpredictable behavior: the final outcome of the throw has no *apparent* reason [86].

It is quite remarkable that such a common concept as "chance" should at the same time be so difficult to fathom. Here is a mathematical example:

Some real numbers have the following property,[2] which we call Property P:

P { the decimal expansion of x contains statistically as many zeros as ones, twos, threes, etc; morevoer, any *finite sequence* of digits (for instance, "123456789" or your birthdate) appears in the decimal expansion and indeed appears statistically as frequently as any other sequence of the same length (thus "12345" is as frequent as "94281").

Mathematicians (in particular, É. Borel) have proved that Property P holds for the *overwhelming majority* of real numbers. More precisely, the probability that P holds for a real number x is equal to 1. The probability that P does not hold is therefore equal to 0.

Now, *can you give me a* single *example of a real number such that Property P holds?*

It should be noted that rational numbers do not have Property P. Indeed, the decimal expansion of a rational number is periodic (for example, $1/7 = 0.142857142857...$ with an infinite repetition of the sequence "142857"), which clearly contradicts Property P. Maybe a number such as π, or e, or $\sqrt{2}$ would work. This is quite possible, but this is a complete mystery at the current time: nobody has been able to show, for instance, that π has Property P, nor indeed that it doesn't.

So, we know that a "randomly chosen" real number must *almost surely* satisfy Property P; yet, (almost) any time we chose a number, either it does

---

[1] E.g., the famous case of "water memory," which is described in detail in [15].

[2] A number for which Property P holds is called a **normal** number.

not have this property, or if it does, we are not able to prove it.[3] The conclusion is that it is in fact very difficult to make a "random" choice of a real number; most of our "choices" are not random at all.[4]

This simple example should be enough to suggest how difficult it may be to express rigorously what is a "random choice." To overcome those difficulties, it has been necessary to formalize and axiomatize probability theory, leaving aside questions of interpretation to concentrate solely on the mathematical content.

In this sense, the forefathers of probability are certainly Blaise PASCAL (see page 606) and Pierre DE FERMAT (1601–1665), during a famous exchange of letters during the summer 1654. The physicist Christiaan HUYGHENS (1629–1695) published this correspondance in the first treatise of probability theory, *De Ratiociniis in ludo aleæ*, where in particular the notion of *mathematical expectation* is introduced.

However, the works of Jacques BERNOULLI (see page 555) and Abraham DE MOIVRE,[5] with the first theorems concerning laws of large numbers, are the real starting point of modern probability theory. LAPLACE and GAUSS studied the theory of errors (with important applications to measurements such as that of the terrestrial meridian).

In the ninetheenth century, Pafnouti TCHEBYCHEV[6] (1821–1894) and his students Andrei MARKOV (1856–1922) and Alexandre LIAPOUNOV (1857–1918) created a formalism suitable for the study of sums of random variables, which marks the beginning of the important Russian school of probability theory. The complete axiomatization of probability is due to Andrei KOLMOGOROV (see below), building on previous work of Sergei BERNSTEIN (1880–1968), Richard VON MISES (1883–1953) and Émile BOREL (1871–1956).

---

[3] To be precise, we know *a few* normal numbers. For instance, the number obtained by first writing successively all positive integers, then concatenating the resulting sequences of digits (Champernowne, 1933):

$$0,1234567891011121314151617\ldots$$

or the number obtained by doing the same with the sequence of prime numbers (Erdös and Copeland, 1945) :

$$0,23571113171923293137\ldots$$

It is quite obvious that those numbers are artificial; in particular, they depend on the fact that we work in base 10. So one may increase the difficulty by asking for an *absolutely normal* number, one which is normal (in an obvious sense) in *any* basis. It seems that today *not a single* absolutely normal number is known, whereas, again, "almost all" numbers must have this property.

[4] Consider also the great difficulty inherent in the writing of a random number generator for a computer.

[5] Abraham DE MOIVRE (1667–1754) left France at eighteen during the repression against Protestants. He discovered mathematics through a chance encounter with Newton's *Principia* [69]. Besides studies of complex numbers, he gave the mathematical definition of independent events, and proved the "Stirling" formula.

[6] The mathematical great-great-great-great-grandfather of the translator.

19.2
Basic definitions

In this section, we introduce the basic vocabulary of probability theory and the axioms stated by Kolmogorov [19] in *Über die analytischen Methoden in der Wahrscheinlichkeitrechnung* (Analytic Methods of Probability Theory, 1931), and also in the historical book *Grundbegriffe der Wahrscheinlichkeitrechnung* (Foundations of Probability Theory, 1933).

The probability of an event that can range only over a discrete set (such as the result of throwing a die) can be captured very intuitively. However, when the space of possible events is continuous (such as the points in space, or an instant in time), a precise mathematical description of the "probability that an event occurs" is required.

Kolmogorov suggested that the probability of an event be the expression of the *measure* of a set.

DEFINITION 19.1 A **probability space** is a pair (Ω, Σ), where

- Ω: a set **(the sample space)**.
- Σ: σ-algebra of events, that is, a set such that

  i) $\Sigma \subset \mathfrak{P}(\Omega)$ ($\mathfrak{P}(\Omega)$ is the set of all subsets of Ω);

  ii) $\Omega \in \Sigma$ and $\varnothing \in \Sigma$;

  iii) if $A \in \Sigma$, then its complement $\complement_\Omega A$ also belongs to Σ;

  iv) if $A_n \in \Sigma$ for any $n \in \mathbb{N}$, then $\bigcup_{n\in\mathbb{N}} A_n \in \Sigma$.

An element of Σ is called an **event**. An element of $\omega \in \Omega$ is sometimes called **the result of the experiment Ω**, or an **atomic** or **elementary event.**

The reader should recognize Definitions 2.7 and 2.10 on page 58. So a measurable space is a probabilistic space.

Example 19.2 When playing (once) heads or tails, one can take $\Omega = \{P, F\}$ and

$$\Sigma = \{\varnothing, \{P\}, \{F\}, \{P, F\}\}.$$

Example 19.3 When rolling once a die with six facets, the space of elementary events can be

$$\Omega = \{1, 2, 3, 4, 5, 6\} \qquad \text{and} \qquad \Sigma = \mathfrak{P}(\Omega).$$

The event "the result of the throw is even" is the element $\{2, 4, 6\} \in \Sigma$.

Example 19.4 Take $\Omega = \mathbb{R}$. Then both $\{\varnothing, \mathbb{R}\}$ and $\{\varnothing, [0,1], \complement_{\mathbb{R}}[0,1], \mathbb{R}\}$ are σ-algebras.

Remark 19.5 Note that it is sometimes useful to take for Ω a "set" which is not an abstract mathematical object, but rather a group of persons, a collection of objects, etc.

DEFINITION 19.6 An event $A \in \Sigma$ is **realized** if the result ω of an experiment is in A.

Working with the σ-algebra Σ, we have at our disposal on the set of events the usual set-theoretic operations of union, intersection, and taking the complementary subset. These have the following intuitive meaning:

- the union of two events A and B, that is, $A \cup B$, corresponds to the realization of A *or* B (an inclusive "or," where A *and* B is possible);
- the intersection of two events A and B, i.e., $A \cap B$, corresponds to the realization of A *and* B;
- taking the complementary subset corresponds to the realization of the opposite of A, or equivalently to the no-realization (the *failure*) of A.

DEFINITION 19.7 Let (Ω, Σ) be a sample probability space $A, B \in \Sigma$. Two events A and B are **simultaneously realized** if the outcome of the experiment is $\omega \in A \cap B$.

The events A and B are **incompatible** if $A \cap B = \varnothing$.

A **complete class of events** is any countable (or finite) sequence of non-empty events $(A_n)_{n \in \mathbb{N}}$, which are mutually incompatible and which together exhaust Ω, that is, such that

i) $A_m \cap A_n = \varnothing$ if $m \neq n$,

ii) $\bigcup_{n \in \mathbb{N}} A_n = \Omega$.

Any elementary event $\{\omega\} \in \mathfrak{P}(\Omega)$ is then contained in one and only one of the elements of the complete class.

DEFINITION 19.8 Let $\mathscr{C} \subset \mathfrak{P}(\Omega)$ be any set of subsets of Ω. The intersection of all σ-algebras containing $\mathscr{C}$ is a σ-algebra, which is called the **σ-algebra generated by** $\mathscr{C}$ and is denoted $\boldsymbol{\sigma}(\mathscr{C})$.

Example 19.9 Consider the experiment of rolling a die once. We take

$$\Omega = \{1, 2, 3, 4, 5, 6\} \quad \text{and we put} \quad \Sigma = \mathfrak{P}(\Omega).$$

Consider now $\mathscr{C} = \{\{1, 2\}\}$. The σ-algebra generated by $\mathscr{C}$ is

$$\sigma(\mathscr{C}) = \{\varnothing, \{1, 2\}, \{3, 4, 5, 6\}, \Omega\}.$$

On the other hand, if we put $\mathscr{C} = \{\{1\}, \{2\}\}$, we have

$$\sigma(\mathscr{C}) = \{\varnothing, \{1\}, \{2\}, \{1, 2\}, \{3, 4, 5, 6\}, \{1, 3, 4, 5, 6\}, \{2, 3, 4, 5, 6\}, \Omega\}.$$

Having defined these notions, which are basically set-theoretic in nature, we can use the σ-algebra for the purpose of defining *measures* of sets in this σ-algebra, as in the definition of the Lebesgue integral. A suitably normalized positive measure will provide the notion of *probability*.

DEFINITION 19.10 (Probability measure) Let (Ω, Σ) denote a sample space with a σ-algebra. A map $\mathsf{P} : \Sigma \to \mathbb{R}^+$ is a **probability (measure)** on (Ω, Σ) if:

(P1) $\mathsf{P}(\Omega) = 1$;

(P2) for any countable (or finite) sequence $(A_n)_{n\in\mathbb{N}}$ of pairwise disjoint events in Σ, we have

$$\mathsf{P}\Big(\bigcup_{n\in\mathbb{N}} A_n\Big) = \sum_{n\in\mathbb{N}} \mathsf{P}(A_n).$$

The triplet $(\Omega, \Sigma, \mathsf{P})$ is also called a **probability space**.

Comparing with Definition 2.11 on page 58, we see that a probability space is simply a measure space where the total measure of the space is supposed to be equal to 1.

The reader is then invited to check the following properties:

PROPOSITION 19.11 *Let* P *be a probability measure. Then the following properties hold:*

i) *if* $A \subset B$, *then we have* $\mathsf{P}(A) \leqslant \mathsf{P}(B)$;

ii) $\mathsf{P}\left(\complement_\Omega A\right) = 1 - \mathsf{P}(A)$;

iii) *if* $A \subset B$, *then we have* $\mathsf{P}(B) = \mathsf{P}(A) + \mathsf{P}(B \setminus A)$;

iv) $\mathsf{P}(A \cup B) = \mathsf{P}(A) + \mathsf{P}(B) - \mathsf{P}(A \cap B)$;

v) *if* $(A_n)_{n\in\mathbb{N}}$ *is an increasing sequence (for inclusion), then we have*

$$\mathsf{P}\Big(\bigcup_{n\in\mathbb{N}} A_n\Big) = \sup_{n\in\mathbb{N}} \mathsf{P}(A_n) = \lim_{n\to\infty} \mathsf{P}(A_n);$$

vi) *if* $(A_n)_{n\in\mathbb{N}}$ *is a decreasing sequence (for inclusion), then we have*

$$\mathsf{P}\Big(\bigcap_{n\in\mathbb{N}} A_n\Big) = \inf_{n\in\mathbb{N}} \mathsf{P}(A_n) = \lim_{n\to\infty} \mathsf{P}(A_n).$$

Example 19.12 Let $a \in \Omega$ be an elementary event. The map

$$\delta_a : \mathfrak{P}(\Omega) \longrightarrow \{0, 1\},$$
$$A \longmapsto \begin{cases} 1 & \text{if } a \in A, \\ 0 & \text{if } a \notin A \end{cases}$$

is a probability measure, which is called the **Dirac measure or mass** at a.

Example 19.13 Let $(p_n)_{n\in\mathbb{N}}$ be an arbitrary sequence of non-negative real numbers such that $\sum_{n\in\mathbb{N}} p_n = 1$. Let Ω be any set and let $(x_n)_{n\in\mathbb{N}}$ be a sequence of elements in Ω. Then $\sum_{n\in\mathbb{N}} p_n\, \delta_{x_n}$ defines (in an obvious way) a probability measure on $(\Omega, \mathfrak{P}(\Omega))$.

Andrei Nikolaievitch KOLMOGOROV (1903–1987) could not receive any education from his parents, as his mother died in childbirth and his father, an agriculturist, was in exile. He was raised by his aunt and became a train conductor. In his free time, Kolmogorov wrote a treatise on Newtonian mechanics. His further works dealt with very varied subjects: trigonometric series, the axiomatization of probability theory, the theory of Markov processes, information theory. In physics, his contributions were important in the theory of turbulence, the study of chaotic systems and dynamical systems (his initial remains in the acronym "KAM theory," with Arnold's and Moser's). He also had deep interests outside mathematics, including the poetic works of Pushkin.

Example 19.14 Let $f : \mathbb{R}^n \to \mathbb{R}^+$ be an integrable function (with respect to Lebesgue measure) such that $\int f \, d\mu = 1$. For any Borel subset $A \in \mathscr{B}(\mathbb{R}^n)$ (recall that the Borel σ-algebra is the σ-algebra on $\mathbb{R}^n$ generated by open sets, or equivalently by "rectangles" $[a_1, b_1] \times \cdots \times [a_n, b_n]$), let

$$\mathsf{P}(A) = \int_A f \, d\mu.$$

The map P thus defined is a probability measure on $(\mathbb{R}^n, \mathscr{B}(\mathbb{R}^n))$.

Example 19.15 Let σ be a positive real number, and let $m \in \mathbb{R}$. The **Gauss distribution** or **normal distribution** with mean m and standard deviation σ is the probability measure on $\mathbb{R}$ (with respect to the Borel σ-algebra) defined by

$$\mathsf{P}(A) = \frac{1}{\sigma\sqrt{2\pi}} \int_A \exp\left\{\frac{(t-m)^2}{2\sigma^2}\right\} dt \qquad \text{for any } A \in \mathscr{B}(\mathbb{R}).$$

DEFINITION 19.16 A subset $N \subset \Omega$ is **negligible** (for a given probability measure P) if it is contained in a measurable subset (and event) with probability zero:

$$\exists N' \in \Sigma \qquad N \subset N' \quad \text{and} \quad \mathsf{P}(N') = 0.$$

A property p that depends on elementary events **holds almost surely** or **is almost sure** if the set of those ω which do not satisfy p, namely, the set

$$\{\omega \in \Omega \, ; \, p(\omega) \text{ is false}\},$$

is negligible.

Example 19.17 Let $\Omega = [0, 1]$, with the Borel σ-algebra $\Sigma = \mathscr{B}([0, 1])$ and the "uniform" probability measure on $[0, 1]$ (i.e., the Lebesgue measure restricted to $[0, 1]$). Then for $x \in [0, 1]$, x is almost surely an irrational number.

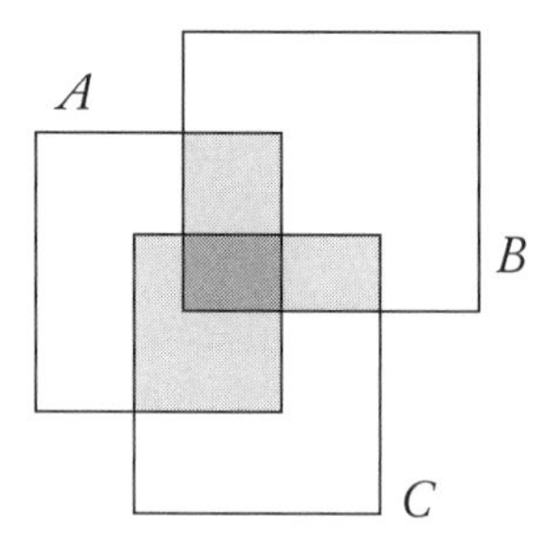

Fig. 19.1 – The Poincaré formula for $n = 3$. In dark gray, $A \cap B \cap C$; in light gray, $A \cap B$, $A \cap C$ and $B \cap C$.

19.3 Poincaré formula

Consider two events A and B. Obviously (just using the definition as the measure of the corresponding events) we have the formula

$$\mathrm{P}(A \cup B) = \mathrm{P}(A) + \mathrm{P}(B) - \mathrm{P}(A \cap B).$$

Can this be generalized to more than two events? Let A, B, C be three events. We can compute

$$\mathrm{P}(A) + \mathrm{P}(B) + \mathrm{P}(C) - \mathrm{P}(A \cap B) - \mathrm{P}(B \cap C) - \mathrm{P}(C \cap A),$$

but as clearly seen in Figure 19.1, this is not the correct result for $\mathrm{P}(A \cup B \cup C)$, because the subset $A \cap B \cap C$ has been counted three times, then subtracted three times. Hence the correct relation is in fact

$$\mathrm{P}(A \cup B \cup C) = \mathrm{P}(A) + \mathrm{P}(B) + \mathrm{P}(C) - \mathrm{P}(A \cap B) \\ - \mathrm{P}(B \cap C) - \mathrm{P}(C \cap A) + \mathrm{P}(A \cap B \cap C).$$

With this done, the general result is easy to guess and prove.

THEOREM 19.18 (Poincaré formula) *Let $n \in \mathbb{N}^*$ and let $A_1, \dots, A_n$ be arbitrary events, not necessarily disjoint. Then we have*

$$\mathrm{P}\left(\bigcup_{i=1}^{n} A_i\right) = \sum_{k=1}^{n} (-1)^{k+1} \sum_{1 \leqslant i_1 < i_2 < \cdots < i_k \leqslant n} \mathrm{P}(A_{i_1} \cap \cdots \cap A_{i_k}).$$

PROOF. A proof using random variables, as defined in the next chapter, is given in the Appendix D. An elementary proof by induction is also easy.

19.4

Conditional probability

♦ **Exercise 19.1** You're taking a plane to Seattle. A colleague tells you: "There is a one in ten thousand chance that there is a bomb in a plane. But there is only one in a hundred million chance that there are *two* bombs. So, for more security, just bring your own bomb."

What should you think of this argument?

DEFINITION 19.19 Let A and B be two events such that $\mathsf{P}(B) \neq 0$. The **conditional probability** of the event A **knowing** the event B is the quantity

$$\mathsf{P}_B(A) = \mathsf{P}(A|B) = \frac{\mathsf{P}(A \cap B)}{\mathsf{P}(B)}.$$

It is denoted either $\mathsf{P}_B(A)$ or $\mathsf{P}(A|B)$.

Example 19.20 The probability that there are two bombs in your plane, *knowing* that there is at least one (yours) is therefore simply one in ten thousand. *Bringing one in your luggage is a useless precaution.*

If $\mathsf{P}(A) \neq 0$ as well as $\mathsf{P}(B) \neq 0$, it is also possible to define the probability of B knowing A. Then we have

$$\mathsf{P}(A \cap B) = \mathsf{P}(B) \cdot \mathsf{P}(A|B) = \mathsf{P}(A) \cdot \mathsf{P}(B|A),$$

hence the following formula, due to the clergyman Thomas Bayes, who published it in 1764 [12]:

THEOREM 19.21 (Bayes formula) *Let A and B be two events in Σ with nonzero probability. Then we have*

$$\mathsf{P}(B|A) = \frac{\mathsf{P}(A|B) \cdot \mathsf{P}(B)}{\mathsf{P}(A)}.$$

Although it is very simple, this is an interesting result since it relates the probability that the event B is realized knowing A to the "reciprocal" probability that A be realized knowing B. One has to be careful because those two quantities are often confused, either knowingly or unknowingly.

For instance, a scary message such as "95% of heroin addicts started by smoking marijuana" may be used with the intent of suggesting (without stating it explicitly) that "95% of smokers of marijuana become heroin addicts." This is the ordinary fare of most political campaign arguments, of advertisements of every kind, and those never-ending polls.

But *the greatest care is necessary,* as the following example shows:

Thomas BAYES (1702–1761), English Presbyterian minister and theologian, studied mathematics, and in particular probability theory, during his free time. He published little of his work during his lifetime and remained unknown to most of his peers. His *Essay towards Solving a Problem in the Doctrine of Chances* was published posthumously in 1764.

Example 19.22 On the planet Zork live two ethnic groups: Ents and Vogons. According to a recent statistical analysis of the distribution of wealth on Zork, it is true that

- 80% among the Vogons are poor
- 80% among the poor are Vogons.

Is it possible to deduce that wealth is inequitably distributed between the two groups?

◊ **Solution**: Certainly not, in the absence of other data such as the proportion of Vogons in the population. It may very well be the case that the Vogons represent 80% of the total population, and that 80% of the total population is poor. If that is the case, wealth is equitably distributed between the two groups (not necessarily *within* each group). Again, this kind of confusion occurs continually, and is sometimes encouraged. (Many examples can be found by reading almost any newspaper, adapting the two words "poor" and "Vogons" to various circumstances.)

We now introduce a minor generalization of a complete system, which allows the events in the system only to cover the sample space up to a set of probability zero.

DEFINITION 19.23 An **almost complete system** is a sequence $(A_n)_{n\in\mathbb{N}}$ of events in Σ such that

i) $A_m \cap A_n = \varnothing$ if $m \neq n$,

ii) $\mathsf{P}\Big(\bigcup_{n\in\mathbb{N}} A_n\Big) = 1$.

THEOREM 19.24 (Bayes formula) *Let $(A_n)_{n\in\mathbb{N}}$ be an almost complete system of events, and let B be an arbitrary event $B \in \Sigma$. Assume that $\mathsf{P}(A_n) > 0$ for any $n \in \mathbb{N}$. Then we have*

i) $\mathsf{P}(B) = \sum_{n\in\mathbb{N}} \mathsf{P}(B|A_n)\mathsf{P}(A_n)$;

ii) if, moreover, $\mathsf{P}(B) > 0$, *then*

$$\forall p \in \mathbb{N} \qquad \mathsf{P}(A_p|B) = \frac{\mathsf{P}(B|A_p)\cdot\mathsf{P}(A_p)}{\sum\limits_{n\in\mathbb{N}} \mathsf{P}(B|A_n)\cdot\mathsf{P}(A_n)}.$$

Remark 19.25 The Collins case is a striking example of the use (or misuse) of probability and conditional probability. In Los Angeles in 1964, a blond woman and a black man with a beard were arrested for a robbery. Despite the absence of convincing evidence, the prosecution argued successfully that they must be guilty because the chance that a random couple corresponded (as they did) to the witness's description was estimated to be one in twelve million. The California Supreme Court reversed the judgment on appeal, since it was shown that the probability that at least *two* couples in the Los Angeles area correspond to the description *knowing that at least one couple does* (namely, the actual thieves) was close to 42 %, and therefore far from negligible (certainly too large to decide guilt "beyond reasonable doubt"!). This shows that there was a high probability that arresting the couple was a mistake.

19.5
Independent events

It is now time to define the notion of *independent* events. Intuitively, to say that two events A and B are independent means that the probability that A is realized is equal to the probability that it is realized *knowing* that B is realized, or indeed to the probability that A is realized *knowing* that "not-B" is realized.

DEFINITION 19.26 Let A and B be two events in Σ. Then **A and B are independent** if $\mathsf{P}(A \cap B) = \mathsf{P}(A)\,\mathsf{P}(B)$, or, equivalently if $0 < \mathsf{P}(B) < 1$, if $\mathsf{P}(A|B) = \mathsf{P}(A|\complement B) = \mathsf{P}(A)$.

Let $\Sigma_1, \ldots, \Sigma_n$ be σ-algebras contained in Σ. They are **independent σ-algebras** if

$$\forall (B_1, \ldots, B_n) \in \Sigma_1 \times \cdots \times \Sigma_n, \qquad \mathsf{P}\Big(\bigcap_{i=1}^{n} B_i\Big) = \prod_{i=1}^{n} \mathsf{P}(B_i).$$

Moreover, **events $A_1, \ldots, A_n$ are independent** if the σ-algebras

$$\sigma(\{A_1\}), \ldots, \sigma(\{A_n\})$$

are independent.

Warning! The pairwise independence of events $A_1, \ldots, A_n$ does not imply their independence. In fact, we have the following result:

THEOREM 19.27 *The events $A_1, \ldots, A_n$ are independent if and only if, for any $p \in \mathbb{N}$, $p \leqslant n$, and for any p-uple of distinct indices $(i_1, \ldots, i_p)$, we have*

$$\mathsf{P}\Big(\bigcap_{k=1}^{p} A_{i_k}\Big) = \prod_{k=1}^{p} \mathsf{P}(A_{i_k}).$$

In particular, for $n = 2$, we recover the first definition: A and B are independent events if and only if $\mathsf{P}(A \cap B) = \mathsf{P}(A)\,\mathsf{P}(B)$.

Example 19.28 Consider the experiment of rolling two dice, and the events:

- **event A:** the sum of the results of both dice is even;
- **event B:** the result of the first die is even;
- **event C:** the result of the second die is even.

It is easy to check that A, B, and C are pairwise independent. Indeed, we have (assuming all 36 outcomes of the experiment have probability $1/36$):

$$\mathsf{P}(A) = \mathsf{P}(B) = \mathsf{P}(C) = \tfrac{1}{2} \qquad \text{whereas} \quad \mathsf{P}(A \cap B) = \mathsf{P}(B \cap C) = \mathsf{P}(C \cap A) = \tfrac{1}{4}.$$

On other hand, A, B, C are *not* independent, since

$$\mathsf{P}(A \cap B \cap C) = \mathsf{P}(B \cap C) = \tfrac{1}{4} \qquad \text{whereas} \quad \mathsf{P}(A)\,\mathsf{P}(B)\,\mathsf{P}(C) = \tfrac{1}{8}.$$

(This is also clear intuitively, since A and B together imply C, for instance.)

One should note that this concept of independence depends only on a *numerical* evaluation of probabilities, and that it may turn out to be in contradiction with common sense or intuition, for which independence may carry deeper meaning. In particular, independence should not be mistaken with *causality* (see Section 20.11, page 550).

Chapter 20

Random variables

FALSTAFF. Prithee, no more prattling; go. I'll hold.
This is the third time; I hope good luck lies in odd numbers.
Away I go. They say there is divinity in odd numbers,
either in nativity, chance, or death. Away!

William SHAKESPEARE
The Merry Wives of Windsor, Act V, Scene I.

As we have noticed in the previous chapter, the sample space Ω is not usually a mathematically simple set – it may well be a population of exotic fishes, a collection of ideas, or representatives of various physical measurements. Worse, this set Ω is often *not known*. Then, instead of working with the probability measure P on Ω, one tries to reduce to various probabilities on $\mathbb{R}$.

For this purpose, we consider a map X from Ω to $\mathbb{R}$ (which is called a *random variable*); $\mathbb{R}$ becomes the new set on which a measure is defined. Indeed, a probability measure denoted P_X, which is the "image" of P by X, and which is called the *probability distribution of* X, is defined.

20.1 Random variables and probability distributions

DEFINITION 20.1 Let (Ω, Σ) be a probability space. A **random variable** is a function X, defined on Ω and with values in $\mathbb{R}$, which is measurable when $\mathbb{R}$

is equipped with the Borel σ-algebra, that is, such that we have $X^{-1}(A) \in \Sigma$ for any Borel set $A \in \mathscr{B}(\mathbb{R})$.

To simplify notation, we will abbreviate **r.v.** instead of writing "random variable."

A "random" function is, in other words, simply a *measurable* function – and, for a physicist not overly concerned with quoting the axiom of choice, this means "any function" that she may think about.

Here is an example. Take as probability space

$$\Omega \stackrel{\text{def}}{=} \{\text{all students at Rutgers university}\},$$

with the σ-algebra $\Sigma = \mathfrak{P}(\Omega)$.

We can define a random variable X corresponding to the "age" of a student:

$$X(\text{student}) = \textit{age of the student} \in \mathbb{R}.$$

Now we can study the distribution of the age in the population (evaluate the average, the standard deviation,...) without knowing too much about Ω, which can be a rather complicated set to envision precisely (if we think that it contains information such as the fact that François prefers to sing Bach in B-flat, or that Winnie's preferred dessert is strawberry cheesecake).

Given a probability space $(\Omega, \Sigma, \mathsf{P})$ and an r.v. X on Ω, the *probability distribution* of X is defined as the *image probability measure* of P by X:

DEFINITION 20.2 (Distribution) Let $(\Omega, \Sigma, \mathsf{P})$ be a probability space and X a random variable. The **probability distribution** of X is the probability measure

$$\begin{aligned} \mathsf{P}_X : \mathscr{B} &\longrightarrow \mathbb{R}^+, \\ B &\longmapsto \mathsf{P}\{X \in B\}, \end{aligned}$$

where

$$\{X \in B\} \stackrel{\text{def}}{=} X^{-1}(B) = \big\{\omega \in \Omega \,;\, X(\omega) \in B\big\}.$$

In particular, we have

$$\begin{aligned} \mathsf{P}_X\big(\,]a,b]\,\big) = \mathsf{P}\{a < X \leqslant b\} &= \mathsf{P}\Big(X^{-1}\big(\,]a,b]\,\big)\Big) \\ &= \mathsf{P}\Big(\big\{\omega \in \Omega \,;\, a < X(\omega) \leqslant b\big\}\Big). \end{aligned}$$

Note that this probability measure is *in general* defined on another set ($\mathbb{R}$) and with respect to another σ-algebra (of Borel sets) than that of the probability space itself (Ω).

The following diagram summarizes this discussion:

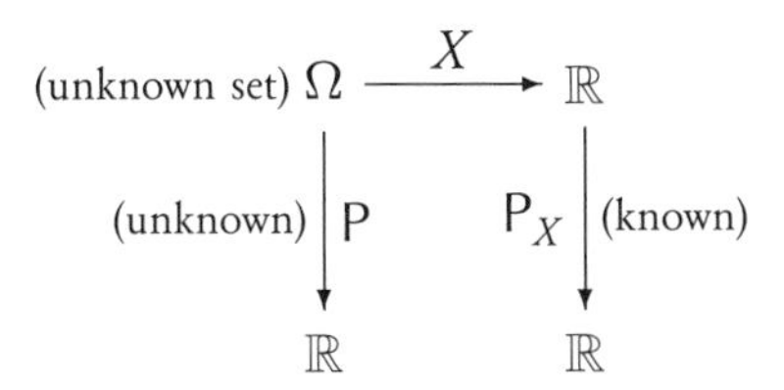

In the example of the Rutgers students, assume moreover that the sample space is given the uniform probability measure (each student is given the same probability, $p = 1/N$, where N is the total number of students). What is the probability distribution of the random variable "age," which we see as integer-valued? This distribution is a probability measure on the set $\mathbb{N}$, an instance of what is often called a **discrete distribution**:

$$\mathsf{P}_X(\{n\}) = \mathsf{P}(\{X = n\}) = \mathsf{P}\Big(\{\text{students}\ ;\ \text{age}(\text{student}) = n\}\Big)$$
$$= \frac{\text{number of students aged } n}{N}.$$

It can very well be the case that two random variables have the same distribution without being closely related to each other. For instance,

THEOREM 20.3 *Let X and Y be two random variables defined on the same sample space which are equal almost everywhere. Then X and Y have the same distribution:* $\mathsf{P}_X = \mathsf{P}_Y$.

But *be warned* that the converse statement is completely false.

Consider, for example, the process of flipping n times a fair (unbiased) coin, using 0 and 1 to denote, respectively, "heads" or "tails." We can take $\Omega = \{0,1\}^n$ and of course $\Sigma = \mathfrak{P}(\Omega)$. Let $\omega = (\mathcal{E}_1, \ldots, \mathcal{E}_n) \in \Omega$, where $\mathcal{E}_i \in \{0,1\}$ for $i = 1, \ldots, n$, be an atomic event. The probability of $\{\omega\} \in \Sigma$, if the successive throws are independent, is equal to

$$\mathcal{P}(\{\omega\}) = (1/2)^n \qquad \text{(uniform probability)}.$$

Consider now the random variable

$$X : \Omega \longrightarrow \mathbb{N},$$
$$\omega \longmapsto \sum_{i=1}^{n} \mathcal{E}_i = \text{number of ``tails.''}$$

We can find the distribution of X by denoting, for instance, A_k the event corresponding to "k tails out of n throws." So $A_k \in \Sigma$ and in fact

$$A_k = \{\omega \in \Omega\ ;\ X(\omega) = k\}.$$

The *distribution of* X is supported, in the case considered, on the set $\{0, \ldots, n\}$, and we have

$$\mathsf{P}_X(\{k\}) = \mathsf{P}(X = k) = \mathsf{P}(A_k) = \binom{n}{k} \cdot \left(\frac{1}{2}\right)^n.$$

This distribution is called the **binomial distribution**, and it is often denoted $\mathscr{B}\left(n, \frac{1}{2}\right)$. Note that it is easily checked that

$$\mathsf{P}_X(\mathbb{N}) = \frac{1}{2^n} \sum_{k=0}^{n} \binom{n}{k} = \frac{1}{2^n}(1+1)^n = 1,$$

as it should be. Moreover, it is easily seen that if X has a binomial distribution, so does $(n - X)$ (corresponding to the number of "heads" in n throws). Of course, X and $(n - X)$ are not equal almost everywhere!

What is clearly visible from this example is that distributions of random variables only provide *some* information on the full sample space, by no means all. This is the price to pay for the simplification that results from their use. Whatever is not encoded in the random variable (all the factors leading to a student of a certain age and not another, the exact trajectory in space of the falling die...) is completely forgotten.

DEFINITION 20.4 (Bernoulli distribution, binomial distribution) Let $n \in \mathbb{N}$ and $p \in [0, 1]$. A random variable X **follows the Bernouilli distribution with parameter** $\boldsymbol{p}$, denoted $\mathscr{B}(\mathbf{1}, \boldsymbol{p})$, if

$$\mathsf{P}_X(\{1\}) = p \qquad \text{and} \qquad \mathsf{P}_X(\{0\}) = 1 - p.$$

A random variable X follows the **binomial distribution** $\mathscr{B}(\boldsymbol{n}, \boldsymbol{p})$ if

$$\mathsf{P}_X(\{k\}) = \binom{n}{k} p^k (1-p)^{n-k} \qquad \text{for } k \in [\![1, n]\!].$$

The binomial distribution $\mathscr{B}(n, p)$ corresponds to the sum of n independent random variables, each of which has a Bernoulli distribution with the same parameter $\mathscr{B}(1, p)$.

20.2

Distribution function and probability density

DEFINITION 20.5 The **distribution function** of an r.v. X is the function defined for $x \geqslant 0$ by

$$F(x) \stackrel{\text{def}}{=} \mathsf{P}(\{X \leqslant x\}) = \mathsf{P}(X \in]-\infty, x]).$$

For all $a, b \in \mathbb{R}$, we have

$$\mathsf{P}(\{a < X \leqslant b\}) = F(b) - F(a).$$

Since the σ-algebra of Borel subsets of the real line is generated by the intervals $]-\infty, a]$, the folowing result follows:

PROPOSITION 20.6 *Let (Ω, Σ) be a probability space. A map $X : \Omega \to \mathbb{R}$ is a random variable if and only if $X^{-1}\big(\,]-\infty, a]\,\big) \in \Sigma$ for all $a \in \mathbb{R}$.*

Moreover, the probability distribution P_X is uniquely determined by the values $\mathsf{P}\big(X^{-1}(]-\infty, a])\big) = F(a)$ for $a \in \mathbb{R}$, that is, by the distribution function.

All information contained in the probability distribution P_X is contained in the distribution function F. **The distribution function is thus a fundamental tool of probability theory.**

♦ **Exercise 20.1** Express $\mathsf{P}\big(\{a < X < b\}\big)$, $\mathsf{P}\big(\{a \leqslant X \leqslant b\}\big)$, and $\mathsf{P}\big(\{a \leqslant X < b\}\big)$ in terms of $F(b)$, $F(a)$, $F(b^-)$, and $F(a^-)$.

From the definition of the distribution function, we immediately derive the following theorem:

THEOREM 20.7 *Let X be a random variable with distribution function F. Then F has the following properties:*

i) F is right-continuous; that is, for all $x \in \mathbb{R}$, we have

$$F(x^+) \stackrel{\text{def}}{=} \lim_{\varepsilon \to 0^+} F(x + \varepsilon) = F(x)\ ;$$

ii) F is not necessarily left-continuous, and more precisely we have

$$F(x) - F(x^-) = \mathsf{P}\{X = x\}\,;$$

iii) for all $x \in \mathbb{R}$, we have $0 \leqslant F(x) \leqslant 1$;

iv) F is increasing;

v) if we define

$$F(-\infty) \stackrel{\text{def}}{=} \lim_{x \to -\infty} F(x) \qquad \text{and} \qquad F(+\infty) \stackrel{\text{def}}{=} \lim_{x \to +\infty} F(x),$$

then

$$F(-\infty) = 0 \qquad \text{and} \qquad F(+\infty) = 1\ ;$$

vi) F is continuous if and only if $\mathsf{P}\{X = x\} = 0$ for all $x \in \mathbb{R}$. Such a probability distribution or random variable is called ***diffuse****.*

PROOF. Only i) and ii) require some care. First, write $F(x + \varepsilon) - F(x) = \mathsf{P}\{x < X \leqslant x + \varepsilon\}$, then notice that, as ε tends to 0 (following an arbitrary sequence), the set $\{x < X \leqslant x + \varepsilon\}$ "tends" to the empty set which has measure zero. This proves i).

Moreover, writing $F(x) - F(x - \varepsilon) = \mathsf{P}\{x - \varepsilon < X \leqslant x\}$, one notices similarly that the set $\{x - \varepsilon < X \leqslant x\}$ "tends" to $\{X = x\}$.

20.2.a Discrete random variables

DEFINITION 20.8 Let $(\Omega, \Sigma, \mathsf{P})$ be a probability space. A random variable X on Ω is a **discrete random variable** if $X(\Omega)$ is almost surely countable, that is, there is a countable set D such that $\mathsf{P}\{X \notin D\} = 0$.

When an r.v. is discrete, it is clear that its distribution function has discontinuities; it is a step function.

Example 20.9 Consider the experience of throwing a weighted die, which falls on the numbers 1 to 6 with probabilities respectively equal to $\frac{1}{4}, \frac{1}{6}, \frac{1}{6}, \frac{1}{6}, \frac{1}{6}, \frac{1}{12}$ (this is the one I hand to my friend Claude when playing with him).

The distribution function of the corresponding random variable is given by the following graph:

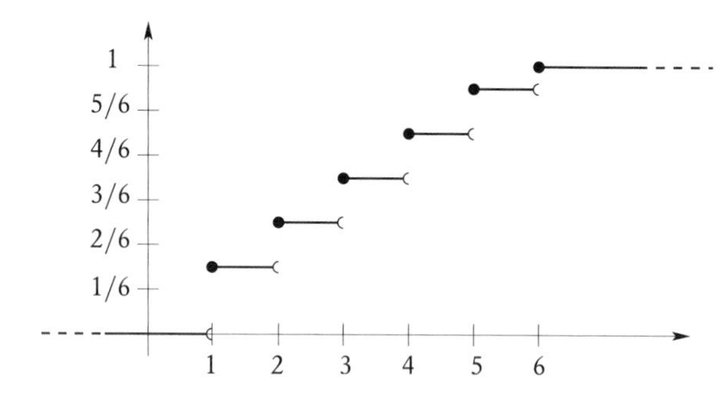

20.2.b (Absolutely) continuous random variables

DEFINITION 20.10 A random variable is **continuous** if its distribution function is continuous.

Remark 20.11 In fact, we should not speak simply of continuity, but rather of a similar but stronger property called **absolute continuity**. The reader is invited to look into a probability or analysis book such as [83] for more details.

Since, for distribution functions, continuity and absolute continuity are in fact equivalent, we will continue writing "continuous" instead of "absolutely continuous" for simplicity.

DEFINITION 20.12 The **probability density** of a continuous r.v. with distribution function F is a non-negative function f such that for all $x \in \mathbb{R}$ we have

$$\mathsf{P}\big(\{X \leqslant x\}\big) = F(x) = \int_{-\infty}^{x} f(t)\,\mathrm{d}t.$$

The density probability is almost everywhere equal to the derivative of F.

For a discrete r.v., the probability density f does not exist as a function, but it may be represented with Dirac distributions, since we want to differentiate a discontinuous function.

Example 20.13 Consider again the weighted die above. The probability density of the r.v. X representing the result of throwing this die is

$$f(x) = \frac{1}{4}\delta(x-1) + \frac{1}{6}\delta(x-2) + \frac{1}{6}\delta(x-3) + \frac{1}{6}\delta(x-4) + \frac{1}{6}\delta(x-5) + \frac{1}{12}\delta(x-6).$$

Remark 20.14 *In many cases,* a random variable has a distribution function which can be expressed as the sum of a continuous distribution function and a step function (which is itself a distribution function). The probability density is then the sum of an ordinary function and at most countably many Dirac distributions.

20.3 Expectation and variance

20.3.a Case of a discrete r.v.

DEFINITION 20.15 Let X be a discrete random variable. Assume that

$$X(\Omega) = \{p_k\ ;\ k \in I\} \cup D,$$

where I is a countable set and $\mathsf{P}\{X \in D\} = 0$, and let $p_k = \mathsf{P}(X = x_k)$ for $k \in I$. The **expectation** of X is the quantity

$$\mathsf{E}(X) = \sum_{k\in I} x_k\,\mathsf{P}(X = x_k) = \sum_{k\in I} x_k\,p_k$$

if this series converges.

Thus, the expectation is the "expected" average of the random variable. We believe that in some ways it represents the typical behavior of X. To make this more precise, we try to characterize how the values of X can be far from the expectation. For instance, during an exam graded on a 200-point scale, the class average may be equal to 100 with grades all contained between 90 and 110, or with grades ranging from 0 to 200; although two such distributions may yield the same expectation, they are of course very different.

To measure the *deviation from the mean*, the first idea might be to consider the average of the differences from the expected value, that is, to consider

$$\mathsf{E}\big(X - \mathsf{E}(X)\big) = \sum_k \big(x_k - \mathsf{E}(x)\big)\,p_k\ ;$$

however, this is always zero, by the definition of the expectation. Another idea is therefore needed, and since we want to *add* all the deviations, irrespective of sign, we can try to use

$$\mathsf{E}\Big(\big|X - \mathsf{E}(X)\big|\Big) = \sum_k \big|x_k - \mathsf{E}(x)\big|\,p_k.$$

However, absolute values are always algebraically delicate to handle. It is better, therefore, to get rid of the signs by taking the *square* of $X - \text{E}(X)$, possibly taking the square root of the result obtained by taking the average of this.[1]

DEFINITION 20.16 The **variance** of the discrete r.v. X is the quantity

$$\begin{aligned}\text{Var}(X) = \sigma^2(X) = \text{E}\Big(\big(X - \text{E}(X)\big)^2\Big) &= \sum_k \big(x_k - \text{E}(X)\big)^2 p_k \\ &= \text{E}(X^2) - \big(\text{E}(X)^2\big)\end{aligned}$$

which may be either a non-negative real number or $+\infty$. The **standard deviation** of X is given by $\sigma(X) = \sqrt{\text{Var}(X)}$.

Consider the example of giving global grades to students having passed four exams, say in Mathematics, Physics, Chemistry, and Medieval French Poetry. All four courses should play a similar role in determining the ranking of the students, even though they may carry different coefficients. For this purpose the graders must ensure that the standard deviation of their own grades is equal to some value fixed beforehand with their colleagues (say, 4). It is then of no consequence that the average in Mathematics is different from the average in Medieval French Poetry. On the other hand, if the standard deviation in Mathematics is equal to 6 while that in Medieval French Poetry is 2, then the Medieval French Poetry exam has little effect in distinguishing the students (they all have more or less the same grade), whereas Mathematics is crucial, even if – this being the entrance exam to the French Academy – the coefficient for Mathematics is a factor of 2 smaller.

20.3.b Case of a continuous r.v.

We will take the following definition as generalizing the discrete case.

DEFINITION 20.17 Let X be a random variable with probability density f. The **expectation E(X)** of the r.v. X is the number

$$\text{E}(X) = \int_{-\infty}^{+\infty} t\, f(t)\, \text{d}t$$

when the integral exists. The **variance Var(X)** and the **standard deviation $\sigma(X)$** are given by

$$\text{Var}(X) = \sigma^2(X) = \int_{-\infty}^{+\infty} \big(t - \text{E}(X)\big)^2 f(t)\, \text{d}t = \int_{-\infty}^{+\infty} t^2 f(t)\, \text{d}t - \big[\text{E}(x)\big]^2$$

when those integrals exist.

---

[1] In order to preserve the unit of a physical quantity, for instance.

Remark 20.18 The reader may check, as an exercise, that if an r.v. X is discrete, and a distribution is used to represent its probability density, then the above definition provides the same definition of the expectation of X as Definition 20.15.

Remark 20.19 Putting $f(x) = |\psi(x)|^2$, the reader can also recognize in the definition of expectation and standard deviation the quantities denoted $\langle x \rangle$ and Δx in quantum mechanics (see Section 13.3 on page 359).

Remark 20.20 The expectation and the variance are special cases of what are called *moments* of random variables.

For a continuous random variable, we write equivalently

$$\int_{\mathbb{R}} x\, f(x)\, \mathrm{d}x = \int_{\Omega} X(\omega)\, \mathrm{dP}(\omega).$$

- The first formula uses the probability density;[2] we can consider that we integrate the function $x \mapsto x$ with respect to the measure $\mathrm{d}\mu = f(x)\, \mathrm{d}x$.
- The second formule uses an integral over the sample space Ω, which is equipped with a σ-algebra and a probability measure, which suffices to defined an integral in the same manner that the Lebesgue integral is constructed.[3] The function considered is this time the random variable X, and it is integrated with respect to the measure $\mathrm{dP}(\omega)$.

Mathematicians usually avoid using the density f (which requires the use of the theory of distributions as soon as F is discontinuous) and appeal instead to another measure denoted $\mathrm{d}F = f(x)\, \mathrm{d}x$ and to the so-called "Stieltjes integrals" (see sidebar 8 on page 552). However, since density probabilities are probably more intuitive to physicists, we will use those instead, it being understood that it may be a distribution, and must be considered accordingly, even if it is written "under an integral sign."

♦ **Exercise 20.2** A random variable X is distributed according to a **Cauchy distribution** if its probability density is of the type

$$f(x) = \frac{a}{\pi} \frac{1}{a^2 + (x - m)^2}$$

for some real number m and some positive real number a. Compute the expectation and standard deviation of a random variable with this distribution *(answer in the table, page 616)*.

---

[2] And, as already noticed, may be generalized to the case of discontinuous F by considering f as a distribution and writing $\langle f, x \mapsto x \rangle$ instead of the integral.

[3] This is indeed what is called the Lebesgue integral with respect to an arbitrary measure on an arbitrary measure space.

20.4
An example: the Poisson distribution

20.4.a Particles in a confined gas

Consider a box with constant volume V, containing N identical, classical (i.e., not quantum) particles, which we assume not to interact. Moreover, assume that the system has reached thermodynamic equilibrium, and is in the situation of *thermodynamic limit* (in particular, N is supposed to be very large).

Fix now a small volume $\delta V \ll V$. What is the probability that *one* given particle is located in this small volume? If the system is ergodic, this probability is $p = \delta V / V$.

What is now the probability that there are two particles in this volume? That there are n particles? More generally, we ask for the probability distribution of the number of particles contained in δV.

To start with, notice that since the particles are *classical* and *do not interact*, there is no correlation between their positions. Thus, the probability for one particle to be in the given volume δV is independent of the number of particles already located there. Hence, the probability distribution for the number of particles in δV is the binomial distribution $\mathsf{P} = \mathscr{B}(N, p)$ with values given by

$$\mathsf{P}(n) = \binom{N}{n} p^n (1-p)^{N-n},$$

since n particles must be in δV (with probability p) and $N - n$ must be outside (with probability $(1-p)$). The expected value of the binomial distribution is

$$\overline{n} = \sum_{n=0}^{N} n \, \mathsf{P}(n) = Np.$$

If N goes to infinity, we will see in the next chapter that the binomial distribution can be approached by a gaussian distribution; however, this is only correct if $Np \gg 1$. For fixed N (which is the case in physics), if we consider a very small volume δV, it is possible to make $\overline{n} = Np = N\,\delta V / V$ very small compared to 1. Then we can write

$$\binom{N}{n} \approx \frac{N^n}{n!} \qquad \text{and} \qquad (1-p)^{N-n} = \exp^{(N-n)\log(1-p)} \approx \mathrm{e}^{-Np} = \mathrm{e}^{-\overline{n}},$$

which gives

$$\mathsf{P}(n) \approx \frac{N^n}{n!} p^n \mathrm{e}^{-\overline{n}} = \frac{\overline{n}^n}{n!} \mathrm{e}^{-\overline{n}}.$$

DEFINITION 20.21 The **Poisson distribution** with parameter λ is the discrete probability measure such that

$$\mathsf{P}_\lambda(n) = \mathrm{e}^{-\lambda}\frac{\lambda^n}{n!} \qquad \text{for all } n \in \mathbb{N}.$$

A random variable X is a Poisson random variable if $\mathsf{P}\{X = n\} = \mathsf{P}_\lambda(n)$ for all $n \in \mathbb{N}$.

The Poisson distribution is useful to describe *rare events*.[4] The maximal value of $n \mapsto \mathsf{P}_\lambda(n)$ is attained for $n \approx \lambda$. Moreover, the reader will easily check the following results:

THEOREM 20.22 *The expectation of a Poisson random with parameter* λ *is* $\mathsf{E}(X) = \lambda$, *and its variance is* $\mathrm{Var}(X) = \lambda$.

20.4.b Radioactive decay

A radiation source randomly emits α-particles. Assume that this radiation is weak enough that it does not perturb the source (i.e., the power which is radiated is statistically constant). Moreover, assume that the probability that a particle is emitted between t and $t + \Delta t$ depends only on Δt, and not on t or on the number of α-particles already emitted before time t (this is called a **Markov hypothesis**). It seems reasonable to postulate that this probability is proportional to Δt when Δt is very small; thus, we take it to be equal to $p\Delta t$.

Let $\mathcal{Q}(t)$ be the probability that the source emits no particle during a length of time of duration t. In the absence of time correlation, we have therefore

$$\mathcal{Q}(t + \mathrm{d}t) = (1 - p\,\mathrm{d}t)\,\mathcal{Q}(t),$$

which, together with the condition $\mathcal{Q}(0) = 1$, implies that $\mathcal{Q}(t) = \mathrm{e}^{-pt}$.

What is the probability that n particles are emitted during an interval of length t? To compute this, partition the interval $[0, t]$ into a large number N of small intervals of length $\Delta t = t/N$. During each of these intervals of length Δt, the probability of emitting an α-particle is $p\Delta t$. Hence the probability of emitting n particles is given by a binomial distribution, which may be approximated by the Poisson distribution with parameter $\lambda = pt$:

$$\mathsf{P}(n, t) = \binom{N}{n}(p\Delta t)^n\,(1 - p\Delta t)^{N-n} \approx \frac{(pt)^n}{n!}\,\mathrm{e}^{-pt}.$$

As might be expected, this probability is maximal when $n \approx pt$; the average number of particles emitted during an interval of length t is given by the expectation of the Poisson distribution, and is also equal to pt.

---

[4] The first recorded use of the Poisson distribution is found in a statistical study of the annual number of deaths caused by horse kicks in the Prussian cavalry in 1898.

20.5

Moments of a random variable

DEFINITION 20.23 The ***k*-th moment** (with $k \in \mathbb{N}$) of a random variable X with distribution function F is the quantity

$$\hat{\mu}_k \stackrel{\text{def}}{=} \int_{-\infty}^{+\infty} x^k f(x)\,\mathrm{d}x = \int_\Omega X^k(\omega)\,\mathrm{dP}(\omega)$$

if it exists.

For $k \geqslant 1$, the k-th moment does not necessarily exist. The 0-th moment always exists, but it is always equal to 1 and is not very interesting. The first moment is simply the expectation of X. One can define variants of the moment for a *centered* random variable, that is, one which expectation equal to zero.

DEFINITION 20.24 A **centered random variable** is a random variable with expectation equal to 0. If X is a random variable such that the expectation of X exists, and $m = \mathrm{E}(X)$, then $(X - m)$ is a centered random variable.

DEFINITION 20.25 The ***k*-th centered moment** of an r.v. X with expectation m is given by

$$\mu_k \stackrel{\text{def}}{=} \int_{-\infty}^{+\infty} (x - m)^k f(x)\,\mathrm{d}x = \int_\Omega (X(\omega) - m)^k\,\mathrm{dP}(\omega).$$

Remark 20.26 For $k \geqslant 1$, the centered k-th moment exists if and only if the original (non-centered) k-th moment exists.

Remark 20.27 The centered second moment is none other than the variance:

$$\mu_2 = \mathrm{Var}(X) = \sigma^2(X).$$

♦ **Exercise 20.3** Show that $\mu^2 = \hat{\mu}_2 - \mu_1^2$.

Show that the kinetic energy of a distribution of mass is the sum of the kinetic energy in the barycentric referential and the kinetic energy of the center of gravity (also called the barycenter), carrying the total mass of the system.

The following definition is sometimes used.

DEFINITION 20.28 A **reduced random variable** is a random variable for which the standard deviation exists and satisfies $\sigma(X) = 1$.

If X is an r.v. with expectation m and standard deviation σ, the random variable $(X - m)/\sigma$ is always centered and reduced.

DEFINITION 20.29 The **median** of a random variable X with distribution function F is the number

$$x_{1/2} = \inf\left\{x \in \mathbb{R}\,;\; F(x) \geqslant \frac{1}{2}\right\}.$$

If F is continuous, then $x_{1/2}$ is a solution of the equation $F(x) = \frac{1}{2}$. If, moreover, F is strictly increasing, this is the unique solution.

Example 20.30 The median occurs frequently in economics and statistics. The *median salary* of a population, for instance, is the salary such that *half* of the population earn less, and *half* earns more. It is therefore different from the *average salary*.

Take for instance the case of six individuals, named Angelo, Bianca, Corin, Dolabella, Emilia, and Fang, with salaries as follows:

Angelo	5000	Bianca	9000	Corin	9000
Dolabella	9000	Emilia	5000	Fang	11000

The average salary is \$8000. However, \$8000 is not the median salary, since only two persons earn less, and four earn more. In this example, the median salary is \$9000.

If three other persons are added to the sample, Grumio, Hamlet and Imogen, earning \$6000, \$7000 and \$8500, respectively, then the median salary becomes \$8500.

Example 20.31 (Half-life and lifetime) Consider a radioactive particle that may disintegrate at any moment. Start the experiment at time $t = 0$ (the particle has not yet disintegrated) and assume that the probability that it does disintegrate during an infinitesimal time interval $\mathrm{d}t$ is equal to $\alpha\,\mathrm{d}t$ (independently of t). As in Section 20.4.b, one can show that the probability that the particle be intact at time t is given by

$$F(t) = \mathrm{e}^{-\alpha t}.$$

The median of the random variable "time before disintegration" is called the **half-life** (the probability of disintegration during this length of time is $\frac{1}{2}$). To compute it, we must solve the equation

$$F(t_{1/2}) = \exp\{-\alpha\, t_{1/2}\} = \frac{1}{2},$$

and therefore we have $t_{1/2} = (\log 2)/\alpha$. Note that the *expectation* of this r.v., which is called the **(average) lifetime** of the particle, is equal to $1/\alpha$. Indeed, the probability density for disintegration at time t is given by

$$f(t) = -F'(t) = \alpha\,\mathrm{e}^{-\alpha t},$$

and the expectation is therefore

$$\tau = \int t\, f(t)\,\mathrm{d}t = \int t\alpha\,\mathrm{e}^{-\alpha t}\,\mathrm{d}t = \frac{1}{\alpha} = \frac{t_{1/2}}{\log 2} > t_{1/2}.$$

It is such that $F(\tau) = 1/\mathrm{e}$.

20.6
Random vectors

20.6.a Pair of random variables

Sometimes it is necessary to consider a random *vector*, or more generally an n-tuple of random variables. We start with the case of a pair (X, Y) of random variables, all defined on the same probability space $(\Omega, \Sigma, \mathrm{P})$.

DEFINITION 20.32 The **distribution function** or **joint distribution function** of a pair (X, Y) of random variables is the function

$$F_{XY}(x, y) \stackrel{\text{def}}{=} \mathrm{P}\{X \leqslant x \text{ and } Y \leqslant y\},$$

defined for $x, y \in \mathbb{R}$, that is, the probability that both X is less than or equal to x, and Y is less than or equal to y.

PROPOSITION 20.33 *Let F_{XY} be the distribution function of the pair (X, Y). Then, for any $x, y \in \mathbb{R}$, we have*

$$F_{XY}(-\infty, y) = F_{XY}(x, -\infty) = 0 \qquad \text{and} \qquad F_{XY}(+\infty, +\infty) = 1.$$

Moreover, F_{XY} in increasing and right-continuous with respect to each variable.

The notion of probability density is also easy to extend:

DEFINITION 20.34 (Joint probability density) The **probability density** (also called **joint probability density**) of the pair (X, Y) is the function (or sometimes the distribution)

$$f_{XY}(x, y) \stackrel{\text{def}}{=} \frac{\partial^2 F_{XY}}{\partial x \, \partial y}(x, y).$$

We then have

$$F_{XY}(x, y) = \int_{-\infty}^{x} \int_{-\infty}^{y} f_{XY}(s, t) \, \mathrm{d}t \, \mathrm{d}s.$$

DEFINITION 20.35 (Marginal distribution) Let (X, Y) be a pair of random variables. The **marginal distribution** of X is the distribution of the random variable X.

Knowing the (joint) distribution function of the pair (X, Y), it is very easy to find the distribution function of X, and hence its marginal distribution. Since $F_X(x)$ is the probability that $X \leqslant x$, with no condition whatsoever imposed on the variable Y, we have

$$F_X(x) = F(x, +\infty),$$

and similarly the marginal distribution of Y is given by

$$F_Y(y) = F(+\infty, y).$$

It follows that the marginal probability densities are

$$f_X(x) = \int_{-\infty}^{+\infty} f_{XY}(x,y)\,\mathrm{d}y \qquad \text{and} \qquad f_Y(y) = \int_{-\infty}^{+\infty} f_{XY}(x,y)\,\mathrm{d}x.$$

We now introduce the moments of a pair:

DEFINITION 20.36 Let (X, Y) be a pair of r.v. and let k and ℓ be integers. The **(k,ℓ)-th moment** or **moment of order (k,ℓ)** of (X, T) is the quantity

$$\hat{\mu}_{k,\ell} \stackrel{\text{def}}{=} \iint_{\mathbb{R}^2} x^k\, y^\ell\, f_{XY}(x,y)\,\mathrm{d}x\,\mathrm{d}y,$$

if it exists. If we denote $m_X \stackrel{\text{def}}{=} \hat{\mu}_{1,0}$ and $m_Y \stackrel{\text{def}}{=} \hat{\mu}_{0,1}$ the expectations of X and Y, the corresponding **centered moment** is given by

$$\mu_{k,\ell} \stackrel{\text{def}}{=} \iint_{\mathbb{R}^2} (x - m_X)^k\, (y - m_Y)^\ell\, f_{XY}(x,y)\,\mathrm{d}x\,\mathrm{d}y.$$

DEFINITION 20.37 The **variances** of (X, Y) are the centered moments of order $(2,0)$ and $(0,2)$, denoted $\mathrm{Var}(X) = \sigma_X^2 = \mu_{20}$ and $\mathrm{Var}(Y) = \sigma_Y^2 = \mu_{02}$. The **covariance** of (X,Y), denoted $\mathrm{Cov}(X,Y)$, is the centered moment of order $(1,1)$, that is,

$$\mathrm{Cov}(X,Y) \stackrel{\text{def}}{=} \mu_{11}.$$

We then have the relation

$$\mathrm{Cov}(X,Y) = \mathrm{E}\Big[\big(X - \mathrm{E}(X)\big)\big(Y - \mathrm{E}(Y)\big)\Big].$$

Remark 20.38 Note also the obvious formula $\mathrm{Cov}(X,X) = \sigma_X^2 = \mathrm{Var}^2(X)$.

♦ **Exercise 20.4** Let X and Y be two random variables with moments of order 1 and 2. Show that

$$\mathrm{Var}(X+Y) = \mathrm{Var}(X) + \mathrm{Var}(Y) + 2\,\mathrm{Cov}(X,Y).$$

Note that the centered moment of order $(1,1)$ is related with the non-centered moment by means of the formula $\mu_{11} = \hat{\mu}_{11} - \mu_{10}\,\mu_{01}$.

Finally, we have the correlation of random variables:

DEFINITION 20.39 The **correlation** of the pair (X,Y) is the quantity

$$r \stackrel{\text{def}}{=} \frac{\mu_{11}}{\sqrt{\mu_{20}\,\mu_{02}}} = \frac{\mathrm{Cov}(X,Y)}{\sqrt{\mathrm{Var}(X)\,\mathrm{Var}(Y)}}$$

if the moments involved exist.

This has the following properties (following from the Cauchy-Schwarz inequality):

THEOREM 20.40 *Let (X, Y) be a pair of random variables with correlation r, assumed to exist. Then we have*

i) $|r| \leqslant 1$;

ii) $|r| = 1$ *if and only if the random variables X and Y are related by a linear transformation, more precisely if*

$$\frac{Y - m_Y}{\sigma_Y} = r\frac{X - m_X}{\sigma_X} \qquad \textit{almost surely.}$$

DEFINITION 20.41 (Uncorrelated r.v.) Two random variables X and Y are **uncorrelated** if their correlation exists and is equal to zero.

The correlation is useful in particular when trying to quantify a link between two statistical series of numbers. Suppose that for N students, we know their size T_n and their weight P_n, $1 \leqslant n \leqslant N$, and we want to show that there exists a statistical link between the size and the weight. Compute the average weight $\overline{P}$ and the average size $\overline{T}$. The standard deviations of those random variables are given by

$$\sigma^2(T) = \frac{1}{N}\sum_{n=1}^{N} T_n^2 - \overline{T}^2 \qquad \text{and} \qquad \sigma^2(P) = \frac{1}{N}\sum_{n=1}^{N} P_n^2 - \overline{P}^2.$$

The correlation between size and weight is then equal to

$$r = \frac{1}{\sigma(T)\,\sigma(P)}\sum_{n=1}^{N}(T_i - \overline{T})(P_i - \overline{P}).$$

If it is close to 1, there is a strong *positive* correlation between size and weight, that is, the weight increases in general when the size increases. If the correlation is close to -1, the correlation is negative (the taller you are, the lighter). Otherwise, there is little or no correlation. If $|r| = 1$, there is even an affine formula relating the two statistical series.

The correlation must be treated with some care. A value $r = 0.8$ may indicate an interesting link, or not, depending on the circumstances. For instance, the size of the dataset may be crucial, as the following example (taken from Jean MEEUS [66]) indicates.

The table below gives the absolute magnitude[5] m of some asteroids and the number of letters ℓ in their name:

---

[5] A logarithmic measure of brightness at fixed distance.

Name	m	ℓ	Name	m	ℓ
Niobe	8.5	5	Frigga	9.6	6
Feronia	10.3	7	Diana	9.1	5
Klytia	10.3	6	Eurynome	9.3	8
Galatea	10.1	7	Sappho	9.3	6
Eurydike	10.0	8	Terpsichore	9.7	11
Freia	9.0	5	Alkmene	9.4	7

The correlation for the first eight asteroids is $r \approx 0.785$, which is quite large. Worse, the correlation for the last five is $r \approx 0.932$. Of course, this is merely an accident, due to the small sample used. There is in fact no correlation between the length of the name and the magnitude of asteroids. With the twelve given, the correlation becomes ≈ 0.4 and with fifty asteroids[6] it is only $r \approx -0.097$.

Remark 20.42 The covariance of (X, Y) may be seen as a kind of "scalar product" of $X - \mathrm{E}(X)$ with $Y - \mathrm{E}(Y)$ and the standard deviation as the associated "norm." The correlation is then the *cosine of the angle* between the two functions in L^2 space:

$$r = \frac{(X|Y)}{\|X\|_2 \, \|Y\|_2}.$$

So there is no surprise in the fact that this number is between -1 and 1. Moreover, $|r| = 1$ if and only if the functions $X - \mathrm{E}(X)$ and $Y - \mathrm{E}(Y)$ are linearly dependent, as we found above.

20.6.b Independent random variables

DEFINITION 20.43 (Independence) Two random variables X and Y are **independent random variables** if the joint distribution function can be expressed as the direct product (or "tensor" product) of their marginal distribution functions $F_{XY} = F_X \otimes F_Y$, that is, if

$$F_{XY}(x, y) = F_X(x)\, F_Y(y) \quad \text{for all } x, y \in \mathbb{R}.$$

It is usual to abbreviate "independent random variables" by **i.r.v.**

There is an equivalent form of the definition which involves the *σ-algebra generated by a random variable*:

DEFINITION 20.44 (σ-algebra generated by an r.v.) Let X be a random variable. The **σ-algebra generated by X**, denoted $\mathscr{T}(X)$, is the set of all subsets of the form $X^{-1}(A)$ for a Borel subset A.

PROPOSITION 20.45 *Two random variables X and Y are independent if and only if the σ-algebras generated by X and Y are independent, that is, if, for all $A \in \mathscr{T}(X)$ and $B \in \mathscr{T}(Y)$, we have* $\mathrm{P}(A \cap B) = \mathrm{P}(A) \cdot \mathrm{P}(B)$.

[6] Numbered 51 to 100 in the *Ephemerids of Minor Planets* (Leningrad, 1979).

If the joint probability density exists, we then have $f_{XY} = f_X \otimes f_Y$, that is,

$$f_{XY}(x,y) = f_X(x)\, f_Y(y) \quad \text{for all } x, y \in \mathbb{R}.$$

We have the following important result:

THEOREM 20.46 *Two independent r.v. X and Y are uncorrelated.*

PROOF. We can assume that the variances σ_X^2 and σ_Y^2 exist; otherwise the correlation is not defined. Then we have

$$\begin{aligned}\operatorname{Cov}(X,Y) &= \iint (x-m_X)(y-m_Y)\, f_{XY}(x,y)\,\mathrm{d}x\,\mathrm{d}y \\ &= \int (x-m_X)\, f_X(x)\,\mathrm{d}x \int (y-m_Y)\, f_Y(y)\,\mathrm{d}y = 0,\end{aligned}$$

since each factor vanishes.

The converse is not true in general, however.

Example 20.47 Let $\Omega = [0, 2\pi]$ with the uniform probability measure $\mathrm{P} = \mathrm{d}\theta/2\pi$. Let $X(\theta) = \cos\theta$ and $Y = \sin\theta$. The r.v. X and Y are not independent (intuitively, because there is a relation between them, namely $X^2 + Y^2 = 1$), but on the other hand the covariance is equal to

$$\operatorname{Cov}(X,Y) = \int_\Omega \cos\theta \sin\theta \,\frac{\mathrm{d}\theta}{2\pi} = 0,$$

so they are uncorrelated.

20.6.c Random vectors

It is now possible to generalize the preceding results to n random variables.

DEFINITION 20.48 An n-dimensional **random vector** is an n-tuple of random variables, or in other words a measurable map

$$\boldsymbol{X} = (X_1, \dots, X_n) : \Omega \longrightarrow \mathbb{R}^n.$$

DEFINITION 20.49 The **distribution function** (or **joint distribution function**) of a random vector $\boldsymbol{X}$ is the function defined on $\mathbb{R}^n$ by

$$F_{\boldsymbol{X}}(\boldsymbol{x}) = \mathrm{P}\{X_i \leqslant x_i \text{ for } i = 1, \dots, n\}, \qquad \text{where} \quad \boldsymbol{x} = (x_1, \dots, x_n).$$

PROPOSITION 20.50 *Let $F_{\boldsymbol{X}}$ be the distribution function of $\boldsymbol{X}$. Then for all $\boldsymbol{x} \in \mathbb{R}^n$ we have*

$$\begin{aligned}F_{\boldsymbol{X}}(-\infty, x_2, \dots, x_n) = \cdots = F_{\boldsymbol{X}}(x_1, \dots, -\infty, \dots, x_n) = \cdots \\ = F_{\boldsymbol{X}}(x_1, \dots, -\infty) = 0\end{aligned}$$

and

$$F_{XY}(+\infty, \dots, +\infty) = 1.$$

Morevoer, $F_{\boldsymbol{X}}$ is increasing and right-continuous with respect to each variable.

DEFINITION 20.51 The **probability density** (or **joint probability density**) of a random vector X is the function (or distribution)

$$f_X(x_1, \ldots, x_n) \stackrel{\text{def}}{=} \frac{\partial^n}{\partial x_1 \ldots \partial x_n} F_X(x_1, \ldots, x_n).$$

We then have

$$F_X(x_1, \ldots, x_n) = \int_{-\infty}^{x_1} \cdots \int_{-\infty}^{x_n} f(s_1, \ldots, s_n) \, \mathrm{d}s_1 \cdots \mathrm{d}s_n.$$

20.7
Image measures

20.7.a Case of a single random variable

♦ **Exercise 20.5** The probability density of the velocity of an atom of a perfect gas is known (it is a maxwellian[7]). What is the probability density of the kinetic energy $mv^2/2$ of this atom?

To treat this kind of problem, we must consider the situation where, given a random variable X, a *second* random variable $Y = \varphi(X)$ is defined. This is called a **change of random variable**. The task is to determine the distribution function of Y, and its probability density, knowing those of X.

When the change of random variable is *bijective*, there is a unique $y = \varphi(x)$ associated to any $x \in \mathbb{R}$, and it is then easy to see that if f denotes the probability density of X and g denotes that of Y, then

$$g(y) = \left. \frac{f(x)}{|\varphi'(x)|} \right|_{x=\varphi^{-1}(y)},$$

provided φ' has no zero.

Proof. Indeed, we have $G(y) = \mathrm{P}\{X \leqslant \varphi^{-1}(y)\} = F \circ \varphi^{-1}(y)$ if φ is increasing, and $G(y) = \mathrm{P}\{X \geqslant \varphi^{-1}(x)\} = 1 - \mathrm{P}\{X < \varphi^{-1}(y)\} = 1 - F \circ \varphi^{-1}(y)$ if φ is decreasing. Then

$$g(y) = G'(y) = \left[F \circ \varphi^{-1}\right] = \pm F' \circ \varphi^{-1} \cdot (\varphi^{-1})' = \pm \frac{f(x)}{\varphi'(x)},$$

as was to be shown.

If the function φ is not a bijection, we have the following more general result (compare with the change of variable formula for a Dirac distribution, Exercise 7.4.d, page 198):

[7] In other words, a gaussian with a physicist's name.

THEOREM 20.52 *Let $\varphi : X \mapsto Y$ be a change of random variable and $y \in \mathbb{R}$ such that the equation*

$$y = \varphi(x)$$

has at most countably many solutions, and let S_y denote the set of these solutions. Assume, moreover, that $\varphi'(x) \neq 0$ for $x \in S_y$. Then the probability density g of Y is related to the probability density f of X by

$$g(y) = \sum_{x \in S_y} \frac{f(x)}{|\varphi'(x)|}.$$

20.7.b Case of a random vector

Theorem 20.52 generalizes to a function $\varphi : \mathbb{R}^n \to \mathbb{R}^n$ as follows:

THEOREM 20.53 *Let $\varphi : \boldsymbol{X} \mapsto Y$ be a change of random vector, and let $\boldsymbol{y} \in \mathbb{R}^n$ be such that the equation $\boldsymbol{y} = \varphi(\boldsymbol{x})$ has at most countably many solutions $\boldsymbol{x}_k$, $k \in I$. Assume, moreover, that the differential $\mathrm{d}\varphi(\boldsymbol{x}_k)$ is invertible for all $i \in I$. Then the probability density g of Y is related to the probability density f of $\boldsymbol{X}$ by*

$$g(\boldsymbol{y}) = \sum_{k \in I} \frac{f(\boldsymbol{x}_k)}{|\det \mathrm{d}\varphi(x_k)|} = \sum_{k \in I} f(\boldsymbol{x}_k) \left| \frac{\mathrm{D}(x_1, \dots, x_n)}{\mathrm{D}(y_1, \dots, y_n)} \right|,$$

where the last factor is the inverse of the jacobian matrix of the change of variable.

20.8 Expectation and characteristic function

20.8.a Expectation of a function of random variables

Let X be a random variable and $Y = \varphi(X)$. The expectation of Y, if it exists, can be computed using the probability density f_X of X by the formula

$$\mathrm{E}(Y) = \mathrm{E}\big(\varphi(X)\big) = \int_\Omega \varphi\big(X(\omega)\big)\,\mathrm{dP}(\omega) = \int_{\mathbb{R}} \varphi(x)\, f_X(x)\,\mathrm{d}x.$$

More generally, if $Y = \varphi(X_1, \dots, X_n)$ is a random variable defined by change of variable from a random vector, the expectation of Y (if it exists) may be computed using the joint probability density $f_{X_1 \cdots X_n}(x_1, \dots, x_n)$ of $(X_1, \dots, X_n)$ by the formula

$$\mathrm{E}(Y) = \iint_{\mathbb{R}^n} \varphi(x_1, \dots, x_n)\, f_{X_1 \cdots X_n}(x_1, \dots, x_n)\,\mathrm{d}x_1 \dots \mathrm{d}x_n.$$

For instance, if X and Y are random variables, the expectation $\mathsf{E}(X\,Y)$ of the product $X\,Y$ can be computed using the joint probability density f_{XY}:

$$\mathsf{E}\{X\,Y\} = \int_\Omega X(\omega)\,Y(\omega)\,\mathrm{d}\mathsf{P}(\omega) = \iint_{\mathbb{R}^2} x\,y\,f_{XY}(x,y)\,\mathrm{d}x\,\mathrm{d}y.$$

THEOREM 20.54 *The expectation has the following properties:*

i) it is linear, that is, we have

$$\mathsf{E}\Big\{\sum_i \lambda_i X_i\Big\} = \sum_i \lambda_i\,\mathsf{E}\{X_i\}$$

whenever the right-hand side is defined;

ii) if X is almost surely equal to a constant a, $\mathsf{E}\{X\} = a$; in particular, we have $\mathsf{E}\{\mathsf{E}\{X\}\} = \mathsf{E}\{X\}$;

iii) we have

$$\big|\mathsf{E}\{X\,Y\}\big|^2 \leqslant \mathsf{E}\{X^2\}\,\mathsf{E}\{Y^2\},$$

by the Cauchy-Schwarz inequality, with equality if and only if X and Y are linearly dependent;

iv) if X and Y are independent, then $\mathsf{E}\{X\,Y\} = \mathsf{E}\{X\}\,\mathsf{E}\{Y\}$.

20.8.b Moments, variance

Using the formulae of the previous section, we see that the moments and centered moments of a random variable X are given by expectations of functions of X, namely,

$$\hat{\mu}_k = \mathsf{E}\{X^k\} \qquad \text{and} \qquad \mu_k = \mathsf{E}\Big\{\big(X - \mathsf{E}\{X\}\big)^k\Big\}.$$

Moreover, the k-th moment (centered or not) exists if and only if the function $x \longmapsto x^k\,f(x)$ is Lebesgue-integrable, where f is the probability density of X, that is, if

$$\int \big|x^k\big|\,f(x)\,\mathrm{d}x < +\infty.$$

As a special case, the variance is given by

$$\sigma^2 = \mathsf{E}\{(X - \mathsf{E}\{X\})^2\} = \mathsf{E}\{X^2\} - \mathsf{E}\{X\}^2.$$

20.8.c Characteristic function

DEFINITION 20.55 The **characteristic function** φ_X of a random variable X is the function defined for $\nu \in \mathbb{R}$ by

$$\boxed{\varphi_X(\nu) = \mathsf{E}\left\{\mathrm{e}^{\mathrm{i}\nu X}\right\}.}$$

The following result is then immediate.

THEOREM 20.56 *The characteristic function φ_X of a random variable X is the Fourier transform*[8] *of its probability density (in the sense of distributions):*

$$\varphi_X(\nu) = \int_\Omega e^{i\nu X(\omega)}\,dP(\omega) = \int_{-\infty}^{\infty} e^{i\nu x} f(x)\,dx.$$

Notice that this quantity is always defined, for any random variable X, because $|e^{i\nu X}| = 1$ and Ω is of total measure equal to 1 (or equivalently, because the density probability f, when it exists as a function, is Lebesgue-integrable).

According to the general theory of the Fourier transform, we have therefore:

PROPOSITION 20.57 *Let X be a random variable and let φ_X be its characteristic function. Then*

i) $\varphi_X(0) = 1$ and $|\varphi_X(\nu)| \leqslant 1$ for all $\nu \in \mathbb{R}$;

ii) if the k-th order moment of X exists, φ_X can be differentiated k times and we have

$$\mu_k = E\{X^k\} = (-i)^k \varphi_X^{(k)}(0);$$

iii) since X is real-valued, φ_X is hermitian, i.e., we have $\varphi_X(-\nu) = \overline{\varphi_X(\nu)}$.

PROOF. Note that $\varphi_X(0) = \int_{-\infty}^{+\infty} f_X(x)\,dx = 1$ or, to phrase it differently, we have $\varphi_X(0) = E\{1\} = 1$. The last two properties are direct translations of known results of Fourier transforms.

THEOREM 20.58 *The characteristic function of a random variable X determines uniquely the distribution function of X.*

In other words, two random variables have the same characteristic function if and only if they have the same distribution function. (This theorem follows from an inversion formula which is the analogue of the Fourier inversion formula and is due to P. Lévy).

Before going to the next section, note also the following important fact:

PROPOSITION 20.59 *The characteristic function of the sum of two* independent *random variables is the product of the characteristic functions of the summands:*

$$\varphi_{X+Y}(\nu) = \varphi_X(\nu) \cdot \varphi_Y(\nu).$$

---

[8] Be careful with the various conventions for the Fourier transform! It is usual in probability to omit the factor 2π which often occurs in analysis.

PROOF. Let f_{XY} be the joint probability density of (X, Y). Since X and Y are independent, we have $f_{XY}(x, y) = f_X(x)\, f_Y(y)$ and hence

$$\begin{aligned}\varphi_{X+Y}(\nu) &= \mathrm{E}\left\{e^{i\nu(X+Y)}\right\} \\ &= \iint e^{i\nu(x+y)} f_{XY}(x, y)\,\mathrm{d}x\,\mathrm{d}y \\ &= \int e^{i\nu x} f_X(x)\,\mathrm{d}x \int e^{i\nu y} f_Y(y)\,\mathrm{d}y = \varphi_X(\nu)\cdot\varphi_Y(\nu)\end{aligned}$$

by the expectation formula for a change of random variables.

20.8.d Generating function

DEFINITION 20.60 The **generating function** G_X of a random variable X is the function of a real variable s defined as the expectation of s^X:

$$\boxed{G_X(s) = \mathrm{E}(s^X)}$$

when it exists.

THEOREM 20.61 *Let $X : \Omega \to \mathbb{N}$ be an integer-valued discrete r.v., and let $p_n = \mathrm{P}(X = n)$ for $n \in \mathbb{N}$. Then G_X is defined at least for $s \in [-1, 1]$ and is of $\mathscr{C}^\infty$ class on $]-1, 1[$. Moreover, we have $G_X(1) = 1$ and $\left|G_X(s)\right| \leqslant 1$ for all $s \in [-1, 1]$.*

In addition, for $s \in [-1, 1]$, we have the series expansion

$$G_X(s) = \sum_{n=0}^{\infty} p_n\, s^n \qquad \textit{and hence} \qquad \mathrm{P}(X = n) = \frac{G_X^{(n)}(0)}{n!}.$$

In particular

$$\mathrm{E}(X) = G_X'(1) \qquad \textit{and} \qquad \mathrm{Var}(X) = G_X''(1) + G_X'(1) - \left[G_X'(1)\right]^2.$$

20.9 Sum and product of random variables

20.9.a Sum of random variables

Let us assume that every day I eat an amount of chocolate which is between 0 and 100 g, and let us model this situation by a random variable X_1 for the amount eaten on the first day, another random variable X_2 for the amount eaten on the second day, and so on, each random variable being assumed to have a uniform distribution on $[0, 100]$. We would like to known the distribution of the total amount of chocolate eaten in two days.

Let us start with a slightly simplistic reasoning, where the amount of chocolate can only take integral values k between 1 and 100 with probability $\mathsf{P}(X_i = k) = 1/100$. There is a very small probability that I will eat only 2 g of chocolate, since this requires that I eat 1 g the first day *and* 1 g the second, with a total probability equal to 1/10 000 (if we assume that the amount eaten the first day is independent of the amount eaten the second). The probability that I eat 5 g is larger, since this may happen if I eat 1 g and then 4 g, or 2 g and then 3 g, or 3 g and then 2 g, or (finally) 4 g and then 1 g; this gives a total probability equal to 4/10 000.

The reader will now have no trouble showing that the probability that I eat n grams of chocolate is given by the formula

$$\begin{aligned}\mathsf{P}(X_1 + X_2 = n) &= \mathsf{P}(X_1 = 1)\,\mathsf{P}(X_2 = n-1)\\ &\quad + \mathsf{P}(X_1 = 2)\,\mathsf{P}(X_2 = n-2) + \cdots + \mathsf{P}(X_1 = n-1)\,\mathsf{P}(X_2 = 1).\end{aligned}$$

With a finer subdivision of the interval $[0, 100]$, a similar formula would hold, and passing to the continuous limit, we expect to find that

$$f_{\text{total}}(x) = \int_0^{100} f(t)\,f(x-t)\,\mathrm{d}t = f * f(x),$$

where f_{total} is the density of $X_1 + X_2$ and f that of X_1 or X_2 (since the support of f is $[0, 100]$). In other words, the probability density of the sum of two independent random variables is the *convolution product* of the probability densities of the arguments.

This result can be established rigorously in two different manners, using a change of variable or the characteristic function. The reader may find either of those two proofs more enlightening than the other.

(Using the characteristic function)

From Proposition 20.59 we know that the characteristic function of the sum of two *independent* random variables is the product of the characteristic functions of the summands. Using the inverse Fourier transform, we obtain:

THEOREM 20.62 (Sum of r.v.) *Let X_1 and X_2 be two independent random variables, with probability densities f_1 and f_2, respectively (in the sense of distributions). Then the probability density of the sum $X_1 + X_2$ is given by*

$$f_{X_1+X_2} = f_1 * f_2,$$

or equivalently

$$f_{X_1+X_2}(x) = \int_{-\infty}^{+\infty} f_1(t)\,f_2(x-t)\,\mathrm{d}t,$$

which is only correct if the random variables involved are continuous, the probability densities then being integrable functions.

If we denote by $\check{f}_2$ the transpose of f_2, then the probability density of the difference $X_1 - X_2$ is given by

$$f_{X_1-X_2} = f_1 * \check{f}_2 \qquad \textit{or} \qquad f_{X_1-X_2}(x) = \int_{-\infty}^{+\infty} f_1(t)\, f_2(x+t)\, \mathrm{d}t.$$

Note that $\int f_{X_1+X_2} = \int f_{X_1} \cdot \int f_{X_2} = 1$, as it should be.

(With change of variables)

Let now $f_2(x_1, x_2)$ denote the joint probability density of the pair of random variables (X_1, X_2), and make the change of variables $Y_1 = X_1$ and $Y_2 = X_1 + X_2$. The jacobian of this transformation is given by

$$|J| = \left| \frac{\mathrm{D}(y_1, y_2)}{\mathrm{D}(x_1, x_2)} \right| = \begin{vmatrix} 1 & 0 \\ 1 & 1 \end{vmatrix} = 1,$$

and the joint probability density of (Y_1, Y_2) is $g(y_1, y_2) = f_2(y_1, y_2 - y_1)$; the probability density of $Y = X_1 + X_2 = Y_2$ is the marginal density, so we get:

THEOREM 20.63 (Sum of r.v.) *Let $Y = X_1 + X_2$ and let $f_2(x_1, x_2)$ denote the joint probability density of the pair (X_1, X_2). Then the probability density of Y is given by*

$$g(y) = \int_{-\infty}^{+\infty} f_2(t, y-t)\, \mathrm{d}t.$$

If the random variables X_1 and X_2 are independent, we have $f_2 = f_{X_1} \otimes f_{X_2}$ and the probability density of Y is the convolution $g = f_{X_1} * f_{X_2}$:

$$g(y) = \int_{-\infty}^{+\infty} f_{X_1}(t)\, f_{X_2}(y-t)\, \mathrm{d}t.$$

Example 20.64 The sum of two independent gaussian random variables X_1 and X_2, with means m_1 and m_2 and variance σ_1^2 and σ_2^2, respectively is another gaussian variable.

Adpating the result of Exercise 10.2 on page 295, which computes the convolution of two centered gaussians, one can show that $X_1 + X_2$ is gaussian with mean $m_1 + m_2$ and variance $\sigma_1^2 + \sigma_2^2$.

The following result is also very useful:

THEOREM 20.65 (Bienaymé identity) *Let $X_1, \ldots, X_n$ be* independent *random variables, each of which has finite variance* $\mathrm{Var}(X_i) = \sigma_i^2$. *Then the variance of the sum exists and is the sum of the variances of the X_i: we have*

$$\mathrm{Var}(X_1 + \cdots + X_n) = \mathrm{Var}(X_1) + \cdots + \mathrm{Var}(X_n).$$

PROOF. Without loss of generality, we may assume that each random variable is centered (otherwise, replace X_i by $X_i - \mathrm{E}(X_i)$ without changing its variance).

Let φ_i denote the characteristic function of the random variable X_i. We have $\varphi_k'(0) = 0$ for $k = 1, \ldots, n$, and by Theorem 20.59 on page 542 we can write

$$\varphi_{X_1+\cdots+X_n}(\nu) = \varphi_1(\nu) \cdot \varphi_2(\nu) \cdots \varphi_n(\nu).$$

On the other hand, the variance of X_k is given by $\mathrm{Var}(X_k) = -\varphi_k''(0)$. It therefore suffices to differentiate the above function twice and then evaluate the resulting expression at 0. Since the reader will easily check that all first derivatives give zero contributions, there only remains the formula

$$\varphi_{X_1+\cdots+X_n}''(0) = \varphi_1''(0) \cdot \varphi_2(0) \cdots \varphi_n(0) + \cdots + \varphi_1(0) \cdots \varphi_{n-1}(0) \cdot \varphi_n''(0).$$

Since in addition we have $\varphi_k(0) = 1$ for all k, the proof is complete.

20.9.b Product of random variables

While we are on this topic, we might as well consider the topic of product of random variables. For this, put $Y_1 = X_1$ and $Y_2 = X_1 X_2$. This change of variable has jacobian

$$|J| = \left| \frac{\mathrm{D}(y_1, y_2)}{\mathrm{D}(x_1, x_2)} \right| = \begin{vmatrix} 1 & 0 \\ x_2 & x_1 \end{vmatrix} = |x_1|,$$

and hence we obtain:

PROPOSITION 20.66 (Product of r.v.) *The probability density of $Y = X_1 X_2$ is given by*

$$g(y) = \int_{-\infty}^{+\infty} f_2\left(t, \frac{y}{t}\right) \frac{1}{|t|} \,\mathrm{d}t,$$

in terms of the joint probability density f_2 of (X_1, X_2). If the random variables X_1 and X_2 are independent, we have

$$g(y) = \int_{-\infty}^{+\infty} f_{X_1}(t) f_{X_2}\left(\frac{y}{t}\right) \frac{1}{|t|} \,\mathrm{d}t.$$

Similarly, the reader will have no difficulty checking the following:

PROPOSITION 20.67 (Ratio of r.v.) *The probability density of $Y = X_1/X_2$ is*

$$g(y) = \int_{-\infty}^{+\infty} f_2(t, ty)\, |t| \,\mathrm{d}t$$

and, if the random variables X_1 and X_2 are independent,

$$g(y) = \int_{-\infty}^{+\infty} f_{X_1}(t) f_{X_2}(ty)\, |t| \,\mathrm{d}t.$$

Irénée Jules Bienaymé (1796–1876), studied in Bruges and then at the lycée Louis-le-Grand, and took part in the defense of Paris in 1814. After going to the École Polytechnique, his scholarly life was perturbed by political turbulence (the school closed for some time). He obtained a position at the military school of Saint-Cyr and then became a general inspector in the Ministry for Finance. After the 1848 revolution, he became professor of probability at the Sorbonne and a member of the Académie des Sciences in 1852. He had a knack for language and translated in French the works of his friend Tchebychev. He also had a knack for disputing with others, such as Cauchy or Poisson.

20.9.c Example: Poisson distribution

The following is easy to check:

THEOREM 20.68 *Let X be an r.v. with a Poisson distribution with parameter λ. Then the characteristic function of X is*

$$\varphi_X(\nu) = \mathsf{E}\left\{e^{i\nu X}\right\} = \exp\left\{\lambda(e^{i\nu} - 1)\right\}.$$

Let X_1 and X_2 be two independent r.v. with Poisson distributions with parameters λ_1 and λ_2, respectively. Then the r.v. $X = X_1 + X_2$ has a Poisson distribution with parameter $\lambda = \lambda_1 + \lambda_2$.

20.10 Bienaymé-Tchebychev inequality

20.10.a Statement

Given a random variable X, its standard deviation mesaures the average (quadratic) distance to the mean, that is, the probability that the actual value of X be far from the expectation of X. The Bienaymé-Tchebychev inequality is a way to quantify this idea, giving an upper bound for the probability that X differs from its expectation by a given amount.

THEOREM 20.69 (Bienaymé-Tchebychev inequality) *Let $(\Omega, \Sigma, \mathsf{P})$ be a probability space and let X be a real-valued random variable on Ω such that $\mathsf{E}(X^2) < +\infty$. Let $m = \mathsf{E}(X)$ be the expectation of X and let σ^2 be its variance. Then, for any $\varepsilon > 0$, we have*

$$\mathsf{P}\Big(\,|X - m| \geqslant \varepsilon\Big) \leqslant \frac{\sigma^2}{\varepsilon^2}, \qquad \textit{or equivalently} \qquad \mathsf{P}\Big(\,|X - m| \geqslant \varepsilon\sigma\Big) \leqslant \frac{1}{\varepsilon^2}.$$

Pafnouti TCHEBYCHEV (1821–1894) entered Moscow University in 1837, and Saint Petersburg University in 1847. He studied prime numbers (proving in particular that there always exists a prime number p such that $n < p \leqslant 2n$ for all $n \geqslant 1$, and giving the first correct estimate for the order of magnitude of the number of primes less a large quantity X, which was refined into the prime number theorem in 1896 by Hadamard and de la Vallée Poussin). He was also interested in mechanics, in quadratic forms (Tchebychev polynomials), and in probability theory, where he published four important papers. His works were described very clearly by Andrei MARKOV. He died while drinking tea, like Euler, which is rigorous proof that mathematicians should drink coffee.[9]

In the other direction, knowing the actual value of X for some experience ω, this inequality gives an estimate of the error in assuming that this value is equal to the average value m.

Example 20.70 Let X be a random variable with expectation m and standard deviation σ. The probability that the result $X(\omega)$ of an experiment differ from m by more than 4σ is at most $6,25\,\%$.

To prove the inequality, we start with the following lemma:

LEMMA 20.71 *Let $(\Omega, \Sigma, \mathrm{P})$ be a probability space and let X be a non-negative random variable such that $\mathrm{E}(X) < +\infty$ (i.e., in the terminology of analysis, we consider a measure space and a non-negative integrable function). For any $t \in \mathbb{R}^{+*}$, we have*

$$\mathrm{P}\Big(\{X \geqslant t\}\Big) \leqslant \frac{1}{t} \int_\Omega X(\omega)\,\mathrm{dP}(\omega).$$

PROOF. This is clear by looking at the following picture:

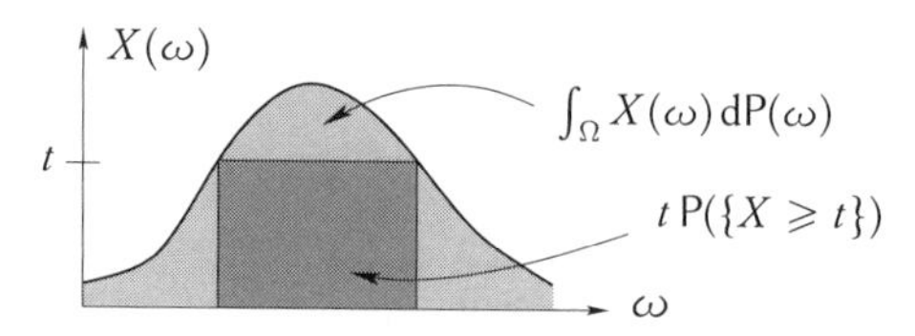

The dark region is a rectangle with height t and width $\mathrm{P}\big(\{X \geqslant t\}\big)$. Of course, it may well be something else than a rectangle (for instance many rectangles, depending on the structure of the event $\{X \geqslant t\}$), but in any case the dark region has measure $t\,\mathrm{P}\big(\{X \geqslant t\}\big)$. It is clear that this measure is smaller than or equal to $\int_\Omega X(\omega)\,\mathrm{dP}(\omega)$, since $X \geqslant 0$.

[9] "Mathematicians are machines for turning coffee into theorems", Paul ERDÖS (or Alfred RÉNYI).

More rigorously, we can write

$$0 \leqslant t\,\chi_{\{X \geqslant t\}}(\omega) \leqslant X(\omega)$$

(since $t\,\chi_{\{X \geqslant t\}}(\omega)$ is equal to t or 0). Integrating over Ω with respect to the probability measure dP, the inequality follows.

Applying this lemma to $\big(X - \mathsf{E}(X)\big)^2/\sigma^2$ and $t = \varepsilon^2$, we derive the Bienaymé-Tchebychev inequality.

20.10.b Application: Buffon's needle

Buffon[10] computed in 1777 that when throwing a needle of length a onto a floor with floorboards of width $b \geqslant a$, the probability that the needle will cross a line between two boards is equal to $\frac{2a}{\pi b}$. Take $a = b$ to simplify, and let $p = 2/\pi$ denote the resulting probability.

Suppose the needle is thrown N times, where N is very large. If we let X denote the random variable corresponding to the number of times the needle has crossed a line between two planks, it follows that X has a binomial distribution $\mathscr{B}(N, p)$. Thus the average number of needles crossing a line is given by the expectation $\mathsf{E}(X) = pN$, and the standard deviation of X is $\sigma^2 = Np(1-p)$. We can expect than when N is very large, the ratio X/N should be close to $p = 2/\pi$, given an "experimental" numerical approximation of π (this experiment has been performed a number of times, and many simulations can be found on the Web).

The question that may be asked is then: to obtain a approximation of π with a given precision, how many times should the needle be thrown? For instance, suppose I want to obtain the value of π with an error of at most $1/1000$, how many times should I throw the needle to be certain that the resulting value will be correct with a probability of 99%?

The simplest estimate for such a question can be obtained simply from the Bienaymé-Tchebychev inequality, which we recall for convenience:

$$\mathsf{P}\Big(\,|X - Np| \geqslant \varepsilon\sigma\Big) \leqslant \frac{1}{\varepsilon^2}.$$

In order for the probability that X satisfies the inequality to be at most 1%, we must take $1/\varepsilon^2 = 1/100$; hence we take $\varepsilon = \sqrt{100} = 10$. Let π' be the value of π obtained experimentally, namely $\pi' = 2N/X$. We want π' to satisfy $\pi - 10^{-3} \leqslant \pi' \leqslant \pi + 10^{-3}$, and therefore for the number X of needles crossing a line to satisfy

$$\frac{2N}{\pi + 10^{-3}} \leqslant X \leqslant \frac{2N}{\pi - 10^{-3}}.$$

---

[10] Georges Leclerc, comte de Buffon (1707–1788), famous naturalist, author of the monumental *Histoire naturelle*, was also interested in the theory of probability.

Writing $|X - Np| = \varepsilon\sigma$ with values

$$X = \frac{2N}{\pi \pm 10^{-3}}$$

(both signs "+" and "−" lead to the same result), we find

$$N = \frac{\pi^2 \varepsilon}{2.10^{-3}} \sqrt{Np(1-p)}, \qquad \text{hence} \qquad N = \frac{\pi^4 \varepsilon^2}{4.\,10^{-6}} p(1-p)$$

in other words, $N = 560$ million, more or less. It is not very reasonable to expect that π can be computed in this manner.

However, this is a very coarse estimate, since the Bienaymé-Tchebychev inequality holds with no assumptions on the distribution of the random variable being considered. If we take into account the precise distribution of X, a more precise result can be derived. But this remains *of the same order of magnitude*, and does not really affect the discussion. For instance, since the binomial distribution may be approximated by a gaussian distribution for N such that $Np(1-p)$ is very large, we can find numerically (using tables or computations of the function $\operatorname{erf}(x)$) a value $N \approx 150$ million throws, which is only better by a factor of 3.

20.11 Independance, correlation, causality

It is very important to distinguish between independance, correlation, and causality. In Example 20.47, we saw that two *dependent* events can be *uncorrelated.*

Two events are *causally linked* if one is the cause of the other. For instance, statistics show that road accidents are more frequent on Saturdays than other days of the week (thus they are correlated). There is causality here: usually more alcohol is consumed on Saturday, leading to a higher rate of accidents.

However, events may be correlated *without* causality. An example is given by Henri Broch [15]: in villages in Alsace, statistics show that the number of births per year is highly correlated with the number of storks (the correlation is close to 1). Should we conclude that storks carry babies home (which would mean causality)?[11]

David Ruelle, in his excellent book [77], mentions events which are *causally linked* but *uncorrelated*, which is more surprising. Consider, on the one hand, the position of the planet Venus in the sky, and on the other hand the weather

[11] There is in fact a hidden "causal" explanation: the more families in a village, the more houses, hence the more chimneys, and the more room for storks. This causality is not because of the two events discussed, but because of a third, linked independently with both.

one month after the date of observation. We know very well that weather [86] is a highly chaotic system, and hence is very sensitive to initial conditions. The position of Venus on a given day is therefore highly likely to influence the weather one month later. There is therefore strong causality (and dependence) between the position of Venus and the weather. However, no correlation can be observed. Indeed, other factors (the day of the year, the weather in previous months, the solar activity, the position of other planets, etc.) are equally or more crucial in the evolution of the weather than the position of Venus. Looking only at Venus and "averaging" over all other variables, no trace of the influence of this fair planet will be noticeable. Causality without correlation.

Sidebar 8 (Riemann-Stieltjes integral) *Let $\alpha : \mathbb{R} \to \mathbb{R}$ be an increasing function (not necessarily continuous). A measure μ_α is defined on open intervals by $\mu_\alpha(\,]a,b[\,) = \alpha(b^-) - \alpha(a^+)$ (where $\alpha(b^-)$ denotes the limit of $\alpha(x)$ as x tends to b on the left, and $\alpha(a^+)$, similarly, is the limit as x tends to a on the right; both are well defined because α is increasing). This measure may be extended to Borel sets in $\mathbb{R}$ uniquely, by defining*

$$\mu_\alpha\left(\bigcup_i B_i\right) = \sum_i \mu_\alpha(B_i)$$

if $B_i \cap B_j = \varnothing$ for $i \neq j$. This yields a Borel measure, that is, a measure on $(X, \mathscr{B})$, with the property that

$$\mu(\,[a,b]\,) = \alpha(b^+) - \alpha(a^-), \qquad \mu(\,[a,b[\,) = \alpha(b^-) - \alpha(a^-), \qquad \text{etc.}$$

The integral of a function f with respect to this measure is defined as described in this chapter, and is denoted $\int f \, \mathrm{d}\mu_\alpha$ or $\int f \, \mathrm{d}\alpha$, and called the ***Riemann-Stieltjes integral*** *of f with respect to α. It satisfies the properties stated in Theorem 2.28 on page 64. There are associated spaces $L^1(\mathbb{R}, \mathrm{d}\alpha)$ and $L^1(\,[a,b]\,, \mathrm{d}\alpha)$, which are complete by the Riesz-Fischer theorem.*

Finally, if α is of $\mathscr{C}^1$ class, one can show that

$$\int f \, \mathrm{d}\alpha = \int f(x)\, \alpha'(x) \, \mathrm{d}x = \int f(x) \frac{\mathrm{d}\alpha}{\mathrm{d}x} \, \mathrm{d}x$$

for any function f integrable with respect to μ_α.

This construction is interesting in part because it gives a definition of certain integrals or "generalized functions" without requiring the theory of distributions.

Example 20.72 Let $\alpha(x) = 1$ for $x \geqslant 0$ and $\alpha(x) = 0$ otherwise. This is an increasing function. The associated measure μ_α is called the **Dirac measure**; it has the property that any Dirac-measurable function is integrable and $\int f \, \mathrm{d}\alpha = f(0)$ for any f. Note that functions in $L^1(\mathbb{R}, \mathrm{d}\alpha)$ are always defined up to equality almost everywhere, but there $\{0\}$ is of measure 1 for μ_α, so it is not a negligible set.

Let $P = \{x \in \mathbb{R} \; ; \; \mu_\alpha(\{x\}) \neq 0\}$ (the set of discontinuities of α). One can define a measure μ_p by $\mu_p(X) = \mu_\alpha(P \cap X)$, and moreover $\mu_c = \mu_\alpha - \mu_p$ is also a Borel measure.

Chapter 21

Convergence of random variables: central limit theorem

Guildenstern *(Flips a coin)*: The law of averages, if I have got this right, means that if six monkeys were thrown up in the air for long enough they would land on their tails about as often as they would land on their –
Rosencrantz: Heads. *(He picks up the coin.)*

Tom Stoppard, *Rosencrantz & Guildenstern are dead* [87]

21.1 Various types of convergence

We are going to define three types of convergence of a sequence $(X_n)_{n\in\mathbb{N}}$ of random variables to a limit X.

The first two are parallel to well-known concepts in the theory of integration.

Let $(X, \mathscr{T}, \mu)$ be a measure space (for instance, $X = \mathbb{R}$, $\mathscr{T} = \mathscr{B}$, the Borel σ-algebra, and μ the Lebesgue measure). A sequence $(f_n)_{n\in\mathbb{N}}$ of measurable functions on X **converges almost everywhere to a measurable function f** if there exists a measurable subset $N \subset X$ such that $\mu(N) = 0$ and, for all $x \notin N$, the sequence $\big(f_n(x)\big)_{n\in\mathbb{N}}$ converges to $f(x)$. In other words, the sequence $(f_n)_{n\in\mathbb{N}}$ converges simply to f for almost all x. In probability language, we have the following analogous definition:

DEFINITION 21.1 (Almost sure convergence) Let $(\Omega, \Sigma, \mathsf{P})$ be a probability space, $(X_n)_{n\in\mathbb{N}}$ a sequence of random variables, and X a random variable. Then **$(X_n)_{n\in\mathbb{N}}$ converges almost surely to X**, denoted $X_n \xrightarrow{\text{C.A.S.}} X$, if there exists a subset $N \subset \Omega$ with probability zero such that, for any $\omega \notin N$, the sequence $\big(X_n(\omega)\big)_{n\in\mathbb{N}}$ converges to X.

In integration theory, another notion is convergence *in measure*: a sequence $(f_n)_{n\in\mathbb{N}}$ of measurable functions **converges in measure to a measurable function f** if

$$\forall \varepsilon > 0 \qquad \lim_{n\to\infty} \mu\Big(\Big\{x \in X \,;\, \big|f(x) - f_n(x)\big| \geqslant \varepsilon\Big\}\Big) = 0.$$

In other words, with a margin of error $\varepsilon > 0$, the subset of those points where f_n is "too far" from f shrinks down to nothingness as n becomes larger and larger.

Similarly, we have the probabilistic analogue:

DEFINITION 21.2 (Convergence in probability) Let $(X_n)_{n\in\mathbb{N}}$ be a sequence of r.v. and let X be a r.v. The sequence **$(X_n)_{n\in\mathbb{N}}$ converges in probability to X**, denoted $X_n \xrightarrow{\text{C.P.}} X$, if

$$\forall \varepsilon > 0 \qquad \lim_{n\to\infty} \mathsf{P}\Big(\Big\{\omega \in \Omega \,;\, \big|X(\omega) - X_n(\omega)\big| \geqslant \varepsilon\Big\}\Big) = 0.$$

In general, almost everywhere convergence does not imply convergence in measure. However, when the measure space has finite total measure ($\mu(X) < +\infty$), this is the case (for instance, this follows from Egorov's theorem, stated on page 69). The case of probability spaces is of this type, since $\mathsf{P}(\Omega) = 1$, so we deduce:

THEOREM 21.3 *Almost sure convergence implies convergence in probability.*

The third type of convergence, which is very important in probability, is the following:

DEFINITION 21.4 (Convergence in distribution) A sequence $(X_n)_{n\in\mathbb{N}}$ of r.v. with distribution functions $(F_n)_{n\in\mathbb{N}}$ **converges in distribution** if there exists a distribution function F such that the sequence $(F_n)_{n\in\mathbb{N}}$ converges simply to F at any point of continuity of F. This function determines the **limit distribution** of the sequence $(X_n)_{n\in\mathbb{N}}$.

If X is an r.v. with distribution function F, then we say that **$(X_n)_{n\in\mathbb{N}}$ converges in distribution to X**, and we denote $X_n \xrightarrow{\text{C.D.}} X$.

Note that if $X_n \xrightarrow{\text{C.D.}} F$, then we have

$$\mathsf{P}(X_n \leqslant x) \xrightarrow[n\to\infty]{} \mathsf{P}(X \leqslant x)$$

for any $x \in \mathbb{R}$ which is a point of continuity of F.

Convergence in distribution is the weakest of the three types of convergence introduced. In particular, we have the implications:

$$(X_n \xrightarrow{\text{C.A.S.}} X) \Longrightarrow (X_n \xrightarrow{\text{C.P.}} X) \Longrightarrow (X_n \xrightarrow{\text{C.D.}} X).$$

Certainly, convergence in distribution does not imply convergence in probability or almost sure, since the random variables X_n and X may have *nothing to do* with each other; in fact, they may even be defined on different probability spaces, since only the distribution function matters! It is perfectly possible to have two r.v. X and Y with the same distribution, but which are never equal (see page 524).

21.2
The law of large numbers

Consider a game of dce. I roll the die, and you must announce the result in advance (with money at stake). If I only roll the die once, there is nothing interesting you can really say (except if my die was false, but I would not say so in advance). However, if I roll the die ten times, and you must guess the *sum* of the results of the ten throws, there is much more you can say. More precisely, you may resonably think that the sum will be *close* to 35, maybe 33 or 38, but most certainly not 6 (there is only one chance in 6^{10}, that is, one in sixty million to obtain a 1 ten times in a row; you may as well play the lottery). And as I roll the die more and more times, you will be able to bet with increasing confidence that the sum will be roughly $n \times 3.5$. This, in effect, is what the weak law of large numbers makes precise[1]:

THEOREM 21.5 (Weak law of large numbers) *Let* $(\Omega, \Sigma, \mathsf{P})$ *be a probability space and let* $(X_n)_{n\in\mathbb{N}^*}$ *be a sequence of square integrable* independent *r.v. with the same expectation m (i.e., for all n, we have* $\mathsf{E}(X_n) = m$*) and the same variance* σ^2*. Define*

$$Y_n = \frac{X_1 + \cdots + X_n}{n}.$$

[1] This theorem was proved (rigorously) for the first time by the Swiss mathematician James (Jakob) Bernoulli (1654–1705) (who also discovered the famous Bernoulli numbers), in his treatise *Ars Conjectandi*, published in 1713. His nephew Daniel Bernoulli (1700–1782) also studied probability theory, as well as hydrodynamics. A generalization of Bernoulli's result was given by the marquis de Laplace, in his *Théorie analytique des probabilités*.

Then the sequence $(Y_n)_{n\in\mathbb{N}^*}$ *converges in probability to the* constant *random variable equal to* m *(i.e,* $Y(\omega) = m$ *for all* $\omega \in \Omega$*). In other words, we have*

$$\forall \varepsilon > 0 \qquad \mathsf{P}\big(\,|Y_n - m| \geqslant \varepsilon\big) \xrightarrow[n\to\infty]{} 0.$$

PROOF. The simplest proof uses the Bienaymé-Tchebychev inequality: notice that $\mathsf{E}(Y_n) = m$ and then write

$$\mathsf{P}\big(\,|Y_n - m| \geqslant \varepsilon\big) \leqslant \frac{\sigma^2(Y_n)}{\varepsilon^2}.$$

Since the random variables $X_1, \dots, X_n$ are independent, the Bienaymé identity 20.65 yields the formula

$$\sigma^2(n\,Y_n) = \sigma^2(X_1 + \cdots + X_n) = n\,\sigma^2,$$

hence

$$\sigma^2(Y_n) = \frac{n\sigma^2}{n^2} = \frac{\sigma^2}{n},$$

from which we derive

$$\mathsf{P}\big(\,|Y_n - m| \geqslant \varepsilon\big) \leqslant \frac{\sigma^2}{n\varepsilon^2} \xrightarrow[n\to\infty]{} 0.$$

Remark 21.6 It is possible to show that this result is still valid without assuming that the variables are square integrable (but they are always integrable). The proof is much more involved, however.

There is a stronger version of the law of large numbers, due to Kolmogorov and Khintchine (see, e.g., [33]):

THEOREM 21.7 (Strong law of large numbers) *Let* $(X_n)_{n\in\mathbb{N}}$ *be a sequence of independent identically distributed r.v. Assume that the* X_n *are integrable, and let* m *denote their common expectation. Then the sequence of Cesàro means converges almost surely to* m*: we have*

$$\frac{X_1 + \cdots + X_n}{n} \xrightarrow{\text{C.A.S.}} m.$$

21.3

Central limit theorem

Let us come back to our game of dice. We know, according to the law of large numbers, that we must bet that the sum is close to the average. It would be interesting to know *how* the sequence $(Y_n)_{n\in\mathbb{N}^*}$ of Cesàro means converges to the constant random variable m. And more precisely, what should we expect if we bet on a value *other* than the average?

[2] He showed in particular that it is possible to cut a cake into 65,537 equal pieces with straightedge, compass, and knife. A very useful trick if you plan a large wedding.

A child prodigy, Carl Friedrich GAUSS (1777–1855), son of an artisan, entered the University of Göttingen at eighteen. He made contributions to astronomy (rediscovering Ceres, which lesser minds had managed to lose in the firmament, by means of orbit computations) and geodesy. With Wilhelm Weber, he created the theory of magnetism. However, his greatest discoveries were in mathematics. At nineteen, he proved that the regular seventeen-sided polygon can be constructed with straightedge and compass, and further generalized this result.[2] One of the greatest among arithmeticians (he introduced congruences, for instance), he also laid the foundation of differential geometry, defining in particular the *Gauss curvature* of a surface. He was named "the prince of mathematicians."

The fundamental result (in particular, in many physical applications) is that the successive sums of the terms of a sequence of independent, identically distributed random variables (square integrable), will *always* converge (in distribution) to a *Gaussian* random variable, after an appropriate scaling. In particular, in statistical physics, this is why many measurable quantities can be expressed by the computation of a moment of Gaussian variables. Problem 8 gives the example of a simple problem of a random walk in the plane.

The theorem stating this fact is called the "central limit theorem."[3] In the case of Bernoulli random variables (heads or tails), it was stated and proved by Abraham de Moivre.

We begin by recalling the main facts conerning the Gaussian or normal distribution, which is so essential in this result.

DEFINITION 21.8 (Normal distribution) The **normal distribution** with mean m and standard deviation $\sigma > 0$, or **Gaussian distribution** with mean m and standard deviation $\sigma > 0$, denoted $\mathscr{N}(m, \sigma^2)$, is the probability distribution on $\mathbb{R}$ with density given by

$$g_{m,\sigma}(x) = \frac{1}{\sqrt{2\pi}\sigma} \exp\left\{-\frac{(x-m)^2}{2\sigma^2}\right\}.$$

A normal distribution of the type $\mathscr{N}(0, \sigma^2)$ is called a **centered normal distribution**.

The reader will easily check that the characteristic function of the normal distribution $\mathscr{N}(m, \sigma^2)$ is given by

$$\Phi_{m,\sigma}(u) = \mathrm{e}^{\mathrm{i}um} \exp\left\{-\frac{u^2\sigma^2}{2}\right\}, \quad \text{in particular,} \quad \Phi_{0,\sigma}(u) = \exp\left\{-\frac{u^2\sigma^2}{2}\right\}.$$

[3] Translated lierally from the German, where a more suitable translation would have been "fundamental limit theorem (of probability theory)."

Paul LÉVY (1886–1971), son and grandson of mathematicians, studied at the École Polytechnique, then at the École des Mines, where he became professor in 1913. Hadamard, one of his teachers, asked him to collect the works of René Gâteaux, a young mathematician killed at the beginning of the First World War. After doing this, Lévy developed the ideas of Gâteaux on differential calculus and functional analysis. At the same time as Khintchine, he studied random variables, and in particular the problem of the convergence of sequences of random variables. He also improved and expanded considerably the results of Wiener concerning Brownian motion. He was also interested in partial differential equations and the use of the Laplace transform.

THEOREM 21.9 (Central limit theorem) *Let $(X_n)_{n\in\mathbb{N}^*}$ be a sequence of independent, identically distributed random variables with finite expectation m and finite variance σ^2. Let*

$$Y_n \stackrel{\text{def}}{=} \frac{X_1 + \cdots + X_n}{n} \qquad \text{and} \qquad Z_n = \sqrt{n}(Y_n - m).$$

Then the sequence $(Z_n)_{n\in\mathbb{N}^*}$ converges in distribution to the centered normal distribution $\mathscr{N}(0,\sigma^2)$. In other words, we have

$$\mathsf{P}\left\{\frac{S_n - n\mu}{\sigma\sqrt{n}} < x\right\} \xrightarrow[n\to\infty]{} \mathscr{N}(x) = \int_{-\infty}^{x} \mathrm{e}^{-t^2/2}\,\mathrm{d}t.$$

for all $x \in \mathbb{R}$.

PROOF. Let $S_n \stackrel{\text{def}}{=} X_1 + \cdots + X_n$. We can compute the characteristic function of Z_n, namely:

$$\Phi_{Z_n}(u) = \mathsf{E}\left(\mathrm{e}^{\mathrm{i}uZ_n}\right) = \mathsf{E}\left(\mathrm{e}^{\mathrm{i}u\frac{S_n}{\sqrt{n}} - \mathrm{i}u\sqrt{n}m}\right)$$

$$= \mathrm{e}^{-\mathrm{i}um\sqrt{n}}\Phi_{S_n}\left(\frac{u}{\sqrt{n}}\right) = \mathrm{e}^{-\mathrm{i}um\sqrt{n}}\left[\Phi_X\left(\frac{u}{\sqrt{n}}\right)\right]^n,$$

where the last relation follows from the independence of the r.v. X_n and from Theorem 20.59.

We now perform a Taylor expansion of $\Phi_X\left(u/\sqrt{n}\right)$ to order 2, using the fact that

$$\Phi'_X(0) = \mathrm{i}\,\mathsf{E}(X) \qquad \text{and} \qquad \Phi''_X(0) = -\mathsf{E}(X^2),$$

and this yields $\quad \Phi_X\left(\dfrac{u}{\sqrt{n}}\right) = 1 + \mathrm{i}\dfrac{u}{\sqrt{n}}m - \dfrac{u^2}{2n}\mathsf{E}(X^2) + o\left(\dfrac{u^2}{n}\right).$

Inserted in the preceding relation, this implies that

$$\Phi_{Z_n}(u) = \exp\left\{-\mathrm{i}\frac{u^2\sigma^2}{2} + o\left(\frac{1}{n}\right)\right\},$$

so the characteristic functions of Z_n converge pointwise to that of the normal distribution. This turns out to be enough to imply the result (which is not too surprising, since the characteristic function of an r.v. determines its distribution uniquely, as recalled earlier), but the proof of this implication involves the continuity theorem of Paul Lévy and is by no means easy.

Remark 21.10 The condition that the variance $\mathrm{Var}(X_n)$ be finite is necessary. Exercise 21.15 on page 562 gives a counterexample to the central limit theorem when this assumption is not satisfied. It should be mentioned that *non-Gaussian processes* are also very much in current fashion in probability theory.

Application

Consider a piece of copper, of macroscopic size (electrical wiring, for instance). Because of thermal fluctuations, the average velocity of a conducting electron in this copper wire is $\sqrt{3k_B T/m}$, that is, of the order of $10^5\ \mathrm{m \cdot s^{-1}}$, or *tens of thousands of kilometers per second!* Yet, the copper wire, on a table, does not move. This is due to the fact that the velocities of the electrons (and of nuclei, although they are much slower) are randomly distributed and "cancel" each other. How does this work?

Let N be the number of conducting electrons in the wire. We assume that the wire also contains exactly N atoms,[4] and we will denote by m and M, respectively, the masses of the electrons and atoms. You should first check that the velocity of atoms is negligible in this problem.

We denote by X_n the random variable giving the velocity (at a given time) of the n-th electron. What is called the "average velocity" is *not* the expectation of X_n, since (in the absence of an electric field), the system is isotropic and $\mathrm{E}(X_n) = 0$. In fact, the "average velocity" is best measured by the *standard deviation* of the velocity $\sigma(X_n)$.

What is then "average velocity" (the standard deviation of the velocity) of the whole wire? Since the number N of atoms and electrons is very large (of the order of 10^{23}), we can write

$$Y = \frac{m}{N(m+M)} \sum_{n=1}^{N} X_n \approx \frac{1}{\sqrt{N}} \frac{m}{M} \left(\frac{1}{\sqrt{N}} \sum_{n=1}^{n} X_n \right),$$

and the term in parentheses is close to a centered gaussian with variance σ^2 by the central limit theorem.

The standard deviation of Y is therefore very close to

$$\sigma(Y) = \frac{1}{\sqrt{N}} \frac{m}{M} \sigma.$$

Notice the occurrence of the famous ratio $1/\sqrt{N}$, which is characteristic of the fluctuations of a system with N particles. In the present case, this reduces the "average velocity" of the set of electrons to $1.5 \cdot 10^{-7}\ \mathrm{m \cdot s^{-1}}$.

[4] Depending on the metal used, it may not be N, but rather $N/2$ or $N/3$; however, the principle of the calculation remains the same.

EXERCISES

Combinatorics

♦ **Exercise 21.1** All 52 cards in a pack of cards are dealt to four players. What is the probability that each player gets an ace? Does the probability change a lot when playing algebraic whist? (Recall that algebraic whist is played with a pack of 57 cards.)

HINT: Compute first, in general, the number of ways of subdividing n objects into k subsets containing $r_1, \ldots, r_k$ objects, respectively.

♦ **Exercise 21.2** Take a pack of n cards, and deal them face up on a table, noting the order the cards come in. Then shuffle the pack thoroughly and deal them again. What is the probability that at least one card occurs at the same position both times? Does this probability have a limit as n goes to infinity?

It is amusing to know that this result was used by con-men on Mississippi steamboats, who exploited the passengers' intuitive (wrong) assumption that the probability would be much smaller than it actually is.

Conditional probabilities

♦ **Exercise 21.3** Mrs. Fitzmaurice and Mrs. Fitzsimmons discuss their respective children during a plane trip. Mrs. Fitzmaurice says: "I have two children, let me show you their pictures." She looks in her bag, but finds only one picture: "This is my little Alicia, playing the piano last summer."

What is the probability that Alicia has a brother?

♦ **Exercise 21.4** A religious sect practices population control by means of the following rules: each family has children (made in the dark) until a boy is born. Then, and only then, they abstain for the rest of their days. If the (small) possibility of twins is neglected, and knowing that without any control, about 51% of babies are boys, what will be the percentage of males in the community?

Same question if each family also limits to twelve the total number of children it can have.

♦ **Exercise 21.5** A new diagnostic test for a lethal disease has been devised. Its reliability is impressive, says the company producing it: clinical trials show that the test is positive for 95% of subjects who have the disease, and is negative for 95% of subjects who do not have it. Knowing that this disease affects roughly 1% of the population, what probability is there that a citizen taking the test and getting a positive result will soon end up in eternal rest in the family plot?

♦ **Exercise 21.6** Another amusing exercise. The other day, my good friend J.-M. told me that he knows someone to whom the most amazing thing happened. This other person was just thinking of one of his former relations, who he had not thought about for 10 years, and the very next day he learned the person had just died, at the same time he was thinking of him! (Or maybe within ten minutes.)

What should one think of this "extraordinary" (rigorously true) fact? Does this kind of thing happen often, or do we have here an example of ESP (Extra-Sensory Perception, for the uninitiated)?

♦ **Exercise 21.7** In 1761, Thomas Bayes, Protestant theologian, leaves for ever this vale of tears. He arrives at the Gates of Heaven. But there is not much room left, and anyway God is a strict Presbyterian and Bayes, unluckily for him, was a Nonconformist minister. However, St. Peter gives him a chance: Bayes is shown three identical doors, two of which lead to Hell and only one to Heaven. He has to choose one. With no information available, Bayes selects one at random. But before he can open it, St. Peter – who is good and likes mathematicians – stops him. "Wait, here is a clue," he says, and opens one of the two remaining doors, which Bayes can see leads to Hell.

What should Bayes do? Stick to his door, or change his mind and select the other unopened door?[5]

Expectation, standard deviation

♦ **Exercise 21.8** When a students hands in an exam paper, it will get a grade corresponding roughly to the objective quality level of the paper. However, because of random perturbations of the grader's environment (did he eat well at lunch, is his cat purring on his knees,...), the actual grade is really a random variable, with expectation equal to the "objective" grade m, and with standard deviation $\sigma = 2$, say, for a paper graded out of 20 points.

To get a more exact result, it is possible to ask N different graders to look at the paper and then average their respective grades, each of which is a random variable with the same distribution, with expectation m and standard deviation σ.

Denote by $X_1, \dots, X_N$ the corresponding random variables for each grader, and let $S_N = (X_1 + \cdots + X_N)/N$ be the average of their grades. What is the expectation of S_N? What is its standard deviation?

How many correctors are needed so that the probability is at least 0.8 that the average grade corresponds to the exact grade, within half a point? Same question if the standard deviation σ is only half a point. Do you have any comments about the results?

♦ **Exercise 21.9** Solveig wants to make a chocolate cake. The recipe states that "two large eggs (100 g)" are needed. But she only has two eggs from two different brands. The first package says the eggs inside are "large" (between 50 and 60 g, with mass uniformly distributed), and the second says the eggs are "medium" (between 40 and 45 g with uniform distribution). What is the probability density of the total mass of the two eggs? What is the average and standard deviation of this mass? What is the probability that the mass is greater than 100 g?

Covariance

♦ **Exercise 21.10** Let $(X_n)_{n\in\mathbb{N}^*}$ be a sequence of random variables, pairwise independent, with the Bernoulli distribution.[6] with parameter p. Let $Y_n = X_n X_{n+1}$ for $n \in \mathbb{N}^*$.

1. Find the distribution of Y_n.
2. Compute $\operatorname{Cov}(Y_n, Y_{n+k})$ for $n, k \in \mathbb{N}^*$.
3. Let

$$S_n = \sum_{i=1}^{n} \frac{Y_i}{n}$$

for $n \in \mathbb{N}^*$. Show that the sequence $(S_n)_{n\in\mathbb{N}^*}$ converges in probability to the determinist (constant) random variable p^2.

---

[5] This version of the classical "Monty Hall paradox" was suggested by Claude Garcia.

[6] Recall that a Bernoulli distribution with parameter p is the distribution of a random variable which takes the value 1 with probability p and 0 with probability $1 - p$.

Sum and product of r.v.

♦ **Exercise 21.11** Show that the ratio of two independent centered normal random variables is a r.v. with a Cauchy distribution; recall that this means that for some $a \in \mathbb{R}$, the probability density is

$$f(x) = \frac{a}{\pi} \frac{1}{a^2 + x^2}.$$

♦ **Exercise 21.12** Let X be Y be two independent random variables, such that X has a Poisson distribution with parameter λ and Y has a Poisson distribution with parameter μ, where $\lambda, \mu > 0$. Let $p_n = \mathrm{P}(X = n)$ and $q_n = \mathrm{P}(Y = n)$ for $n \in \mathbb{N}$.

i) Show that $Z = X + Y$ has a Poisson distribution with parameter $\lambda + \mu$.

ii) Let $k \in \mathbb{N}$. Compute the conditional probability distribution of X knowing that $Z = k$.

iii) Let $W = X - Y$. Compute the expectation and the variance of W.

Convergence

♦ **Exercise 21.13** A player flips two coins B_1 and B_2 repeatedly according to the following rule: if flipping whichever coin is in his hand yields tails, switch to the other coin for the next toss, otherwise continue with the same. Assume that tails is the outcome of flipping B_1 with probability $p_1 = 0.35$, and of flipping B_2 with probability $p_2 = 0.4$. Let X_k be the random variable equal to 1 if the player is flipping coin B_1 at the k-th step, and equal to 0 otherwise. Let

$$h_1(k) = \mathrm{P}(X_k = 1), \qquad h_2(k) = \mathrm{P}(X_k = 0), \qquad \text{and} \qquad H(k) = \begin{pmatrix} h_1(k) \\ h_2(k) \end{pmatrix}.$$

i) Show that $H(k+1) = M \cdot H(k)$, where M is a certain 2×2 matrix. Deduce that $H(k) = M^k H(0)$, where $H(0)$ is the vector corresponding to the start of the game.

ii) Diagonalize the matrix M. Show that as k tends to infinity, $H(k)$ converges to a vector which is independent of the initial conditions. Deduce that the sequence $(X_k)_{k \in \mathbb{N}}$ converges almost surely to a random variable, and write down the distribution of this random variable.

♦ **Exercise 21.14** Let (U_n) be a sequence of identically distributed independent random variables such that

$$\mathrm{P}(U_n = 1) = p, \qquad \mathrm{P}(U_n = -1) = q = 1 - p \quad \text{with } p \in [0, 1].$$

Define a sequence of random variables V_n by $V_n = \prod_{i=1}^{n} U_i$. Find the distribution of V_n, then show that (V_n) converges in distribution to a random variable V, and find the corresponding distribution (it may be helpful to compute the expectation of V_n).

♦ **Exercise 21.15 (Counterexample to the central limit theorem)** Show that if X_1 and X_2 are Cauchy random variables with parameters a and b, respectively (see Exercise 21.11), then the random variable $X_1 + X_2$ is a Cauchy random variable with parameter $a + b$ (compute the characteristic function of a Cauchy random variable). What conclusion can you draw regarding the central limit theorem?

PROBLEMS

♦ **Problem 7 (Chain reaction)** Let X and N be two square integrable random variables taking values in $\mathbb{N}$. Let $(X_k)_{k\geqslant 1}$ be a sequence of independent random variables with the same distribution as X. Assume further that N is independent of the sequence $(X_k)_{k\geqslant 1}$. Let G_X and G_N denote the generating functions of X and N. For $n \geqslant 1$, let

$$S_n = \sum_{k=1}^{n} X_k.$$

Finally, let S be defined by

$$\forall \omega \in \Omega \qquad S(\omega) = \sum_{k=1}^{N(\omega)} X_k(\omega).$$

i) For $n \in \mathbb{N}$, compute the generating function G_n of S_n, in terms of G_X.

ii) Show that $G_S(t) = G_N\big(G_X(t)\big) = G_N \circ G_X(t)$ for any $t \in [-1,1]$.

iii) Compute $\mathsf{E}(S)$ and $V(S)$ in terms of the expectation and the variance of X and N.

—

Consider now a reaction (for instance, a nuclear reaction) occurring in a certain environment. Time is measured by an integer $n \geqslant 0$ for simplicity. A particle (time 0) comes into the environment, is absorbed, and produces new particles (time 1). In turn, each new particle generates new particles at the next instant.

We denote by p_0 the (constant) probability that a particle creates no other particle, and for $k \in \mathbb{N}^*$, we let p_k be the probability that it creates exactly k particles. Let

$$\forall t \in \mathbb{R} \qquad G(t) = \sum_{k=0}^{\infty} p_k\, t^k.$$

Finally, let Z_k be the random variable which is the total number of particles at time k, and let x_n be the probability that no particle survives at time n (end of the reaction):

$$\forall n \in \mathbb{N}^* \qquad x_n = \mathsf{P}(Z_n = 0).$$

iv) Find G when $p_0 = 1$. We now assume that $p_0 < 1$. Show that G is strictly increasing on $[0,1]$.

v) Find G if $p_0 + p_1 = 1$. Assuming now that $p_0 + p_1 < 1$, show that G is convex.

vi) In the following cases, find the number of solutions to the equation $G(x) = x$:

$$a)\ p_0 = 0\,; \quad b)\ 0 < p_0 < 1 \text{ and } G'(1) \leqslant 1\,; \quad c)\ 0 < p_0 < 1 \text{ and } G'(1) > 1$$

vii) Find a recurrence relation expressing the generating function G_{n+1} of Z_{n+1} in terms of the generating function G_n of Z_n and of G.

viii) Still under the condition $0 < p_0 < 1$, find a recurrence relation between x_n and x_{n+1}. Show that the sequence $(x_k)_{k\in\mathbb{N}}$ is increasing, and deduce that it converges to a limit ℓ such that $0 < \ell < 1$.

ix) Show that $\mathsf{E}(Z_n) = \mathsf{E}(Z_1)^n$.

x) Show from the preceding results that:

  (a) if $\mathsf{E}(Z_1) \leqslant 1$, then the probability that the reaction stop before time n converges to 1 as n tends to infinity;

  (b) if $\mathsf{E}(Z_1) > 1$, then this probability converges to a limite $0 < \ell < 1$. What should we expect from the reaction? What are your comments?

♦ **Problem 8 (Random walk in the plane)** Consider a plane named P, a point named O which serves as origin for some cartesian coordinate system on P, and a drunk mathematician named Pierre who, having profited extensively from his pub's extensive selection of single malt whiskey, leaves the afore mentioned pub (located at the origin O) at time $t = 0$ to walk home. Each second, he make a single step of fixed unit length, in a *random* direction (obstacles, constables, and other mishaps are considered negligible).

Let X_n and Y_n be the r.v. corresponding to the x- and y-coordinates of Pierre after the n-th step, and let θ_n denote the angle (measured from the x-axis) giving the direction following during the n-th step. Thus we have

$$X_n = \sum_{k=1}^{n} \cos\theta_k \qquad \text{and} \qquad Y_n = \sum_{k=1}^{n} \sin\theta_k .$$

We assume that the θ_k are independent random variables *uniformly* distributed between 0 and 2π.

The objective is to study the r.v.

$$R_n = \sqrt{X_n^2 + Y_n^2}$$

representing the distance from the pub.

i) Show that the r.v. X_n and Y_n have expectation zero, and compute their variance.

ii) Are X and Y correlated? Are they independent?

iii) What is the expectation of R_n^2?

iv) Explain why it is reasonable to assume that for large n, the joint distribution of (X_n, Y_n) may be approximated by the joint distribution of a pair of independent centered normal distributions. (Use the fact that uncorrelated gaussian variables are in fact independent.)

What is the (approximate) probability density of the pair (X_n, Y_n)?

v) Deduce the (approximate) distribution function, and then the probability density, of R_n.

vi) Finally, find an approximation for the expectation $\mathsf{E}(R_n)$.

SOLUTIONS

♦ **Solution of exercise 21.3.** The probability that Alicia has a brother is $\frac{2}{3}$, and not $\frac{1}{2}$.

Indeed, among families with two children, about 25% have two boys, 25% have two girls, and therefore 50% have one each. Since we know that there is at least one girl, the probability of having two girls is $\frac{1}{4}$ divided by $\frac{1}{2} + \frac{1}{4}$, which is $\frac{1}{3}$.

If Mrs. Fitzmaurice had distinguished between her children (saying, for instance, "Here is my older child at the piano"), the probability would have been $\frac{1}{2}$. (Among those families whose first child is a girl, there are 50% where the second is a girl.)

♦ **Solution of exercise 21.2.** Let A_i denote the event which is "the i-th card is at the same place both times." It is easy to see that $\mathsf{P}(A_i) = 1/n$, that $\mathsf{P}(A_i \cap A_j) = 1/n(n-1)$ for $i \neq j$ and that more generally

$$\mathsf{P}(A_{i_1} \cap \cdots \cap A_{i_k}) = \frac{(n-k)!}{n!}$$

when all indices are distinct. Using the Poincaré formula, the probability of the event B: "at least one of the cards is at the same place both times" is

$$\mathsf{P}(B) = \mathsf{P}\left(\bigcup_{i=1}^{n} A_i\right),$$

and is therefore equal to

$$P(B) = \sum_{k=1}^{n} (-1)^{k+1} \sum_{1 \leqslant i_1 < i_2 < \cdots < i_k \leqslant n} \frac{(n-k)!}{n!}.$$

There are $\binom{n}{k}$ distinct k-tuples in the sum, and we obtain

$$P(B) = \sum_{k=1}^{n} (-1)^{k+1} C_n^k \frac{(n-k)!}{n!} = 1 - \sum_{k=0}^{n} \frac{(-1)^k}{k!}.$$

This probability converges to $1 - 1/e$ (this is the beginning of the power series expansion for e^{-1}), or roughly 63%.

♦ **Solution of exercise 21.4.** The probability hasn't changed: there are still 51% of baby boys in the community.

This is easy to check: notice that every family has exactly *one* boy; those with two children have exactly one boy and one girl, those with three have one boy and two girls, etc. The families with n children have *one* boy and $n-1$ girls.

Let $P(n)$ be the probability that a family has n children. We have of course $P(n) = pq^{n-1}$, with $p = 0.51$ the probability that a baby is a boy, and with $q = 1 - p$.

The proportion of boys is then given by

$$t = \frac{P(1) + P(2) + \cdots + P(n) + \cdots}{P(1) + 2P(2) + \cdots + nP(n) + \cdots} = \frac{p(1 + q + q^2 + \cdots)}{p(1 + 2q + 3q^2 + \cdots)} = \frac{p/(1-q)}{p/(1-q)^2} = p.$$

♦ **Solution of exercise 21.5.** Let P denote the event "the result of the test is positive," and let $\overline{P}$ be the opposite event. Let M denote the event "the person taking the test has the disease," and let $\overline{M}$ be its opposite.

From the statement of the problem, we know that $P(P|M) = 0.95$ and that $P(\overline{P}|\overline{M}) = 0.95$. Since $(M, \overline{M})$ is a partition of the sample space (either a person has the disease, or not), by Bayes's formula, we have

$$P(M|P) = \frac{P(P|M) \cdot P(M)}{P(P|M) \cdot P(M) + P(P|\overline{M}) \cdot P(\overline{M})}.$$

With $P(P|\overline{M}) = 1 - P(\overline{P}|\overline{M}) = 0.05$, the probability of having the disease when the test is positive is "only" around 16%.

♦ **Solution of exercise 21.6.** Before making computations, two comments are in order. Saying that the story is true does not mean that premonition really occurred, but *that this is how the story was told*. Moreover, you are entitled to disbelief – it is not because it is printed, with "True Story" explicitly stated, even in Bold Face, that it *is* really true. For this type of debate, this is an important remark.

What is the probability for *one* given person, whom I think about once in ten years, that I will do so within ten minutes of his death? Let's spread out death on an interval of 80 years, or roughly 42 million minutes, so that we can guess that the probability is roughly one out of 4.2 million.

I estimate that I know, more or less, roughly a thousand persons (including those I only think about only once ten or twenty years); there may therefore be about one million people who know someone I know.[7] Hence it is not only possible, but in fact rather unsurprising, that someone could tell me such a story!

[7] This may be of the wrong order of magnitude, since my acquaintances each have, among the people they "know" (those they only think about once in ten years, including people known only by name, actors, and so on) rather more than a thousand individuals. On the other hand, the sets are certainly not disjoint. Any finer estimate would be interesting.

This does not mean that ESP does not exist; only that there is no need here to invoke a paranormal phenomenon in order to explain what happened.

♦ **Solution of exercise 21.7.** Let us assume that door A leads to Heaven, and doors B and C to Hell. If Bayes had chosen A, Saint Peter will open B or C for him; changing his mind means Bayes will go to Hell.

If Bayes had chosen door B, Saint Peter, necessarily, must open door C for him (since he opens a door leading to Hell), so if Bayes changes his mind, he will go to Heaven. The same is obviously true if Bayes had chosen door C.

Therefore, independently of his original choice, if Bayes changes his mind after Saint Peter has shown him one door, this change of mind will *switch* his destination from the one first chosen. This means the probability that Bayes will go to Heaven is now $2/3$ and the probability of going to Hell is now $1/3$.

Conclusion: well, this rather depends on what Bayes really wants – does he want to spend the rest of his (eternal) life hearing angels play the harp, or is he more interested in playing roulette with the devils?

♦ **Solution of exercise 21.8.** This exercise is a simple application of the Bienaymé-Tchebychev inequality. The goal is that the probability of the grade being too far from m be less than 0.2. Consider the random variable $X_1 + \cdots + X_N$. It has expectation Nm, and S_N has expectation m. The variance of $X_1 + \cdots + X_N$, on the other hand, is equal to $N\sigma^2$, as can be seen by looking at the centered random variables $X_i' = X_i - m$, with zero expectation and standard deviation σ. The variance of the sum $X_1 + X_2$ is simply $E((X'^1 + X'^2)^2) = E((X_1')^2) + E((X_2')^2) = 2\sigma^2$, and similarly by induction for the sum of N random variables (since every crossed term has expectation equal to zero). Hence the standard deviation of $X_1 + \cdots + X_N$ is $\sqrt{N}\sigma$ and that of $S_N = (X_1 + \cdots + X_n)/N$ is $\sigma/\sqrt{N}$.

We are looking for a value of N such that

$$\mathrm{P}\left(|S_N - m| \geqslant \frac{1}{2}\right) \leqslant 0.2.$$

The Bienaymé-Tchebychev inequality states that we always have

$$\mathrm{P}\left(|S_N - m| \geqslant \frac{\varepsilon\sigma}{\sqrt{N}}\right) \leqslant \varepsilon^2.$$

So we need $1/\varepsilon^2 = 0{,}2$, or $\varepsilon = \sqrt{5}$, and it then suffices that

$$\frac{\sqrt{5}\sigma}{\sqrt{N}} = \frac{1}{2},$$

or $N = 4 \times 5 \times \sigma^2 = 20\,\sigma^2$ to be certain that the desired inequality holds.

For a value $\sigma = 2$, one finds $N = 80$!!! This is rather unrealistic, considering the average budget of a university. If the graders manage to grade very consistently, so that their individual standard deviation is $\sigma = \frac{1}{2}$, it will be enough to take $N = 5$. This is still a lot but will provoke fewer cardiac incidents among university administrators.

However, notice that those estimates are *not optimal*, because the Bienaymé-Tchebychev inequality can be refined significantly if the distribution of the random variables X_i is known (it is only in itself a worst-case scenario). We know that 80 graders will achieve the desired goal, independently of the distribution (given the expectation and standard deviation), but it is very likely that a smaller number suffices.

♦ **Solution of exercise 21.9.** The probability densities of the masses of the first and second egg, respectively, are

$$f_1(x) = \begin{cases} 0 & \text{if } x \notin [50, 60], \\ \dfrac{1}{10} & \text{if } x \in [50, 60], \end{cases} \qquad \text{and} \qquad f_2(x) = \begin{cases} 0 & \text{if } x \notin [40, 45], \\ \dfrac{1}{5} & \text{if } x \in [40, 45]. \end{cases}$$

The expectation for the mass of the first egg is of course 55 g, and the expectation for the mass of the second is 42.5 g. Thus the expectation for the total mass is 97.5 g.

The probability density of the mass of the two eggs together is the convolution

$$f(x) = f_1 * f_2(x) = \int_{-\infty}^{+\infty} f_1(t)\, f_2(x-t)\,\mathrm{d}t,$$

(because, obviously, we can assume that the two masses are independent), which is the function with the following graph:

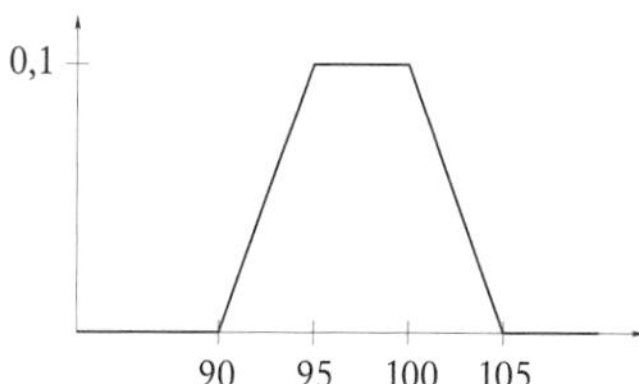

The expectation may also be computed from this formula: it is equal to

$$E = \int_0^{+\infty} x\, f(x)\,\mathrm{d}x = 97.5,$$

as it should be. The standard deviation satisfies

$$\sigma^2 = \int_{-\infty}^{+\infty} (x-E)^2 f(x)\,\mathrm{d}x = \sigma_1^2 + \sigma_2^2.$$

Since $\sigma_1^2 = 25/3$ and $\sigma_2^2 = 25/6$ by immediate computations, we obtain $\sigma^2 = 25/2$. So the standard deviation is $\sigma = 5/\sqrt{2}$. From the graph, the probability that the mass is at least 100 g is $1/4$.

♦ **Solution of exercise 21.11.** Let X_1 and X_2 be two normal centered random variables, with variances σ_1^2 and σ_2^2, respectively. By Proposition 20.67, the probability density of the ratio X_1/X_2 is given by

$$\begin{aligned} g(y) &= \frac{1}{2\pi\sigma_1\sigma_2} \int \exp\left\{-\frac{t^2}{2\sigma_1^2}\right\} \exp\left\{-\frac{t^2y^2}{2\sigma_2^2}\right\} |t|\,\mathrm{d}t \\ &= \frac{1}{2\pi\sigma_1\sigma_2} \int \exp\left\{-\frac{t^2}{2\sigma_2^2}\left(\frac{\sigma_2^2}{\sigma_1^2}+y^2\right)\right\} |t|\,\mathrm{d}t \\ &= \frac{1}{\pi\sigma_1\sigma_2}\,\frac{1}{\sigma_2^2/\sigma_1^2+y^2}, \end{aligned}$$

which is the distribution function of a Cauchy random variable. (To obtain the last line, notice that the integral to compute is of the type $\int u(t)\,u'(t)dt$.)

♦ **Solution of exercise 21.15.** The characteristic function of a Cauchy r.v. is given by

$$f(u) = \int \frac{a}{\pi}\frac{\mathrm{e}^{\mathrm{i}ux}}{a^2+x^2}\,\mathrm{d}x = \mathrm{e}^{-a|u|}.$$

Since the characteristic function of the sum of independent random variables is the product of the characteristic functions of the arguments, we get that the characteristic function of $X_1 + \cdots + X_n$ is

$$\exp\Big((a_1+\cdots+a_n)\,|x|\Big),$$

which does not converge pointwise to the characteristic function of a normal random variable.

♦ **Solution of problem 7**

i) From the independence of the r.v. X_k, we deduce that

$$G_n = \prod_{k=1}^{n} G_{X_k} = (G_X)^n.$$

ii) Using the fact that

$$\mathsf{P}(S=n) = \sum_{k=0}^{\infty} \mathsf{P}(S=n \text{ and } N=k) = \sum_{k=0}^{\infty} \mathsf{P}(S_k=n \text{ and } N=k),$$

we derive

$$\begin{aligned} G_S(t) &= \sum_{n=0}^{\infty} \mathsf{P}(S=n)\, t^n = \sum_{n=0}^{\infty}\sum_{k=0}^{\infty} \mathsf{P}(S_k=n \text{ and } N=k)\, t^n \\ &= \sum_{k=0}^{\infty}\sum_{n=0}^{\infty} \mathsf{P}(S_k=n)\cdot \mathsf{P}(N=k)\, t^n = \sum_{k=0}^{\infty} \mathsf{P}(N=k)\cdot \underbrace{\sum_{n=0}^{\infty} \mathsf{P}(S_k=n)\, t^n}_{G_k(t)} \\ &= \sum_{k=0}^{\infty} \mathsf{P}(N=k)\,\bigl(G_k(t)\bigr)^n = G_N\bigl(G_X(t)\bigr). \end{aligned}$$

iii) $\mathsf{E}(S) = G_S'(1) = G_N'\left(G_X(1)\right)\cdot G_X'(1) = \mathsf{E}(N)\cdot \mathsf{E}(X)$. An easy computation yields

$$\mathrm{Var}(S) = G_S''(1) + G_S'(1) - G_S'(1)^2 = \mathsf{E}^2(X)\, V(N) + \mathsf{E}(N)\, V(X).$$

iv) If $p_0 = 1$, we have $G \equiv 1$. If $p_0 < 1$, G is a sum of increasing functions, one of which at least is strictly increasing.

v) If $p_0 + p_1 = 1$, then G is affine, with $G(0) > 0$ and $G(1) = 1$. If $p_0 + p_1 < 1$, G is the sum of convex functions, one of which at least is strictly convex, so it is strictly convex.

vi) The graph of G may be of one of the following types:

a) : 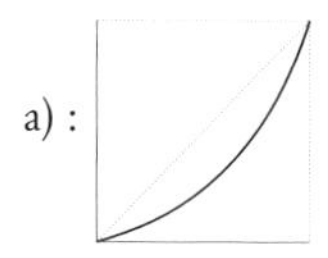 b) : 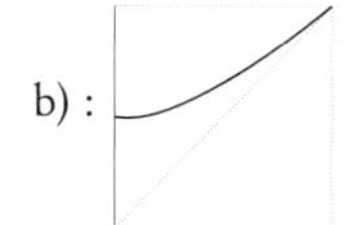 c) : 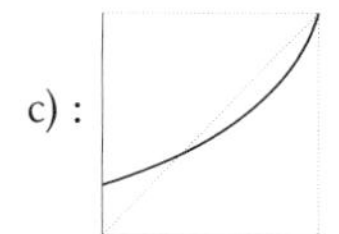

The equation $G(x) = x$ has two, one, and two solutions, respectively, in the three cases.

vii) Using question ii), we find that $G_{Z_2} = G \circ G$ and then, by induction, that

$$G_{Z_{n+1}} = G_{Z_n} \circ G = \underbrace{G \circ \cdots \circ G}_{n+1 \text{ times}}.$$

viii) We have $x_{n+1} = G \circ G \circ \cdots \circ G(0) = G(x_n)$. In all cases above, there exists an interval $[0, \ell]$ on which $G(x) \geqslant x$, and this interval is stable by G, so that the sequence $(x_n)_{n\in\mathbb{N}}$ is increasing and bounded from above by ℓ. Hence, it converges.

ix) We have $\mathsf{E}(Z_{n+1}) = G_{Z_{n+1}}'(1) = G_{Z_n}'\left(G(1)\right)\cdot G'(1) = \mathsf{E}(Z_n)\cdot \mathsf{E}(Z)$ since $G(1) = 1$ and hence, by induction again, we get $\mathsf{E}(Z_n) = \mathsf{E}(Z)^n$.

x) (a) There are two cases.

If $\mathsf{E}(Z) < 1$: the curve is as in Figure b) above, which shows that $x_n \to 1$: there is certain extinction, and $\mathsf{E}(Z_k) \to 0$.

If $\mathsf{E}(Z) = 1$ and $p_0 > 0$: the curve is again of type b), so $x_n \to 1$: there is again certain extinction and of course $\mathsf{E}(Z_n) = 1$.

(b) We assume now that $\mathsf{E}(Z_1) > 1$. The graph of G is of type c), so that $x_n \to \ell \in]0,1]$. The probability of extinction is not 1 (there is positive probability because the particles may all disappear during the first steps). Moreover, $\mathsf{E}(Z_n) \to +\infty$: the reaction is explosive.

◆ **Solution of problem 8**

i) We compute the expectation of X_n:

$$\mathsf{E}(X_n) = \mathsf{E}\left(\sum_{k=1}^{n} \cos\theta_k\right) = \sum_{k=1}^{n} \mathsf{E}(\cos\theta_k) = \sum_{k=1}^{n} \frac{1}{2\pi}\int_0^{2\pi} \cos\theta \,\mathrm{d}\theta = 0,$$

so X_n is centered, and of course so is Y_n.

There only remains to compute $\sigma^2(X_n) = \mathsf{E}(X_n^2) - \mathsf{E}(X_n)^2 = \mathsf{E}(X_n^2)$, which is given by

$$\sigma^2(X_n) = \mathsf{E}\left\{\left(\sum_{k=1}^{n} \cos\theta_k\right)^2\right\} = \mathsf{E}\left(\sum_{k=1}^{n} \cos^2\theta_k + \sum_{k\neq l} \cos\theta_k \cdot \cos\theta_l\right)$$
$$= n\,\mathsf{E}(\cos^2\theta) + n(n-1)\,\mathsf{E}(\cos\theta_k \cdot \cos\theta_l).$$

Since the variables θ_k are independent, we have

$$\mathsf{E}(\cos\theta_k \cdot \cos\theta_l) = \mathsf{E}(\cos\theta_k)\cdot \mathsf{E}(\cos\theta_l) = 0,$$

so we obtain finally

$$\sigma^2(X_n) = n\,\mathsf{E}(\cos^2\theta) = \frac{n}{2\pi}\int_0^{2\pi} \cos^2\theta\,\mathrm{d}\theta = \frac{n}{2}.$$

Similarly, we have $\sigma^2(Y_n) = n/2$.

ii) To determine whether the variables X_n and Y_n are correlated or not, we compute the covariance. We have

$$\mathrm{Cov}(X_n, Y_n) = \mathsf{E}\Big[(X_n - \mathsf{E}(X_n))\,(Y_n - \mathsf{E}(Y_n))\Big] = \mathsf{E}(X_n \cdot Y_n)$$
$$= \mathsf{E}\left\{\sum_{k=1}^{n}\sum_{l=1}^{n} \cos\theta_k \cdot \sin\theta_l\right\}$$
$$= n\,\mathsf{E}(\cos\theta \cdot \sin\theta) + n(n-1)\,\mathsf{E}(\cos\theta \cdot \sin\theta')$$
$$= n\,\mathsf{E}(\cos\theta \cdot \sin\theta) + n(n-1)\,\mathsf{E}(\cos\theta)\,\mathsf{E}(\sin\theta') = 0.$$

Hence X_n and Y_n are uncorrelated.

However, they are certainly not independent. For instance, if $X_n = n$, we necessarily have $Y_n = 0$ (Pierre has walked due east during the first n steps), so

$$\mathsf{P}(X_n = n \text{ and } Y = 0) = \mathsf{P}(X_n = n) \neq \mathsf{P}(X_n = n)\cdot \mathsf{P}(Y_n = 0).$$

iii) The expectation of R_n^2 is

$$\mathsf{E}(R_n^2) = \mathsf{E}(X_n^2 + Y_n^2) = \mathsf{E}(X_n^2) + \mathsf{E}(Y_n^2) = n.$$

So the average of the square of the distance after n steps is n. Of course, one can not deduce that the average distance is $\sqrt{n}$ (which is in fact false).

iv) Since the r.v. $\cos\theta_n$ are independent, have zero expectation and finite variance, we may use the central limit theorem to deduce that the distribution of the r.v. X_n is close to that of a normal centered r.v. with the same variance $n/2$. The density probability for such a random variable $\tilde{X}_n$ is

$$f_n(x) = \frac{1}{\sqrt{n\pi}} \exp\left(-\frac{x^2}{n}\right).$$

Of course, the same is true for Y_n, which can be approximated by $\tilde{Y}_n$ which also has probability density f_n.

Since X_n and Y_n are uncorrelated, it is natural to assume that $\tilde{X}_n$ and $\tilde{Y}_n$ are also uncorrelated, and therefore they are independent, so that the joint probability density is

$$f_{X_n,Y_n}(x,y) \approx \frac{1}{n\pi} \exp\left(-\frac{x^2+y^2}{n}\right).$$

v) The distribution function $H_n(r)$ of the r.v. R_n is given by

$$H_n(r) = \mathsf{P}(R \leqslant r) = \iiint_{\overline{\mathcal{B}}(0\,;\,r)} f_{X_n,Y_n}(x,y)\,\mathrm{d}x\,\mathrm{d}y,$$

where $\overline{\mathcal{B}}(0;r)$ is the closed disc of radius r centered at the origin. Using the approximation above for the density and polar coordinates, we obtain

$$H_n(r) \approx \frac{1}{n\pi} \int_0^{2\pi} \int_0^r \mathrm{e}^{-\rho^2/n} \rho\,\mathrm{d}\rho\,\mathrm{d}\theta = \frac{2}{n} \int_0^r \mathrm{e}^{-\rho^2/n} \rho\,\mathrm{d}\rho$$
$$= \int_0^{r^2} \mathrm{e}^{-u/n} \frac{\mathrm{d}u}{n} = 1 - \mathrm{e}^{-r^2/n}.$$

The probability density of R_n is therefore

$$h_n(r) = H_n'(r) \approx \frac{2r}{n}\,\mathrm{e}^{-r^2/n}.$$

vi) Finally, the expectation of R_n is computed by

$$\mathsf{E}(R_n) = \int_0^{+\infty} r\,h_n(r)\,\mathrm{d}r \approx \frac{2}{n} \int_0^{+\infty} r^2\,\mathrm{e}^{-r^2/n}\,\mathrm{d}r = \frac{\sqrt{n\pi}}{2}.$$

Notice that the distance to the origin increases with time (rather slowly, compared with n).

The following heuristic reasoning suggests an explanation: the probability of getting *farther* from the pub is larger than the probability of getting *closer* to the pub, as can be seen in the picture below: the circular arc describing the values of θ leading to an increase of the distance is larger than the other (in fact, a simple computation shows that the difference between the length of the two arcs if of order $1/\sqrt{r}$, which is consistent with the result obtained).

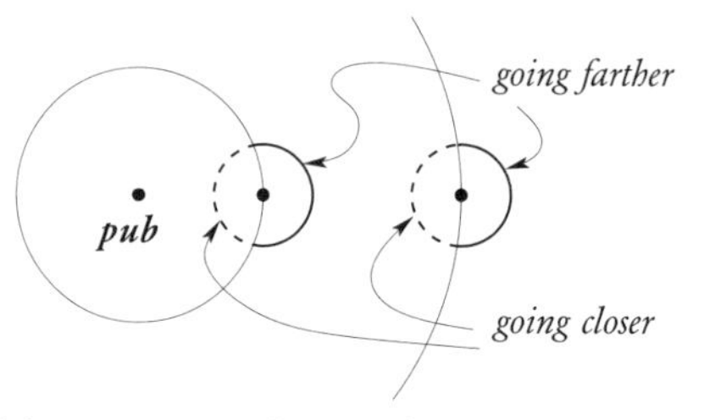

What a beautiful argument! Unfortunately, it comes somewhat after the fact, and it has limits. Indeed, a one-dimensional random walk is also characterized by a distance to the origin $E(R_n)$ proportional to $\sqrt{n}$, whereas the probability of getting closer is *the same* as the probability of going further from the origin.

The conclusion to draw from this tale: beware nice heuristic reasonings; sometimes, they are but webs of lies and deceit.

Appendix

A

Reminders concerning topology and normed vector spaces

A.1

Topology, topological spaces

Since topology is probably not well known to many physicists, we recall a few elementary definitions. At the very least, they are very useful for stating precisely properties of continuity and convergence in sets which are more complicated than $\mathbb{R}$.

DEFINITION A.1 (Open sets, neighborhoods) Let E be an arbitrary set. A **topology** on E is the data of a set $\mathscr{O}$ of subsets of E, called **the open subsets of E** (for the given topology), which satisfy the following three properties:

i) $\varnothing \in \mathscr{O}$ and $E \in \mathscr{O}$;

ii) for any set I, and any family $(A_i)_{i \in I}$ of open subsets, the union $\bigcup_{i \in I} A_i$ is also an open set *(any union of open sets is open)*;

iii) for any $n \in \mathbb{N}$ and any finite family $(A_1, \ldots, A_n) \in \mathcal{O}^n$ of open sets, the intersection $\bigcap_{i=1}^{n} A_i \in \mathcal{O}$ is open *(any finite intersection of open sets is open).*

Finally, a **neighborhood V of a point $a \in E$** is any subset $V \subset E$ containing an open set $\Omega \subset V$ such that $a \in \Omega$.

If a set E is given a topology $\mathcal{O}$, then the pair $(E, \mathcal{O})$, or E itself by abuse of notation, is called a **topological space**.

Remark A.2 If E is a metric space, that is, a set on which a *distance* function is defined, the distance can be used to define a *metric topology*, using the open balls defined by the distance. In particular this is the case for normed vector spaces, which we will discuss separately below.

Using the notion of open set, the continuity of a map can be defined.

DEFINITION A.3 (Continuity of a map) Let $(E, \mathcal{O})$ and $(F, \mathcal{O}')$ be two topological spaces. A map $f : E \to F$ is **continuous** if, for any open set $\Omega \subset F$, the inverse image $f^{-1}(\Omega) = \{x \in E \; ; \; f(x) \in \Omega\}$ is an open set in E, tha is, if $f^{-1}(\Omega) \in \mathcal{O}$ for any $\Omega \in \mathcal{O}'$.

Remark A.4 This definition should be compared with that of a *measurable* map between two measurable spaces (see Definition 2.21 on page 61).

DEFINITION A.5 (Dense subset) Let $(E, \mathcal{O})$ be a topological space. A subset $X \subset E$ is **dense in E** if, for any $a \in E$ and any V which is a neighborhood of a in E, V meets X, that is, $V \cap X \neq \varnothing$.

DEFINITION A.6 (Connected sets) A topological space $(E, \mathcal{O})$ is *connected* if it is not the union of two nonempty disjoint open sets: it is not possible to find A and B such that

$$X = A \cup B \qquad \text{with } A \in \mathcal{O}, \quad B \in \mathcal{O}, \quad A \neq \varnothing, \quad B \neq \varnothing, \quad A \cap B = \varnothing.$$

A subset $X \subset E$ is **connected** if $(X, \mathcal{O}_X)$ is connected, where the **induced topology** $\mathcal{O}_X$ on X is the topology with open sets of the form $U = A \cap X$ for some $A \in \mathcal{O}$.

In other words, the topological space is connected if it is "in one piece."

Example A.7 Below are pictures of two sets in the plane; the set (a) is connected, while (b) is not.

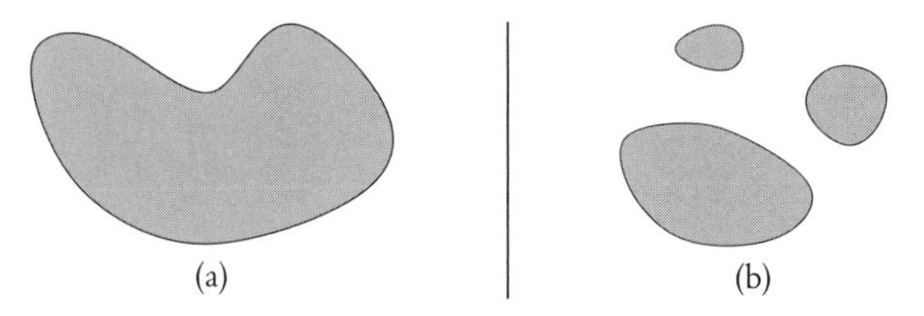

Be careful not to mix *connected* and *convex*. The first is a purely topological notion, the second is geometric.

DEFINITION A.8 (Convexity) A subset X in a vector space or in an affine space is **convex** if, for any points A, B in X, the line segment $[A, B]$ is contained in X. (The segment $[A, B]$ is defined as the set of points $\lambda A + (1 - \lambda)B$, for $\lambda \in [0, 1]$, and this definition requires a vector space, or affine space, structure.)

Example A.9 All vector spaces are convex. Below are examples (in the plane) of a convex set (a) and a nonconvex set (b); note that both are connected.

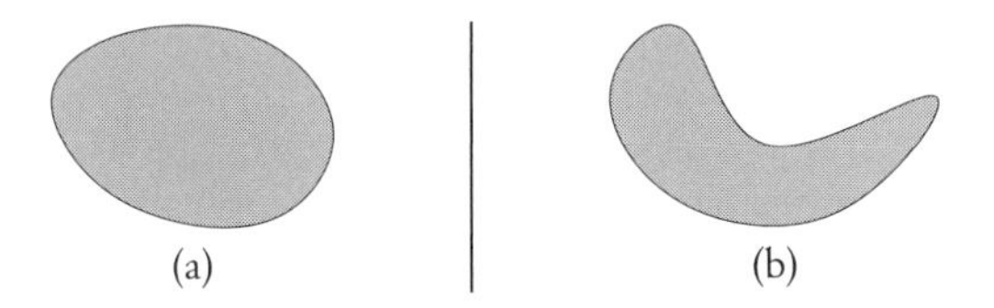

DEFINITION A.10 (Simply connected set) A topological space $(E, \mathscr{O})$ is **simply connected** if any closed loop inside E may be deformed continuously to a trivial constant loop (i.e., all loops are **homotopic to zero**).

A subset $X \subset E$ is simply connected if the topological space $(X, \mathscr{O}_X)$ is simply connected.

In the special case $E = \mathbb{R}^2$ (or $E = \mathbb{C}$), a subset $X \subset \mathbb{R}^2$ is simply connected if "there is no hole in X."

Example A.11 Still in the plane, the set in (a) is *simply connected*, whereas the set in (b) is not. Again both are connected.

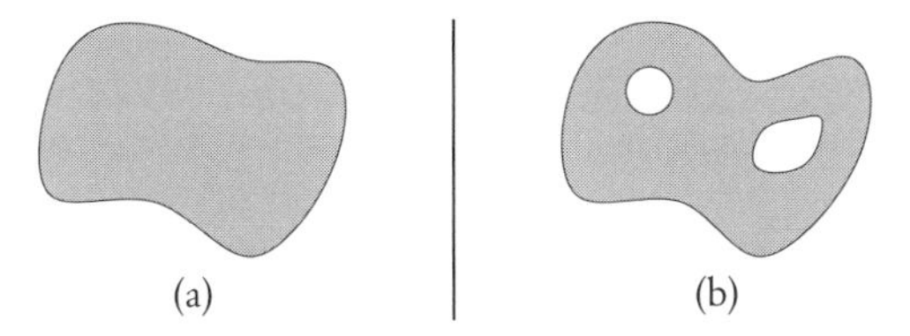

Example A.12 Let X be the complex plane minus the origin 0; then X is not simply connected. Indeed, any loop turning once completely around the origin *cannot be* retracted to a single point (since 0 is not in X!).

On the other hand, the complex plane minus a half-line, for instance, the half-line of negative real numbers (i.e., $\mathbb{C} \setminus \mathbb{R}^-$) is simply connected. Below are some illustrations showing how a certain loop is deformed to a point.

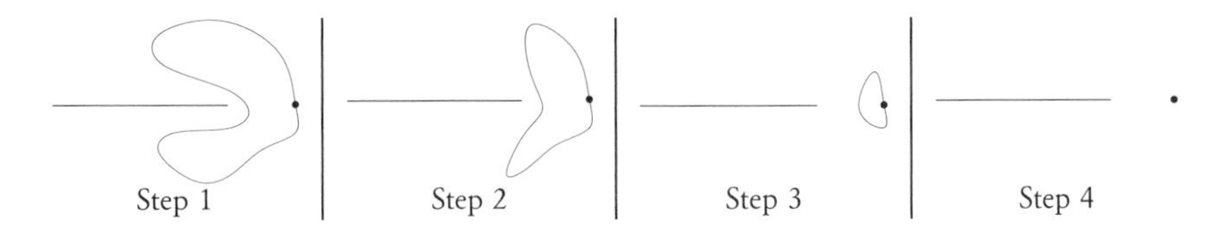

Example A.13 A torus (the surface of a doughnut) is not simply connected. In the picture below one can see a closed loop which can not be contracted to a single point.

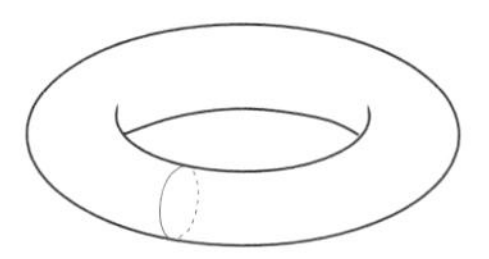

Example A.14 The unit circle $\mathbf{U} = \{z \in \mathbb{C} \; ; \; |z| = 1\}$ in the complex plane is not simply connected.

An important point is that connected and simply-connected sets are preserved by continuous maps.

THEOREM A.15 *The image of a connected set (resp. simply connected set) by a continuous map is connected (resp. simply connected).*

DEFINITION A.16 (Compact sets) An **open covering** of a topological space K is any family of open sets $(U_i)_{i \in I}$ such that K is the union of the sets U_i:

$$\bigcup_{i \in I} U_i = K.$$

The topological space K is **compact** if, from every open covering of K, one can extract a *finite* subcovering, i.e., there exist finitely many indices $i_1, \ldots, i_n$, such that

$$\bigcup_{j=1}^{i_n} U_{i_j} = K.$$

A subset $X \subset K$ is compact if it is compact with the induced topology.

The definition of compact spaces by this property is mostly of interest to mathematicians. In metric spaces (in particular, in normed vector spaces), this property is equivalent with the following: *from any sequence* $(x_n)_{n \in \mathbb{N}}$ *with values in* K*, one can extract a subsequence that converges in* K.

Moreover, in any finite-dimensional normed vector space, compact sets are exactly those which are closed and bounded (see Theorem A.42).

A.2

Normed vector spaces

A.2.a Norms, seminorms

In this section, $\mathbb{K}$ is either $\mathbb{R}$ or $\mathbb{C}$, and E is a $\mathbb{K}$-vector space, not necessarily of finite dimension.

DEFINITION A.17 (Norm) Let E be a $\mathbb{K}$-vector space. A **norm on E** is a map $N : E \to \mathbb{R}$ which satisfies the following properties, for all $x, y \in E$ and $\lambda \in \mathbb{K}$:

N1 : we have $N(x) \geqslant 0$;

N2 : we have $\big(N(x) = 0 \iff x = 0\big)$;

N3 : we have $N(\lambda x) = |\lambda|\, N(x)$;

N4 : we have $N(x + y) \leqslant N(x) + N(y)$ *(triangular inequality)*

If is customary to write $\|x\|$ for the norm of an element x, and to denote by $\|\cdot\|$ the norm map itself.

A **normed vector space** is a pair (E, N) where E is a vector space and N is a norm on E.

DEFINITION A.18 (Seminorm) Let E be a $\mathbb{K}$-vector space. A **seminorm** on E is a map $N : E \to \mathbb{R}$ satisfying properties N1, N3, and N4 above. Note that N3 still implies $N(0) = 0$.

Example A.19 Let $a, b \in \mathbb{R}$, $a < b$. Let $\mathscr{C}([a, b])$ be the vector space of continuous functions defined on $[a, b]$, with values in $\mathbb{K} = \mathbb{R}$ or $\mathbb{C}$. For $f \in \mathscr{C}([a, b])$, let

$$\|f\|_1 = \int_a^b |f(x)|\,\mathrm{d}x \qquad \textbf{norm of convergence in mean,}$$

$$\|f\|_2 = \sqrt{\int_a^b |f(x)|^2\,\mathrm{d}x} \qquad \textbf{norm of convergence in mean-square,}$$

$$\|f\|_\infty = \sup_{[a,b]} |f| \qquad \textbf{norme of uniform convergence.}$$

Then $\|\cdot\|_1$, $\|\cdot\|_2$ and $\|\cdot\|_\infty$ are norms on $\mathscr{C}([a, b])$.

On the space of *piecewise continuous functions*, $\|\cdot\|_1$ and $\|\cdot\|_2$ are seminorms (a function which is zero almost everywhere satisfies $\|f\|_1 = \|f\|_2 = 0$ but is not necessarily zero everywhere) and $\|\cdot\|_\infty$ is a norm.

If $a, b \in \overline{\mathbb{R}}$, it is necessary to restrict attention to integrable (resp. square integrable, resp. bounded continuous) functions in order that $\|\cdot\|_1$, $\|\cdot\|_2$, or $\|\cdot\|_\infty$ be defined; then they are norms.

DEFINITION A.20 (Distance) Let $(E, \|\cdot\|)$ be a normed vector space. The **distance associated to the norm** $\|\cdot\|$ is the map

$$\begin{aligned} d: \quad & E^2 \longrightarrow \mathbb{R}, \\ & (x,y) \longmapsto d(x,y) \stackrel{\text{def}}{=} \|x-y\|. \end{aligned}$$

Let $A \subset E$ be a non-empty subset of E and let $x \in E$. The **distance from x to A** is the real number

$$d(x;A) \stackrel{\text{def}}{=} \inf_{y \in A} \|x-y\|.$$

A.2.b Balls and topology associated to the distance

The distance associated to a norm gives a normed vector space the structure of a metric space. There is then a natural topology associated with this metric; it is of course a (very) special case of the general topological notions described in Section A.1.

DEFINITION A.21 Let $(E, \|\cdot\|)$ be a normed vector space. Let $a \in E$ and let $r \in \mathbb{R}^+$. The **open ball** centered at a with radius r (for the norm $\|\cdot\|$) is the set defined by

$$\mathcal{B}(a\,;\,r) \stackrel{\text{def}}{=} \{x \in E\,;\, \|x-a\| < r\}.$$

The **closed ball** centered at a with radius r is the set defined by

$$\overline{\mathcal{B}}(a\,;\,r) \stackrel{\text{def}}{=} \{x \in E\,;\, \|x-a\| \leqslant r\}.$$

DEFINITION A.22 (Open set) A subset $A \subset E$ is **open** in E if and only if, for any $x \in A$, there exists a ball centered at x with positive radius $r > 0$ contained in A:

$$\forall x \in A \quad \exists r > 0 \qquad \mathcal{B}(x\,;\,r) \subset A.$$

DEFINITION A.23 (Closed) A subset $A \subset E$ is **closed** in E if and only if the complement of A in E, denoted $\complement_E A$, is open.

Example A.24 A well-known French saying is that "Il faut qu'une porte soit ouverte ou fermée" (Musset; "A door must be either open or closed"); this is quite true in its way, but not in the setting of topological spaces and normed vector spaces in particular. For instance, in $\mathbb{R}^2$, the set $]0,1[\times\{0\}$ is neither open nor closed. Similarly, in $\mathbb{R}$, $\mathbb{Q}$ is neither open nor closed. More surprising maybe, $\mathbb{R}$ (in $\mathbb{R}$) is *both* open and closed.

DEFINITION A.25 (Interior) Let A be a subset of E. A point $a \in E$ is **in the interior of A** if there exists $r > 0$ such that the open ball centered at a with radius r is contained in A:

$$\exists r > 0 \qquad \mathcal{B}(a\,;\,r) \subset A.$$

The **interior of A**, denoted $\overset{\circ}{A}$, is the set of all interior points of A.

DEFINITION A.26 (Closure) A point $a \in E$ is **in the closure of** A if, for any positive $r > 0$, the open ball centered at a with radius r has non-empty intersection with A:

$$\forall r > 0 \qquad \mathcal{B}(a\,;\,r) \cap A \neq \varnothing.$$

The **closure** of A, denoted $\overline{A}$, is the set of all a satisfying this property.

PROPOSITION A.27 *For any $A \subset E$, we have $\overset{\circ}{A} \subset A \subset \overline{A}$.*

Example A.28 Let A be a non-empty subset of $\mathbb{R}$ which has an upper bound. Then $\sup A \in \overline{A}$.

PROPOSITION A.29 *A subset $A \subset E$ is closed if and only if A is equal to its closure: $\overline{A} = A$. A subset $B \subset E$ is open if and only if B is equal to its interior: $\overset{\circ}{B} = B$.*

The definition of a dense set may then be rephrased in this manner:

DEFINITION A.30 A subset $A \subset E$ is **dense in E** if and only if the closure of A is the whole space E.

Example A.31 The set $\mathbb{Q}$ of rationals is dense in $\mathbb{R}$; the set of invertible matrices $\mathrm{GL}_n(\mathbb{K})$ is dense is the space $\mathfrak{M}_n(\mathbb{K})$ of all matrices. The set of diagonalizable matrices is dense in $\mathfrak{M}_n(\mathbb{C})$.

THEOREM A.32 *Let E and F be normed vector spaces, $D \subset E$ a dense subset of E, and $f : E \to F$ a continuous map. If $f(x) = 0$ for all $x \in D$, then f is zero on E.*

Example A.33 Here is an application: for any matrix $A \in \mathfrak{M}_n(\mathbb{C})$, we have $\det(\exp A) = \exp(\operatorname{tr} A)$. This property is obvious in the case of a diagonalizable matrix: indeed, assume that $A = P \operatorname{diag}(\lambda_1, \dots, \lambda_n) P^{-1}$ (with $P \in \mathrm{GL}_n(\mathbb{C})$); then we have

$$\exp A = P \begin{pmatrix} \exp(\lambda_1) & & \\ & \ddots & \\ & & \exp(\lambda_n) \end{pmatrix} P^{-1} \qquad \text{and} \qquad \prod_{i=1}^{n} \exp(\lambda_i) = \exp\left(\sum_{i=1}^{n} \lambda_i\right).$$

Since the set of diagonalizable matrices is dense in $\mathfrak{M}_n(\mathbb{C})$, and since the determinant, the trace and the exponential function for matrices are continuous (the last being not quite obvious), it follows that the identity is also valid for an arbitrary $A \in \mathfrak{M}_n(\mathbb{C})$.

Example A.34 Let $f : [0,1] \to \mathbb{R}$ be a continuous function such that

$$f\left(\frac{x+y}{2}\right) = \frac{f(x) + f(y)}{2}$$

for all $(x, y) \in \mathbb{R}^2$. We want to prove that f is affine, i.e., there exist $a, b \in R$ such that $f(x) = ax + b$ for all $x \in [0,1]$. For this purpose, let $b = f(0)$ and $a = f(1) - f(0)$, and let $g : x \mapsto f(x) - (ax + b)$. The function g is then continuous, and satisfies $g(0) = g(1) = 0$. Moreover, g satisfies the same relation

$$g\left(\frac{x+y}{2}\right) = \frac{g(x) + g(y)}{2}$$

Edmund Georg Hermann LANDAU (1877–1938), German mathematician, was famous in particular for his many works of analytic number theory (in particular concerning the Riemann zeta function) and his treatise on number theory. From his youngest days, he was also interested in mathematical problems and games, and had published two books of chess problems even before finishing his thesis. He taught at the University of Berlin from 1899 to 1919, then at the University of Göttingen (replacing Minkowski) until he was forced to resign by the Nazis in 1934 (the SS mathematician Teichmüller organized a boycott of his classes).

as before. From this we deduce immediately that $g(\frac{1}{2}) = 0$, then, repeating the process, that

$$g(\tfrac{1}{4}) = g(\tfrac{2}{4}) = g(\tfrac{3}{4}) = 0,$$

and by an obvious induction we obtain

$$g\left(\frac{k}{2^n}\right) = 0 \qquad \text{for all } n \in \mathbb{N}^* \text{ and all } k \in [\![0, 2^n]\!].$$

Since the set $E = \{k \cdot 2^{-n} \; ; \; n \in \mathbb{N}^*, \; k \in [\![0, 2^n]\!]\}$ is dense in $[0, 1]$, and since g is continuous, we can conclude that $g = 0$.

THEOREM A.35 (Properties of open sets) *The union of any family of open sets is an open set.*

The intersection of finitely many *open sets is an open set.*

Example A.36 $\bigcup_{n \in \mathbb{N}} \left]\frac{1}{n}, 1\right[=]0, 1[$ is open in $\mathbb{R}$.

Counterexample A.37 The intersection of *infinitely many* open sets has no reason to be open: for instance, $\bigcap_{n \in \mathbb{N}} \left]0, 1 + \frac{1}{n}\right[=]0, 1]$ is not open in $\mathbb{R}$.

THEOREM A.38 (Properties of closed sets) *The union of* finitely many *closed sets is a closed set.*

The intersection of any family of closed sets is a closed set.

Counterexample A.39 As before, the union of infinitely many closed sets has no reason to be closed: for instance, $\bigcup_{n \in \mathbb{N}} \left[0, 1 - \frac{1}{n}\right] = [0, 1[$ is not closed.

A.2.c Comparison of sequences

DEFINITION A.40 Let E be a normed vector space.

A sequence $(u_n)_{n \in \mathbb{N}}$ with values in E **is dominated by** a sequence $(\alpha_n)_{n \in \mathbb{N}}$ of non-negative real numbers, denoted $\boldsymbol{u_n = \mathrm{O}(\alpha_n)}$ and pronounced "u_n is big-Oh of α_n," if there exist a real number A and an integer N such that $\|u_n\| \leqslant A\,|\alpha_n|$ for all $n \geqslant N$.

The Czech priest Bernhard BOLZANO (1781–1848), was born and died in Prague, in a German-speaking family of Italian origin. He studied mathematics during his free time. Because of his criticism of the Austro-Hungarian Empire, he was forbidden to publish his work and was closely watched. He sought to introduce rigor in the foundations of mathematics: function theory, set theory, logic. His impressive works (on cardinal theory, continuity, real analysis) were in advance on their time and were neglected during his lifetime (his discovery of the Cauchy criterion for convergence of sequences was not known). Finally, as a philosopher of science, he exerted an important influence (notably on Husserl), and he founded epistemology on formal logic.

The sequence (u_n) is **negligible compared to** $(\alpha_n)_{n\in\mathbb{N}}$ if, for any positive real number $\varepsilon > 0$, there exists an index $N \in \mathbb{N}$, depending on ε, such that $\|u_n\| \leqslant \varepsilon\,|\alpha_n|$ for all $n \geqslant N$. This is denoted $\boldsymbol{u_n = \mathrm{o}(\alpha_n)}$, pronounced "$u_n$ is little-Oh of α_n."

Two real- or complex-valued sequences $u = (u_n)_{n\in\mathbb{N}}$ and $v = (v_n)_{n\in\mathbb{N}}$ are **equivalent**, denoted $\boldsymbol{u_n \sim v_n}$, if $u_n - v_n = \mathrm{o}(u_n)$ or, equivalently (and it shows that the relation "$\sim$" is an equivalence relation), if $u_n - v_n = \mathrm{o}(v_n)$.

The notation above is due to E. Landau.

A.2.d Bolzano-Weierstrass theorems

THEOREM A.41 (Bolzano-Weierstrass) *Any bounded complex- or real-valued sequence has a convergent subsequence.*

THEOREM A.42 (Bolzano-Weierstrass) *Any subset of a finite-dimensional normed vector space which is bounded and closed is compact, and in particular, any sequence with values in such a subset has a convergent subsequence.*

A.2.e Comparison of norms

THEOREM A.43 *Let E be a vector space and let N, N' be two norms on E. Then any sequence converging to 0 in the sense of N also converges to 0 in the sense of N' if and only if there exists a real number $\alpha \geqslant 0$ such that*

$$N' \leqslant \alpha N,$$

that is, $N'(x) \leqslant \alpha N(x)$ for all $x \in E$.

DEFINITION A.44 Two norms N and N' on E are **equivalent** if there exist real numbers $\alpha, \beta \geqslant 0$ such that

$$\alpha N \leqslant N' \leqslant \beta N.$$

Karl Theodor Wilhelm WEIERSTRASS (1815–1897), German mathematician (from Westphalia), a famous teacher, was the first to give a convincing construction of the set $\mathbb{R}$ of real numbers (such a construction was lacking in Bolzano's work). He also constructed an example of a continuous function on $\mathbb{R}$ which is nowhere differentiable (ignorant of Bolzano's prior work). The famous theorem stating that any continuous function on an interval $[a,b]$ can be uniformly approximated by polynomials is also due to him. Finally, wishing to provide rigorous and unambiguous definitions of the concepts of analysis, he introduced the definition of continuity based on "epsilons and deltas," which are the cause of such happy moments and memories in the lives of students everywhere.

THEOREM A.45 *If the norms N and N' are equivalent, then any sequence that converges for N also converges for N' and conversely, and moreover, the limits are equal.*

Example A.46 Let $E = \mathbb{K}^n$. We can easily compare the norms N_1, N_2, and N_∞ defined by $N_1(\boldsymbol{x}) = \sum |x_i|$, $N_2(\boldsymbol{x}) = \left(\sum |x_i|^2\right)^{1/2}$ and $N_\infty(\boldsymbol{x}) = \max |x_i|$. Indeed, we have

$$\left.\begin{array}{l} N_\infty \leqslant N_1 \leqslant n\, N_\infty \\ N_\infty \leqslant N_2 \leqslant \sqrt{n}\, N_\infty \end{array}\right\} \quad \text{hence} \quad \frac{1}{\sqrt{n}} N_2 \leqslant N_1 \leqslant n N_2.$$

Thus, those three norms are equivalent.

The previous example illustrates an important theorem in finite-dimensional vector spaces:

THEOREM A.47 (Equivalence of norms) *Let E be a finite-dimensional vector space. Then all norms on E are equivalent. In particular, all notions defined in terms of the topology associated to a norm on E are identical whatever the norm used in the definition (open sets, closed sets, compact sets, convergence of sequences, limits of sequences, boundednes,[1] Cauchy sequences, Lipschitz functions,..).*

Counterexample A.48 This is false for infinite-dimensional vector spaces. For instance, consider the space $E = \mathscr{C}([0,1],\mathbb{R})$ of real-valued continuous functions on $[0,1]$, with the two norms

$$N_1(f) = \sup_{t\in[0,1]} \left|f(t)\right| \qquad \text{and} \qquad N_2(f) = \int_0^1 \left|f(t)\right| \mathrm{d}t.$$

It is clear that $N_2 \leqslant N_1$, so any sequence of functions that converges uniformly to 0 also converges in mean to 0. However, the converse is false: for instance, let $(f_n)_{n\in\mathbb{N}}$ be the sequence of functions defined by $f_n(x) = x^n$. Then $(f_n)_{n\in\mathbb{N}}$ converges to 0 in means ($N_2(f_n) = 1/(n+1)$), whereas $\|f_n\|_\infty = 1$ for all $n \in \mathbb{N}$.

Counterexample A.49 On $\mathbb{R}$, the norms of convergence in mean, in quadratic mean, and of uniform convergence, are pairwise non-equivalent. To see this, let Λ be the function defined by $\Lambda(x) = 1 - |x|$ if $x \in [-1,1]$ and $\Lambda(x) = 0$ otherwise.

[1] The *property* of being bounded; of course, a bound for a set may depend on the norm.

- The sequence of continuous functions on $\mathbb{R}$ defined by $f_n(x) = \Lambda(x/n)/n$ converges uniformly and in quadratic mean to 0, but it does not converge in mean to 0.
- The sequence of continuous functions on $\mathbb{R}$ defined by $g_n(x) = \sqrt{n}\,\Lambda(nx)$ converges in mean to 0, but does not converge in quadratic mean, or uniformly.
- The sequence of continuous functions on $\mathbb{R}$ defined by $h_n(x) = \Lambda(x/n)/\sqrt{n}$ converges uniformly to 0, but converges neither in mean nor in quadratic mean.

The reader will now easily find examples illustrating the other cases, using the same ideas.

Another fundamental result specific to finite-dimensional vector spaces is the following:

THEOREM A.50 (Finite-dimensional normed vector spaces are complete) *The vector spaces $\mathbb{R}$, $\mathbb{C}$, and more generally all finite-dimensional vector spaces are complete, that is, any Cauchy sequence in a finite-dimensional vector space is convergent.*

A.2.f Norm of a linear map

The continuity of a linear map between normed vector spaces is characterized by the following theorem:

THEOREM A.51 *Let E and F be normed vector spaces and let $f \in L(E,F)$ be a linear map from E to F. Then f is continuous if and only if there exists a real number $k \in \mathbb{R}$ such that $\|f(x)\|_F \leqslant k\,\|x\|_E$ for all $x \in E$.*

If $f \in L(E,F)$ is continuous, the real number defined by

$$|||f||| = \sup_{\|x\|_E=1} \|f(x)\|_F = \sup_{x\in E\setminus\{0\}} \frac{\|f(x)\|_F}{\|x\|_E}$$

*is called the **norm of f**. Hence, by definition, we have*

$$\|f(x)\|_F \leqslant |||f||| \cdot \|x\|_E$$

for $x \in E$.

The map $f \mapsto |||f|||$ is a norm on the vector space of continuous linear maps from E to F.

EXERCISE

♦ **Exercise A.1** Let $(E, \|\cdot\|)$ be a normed vector space and let $\mathscr{L}(E)$ denote the vector space of continuous maps from E to itself with the associated norm as above. Assume that $(E, \|\cdot\|)$ is complete. Prove that $(\mathscr{L}(E), |||\cdot|||)$ is also complete.

SOLUTION

♦ **Solution of exercise A.1.** Let $(u_n)_{n\in\mathbb{N}}$ be a Cauchy sequence in $\mathscr{L}(E)$. For all $x \in E$, we have

$$\left\|u_p(x) - u_q(x)\right\| \leqslant \|x\| \cdot \left|\!\left|\!\left|u_p - u_q\right|\!\right|\!\right|,$$

and this proves that the sequence $\big(u_n(x)\big)_n$ is a Cauchy sequence in E. By assumption, it converges; we then let

$$u(x) = \lim_{n\to\infty} u_n(x) \qquad \forall x \in E.$$

This defines the map $u : E \to E$, which a moment's thought shows to be linear. Obviously, we want to prove that u is continuous, and is the limit of $(u_n)_{n\in\mathbb{N}}$.

i) *The map u is in $\mathscr{L}(E)$.*

By the Cauchy property, there exists $N \in \mathbb{N}$ such that $\left|\!\left|\!\left|u_p - u_q\right|\!\right|\!\right| \leqslant 1$ for all $p, q \geqslant N$.

□ Let $x \in E$ be such that $\|x\| = 1$.

For all $p, q \geqslant N$, we have $\left\|u_p(x) - u_q(x)\right\| \leqslant 1$. Letting q go to $+\infty$, we obtain $\left\|u_p(x) - u(x)\right\| \leqslant 1$ for all $p \geqslant N$. Hence,

$$\left\|u(x)\right\| \leqslant \left\|u(x) - u_N(x)\right\| + \left\|u_N(x)\right\| \leqslant 1 + |\!|\!|u_N|\!|\!|. \;\square$$

This proves that

$$\sup_{\|x\|=1} \left\|u(x)\right\| \leqslant 1 + |\!|\!|u_N|\!|\!|$$

and in particular this is finite, which means that u is continuous.

ii) *We have $\lim_{n\to\infty} u_n = u$ in the sense of the norm $|\!|\!|\cdot|\!|\!|$ on $\mathscr{L}(E)$.*

□ Let $\varepsilon > 0$. There exists $N \in \mathbb{N}$ such that $\left|\!\left|\!\left|u_p - u_q\right|\!\right|\!\right| \leqslant \varepsilon$ for all $p, q \geqslant N$. Let $x \in E$ be such that $\|x\| = 1$. Then

$$\forall p, q \geqslant N \qquad \left\|u_p(x) - u_q(x)\right\| \leqslant \varepsilon,$$

and letting q go to $+\infty$ again, we obtain

$$\forall p \geqslant N \qquad \left\|u_p(x) - u(x)\right\| \leqslant \varepsilon$$

(we use the fact that the norm on E is itself a continuous map). Since this holds for all $x \in E$ such that $\|x\| = 1$, we have proved that

$$\forall p \geqslant N \qquad \left|\!\left|\!\left|u_p - u\right|\!\right|\!\right| \leqslant \varepsilon. \;\square$$

And of course, since $\varepsilon > 0$ was arbitrary, this is precisely the definition of the fact that $(u_n)_{n\in\mathbb{N}}$ converges to u in $\mathscr{L}(E)$.

Appendix B

Elementary reminders of differential calculus

B.1 Differential of a real-valued function

B.1.a Functions of one real variable

Let $f : \mathbb{R} \to \mathbb{R}$ be a function which is differentiable at a point $a \in \mathbb{R}$. By definition, this means that the limit

$$\ell \stackrel{\text{def}}{=} \lim_{x \to a} \frac{f(x) - f(a)}{x - a} = \lim_{h \to 0} \frac{f(x+h) - f(x)}{h}$$

exists. This also means that there exists a unique real number $q \in \mathbb{R}$ such that

$$f(x+h) = f(x) + q \cdot h + \underset{h \to 0}{\mathrm{o}}(h).$$

In fact, we have $q = \ell$, and this is the derivative of f at a, that is,

$$f'(a) = q = \ell.$$

An equivalent rephrasing is that there exists a unique linear map $\Theta \in \mathscr{L}(\mathbb{R})$ such that

$$f(x+h) = f(x) + \Theta.h + \underset{h \to 0}{\mathrm{o}}(h),$$

since a linear map of $\mathbb{R}$ into itself is simply multiplication by a constant value. (We use the notation $\Theta.h$ for the value of Θ evaluated at h, instead of the more usual $\Theta(h)$; this is customary in differential calculus.) This new formulation has the advantage of generalizing easily to a function of many variables, that is, to functions defined on $\mathbb{R}^n$.

B.1.b Differential of a function $f : \mathbb{R}^n \to \mathbb{R}$

Let $f : \mathbb{R}^n \to \mathbb{R}$ be a real-valued function of n real variables, or one "vector" variable. We want to generalize the notion of derivative of f, at a point $\boldsymbol{a} \in \mathbb{R}^n$. First, the **$i$-th partial derivative** of f at the point $\boldsymbol{a} \in \mathbb{R}^n$ is defined as the limit

$$\frac{\partial f}{\partial x_i}(\boldsymbol{a}) \stackrel{\text{def}}{=} \lim_{h \to 0} \frac{f(\boldsymbol{a} + h\,\mathbf{e}_i) - f(\boldsymbol{a})}{h}$$

(if it exists), where $\mathbf{e}_i$ is the i-th vector of the canonical basis of $\mathbb{R}^n$. A classical result of differential calculus states that if all partial derivatives $\partial f / \partial x_i$ exist and are continuous functions of $\boldsymbol{a}$, then we have

$$f(\boldsymbol{a} + \boldsymbol{h}) = f(\boldsymbol{a}) + \frac{\partial f}{\partial x_i}(\boldsymbol{a}) \cdot h_i + \underset{\boldsymbol{h} \to \boldsymbol{0}}{\mathrm{o}}(\boldsymbol{h}). \tag{b.1}$$

for $\boldsymbol{h} \in \mathbb{R}^n$. Denote by $\mathrm{d}x_i$ the linear form "i-th coordinate" on $\mathbb{R}^n$, that is, the linear map associating $\mathrm{d}x_i.\boldsymbol{h} = h_i$ to a vector $\boldsymbol{h} = (h_1, \ldots, h_n)$. Then we can state the following definition:

DEFINITION B.1 (Differential) Let $f : \mathbb{R}^n \to \mathbb{R}$ be a map which has continuous partial derivatives at $\boldsymbol{a} \in \mathbb{R}^n$. The **differential of f at $\boldsymbol{a}$**, denoted $\mathrm{d}f_{\boldsymbol{a}}$, is the linear map $\mathrm{d}f_{\boldsymbol{a}} : \mathbb{R}^n \to \mathbb{R}$ defined by

$$\mathrm{d}f_{\boldsymbol{a}}.\boldsymbol{h} = \sum_{i=1}^{n} \frac{\partial f}{\partial x_i}(\boldsymbol{a}) \cdot h_i,$$

or equivalently

$$\mathrm{d}f_{\boldsymbol{a}} = \sum_{i=1}^{n} \frac{\partial f}{\partial x_i}(\boldsymbol{a})\, \mathrm{d}x_i.$$

We can then write (b.1) as follows:

$$f(\boldsymbol{a} + \boldsymbol{h}) = f(\boldsymbol{a}) + \mathrm{d}f_{\boldsymbol{a}}.\boldsymbol{h} + \underset{\boldsymbol{h} \to \boldsymbol{0}}{\mathrm{o}}(\boldsymbol{h}).$$

This turns out to be the right way of defining differentiable functions: f is **differentiable at $\boldsymbol{a}$** if and only if there exists a linear map $\Theta : \mathbb{R}^n \to \mathbb{R}$ such that

$$f(\boldsymbol{a} + \boldsymbol{h}) = f(\boldsymbol{a}) + \Theta.\boldsymbol{h} + \underset{\boldsymbol{h} \to \boldsymbol{0}}{\mathrm{o}}(\boldsymbol{h}),$$

and the linear map Θ is the same as the differential defined above using partial derivatives.

B.1.c Tensor notation

In relativity theory, physicists usually use a single letter to denote a vector, and use exponents over this letter to indicate the coordinates. Using this tensor notation (see Chapter 16), we can express the relations above as follows:

$$f(x+h) = f(x) + h^\mu\,\partial_\mu f + o(h) \qquad \text{and} \qquad \mathrm{d}f = \partial_\mu f\,\mathrm{d}x^\mu.$$

Writing $f_{,\mu} \stackrel{\text{def}}{=} \partial_\mu f$, we also have

$$f(x+h) = f(x) + h^\mu f_{,\mu} \qquad \text{and} \qquad \mathrm{d}f = f_{,\mu}\,\mathrm{d}x^\mu.$$

B.2 Differential of map with values in $\mathbb{R}^p$

It is very easy to generalize the previous results for vector-valued functions of a vector variable. Let thus $\boldsymbol{f} : \mathbb{R}^n \to \mathbb{R}^p$ be a map with vector values. All previous notions and results apply to each coordinate of $\boldsymbol{f}$. We therefore denote

$$\boldsymbol{f}(\boldsymbol{a}) = \begin{pmatrix} f_1(\boldsymbol{a}) \\ \vdots \\ f_p(\boldsymbol{a}) \end{pmatrix} \qquad \text{for all } \boldsymbol{a} \in \mathbb{R}^n.$$

Then, if each coordinate of $\boldsymbol{f}$ has continuous partial derivatives at $\boldsymbol{a} \in \mathbb{R}^n$, we can write down (b.1) for each component:

$$f_i(\boldsymbol{a}+\boldsymbol{h}) = f_i(\boldsymbol{a}) + \sum_{j=1}^{n} \frac{\partial f_i}{\partial x_j}(\boldsymbol{a})\cdot h_j + o(\boldsymbol{h}), \qquad i = 1,\dots,p.$$

These relations can be written in matrix notation as follows:

$$\begin{pmatrix} f_1(\boldsymbol{a}+\boldsymbol{h}) \\ \vdots \\ f_p(\boldsymbol{a}+\boldsymbol{h}) \end{pmatrix} = \begin{pmatrix} f_1(\boldsymbol{a}) \\ \vdots \\ f_p(\boldsymbol{a}) \end{pmatrix} + \begin{pmatrix} \dfrac{\partial f_1}{\partial x_1}(\boldsymbol{a}) & \cdots & \dfrac{\partial f_1}{\partial x_n}(\boldsymbol{a}) \\ \vdots & & \vdots \\ \dfrac{\partial f_p}{\partial x_1}(\boldsymbol{a}) & \cdots & \dfrac{\partial f_p}{\partial x_n}(\boldsymbol{a}) \end{pmatrix} \cdot \begin{pmatrix} h_1 \\ \vdots \\ h_p \end{pmatrix} + o(\boldsymbol{h}). \tag{b.2}$$

Hence the following definition:

DEFINITION B.2 (Differential) Let $\boldsymbol{f} : \mathbb{R}^n \to \mathbb{R}^p$ be a map such that each coordinate has continuous partial derivatives at $\boldsymbol{a} \in \mathbb{R}^n$. The **differential**

of f at a is the linear map $\mathrm{d}f_a : \mathbb{R}^n \to \mathbb{R}^p$ such that

$$\mathrm{d}f_a \begin{pmatrix} h_1 \\ \vdots \\ h_n \end{pmatrix} = \begin{pmatrix} \frac{\partial f_1}{\partial x_1}(a) & \cdots & \frac{\partial f_1}{\partial x_n}(a) \\ \vdots & & \vdots \\ \frac{\partial f_p}{\partial x_1}(a) & \cdots & \frac{\partial f_p}{\partial x_n}(a) \end{pmatrix} \cdot \begin{pmatrix} h_1 \\ \vdots \\ h_p \end{pmatrix}.$$

The **differential** of f, denoted $\mathrm{d}f$, is the map

$$\mathrm{d}f : \mathbb{R}^n \longrightarrow \mathscr{L}(\mathbb{R}^n, \mathbb{R}^p), \qquad a \longmapsto \mathrm{d}f_a.$$

Thus, the matrix representing the differential $\mathrm{d}f_a$ at a is the matrix[1] of partial derivatives of f. The relation (b.2) may also be written in the form

$$f(a+h) = f(a) + \mathrm{d}f_a.h + \underset{h\to 0}{\mathrm{o}}(h).$$

Using tensor notation, this becomes

$$f^\mu(x+h) = f^\mu(x) + h^\nu \partial_\nu f^\mu + \underset{h\to 0}{\mathrm{o}}(h) \qquad \text{and} \qquad \mathrm{d}f^\mu = \partial_\nu f^\mu \, \mathrm{d}x^\nu.$$

B.3

Lagrange multipliers

Using linear forms, it is easy to establish the validity of the *method of Lagrange multipliers.*

We start with an elementary lemma of linear algebra.

LEMMA B.3 *Let ψ and $\varphi_1, \dots, \varphi_k$ be linear forms defined on a vector space E of dimension n. If the forms $(\varphi_1, \dots, \varphi_k)$ are linearly independent and if*

$$\bigcap_{i=1}^{k} \operatorname{Ker} \varphi_i \subset \operatorname{Ker} \psi,$$

then ψ is a linear combination of the φ_i.

♦ **Exercise B.1** Prove this lemma *(see page 591).*

Now consider the following problem. We wish to find the extrema of a real-valued function $f(x_1, \dots, x_n)$ on $E = \mathbb{R}^n$. If is differentiable, we know that a *necessary* condition (which is not sufficient) for f to be extremal at $a = (a_1, \dots, a_n)$ is that

[1] If $p = n$, it is called the **Jacobian matrix** of f.

$$\mathrm{d}f_{\boldsymbol{a}} = 0, \qquad \text{that is,} \qquad \frac{\partial f}{\partial x_i}(\boldsymbol{a}) = 0 \quad \forall i \in [\![1,n]\!].$$

Now let us make the problem more complicated: we are looking for extrema of f on a (fairly regular) subset of $\mathbb{R}^n$, such as a curve in $\mathbb{R}^n$, or a surface, or some higher-dimensional analogue (what mathematicians call a **differentiable subvariety** of E. Assume that this set is of dimension $n-k$ and is defined by k equations (represented by differentiable functions), namely,

$$\mathscr{S} : \begin{cases} C^{(1)}(x_1,\dots,x_n) = 0 \\ \vdots \\ C^{(k)}(x_1,\dots,x_n) = 0. \end{cases} \tag{b.3}$$

In order that the set $\mathscr{S}$ defined by those equations be indeed "an $(n-k)$-dimensional subvariety," it is necessary that the differentials of the equations $C^{(1)},\dots,C^{(k)}$ be linearly independent linear forms at every point in $\mathscr{S}$:

$$\left(\mathrm{d}C_{\boldsymbol{x}}^{(1)},\dots,\mathrm{d}C_{\boldsymbol{x}}^{(k)}\right) \quad \text{are linearly independent for all } \boldsymbol{x} \in \mathscr{S}.$$

We are now interested in finding the points on a subvariety $\mathscr{S}$ where f has an extremum.

THEOREM B.4 (Lagrange multipliers) *Let f be a real-valued differentiable function on $E = \mathbb{R}^n$, and let $\mathscr{S}$ be a differential subvariety of E defined by equations* (b.3)*. Then, in order for $\boldsymbol{a} \in \mathscr{S}$ to be an extremum for f restricted to $\mathscr{S}$, it is* necessary *that there exist real constants $\lambda_1,\dots,\lambda_k$ such that*

$$\mathrm{d}f_{\boldsymbol{a}} = \lambda_1 \cdot \mathrm{d}C_{\boldsymbol{a}}^{(1)} + \cdots + \lambda_k \cdot \mathrm{d}C_{\boldsymbol{a}}^{(k)}.$$

PROOF. Assume that $\boldsymbol{a} \in \mathscr{S}$ is an extremum of f restricted to $\mathscr{S}$, and let U be the tangent space to $\mathscr{S}$ at $\boldsymbol{a}$. This tangent space is an $(n-k)$-dimensional vector space, where $\boldsymbol{a} \in U$ is the origin.

If $\boldsymbol{a}$ is an extremum of f on $\mathscr{S}$, we necessarily have $\mathrm{d}f_{\boldsymbol{a}}.\boldsymbol{h} = 0$ for any vector $\boldsymbol{h}$ tangent to $\mathscr{S}$ (if we move infinitesimally on $\mathscr{S}$ from $\boldsymbol{a}$, the values of f cannot increase or decrease in any direction). Thus we have $U \subset \operatorname{Ker} \mathrm{d}f_{\boldsymbol{a}}$.

On the other hand, U is defined by the intersection of the tangent planes to the subvarieties with equations $C^{(i)}(\boldsymbol{x}) = 0$ for $1 \leqslant i \leqslant k$ (because $\mathscr{S}$ is the intersection of those). Hence we have $U = \bigcap_{i=1}^k \operatorname{Ker} \mathrm{d}C_{\boldsymbol{a}}^{(i)}$, and therefore

$$\bigcap_{i=1}^{k} \operatorname{Ker} \mathrm{d}C_{\boldsymbol{a}}^{(i)} \subset \operatorname{Ker} \mathrm{d}f_{\boldsymbol{a}}.$$

There only remains to apply Lemma B.3.

In order to find the points $\boldsymbol{a} \in \mathscr{S}$ where f may have an extremum, one solves the system of linear equations $\mathrm{d}f_{\boldsymbol{a}} = \sum \lambda_i \cdot \mathrm{d}C_{\boldsymbol{a}}^{(i)}$, where λ_i are unknowns, together with the equations expressing that $\boldsymbol{a} \in \mathscr{A}$. In other words, the system to solve is

$$\begin{cases} \dfrac{\partial f}{\partial x_1}(\boldsymbol{a}) - \displaystyle\sum_{i=1}^{k} \lambda_i \frac{\partial C^{(i)}}{\partial x_1}(\boldsymbol{a}) = 0, \\ \qquad \vdots \qquad\qquad \vdots \\ \dfrac{\partial f}{\partial x_n}(\boldsymbol{a}) - \displaystyle\sum_{i=1}^{k} \lambda_i \frac{\partial C^{(i)}}{\partial x_n}(\boldsymbol{a}) = 0 \end{cases} \quad \text{and} \quad \begin{cases} C^{(1)}(\boldsymbol{a}) = 0, \\ \quad \vdots \\ C^{(k)}(\boldsymbol{a}) = 0, \end{cases} \tag{b.4}$$

where there are $n+k$ equations, with $n+k$ unknowns, the coordinates of $\boldsymbol{a} = (a_1, \ldots, a_n)$ and the auxiliary unknowns $\lambda_1, \ldots, \lambda_k$, called the **Lagrange multipliers**. Those auxiliary unknowns have the advantage of preserving symmetry among all variables. Moreover, they sometimes have physical significance (for instance, temperature in thermodynamics is the inverse of a Lagrange multiplier).

Physicists usually proceed as follows: introduce k multipliers $\lambda_1, \ldots, \lambda_k$ and construct an auxiliary function

$$F(\boldsymbol{x}) \stackrel{\text{def}}{=} f(\boldsymbol{x}) - \sum_{i=1}^{k} \lambda_i \, C^{(i)}(\boldsymbol{x}).$$

Look then for points $\boldsymbol{a} \in \mathbb{R}^n$ giving *free* extrema of F, that is, where the n variables $x_1, \ldots, x_n$ are independent (do not necessarily lie on $\mathscr{S}$). The condition that the differential of F vanishes gives the first set of equations in (b.4), and of course the solutions depend on the additional parameters λ_i. The values for those are found at the end, using the constraints on the problem, namely, the second set of equations in (b.4).

Example B.5 Let $\mathscr{S}$ be a surface in $\mathbb{R}^3$ with equation $C(x,y,z) = 0$. Let $\boldsymbol{r}_0 \in \mathbb{R}^3$ be a point not in $\mathscr{S}$, and consider the problem of finding the points of $\mathscr{S}$ closest to $\boldsymbol{r}_0$, that is, those minimizing the function $f(\boldsymbol{x}) = \|\boldsymbol{x} - \boldsymbol{r}_0\|^2$, with the constraint $C(\boldsymbol{x}) = 0$. The auxiliary function is $F(\boldsymbol{x}) = \|\boldsymbol{x} - \boldsymbol{r}_0\|^2 - \lambda C(\boldsymbol{x})$, with differential (at a point $\boldsymbol{a} \in \mathbb{R}^3$) given by

$$\mathrm{d}F_{\boldsymbol{a}} = 2(\boldsymbol{a} - \boldsymbol{r}_0|\cdot) - \lambda(\mathbf{grad}\, C(\boldsymbol{a})|\cdot),$$

in other words,

$$\mathrm{d}F_{\boldsymbol{a}} . \boldsymbol{b} = 2(\boldsymbol{a} - \boldsymbol{r}_0) \cdot \boldsymbol{b} - \lambda\, \mathbf{grad}\, C(\boldsymbol{a}) \cdot \boldsymbol{b} \qquad \text{for all } \boldsymbol{b} \in \mathbb{R}^3.$$

The free extrema of F are the points $\boldsymbol{a}$ such that $(\boldsymbol{a} - \boldsymbol{r}_0)$ is colinear with the gradient of C at $\boldsymbol{a}$. Using the condition $C(\boldsymbol{a}) = 0$, the values of λ are found. Since $\mathbf{grad}\, C(\boldsymbol{a})$ is a vector perpendicular to the surface $\mathscr{S}$, the geometric interpretation of the necessary condition is that points of $\mathscr{S}$ closest to $\boldsymbol{r}_0$ are those for which $(\boldsymbol{a} - \boldsymbol{r}_0)$ is perpendicular to the surface at the point $\boldsymbol{a}$.

Of course, this is a necessary condition, but not necessarily a sufficient condition (the distance may be maximal at the points found in this manner, or there may be a saddle point).

Remark B.6 Introducting the constraint multiplied by an additional parameter in the auxiliary function in order to satisfy this constraint is a trick used in electromagnetism in order to fix the gauge. In the Lagrangian formulation of electromagnetism, the goal is indeed to minimize the action given by the time integral of the Lagrangian under the constraint that the potential four-vector has a fixed gauge. A gauge-fixing term, for instance $\lambda(\partial \cdot A)^2$ in Lorentz gauge, is added to the Lagrangian density (see [21,49]).

SOLUTION

♦ **Solution of exercise B.1 on page 588.** By linear algebra, there exist additional linear forms $\varphi_{k+1}, \dots, \varphi_n$ such that $(\varphi_1, \dots, \varphi_n)$ is a basis of E^*, since the given forms are linearly independent. Let $(e_1, \dots, e_n)$ be the dual basis. By the definition of a basis, there are constants λ_i, $1 \leqslant i \leqslant n$, such that $\psi = \sum_{i=1}^{n} \lambda_i \varphi_i$. Indeed, we have $\lambda_i = \psi(e_i)$ using the formula $\varphi_m(e_\ell) = \delta_{m\ell}$ defining the dual basis. Let $j = i + 1, \dots, n$. The same formula gives $e_j \in \bigcap_{1 \leqslant i \leqslant k} \operatorname{Ker} \varphi_i \subset \operatorname{Ker} \psi$ by assumption, hence $\lambda_j = 0$ for $j > i$. This implies $f \in \operatorname{Vect}(\varphi_1, \dots, \varphi_k)$, as desired.

Appendix C

Matrices

In this short appendix, we recall the formalism of duality (duality bracket or pairing) and show how it can be used easily to recover the formulas of change of basis, and to better understand their origin; this can be considered as a warm-up for the chapter on the formalism of tensors, which generalizes these results.

C.1 Duality

Let E be a finite-dimensional vector space over the field $\mathbb{K} = \mathbb{R}$ or $\mathbb{C}$, of dimension n.

DEFINITION C.1 A **linear form** on E is a linear map from E to $\mathbb{K}$. The **dual space** of E is the space $E^* = \mathscr{L}(E, \mathbb{K})$ of linear forms on E. For $\varphi \in E^*$ a linear form and $x \in E$ a vector, we denote

$$\langle \varphi, x \rangle \stackrel{\text{def}}{=} \varphi(x).$$

The map $\langle \cdot, \cdot \rangle : E \times E^* \to \mathbb{K}$ is called the **duality bracket** or **duality pairing**.

DEFINITION C.2 (Dual basis) Let $\mathcal{E} = (e_1, \dots, e_n)$ be a basis of E. There is a unique family $\mathcal{E}^* = (e_1^*, \dots, e_n^*)$ of linear forms on E such that

$$\forall i, j \in [\![1, n]\!] \qquad \langle e_i^*, e_j \rangle = \delta_{ij}.$$

The family $\mathcal{E}^*$ is a basis of E^*, and is called the **dual basis** of $\mathcal{E}$.

For $x = \sum_{i=1}^n x_i\, e_i$, we have $x_i = \langle e_i^*, x \rangle$ for all $i \in [\![1, n]\!]$. The elements of the dual basis $\mathcal{E}^*$ are also called **coordinate forms**.

C.2
Application to matrix representation

We will now explain how the formalism of linear forms can be used to clarify the use of matrices to represent vectors, and how to use them to remember easily the formulas of change of basis.

C.2.a Matrix representing a family of vectors

DEFINITION C.3 Let $\mathcal{E} = (e_1, \dots, e_n)$ be a basis of E and let $(x_1, \dots, x_q)$ be q vectors in E. The **matrix representing the family $(x_j)_{1\leqslant j\leqslant q}$ in the basis $\mathcal{E}$** is the $(n \times q)$-matrix $M = (a_{ij})_{\substack{1\leqslant i\leqslant n\\ 1\leqslant j\leqslant q}}$ such that

> for all $j \in [\![1,q]\!]$, the column coefficients $a_{1j}, \dots, a_{nj}$ are the coefficients of x_j expressed in the basis $\mathcal{E}$.

In other words, the vector x_j is expressed in the basis $\mathcal{E}$ by the linear combination

$$x_j = \sum_{i=1}^{n} a_{ij}\, e_i.$$

$$M = \begin{pmatrix} a_{11} & \cdots & \boxed{a_{1j}} & \cdots & a_{1q} \\ a_{21} & \cdots & a_{2j} & \cdots & a_{2q} \\ \vdots & & \vdots & & \vdots \\ a_{n1} & \cdots & a_{nj} & \cdots & a_{nq} \end{pmatrix}$$

↑ vector x_j "seen in $\mathcal{E}$"

Using the dual basis, we can state the formula

$$a_{ij} = e_i^*(x_j) = \langle e_i^*, x_j \rangle$$

for all $i \in [\![1,n]\!]$ and $j \in [\![1,q]\!]$. We can express the definition $M = \big(\langle e_i^*, x_j\rangle\big)_{ij}$ symbolically in the form

$$\boxed{M = \mathrm{mat}_{\mathcal{E}}(x_1, \dots, x_q) = \langle e^* | x \rangle}.$$

C.2.b Matrix of a linear map

As before, let E be a $\mathbb{K}$ vector space with a basis $\mathcal{B} = (b_1, \dots, b_n)$; in addition, let F be another $\mathbb{K}$-vector space with basis $\mathcal{C} = (c_1, \dots, c_p)$. We denote $\mathcal{C}^* = (c_1^*, \dots, c_p^*)$ the basis dual to $\mathcal{B}$.

Let $f \in \mathscr{L}(E, F)$ be a linear map from E to F, and let $A = \text{mat}_{\mathcal{B},\mathcal{C}}(f)$ be the matrix representing the map f in the basis $\mathcal{B}$ and $\mathcal{C}$. Recall that, by definition, this means that the first column of A contains the coefficients expressing the image $f(b_1)$ of the first basis vector of $\mathcal{B}$ in the basis $\mathcal{C}$, and so on: the j-th column contains the coefficients of $f(b_j)$ in the basis $\mathcal{C}$.

It follows that for $i \in [\![1, p]\!]$ and $j \in [\![1, n]\!]$, the coefficient A_{ij} is the i-th coordinate of $f(e_j)$ in the basis $\mathcal{C}$, that is, we have

$$A_{ij} = c_i^*\big(f(b_j)\big) = \big\langle c_i^*, f(b_j)\big\rangle.$$

This is denoted also

$$\boxed{A = \big\langle c^* \big| f \cdot b\big\rangle.} \tag{$*$}$$

C.2.c Change of basis

Now let $\mathcal{B}' = (b'_1, \ldots, b'_n)$ be another basis of the space E. The **change of basis matrix** from $\mathcal{B}$ to $\mathcal{B}'$, denoted $P = \text{Pass}(\mathcal{B}, \mathcal{B}')$, is defined as follows: the columns of P represent the vectors of $\mathcal{B}'$ expressed in the basis $\mathcal{B}$. In other words, we have

$$P_{ij} = i\text{-th coordinate of } b'_j = b_i^*(b'_j).$$

for all $(i, j) \in [\![1, n]\!]$.

We denote

$$\boxed{P = \text{Pass}(\mathcal{B}, \mathcal{B}') = \big\langle b^* \big| b'\big\rangle}$$

Notice that the inverse of the matrix P, which is simply the change of basis matrix from $\mathcal{B}'$ to $\mathcal{B}$, is

$$\boxed{\big\langle b^* \big| b'\big\rangle^{-1} = \big\langle b'^* \big| b\big\rangle.}$$

C.2.d Change of basis formula

Let again $\mathcal{B} = (b_1, \ldots, b_n)$ and $\mathcal{B}' = (b'_1, \ldots, b'_n)$ be two bases of E, and let $x \in E$ be a vector. The change of basis formula

$$\text{mat}_{\mathcal{B}}(x) = \text{Pass}(\mathcal{B}, \mathcal{B}') \cdot \text{mat}_{\mathcal{B}'}(x)$$

expresses the relation between the coordinates for x in each basis. In abbreviated form, we write

$$X = PX'$$

or, with the previous notation,

$$\boxed{\langle b^* | x\rangle = \big\langle b^* \big| b'\big\rangle \cdot \big\langle b'^* \big| x\big\rangle.}$$

To remember this formula, it suffices to use the relation

$$\mathrm{Id} = \sum_{k=1}^{n} \left|b'_k\right\rangle \cdot \left\langle b'^*_k\right|$$

or its short form

$$\mathrm{Id} = \left|b'\right\rangle \cdot \left\langle b'^*\right| \tag{c.1}$$

to recover the formula.

Let now $f : E \to F$ be a linear map. In addition to the basis $\mathcal{B}$ and $\mathcal{B}'$ of E, let $\mathcal{C}$ and $\mathcal{C}'$ be basis of F, and let

$$M = \mathrm{mat}_{\mathcal{B},\mathcal{C}}(f) \qquad \text{and} \qquad M' = \mathrm{mat}_{\mathcal{B}',\mathcal{C}'}(f).$$

If we denote by P and Q respectively the change of basis matrices for E and F, i.e.,

$$P = \mathrm{Pass}(\mathcal{B},\mathcal{B}') = \left\langle b^* \middle| b'\right\rangle \qquad \text{and} \qquad Q = \mathrm{Pass}(\mathcal{C},\mathcal{C}') = \left\langle c^* \middle| c'\right\rangle,$$

the change of basis formula relating M and M' can be recovered by writing

$$M' = \left\langle c'^* \middle| f \cdot b'\right\rangle = \underbrace{\left\langle c'^* \middle| c\right\rangle}_{Q^{-1}} \underbrace{\left\langle c^* \middle| f \cdot b\right\rangle}_{M} \underbrace{\left\langle b^* \middle| b'\right\rangle}_{P} = Q^{-1} M P \tag{c.2}$$

One should remember that (c.1) provides a way to recover without mistake the change of basis formula for matrices, in the form

$$\boxed{\left\langle c'^* \middle| f \cdot b'\right\rangle = \left\langle c'^* \middle| c\right\rangle \left\langle c^* \middle| f \cdot b\right\rangle \left\langle b^* \middle| b'\right\rangle.}$$

♦ **Exercise C.1** The trace of an endomorphism $f \in \mathscr{L}(E)$ is defined by the formula

$$\mathrm{tr}(f) = \sum_{i=1}^{n} \left\langle e_i^*, f(e_i)\right\rangle$$

in a basis $\mathcal{E} = (e_1, \dots, e_n)$. Show that this definition is independent of the chosen basis.

C.2.e Case of an orthonormal basis

When E has a euclidean or Hilbert space structure and the basis $\mathscr{E} = (e_1, \dots, e_n)$ is an orthonormal basis, the coordinate forms e_i^* are associated with the orthogonal projections on the lines spanned by the vectors e_i: we have

$$\forall x \in E \quad \forall i \in [\![1,n]\!] \qquad e_i^*(x) = (e_i|x).$$

The formula of the previous section, provided all bases involved are orthonormal, becomes

$$\left(c' \middle| f \cdot b'\right) = \left(c' \middle| c\right)\left(c \middle| f \cdot b\right)\left(b \middle| b'\right).$$

or, with all indices spelled out precisely

$$\left(c'_i \middle| f(b'_j)\right) = \sum_{k=1}^{n} \sum_{\ell=1}^{n} \left(c'_i \middle| c_k\right) \left(c_k \middle| f(b_\ell)\right) \left(b_\ell \middle| b'_j\right).$$

A few proofs

This appendix collects a few proofs which were too long to insert in the main text, but are interesting for various reasons; they may be intrinsically beautiful, or illustrate important mathematical techniques which physicists may not have other opportunities to discover.

THEOREM 4.27 on page 96 (Winding number) *Let γ be a closed path in $\mathbb{C}$, U the complement of the image of γ in the complex plane. The* ***winding number of γ around the point z***,

$$\operatorname{Ind}_\gamma(z) \stackrel{\text{def}}{=} \frac{1}{2\pi i} \int_\gamma \frac{d\zeta}{\zeta - z} \qquad \text{for all } z \in U \tag{d.1}$$

is an integer, and as a function of z it is constant on each connected component of U, and is equal to zero on the unbounded connected component of U.

Let $z \in \Omega$. Assume that γ is parameterized by $[0,1]$, so that

$$\operatorname{Ind}_\gamma(z) = \frac{1}{2i\pi} \int_0^1 \frac{\gamma'(t)}{\gamma(t) - z}\, dt. \tag{$*$}$$

If w is a complex number, it is equivalent to show that w is an integer or to show that $e^{2i\pi w} = 1$. Hence, to prove that the winding number $\operatorname{Ind}_\gamma(z)$ is an integer, it suffices to prove that $\varphi(1) = 1$, where

$$\varphi(t) \stackrel{\text{def}}{=} \exp\left\{ \int_0^t \frac{\gamma'(s)}{\gamma(s) - z}\, ds \right\} \qquad (0 \leqslant t \leqslant 1).$$

Now, we can compute easily that $\varphi'(t)/\varphi(t) = \gamma'(t)/\big(\gamma(t) - z\big)$, for all $t \in [0,1]$, except possibly at the finitely many points $\mathscr{S} \subset [0,1]$ where the curve γ is not differentiable. Hence we find

$$\left(\frac{\varphi(t)}{\gamma(t)-z}\right)' = \frac{\varphi'(t)}{\gamma(t)-z} - \frac{\varphi(t)\gamma'(t)}{\big(\gamma(t)-z\big)^2} = \frac{\varphi(t)\gamma'(t)}{\big(\gamma(t)-z\big)^2} - \frac{\varphi(t)\gamma'(t)}{\big(\gamma(t)-z\big)^2} = 0$$

on $[0,1] \setminus \mathscr{S}$. Because the set $\mathscr{S}$ is *finite*, it follows that $\varphi(t)/\big(\gamma(t)-z\big)$ is a constant. In addition, we have obviously $\varphi(0) = 1$, and hence

$$\varphi(t) = \frac{\gamma(t)-z}{\gamma(0)-z}.$$

Finally, since $\gamma(1) = \gamma(0)$, we obtain $\varphi(1) = 1$ as desired.

It is clear from the definition of Ind_γ as function of z that it is a continuous function on Ω. But a continuous function defined on Ω which takes integral values is well known to be constant on each connected component of Ω.

Finally, using the definition $(*)$ it follows that $\mathrm{Ind}_\gamma(z) < 1$ if $|z|$ is large enough, and therefore the integer Ind_γ must be zero on the unbounded component of Ω.

THEOREM 4.31 on page 97 (Cauchy for triangles) *Let Δ be a closed triangle in the plane (by this is meant a "filled" triangle, with boundary $\partial\Delta$ consisting of three line segments), entirely contained in an open set Ω. Let $p \in \Omega$, and let $f \in \mathscr{H}(\Omega \setminus \{p\})$ be a function continuous on Ω and holomorphic on Ω except possibly at the point p. Then*

$$\int_{\partial\Delta} f(z)\,\mathrm{d}z = 0.$$

Let $[a,b]$, $[b,c]$, and $[c,a]$ denote the segments which together form the boundary $\partial\Delta$ of the filled triangle Δ. We can assume that a,b,c are not on a line, since otherwise the result is obvious. We now consider three cases: ① $p \notin \Delta$, ② p is a vertex, ③ p is inside the triangle, but not a vertex.

① *p is not in* Δ. Let c' be the middle of $[a,b]$, and a' and b' the middle, respectively, of $[b,c]$ and $[c,a]$. We can subdivide Δ into four smaller similar triangles Δ_i, $i = 1,\dots,4$, as in the picture, and we have

$$J \stackrel{\text{def}}{=} \int_{\partial\Delta} f(z)\,\mathrm{d}z = \sum_{i=1}^{4} \int_{\partial\Delta_i} f(z)\,\mathrm{d}z$$

(the integrals on $[a',b']$, $[b',c']$, and $[c',a']$ which occur on the right-hand side do so in opposite pairs because of orientation). It follows that for one index $i \in \{1,2,3,4\}$ at least, we have $\left|\int_{\Delta_i} f(z)\,\mathrm{d}z\right| \geqslant |J/4|$. Let $\Delta^{(1)}$ denote one triangle (any will do) for which this inequality holds. Repeating the reasoning above with this triangle, and so on, we obtain a sequence $(\Delta^{(n)})_{n\in\mathbb{N}}$ of triangles such that $\Delta \supset \Delta^{(1)} \supset \Delta^{(2)} \supset \cdots \supset \Delta^{(n)} \supset \cdots$ and

$$\left|\int_{\partial\Delta^{(n)}} f(z)\,\mathrm{d}z\right| \geqslant \frac{|J|}{4^n}.$$

at each step. Let L be the length of the boundary $\partial\Delta$. Then it is clear that the length of $\partial\Delta^{(n)}$ is equal to $L/2^n$ for $n \geqslant 1$.

Finally, because the sequence of triangles is a decreasing (for inclusion) sequence of compact sets with diameter tending to 0, there exists a unique intersection point:

$$\bigcap_{i\geqslant 1} \Delta^{(i)} = \{z_0\}.$$

By definition, we have $z_0 \in \Delta$ and by assumption, f is therefore holomorphic at z_0.

Now we can show that $J = 0$.

Let $\varepsilon > 0$. There exists a positive real number $r > 0$ such that

$$|f(z) - f(z_0) - f'(z_0)(z - z_0)| \leqslant \varepsilon\,|z - z_0|$$

for all $z \in \mathbb{C}$ such that $|z - z_0| < r$. Since the diameter of $\Delta^{(n)}$ tends to zero, we have $|z - z_0| < r$ for all $z \in \Delta^{(n)}$ if n is large enough. Using Proposition 4.30, we have

$$\int_{\partial\Delta^{(n)}} f(z_0)\,\mathrm{d}z = f(z_0)\int_{\partial\Delta^{(n)}} \mathrm{d}z = 0 \qquad \text{and} \qquad \int_{\partial\Delta^{(n)}} f'(z_0)\,(z - z_0)\,\mathrm{d}z = 0.$$

Hence

$$\int_{\partial\Delta^{(n)}} f(z)\,\mathrm{d}z = \int_{\partial\Delta^{(n)}} [f(z) - f(z_0) - f'(z_0)\,(z - z_0)]\,\mathrm{d}z,$$

which proves that

$$\left|\int_{\partial\Delta^{(n)}} f(z)\,\mathrm{d}z\right| \leqslant \varepsilon\frac{L}{2^n}\frac{L}{2^n} = \varepsilon\frac{L^2}{4^n},$$

and therefore

$$J \leqslant 4^n \left|\int_{\partial\Delta^{(n)}} f(z)\,\mathrm{d}z\right| \leqslant \varepsilon L^2.$$

This holds for all $\varepsilon > 0$, and consequently we have $J = 0$ in the case where $p \notin \Delta$.

② *p is a vertex, for instance $p = a$.* Let $x \in [a,b]$ and $y \in [a,c]$ be two arbitrarily chosen points. According to the previous case, we have

$$\int_{[xyb]} f(z)\,\mathrm{d}z = \int_{[ybc]} f(z)\,\mathrm{d}z = 0$$

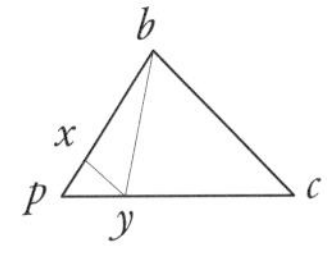

and hence

$$\int_{\partial\Delta} f(z)\,\mathrm{d}z = \int_{[a,x]} + \int_{[x,y]} + \int_{[y,a]} f(z)\,\mathrm{d}z,$$

which tends to 0 as $[x \to a]$ and $[y \to a]$, since f is bounded, and the length of the path of integration tends to 0.

③ *p is inside the triangle, but is not a vertex.* It suffices to use the previous result applied to the three triangles $[a,b,p]$, $[b,p,c]$, and $[a,p,c]$.

THEOREM 7.19 on page 187 (Cauchy principal value) *Let $\varphi \in \mathscr{D}(\mathbb{R})$ be a test function. The quantity*

$$\int_{-\infty}^{-\eta} \frac{\varphi(x)}{x}\,\mathrm{d}x + \int_{+\eta}^{+\infty} \frac{\varphi(x)}{x}\,\mathrm{d}x$$

has a finite limit as $[\eta \to 0^+]$. Moreover, the map

$$\varphi \longmapsto \lim_{\eta\to 0^+} \int_{|x|>\eta} \frac{\varphi(x)}{x}\,\mathrm{d}x \tag{d.2}$$

is linear and continuous on $\mathscr{D}(\mathbb{R})$.

Let $\varphi \in \mathscr{D}$, so φ is of $\mathscr{C}^\infty$ class and has bounded support. In particular the integral $\int_{|x|>\eta} \varphi(x)/x\,\mathrm{d}x$ exists for all $\eta > 0$. Let $[-M, M]$ be an interval containing the support of φ. For all $x \in [-M, M]$, we have $\varphi(x) - \varphi(0) = x\big[\varphi'(0) + \omega(x)\big]$, where ω is a function such that $\lim_{x\to 0} \omega(x) = 0$.

Notice that
$$\int_{\eta<|x|<M} \frac{\varphi(0)}{x}\,\mathrm{d}x = 0 \qquad \text{for all } \eta > 0,$$

so that, by parity, we have

$$f(\eta) \stackrel{\text{def}}{=} \int_{\eta<|x|<M} \frac{\varphi(x)}{x}\,\mathrm{d}x = \int_{\eta<|x|<M} \frac{x\big[\varphi'(0)+\omega(x)\big]}{x}\,\mathrm{d}x$$
$$= \int_{\eta<|x|<M} \big[\varphi'(0)+\omega(x)\big]\,\mathrm{d}x,$$

and as $[\eta \to 0^+]$ this last expression obviously has a limit.

We know must show that the linear map defined for test functions by (d.2) is continuous on $\mathscr{D}(\mathbb{R})$.

Let $(\varphi_n)_{n\in\mathbb{N}}$ be a sequence of test functions converging to 0. By definition, there exists a fixed interval $[-M, M]$ containing the support of *all* the φ_n.

□ Let $\varepsilon > 0$. There exists an integer N such that, if $n \geqslant N$, we have $\|\varphi_n\|_\infty \leqslant \varepsilon/M$, $\|\varphi_n'\|_\infty \leqslant \varepsilon$, and $\|\varphi_n''\|_\infty \leqslant \varepsilon$. Moreover, the Taylor-Lagrange inequality shows that for $n \in \mathbb{N}$ and $x \in \mathbb{R}$, we have

$$|\varphi_n(x) - \varphi_n(0) - x\,\varphi_n'(0)| \leqslant \frac{x^2}{2}\,\|\varphi''\|_\infty.$$

Let $n \geqslant N$ be an integer. Choose, arbitrarily for the moment, a real number $\eta \in\,]0,1]$, and consider

$$\int_{|x|>\eta} \frac{\varphi_n(x)}{x}\,\mathrm{d}x = \int_{\eta<|x|<1} \frac{\varphi_n(x)}{x}\,\mathrm{d}x + \int_{|x|>1} \frac{\varphi_n(x)}{x}\,\mathrm{d}x.$$

The second integral is bounded by $2M \times \|\varphi_n\|_\infty \leqslant 2\varepsilon$. Using the symmetry of the interval in the first integral, we can write

$$\int_{\eta<|x|<1} \frac{\varphi_n(x)}{x}\,\mathrm{d}x = \int_{\eta<|x|<1} \frac{\varphi_n(x)-\varphi_n(0)}{x}\,\mathrm{d}x.$$

But we have

$$\left|\int_{\eta<|x|<1} \frac{\varphi_n(x)-\varphi_n(0)}{x}\,\mathrm{d}x\right| \leqslant \left|\int_{\eta<|x|<1} \frac{\varphi_n(x)-\varphi_n(0)-x\varphi_n'(0)}{x}\,\mathrm{d}x\right| + \left|\int_{\eta<|x|<1} \frac{x\varphi_n'(0)}{x}\,\mathrm{d}x\right|$$
$$\leqslant \|\varphi_n''\|_\infty \int_{\eta<|x|<1} \left|\frac{x}{2}\right|\,\mathrm{d}x + \|\varphi_n'\|_\infty \leqslant 2\varepsilon.$$

This is valid for any $\eta \in \,]0,1]$. Letting then $[\eta \to 0^+]$ for fixed n, we find that

$$\left| \lim_{\eta \to 0^+} \int_{|x|>\eta} \frac{\varphi_n(x)}{x}\,\mathrm{d}x \right| \leqslant 4\varepsilon \qquad \forall n \geqslant N. \ \square$$

This now proves that the map (d.2) is continuous.

THEOREM 8.18 on page 232 *Any Dirac sequence converges weakly to the Dirac distribution δ in $\mathscr{D}'(\mathbb{R})$.*

Let $(f_n)_{n\in\mathbb{N}}$ be a Dirac sequence and let $A > 0$ be such that f_n is non-negative on $[-A, A]$ for all n.

Let $\varphi \in \mathscr{D}$ be given. We need to show that $\langle f_n, \varphi \rangle \longrightarrow \varphi(0)$. First, let $\psi \stackrel{\text{def}}{=} \varphi - \varphi(0)$, so that ψ is a function of $\mathscr{C}^\infty$ class, with bounded support, with $\psi(0) = 0$. Let K be the support of φ.

$\square$ Let $\varepsilon > 0$. Since ψ is continuous, there exists a real number $\eta > 0$ such that

$$|\psi(x)| \leqslant \varepsilon$$

for all real numbers x with $|x| < \eta$. We can of course assume that $\eta < A$. Because of Condition ② in the definition 8.15 on page 231 of a Dirac sequence, there exists an integer N such that

$$\left| \int_{|x|>\eta} f_n(x)\,\psi(x)\,\mathrm{d}x \right| \leqslant \varepsilon$$

for all $n \geqslant N$. Because f_n is non-negative on $[-\eta, \eta]$, we can also write

$$\left| \int_{|x|\leqslant\eta} f_n(x)\,\psi(x)\,\mathrm{d}x \right| \leqslant \varepsilon \int_{|x|\leqslant\eta} |f_n(x)|\,\mathrm{d}x = \varepsilon \int_{|x|\leqslant\eta} f_n(x)\,\mathrm{d}x.$$

Again from the definition, the last integral tends to 1, so that if n is large enough we have

$$\left| \int_{|x|\leqslant\eta} f_n(x)\,\psi(x)\,\mathrm{d}x \right| \leqslant 2\varepsilon, \qquad \text{hence} \qquad \left| \int f_n(x)\,\psi(x)\,\mathrm{d}x \right| \leqslant 3\varepsilon.$$

Now, there only remains to notice that $\int_K f_n(x)\,\mathrm{d}x \xrightarrow[n\to\infty]{} 1$ (since K is a bounded set), and then for all n large enough, we obtain

$$\left| \int f_n(x)\,\varphi(x)\,\mathrm{d}x - \varphi(0) \right| \leqslant 4\varepsilon. \ \square$$

Since φ is an arbitrary test function, this finally shows that $f_n \xrightarrow{\mathscr{D}'} \delta$.

THEOREM 9.28 on page 261 *The space ℓ^2, with the hermitian product*

$$(a|b) = \sum_{n=0}^{\infty} \overline{a_n}\, b_n,$$

is a Hilbert space.

The inequality $|\alpha\beta| \leqslant \frac{1}{2}(\alpha^2 + \beta^2)$ shows that ℓ^2 is a vector space. It is also immediate to check that the formula above defines a hermitian product on ℓ^2. The hard part is to prove that ℓ^2 is a complete space for the norm $\|\cdot\|_2$ associated to this hermitian product.

Let $(u^{(p)})_{p\in\mathbb{N}}$ be a Cauchy sequence in ℓ^2. We need to prove that it converges in ℓ^2. There are three steps: first a candidate u is found that should be the limit of $(u^{(p)})_{p\in\mathbb{N}}$; then it is proved that u is an element of ℓ^2; and finally, the convergence of $(u^{(p)})_{p\in\mathbb{N}}$ to u in ℓ^2 is established.

• **Looking for a limit** Since the sequence $(u^{(p)})_{p\in\mathbb{N}}$ is a Cauchy sequence, it is obvious that for any $p \in \mathbb{N}$, the sequences of complex numbers $(u^{(p)}{}_n)_{n\in\mathbb{N}}$ is a Cauchy sequence in $\mathbb{C}$. Since $\mathbb{C}$ is complete, this sequence converges, and we may define

$$u_n \stackrel{\text{def}}{=} \lim_{p\to\infty} u_n^{(p)}.$$

Obviously, the sequence $(u_n)_{n\in\mathbb{N}}$ of complex numbers should be the limit of $(u^{(p)})_p$.

• **The sequence u is in ℓ^2** The Cauchy sequence $(u^{(p)})_{p\in\mathbb{N}}$, like any other Cauchy sequence, is bounded in ℓ^2: there exists $A > 0$ such that

$$\left\|u^{(p)}\right\|_2 = \sum_{n=0}^{\infty} \left|u_n^{(p)}\right|^2 \leqslant A$$

for all $p \in \mathbb{N}$. Let N be an integer $N \in \mathbb{N}$. We obtain in particular

$$\forall p \in \mathbb{N}, \quad \sum_{n=0}^{N} \left|u_n^{(p)}\right|^2 \leqslant \left\|u^{(p)}\right\|_2 \leqslant A,$$

and as $[p \to \infty]$, we find

$$\sum_{n=0}^{N} \left|\lim_{p\to\infty} u_n^{(p)}\right|^2 = \sum_{n=0}^{N} |u_n|^2 \leqslant A$$

(since the sum over n is finite, exchanging limit and sum is obviously permitted). Now this holds for all $N \in \mathbb{N}$, and it follows that $u \in \ell^2$ and in fact $\|u\|_2 \leqslant A$.

• **The sequence $u^{(p)}$ tends to u in ℓ^2** that is, we have $\left\|u^{(p)} - u\right\|_2 \xrightarrow[p\to\infty]{} 0$.

□ Let $\varepsilon > 0$. There exists an integer $N \in \mathbb{N}$ such that $\left\|u^{(p)} - u^{(q)}\right\|_2 \leqslant \varepsilon$ if $p, q \geqslant N$. Let $p \geqslant N$. Then for all $q \geqslant p$ and all $k \in \mathbb{N}$, we have

$$\sum_{n=0}^{k} \left|u_n^{(p)} - u_n^{(q)}\right|^2 \leqslant \left\|u^{(p)} - u^{(q)}\right\|_2 \leqslant \varepsilon.$$

We fix k, and let $[q \to \infty]$, deriving that

$$\sum_{n=0}^{k} \left|u_n^{(p)} - u_n\right|^2 \leqslant \varepsilon$$

(as before, the sum is finite, so exchanging limit and sum is possible). Again, this holds for all $k \in \mathbb{N}$, so in the limit $[k \to \infty]$, we obtain $\left\|u^{(p)} - u\right\|_2 \leqslant \varepsilon$. □

All this shows that $\left\|u^{(p)} - u\right\|_2 \xrightarrow[p\to\infty]{} 0$, and concludes the proof.

THEOREM 12.36 on page 349

$$\mathscr{F}^{-1}\left[\frac{\sin kct}{kc}\right] = \frac{1}{4\pi cr}\Big[\delta(r-ct) - \delta(r+ct)\Big],$$

$$\mathscr{F}^{-1}\left[\cos kct\right] = \begin{cases} \dfrac{1}{4\pi r}\delta'(r-ct) & \text{if } t > 0, \\ \delta(\boldsymbol{r}) & \text{if } t = 0. \end{cases}$$

We prove the second formula. It is obviously correct for $t = 0$, and if $t > 0$, we compute

$$\begin{aligned}
\mathscr{F}^{-1}\left[\cos kct\right](\boldsymbol{r}) &= \frac{1}{(2\pi)^3}\iiint \cos(kct)\,\mathrm{e}^{\mathrm{i}\boldsymbol{k}\cdot\boldsymbol{r}}\,\mathrm{d}^3\boldsymbol{k} \\
&= \frac{1}{(2\pi)^3}\int_0^{2\pi}\mathrm{d}\varphi\int_0^{\pi}\mathrm{d}\theta\int_0^{+\infty}\mathrm{d}k\,\cos(kt)\,\mathrm{e}^{\mathrm{i}kr\cos\theta}k^2\sin\theta \\
&= \frac{1}{(2\pi)^2}\int_0^{+\infty}\cos(kt)\frac{(\mathrm{e}^{\mathrm{i}kr} - \mathrm{e}^{-\mathrm{i}kr})}{\mathrm{i}kr}k^2\,\mathrm{d}k \\
&= \frac{1}{8\pi^2 r}\int_0^{+\infty}\left(\mathrm{e}^{\mathrm{i}k(t-r)} - \mathrm{e}^{-\mathrm{i}k(t-r)} + \mathrm{e}^{\mathrm{i}k(t+r)} - \mathrm{e}^{-\mathrm{i}k(t+r)}\right)\mathrm{i}k\,\mathrm{d}k \\
&= \frac{1}{8\pi^2 r}\int_{-\infty}^{+\infty}\left(\mathrm{e}^{\mathrm{i}k(t-r)} + \mathrm{e}^{\mathrm{i}k(t+r)}\right)\mathrm{i}k\,\mathrm{d}k \qquad \text{by parity} \\
&= \frac{1}{4\pi r}\mathscr{F}^{-1}\left[\mathrm{i}k\widetilde{\delta(r-t)} + \widetilde{\delta(r+t)}\right] = \frac{1}{4\pi r}\left[\delta'(r-t) + \delta'(r+t)\right]
\end{aligned}$$

by the properties of Fourier transforms of derivatives. Since $t > 0$, the variable r is positive and the distribution $\delta'(r+t)$ is identically zero, leading to the result stated.

The first formula is obtained in the same manner.

THEOREM 16.18 on page 439 (Existence and uniqueness of the tensor product) *Let E and F be two finite-dimensional vector spaces over $\mathbb{K} = \mathbb{R}$ or $\mathbb{C}$. There exists a $\mathbb{K}$-vector space $E \otimes F$ such that, for any vector space G, the space of* ***linear*** *maps from $E \otimes F$ to G is isomorphic to the space of* ***bilinear*** *maps from $E \times F$ to G, that is,*

$$\mathscr{L}(E \otimes F, G) \simeq \mathcal{B}il(E \times F, G).$$

More precisely, there exists a bilinear map

$$\varphi : E \times F \longrightarrow E \otimes F$$

such that, for any vector space G and any bilinear map f from $E \times F$ to G, there exists a unique linear map f^* from $E \otimes F$ into G such that $f = f^* \circ \varphi$, which is summarized by the following diagram:

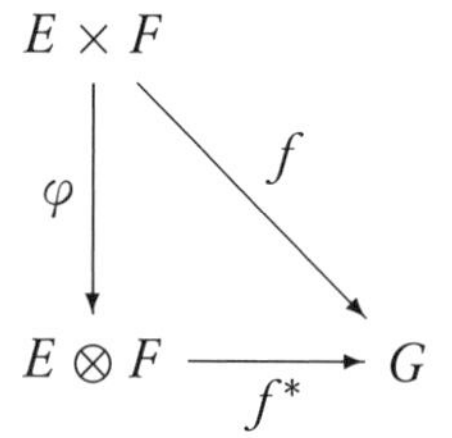

The space $E \otimes F$ is called the ***tensor product*** *of E and F. It is only unique up to (unique) isomorphism.*

The bilinear map φ from $E \times F$ to $E \otimes F$ is denoted "$\otimes$" and also called the ***tensor product****. It is such that if $(\mathbf{e}_i)_i$ and $(\mathbf{f}_j)_j$ are bases of E and F, respectively, the family $(\mathbf{e}_i \otimes \mathbf{f}_j)_{i,j}$ is a basis of $E \otimes F$.*

All this means that the tensor product "linearizes what was bilinear," or also that "to any bilinear map on $E \times F$, one can associate a unique linear map on the space $E \otimes F$."

• **Construction of the tensor product** First, consider the set $E \times F$ of pairs (x, y) with $x \in E$, $y \in F$. Then consider the *free vector space* M generated by $E \times F$, that is, the set of "abstract" linear combinations of the type

$$\sum_{(x,y)\in E\times F} \lambda_{(x,y)}[(x,y)],$$

where $[(x,y)]$ is just some *symbol* associated with the pair (x,y) and $\lambda_{(x,y)}$ is in $\mathbb{K}$, with the additional condition that $\lambda_{(x,y)} = 0$ except for finitely many pairs (x,y).

The set M has an obvious structure of a vector space (infinite-dimensional), with basis the symbols $[(x,y)]$ for $(x,y) \in E \times F$: indeed we add up elements of M "term by term", and multiply by scalars in the same way:

$$\sum_{(x,y)\in E\times F} \lambda_{(x,y)}[(x,y)] + \sum_{(x,y)\in E\times F} \mu_{(x,y)}[(x,y)] = \sum_{(x,y)\in E\times F} (\lambda_{(x,y)} + \mu_{(x,y)})[(x,y)],$$

$$\mu \cdot \Big(\sum_{(x,y)\in E\times F} \lambda_{(x,y)}[(x,y)] \Big) = \sum_{(x,y)\in E\times F} \mu\lambda_{(x,y)}[(x,y)].$$

Now, in order to transform *bilinear* maps on $E \times F$ into *linear* maps, we will use M and identify in M certain elements (for instance, we want the basis elements $[(x,2y)]$ and $[(2x,y)]$ to be equal, and to be equal to $2[(x,y)]$).

For this purpose, let N be the subspace of M generated by all elements of the following types, which are obviously in M:

- $[(x+x',y)] - [(x,y)] - [(x',y)]$;
- $[(x,y+y')] - [(x,y)] - [(x,y')]$;
- $[(ax,y)] - a[(x,y)]$;
- $[(x,ay)] - a[(x,y)]$;

where $x \in E$, $y \in F$ and $a \in \mathbb{K}$ are arbitrary.

Consider now $T = M/N$, the quotient space, and let π be the canonical projection map $M \to M/N$, that is, the map which associates to $z \in M$ the *equivalence class* of z, namely, the set of elements of the form $z+n$ with $n \in N$. This map is surjective and linear.

Now, $E \times F$ can be identified with the "canonical" basis of M, and we can therefore construct a map

$$\varphi : E \times F \longrightarrow T = M/N$$

by sending (x,y) to the class of the basis vector $[(x,y)]$.

We now claim that this map φ and T as $E \otimes F$ satisfy the properties stated for the tensor product.

Indeed, let us first show that φ is *bilinear*. Let $x,x' \in E$ and $y \in F$. Then $\varphi(x+x',y)$ is the class of $[(x+x',y])$. But notice that the element $[(x+x',y)] - [(x,y)] - [(x',y')]$ is in N by definition, which means that the class of $[(x+x',y)]$ is the class of $[(x,y)] + [(x',y)]$. This means that $\varphi(x+x',y) = \varphi(x,y) + \varphi(x',y)$. Exactly in the same manner, the other elements we have put in N ensure that all properties required for the bilinearity of φ are valid.

We now write $E \otimes F = T$, and we simply write $\varphi(x,y) = x \otimes y$.

Consider a $\mathbb{K}$-vector space G. If we have a linear map $T = E \otimes F \to G$, then the composite

$$E \times F \longrightarrow E \otimes F \longrightarrow G$$

is a bilinear map $E \times F \to G$. Conversely, let $B : E \times F \to G$ be a bilinear map. We construct a map $E \otimes F \to G$ as follows. First, let $\tilde{B} : M \to G$ be the map defined by

$$\sum_{(x,y)\in E\times F} \lambda_{(x,y)}[(x,y)] \longmapsto \sum_{(x,y)\in E\times F} \lambda_{(x,y)} B(x,y).$$

It is obvious that $\tilde{B}$ is linear. Moreover, it is immediate that we have $\tilde{B}(n) = 0$ for all $n \in N$ *because* B is bilinear; for instance,

$$\tilde{B}([(x+x',y)] - [(x,y)] - [(x',y)]) = B(x+x',y) - B(x,y) - B(x',y) = 0.$$

This means that we can unambiguously use $\tilde{B}$ to "induce" a map $T \to G$ by $x \otimes y \mapsto B(x,y)$, because if we change the representative of $x \otimes y$ in M, the value of $B(x,y)$ does not change.

Thus we have constructed a *linear* map $E \otimes F \to G$.

It is now easy (and an excellent exercise) to check that the applications just described,

$$\mathscr{L}(E \otimes F, G) \longrightarrow \mathcal{B}il(E \times F, G) \quad \text{and} \quad \mathcal{B}il(E \times F, G) \longrightarrow \mathscr{L}(E \otimes F, G),$$

are reciprocal to each other.

- **Uniqueness** Let $\varphi : E \times F \to H$ and $\varphi' : E \times F \to H'$ be two maps satisfying the property stated in the theorem, where H and H' are $\mathbb{K}$-vector spaces.

Since φ' is itself bilinear, there exists (by the property of φ) a unique linear map $\psi : H \to H'$ such that $\varphi' = \psi \circ \varphi$. Similarly, exchanging the roles of φ and φ', there exists a unique $\psi' : H' \to H$ such that $\varphi = \psi' \circ \varphi'$. Now notice that we have both $\varphi = \mathrm{Id}_H \circ \varphi$ (a triviality) and $\varphi = \psi' \circ \psi \circ \varphi$. The universal property of φ again implies that $\psi' \circ \psi = \mathrm{Id}_H$. Similarly, we find $\psi \circ \psi' = \mathrm{Id}_{H'}$, and hence ψ is an isomorphism from H to H'.

This shows that the tensor product is unique up to isomorphism, and in fact "up to unique isomorphism."

THEOREM 19.18 on page 516 (Poincaré formula) *Let $n \in \mathbb{N}^*$ and let $A_1, \dots, A_n$ be arbitrary events, not necessarily disjoint. Then we have*

$$\mathsf{P}\left(\bigcup_{i=1}^{n} A_i\right) = \sum_{k=1}^{n} (-1)^{k+1} \sum_{1 \leqslant i_1 < i_2 < \cdots < i_k \leqslant n} \mathsf{P}(A_{i_1} \cap \cdots \cap A_{i_k}).$$

Following Pascal,[1] consider the random variables $\mathbf{1}_{A_i}$ which are characteristic functions of the events A_i, that is,

$$\mathbf{1}_{A_i} : \Omega \longrightarrow \mathbb{R},$$
$$\omega \longmapsto \begin{cases} 0 & \text{if } \omega \in A, \\ 1 & \text{if } \omega \notin A. \end{cases}$$

Then we have $E(\mathbf{1}_{A_i}) = \mathsf{P}(A_i)$. Moreover, these random variables satisfy

$$\mathbf{1}_{A_i \cap A_j} = \mathbf{1}_{A_i} \cdot \mathbf{1}_{A_j} \qquad \text{and} \qquad \mathbf{1}_{A_i^c} = \mathbf{1}_\Omega - \mathbf{1}_{A_i}.$$

Using both properties, we find

$$\mathsf{P}(A_1 \cup \cdots \cup A_n) = E\left[\mathbf{1}_\Omega - \prod_{i=1}^{n} (\mathbf{1}_\Omega - \mathbf{1}_{A_i})\right].$$

Now expand the product on the right-hand side, noting that $\mathbf{1}_\Omega \cdot \mathbf{1}_{A_i} = \mathbf{1}_{A_i}$; we get[2]

$$\prod_{i=1}^{n} (\mathbf{1}_\Omega - \mathbf{1}_{A_i}) = \mathbf{1}_\Omega + \sum_{k=1}^{n} (-1)^k \sum_{1 \leqslant i_1 < i_2 < \cdots < i_k \leqslant n} \mathbf{1}_{A_{i_1}} \cdot \mathbf{1}_{A_{i_2}} \dots \mathbf{1}_{A_{i_k}}.$$

It only remains to compute the expectation of both sides to derive the stated formula.

[1] Blaise Pascal (1623–1662) studied mathematics when he was very young, and investigated the laws of hydrostatics and hydrodynamics (having his brother in law climb to the top of the Puy de Dôme with a barometer in order to prove that atmospheric pressure decreases with altitude). He found many results in geometry, arithmetics, and infinitesimal calculus, and was one of the very first to study probability theory. In addition, he invented and built the first mechanical calculator. In 1654, Pascal abandoned mathematics and science for religion. In his philosophy there remain, however, some traces of his scientific mind.

[2] This may be proved by induction, for instance, in the same manner that one shows the that

$$\prod_{i=1}^{n} (1 - x_i) = 1 + \sum_{k=1}^{n} (-1)^k \sum_{1 \leqslant i_1 < i_2 < \cdots < i_k \leqslant n} x_{i_1} x_{i_2} \dots x_{i_k}.$$

Tables

Fourier Transforms

The convention used here is the following:

$$\tilde{f}(\nu) \stackrel{\text{def}}{=} \int_{-\infty}^{+\infty} f(x)\,\mathrm{e}^{-2\mathrm{i}\pi\nu x}\,\mathrm{d}x \qquad \text{and} \qquad f(x) = \int_{-\infty}^{+\infty} \tilde{f}(\nu)\,\mathrm{e}^{2\mathrm{i}\pi\nu x}\,\mathrm{d}\nu.$$

For other standard conventions, see the table on page 612.

$f(x)$	$\tilde{f}(\nu)$	
$f(ax)$	$\dfrac{1}{\lvert a\rvert}\tilde{f}\left(\dfrac{\nu}{a}\right)$	
$f^{(n)}(x)$	$(2\mathrm{i}\pi\nu)^n\,\tilde{f}(\nu)$	
$(-2\mathrm{i}\pi x)^n f(x)$	$\tilde{f}^{(n)}(\nu)$	
$\Pi(x)$	$\dfrac{\sin\pi\nu}{\pi\nu} = \operatorname{sinc}\pi\nu$	
$\Pi\left(\dfrac{x}{a}\right)$	$\dfrac{\sin\pi\nu\,\lvert a\rvert}{\pi\nu}$	
$\dfrac{1}{b-a}\chi_{[a,b]}$	$\dfrac{\sin\big(\pi(b-a)\nu\big)}{\pi\nu}\,\mathrm{e}^{-\mathrm{i}\pi(a+b)\nu}$	$(b > a)$
$\mathrm{e}^{-\mu\lvert x\rvert}$	$\dfrac{2\mu}{\mu^2+4\pi^2\nu^2}$	$(\mu > 0)$
$\dfrac{2a}{a^2+4\pi^2\nu^2}$	$\mathrm{e}^{-a\lvert\nu\rvert}$	$(a > 0)$
$\mathrm{e}^{-\sigma x^2}$	$\sqrt{\dfrac{\pi}{\sigma}}\,\mathrm{e}^{-\pi^2\nu^2/\sigma}$	$(\sigma > 0)$
1	δ	
x	$-\dfrac{1}{2\mathrm{i}\pi}\,\delta'$	
x^k	$\dfrac{1}{(-2\mathrm{i}\pi)^k}\,\delta^{(k)}$	

$f(x)$	$\widetilde{f}(\nu)$
$H(x)$	$\dfrac{\delta}{2} + \dfrac{1}{2\mathrm{i}\pi}\,\mathrm{pv}\left(\dfrac{1}{\nu}\right)$
$H(-x)$	$\dfrac{\delta}{2} - \dfrac{1}{2\mathrm{i}\pi}\,\mathrm{pv}\left(\dfrac{1}{\nu}\right)$
$\mathrm{sgn}(x)$	$\dfrac{1}{\mathrm{i}\pi}\,\mathrm{pv}\left(\dfrac{1}{\nu}\right)$
$\mathrm{pv}\left(\dfrac{1}{x}\right)$	$-\mathrm{i}\pi\,\mathrm{sgn}\,\nu$
$\sin(2\pi\nu_0 x)$	$\dfrac{1}{2\mathrm{i}}\left[\delta(\nu-\nu_0) - \delta(\nu+\nu_0)\right]$
$\cos(2\pi\nu_0 x)$	$\dfrac{1}{2}\left[\delta(\nu-\nu_0) + \delta(\nu+\nu_0)\right]$
$\delta(x)$	1
$\delta(x-x_0)$	$\mathrm{e}^{-2\mathrm{i}\pi\nu x_0}$
$Ш(x)$	$Ш(\nu)$
$Ш(ax)$	$\dfrac{1}{\lvert a\rvert}\,Ш(\nu/a)$

Three-dimensional Fourier transforms

For the following three-dimensional Fourier transforms, the convention chosen is

$$F(\boldsymbol{k}) = \iiint f(\boldsymbol{x})\,\mathrm{e}^{-\mathrm{i}\boldsymbol{k}\cdot\boldsymbol{r}}\,\mathrm{d}^3\boldsymbol{r}, \qquad f(\boldsymbol{r}) = \frac{1}{(2\pi)^3}\iiint F(\boldsymbol{k})\,\mathrm{e}^{\mathrm{i}\boldsymbol{k}\cdot\boldsymbol{r}}\,\mathrm{d}^3\boldsymbol{k}.$$

Moreover, we put $r = \|\boldsymbol{r}\|$, $\boldsymbol{n} = \boldsymbol{r}/r$ and $k = \|\boldsymbol{k}\|$.

$f(\boldsymbol{r})$	$F(\boldsymbol{k})$	
$\triangle f(\boldsymbol{r})$	$-k^2 F(\boldsymbol{k})$	
$1/r$	$4\pi/k^2$	
$\mathrm{e}^{-\mu r}/r$	$4\pi/(\mu^2+k^2)$	$(\mu > 0)$
$\dfrac{1}{r^2+a^2}$	$\mathrm{e}^{-ak}/2\pi^2 k$	$(a > 0)$
$1/r^2$	$1/2\pi^2 k$	
$\dfrac{1}{r}(\boldsymbol{p}\cdot\boldsymbol{n})(\boldsymbol{q}\cdot\boldsymbol{n})$	$\dfrac{4\pi}{k^2}\left[\boldsymbol{p}\cdot\boldsymbol{q} - 2\dfrac{(\boldsymbol{p}\cdot\boldsymbol{k})(\boldsymbol{q}\cdot\boldsymbol{k})}{k^2}\right]$	$\boldsymbol{p},\ \boldsymbol{q} \in \mathbb{R}^3$
$\mathrm{e}^{-\alpha r^2}$	$\left(\dfrac{\pi}{\alpha}\right)^{3/2}\mathrm{e}^{-k^2/4\alpha}$	$(\alpha > 0)$
$\delta(\boldsymbol{r})$	1	
1	$(2\pi)^3\delta$	

Table of the usual conventions for the Fourier transform

	Fourier transform	Inverse transform	$\mathscr{F}[1]$	$\mathscr{F}[\delta]$	Parseval-Plancherel
	$\tilde{f}(\nu) = \int f(x)\,\mathrm{e}^{-2\mathrm{i}\pi\nu x}\,\mathrm{d}x$	$f(x) = \int \tilde{f}(\nu)\,\mathrm{e}^{2\mathrm{i}\pi\nu x}\,\mathrm{d}\nu$	δ	1	$\int \overline{f}\,g = \int \overline{\tilde{f}}\,\tilde{g}$
	$\tilde{f}(\omega) = \dfrac{1}{\sqrt{2\pi}} \int f(x)\,\mathrm{e}^{-\mathrm{i}\omega x}\,\mathrm{d}x$	$f(x) = \dfrac{1}{\sqrt{2\pi}} \int \tilde{f}(\omega)\,\mathrm{e}^{\mathrm{i}\omega x}\,\mathrm{d}\omega$	$\sqrt{2\pi}\,\delta$	$\dfrac{1}{\sqrt{2\pi}}$	$\int \overline{f}\,g = \int \overline{\tilde{f}}\,\tilde{g}$
	$\tilde{f}(\omega) = \int f(x)\,\mathrm{e}^{-\mathrm{i}\omega x}\,\mathrm{d}x$	$f(x) = \dfrac{1}{2\pi} \int \tilde{f}(\omega)\,\mathrm{e}^{\mathrm{i}\omega x}\,\mathrm{d}\omega$	$2\pi\delta$	1	$\int \overline{f}\,g = \dfrac{1}{2\pi} \int \overline{\tilde{f}}\,\tilde{g}$
3D	$F(\boldsymbol{k}) = \iiint f(\boldsymbol{r})\,\mathrm{e}^{-\mathrm{i}\boldsymbol{k}\cdot\boldsymbol{r}}\,\mathrm{d}^3\boldsymbol{r}$	$f(\boldsymbol{r}) = \dfrac{1}{(2\pi)^3} \iiint F(\boldsymbol{k})\,\mathrm{e}^{\mathrm{i}\boldsymbol{k}\boldsymbol{r}}\,\mathrm{d}^3\boldsymbol{k}$	$(2\pi)^3\delta$	1	$\int \overline{f}\,g = \dfrac{1}{(2\pi)^3} \int \overline{\tilde{f}}\,\tilde{g}$
QM (1D)	$\tilde{f}(p) = \int f(x)\,\mathrm{e}^{-\mathrm{i}px/\hbar}\,\dfrac{\mathrm{d}x}{\sqrt{2\pi\hbar}}$	$f(x) = \int \tilde{f}(p)\,\mathrm{e}^{\mathrm{i}px/\hbar}\,\dfrac{\mathrm{d}p}{\sqrt{2\pi\hbar}}$	$\sqrt{2\pi\hbar}\,\delta$	$\dfrac{1}{\sqrt{2\pi\hbar}}$	$\int \overline{f}\,g = \int \overline{\tilde{f}}\,\tilde{g}$
QM (3D)	$\tilde{f}(\boldsymbol{p}) = \iiint f(\boldsymbol{x})\,\mathrm{e}^{-\mathrm{i}\boldsymbol{p}\cdot\boldsymbol{x}/\hbar}\,\dfrac{\mathrm{d}^3\boldsymbol{x}}{(2\pi\hbar)^{3/2}}$	$f(\boldsymbol{x}) = \iiint \tilde{f}(\boldsymbol{p})\,\mathrm{e}^{\mathrm{i}\boldsymbol{p}\cdot\boldsymbol{x}/\hbar}\,\dfrac{\mathrm{d}^3\boldsymbol{p}}{(2\pi\hbar)^{3/2}}$	$(2\pi\hbar)^{3/2}\,\delta$	$(2\pi\hbar)^{-3/2}$	$\int \overline{f}\,g = \int \overline{\tilde{f}}\,\tilde{g}$

Various definitions used for the Fourier transforms. The formulas for those different from the ones used in this book can be recovered by elementary changes of variables. The convention named "QM" are used in quantum mechanics. Note that the simplest definition in terms of memorizing the formulas is the first one. The definition for quantum mechanics is the one in the book by Cohen-Tannoudji et al. [20]

Laplace transforms

In some problems (in particular, to solve differential equations while taking initial conditions at $t = 0$ into account), it is useful to go through the Laplace transform, which transforms the differential operations into algebraic operations, which can easily be inverted. The table of Laplace transforms is therefore usually read from right to left: one looks for an "original" $t \mapsto f(t)$ for the function $p \mapsto F(p)$ that has been calculated.

$f(t)$	$F(p) = \int_0^{+\infty} e^{-pt} f(t)\,dt$	
$\dfrac{1}{2i\pi} \int_{c-i\infty}^{c+i\infty} e^{tz} F(z)\,dz$	$F(p)$	inversion
$f'(t)$	$p\,F(p) - f(0^+)$	
$\int_0^t f(s)\,ds$	$\dfrac{F(p)}{p}$	
$(-1)^n t^n f(t)$	$F^{(n)}(p)$	$(n \in \mathbb{N})$
$\dfrac{f(t)}{t}$	$\int_p^{\infty} F(z)\,dz$	$(*)$
$e^{at} f(t)$	$F(p-a)$	$(a \in \mathbb{C})$

$(*)$ The integral "$\int_p^\infty$" is defined on any path joining p to infinity "on the right-hand side" ($\mathrm{Re}(z) \to +\infty$), which is entirely contained in the half-plane of integrability where F is defined. Because F is analytic, this integral is independent of the chosen path.

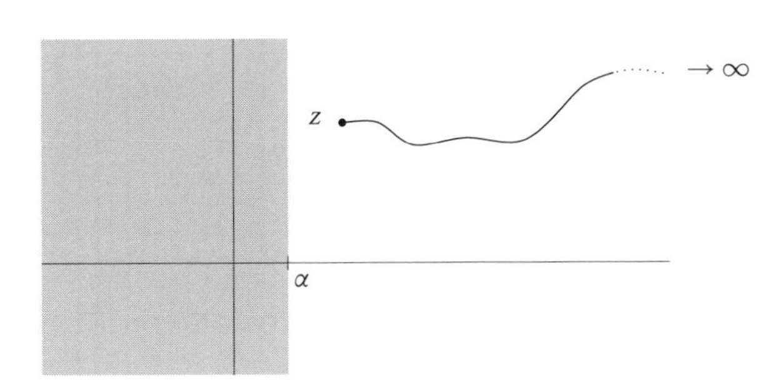

$f(t)$	$F(p)$	
$\delta(t)$	1	
$\delta^{(n)}(t)$	p^n	$(n \in \mathbb{N})$
$\text{III}(t)$	$\dfrac{1}{1-\mathrm{e}^{-p}}$	
1	$1/p$	
t	$1/p^2$	
$1/\sqrt{\pi t}$	$1/\sqrt{p}$	
e^{at}	$1/(p-a)$	$(a \in \mathbb{C})$
$t^n\,\mathrm{e}^{at}/n!$	$(p-a)^{-(n+1)}$	$(n \in \mathbb{N}, \quad a \in \mathbb{C})$
$\cos\omega t$	$\dfrac{p}{p^2+\omega^2}$	$(\omega \in \mathbb{R}$ or $\mathbb{C})$
$\sin\omega t$	$\dfrac{\omega}{p^2+\omega^2}$	
$\cosh\omega t$	$\dfrac{p}{p^2-\omega^2}$	
$\sinh\omega t$	$\dfrac{\omega}{p^2-\omega^2}$	
$\dfrac{1}{\omega^2}(1-\cos\omega t)$	$\dfrac{1}{p(p^2+\omega^2)}$	$(\omega \in \mathbb{C})$
$\dfrac{1}{\omega^3}(\omega t-\sin\omega t)$	$\dfrac{1}{p^2(p^2+\omega^2)}$	
$\dfrac{1}{2\omega^3}(\sin\omega t-\omega t\cos\omega t)$	$\dfrac{1}{(p^2+\omega^2)^2}$	
$\dfrac{t}{2\omega}\cos\omega t$	$\dfrac{p}{(p^2+\omega^2)^2}$	
$\dfrac{1}{2\omega}(\sin\omega t+\omega t\cos\omega t)$	$\dfrac{p^2}{(p^2+\omega^2)^2}$	

$f(t)$	$F(p)$	
$t\cos\omega t$	$\dfrac{p^2-\omega^2}{(p^2+\omega^2)^2}$	
$J_0(\omega t)$	$\dfrac{1}{\sqrt{p^2+\omega^2}}$	
$\omega\, J_1(\omega t)$	$1-\dfrac{\omega}{\sqrt{p^2+\omega^2}}$	
$\omega\,\dfrac{J_1(\omega t)}{t}$	$\sqrt{p^2+\omega^2}-p$	
$\log t$	$-\dfrac{\log p}{p}-\gamma$	$(\gamma\approx 0,577...)$
$\dfrac{\sqrt{\pi}}{\Gamma(k)}\left(\dfrac{t}{2\omega}\right)^{k-\frac{1}{2}} J_{k-\frac{1}{2}}(\omega t)$	$\dfrac{1}{(p^2+a^2)^k}$	$(k>0)$

Tables of probability laws

Discrete laws

Name	Set of values	Law	Expectation	Variance
Uniform law	$[\![1, n]\!]$	$\mathrm{P}(X = k) = \dfrac{1}{n}$	$\dfrac{n+1}{2}$	$\dfrac{n^2 - 1}{12}$
Bernoulli law $\mathscr{B}(1, p)$	$\{0, 1\}$	$\begin{cases} \mathrm{P}(X = 1) = p \\ \mathrm{P}(X = 0) = 1 - p \end{cases}$	p	$p(1 - p)$
Binomial law $\mathscr{B}(n, p)$	$[\![0, n]\!]$	$\mathrm{P}(X = k) = \dbinom{n}{k} p^k (1 - p)^{n-k}$	np	$np(1 - p)$
Poisson law	$\mathbb{N}$	$\mathrm{P}(X = n) = \mathrm{e}^{-\lambda} \cdot \dfrac{\lambda^n}{n!}$	λ	λ

Continuous laws

Name	Set of values	Density $f(x)$	Expectation $E(X)$	Variance σ^2	Char. function $\varphi(\nu)$
Uniform law	$[a, b]$	$\dfrac{1}{b - a} \chi_{[a,b]}$	$\dfrac{a + b}{2}$	$\dfrac{(b - a)^2}{12}$	$\dfrac{\mathrm{e}^{\mathrm{i}\nu b} - \mathrm{e}^{\mathrm{i}\nu a}}{\mathrm{i}\nu(b - a)}$
Normal law	$\mathbb{R}$	$\dfrac{1}{\sigma\sqrt{2\pi}} \exp\left\{-\dfrac{(x - m)^2}{2\sigma^2}\right\}$	m	σ^2	$\mathrm{e}^{\mathrm{i}m\nu} \mathrm{e}^{-\sigma^2\nu^2/2}$
Exponential law	$\mathbb{R}^+$	$\mu^{-1} \mathrm{e}^{-x/\mu}$	μ	μ^2	$\dfrac{1}{1 - \mathrm{i}\mu\nu}$
Cauchy law	$\mathbb{R}$	$\dfrac{a}{\pi} \dfrac{1}{a^2 + (x - m)^2}$	not defined	not def.	$\mathrm{e}^{\mathrm{i}m\nu - \lvert a\nu \rvert}$
χ^2 law	$\mathbb{R}^+$	$\dfrac{1}{2\Gamma(n/2)} \left(\dfrac{x}{2}\right)^{\frac{n}{2}-1} \mathrm{e}^{-x/2}$	n	$2n$	$(1 - 2\mathrm{i}\nu)^{-n/2}$

Further reading

General books

- ***Methods of theoretical physics***
  P. M. MORSE and H. FESHBACH (McGraw-Hill, 1953)

  A great classic, very complete.

- ***Mathematical methods for physicists***
  G. B. ARFKEN and H. J. WEBER (Academic Press, 4th edition, 1995)

  Many subjects are covered, in particular, differential equations, Bessel and Legendre functions, other special functions, spherical harmonics.

Integration

- ***Lebesgue measure and integration, an introduction***
  F. BURK (Wiley Interscience, 1998)

  A very clear book, with many examples guiding the reader step by step. Also a very good historical introduction.

Complex analysis

- ***Function theory of one complex variable***
  R. E. GREENE and S. G. KRANTZ (Wiley Interscience, 1997)

  Very pedagogical.

Distributions and Fourier analysis

- ***Fourier analysis: an introduction***
  E. M. STEINZ and R. SHAKARCHI (Princeton University Press, 2003)

  A very clear introduction.

- ***A wavelet tour of signal processing***
  S. MALLAT (Academic Press, 1999)

  A nice and clear treatment of a difficult subject, with many illustrations and examples of practical applications of wavelets.

Integral transforms, operator theory, Hilbert spaces

- *Theory of linear operators in Hilbert space*
  N. I. AKHIEZER and I. M. GLAZMAN (2 volumes, Dover, 1993)

  A very complete book, rather difficult to read.

- *Applied functional analysis*
  E. ZEIDLER (Springer-Verlag, 1995, Applied Mathematical Sciences 108)

  This book is oriented toward physical applications, but remains quite difficult.

- *Quantum mechanics in Hilbert space*
  E. Prugovečki (Academic Press, 1971, Pure and Applied Mathematics 41)

  Another very complete treatement, mathematically rigorous, using powerful tools which go far beyond what we have seen in the present text.

Tensors, geometry

- *Riemannian geometry*
  S. GALLOT, D. HULIN and J. LAFONTAINE (Springer-Verlag, 1990)

- *Leçon sur la théorie des spineurs*
  É. CARTAN (Hermann, 1966)

  An English version is published by Dover with the title *The theory of spinors*, 1981.

- *Introduction to differentiable manifolds*
  L. AUSLANDER and R. E. MACKENZIE (Dover, 1977)

  A clear introduction to differential geometry.

- *Modern geometry—Methods and applications*
  B. DUBROVIN, S. NOVIKOV and A. FOMENKO (Springer, 1991)

  Three volumes: I. *Geometry of surfaces, of transformation groups and fields*; II. *Geometry and topology of varieties*; III. *Methods of homology theory*. The first two volumes are most likely to be of interest to physicists. The geometrical intuition is emphasized throughout. The English edition published by Springer-Verlag is much more expensive than the original Russian or French editions.

- *Geometry, topology and physics*
  M. NAKAHARA (Adam Hilger, 1990)

  A unique book that makes the link between theoretical physics and high-level mathematics.

Groups

- *Symmetries in quantum mechanics: from angular momentum to supersymmetry*
  M. CHAICHIAN and R. HAGEDORN (Institute of Physics, 1998)

  A very interesting book, with increasing difficulty, giving a vast overview of symmetries, and explaining clearly the relation between symmetries and dynamics.

- ***Lie groups and algebra, with applications to physics, geometry and mechanics***
  D. H. SATTINGER and O. L. WEAVER (Springer-Verlag, 1986, Applied Mathematical Sciences 61)

Probabilities, statistics

- ***An introduction to probability theory***
  W. FELLER (Wiley, 1968 (vol. I) and 1971 (vol. II))

  The first volume, with little formalism, is a remarkable book directed toward concrete applications, analyzed in great detail and depth.

- ***Stochastik***
  A. ENGEL (Ernst Klett Verlag, 2000)

  This book has a fascinating approach to probability theory, full of applications and with little formalism. Moreover, this is one of the rare books dealing with probabilities and statistics simultaneously. (A French translation of the original German edition is available: *Les certitudes du hasard*, ALÉAS Éditeur, Lyon, 1990).

References

[1] Edwin A. Abbott. *Flatland.* Dover, 1992.

[2] Milton Abramowitz and Irene A. Stegun, editors. *Handbook of mathematical functions.* Dover, 1972.

[3] Sir George Airy. On the intensity of light in the neighbourhood of a caustic. *Camb. Phil. Trans.*, **6**, 379–402, 1838.

[4] Naum Ilitch Akhiezer and Israël Markovitch Glazman. *Theory of linear operators in Hilbert space.* Dover, 1993. (two volumes bound as one).

[5] André Angot. *Compléments de mathématiques à l'usage des ingénieurs de l'électrotechnique et des télécommunications.* Masson, 1982.

[6] Walter Appel and Angel Alastuey. Thermal screening of Darwin interactions in a weakly relativistic plasma. *Phys. Rev. E*, **59**(4), 4542–4551, 1999.

[7] Walter Appel and Michael K.-H. Kiessling. Mass and spin renormalization in Lorentz electrodynamics. *Annals of Physics*, **289**(1), 24–83, April 2001. xxx.lanl.gov/math-ph/00090003.

[8] George B. Arfken and Hans J. Weber. *Mathematical methods for physicists.* Academic Press, fourth edition, 1995.

[9] Vladimir I. Arnold. *Mathematical methods of classical mechanics.* Springer-Verlag, 1989.

[10] Gaston Bachelard. *Le nouvel esprit scientifique.* Quadrige / P.U.F., 1987.

[11] Gernot Bauer and Detlef Dürr. The Maxwell-Lorentz system of a rigid charge distribution. *Preprint Universität München*, 1999.

[12] Thomas Bayes. An essay towards solving a problem in the doctrine of chances. *Phil. Trans.*, **53**, 370–418, 1764.

[13] Émile Borel. *Leçons sur les séries divergentes.* Gauthier-Villars, Paris, second edition, 1928.

[14] Max Born and Emil Wolf. *Principles of optics.* Cambridge University Press, seventh edition, 1999.

[15] Henri BROCH. *Au cœur de l'extraordinaire.* Collection Zététique, 2002. A book about "the spirit of doubt a priori," which should be always present to the mind of scientists...

[16] Frank BURK. *Lebesgue measure and integration.* Wiley-Interscience, 1998.

[17] Henri CARTAN. *Cours de calcul différentiel.* Hermann, 1979. (*Differential forms*, Dover, 2006).

[18] Masud CHAICHIAN and Rolf HAGEDORN. *Symmetries in quantum mechanics.* Institute of Physics Publishing, 1998.

[19] Éric CHARPENTIER, Annick LESNE, and Nikolaï NIKOLSKI. *L'héritage de Kolmogorov en mathématiques.* Belin, 2004.

[20] Claude COHEN-TANNOUDJI, Bernard DIU, and Frank LALOË. *Quantum Mechanics.* Wiley-Interscience, 2001.

[21] Claude COHEN-TANNOUDJI, Jacques DUPONT-ROC, and Gilbert GRYNBERG. *Photons and atoms, Introduction to quantum electrodynamics.* Wiley-Interscience, 1997.

[22] Jean-François COLOMBEAU. *Elementary introduction to new generalized functions.* North Holland, 1984.

[23] Jean-François COLOMBEAU. *New generalized functions and multiplication of distributions.* North Holland Mathematical Studies 84, 1984.

[24] Robert DAUTRAY and Jacques-Louis LIONS. *Analyse mathématique et calcul numérique pour les sciences et les techniques.* Collection du CEA. Masson, 1984.

[25] Jean DE PRINS. Couvrez ce sein que je ne saurais voir... In *Cahiers d'histoire & de philosophie des sciences*, volume 40 of *Actes du colloque de Lille, 11-13 avril 1991*, 165-174. Société française des sciences et des techniques, 1992.

[26] Petrus DEBYE and Erich HÜCKEL. Zur theorie der elektrolyte. *Phys. Z.*, **24**, 185-206, 1923.

[27] André DELACHET. *Le calcul tensoriel*, volume 1336 of *Que sais-je ?* Presses Universitaires de France, 1974.

[28] Bertrand DELAMOTTE. Un soupçon de théorie des groupes: groupe des rotations et groupe de Poincaré. (Electronic document which can be found at www.lpthe.jussieu.fr/DEA/delamotte.html), 1997.

[29] Pierre DOLBEAULT. *Analyse complexe.* Masson, 1990.

[30] Boris DUBROVIN, Sergeï NOVIKOV, and Anatoli FOMENKO. *Modern geometry – Methods and applications*, volume 1. Springer, 1982. Old-fashioned Russian notation with indices everywhere, but the text is very clear; the first two volumes are especially recommended for physicists.

[31] Phil P. G. DYKE. *An introduction to Laplace transforms and Fourier series*. Springer, 2001.

[32] Freeman J. DYSON. Divergences of perturbation theory in quantum electrodynamics. *Phys. Rev.*, **85**, 631–632, 1952.

[33] William FELLER. *An introduction to probability theory*. Wiley, 1968, 1971. 2 volumes.

[34] Richard P. FEYNMAN and A. R. HIBBS. *Quantum mechanics and integrals*. McGraw-Hill, 1965.

[35] Richard P. FEYNMAN, Robert B. LEIGHTON, and Matthew SANDS. *The Feynman lectures on physics*, volume 2. Addison Wesley, 1977.

[36] Richard P. FEYNMAN, Robert B. LEIGHTON, and Matthew SANDS. *The Feynman lectures on physics*, volume 3. Addison Wesley, 1977.

[37] Claude GASQUET and Patrick WITOMSKI. *Fourier analysis and applications: filtering, numerical computation, wavelets*. Springer, 1998.

[38] Israël M. GELFAND and G. E. SHILOV. *Generalized functions*. Academic Press, 1964.

[39] Josiah W. GIBBS. Letters to the editor. *Nature*, 200–201, December 1898.

[40] Josiah W. GIBBS. Letters to the editor. *Nature*, 606, April 1899.

[41] François GIERES. Mathematical surprises and Dirac's formalism in quantum mechanics. *Rep. Prog. Phys*, **63**, 1893–1931, 2000.

[42] I. S. GRADSTEIN and I. M. RYSHIK. *Summen-, Produkt- und Integral-Tafeln*. Deutscher Verlag der Wissenschaften, 1963.

[43] Robert E. GREENE and Steven G. KRANTZ. *Function theory of one complex variable*. Wiley Interscience, 1997. Very pedagogical.

[44] Étienne GUYON, Jean-Pierre HULIN, and Luc PETIT. *Physical hydrodynamics*. Oxford University Press, 2001.

[45] Paul HALMOS. *Naive set theory*. Springer-Verlag, 1998. A very clear introduction.

[46] Godfrey H. HARDY. *A mathematician's apology*. Cambridge University Press, 1992. Mathematics as seen by a profound and very appealing mathematician. With an account of Ramanujan.

[47] Harry HOCHSTADT. *The functions of mathematical physics.* Dover, 1986.

[48] Bruce J. HUNT. Rigorous discipline: Oliver Heaviside versus the mathematicians. In Peter DEAR, editor, *The literary structure of scientific argument*, chapter 3, 72–95. University of Pennsylvania Press, 1991.

[49] Claude ITZYKSON and Jean-Bernard ZUBER. *Quantum field theory.* McGraw Hill, 1980.

[50] John D. JACKSON. *Classical electrodynamics.* John Wiley and Sons, 1962.

[51] Winfried JUST and Martin WEESE. *Discovering modern set theory*, volume I. American Mathematical Society, 1996.

[52] Jean-Pierre KAHANE. *Nécessité et pièges des définitions mathématiques*, chapter 174. *In* Qu'est-ce que l'Univers ? Odile Jacob, 2001. (Université de tous les savoirs).

[53] Jean-Pierre KAHANE, Pierre CARTIER, and Vladimir I. ARNOLD. *Leçons de mathématiques d'aujourd'hui.* Cassini, 2000. Some of the lectures (in particular, the one by P. Cartier) are particularly relevant to physicists and give examples of contemporary researches in pure and applied mathematics.

[54] Jean-Pierre KAHANE and Pierre Gilles LEMARIÉ-RIEUSSET. *Séries de Fourier et ondelettes.* Cassini, 1998. In the first part of the book, there is a remarkable description of Fourier series and their history, from the origin to the present days. The second part deals with wavelet analysis and is written by a specialist in this topic.

[55] J. I. KAPUSTA. *Finite-temperature field theory.* Cambridge University Press, 1989.

[56] Michael K.-H. KIESSLING. The "Jeans swindle". A true story – mathematically speaking. *Adv. Appl. Math.*, **31**, 132–149, 2003.

[57] Lev LANDAU and Evgueni LIFCHITZ. *The classical theory of fields.* Butterworth-Heinemann, 1995.

[58] Lev LANDAU and Evgueni LIFCHITZ. *Quantum mechanics.* Butterworth-Heinemann, 1996.

[59] Serge LANG. *Complex analysis.* Springer-Verlag, 1993.

[60] Michel LE BELLAC. *Quantum and statistical field theory.* Oxford University Press, 1992.

[61] Jean-Claude LE GUILLOU and Jean ZINN-JUSTIN, editors. *Large-order behaviour of perturbation theory.* Current physics sources and comments 7. North-Holland, 1990.

[62] André LICHNEROWICZ. *Éléments de calcul tensoriel*. Librairie Armand Colin, 1958. With applications to classical dynamics.

[63] Eliott H. LIEB. The stability of matter. *Rev. Mod. Phys.*, **48**(4), 553–569, 1976.

[64] Eliott H. LIEB and Joel L. LEBOWITZ. The constitution of matter: existence of thermodynamical systems composed of electrons and nuclei. *Adv. Math.*, **9**, 316–398, 1972.

[65] Stéphane MALLAT. *A wavelet tour of signal processing*. Academic Press, 1999.

[66] Jean MEEUS. *Calculs astronomiques à l'usage des amateurs*. Société astronomique de France, 1986.

[67] Charles MISNER, Kip THORNE, and John Archibald WHEELER. *Gravitation*. Freeman, 1973. Tensor calculus is explained in detail, patiently, on affine spaces as well as in Riemannian geometry. The mathematical formalism is very well presented.

[68] Mikio NAKAHARA. *Geometry, topology and physics*. Adam Hilger, 1990.

[69] Sir Isaac NEWTON. *Philosophiæ Naturalis Principia Mathematica (1687)*. Prometheus Books, 1995.

[70] Henri POINCARÉ. *Les méthodes nouvelles de la mécanique céleste*, volume 2 (sur 3). Gauthier-Villars, 1892–99. (Distributed by the Librairie Blanchard in Paris).

[71] Eduard PRUGOVEČKI. *Quantum mechanics in Hilbert spaces*. Pure and Applied mathematics 41. Academic Press, 1971.

[72] Jean-Pierre RAMIS. Séries divergentes et théories asymptotiques. Preprint Strasbourg, I.R.M.A., 1992–1993.

[73] Tanguy RIVOAL. Irrationalité d'au moins un des neuf nombres $\zeta(5)$, $\zeta(7), \ldots, \zeta(21)$, 2002.

[74] R. M. ROBINSON. On the decomposition of spheres. *Fundamenta Mathematica*, **34**, 246–266, 1947.

[75] François ROUVIÈRE. *Petit guide de calcul différentiel à l'usage de la licence et de l'agrégation*. Cassini, 1999.

[76] Walter RUDIN. *Real and complex analysis*. McGraw Hill, 1986. Rather mathematical.

[77] David RUELLE. *Chance and chaos*. Princeton University Press, 1991.

[78] Peter SARNAK. Arithmetic quantum chaos. The Schur lectures (1992). In *Israel Math. Conf. Proc.*, **8**, 183–236. Bar-Ilan Univ., 1995.

[79] Laurent SCHWARTZ. Causalité et analyticité. *Anais da Academia Brasileira de Ciências*, **34**, 13–21, 1962.

[80] Laurent SCHWARTZ. *Théorie des distribution*. Hermann, second edition, 1966.

[81] Laurent SCHWARTZ. *Mathematics for the physical sciences*. Addison-Wesley, 1967. An important reference, by the father of distributions.

[82] Laurent SCHWARTZ. *A mathematician grappling with his century*. Birkhauser, 2000. The scientific and personal autobiography of the inventor of the theory of distributions, pacifist, anti-colonialist, human-rights activist.

[83] Albert Nicolaievitch SHIRYAEV. *Probability*. Springer-Verlag, 1996.

[84] Herbert SPOHN. Dynamics of charged particles and their radiation field. xxx.lanl.gov/math-ph/9908024, 1999.

[85] Elias M. STEIN and Rami SHAKARCHI. *Fourier analysis: an introduction*. Princeton University Press, 2003.

[86] Ian STEWART. *Does God play dice? The new mathematics of chaos*. Blackwell, 2001.

[87] Tom STOPPARD. *Rosencrantz & Guildenstern are dead*. Faber, 1991.

[88] Tom STOPPARD. *Arcadia*. Faber, 1993.

[89] Craig THOMPSON. *Good bye, Chunky Rice*. Top Shelf, 1999.

[90] Craig THOMPSON. *Blankets*. Top Shelf, 2004.

[91] Sin-Itiro TOMONAGA. *The story of spin*. University of Chigago Press, 1998.

[92] Steven WEINBERG. *Gravitation and cosmology*. John Wiley, 1972. Notation "with subscripts"; very deep physically.

[93] Lee Peng YEE and Rudolf VÝBORNÝ. *The integral: an easy approach after Kurzweil and Henstock*. Cambridge University Press, 2000.

[94] Kôsaku YOSIDA. *Functional Analysis*. Classics in Mathematics. Springer, 1995.

[95] Eberhard ZEIDLER. *Applied functional analysis: applications to mathematical physics*. Applied Mathematical Sciences 108. Springer, 1997.

Portraits

Sidebars

Index

The **boldface** page numbers refer to the main definitions, the *italic* ones to exercises. The family names in SMALL CAPITALS refer to short biographies.

Symbols

D

E

F

R

S

T

U

V

W

X

Y

Z